To convert		
from	to	multiply by
Joules	calories	0.2390
Calories	joules	4.184
Nanometers	angstroms	10
Angstroms	nanometers	0.1
Atmospheres	mmHg (Torr)	760
mmHg (Torr)	atmospheres	0.001316
Pascal (Pa)	mmHg (Torr)	0.00750
mmHg (Torr)	pascal (Pa)	133.3
Electron volts	kilojoules per mole	96.488
Liter atmospheres	joules	101.32

Physical quantities

Atomic mass unit	amu	1.6605×10^{-24} g
Avogadro number	N	6.0222×10^{23} molecules per mole
Boltzmann constant	k	1.3806×10^{-23} J/deg
Electron charge	e	-1.6022×10^{-19} C
Electron mass	m	9.1096×10^{-28} g
		0.0005486 amu
Electron volt	eV	1.6022×10^{-19} J
Faraday constant	$\mathcal{F}$	9.6487×10^{4} C/equiv
Gas constant	R	0.082057 liter atm mol^{-1} deg^{-1}
		8.3143 J mol^{-1} deg^{-1}
Neutron mass	n	1.6749×10^{-24} g
		1.00866 amu
Planck's constant	h	6.6262×10^{-34} J sec
Proton mass	p	1.6726×10^{-24} g
		1.00728 amu
Speed of light	c	2.9979×10^{10} cm/sec

CHEMISTRY
PRINCIPLES AND APPLICATIONS

CHEMISTRY
PRINCIPLES AND APPLICATIONS

Michell J. Sienko
Professor of Chemistry
Cornell University

Robert A. Plane
President and Professor of Chemistry
Clarkson College of Technology

McGraw-Hill Book Company

New York St. Louis San Francisco Auckland Bogotá Düsseldorf
Johannesburg London Madrid Mexico Montreal New Delhi
Panama Paris São Paulo Singapore Sydney Tokyo Toronto

Library of Congress Cataloging in Publication Data

Sienko, Michell J
    Chemistry, principles and applications.

    Published in 1974 under title: Chemical
principles and properties.
    Includes bibliographical references and index.
    1. Chemistry. I. Plane, Robert A., joint
author. II. Title.
QD31.2.S558 1979   540   78-11169
ISBN 0-07-057321-2

CHEMISTRY: Principles and Applications

A major revision and reorganization of *Chemical Principles and Properties.*

1234567890 VHVH 7832109

This book was set in Baskerville by A Graphic Method Inc.
The editors were Donald C. Jackson, Anne T. Vinnicombe,
and Laura D. Warner;
the designer was Merrill Haber;
the production supervisor was Joe Campanella.
The new drawings were done by J & R Services, Inc.
Von Hoffmann Press, Inc., was printer and binder.

Cover Photograph: Photomicrograph of recrystallization of molten sulfur taken
under polarized light, ×283. (*Photographer Manfred Kage, Peter Arnold, Inc.*)

CONTENTS

v

vii

22 THE ATOMIC NUCLEUS

Nuclear Stability. Types of Radioactivity. Applications of Radioactivity. Nuclear Energy. Energy Crisis. *Important Concepts. Exercises.*

APPENDIX

PREFACE

In preparing this edition, we came to the conclusion that a new title was necessary to emphasize that this is a major reorganization and revision of *Chemical Principles and Properties.* So, we have chosen the title *Chemistry: Principles and Applications.*

Unlike preceding editions, which assumed that high school chemistry courses would become more mathematical and more abstract with time, this edition recognizes that present-day college students need more help than ever to cope with both the theoretical and experimental sides of freshman chemistry. As a result, there has been a major recasting of our approach. Although the basic sequence of topics—atoms to molecules to more complex systems—is retained, the development is more gradual and starts with a broader base. We begin in Chapter 1 with a description of what chemistry is and what vocabulary and notation are associated with it. Then, in Chapter 2, we move on to stoichiometry, partly to increase students' abilities to handle chemical notation and partly to allow instructors to begin meaningful related laboratory work early in the course. Chapter 3 provides an overview of some descriptive chemistry based on the periodic table. This sets the stage for Chapters 4 and 5, where we start to examine the details of atomic structure, the nature of molecules, and chemical bonding. From these chapters, the student should develop sufficient ease in handling the ideas of molecular shapes and molecular reactivity to go rather extensively into states of matter, solutions, solution reactions, kinetics, and equilibrium (Chapters 6 to 15). Chemical thermodynamics is introduced in Chapter 10, considerably later than before, and used subsequently to tie in with electrochemistry (Chapter 11) and equilibrium (Chapter 13). Chemical equilibrium is handled in two chapters. First, Chapter 13 develops the basic ideas of equilibria using gaseous systems. Chapter 14, devoted to the chemistry of hydrogen, oxygen, and water, provides a bridge to Chapter 15, where we undertake a consideration of the more complex equilibria in aqueous solutions. In Chapter 16 we return to take a harder look at the problem of describing electrons in atoms and molecules, first explored in Chapter 5.

There has been considerable revision of the descriptive chemistry that constituted Part II of the second edition. Some has been brought forward and integrated with the discussion of chemical principles; some, particularly that of the less common elements, has been omitted entirely; the rest has been rearranged to make a more practical, coherent package. The typical metals are discussed in Chapter 17; the transition metals, in Chapters 18 and 19; and the nonmetals, in Chapter 21. The material on carbon, Chapter 20, has been expanded to give a better idea of the scope of organic chemistry: We have more on organic functional groups, more on mechanisms of reaction, and new material on carbohydrates, proteins, and nucleic acids.

Freshman chemistry texts tend to add or expand topics without dropping any. In this edition, pruning has been necessary to keep the depth and breadth of coverage manageable. Although new subjects have been added, such as valence-shell–electron-pair repulsion (Chapter 5), osmotic pressure (Chapter 8), photoelectrolysis (Chapter 11), polymers (Chapter 20), and applications of radioactivity (Chapter 22), a page has been deleted for each one added.

The number of worked-out problems has been increased, and they have been tied more closely to the development in the text. The exercises at the ends of chapters—more than 900 of them at varying levels of difficulty—are all new. As in the second edition, asterisks are used to designate the level of difficulty: easy*, moderate**, hard***. The relative number of one-asterisk problems has been increased, and the relative number of three-asterisk problems has been reduced. Answers to some of the exercises have been included. The remaining answers as well as worked-out solutions are included in a *Student/Instructor Solution Supplement* available from the publisher.

This book is designed for careful study, and a student who makes such effort will be repaid with a comprehensive grasp of chemistry. The book is intended for college students who have had one year of high school chemistry. A knowledge of calculus is not assumed, and none is used. The text is designed mainly for those interested in science or science-related subjects. It is directed toward premedical students, engineers, students of agriculture and biological science, and all those who may have a need for knowledge of chemistry in their careers. It can also be used by students majoring in chemistry.

For students whose high school chemistry preparation is not strong or who simply wish to refresh their memories or review material covered in this text, a *Study Guide* is available from the publisher. It includes a complete glossary of the key terms and concepts, skills to master, step-by-step worked problems covering the skills, and chapter self-tests.

For the preparation of this edition, we wish to express our deep appreciation to reviewers David L. Adams, Robert C. Atkins, Jon M. Bellama, Ellen A. Keiter, and Fred H. Redmore for their perceptive criticism; to the McGraw-Hill staff for their help and guidance; to Ann Lemley for checking the problems; and especially to Christina Fuiman for expertly and cheerfully working her way through typescripts of the several versions that ended up as the chapters that follow.

Michell J. Sienko
Robert A. Plane

CHEMISTRY
PRINCIPLES AND APPLICATIONS

Chapter 1

CHEMISTRY AND CHEMICAL NOMENCLATURE

In the irrigation ditches of North Africa and the jungle swamps of Brazil lives a fresh-water snail that is the intermediate host for a debilitating human parasite, the schistosome, or blood-fluke. The disease that results, *schistosomiasis,* is widespread throughout Africa and Brazil. It takes a fearful economic toll because it drains all a person's energy so that just living becomes an ultimate chore. How can this disease be cured? Early diagnosis is important. Chemotherapy, the treatment of disease by chemical agents, has been somewhat effective, but the compounds used, such as those of the element antimony, are exceedingly toxic, and there are frequently unwanted side effects. One of the desperate calls currently sounded by the World Health Organization is for a systematic search for a cure to this disease. Specifically, what is needed is a replacement of older, dangerous drugs by a safe, effective remedy.

On the other side of the world, in the sunlit deserts of Arizona, experts gather for a symposium to discuss the utilization of solar energy. We all know that the time of fossil fuels, petroleum and natural gas, is running out. New sources of energy have to be developed. Can we perhaps design an electrochemical cell that will directly use solar energy to split water into its component elements, hydrogen and oxygen, and thus provide the cleanest fuel of all, hydrogen? What materials might be used? What new compounds should be invented to exploit this mode of energy conversion and make it most efficient?

These are but two of the problems that chemistry is currently involved in. They may appear to be very far from questions such as why the bond angle in the water molecule is only 104.5° or why potassium chloride spontaneously dissolves in water even though the process is energetically unfavorable, but the experience of the past tells us that knowledge of fundamentals is generally the key to successful solution of an applied problem.

In this chapter we concern ourselves with an introduction to the language of chemistry. We will need it before undertaking the systematic study of chemistry in later chapters. Precision in language is a distinctive feature of science. Although much of the material in these sections will be familiar to most readers, importance of its mastery cannot be overemphasized.

1.1 CHEMISTRY

Chemistry concerns itself with the composition of matter, the changes that matter undergoes, and the relation between changes in composition and changes in energy. The scope of the subject is vast, as vast as the universe.

Astrochemistry relies for its information about the composition of the stars on analysis of the radiation that comes to us through space and time. Even for the closest star, the sun, it takes 8 min for the information to reach us. What changes have occurred in those 8 min? How much greater are the changes for the next closest star where it takes 4 years for the light to reach us? How much do we really know at the present time about other stars, so far distant that light, traveling at its constant speed of 3×10^{10} cm/sec through the centuries, has barely had time to reach us? Let us confine our study of chemistry to the earth.

Geochemistry concerns itself with the composition of the earth. Much can be learned by gathering pebbles on the surface of the earth and analyzing them for chemical content, but what about the earth's interior? Most people believe that the core, almost as inaccessible as the farthest star, is composed of iron and nickel. How do we know? We get information indirectly. We observe echo waves of earthquakes that pass through the earth's interior; we observe the behavior of liquid iron and nickel in the laboratory; we calculate how properties are affected by the enormous pressures under the earth's surface. We proceed indirectly, and eventually we are led to a belief that we all accept because our hypothesis, our suggested assumption, turns out to predict so many things that are consistent with observation.

This is generally the way science proceeds—collection of facts, interpretation of facts in terms of a proposed model, and then acceptance of the model as "fact." Few of us realize how great a portion of our thinking is not really fact but only a model that was suggested to explain some facts. Consider the atom. For most chemists, it is a real entity that is the building block of all materials. Yet no one has ever seen an atom. Even the most sophisticated, modern, high-resolution electron microscope gives only indirect evidence of an atom's existence. Why then do we have so much faith in it? The answer is that the body of knowledge we can account for is so vast and so elegantly simplified by thinking in terms of atoms that we accept the atom as a physical actuality.

Chemistry is traditionally divided into four broad areas: analytical, physical, organic, and inorganic. *Analytical chemistry* emphasizes the techniques that are used to find out about the composition of matter: What are the atoms that make up a given substance? How are they arranged in clusters? What is their relative arrangement in space? *Physical chemistry* is concerned rather with the general laws that govern the behavior of all matter: What governs the particular arrangements of atoms in space? How do these generally change with time? What are the great driving forces that generate the particular material or combination of materials that we have at hand? *Organic chemistry* is special only in that it concentrates on the compounds of the element carbon. *Inorganic chemistry* has for its province the compounds of all the other 100 or so elements.

When chemistry started some 500 years ago in the dim mysterious reaches of alchemy, it was primarily an experimental science. Main interest then was in discovering new recipes: How to extract metals from the earth; how to make medicines from plants; how to convert lead into gold. As the recipes patiently accumulated, chemistry became more abstract. Venturesome thinkers wondered why certain general patterns of observation existed and began to create models of the underlying structure. At present, chemistry stands on the scale of abstraction about midway between biology and physics. It shares with biology the need to systematize an extraordinarily large body of observations on complex systems. It shares with physics the effort to explain observed laws of behavior in terms of a few principles applied to simple particles.

Recently, as with most sciences, chemistry has spread out to overlap neighboring disciplines and thereby produce challenging new areas of interdisciplinary study. *Biochemistry,* concerned with the chemical reactions that occur in living tissues, came from a fusion of biology and chemistry; it has long been recognized as an important adjunct for the study of medicine. *Molecular biology,* which is closely related to biochemistry, concentrates on biological events such as growth and reproduction in terms of molecules in the cell; it came from a fusion of biochemistry and genetics. It is particularly concerned with the replication of giant molecules (macromolecules) in living systems. Toward the other end of the scale of abstraction lies *theoretical chemistry,* where generalized statements of chemical phenomena are subjected to the most powerful methods of mathematics for rigorous solution. Because mathematics is highly abstract and divorces itself from specific materials or reactions, it is capable of giving great insight for predicting the general behavior of chemical substances. On a more practical side, chemistry has merged with engineering to give, for example, *chemical engineering;* it concentrates on the problem of translating a basic discovery in a research laboratory into plant production of a real product in a real economic world. Another example is *materials engineering,* where synthetic and analytical aspects of chemistry have been applied to understand why there are limits to desirable properties in specific materials and thereby lay the basis for rational improvement of these properties. Noteworthy also is a relatively new but rapidly growing field, *solid-state chemistry;* it straddles chemistry and physics and addresses itself to the problem of how physical properties, i.e., those that can be described without reference to interaction with other substances (e.g., electric conductivity), are related to composition and arrangement of atoms in the solid state.

All these interdisciplinary fields have one thing in common. They demand equally comprehensive knowledge of the two disciplines that are being merged. For basis, they all go back to chemistry. The fact that schools of agriculture, medicine, life sciences, engineering, etc., all require knowledge of chemistry underlines its widespread importance.

1.2
ELEMENTS, COMPOUNDS, SOLUTIONS, AND PHASES

One of the most promising areas of current technological research is the development of *composite materials.* These are usually made as mixtures of materials in which a binder, e.g., epoxy resin, has some type of reinforcement, e.g., boron fibers, dispersed through it. The goal is to optimize combinations of high strength, high stiffness, and light weight that are required for applications such as rocket-motor casings and structural components of aircraft. Clearly the properties of a composite are a complex blend of the properties of binder, fibers, and interfacial surface between the two. One can well imagine how difficult it would be to work back from the observed properties of the mixture to deduce the properties of pure boron alone. Yet this is the problem constantly faced in chemistry. We make a new compound; we wish to catalog its properties. How do we know we have a pure compound? How do we know the properties we are measuring are not due to an impurity or, worse yet, to some cooperative effect in which impurity and compound work together to produce an observed effect? The constant fussing of the chemist to establish clearly the distinction between *mixture, compound,* and *phase* stems from a desire to avoid such traps.

First, let us get together on some fundamental terms. *Matter* is anything that has mass and occupies space. It can occur in any of three states: solid, liquid, or gas. The *solid state* is characterized by retention of shape, of form, and of volume, no matter what the shape of the container. The *liquid state* retains volume when a sample is transferred from container to container, but the shape of the sample adjusts itself to take on the shape of the bottom of the container in which the sample is confined. The *gas state* is characterized by retention of neither shape nor volume since the sample expands to fill any container accessible to it. As an example of the foregoing, water is a sample of matter. In the solid state, it exists as *ice;* in the liquid state, as *liquid water;* in the gaseous state, as *water vapor.* Note that, in describing the gaseous state of water, we have not said steam. The reason for this is that the term "steam" is ambiguous. Sometimes it is used to refer to the invisible vapor into which water is converted when heated to the boiling point; more often it is used to describe the visible mist formed by condensation on cooling of water vapor (as at the spout of a teakettle). When terms are ambiguous, it is better to avoid them unless the meaning is not critical.

By *element* we mean one of the 100 or so fundamental chemical substances, composed of atoms of only one kind, into which all matter can be resolved. By *compound* we mean a complex substance, composed of two or more elements, which has the property that no change in composition (i.e., no change in the percent of the different elements) occurs when a sample is partly put through a change of state. Why do we include the restriction as to change of state? Why make such an involved definition? It is necessary if we want to make a clear distinction between a *compound* and a *solution.* Some people differentiate the two by saying *a compound has a fixed composition,* whereas *the composition of a solution is variable.* It is important to note that "fixed" and "variable" are being used here in two contexts: going from sample to sample and going from one state of matter to another. The latter is the more significant. Given a single sample of a compound and a single sample of a solution, both will have fixed compositions. The compound, however, will not change in percent composition when the sample is partly put through a change of state; the solution will. As a specific illustration, when a sample of pure liquid water is partly boiled, the vapor (gas) that is formed consists of 11% by weight hydrogen and 89% by weight oxygen just as does the starting liquid water; this is typical of a compound's behavior. On the contrary, if a 10% salt-90% water solution is partially boiled, the gaseous substance that first forms is practically pure water; i.e., it is much richer in the water component than is the starting water-salt solution. This is typical of a solution's behavior. When a solution starts to go through a change of state, one of the components of the solution generally tends to separate out more than the other.

Another useful but sometimes confused term is *phase*. It applies to the region (or regions) of a sample characterized by the same set of properties. As a simple example we can consider several ice cubes floating in liquid water. All the ice cubes taken together represent the ice phase; all the liquid water represents another phase. This is a two-phase system, no matter what the relative amounts of ice and liquid water are. The distinctive feature about a phase is that properties change when one crosses a *phase boundary,* i.e., when one goes out from one phase into another. Thus, for example, in going from the ice phase into the liquid-water phase there is a change in density from 0.9 g/cm³ (gram per cubic centimeter) to 1.0 g/cm³. Confusion in the use of the word *phase* comes from the fact that occasionally the word *phase* can be substituted for the word *compound* without saying anything wrong. For instance, when a solution of sodium chloride dissolved in water is mixed with a solution of silver nitrate in water, the *compound* that separates out as a white precipitate is silver chloride. One could equally well say the *phase* that separates out is silver chloride. In the former case, the emphasis is on formation of a complex material composed of the elements silver and chlorine; in the latter, emphasis is on a new region that has appeared in which properties are different from those in the starting system.

1.3
SYMBOLS AND FORMULAS

The chemical elements are generally represented by *symbols,* such as H for hydrogen and O for oxygen. A symbol represents not only an element in general but also a specific amount of that element, either one atom or, as described in the next chapter, a standard large number (6×10^{23}) of such atoms. The latter is called a *mole* of atoms. Thus, N stands for nitrogen, one atom of nitrogen, and also for 1 mol of nitrogen atoms.

In many cases the symbol for an element is just the capitalized first letter of the name, for example, P for phosphorus. If several elements have the same initial letter, a second small letter may be included in the symbol, for example, Pt for platinum, Pu for plutonium. In some cases, particularly for elements known since antiquity, the symbols are derived from Latin names, for example, Ag for silver from *argentum,* Au for gold from *aurum,* Cu for copper from *cuprum,* Fe for iron from *ferrum.* A complete list of element names and symbols is given on the inside back cover of this book.

Chemical compounds are represented by *formulas,* such as H_2O, which are combinations of the symbols of the constituent elements. The subscript numbers, 2 for H and 1 (understood) for O, designate the relative numbers of atoms of the different kinds. Again the formula does triple duty: It represents the compound in general, it represents 1 formula-unit of the compound (a formula-unit is the set of atoms shown by the subscripts in the formula), and it represents a standard large number (6×10^{23}) of such formula-units. The term *mole* is used for 6×10^{23} formula-units. Thus, H_2O stands for water, 1 formula-unit of water, and also for 1 mol of water. In some cases, as with water, the formula-unit exists as a separate discrete entity, in which case it may be called a *molecule*. In other cases, as with SiO_2, there is no simple molecule and only the ratio of atoms is indicated.

Chemical formulas are usually determined from experimental analysis of a compound to determine what atoms make up the compound and how the atoms are connected. One technique, which is discussed in more detail in Sec. 2.3, uses weight relations in chemical reactions to fix the ratio of the number of atoms of each kind. This gives us what is called the *empirical formula,* or the *simplest formula;* it tells nothing about how the atoms are joined together. Another technique, which is particularly applicable to organic compounds, is to break the compound apart and analyze for the masses of the various fragments. Computer analysis of the fragments can often tell us which atoms are connected to which and how many atoms of each kind there are in the molecule. The term *molecule* is used to designate the discrete unit which includes all the atoms that are strongly joined together. Thus, we can get not only the *molecular formula* but also some idea of the specific atomic groupings that occur in the molecule. Common groupings that we will frequently encounter are hydroxyl (OH), methyl (CH_3), and ethyl (C_2H_5). The compound methyl alcohol shows the groupings CH_3 and OH and is usually written CH_3OH. There are various other techniques for finding out which atoms are connected to which and how they are arranged in space, but we shall postpone consideration of them until after we have developed more background information.

Names of chemical compounds are usually described as being trivial or systematic. *Trivial names,* such as wood alcohol for CH_3OH, are like nicknames; they generally reflect some characteristic of the compound, and having grown to be accepted through repeated usage, they are likely to be commonly used in informal conversation. Thus, the name "wood alcohol" for CH_3OH reflects the fact that the compound was originally obtained by destructive distillation of wood, i.e., by heating wood in a closed container and collecting the volatile products of the decomposition. *Systematic names,* as the term implies, are more formal; they generally reflect the classification of the compound and often tell something about the atomic arrangement. For example, CH_3OH is called *methanol.* The ending *-ol* indicates that the compound is classified as an alcohol (characterized by OH groups attached to C). The root *methan-* is derived from the parent compound methane, CH_4. Another systematic name for CH_3OH would be hydroxymethane, emphasizing that CH_3OH is derived from CH_4 by replacing an OH group for one of the H's.

The rules for giving systematic names of organic compounds are rather involved, because there are over a million such compounds and it requires some care to spell out the rules so as to cover all possibilities. A brief summary of the rules is given in Appendix A2.2.

Systematic naming of inorganic compounds is a bit simpler. A summary of the rules is given in Appendix A2.1. Here we simply mention a few of the rules that will be most useful for the early part of this book. Compounds composed of but two elements have names derived directly from the elements, with the second element generally given an *-ide* ending. Thus, CaO, for which the trivial name is *lime,* is called calcium oxide. If we want to emphasize the number of atoms of a given element, prefixes such as *mono* (for 1), *di* (for 2), *tri* (for 3), *tetra* (for 4), *penta* (for 5), and *hexa* (for 6) can be used to precede the name of the element. Thus, carbon monoxide (CO) is distinguished from carbon dioxide (CO_2).

Compounds containing more than two elements are usually named to reflect classification of the compound as an acid, a base, or a salt, or to emphasize certain characteristic groupings that commonly occur. We shall discuss acids, bases, and salts in considerable detail later in this text, but here we give a few preliminary definitions. *Acids* are a class of substances that have the following characteristic properties: (1) sour taste, (2) ability to turn litmus red, (3) ability to react with most metals to liberate hydrogen, and (4) ability to neutralize, or destroy, bases. *Bases* are substances that have the following properties: (1) bitter taste, (2) ability to turn litmus blue, and (3) ability to neutralize acids. *Salts* are the products of neutralization reactions between acids and bases. Two hundred years ago, the French chemist Lavoisier suggested that to have acid properties, a compound needs to contain oxygen. A hundred years later, the Swedish chemist Arrhenius argued that to have acid properties, a compound needs to contain hydrogen. The Arrhenius view is more in keeping with modern ideas. However, as we shall see later, it will need to be broadened considerably. In the Arrhenius view, acid properties are attributable to the presence of an easily detachable hydrogen ion, H^+. The superscript + indicates the electric charge of the particle, and the word *ion* describes in general any grouping of one or more atoms that carries a charge. The formulas of acids are generally written to show the detachable hydrogen first. Thus, as common examples we have the following:

HCl	hydrochloric acid, or hydrogen chloride
HNO_3	nitric acid, or hydrogen nitrate
H_2SO_4	sulfuric acid, or dihydrogen sulfate
H_3PO_4	phosphoric acid, or trihydrogen phosphate
$HC_2H_3O_2$	acetic acid, or hydrogen acetate

The first of the names are trivial names; the second, systematic. The systematic names illustrate the special names that go with certain characteristic groupings: nitrate for NO_3^-, sulfate for SO_4^{2-}, phosphate for PO_4^{3-}, and acetate for $C_2H_3O_2^-$.

Bases can be interpreted as owing their characteristic properties to action of the hydroxide ion OH^-. Typical bases are:

NaOH	sodium hydroxide
$Ca(OH)_2$	calcium hydroxide

When an acid such as HCl neutralizes a base such as NaOH, it is believed that the H^+ of the acid combines with the OH^- of the base to form the compound HOH, that is, water. If the water is now boiled off, the Na^+ of the base and the Cl^- of the acid are left to combine with each other as the salt sodium chloride, NaCl. The formulas of salts are generally written with the residue from the parent base first and the residue from the parent acid second. The systematic name is then simply the combination of the names of the two groupings. As a further example, neutralization of H_2SO_4 by NaOH produces the salt Na_2SO_4, for which the name is sodium sulfate.

1.4
CHEMICAL REACTIONS AND CHEMICAL EQUATIONS

The simplest kind of chemical reaction is one in which one chemical element combines with another chemical element to form a compound. In shorthand form, this is usually written as a *chemical equation,* which is simply a symbolic statement of observed chemical facts. As an example, let us consider the reaction of the element zinc with the element sulfur to produce the compound zinc sulfide. We write this as

$$Zn + S \longrightarrow ZnS$$

where the arrow is read as "reacts to give." The symbols $Zn + S$ to the left of the arrow represent *reactants,* i.e., starting materials; the formula ZnS to the right of the arrow represents the *product.* If the direction of the arrow were reversed, the role of starting reactants and final products would be interchanged. Specifically,

$$Zn + S \longleftarrow ZnS$$

stands for a chemical reaction in which the starting material ZnS is decomposed (broken down) into constituent elements Zn and S.

To be valid, a chemical equation needs to satisfy two conditions: It must be consistent with the observed facts, and it must be balanced. The word "balanced" means that atoms must be conserved in a chemical reaction; i.e., each atom on the left side of a chemical equation must also show up on the right side of the equation.

There are systematic ways to go about balancing a chemical equation which we shall consider in later chapters, but the essence of the procedure is to adjust the coefficients of the equation (the numbers that go in front of each symbol and each formula) to ensure that all the atoms are accounted for. It should be emphasized that the *subscripts* in a formula cannot be changed to achieve balance, because that would imply a different chemical species.

Suppose, for example, we are asked to write a balanced equation for the reaction of calcium (Ca) with oxygen (O_2) to form calcium oxide (CaO). We first write

Step (1) $\qquad ?Ca + ?O_2 \longrightarrow ?CaO$

to represent the chemical species involved. We then select coefficients to balance the equation. Specifically, we could note that there are two oxygen atoms to the left of the arrow, and so there should be two to the right. We get these by putting a 2 before the CaO. The equation now reads

Step (2) $\qquad\qquad$ $?Ca + O_2 \longrightarrow 2CaO$

We now note that there are two calcium atoms to the right of the arrow, and so there should be two to the left. We can get these by putting 2 before the Ca. The final result is

Step (3) $\qquad\qquad$ $2Ca + O_2 \longrightarrow 2CaO$

It would have been entirely wrong in step (1) to have changed the subscript of oxygen in CaO from 1 (understood) to 2. This would have given us

$$Ca + O_2 \longrightarrow CaO_2$$

which is a perfectly valid, balanced equation, but *it is not the correct equation for the specific reaction that was asked for.* This is an entirely different chemical reaction.

Equations for reactions between compounds, such as neutralization, follow the same principles. For the reaction between HCl and NaOH the balanced equation is

$$HCl + NaOH \longrightarrow HOH + NaCl$$

This is sometimes called double-substitution, or *metathesis,* since the pieces of the compounds have simply changed partners. For the neutralization between HCl and $Ca(OH)_2$ the balanced equation is

$$2HCl + Ca(OH)_2 \longrightarrow 2HOH + CaCl_2$$

We may ask two questions here: (1) How do we know we should write $CaCl_2$ for the second product? Why not CaCl? (2) How do we know that both hydroxides of the $Ca(OH)_2$ are neutralized? The answer to the first question, as we shall learn in subsequent chapters, is that calcium in a compound occurs almost invariably as a dipositive ion; i.e., it exists as a charged particle carrying an electric charge of +2. Hence, Ca^{2+} needs to be paired with two negative charges to give electrical neutrality. The hydroxide group, OH^-, carries a single negative charge, and so, as in $Ca(OH)_2$, there must be two hydroxides per doubly charged calcium. The situation may perhaps be more clearly understood if one thinks of $Ca(OH)_2$ as $(Ca^{2+})(OH^-)_2$, where the subscript 2 applies to everything inside the parentheses enclosing the OH^-. Similarly, chloride carries a single negative charge, and so, as in $(Ca^{2+})(Cl^-)_2$, there must be two singly negative chlorides per doubly positive calcium. Commonly observed charges for ions should be remembered as they are encountered, since they are useful for checking correctness of formulas. Figure 1.1 gives a brief listing of frequently encountered ions.

The answer to the second question is more difficult. Partial neutralization of $Ca(OH)_2$ can indeed occur, and it is possible to produce a mixed compound of the type $Ca(OH)Cl$ in which only half the hydroxide has been neutralized.

**FIG. 1.1   Names and Symbols of Commonly
Encountered Ions**

Name	Charge	Symbol
Sodium	+1	Na^+
Potassium	+1	K^+
Ammonium	+1	NH_4^+
Silver	+1	Ag^+
Calcium	+2	Ca^{2+}
Magnesium	+2	Mg^{2+}
Zinc	+2	Zn^{2+}
Aluminum	+3	Al^{3+}
Hydroxide	−1	OH^-
Fluoride	−1	F^-
Chloride	−1	Cl^-
Bromide	−1	Br^-
Iodide	−1	I^-
Nitrate	−1	NO_3^-
Oxide	−2	O^{2-}
Sulfide	−2	S^{2-}
Sulfate	−2	SO_4^{2-}
Carbonate	−2	CO_3^{2-}
Phosphate	−3	PO_4^{3-}

For getting this material from HCl and $Ca(OH)_2$, the balanced equation would be

$$HCl + Ca(OH)_2 \longrightarrow HOH + Ca(OH)Cl$$

However, this reaction, where 1 formula-unit of HCl reacts with 1 formula-unit of $Ca(OH)_2$, does not correspond to complete neutralization. It occurs only under rather special conditions, i.e., when the amount of $Ca(OH)_2$ relative to the amount of HCl is severely restricted and when the temperature of the subsequent evaporation is delicately controlled. Under normal conditions of neutralization, $Ca(OH)Cl$ is not formed. That the nature of a chemical product frequently depends on the relative amounts of starting reagents will be made clearer by the discussion in the next chapter. The problem of partial neutralization is considered in greater detail in Chap. 15.

1.5
ENERGY, HEAT, AND TEMPERATURE

When a chemical reaction occurs, there are frequently visual signals that something has happened. Colors may change; gases may evolve; precipitates may form. Less obvious are changes in energy which almost invariably accompany chemical reactions.

Energy is a powerful but abstract physical concept that describes the *ability to do work*.* It is of great public interest at the present time because there is a worldwide shortage of energy and there are problems of safe, efficient, and economical conversion from one form of energy to another. Also there is a tragic imbalance in worldwide distribution: Some countries have too much energy resources for their population, and others have too little.

In its broadest sense, energy can be classified as potential or kinetic. *Potential energy* is that which arises from position; *kinetic energy* is that which arises from motion. It is easier to grasp the concept of kinetic energy since it is simple to visualize the motion of an object, a particle, or any entity having mass. By definition, kinetic energy is equal to $\frac{1}{2}mv^2$, where m is the mass of the moving object and v is its velocity or speed of motion. Potential energy is more subtle. It comes about because there are forces between particles, and it requires work (or work is done for us) when we go from one relative arrangement of particles to another. Let us consider several examples.

A boulder poised on the edge of a cliff is a certain distance from the center of the earth. Corresponding to this, there is a certain amount of *potential energy* in the earth-boulder system. Let the boulder now fall to the foot of the cliff because of the well-known attractive force of gravity. As shown in Fig. 1.2, the earth-boulder system goes to a state of lesser potential energy, since the boulder has moved closer to the center of the earth. It would take work to lift the boulder back to the top of the cliff, the same amount of work that was done when the boulder fell from top to bottom. What happens to the *kinetic energy* in the process of falling? The boulder at the top of the cliff is at rest; its velocity is zero, and so the kinetic energy is zero. As the boulder falls, its velocity increases. Its store of kinetic energy increases as the potential energy of the earth-boulder system decreases. At the foot of the cliff, just before impact,

*Appendix 5 at the back of the book has a discussion of energy and other concepts from physics.

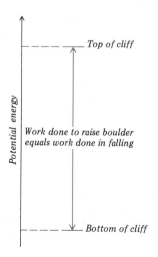

FIG.1.2   Potential energy of earth-boulder system.

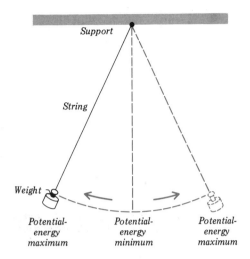

Support

String

Weight

Potential-
energy
maximum

Potential-
energy
minimum

Potential-
energy
maximum

FIG. 1.3 Weight-on-string pendulum showing energy relations.

the kinetic energy is a maximum, just equal in measure to the distance between the dashed lines of Fig. 1.2. At impact, the boulder again comes to rest.
It gives up its kinetic energy by doing work, probably by demolishing some
object at the foot of the cliff. In going from the initial state (boulder at rest on
top of cliff) to the final state (boulder at rest at foot of cliff), the earth-boulder
system has decreased in potential energy. The difference has gone into work
demolishing the object, raising its potential energy by breaking it apart and
raising its kinetic energy by heating up the fragments. We are considering
here an observed general regularity in natural behavior, *the law of conservation
of energy:* The energy of the universe is constant; it simply changes from one
manifestation to another.

As we shall see in Chap. 7, the atoms that make up a solid are not completely at rest but are in constant vibratory motion. Each atom behaves somewhat like a pendulum oscillating back and forth with respect to its neighbors.
Such a pendulum, which is more generally typified by a weight hung from a
string, is a good example of a system showing continual interchange between
kinetic and potential energy. As shown in Fig. 1.3, the potential energy is a
maximum at the ends of the swing, where the weight is farthest from the
center of the earth. This, of course, is where the weight comes to a stop, and
so the kinetic energy is zero. In the middle of the swing, where the weight is
closest to the center of the earth, the potential energy is a minimum, but the
velocity of motion is greatest, and so the kinetic energy is a maximum. At
every point during the swing, the sum (potential energy plus kinetic energy)
stays constant, but the apportionment between the two kinds of energy is
changing continuously. Figure 1.4 shows the atomic analog where, as in
solid NaCl, the Cl^- vibrates back and forth with respect to the neighboring
Na^+. The potential energy is lowest when the Cl^- is midway between its Na^+
neighbors, but it rapidly rises when the Cl^- moves closer to one Na^+ or
another.

Besides vibratory motion of atoms, which is always present in all samples
of matter, atoms can change positions relative to other atoms in a major way

12

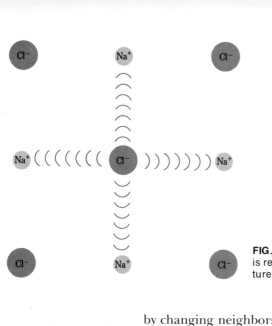

FIG. 1.4 Schematic representation of vibratory motion in NaCl. (What is represented is but a small portion of the repeating solid-state structure.)

by changing neighbors. This is what happens in a chemical reaction. As an example, when zinc reacts with sulfur to form zinc sulfide,

$$Zn + S \longrightarrow ZnS$$

the zinc atoms change from an environment where they are completely surrounded by other zinc atoms to one where they are surrounded by sulfur atoms. Similarly, the sulfur atoms go from a state where they are joined to other sulfur atoms to a state where they are joined to zinc atoms. Clearly we expect a change in the potential energy of the system in going from Zn + S to ZnS. This change in potential energy, which may be up or down, usually shows up as heat, but it may also appear as light, sound, shock waves, or any other of the numerous manifestations of energy. Most commonly, the energy change appears as heat, either *evolved to* the surroundings, in which case the chemical change is said to be *exothermic,* or *absorbed from* the surroundings, in which case the chemical change is said to be *endothermic.* How does one tell the difference? If the change is exothermic, the reaction mixture generally gets warmer; if endothermic, it gets colder. There may be a complication, however, in that some reagents do not react when they are just mixed together. Often the reaction mixture must be heated to get the reaction started. As an example, when powdered zinc is mixed with powdered sulfur, nothing happens. The mixture has to be heated, at which point it takes off and gets quite a bit hotter without input of more heat. The heat that is required to get a reaction going has nothing to do with whether the reaction is exothermic or endothermic. What counts is whether there is *net* evolution or absorption of heat when reactants convert to products. The net heat liberated or absorbed, when referred to standard quantities (moles) of material, is called the *heat of the reaction.*

Temperature is often confused with heat, but the terms are not synonymous. *Heat* is a *quantity of energy,* whereas *temperature* describes the *intensity of heat.* The distinction may be made clear by noting that a burning

13

100°C — *Boiling-water point*

0°C — *Melting-ice point*

FIG. 1.5 Mercury-in-glass thermometer.

match and a bonfire can be at the same temperature, but there is more heat in a bonfire. The difference can also be illustrated by observing that a given quantity of heat will raise the temperature of a teaspoon of water more than it will raise the temperature of a bucket of water.

Temperature can also be defined as a property that fixes the direction of heat flow. Heat always tends to flow from a body at high temperature to a body at low temperature. Temperature can be measured by taking advantage of the fact that most substances expand as they get hot. The mercury *thermometer* is a temperature-measuring device which works because the volume occupied by a given mass of mercury increases as its temperature increases. As shown in Fig. 1.5, the sensing bulb contains a large volume of mercury which on expansion can move into a capillary. The volume of the capillary is much less than the volume of the bulb, so that a small expansion of the mercury causes a large movement of the mercury thread. The *Celsius* scale of temperature, often called the *centigrade* scale, can be reproduced by first placing the thermometer in melting ice and marking the position of the mercury thread as 0°C, and then placing it in boiling water and marking the position as 100°C. The space between is divided into 100 equal parts, assuming the bore of the capillary is constant. The *Fahrenheit* scale differs in that the ice point is 32°F and the boiling point 212°F, with 180 equal divisions between them. To convert from Fahrenheit to Celsius, subtract 32 from the Fahrenheit reading and multiply the result by $\frac{100}{180}$, or $\frac{5}{9}$.

Another common temperature-sensing device is a *thermocouple*. This consists of a junction between two dissimilar metals and has the property that the electric voltage across the junction changes as the temperature changes. Thus, by measuring the voltage, one can easily determine the temperature. Most household thermostats use thermocouples for sensing changes in temperature.

1.6
SCIENTIFIC UNITS

As recommended by the International Committee on Weights and Measures, most of the units used in this text are SI units. The designation SI comes from the French label *Système International*. In the system there are seven base units: meter, kilogram, second, ampere, kelvin, candela,* and mole. Figure 1.6 summarizes the nature of these units.

The *meter,* for which the symbol is *m,* is the unit of length. It was originally defined at the time of the French Revolution as equal to one ten-millionth of the quarter circle around the earth passing through Paris. Standard platinum bars were made of this distance, and when, subsequently, it was found that the earth's quadrant had been erroneously surveyed, the defini-

---

*The *candela,* for which the symbol is *cd,* is the basic unit of light intensity. We do not have much use for this unit in chemistry. It corresponds to one-sixtieth of the luminous intensity per square centimeter of a glowing white-hot solid at the temperature of solidification of platinum. An ordinary candle corresponds roughly to a one-candela light source.

FIG. 1.6 SI Base Units

Name of unit	Physical quantity	Symbol for unit
Meter	Length	m
Kilogram	Mass	kg
Second	Time	s (sec)
Ampere	Electric current	A (amp)
Kelvin	Temperature	K ($°$K)
Candela	Light intensity	cd
Mole	Amount of substance	mol

tion was changed to the distance between two marks on a standard platinum bar preserved in Paris. It had been suggested early that the wavelength of light would provide a more natural standard unit of length, but this suggestion has been adopted only recently. At present, the meter is defined in terms of a specific orange spectral line emitted by an electric discharge in krypton at a specified temperature. One meter is represented by 1,650,763.73 wavelengths of this light. It is equivalent to 39.37 inches, which is about the distance from your outstretched fingertips to the middle of your chest.

The *kilogram,* for which the symbol is *kg,* is the unit of mass.* It was originally supposed to be the mass of 1000 cubic centimeters of water at the temperature of its maximum density, which is 3.98°C. Again, a small error was made in construction of the standard, so that at present a kilogram is defined as the mass of a certain platinum-iridium cylinder kept at the International Bureau of Weights and Measures near Paris and very nearly equal to 1000 cubic centimeters of water at the temperature of its maximum density. One kilogram is equivalent to 2.2 pounds and is roughly the mass of two quarts of milk.

The *second* is the fundamental unit of time. It was originally one-sixtieth of a minute or 1/86,400th of the mean solar day, but it is now defined as the duration of a specific number of pulses (9,192,631,770) involving transition between two energy levels of a particular cesium atom. The recommended symbol for second is *s,* but *sec* is often used instead and will be used in this book.

The *ampere* is the unit of electric current. It is defined as the intensity of current which, when maintained in two parallel straight wires one meter apart in a vacuum, produces a force between the wires equal to 2×10^{-7} newton per meter of length. (A newton, which is the force needed to accelerate one kilogram by one meter per second per second, is approximately the force exerted by an apple in the earth's gravity.) The recommended symbol for ampere is

---

*Although often used interchangeably by nonscientists, mass is not precisely the same as weight. Mass is a quantitative measure of the intrinsic inertial properties of an object, i.e., the tendency of an object to remain at rest if stationary or to continue in motion if already moving. Weight, on the other hand, is a measure of the attractive force that a mass experiences in a particular gravitational field. Because the force of gravity can change (e.g., it goes to zero in an orbiting satellite), the weight of an object is not constant. However, the mass of an object is constant and can be determined by comparing its weight with that of a known mass. Chemists often use mass and weight interchangeably because in the typical chemistry laboratory gravity does not change.

A, but *amp* is often used instead. An ampere is about the amount of electric current it takes to light a 100-watt light bulb.

The *kelvin* is the unit of thermodynamic temperature. It has the same size as the Celsius degree mentioned in Sec. 1.5, but it is more rigorously defined as 1/273.16 of the thermodynamic temperature at which liquid water, ice, and water vapor coexist (so-called "triple point"). The international committee has recommended that kelvin be symbolized by *K*, not by °*K* as one might expect from comparison with other temperature scales. There is a great deal of resistance to accepting the committee's recommendation. We will use °*K* in this book.

The *mole*, for which the symbol is *mol*, is the SI unit for the amount of a substance. It is the amount of material which has as many elementary entities of specified composition as there are atoms in 0.012 kg of carbon 12, the most abundant variety of carbon atom. The number of atoms in a mole of atoms is referred to as the *Avogadro number*. Its numerical value is 6.0222×10^{23}.

Decimal fractions or multiples of units are indicated by prefixes as in the following listing:

Fraction	Prefix	Symbol	Multiple	Prefix	Symbol
10^{-1}	Deci-	d	10^{1}	Deka-	da
10^{-2}	Centi-	c	10^{2}	Hecto-	h
10^{-3}	Milli-	m	10^{3}	Kilo-	k
10^{-6}	Micro-	μ	10^{6}	Mega-	M
10^{-9}	Nano-	n	10^{9}	Giga-	G
10^{-12}	Pico-	p	10^{12}	Tera-	T
10^{-15}	Femto-	f			
10^{-18}	Atto-	a			

To appreciate how these work, we consider fractions and multiples of a meter. A tenth of a meter, corresponding to 10^{-1}, is a *decimeter*, for which the symbol is dm. A hundredth of a meter, corresponding to 10^{-2}, is a *centimeter* (cm); a thousandth, 10^{-3}, is a *millimeter* (mm); etc. One *micrometer* (μm), equal to 10^{-6} m, is often called a *micron*. A *kilometer* (km), which is equal to 10^{3} m, is about equal to five-eighths of a mile.

Besides the SI base units, there are so-called "derived units" which are combinations of base units. Simple examples would be the *square meter*, m², which is the unit for area, and the *cubic meter*, m³, which is the unit for volume. A special name, *liter*, for which the symbol is 1, is given to one-thousandth of a cubic meter, 10^{-3} m³. One *milliliter*, symbol ml, is the same as one cubic centimeter, cm³. Some derived units have special names which have been chosen to honor scientists who contributed greatly to increasing the understanding of the physical quantities involved. In such cases the name of the unit is always written lowercase, whereas the symbol or abbreviation uses a capital letter. Thus, the derived unit for force is the *newton*, after Sir Isaac Newton (1642–1727), and has the symbol N; the derived unit for energy is the *joule,* after James Prescott Joule (1818–1889), and has the symbol J. Figure 1.7 summarizes some derived units and shows how they are defined in terms of base units.

In reading the last column of Fig. 1.7, it is important to note that negative exponents mean division by that unit to the power indicated. Thus, for newton we have kg m sec^{-2}, which means kilogram meter per second squared, or

$$\text{Newton} = \text{kg m sec}^{-2} = \frac{\text{kg m}}{\text{sec}^2} = \frac{(\text{kilogram})(\text{meter})}{\text{second}^2}$$

The rationale for this set of units to these powers is as follows: According to the simple laws of physics, $F = ma,$ which states that the force F is equal to the mass m times the acceleration $a.$ For unit mass, $m = 1$ kilogram, and for unit acceleration, $a = 1$ meter per second per second. This gives

$$\text{F (in newtons)} = (m)(a) = (\text{kilogram})\left(\frac{\text{meter/second}}{\text{second}}\right) = \text{kg m sec}^{-2}$$

As will be noted in the following section, careful attention to units is a necessity if we want to make successful use of the problem-solving technique known as *dimensional analysis.* In particular, cancellation of units between numerator and denominator is often required. In Fig. 1.7, we have used positive and negative exponents on the units rather than a diagonal line (/) to distinguish what is in the numerator from what is in the denominator of a fraction. The reason for this is that the diagonal-line notation sometimes leads to confusion. For example, g/cm/sec can be interpreted to mean grams per centimeter divided by seconds or to mean grams divided by centimeters per second. If the former is intended, it would be better written g cm^{-1} sec^{-1}; if the latter,

FIG. 1.7 SI Derived Units

Name of unit	Physical quantity	Symbol for unit	Definition in base units
Newton	Force	N	kg m sec^{-2}
Pascal	Pressure	Pa	N m^{-2} = kg m^{-1} sec^{-2}
Joule	Energy	J	kg m^2 sec^{-2}
Watt	Power	W	J sec^{-1} = kg m^2 sec^{-3}
Coulomb	Electric charge	C	A sec
Volt	Electrical potential difference	V	J A^{-1} sec^{-1} = kg m^2 sec^{-3} A^{-1}
Ohm	Electric resistance	Ω	V A^{-1} = kg m^2 sec^{-3} A^{-2}
Siemens	Electric conductance	S	Ω^{-1} = A V^{-1} = sec^3 A^2 kg^{-1} m^{-2}
Farad	Electric capacitance	F	A sec V^{-1} = A^2 sec^4 kg^{-1} m^{-2}
Hertz	Frequency	Hz	sec^{-1} (cycle per second)

g cm^{-1} sec. In this text we use diagonal-line notation when only two quantities are involved; otherwise, we use negative exponents.

The other important entry in Fig. 1.7 is the third one, joule (symbolized by J) for the unit of energy. It is defined as kg m^2 sec^{-2}, which comes about as follows: Energy can be measured as work. Work equals force times the distance through which the force operates. Hence, we expend 1 unit of energy (1 J) by working a unit of force (1 N) through a unit distance (1 m). As noted above, 1 N = 1 kg m sec^{-2} so that we can write

$$1\ J = \text{(force) (distance)}$$
$$= (1\ N)\ (1\ m)$$
$$= \left(\frac{1\ \text{kg m}}{\text{sec}^2}\right) (m) = \text{kg m}^2\,\text{sec}^{-2}$$

As a rough illustration, 1 J amounts to about $\frac{1}{2000}$ of the energy liberated in the burning of a match.

Until recently, chemists traditionally measured energy in units of calorie, for which the symbol is *cal*. One *calorie* was defined as the energy required to heat a gram of water by 1°C—more precisely, from 14.5 to 15.5°C. The conversion factors are

$$1\ \text{cal} = 4.184\ J$$
$$1\ J = 0.2390\ \text{cal}$$

To add to the confusion, there is another Calorie, the so-called "big calorie" used in nutrition. One nutritional Calorie equals 1000 small calories:

$$1\ \text{Calorie (nutrition)} = 1000\ \text{cal} = 1\ \text{kcal}$$

The International Union of Pure and Applied Chemistry (IUPAC) has recommended that the calorie be abandoned as a unit of energy, but the calorie refuses to die. In this text, we will use joule as the unit of energy, but the reader should be prepared to encounter either joules or calories.

To complete this section on units, we need to recognize that there are certain decimal fractions and multiples of SI units, having special names, which do not belong to the International System of Units. Their use is to be *progressively discouraged*. Among such are the following:

Name of unit	Physical quantity	Symbol for unit	Definition of unit
Angstrom	Length	Å	10^{-10} m = 10^{-8} cm
Dyne	Force	dyn	10^{-5} N
Bar	Pressure	bar	10^{5} N/m^2
Erg	Energy	erg	10^{-7} J

The first of these units, the angstrom, is a favorite of chemists because it is about the size of an atom. Atomic sizes and distances between atoms are frequently quoted in angstrom units. In SI units, we are discouraged from talking in terms of angstroms, and we are supposed to use *nanometers* (nm) or *picometers* (pm). Given that 1 nm = 10^{-9} m and 1 pm = 10^{-12} m, we can write

$$1 \text{ Å} = 0.10 \text{ nm} = 100 \text{ pm}$$

$$1 \text{ nm} = 10 \text{ Å} \qquad 1 \text{ pm} = 0.01 \text{ Å}$$

In this text, we shall mainly use nanometers for atomic dimensions. Hence, the radius of the sodium ion, Na^+, will be given as 0.097 nm, not as 0.97 Å. We could just as well use picometers, in which case the radius would be 97 pm.

Besides the above units, there are other frequently encountered units not in the International System which can now be defined exactly in terms of SI units. These units, which are supposed to be abandoned as quickly as possible, include the following:

Name of unit	Physical quantity	Symbol for unit	Definition of unit
Inch	Length	in	2.54×10^{-2} m
Pound	Mass	lb	0.453502 kg
Atmosphere	Pressure	atm	101,325 N/m²
Torr	Pressure	torr	(101,325/760) N/m²
Millimeter of mercury	Pressure	mmHg	$13.5951 \times 980.665 \times 10^{-2}$ N/m²
Calorie	Energy	cal	4.184 J

Some of these units, such as the atmosphere, are showing great ability to survive, and their continued use has been sanctioned for a limited time.

1.7
DIMENSIONAL ANALYSIS

Dimensional analysis is a problem-solving technique in which the units used to express physical quantity measurements are monitored to test whether physical quantities have been put together correctly to produce a desired result. So, for example, to calculate the mass in grams of a given sample from its given density (in grams per cubic centimeter) and its given volume (in cubic centimeters), dimensional analysis says that

$$\text{Mass (g)} = \text{density} \left(\frac{g}{cm^3} \right) \times \text{volume} (cm^3) = g$$

Since the cubic centimeters in numerator and denominator cancel on the right side of the equation, the two sides of the equation have the same units—an obvious requisite for equality. It would be wrong to calculate mass by dividing density by volume since dimensional analysis would then show

$$\frac{\text{Density (g/cm}^3)}{\text{Volume (cm}^3)} = \frac{\text{g}}{\text{cm}^6} \neq \text{mass (g)}$$

(The symbol $\neq$ means "is not equal to.")

As another example, suppose you need to calculate an amount of material (in moles), given concentration (in moles per liter) and volume (in liters). The correct way to do this would be as follows:

$$\text{Amount of material} = \text{concentration} \times \text{volume}$$

$$\text{mol} = \left(\frac{\text{mol}}{\cancel{\text{liter}}}\right) (\cancel{\text{liter}})$$

If, by accident, we should write

$$\text{Amount of material} = \frac{\text{concentration}}{\text{volume}} \qquad \textit{wrong!}$$

then we could easily prove it wrong by writing in the dimensions:

$$\text{mol} \neq \frac{\text{mol/liter}}{\text{liter}} = \frac{\text{mol}}{\text{liter}^2}$$

Dimensional analysis is a very powerful tool. It can be used to check equations, since a necessary condition for the correctness of any equation is that the two sides have the same dimensions. Thus, the famous equation that describes the state of a gas

$$PV = nRT$$

is valid if P is in atmospheres, V is in liters, n is in number of moles, R is in **liter atmospheres** per mole degree, and T is in degrees. Dimensionally, this would look as follows:

$$(\text{atm})(\text{liter}) = (\cancel{\text{mol}})\left(\frac{\text{liter atm}}{\cancel{\text{mol}}\,\cancel{\text{deg}}}\right)(\cancel{\text{deg}})$$

If an equation does not check dimensionally, then either the quantities are not being combined correctly or the units used are not self-consistent.

Another powerful use of dimensional analysis is in conversion of units. If, for example, we want to convert miles per hour to centimeters per second, we could proceed as follows:

$$1\frac{\text{mi}}{\text{h}} = \frac{\text{mi}\,(?\,\text{cm/mi})}{\text{h}\,(?\,\text{sec/h})} = ?\,\text{cm/sec}$$

As can be seen, in the numerator we want to multiply miles by centimeters per mile to get centimeters; in the denominator we want to multiply hours by seconds per hour to get seconds. First we find out how many centimeters there are per mile:

$$?\,\frac{\text{cm}}{\text{mi}} = \left(100\,\frac{\text{cm}}{\cancel{\text{m}}}\right)\left(\frac{1000\,\cancel{\text{m}}}{1\,\cancel{\text{km}}}\right)\left(\frac{1\,\cancel{\text{km}}}{0.621\,\text{mi}}\right)$$

$$= 1.61 \times 10^5\,\text{cm/mi}$$

Then we find out how many seconds there are per hour:

$$? \; \frac{\text{sec}}{\text{h}} = \left(\frac{60 \; \text{sec}}{1 \; \text{min}}\right)\left(\frac{60 \; \text{min}}{1 \; \text{h}}\right)$$

$$= 3.60 \times 10^3 \; \text{sec/h}$$

Finally, by putting these together, we get

$$1 \; \frac{\text{mi}}{\text{h}} = \frac{\text{mi}}{\text{h}} \; \frac{(1.61 \times 10^5 \, \text{cm/mi})}{(3600 \, \text{sec/h})}$$

$$= 44.7 \; \text{cm/sec}$$

In words, given the speed in miles per hour, we multiply by 44.7 to get centimeters per second.

We shall use dimensional analysis frequently throughout this book. The key to its successful use is to be careful to write out all the units and keep straight what is in the numerator and what is in the denominator.

Although not explicitly stated above, a measured physical quantity consists of two parts: a numerical coefficient and a set of units. Either is meaningless without the other. Thus, it makes no sense to say that the density of gold is 19.3 or that it is grams per cubic centimeter; the correct statement is that the density of gold is 19.3 g/cm³. In carrying out the multiplications or divisions as indicated above, the numerical coefficients are multiplied or divided just as are the units.

Finally, a word of caution needs to be added. There are some quantities that do not have units; they are dimensionless. An example is *specific gravity*, which is defined as the ratio of the density of a material to the density of water. Because the density of water is very nearly 1.0 g/cm³, division by 1.0 g/cm³ gets rid of the units without changing the numerical coefficient. Thus we have for the specific gravity of gold:

$$\frac{\text{Density of gold}}{\text{Density of water}} = \frac{19.3 \, \text{g/cm}^3}{1.0 \, \text{g/cm}^3} = 19.3$$

1.8
PRECISION AND ACCURACY

To the nonscientist the words "precision" and "accuracy" seem to mean the same thing. There is, however, a subtle difference. *Accuracy* refers to conformity to the truth, no matter how elegant or fancily marked the measuring instrument. If, for example, the measuring scale is put on wrong, it will be impossible to get accurate readings from it. *Precision,* on the other hand, refers to the ability to reproduce a reading or measurement from one trial to the next, relying only on the fineness with which the measuring scales have been put on the instrument and ignoring whether the scale conforms to the truth. The following example may help to illustrate the difference between the two terms: Suppose we have two thermometers, A and B, which we are to use to measure the boiling point of water. At normal atmospheric pressure, pure

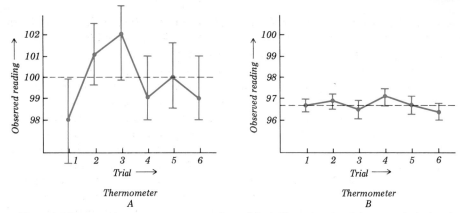

FIG. 1.8 Difference between accuracy and precision. Thermometer A is accurate but not precise; thermometer B is precise but not accurate. (The vertical bar through each data point indicates the uncertainty in each measurement.)

water boils at 100.00°C, and so we have in this case a known quantity that we are trying to measure. Let us say that thermometer A is graduated with markings 10 degrees apart. The best we can do in taking a reading is to get within 10 degrees and make a guess as to the next subdivision. In a typical run of six successive readings we might get the following: 98°, 101°, 102°, 99°, 100°, 99°. The precision of these readings is not very good. They can be off by a couple of degrees, too high or two low, i.e., ±2°. However, the average of the six readings is 100°; therefore, you report the temperature as 100 ± 2°C. You have good accuracy, poor precision.

Let us now look at thermometer B. It is very elegant, with degree markings every degree, and so you can read the temperature to the nearest degree and even estimate the next decimal place to the nearest tenth of a degree. (Unfortunately, unknown to you, the fancy markings were put on in the wrong place, too high up on the thermometer stem.) You take six successive readings. You observe: 96.6°, 96.7°, 96.5°, 96.7°, 96.6°, 96.5°. The average is 96.6°. You report the temperature as 96.6±0.1°C. You have poor accuracy, good precision.

As shown in Fig. 1.8, the second thermometer is actually a better thermometer, since it is capable of measuring to a smaller deviation. However, it is flawed by incorrect calibration, i.e., checking of the markings to see how close they conform to the truth. We could, for example, check the calibration of thermometer B by putting it in water known to be at 100°C and noting that 3.4°C has to be added to any observed reading to get a true value. With this correction, all the readings of thermometer B would be not only more accurate but also more precise than the readings on thermometer A.

Generally, when we buy an instrument with a scale on it, we assume the scale has been accurately placed. Unfortunately, this is not always so. Ordinary thermometers, for example, can often be off by 1°C. The only sure way to tell is by calibrating against a known temperature. Fever thermometers for use on humans, however, are surprisingly accurate. Most of them have been calibrated by putting on the markings so that 37.0°C (or 98.6°F) matches a known temperature bath of average body temperature.

The above discussion illustrates another important point encountered in the handling of physical measurements: The digits used to record the measurement should conform to the precision of the measuring instrument. The convention normally followed is that all the digits are assumed to be known with certainty (assuming the measuring scale has been put on with complete accuracy) except for the last digit, which is assumed to be uncertain to the extent of ±1 in that decimal place. Thus, if a mass of sodium nitrate is reported to be 136.2 g, then it is assumed to be certainly between 136 and 137 and approximately two-tenths of the way between. The implication is that the mass has been determined on a balance which "weighs to the nearest tenth of a gram," which means the decimal place for grams is known with certainty, but the decimal place corresponding to tenths of a gram has been estimated. If, on the other hand, the mass of the same sample has been reported as 136.2684 g, then we infer that it has been measured on a much more precise instrument, that the mass is greater than 136.268 g, less than 136.269 g, and about four-tenths of the way between 136.268 and 136.269 g.

Numbers which express the result of a measurement such that all digits except the last one are known with certainty are known as *significant figures*. In the measurement 136.2 g, there are four significant figures: the 1, 3, and 6 are known with certainty, and the 2 is in doubt. In the measurement 136.2684 g, there are seven significant figures, all except the 4 being known with certainty.

Understanding significant figures is important in calculations involving measured quantities. Consider, for example, the experiment illustrated in Fig. 1.9. We have two graduated cylinders: The large one on the left, calibrated with divisions 10 cm³ apart, contains a volume of water which we estimate to be equal to 168 cm³. The meniscus is certainly between the 160- and the 170-cm³ marks, and it is estimated to be about eight-tenths of the way from 160 to 170 cm³. The volume reading, $V_1 = 168$ cm³, has three significant figures. The 1 and 6 are known with certainty; the 8 is in doubt. The middle cylinder, a small one, is calibrated with divisions 1 cm³ apart. We measure the volume of water contained in it and find it to be 2.6 cm³. The meniscus lies be-

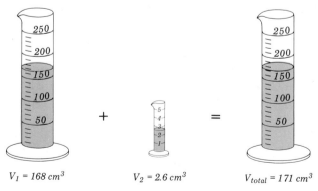

FIG. 1.9 Addition of volumes of water to show limitation of significant figures.

$V_1 = 168$ cm³ $V_2 = 2.6$ cm³ $V_{total} = 171$ cm³

tween the 2- and 3-cm³ markings and is estimated to be about six-tenths of the way from 2 to 3. The volume reading, $V_2 = 2.6$ cm³, is known to two significant figures, the 2 being certain and the 6 in doubt. Now suppose we pour volume V_2 into V_1. What will be the total volume? The first impulse is to set up the problem as a simple addition:

$$
\begin{aligned}
V_1 &= 168 \quad \text{cm}^3 \\
+ V_2 &= \quad 2.6 \text{ cm}^3 \\
\hline
\text{Total} \quad V &= 170.6 \text{ cm}^3
\end{aligned}
$$

However, the answer is wrong. It implies more information about the total volume than we really have. It implies that in the total we know the digits 1, 7, and 0 with certainty, and that only the 6 is in doubt, in other words, that we know the final answer to four significant figures. That this is not right can be seen by setting up the problem again and underlining the doubtful digits:

$$
\begin{aligned}
V_1 &= 16\underline{8} \quad \text{cm}^3 \\
+ V_2 &= \quad 2.\underline{6} \text{ cm}^3 \\
\hline
\text{Total} \quad V &= 17\underline{0}.\underline{6} \text{ cm}^3
\end{aligned}
$$

This clearly shows that, because the 8 is doubtful, anything added to it gives a doubtful result in that decimal place. Hence, in the total the 0 is a doubtful digit, as is the 6. The convention is that only one doubtful digit is to be retained, corresponding to the first decimal column in which there is uncertainty. All digits to the right of that are doubtful and should be discarded. So we throw away the 6, stepping up the preceding digit from 0 to 1. This is in accord with the general convention that dropping digits greater than 5 increases the preceding decimal place by one unit. Dropping digits less than 5 has no effect on the preceding digit. If we drop a 5, then the commonly accepted practice is to step up the preceding digit if it is odd, leave it alone if it is even. This process is known as *rounding off*. The number $17\underline{0}.\underline{6}$ is properly rounded off to $17\underline{1}$. The number $17\underline{0}.\underline{4}$ would be rounded off to $17\underline{0}$.

The above exercise illustrates an important rule for handling physical quantities. When physical quantities *are added together (or subtracted), the sum (or difference) should be expressed only to the first doubtful decimal place*. All digits to the right of that should be discarded.

EXAMPLE 1

What is the sum of 1.562 cm and 3.08 cm?

Solution

$$
\begin{aligned}
1.56\underline{2} \text{ cm} \\
+ 3.0\underline{8} \quad \text{cm} \\
\hline
4.6\underline{4}\underline{2} \text{ cm}
\end{aligned}
$$

Sum equals 4.64 cm

What do we do about multiplication and division? Here the rule is somewhat simpler. *The product or quotient should contain no more significant figures than the least number of significant figures in any number going into the multiplication or division calculation.* For example, in calculating the area of a rectangle that is 8.3 cm on one side and 4.358 cm on another, the product should have no more than two significant figures:

$$8.3 \text{ cm} \times 4.358 \text{ cm} = 36.17 \text{ cm}^2$$
(2 significant   (4 significant   (2 significant
figures)   figures)   figures)

The area should be reported as 36 cm², not 36.17 cm². The rationale for this rule is that the quantity 8.3 implies 8 is known with certainty, the 3 being in doubt. Allowing ± 1 leeway in the 3, we know the number 8.3 to about 0.1 part in 8.3, or 1 part in 83, or approximately to 1 percent precision. No matter what we multiply this number 8.3 by, the product cannot be known better than to about 1 percent, i.e., to two significant figures.

Zeros often cause special difficulties since they can be used to serve two functions: (1) to indicate that a particular decimal place has been measured and found to be zero; (2) to fill in a decimal place as a placeholder so that the decimal point can be properly set. An illustration of the first function is in the quantity 1.003 g. An illustration of the second function is in the quantity 1300 g. In the first quantity, 1.00<u>3</u> g, the zeros are counted as significant figures; these decimal places are known with certainty to be represented by the digits zero, since the convention is to retain only one doubtful digit, namely, the 3. In the second quantity, 1300 g, we do not really know what the zeros mean. They could mean that those decimal places were examined and truly found to be represented by the digits zero, or the zeros could simply be there to indicate that the order of magnitude of the quantity was 1300 g, not 130 g, nor 13 g. Zeros at the end of a number have this ambiguity, and so it is best to avoid using them. The way to do this is to use *scientific notation*, which means expressing numbers as powers of 10, usually with coefficients between 1 and 10. (See Appendix A4.1.) For example, 1300 g can be written in several ways, depending on what information is being conveyed:

$$1.300 \times 10^3 \quad \text{means} \quad 1300 \pm 1$$
$$1.30 \ \times 10^3 \quad \text{means} \quad 1300 \pm 10$$
$$1.3 \ \ \times 10^3 \quad \text{means} \quad 1300 \pm 100$$

The first number 1.300×10^3 has four significant figures, 1.30×10^3 has three significant figures, and 1.3×10^3 has two. The exponential part 10^3 simply gives the order of magnitude of the quantity; it has nothing to do with significant figures. The burden of conveying information about significant figures is in the coefficient that precedes the multiplier 10^3. Since zeros written after a decimal point are not needed to indicate where the decimal point goes, i.e., they are not placeholders, the only reason for writing them to the right of a decimal is to indicate information about the digit value of that decimal place. Hence, zeros to the right of a decimal point are counted as significant figures even though they may be at the end of a number.

There is another way to escape the ambiguity of zeros at the end of a number and that is to change the scale of the unit. For example, 1300 g known to four significant figures can be written 1.300 kg. Similarly, 1300 g known to two significant figures can be written 1.3 kg.

A final word needs to be said about *exact* numbers that are not the result of measurement. For example, the atomic weight of the carbon-12 isotope is set as 12 amu (atomic mass units). This number 12 has complete certainty and can be written as 12.00000 ad infinitum. If such an exact number is multiplied by another number, it itself is regarded to have an infinite number of significant figures, and the ultimate accuracy of the product is limited by the other number.

EXAMPLE 2

What is the total mass of 6.0222×10^{23} carbon-12 atoms?

Solution

$$(6.0222 \times 10^{23} \text{ atoms}) \left(12 \, \frac{\text{amu}}{\text{atom}}\right) = 7.2266 \times 10^{24} \text{ amu}$$

(5 significant figures)   (Infinite number of significant figures)   (5 significant figures)

Exact numbers occasionally arise even from measurement, as in counting discrete objects. For example, the number of men on a football team is exactly 11, not 11.62 or some other fractional number.

Important Concepts

fields of chemistry	symbols	kinetic vs. potential	newton
elements	formulas	heat	joule
compounds	molecule	temperature	dimensional analysis
mixtures	names of compounds	exothermic vs. endothermic	precision vs. accuracy
phase	energy	SI units	significant figures

Exercises

Here and at the end of succeeding chapters are given questions and problems designed to test and amplify your mastery of the foregoing material. Each exercise is marked as to its level of difficulty: *easy, **moderate, ***hard. Answers are provided for some but not all the problems. Special care should be given to understanding each step toward the final numerical answer. Blind substitution in a formula is not encouraged as a way to develop mastery of fundamental concepts. Needed constants and conversion factors are given in the appendix section of this book or can be found in most handbooks of chemistry and physics.

1.1 Composition of matter If it takes so very long for light to reach the earth from distant stars, how might one use the analysis of such light to tell how the composition of stars changes with time?

1.2 Composition of earth's crust Suppose you go into a Natural History Museum and run a chemical analysis on all the mineral specimens in that museum. Discuss the validity of using the result as a fair estimate of the composition of the earth's crust.

*1.3 **Fields of chemistry** If you were to select one characteristic to distinguish each area of chemistry—analytical, physical, organic, inorganic, or theoretical—what phrase would you use to describe each?

*1.4 **Compound** What is wrong with defining a compound as a material that has complex chemical composition?

*1.5 **Phase** What is the difference between a phase and a compound? Under what circumstances might you be correct to substitute the word *phase* for *compound*?

*1.6 **Symbols** Suppose I have just discovered a new element which I choose to call cornellium. What symbol should I pick for this new element? Indicate three things that the symbol would stand for.

*1.7 **Formulas** Write the simplest formula for each of the following compounds:

a Two atoms of nitrogen per one of oxygen

b Two atoms of nitrogen per two of oxygen
c Two atoms of nitrogen per three of oxygen
d Two atoms of nitrogen per four of oxygen

*1.8 **Nomenclature** Name each of the following compounds: (*a*) Na_2O, (*b*) $CaBr_2$, (*c*) Al_2S_3, (*d*) P_4S_3, (*e*) SF_6.

*1.9 **Nomenclature** Distinguish between systematic and trivial names for each of the following: CH_3OH, CaO, H_2O, H_2SO_4.

*1.10 **Acids, bases, salts** Classify each of the following as an acid, base, or salt: HF, KOH, H_2SO_3, $BaSO_3$, $Ba(OH)_2$, H_3PO_4, KNO_3.

*1.11 **Chemical equations** Write balanced chemical equations for each of the following reactions:

a N_2 plus H_2 to give NH_3
b NH_3 plus O_2 to give NO plus H_2O
c NO plus O_2 to give NO_2
d NO_2 plus O_2 to give N_2O_5
e N_2O_5 plus H_2O to give HNO_3

*1.12 **Chemical equations** Write balanced chemical equations for the complete neutralization reaction in each of the following:

a $Ba(OH)_2$ plus HCl
b $Ba(OH)_2$ plus H_2SO_4
c $Ba(OH)_2$ plus H_3PO_4

1.13 **Energy What objection might you raise to defining potential energy as "stored energy"?

***1.14 **Energy** What is the difference between potential energy and kinetic energy? Trace the interconversion of one to the other in the following sequence: a hydrogen atom attached to the hydroxy part of a methanol molecule (CH_3OH) is pulled out beyond its normal distance and then let go to be attracted back to the molecule.

1.15 **Potential energy Why is it strictly arbitrary as to where, on a potential-energy scale, the zero is placed? Show that the zero on a kinetic-energy scale is also somewhat arbitrary.

*1.16 **Kinetic energy** Which has more kinetic energy: 1 g moving at 1 cm/sec or 0.5 g moving at 2 cm/sec? What should one do to the velocity to double the kinetic energy of a moving mass? to triple it?

**1.17 Potential and kinetic energy On the same axes make a graph showing how (a) the potential energy and (b) the kinetic energy change as a pendulum oscillates from one end of the swing to the other. Show on the same graph what the total energy looks like.

**1.18 Heat and temperature Comment critically on the following statement: A high-temperature system is a high-energy system.

*1.19 Celsius scale Data from U.S. laboratories are often standardized at 25°C; from European laboratories, at 18°C. What would be the equivalent Fahrenheit readings?

*1.20 Units The cost of 99.9% europium is quoted as $300/10 g. What would be the cost of each of the following:
a 0.166 kg
b 16.6 dg
c 1.66 × 10⁻⁴ Mg
d 1.66 × 10⁷ μg
e 1.66 cg   *Ans. (d) $498*

**1.21 Units Convert each of the following:
a 125 hl to liters
b 47.3 ml to microliters
c 1.86 × 10⁻⁸ cm to nanometers
d 1.86 × 10⁻⁸ cm to picometers
e 6.7 ps to seconds   *Ans. (d) 186 pm*

**1.22 Units A fly weighing 10.0 mg and moving at 454 cm/sec smashes into a window in a collision that takes but 0.0010 sec. Calculate in joules the energy expended.   *Ans. 0.00010 J*

*1.23 Units An average person's minimum daily food requirement is estimated to be 2400 nutritional Calories. Assuming all these Calories come from sugar (which liberates 15.5 kJ/g), how many grams of sugar do you need per day?

**1.24 Dimensional analysis In the equation

$$Q = \left(\frac{2\pi mkT}{h^2}\right)^{3/2} V + \frac{8\pi^2 IkT}{h^2}$$

the units are kilograms for m, joules per degree for k, degrees for T, cubic meters for V, and joule seconds for h. What must the units of Q be? What must the units of I be?

*1.25 Dimensional analysis Chemists usually report density in grams per cubic centimeter, but the more systematic SI unit would be kilograms per cubic meter. To convert from one to the other, what would be the conversion factor?

**1.26 Dimensional analysis The ideal equation for a gas $PV = nRT$ is often not followed, and so van der Waals developed an alternative equation which reads $\left(P + \dfrac{a}{V^2}\right)(V - b) = nRT$. Given the units shown on page 20, what must be the units for a? for b?

*1.27 Dimensional analysis Give the units of the final answer obtained after performing each of the following operations:
a Divide a volume in cubic centimeters by an area in square centimeters
b Divide an area in square centimeters by a volume in cubic centimeters
c Multiply a density in grams per cubic centimeter by a volume in cubic centimeters
d Divide a mass in grams by a density in grams per cubic centimeter
e Multiply a mass in kilograms by the square of a velocity in meters per second and divide by an energy (in joules).

**1.28 Dimensional analysis Figure out the conversion factor for going from pounds per cubic foot to grams per cubic centimeter.
                                   Ans. 0.0160

**1.29 Dimensional analysis For an oil droplet of mass m the force F due to gravity g is given by $F = mg$. For a spherical droplet of radius 0.010 cm and density 0.98 g/cm³, what would be the force in newtons? The value of g is 980 cm/sec².

*1.30 Units What is wrong with saying that a given sample of sulfuric acid has a specific gravity of 1.8305 kg/liter? Suggest a reason why handbooks of chemistry tabulate specific gravity of sulfuric acid under the heading "at 20°/4°."

**1.31 Precision Suppose you have four timers measuring the elapsed time of a 100-m race. The respective reports are 58.6 sec, 58.6 sec, 55.48 sec, 55.52 sec. What might you conclude about the timers?

**1.32 Significant figures An accurately calibrated pipette delivers 10.00 ml of water into an accurately calibrated graduated cylinder initially containing water just even with the 90-ml mark. Markings on the cylinder are at intervals of 10 ml. What is the final total volume of water in the cylinder?

*1.33 **Significant figures** Round off the following as indicated:

a 1.86 liters to two significant figures
b 1.85 liters to two significant figures
c 1.855 liters to two significant figures
d 1855 ml to two significant figures
e 0.186 liter plus 0.0186 liter to two significant figures

*1.34 **Significant figures** Express each of the following to the proper number of significant figures:

a $\left(1.35 \times 10^{-3} \dfrac{\text{mol}}{\text{liter}}\right) (0.20 \text{ liter})$

b $\left(0.200 \dfrac{\text{mol}}{\text{liter}}\right) (1.35 \times 10^{-3} \text{ liter})$

c $0.872 \text{ mol} + \left(2.35 \times 10^{-2} \dfrac{\text{mol}}{\text{liter}}\right) (40.78 \text{ liters})$

d $\left(\dfrac{1.253 \text{ mol}}{0.8694 \text{ liter}}\right) (1.254 \text{ liters}) + 0.86 \text{ mol}$

Ans. (d) 2.67 mol

1.35 **Scientific notation Write out each of the following in scientific notation:

a $22{,}400 \pm 100$ b $22{,}400 \pm 1$
c 0.00186 ± 0.0001 d 0.00186 ± 0.00001
e 200.0 ± 0.1

Chapter 2

STOICHIOMETRY

What is stoichiometry?* The word comes from the Greek word *stoicheon,* meaning "element," and *metron,* meaning "measure." Stoichiometry concerns itself with the weight relations of elements in chemical reactions. There are two aspects: (1) description of compounds in terms of the elements that combine to form them, and (2) quantitative consideration of the relative amounts of reagents that are consumed or produced in chemical reactions. In this chapter we consider in detail two prime aspects of stoichiometric calculations: chemical formulas and chemical equations.

2.1
ATOMIC WEIGHT

One of the most important ideas that came out of early atomic theory (ca. 1800) was the concept of atomic weight, the idea that all the atoms of a given element were characterized by identical mass. At that time, there was no clear way to determine what this absolute mass was, but it was recognized that if relative atomic weights could be assigned to the various elements, then it would be possible to exercise precise control over the relative number of atoms going into a reaction by manipulating the relative macroscopic (large-scale or laboratory-scale) weights of the elements chosen for reaction. If the F atom is 19 times as heavy as the H atom, then 19 g of F contains just as many atoms as does 1 g of H.

John Dalton, who set up the first practical scale of atomic weights, chose as his reference the lightest element, hydrogen. He arbitrarily assigned it a value of unity and expressed other elements with higher values, depending on how many times heavier their atoms were than hydrogen. To fix the relative weights, compounds consisting of two elements were analyzed to see how much of the total weight was contributed by each element. Then a decision was made what the formula of the compound was, i.e., how many atoms there were, relatively, of each kind in the compound. The relative weights of individual atoms could then be assigned. In retrospect we can see that, in setting up his original scale, Dalton made two mistakes: First, his analysis of water was faulty. He found that decomposition of water into its component elements hydrogen and oxygen gave 12.5% by weight of hydrogen and 87.5% oxygen. (He should have found 11.19% hydrogen and 88.81% oxygen.) Therefore,

*This is probably the most mispronounced word in chemistry. The first syllable *stoi-* should rhyme with *boy,* but it is frequently heard to rhyme with *crow,* especially from those who try to make a six-syllable word out of a five-syllable one. It is not correct to say stow-icky-om-etry.

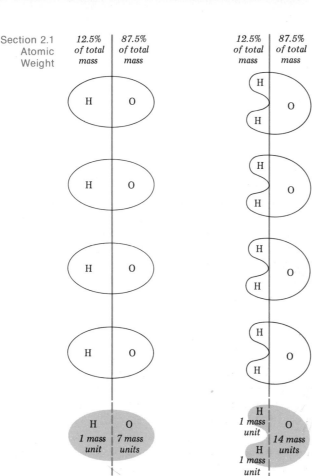

FIG. 2.1 Distribution of mass in water as envisaged by Dalton, on the left; what he should have concluded, on the right. Even the conclusion on the right is wrong because Dalton's analysis of water was faulty; it is actually 11.19% H and 88.81% O.

he concluded that in water the oxygen contributed 7 times as much weight as did the hydrogen. (He should have found 8 times.) His second mistake was more serious. He assumed that the molecule of water consists of one H atom and one O atom. Therefore, the obvious conclusion was that the O atom was 7 times as heavy as the H atom. If H is assigned a mass of 1 unit, O must have a mass of 7 units on the same scale. The situation is depicted on the left side of Fig. 2.1. Actually, as shown on the right side of Fig. 2.1, the water molecule contains two H atoms per O atom, not one as Dalton assumed. Consequently, the mass attributable to the H portion of the molecule must be spread over twice as many atoms, hence making each H atom half as heavy. If again we assign to H a reference value of unity, then the O atom would have to be 14 times as heavy. In other words, even using his faulty chemical analysis, if Dalton had correctly guessed the formula for water, he would have assigned O an atomic weight of 14 on a scale where H was 1. If we go further and correct Dalton's faulty analysis of water from 1:7 parts by weight H:O to the proper ratio 1:8, then we would get 16 for the atomic weight of O, not far from the value presently assigned.

Shortly after Dalton had proposed his list of atomic weights relative to H = 1 as standard, the Swedish chemist Berzelius proposed that the standard should be based on oxygen, not hydrogen. His reasoning was that more elements combine with oxygen than with hydrogen (so presumably it would be easier to prepare compounds containing the standard element for analysis); also, hydrogen is such a light element that its contribution to the total mass is hard to determine accurately. Therefore, Berzelius proposed that oxygen be assigned an atomic weight of 100 and all other elements be referred to it (for example, H, being one-sixteenth as heavy as O would be $\frac{1}{16} \times 100$, or 6.25). Unfortunately, the Berzelius results, especially for the atomic weights of the heavy elements, came out to be such large, unwieldy numbers (e.g., lead came out to have an atomic weight of 1295), that the Berzelius scheme never really gained acceptance. However, the idea of using oxygen as a reference standard was adopted.

Until 1960, atomic weights were expressed on a scale on which oxygen was assumed to have a value of 16 units. However, there were, in practice, two scales. Chemists gave a reference value of 16 to a "natural" oxygen atom—that is, a hypothetical "average" atom that takes account of the fact that there are several kinds of oxygen atoms present in a "natural" mixture. These different kinds of oxygen atoms, which all behave chemically the same but differ slightly in mass, are called *isotopes*. Physicists, on the other hand, singled out the most abundant of the isotopes, so-called "oxygen 16," and gave it an exact reference value of 16. Thus, on the chemists' scale, the unit for expressing atomic weights was one-sixteenth the mass of an "average" oxygen atom; on the physicists' scale, the unit was one-sixteenth the mass of an oxygen-16 atom. Because the chemists' unit was slightly larger, values on the chemists' scale of atomic weight were slightly smaller than those on the physicists' scale. As an example, the *chemical* atomic weight of natural oxygen was given as 16.0000; the *physical* atomic weight of the same natural oxygen was given as 16.0044.

As the determination of atomic weights became more precise, the small differences between chemical and physical values became more irksome. The result, finally, was a unified new scale, adopted by both disciplines in 1961, which abandoned the oxygen standard and moved to the carbon-12 isotope. *The carbon-12 isotope was arbitrarily assigned as having an atomic weight of 12 amu.* Accordingly, at the present time, *one atomic mass unit (amu) is defined as one-twelfth the mass of a carbon atom that is a carbon-12 isotope.* As we shall see in Chap. 4, not all the atoms of a given element are identical. Some are heavier than others because of extra neutral particles (so-called "neutrons") in the nuclei or centers of the atoms. Thus, C 12 is believed to consist of six neutrons, six protons, and six electrons, the whole assembly having an assigned mass of 12.0000 amu, whereas the C-13 isotope has seven neutrons, six protons, and six electrons and has a total mass of 13.00335 amu on the same scale. The natural mixture of carbon atoms has a relative abundance by number of atoms of 98.98% C-12 isotope and 1.11% C-13 isotope. Therefore, the natural mixture is described as having an atomic weight that is the weighted mean of the above two values, that is, 12.011 amu. Actually, different samples of carbon may vary slightly in their isotopic composition.

EXAMPLE 1

A sample of carbon was enriched with C-14 isotope so that the final abundance by number of atoms in a particular sample was 95.36% C 12, 1.07% C 13, and 3.57% C 14. Given that the mass of the C-14 isotope is 14.0044 amu, what is the average atomic weight in the mixture?

Solution

Out of 10,000 atoms

9536	have a mass each of 12.0000 amu =	114,400 amu
107	have a mass each of 13.00335 amu =	1,390 amu
357	have a mass each of 14.0044 amu =	5,000 amu
10,000 atoms		120,790 amu

The average weight is 120,790/10,000, or 12.08 amu.

How are atomic weights determined? In the old days, it was a question of analyzing a compound that contained the unknown combined with the standard reference element and guessing the relative number of atoms of unknown and standard in the formula. Modern procedure relies instead on direct determination of absolute masses of isotopes and on measurement of their relative abundance in a given sample. The device used, a mass spectrometer, deflects beams of atoms into curved paths, with the more massive atoms collecting on the outside because of greater inertia to being deviated from straight-line motion. A beam of atoms composed of several isotopes can thus be split into several beams, each composed of only one isotope. From the deflection of the beams and their relative intensities one can deduce the masses and the number of atoms of each type. The following example shows how mass-spectrometric data can be used to calculate the atomic weight of an element.

EXAMPLE 2

Natural chlorine consists of a mixture of two isotopes, designated, respectively, as chlorine 35 and chlorine 37. Mass-spectrometric analysis indicates that 75.53% of the atoms are chlorine 35, having a mass of 34.968 amu, and 24.47% are chlorine 37, having a mass of 36.956 amu. Calculate the chemical atomic weight of natural chlorine.

Solution

Take 10,000 atoms at random. Of these 10,000 atoms

7553	have a mass of 34.968 amu =	264,100 amu
2447	have a mass of 36.956 amu =	90,430 amu
10,000		354,530 amu

The average weight is 354,530/10,000, or 35.45 amu.

On the inside back cover of the book, the current best values of the atomic weights of the elements are listed on a scale based on the assumption that the carbon-12 isotope has a mass of exactly 12 amu. In the list, there are some atomic-weight entries that are given as simple whole numbers in parentheses. These have not been determined by a mass-spectrometric analysis of the element, but, because the elements concerned are unstable, the numbers simply represent the mass number (i.e., neutrons plus protons) of the longest-lived or best-known isotope. As can be seen from the values in the table, the precision with which atomic weights are known vary from element to element. Thus, lead is given as 207.2 amu, whereas fluorine is given as 18.998403 amu. The reasons for such discrepancies are that some elements, such as lead, differ widely in their isotopic composition, depending on their source. Other elements, such as fluorine, consist in nature of only one stable isotope so that only its mass needs to be determined precisely.

2.2
THE MOLE

One of the central concepts of chemical stoichiometry is the *mole* (symbol *mol*). It is defined as the *amount of material that contains as many elementary units as there are atoms in 12 g of the isotope carbon 12*. This number turns out to be 6.0222×10^{23}, which is known as the *Avogadro number*. Thus, a mole can be defined as the Avogadro number of particles. The mole is a useful concept because it provides a way of talking about a number of particles without worrying about mass. An atom or a molecule is an extremely small entity. In the space of 1 mm one can align between 2 million and 10 million atoms. It would be hopeless to think of weighing out individual atoms for chemical reactions. Even the smallest of weighable amounts would involve millions or billions of atoms. Yet, in the course of a chemical reaction, it is often necessary to maintain precise control on the relative number of atoms going into the reaction. Otherwise, a product other than the desired one might be obtained. Hence, there is obvious need to control relative numbers of atoms even when large amounts of materials are involved. The mole concept provides a way to do this.

The great power of the mole concept lies in the fact that if we take weights of different elements in the same ratio as their atomic weights, then we have identical numbers of atoms. For example, suppose we take 12.0 g of carbon. We would then have 6.02×10^{23} atoms of carbon in our collection. Suppose, further, that we take 16.0 g of oxygen. Since the atomic weight of oxygen is 16.0 amu (the oxygen atom being 16.0/12.0 as heavy as the carbon atom), we also have 6.02×10^{23} atoms of oxygen. In other words, 12.0 g of carbon and 16.0 g of oxygen *have the same number of atoms*. Thus, if we want to make carbon monoxide, CO, where the number of C atoms must match the number of O atoms, 12.0 g of C would just provide the C atoms to match each of the O atoms in 16.0 g of O.

As can be seen, 12.0 g of carbon is a reasonably easy amount of material to weigh out, and so the number of carbon atoms can be precisely controlled at 6.02×10^{23} with a rather easy measurement. Similarly, 16.0 g of oxygen is easy to weigh out, and so the number of oxygen atoms can be precisely controlled at 6.02×10^{23}.

So, this is the first important point to note: *If we take a weight of element that is equal to the number of grams corresponding to the number shown in the atomic-weight table, then we have 1 mol, or 6.02×10^{23}, of atoms of that element.* Specifically, 1.01 g of H, 16.0 g of O, or 19.0 g of F would each contain 6.02×10^{23} atoms. We can use this information in diverse ways, for example, to calculate the weight of an individual atom, the number of atoms in a given weight, or the weight of a specified number of atoms.

EXAMPLE 3

What is the mass of one individual fluorine atom?

Solution

1 mol of fluorine atoms weighs 19.0 g
1 mol of any element contains 6.02×10^{23} atoms

Therefore, $\dfrac{19.0 \text{ g}}{6.02 \times 10^{23} \text{ atoms}} = 3.16 \times 10^{-23}$ g/atom

---

We can also use the concept to calculate how many atoms there are in a given weight of element.

EXAMPLE 4

How many atoms of hydrogen are there in 2.57×10^{-6} g of hydrogen?

Solution

1 mol of hydrogen atoms weighs 1.01 g
1 mol of any element contains 6.02×10^{23} atoms

Hence, $\dfrac{6.02 \times 10^{23} \text{ atoms}}{1.01 \text{ g}}$ gives the number of atoms per gram of hydrogen

In 2.57×10^{-6} g of hydrogen we would have

$(2.57 \times 10^{-6} \text{ g}) \left(\dfrac{6.02 \times 10^{23} \text{ atoms}}{1.01 \text{ g}} \right) = 1.53 \times 10^{18}$ atoms

---

Finally, we can use the concept to calculate the weight of any specified number of atoms.

EXAMPLE 5

A sample of oxygen contains 1.87×10^{27} atoms of oxygen. What would be the weight of this collection of atoms?

Solution

We know that 1 mol of O atoms weighs 16.0 g. All we have to do is to figure out how many moles of O atoms we have.

$$\frac{1.87 \times 10^{27} \text{ atoms}}{6.02 \times 10^{23} \text{ atoms/mol}} = 3.11 \times 10^3 \text{ mol of O atoms}$$

$$(3.11 \times 10^3 \text{ mol of O}) \left(16.0 \frac{\text{g}}{\text{mol of O}} \right) = 4.97 \times 10^4 \text{ g}$$

As implied above, one of the most important applications of the mole concept is to manipulate the relative number of atoms brought together in a reaction. Specifically, if we take 1 mol of X atoms (equal to atomic weight of X in grams) and 1 mol of Y atoms (equal to atomic weight of Y in grams), then we have exactly one atom of X for each atom of Y. By weighing out quantities in the same ratio as the atomic weights, we ensure equality in number of atoms. From this, it is a simple extension to match any ratio of atoms, as the following example shows.

EXAMPLE 6

In a chemical reaction requiring three atoms of magnesium (Mg) for two atoms of nitrogen (N), how many grams of N are required by 4.86 g of Mg? (The atomic weight of Mg is 24.3 amu, and the atomic weight of N is 14.0 amu.)

Solution

Figure out how many moles of Mg atoms we have, calculate from this how many moles of N atoms are required, then convert this to grams.

$$\frac{4.86 \text{ g of Mg}}{24.3 \text{ g/mol of Mg}} = 0.200 \text{ mol of Mg}$$

$$(0.200 \text{ mol of Mg}) \left(\frac{2 \text{ mol of N}}{3 \text{ mol of Mg}} \right) = 0.133 \text{ mol of N}$$

$$(0.133 \text{ mol of N}) \left(\frac{14.0 \text{ g of N}}{1 \text{ mol of N}} \right) = 1.87 \text{ g of N}$$

Note how the dimensions cancel out to give the units shown.

The mole concept is equally powerful when applied to units that are more complicated than atoms. For example, in the case where atoms are known to form specific independent clusters, like the molecule H_2O, we can talk about 1 mol of molecules. One mole of molecules contains 6.02×10^{23}

molecules, just like 1 mol of atoms contains 6.02×10^{23} atoms.* Because each molecule of H_2O contains 2 H atoms and 1 O atom, we can state the following identity:

$$1 \text{ mol of } H_2O \text{ molecules} = 6.02 \times 10^{23} \text{ } H_2O \text{ molecules}$$
$$= 2(6.02 \times 10^{23}) \text{ H atoms plus}$$
$$1(6.02 \times 10^{23}) \text{ O atoms}$$
$$= 2 \text{ mol of H atoms plus 1 mol of O atoms}$$

In some cases, we may not have a simple molecule. We may know only the relative number of the different kinds of atoms in the compound. This happens, for example, in the case of a solid such as calcium dichloride, $CaCl_2$. There are no discrete molecules of $CaCl_2$ in the solid but a complex interlocked network of Ca^{2+} and Cl^- ions in the ratio 1:2. Even in such cases, the mole is a useful concept. We speak then of a mole as being equal to 6.02×10^{23} formula-units, where the formula-unit is a fictitious entity made up of the component atoms in the ratio shown by the chemical composition. As in the case of water shown above, we can make the following identity:

$$1 \text{ mol of } CaCl_2 = 6.02 \times 10^{23} \text{ formula-units of } CaCl_2$$
$$= 1(6.02 \times 10^{23})Ca^{2+} \text{ plus } 2(6.02 \times 10^{23})Cl^-$$
$$= 1 \text{ mol of } Ca^{2+} \text{ ions plus 2 mol of } Cl^- \text{ ions}$$

2.3
SIMPLEST FORMULAS

Elements are denoted by symbols; compounds are denoted by combinations of symbols representing the elements that go into making up the compound. How are these formulas determined? Just exactly what do they represent?

The *simplest formula*, also called the *empirical formula*, gives the bare minimum of information about a compound. It simply states the *ratio* or relative number of moles of atoms of the different kinds in the compound. The convention used in writing the simplest formula is to write the symbols of the elements with subscripts to designate the relative number of moles of atoms of these elements. The formula A_xB_y represents a compound in which there are *x mol of A atoms* for every *y mol of B atoms*. Because of the relationship between 1 mol of atoms and 6.02×10^{23} atoms, the simplest formula also gives

---

*Until recently there used to be a special name, *gram-atom*, applied to 6.02×10^{23} atoms. Thus, 1 gram-atom of hydrogen, which weighed 1.01 g, contained the same number of atoms (6.02×10^{23}) as 1 gram-atom of chlorine, which weighed 35.5 g. In forming HCl, 1 gram-atom of hydrogen just combined with 1 gram-atom of chlorine. The term "gram-atom" fell into disfavor because the prefix *gram* somehow suggested a weight of material, whereas it was the *number of atoms* that was being controlled. The term gram-atom is now being abandoned, as recommended by the IUPAC. There is risk of confusion in using the substitute word mole unless the identity of the counted particle is specified. For example, "1 mol of nitrogen" can mean 1 mol of nitrogen *atoms* or 1 mol of nitrogen *molecules*. The latter are diatomic as N_2, and so 1 mol of N_2 contains twice as many N atoms as 1 mol of N. To avoid possible misunderstanding, the term "mole" should always be used with the formula of the particle that is being counted.

information about the relative number of atoms in the compound. In A_xB_y there are x *atoms of element A* for every y *atoms of element B*. Nothing is implied about the nature of this association—in particular, nothing about the absolute size of the molecular aggregate—except the *relative* number of atoms in it.

EXAMPLE 7

When a weighed piece of metal M (atomic weight 121.75 amu) is heated with excess sulfur (atomic weight 32.06 amu), a chemical reaction occurs between the metal and sulfur. The excess sulfur is then driven off, leaving only the compound, consisting of combined metal and sulfur. From the weight of M and the weight of the compound, the weight of sulfur in the compound can be deduced. Here are some sample figures from an experiment of this type:

Weight of metal   2.435 g ⎫ determined experimentally
Weight of compound   3.397 g ⎭
Weight of sulfur   0.962 g } determined by difference

What is the simplest formula of the compound?

Solution

Since the simplest formula gives the relative number of moles of atoms of the different kinds in the compound, we must calculate how many moles of M atoms and how many moles of S atoms have combined.

$$\frac{2.435 \text{ g of M}}{121.75 \text{ g of M/mol of M}} = 0.0200 \text{ mol of M}$$

$$\frac{0.962 \text{ g of S}}{32.06 \text{ g of S/mol of S}} = 0.0300 \text{ mol of S}$$

The simplest formula shows this ratio as subscripts, and so we write $M_{0.0200}S_{0.0300}$. We can simplify the subscripts by dividing all of them by the smallest subscript to get $M_1S_{1.5}$, and then doubling to get M_2S_3.

The chemical analysis of a compound is often reported in terms of the percent by weight composition. In such cases, a common type of problem is to derive the simplest formula.

EXAMPLE 8

Caffeine, the main stimulant in coffee, can be analyzed by burning it in a stream of oxygen and then collecting and weighing the oxides formed. Analysis by such a method shows that caffeine consists of 49.48% by weight C, 5.19% H, 28.85% N, and 16.48% O. What is its simplest formula?

Solution

To get the simplest formula, we need to know the relative number of moles of C, H, N, and O atoms in some fixed weight of compound. We can take any convenient weight of compound, 1 g, 12.011 g, or any other weight. In practice it is most convenient to solve this problem by taking *100.0 g of compound* since then the numbers giving percentage by weight correspond directly to the grams of each element in the sample. All we need to do then is to convert each of the weights into moles of atoms. We do this by dividing each weight of element by the number of grams needed to make 1 mol of atoms. Thus, in *100.0 g of compound* there would be 49.48 g of C, 5.19 g of H, 28.85 g of N, and 16.48 g of O, or

$$\frac{49.48 \text{ g of C}}{12.011 \text{ g of C/mol of C atoms}} = 4.120 \text{ mol of C atoms}$$

$$\frac{5.19 \text{ g of H}}{1.0079 \text{ g of H/mol of H atoms}} = 5.15 \text{ mol of H atoms}$$

$$\frac{28.85 \text{ g of N}}{14.0067 \text{ g of N/mol of N atoms}} = 2.060 \text{ mol of N atoms}$$

$$\frac{16.48 \text{ g of O}}{15.999 \text{ g O/mol of O atoms}} = 1.030 \text{ mole of O atoms}$$

The simplest formula can be written as $C_{4.120}H_{5.15}N_{2.060}O_{1.030}$ or, dividing all the subscripts by 1.03, $C_4H_5N_2O$.

In general, simplest formulas are reduced so that the subscripts become small whole numbers. The justification for this is that at one time it was believed that the ratio of atoms in a compound had to be expressible by small whole numbers. In the gaseous state, this is still true, since the number of atoms of each kind in a gas molecule must be a discrete small number. In the solid state, it is now recognized that compounds can exist in which the ratio of atoms is not simple. Such compounds are said to be *nonstoichiometric*. For example, the compound obtained by heating copper and sulfur in an open crucible never conforms to the expected exact stoichiometry Cu_2S but almost invariably shows a deficit of copper. Generations of freshman chemistry students have gotten $Cu_{1.87}S$, often believing that if they had done the experiment more carefully they would have gotten Cu_2S. Not so! The compound formed here is a typical example of a nonstoichiometric compound.

2.4
MOLECULAR FORMULAS

Besides the simplest formula, another common type of formula is the molecular formula. In contrast to the simplest formula, where the subscripts indicate the *relative* number of atoms of each kind in the compound, the molecular formula tells in its subscripts the *actual* number of atoms in a molecule of the compound. The molecule is an aggregate of atoms joined together tightly enough to be conveniently treated as a recognizable unit. In order to be able to write a molecular formula, we need to know how many atoms constitute the molecular unit. There are various experimental ways to get at this information, such as X-ray analysis of positions of atoms in a solid or determination of

FIG. 2.2 Molecular and Simplest Formulas of Some Common Chemicals

Substance	Molecular formula	Simplest formula
Gaseous water	H_2O	H_2O
Sucrose (a sugar)	$C_{12}H_{22}O_{11}$	$C_{12}H_{22}O_{11}$
Glucose (a sugar)	$C_6H_{12}O_6$	CH_2O
Ethane	C_2H_6	CH_3
Ethylene	C_2H_4	CH_2
Acetylene	C_2H_2	CH
Benzene	C_6H_6	CH
Morphine	$C_{17}H_{19}NO_3$	$C_{17}H_{19}NO_3$
Caffeine	$C_8H_{10}N_4O_2$	$C_4H_5N_2O$

gas properties or solution properties that are dependent on relative numbers of molecules. Thus, for example, from the observed lowering of the freezing point of water when caffeine is dissolved in it, one can conclude that the caffeine molecule is not given by the formula $C_4H_5N_2O$ but is twice as big, corresponding to the molecular formula $C_8H_{10}N_4O_2$. Examples of other molecular formulas are shown in Fig. 2.2. In some cases, as with gaseous water, sucrose, and morphine, the molecular and simplest formulas are identical. In other cases, e.g., glucose and benzene, they are not. Occasionally, as with benzene (C_6H_6) and acetylene (C_2H_2), the simplest formulas (in this case, CH for each) are identical. We cannot always tell from a formula whether it is molecular or simplest. However, if the subscripts have a common divisor other than 1, then the chance is good that the formula is molecular. The molecular formula gives all the information that the simplest formula gives, i.e., the relative number of atoms, and it also tells the makeup of the molecular complex.

2.5
FORMULA WEIGHTS AND MOLECULAR WEIGHTS

The *formula weight of a compound* is defined as *the sum of the weights of all the atoms in the formula*. For NaCl, the formula weight is the atomic weight of sodium, 23.0 amu, plus the atomic weight of chlorine, 35.5 amu, for a total of 58.5 amu. For caffeine, $C_8H_{10}N_4O_2$, the formula weight is equal to 8 times the atomic weight of carbon plus 10 times the atomic weight of hydrogen plus 4 times the atomic weight of nitrogen plus 2 times the atomic weight of oxygen, or 194.1 amu. In all cases, the formula weight depends on which formula is written. If the molecular formula is used, the formula weight is called the *molecular weight*. For example, 194.1 amu is the molecular weight of caffeine.

As noted, the formula weight is given in atomic mass units. As such, it is an exceedingly small amount. If we take the amount of substance whose mass in grams numerically equals the formula weight, then we have what is called a *gram-mole*. In the case of caffeine, where the formula weight is 194.1 amu, 1 gram-mole weighs 194.1 g. The formula weight can also, when convenient, be taken in tons, ounces, pounds, or any other units of mass. The result then is a ton-mole, an ounce-mole, or a pound-mole. As an example, 194.1 tons of sugar is 1 ton-mole of sugar. Because in chemistry we are primarily concerned with gram-moles, the prefix *gram* is often omitted. When one uses the word *mole,* the term is taken to mean gram-mole. We shall use the term gram-mole for a while to emphasize that we mean formula weight in grams, but we shall gradually simplify to just talking about a mole.

There is an important relationship that exists between the number of gram-moles and the number of fundamental units. Let us consider again caffeine, for which the molecular formula is $C_8H_{10}N_4O_2$. One gram-mole of $C_8H_{10}N_4O_2$ weighs 194 g and contains 96 g of C, 10 g of H, 56 g of N, and 32 g of O. In 96 g of C (atomic weight 12), there are $\frac{96}{12}$, or 8, mol of C atoms; in 10 g of H (atomic weight 1), there are $\frac{10}{1}$, or 10, mol of H atoms; in 56 g of N (atomic weight 14), there are $\frac{56}{14}$, or 4, mol of N atoms; in 32 g of O (atomic weight 16), there are $\frac{32}{16}$, or 2, mol of O atoms. Since 1 mol of any kind of atom contains the Avogadro number of atoms (6.02×10^{23}), we can see that 1 mol of $C_8H_{10}N_4O_2$ contains

$$8 \times 6.02 \times 10^{23} \text{ atoms of C}$$

$$10 \times 6.02 \times 10^{23} \text{ atoms of H}$$

$$4 \times 6.02 \times 10^{23} \text{ atoms of N}$$

$$2 \times 6.02 \times 10^{23} \text{ atoms of O}$$

Furthermore, since each molecule of $C_8H_{10}N_4O_2$ contains 8 atoms of C, 10 atoms of H, 4 atoms of N, and 2 atoms of O, we have 6.02×10^{23} molecules of $C_8H_{10}N_4O_2$ per mole. *For any substance for which the formula is known to be the molecular formula, 1 gram-mole contains the Avogadro number of molecules.*

EXAMPLE 9

What is the weight in grams of one molecule of caffeine, $C_8H_{10}N_4O_2$?

Solution

1 mol of $C_8H_{10}N_4O_2$ weighs 194.1 g
1 mol of any molecular substance contains 6.02×10^{23} molecules

$$\frac{194.1 \text{ g}}{6.02 \times 10^{23} \text{ molecules}} = 3.22 \times 10^{-22} \text{ g/molecule}$$

What about the relationship between number of moles and number of particles, for a compound whose molecular formula is not known? Such a compound is nitrogen tri-iodide, for which the simplest formula is NI_3. This compound is notorious for the fact that a fly lighting on it can cause it to explode. One mole of NI_3 weighs 394.7 g and contains 14.0 g of N and 380.7 g of I. In 14.0 g of N (atomic weight 14.0), there is 1.00 mol of N atoms; in 380.7 g of I (atomic weight 126.9), there are 3.00 mol of I atoms. One mol of

N atoms contains 6.02×10^{23} atoms of N; 3 mol of I atoms contains $3 \times 6.02 \times 10^{23}$ atoms of I. Since the *molecular* formula of NI_3 is not known, the actual number of atoms in the molecule is not known. Therefore, we cannot state the number of *molecules* in one mole of NI_3. However, we can define a formula-unit of NI_3 as a cluster consisting of one N atom and three I atoms. This cluster may not really exist but there would be 6.02×10^{23} of them in 1 mol of NI_3. *For any substance, 1 gram-mole contains the Avogadro number of formula-units.* If the formula is known to be a molecular formula, then we can say that the formula-unit is the same as the molecule.

Even though formulas are generally derived from experimentally determined percentage compositions, it is frequently necessary to go the other way—i.e., to calculate percentage composition from a formula.

EXAMPLE 10

Given the atomic weights Al 26.98, S 32.06, and O 16.00 amu, what is the percent composition of $Al_2(SO_4)_3$?

Solution

One gram-mole of $Al_2(SO_4)_3$ contains

$$
\begin{aligned}
&2 \text{ mol of Al atoms, or } 2 \times 26.98 \text{ g of Al} = &53.96 \text{ g of Al}\\
&3 \text{ mol of S atoms, or } 3 \times 32.06 \text{ g of S} = &96.18 \text{ g of S}\\
&12 \text{ mol of O atoms, or } 12 \times 16.00 \text{ g of O} = &\underline{192.00 \text{ g of O}}\\
& &342.14 \text{ g total mass}
\end{aligned}
$$

$$\text{Percent Al} = \frac{53.96 \text{ g}}{342.14 \text{ g}} \times 100 = 15.77\%$$

$$\text{Percent S} = \frac{96.18 \text{ g}}{342.14 \text{ g}} \times 100 = 28.11\%$$

$$\text{Percent O} = \frac{192.00 \text{ g}}{342.14 \text{ g}} \times 100 = 56.12\%$$

2.6
CHEMICAL EQUATIONS

Besides the chemical formula as discussed above, the other important aspect of stoichiometry is the chemical equation (Sec. 1.4). It states directly what substances and how many formula-units of each are involved in a chemical reaction, and it also indicates indirectly what weight changes occur in the reaction. Thus, a chemical equation is a shorthand statement about chemical change. We shall generally use *net equations, which specify only the substances used up and the substances formed in the chemical reaction.* Net equations omit anything that remains unchanged. The convention used in writing equations is to place what disappears (the reactants) on the left side and what appears (the products) on the right side. The reactants and products are separated by a single arrow $\longrightarrow$, an equals sign $=$, or a double arrow $\rightleftharpoons$, depending on what aspect of the chemical reaction is being emphasized. An example of a net equation is

$$Cl_2(g) + H_2O + Ag^+ \longrightarrow AgCl(s) + HOCl + H^+$$

The reactants and products are designated by symbols or formulas. The symbol can be read as representing one atom or 1 mol of atoms; the formula represents either 1 formula-unit or 1 mol. The notation (g) indicates the gas phase, (l) the liquid phase, and (s) the solid phase. Most of the reactions that we shall consider take place in aqueous solution, in which case we can use the notation (aq). When no phase notation appears, the aqueous phase is understood.

To be valid, a chemical equation must be consistent with the experimental facts—that is, it must state what chemical species disappear and appear. Second, it must be consistent with the conservation of mass. Since we do not destroy mass, we must account for it. If an atom disappears from one substance, it must appear in another. Third, the chemical equation must be consistent with the conservation of electric charge. Since we cannot destroy electric charge, we must account for it. Conditions two and three are expressed by saying that the equation must be balanced. A *balanced* equation contains the same numbers of atoms of the different kinds on the left- and right-hand sides; furthermore, the net charge on the two sides must be the same.

How do we go about writing balanced equations? One method, usually reserved for simple reactions, is to balance the equation "by inspection." This entails juggling the coefficients that appear before the symbols and formulas so that respective numbers of atoms of each element come out to be equal on the left and right sides of the equation. For example, when gaseous hydrogen (H_2) is exploded with gaseous chlorine (Cl_2), the product is gaseous hydrogen chloride (HCl). The balanced equation for the reaction is

$$H_2(g) + Cl_2(g) \longrightarrow 2HCl(g)$$

where a 2 has been inserted as coefficient for the formula HCl to ensure that the 2 H atoms on the left show up as 2 H atoms on the right, and similarly for the Cl. All the species here are electrically neutral, so that electric charge is conserved when the reactants $(H_2 + Cl_2)$ convert to the product (HCl). As another example, when a solution of barium chloride $(BaCl_2)$ is mixed with a solution of sodium sulfate (Na_2SO_4), the balanced net reaction is

$$Ba^{2+} + SO_4^{2-} \longrightarrow BaSO_4(s)$$

Barium ions (Ba^{2+}) and sulfate ions (SO_4^{2-}) disappear, and a white precipitate of solid barium sulfate appears. We could also have written the reaction

$$Ba^{2+} + 2Cl^- + 2Na^+ + SO_4^{2-} \longrightarrow BaSO_4(s) + 2Na^+ + 2Cl^-$$

in keeping with the fact that the reactants are a solution of barium chloride (therefore, containing Ba^{2+} and Cl^-) and a solution of sodium sulfate (containing Na^+ and SO_4^{2-}). However, as is clear from the above equation, the final mixture contains Na^+ and Cl^- just as did the initial mixture. Na^+ and Cl^- are sometimes called "spectator" ions and can be canceled from both sides of the equation. It is a general principle that, in writing net equations, duplications on the left and right side of the equation are canceled out.

Even rather complicated reactions can be balanced "by inspection" provided we are careful to be systematic. For example, suppose we are told to write a balanced equation for the burning of solid caffeine ($C_8H_{10}N_4O_2$) in gaseous oxygen (O_2) to produce the gaseous products CO_2, H_2O, and NO_2. We start out by writing

$$?C_8H_{10}N_4O_2(s) + ?O_2(g) \longrightarrow ?CO_2(g) + ?H_2O(g) + ?NO_2(g) \qquad \text{not balanced}$$

We note that for each molecule of $C_8H_{10}N_4O_2$ there are 8 carbon atoms on the left, and so we put 8 in front of CO_2 on the right. We also note there are 10 hydrogen atoms on the left, and so we put a 5 in front of H_2O on the right. Finally, there are 4 nitrogen atoms on the left, and so we put a 4 in front of NO_2 on the right. The display now looks like this:

$$lC_8H_{10}N_4O_2(s) + ?O_2(g) \longrightarrow$$
$$8CO_2(g) + 5H_2O(g) + 4NO_2(g) \qquad \text{still not balanced}$$

All the atoms are balanced except the oxygen. Checking the right side of the equation, where all the coefficients have been decided, we count up 8×2 for $8CO_2$ plus 5×1 for $5H_2O$ plus 4×2 for $4NO_2$, or a total of 29 oxygen atoms. On the left, where do we get these 29 oxygen atoms? Two of the 29 are already provided in $C_8H_{10}N_4O_2$, and so we need 27 more from the $O_2(g)$. Since each O_2 molecule supplies us with 2 oxygen atoms, we need only $\frac{27}{2}O_2$ to balance the equation:

$$C_8H_{10}N_4O_2(s) + \tfrac{27}{2}O_2(g) \longrightarrow 8CO_2(g) + 5H_2O(g) + 4NO_2(g)$$

This is a perfectly valid equation and can be used for calculation, although some people would prefer to get rid of the fraction by doubling the whole equation. The final result would then be

$$2C_8H_{10}N_4O_2(s) + 27O_2(g) \longrightarrow 16CO_2(g) + 10H_2O(g) + 8NO_2(g)$$

There are other systematic ways to balance equations that we shall consider later. For the time being, it will be sufficient to know what a balanced equation is and how it can be translated to weight relations.

2.7
CALCULATIONS USING CHEMICAL EQUATIONS

A chemical equation is valuable from two standpoints. It gives information on an atomic scale (i.e., atoms and formula-units that are not visible) and also on a laboratory scale (i.e., moles of atoms and moles of formula-units that correspond to visible, convenient weights of chemicals). For example, the chemical equation for the burning of caffeine deduced in the preceding section:

$$2C_8H_{10}N_4O_2 + 27O_2 \longrightarrow 16CO_2 + 10H_2O + 8NO_2$$

tells us the following information:

On an atomic scale:

2(molecules of $C_8H_{10}N_4O_2$) + 27(molecules of O_2) $\longrightarrow$
16(molecules of CO_2) + 10(molecules of H_2O) + 8(molecules of NO_2)

On a laboratory scale:

2(moles of $C_8H_{10}N_4O_2$) + 27(moles of O_2) $\longrightarrow$
16(moles of CO_2) + 10(moles of H_2O) + 8(moles of NO_2)

To go from the atomic scale (invisible, individual molecules) to the laboratory scale (visible, large collections of molecules) we simply need to multiply by 6.02×10^{23}, the Avogadro number. The laboratory scale is roughly the amount of chemical taken for a normal chemical reaction.

We can go one step further. Once we have the chemical equation expressed in number of moles, we can translate it into grams. Thus, the above equation states not only how many moles of each kind are involved but also, provided we supply further atomic-weight information, how many grams of each reagent are participating in the reaction. Specifically, we can say that 1 mol of $C_8H_{10}N_4O_2$ weighs 8×12.0 plus 10×1.01 plus 4×14 plus 2×16.0, or 194.1 g. Two mol would weigh twice as much, or 388.2 g. Similarly, 27 mol of O_2 weighs $(27)(32.0)$, or 864, g; 16 mol of CO_2 weighs $(16)(44.0)$, or 704, g; 10 mol of H_2O weighs $(10)(18.0)$, or 180, g; 8 mol of NO_2 weighs $(8)(46.0)$, or 368, g. Putting the whole thing together, we have

2 mol $C_8H_{10}N_4O_2$ + 27 mol O_2 $\longrightarrow$
16 mol CO_2 + 10 mol H_2O + 8 mol NO_2
(2)(194.1 g) + (27)(32.0 g) $\longrightarrow$
(16)(44.0 g) + (10)(18.0g) + (8)(46.0 g)
$\underbrace{388 \text{ g} \qquad\qquad + 864 \text{ g}}$ $\qquad$ $\underbrace{704 \text{ g} \qquad + 180 \text{ g} \qquad + 368 \text{ g}}$
1252 g $\longrightarrow$ 1252 g

As indicated, 388 g of $C_8H_{10}N_4O_2$ reacts with 864 g of O_2 to give 704 g of CO_2 plus 180 g of H_2O plus 368 g of NO_2. The total mass of reactants on the left side of the equation, 1252 g, is equal to the total mass of products on the right side of the equation, 1252 g. This is in accord with the law of conservation of matter, as it should be, and provides a useful check on the mathematics involved.

Once a balanced chemical equation is obtained, it can be used to calculate various weight relations between reactants and products. This is illustrated by the following examples.

EXAMPLE 11

Given the equation above, how many grams of O_2 are required to burn 1.00 g of caffeine?

Solution

The equation states

$$2C_8H_{10}N_4O_2 + 27O_2 \longrightarrow 16CO_2 + 10H_2O + 8NO_2$$

or, in words, that 2 mol of $C_8H_{10}N_4O_2$ requires 27 mol of O_2 for the reaction. All we have to do is to calculate how many moles of $C_8H_{10}N_4O_2$ we have and from this how much O_2 is required.

How many moles of $C_8H_{10}N_4O_2$ are there in 1.00 g of $C_8H_{10}N_4O_2$? We have already seen that the formula weight of $C_8H_{10}N_4O_2$ is 194.1 amu—in other words, 1 mol weighs 194.1 g.

$$\frac{1.00 \text{ g of } C_8H_{10}N_4O_2}{194.1 \text{ g of } C_8H_{10}N_4O_2/\text{mol of } C_8H_{10}N_4O_2} = 0.00515 \text{ mol of } C_8H_{10}N_4O_2$$

The equation tells us we need 27 mol of O_2 per 2 mol of $C_8H_{10}N_4O_2$, and so we can write

$$(0.00515 \text{ mol of } C_8H_{10}N_4O_2) \left(\frac{27 \text{ mol } O_2}{2 \text{ mol } C_8H_{10}N_4O_2} \right) = 0.0695 \text{ mol of } O_2$$

To convert this number of moles of O_2 to mass, we need to multiply by the number of grams per mole of O_2, 32.0 g.

$$(0.0695 \text{ mol of } O_2) \left(\frac{32.0 \text{ g of } O_2}{1 \text{ mol of } O_2} \right) = 2.22 \text{ g}$$

In the above example, we have calculated how much of another reactant (i.e., same side of the equation) is involved with a given reactant. We can equally well calculate across the arrow, i.e., how much product results from a given reactant.

EXAMPLE 12

Given the same equation above, how many grams of NO_2 would be produced from burning 1.00 g of caffeine?

Solution

$$\frac{1.00 \text{ g of } C_8H_{10}N_4O_2}{194.1 \text{ g of } C_8H_{10}N_4O_2/\text{mol of } C_8H_{10}N_4O_2} = 0.00515 \text{ mol of } C_8H_{10}N_4O_2$$

The equation tells us that there are 8 mol of NO_2 produced when 2 mol of $C_8H_{10}N_4O_2$ are consumed. Therefore, we can write

$$(0.00515 \text{ mol of } C_8H_{10}N_4O_2) \left(\frac{8 \text{ mol of } NO_2}{2 \text{ mol of } C_8H_{10}N_4O_2} \right) = 0.0206 \text{ mol of } NO_2$$

As noted above, 1 mol of NO_2 weighs 46.0 g. The final answer is given by

$$(0.0206 \text{ mol of } NO_2) \left(\frac{46.0 \text{ g of } NO_2}{1 \text{ mol of } NO_2} \right) = 0.948 \text{ g of } NO_2$$

The third variation of the above kind of problem occurs when we are told how much of one product is formed and asked to calculate how much of another product is produced at the same time.

EXAMPLE 13

Suppose in the burning of caffeine, according to the equation given above, 1.00 g of CO_2 is produced. How many grams of H_2O must be produced simultaneously?

Solution

The equation tells us there are 10 mol of H_2O produced per 16 mol of CO_2, and so all we have to do is figure out how many moles of CO_2 we have, multiply that by $\frac{10}{16}$, and then convert to grams of H_2O.

$$\frac{1.00 \text{ g of } CO_2}{44.0 \text{ g of } CO_2/\text{mol of } CO_2} = 0.0227 \text{ mol of } CO_2$$

$$(0.0227 \text{ mol of } CO_2) \left(\frac{10 \text{ mol of } H_2O}{16 \text{ mol of } CO_2} \right) = 0.0142 \text{ mol of } H_2O$$

$$(0.0142 \text{ mol of } H_2O) \left(\frac{18.0 \text{ g of } H_2O}{1 \text{ mol of } H_2O} \right) = 0.256 \text{ g of } H_2O$$

The hardest kind of problem to solve involving chemical equations is the so-called "excess" problem or "deficient" reagent problem, where quantities of reagents are given but not in the exact proportions required by the equation. In other words, one of the reagents is in excess over the amount required. In such cases, one has to be careful to work with the *limiting reagent,* the one that is present in least stoichiometric amount and therefore controls the amount of product.

EXAMPLE 14

How much magnesium sulfide can be made from 1.00 g of magnesium and 1.00 g of sulfur by the reaction $Mg + S \longrightarrow MgS$?

Solution

The first step is to convert the weight data into moles:

$$\frac{1.00 \text{ g of Mg}}{24.3 \text{ g/mol}} = 0.0412 \text{ mol of Mg}$$

$$\frac{1.00 \text{ g of S}}{32.1 \text{ g/mol}} = 0.0312 \text{ mol of S}$$

The equation $Mg + S \longrightarrow MgS$ tells us that we need 1 mol of Mg per 1 mol of S. We are given more moles of Mg than of S. Therefore, the Mg is in excess, and not all of it will be used up. The given moles of S control the amount of product. We get 1 mol of MgS per 1 mol of S, and so we can write

$$(0.0312 \text{ mol of S}) \left(\frac{1 \text{ mol of MgS}}{1 \text{ mol of S}} \right) = 0.0312 \text{ mol of MgS}$$

To get the weight of MgS produced, we multiply by the weight of 1 mol of MgS:

$$(0.0312 \text{ mol of MgS}) \left(\frac{56.4 \text{ g of MgS}}{1 \text{ mol of MgS}} \right) = 1.76 \text{ g of MgS}$$

As a further example of an "excess" problem we consider the burning of caffeine in a limited amount of oxygen. Not only are the products different, but the quantity formed is controlled by only one of the reactants, though amounts of both reactants are specified.

EXAMPLE 15

When caffeine $C_8H_{10}N_4O_2$ is burned in a limited supply of O_2, the products are CO, H_2O, and NO. How many grams of H_2O would be produced in such a reaction if the starting amount of $C_8H_{10}N_4O_2$ is limited to 1.00 g and the supply of O_2 is limited to 2.00 g?

Solution

The products here are different, and so we have a new equation to balance. Given

$$C_8H_{10}N_4O_2 + ?O_2 \longrightarrow ?CO + ?H_2O + ?NO$$

Starting with 1 mol of $C_8H_{10}N_4O_2$ on the left, we note that 8 C on the left requires 8 C on the right, and so we put the coefficient 8 in front of the CO; 10 H's on the left means 10 H's on the right, and so we put a 5 in front of the H_2O; 4 N's on the left means 4 N's on the right, and so we put a 4 in front of the NO. The display now looks like this:

$$C_8H_{10}N_4O_2 + ?O_2 \longrightarrow 8CO + 5H_2O + 4NO$$

On the right side we count up the oxygen as $8 + 5 + 4 = 17$, and so we need 17 O atoms on the left. We already have 2 O atoms in $C_8H_{10}N_4O_2$, and so we need 15 more from O_2. We can get these by taking $\frac{15}{2}O_2$. The final balanced equation reads

$$C_8H_{10}N_4O_2 + \tfrac{15}{2}O_2 \longrightarrow 8CO + 5H_2O + 4NO$$

Now to solve the main problem. We are given a quantity of $C_8H_{10}N_4O_2$ and a quantity of O_2. Only one of these will control the reaction; the other reagent will be in excess. The problem is to decide which reagent will control the amount of reaction, i.e., which is the limiting reagent. The way to tell is to convert each of the given weights to moles and compare with the desired ratio of $1:\frac{15}{2}$.

$$\frac{1.00 \text{ g of } C_8H_{10}N_4O_2}{194.1 \text{ g of } C_8H_{10}N_4O_2/\text{mol of } C_8H_{10}N_4O_2} = 0.00515 \text{ mol of } C_8H_{10}N_4O_2$$

$$\frac{2.00 \text{ g of } O_2}{32.0 \text{ g of } O_2/\text{mol of } O_2} = 0.0625 \text{ mol of } O_2$$

According to the equation, we need $\frac{15}{2}$ mol of O_2 per mole of $C_8H_{10}N_4O_2$, and so 0.00515 mol of $C_8H_{10}N_4O_2$ would require

$$(0.00515 \text{ mol of } C_8H_{10}N_4O_2) \left(\frac{\frac{15}{2} \text{ mol of } O_2}{1 \text{ mol of } C_8H_{10}N_4O_2} \right) = 0.0386 \text{ mol of } O_2$$

Clearly, 0.0386 mol of O_2 is less than the 0.0675 mol of O_2 we are being provided with, and so the given quantity of O_2 is in excess over that required for reaction. Some of the given O_2 will be left over, and calculations based on the quantity of O_2 given will not tell us how much product actually forms. To find how much product is generated, we need to base our calculation on the limiting reagent, the $C_8H_{10}N_4O_2$. Actually, once we decide which is the limiting reagent, we have practically solved the problem. We know how many moles of $C_8H_{10}N_4O_2$ are provided, 0.00515. We can also see from the balanced equation that 5 mol of H_2O are produced per mole of $C_8H_{10}N_4O_2$.

$$(0.00515 \text{ mol of } C_8H_{10}N_4O_2) \left(\frac{5 \text{ mol of } H_2O}{1 \text{ mol of } C_8H_{10}N_4O_2} \right) = 0.0258 \text{ mol of } H_2O$$

To convert this answer to grams, we multiply by the weight of 1 mol of H_2O:

$$(0.0258 \text{ mol of } H_2O) \left(\frac{18.0 \text{ g of } H_2O}{1 \text{ mol of } H_2O} \right) = 0.464 \text{ g of } H_2O$$

Why are "excess" problems so important? One reason is that they represent very practical problems. Chemical reagents are often mixed in quantities that are not in the precise ratio required by the stoichiometric equation. Therefore, we have to know the concept of the limiting reagent to be able to figure out how much product to expect. The other reason for stressing "excess" problems is that they are a good test as to whether we understand the fundamental concepts of moles and stoichiometric reactions.

A tricky variant of the "excess" problem is to specify quantities of two products and ask what is the minimum amount of reaction needed to produce at least so much of the two products. The following example illustrates this.

EXAMPLE 16
When you see the tip of a match flare, the chemical reaction is likely to be $P_4S_3 + 8O_2 \longrightarrow P_4O_{10} + 3SO_2$. What is the minimum amount of P_4S_3 that would have to be burned by this reaction to produce at least 1.00 g of P_4O_{10} and at least 1.00 g of SO_2?

Solution
The trick about this question lies in the implication that more than 1.00 g of P_4O_{10} or of SO_2 may be produced. However, we do not know at first which product will be produced in excess over that required. The most direct way to solve the problem is to calculate how much P_4S_3 would be required to produce 1.00 g of P_4O_{10} and how much would be needed to produce 1.00 g of SO_2. The larger of the two answers would satisfy both requirements.

To solve, we convert the desired weights into moles and use the chemical equation to determine how many moles of P_4S_3 are needed.

a To make the 1.00 g of P_4O_{10}:

$$\frac{1.00 \text{ g of } P_4O_{10}}{283.9 \text{ g of } P_4O_{10}/\text{mol of } P_4O_{10}} = 0.00352 \text{ mol of } P_4O_{10}$$

The equation tells us 1 mol of P_4S_3 is required to produce 1 mol of P_4O_{10}:

$$(0.00352 \text{ mol of } P_4O_{10}) \left(\frac{1 \text{ mol of } P_4S_3}{1 \text{ mol of } P_4O_{10}} \right) = 0.00352 \text{ mol of } P_4S_3 \text{ required}$$

b To make the 1.00 g of SO_2:

$$\frac{1.00 \text{ g of } SO_2}{64.06 \text{ g of } SO_2/\text{mol of } SO_2} = 0.0156 \text{ mol of } SO_2$$

The equation tells us that 1 mol of P_4S_3 is required to produce 3 mol of SO_2:

$$(0.0156 \text{ mol of } SO_2)\left(\frac{1 \text{ mol of } P_4S_3}{3 \text{ mol of } SO_2}\right) = 0.00520 \text{ mol of } P_4S_3 \text{ required}$$

As we can see, it takes more P_4S_3 to make 1.00 g of SO_2 than it does to make 1.00 g of P_4O_{10}. To satisfy both required products, we need to take the larger of the two amounts (the amount of P_4S_3) required:

$$(0.00520 \text{ mol of } P_4S_3)\left(\frac{220.1 \text{ g of } P_4S_3}{1 \text{ mol of } P_4S_3}\right) = 1.14 \text{ g of } P_4S_3$$

Two things should be evident from the above examples: (1) We need the coefficients in the balanced equation to tell us how many moles of each reactant and each product are involved when the reaction takes place as written. (2) We need only to work with the reactants and products asked about in the calculation. It is generally assumed that the other reactants are available as needed.

2.8
STOICHIOMETRY INVOLVING SOLUTIONS

In the preceding section, most of the discussion implied that reactants and products were being monitored by mass; i.e., they were being weighed out on a chemical balance. Actually, when a chemist works, he or she is often likely to pour out a volume of solution to get a desired amount of reagent rather than weigh out a mass of material. The reasons for this are that (1) many reagents routinely come as solutions, and (2) it is considerably quicker and easier to measure out a volume of solution than it is to weigh out a reagent and then dissolve it.

To understand how to calculate with reagents in solution, the most important point to keep clear is the distinction between *concentration* of a reagent and total *amount of reagent* in a given sample. *Concentration* tells us the *strength of a solution; concentration × size of sample* tells us the *total amount of reagent.*

There are many ways to express concentration, some of which we shall consider in detail in Chap. 8. Here we content ourselves with but one concentration unit, molarity, probably the most commonly used of all concentration units in chemistry. *Molarity* is defined as the *number of moles of solute per liter of solution. Solute* generally refers to the substance present in a smaller amount, but in some solutions, for example, concentrated sulfuric acid, H_2SO_4 is referred to as the solute and H_2O as the solvent even when the H_2O is present in much the smaller amount. We shall generally restrict ourselves to aqueous solutions, i.e., those in which water is the solvent. If no solvent is specified, it is generally assumed to be water.

Molarity as a concentration unit is generally designated by M. Since we assume the solvent is water, to know the molarity we need only a number which tells how many moles of solute were used per liter to make the solution and a formula to tell what the solute is. Thus, 6.0 M HCl (which is read "6.0-molar HCl") describes a solution in which 6.0 mol of the solute HCl were dissolved in enough water to make a liter of solution. Obviously, if we want 6.0 mol of HCl, we need to take the full liter of the solution. If we take any fraction of a liter, then we have that fraction of the moles contained per liter. The pertinent formula is

Number of moles of solute = (molarity of the solution) (fraction of a liter taken)

or
$$n = MV$$

where n = number of moles
M = molarity, mol/liter
V = volume, liter

Dimensionally, this checks out as follows:

$$\text{mol} = \left(\frac{\text{mol}}{\text{liter}}\right)\left(\text{liter}\right)$$

The application of this most-useful relation is illustrated in the following problems.

EXAMPLE 17

What is the molarity of a solution made by dissolving 15.0 g of HNO_3 in enough water to make 0.200 liter of solution?

Solution

Convert the mass of HNO_3 into moles and divide by the number of liters of solution:

$$\frac{15.0 \text{ g of } HNO_3}{63.0 \text{ g/mol}} = 0.238 \text{ mol of } HNO_3$$

$$\frac{0.238 \text{ mol of } HNO_3}{0.200 \text{ liter of solution}} = 1.19 \ M \ HNO_3$$

EXAMPLE 18

Suppose you need 0.837 mol of HNO_3 for a particular reaction. You have a solution that is labeled 6.00 M HNO_3. How much of the solution should you take?

Solution

Divide the moles needed by the strength of the solution, and that will tell you how much solution to take:

$$\frac{0.837 \text{ mol}}{6.00 \text{ mol/liter}} = 0.140 \text{ liter}$$

It is interesting to note in these problems how useful dimensional analysis is to check on whether the quantities are being combined correctly. To save time, there is usually a tendency to skimp on writing out units, but even professional chemists will testify that a second or two devoted to this may save hours hunting down an error in calculation.

EXAMPLE 19

A sample consisting of 25.0 ml of 3.00 M HNO_3 is mixed with 75.0 ml of 4.00 M HNO_3. Assuming volumes are additive, what would be the concentration of the final mixture?

Solution

Calculate the moles of solute contributed by each sample and divide the total number of moles by the total volume.

First find the moles of HNO_3 in 25.0 ml of 3.00 M HNO_3:

$$(0.0250 \text{ liter}) \left(\frac{3.00 \text{ mol}}{\text{liter}} \right) = 0.0750 \text{ mol}$$

Then find the moles of HNO_3 in 75.0 ml of 4.00 M HNO_3:

$$(0.0750 \text{ liter}) \left(\frac{4.00 \text{ mol}}{\text{liter}} \right) = 0.300 \text{ mol}$$

Total moles of HNO_3 is 0.0750 plus 0.300 = 0.375 mol
Total final volume is 0.0250 liter plus 0.0750 liter, or 0.100 liter.

$$\text{Final concentration} = \frac{0.375 \text{ mol of } HNO_3}{0.100 \text{ liter of solution}} = 3.75 \ M \ HNO_3$$

EXAMPLE 20

How much water should be added to 1.38 liters of 3.00 M HNO_3 to make a **solution that is** 1.40 M HNO_3? Assume volumes are additive.

Solution

Figure out how many moles of HNO_3 you have, then calculate over what volume you need to spread it out to get a final concentration of 1.80 M:

$$(1.38 \text{ liters}) \left(\frac{3.00 \text{ mol of } HNO_3}{\text{liter}} \right) = 4.14 \text{ mol of } HNO_3$$

$$\frac{4.14 \text{ mol of } HNO_3}{x \text{ liter}} = 1.40 \text{ mol/liter}$$

$$x = \frac{4.14}{1.40} = 2.96$$

But this is the total volume, of which 1.38 liters is contributed by the original 3.00 M HNO_3. To get the volume of water needed to be added, subtract 1.38 liters from 2.96 liters. Final answer: 1.58 liters of water.

In the last two examples, it was assumed that volumes were additive; i.e., when water is added to a solution, the volume of the initial solution plus the volume of water added equals the final total volume, or, when two solutions are mixed, the final volume equals the sum of the initial volumes that are being mixed to give the final solution. This is actually not a bad assumption, but it is not precisely true because on mixing of solutions there is generally some contraction or some expansion. The effect is usually rather small, as can be gathered from the following example.

EXAMPLE 21

When 10.00 ml of ethyl alcohol (C_2H_5OH, density 0.7893 g/ml) is mixed with 20.00 ml of water (density 0.9971 g/ml) at 25°C, the final solution has a density of 0.9571 g/ml. Calculate the percentage change in total volume on mixing, and calculate also the molarity of the final solution.

Solution

We start with 10.00 ml of one component plus 20.00 ml of the other component, so that the initial total volume is 30.00 ml. To find the final total volume, we can use the given density, but we need to know the final mass of solution. Masses are additive, and so we need to know the mass of each component.

$$(10.00 \text{ ml of ethyl alcohol}) \left(0.7893 \frac{g}{ml}\right) = 7.893 \text{ g of ethyl alcohol}$$

$$(20.00 \text{ ml of water}) \left(0.9971 \frac{g}{ml}\right) = 19.94 \text{ g of water}$$

$$\text{Total mass} = 7.893 \text{ g} + 19.94 \text{ g} = 27.83 \text{ g of solution}$$

$$\text{Final volume} = \frac{27.83 \text{ g}}{0.9571 \text{ g/ml}} = 29.08 \text{ ml}$$

$$\% \text{ volume change} = \left(\frac{30.00 - 29.08}{30.00}\right)100 = 3.1\% \text{ contraction}$$

To get the final molar concentration of ethyl alcohol, we divide the moles of ethyl alcohol by the final total volume.

$$\text{Moles of ethyl alcohol} = \frac{7.893 \text{ g}}{46.07 \text{ g/mol}} = 0.1713 \text{ mol}$$

$$\text{Molarity} = \frac{0.1713 \text{ mol}}{0.02908 \text{ liter}} = 5.891 \text{ M } C_2H_5OH$$

The reason for specifying the temperature in the preceding example is that substances usually expand when heated, and so densities generally decrease with rising temperature. Around room temperature ($\sim$ 25°C), the influence of changing temperature on molarity is usually quite small, and so we shall ignore it. For precise work, the temperature at which molarity is determined should be specified. If not given, it is generally assumed to be 25°C.

How are the above ideas applied to the stoichiometry of chemical reactions? One of the most common kinds of reaction is that involving the neutralization of acids and bases. Although the equations for neutralization are usually written out in terms of moles, the practical procedure in the laboratory most often entails mixing a solution of the acid with a solution of the base. The following example illustrates a typical computation.

EXAMPLE 22

If it takes 25.00 ml of 0.198 M HNO_3 to neutralize 37.50 ml of a solution of NaOH, what is the molarity of the NaOH solution?

Solution

Figure out how many moles of HNO_3 we have; use the chemical equation to calculate how many moles of NaOH are needed. Since we know the volume over which the NaOH moles are distributed, we can calculate the concentration:

$$\text{Moles of } HNO_3 = \left(\frac{25.00 \text{ ml}}{1000 \text{ ml/liter}}\right)\left(\frac{0.198 \text{ mol of } HNO_3}{\text{liter}}\right) = 0.00495$$

The balanced equation

$$NaOH + HNO_3 \longrightarrow NaNO_3 + H_2O$$

shows that 1 mol of NaOH is required per mole of HNO_3, and so the 0.00495 mol of HNO_3 requires 0.00495 mol of NaOH. This is distributed over 37.50 ml of solution so that we can write

$$\text{NaOH concentration} = \frac{0.00495 \text{ mol of NaOH}}{0.0375 \text{ liter}} = 0.132 \text{ } M \text{ NaOH}$$

EXAMPLE 23

How many ml of 0.0150 M $Ca(OH)_2$ would be required to neutralize 35.0 ml of 0.0360 M H_3PO_4?

Solution

$$\text{Moles of } H_3PO_4 = (0.0350 \text{ liter})\left(0.0360 \frac{\text{mol}}{\text{liter}}\right) = 0.00126$$

The balanced equation is

$$3Ca(OH)_2 + 2H_3PO_4 \longrightarrow Ca_3(PO_4)_2 + 6H_2O$$

and tells us we need 3 mol of $Ca(OH)_2$ per 2 mol of H_3PO_4.

$$(0.00126 \text{ mol of } H_3PO_4)\left(\frac{3 \text{ mol of } Ca(OH)_2}{2 \text{ mol of } H_3PO_4}\right) = 0.00189 \text{ mol of } Ca(OH)_2 \text{ required}$$

The $Ca(OH)_2$ solution is given to be 0.0150 M, so that we can write

$$\text{Volume required} = \frac{0.00189 \text{ mol}}{0.0150 \text{ mol/liter}} = 0.126 \text{ liter, or } 126 \text{ ml}$$

Acid-base stoichiometry does not necessarily involve mixing of volumes of both solutions. In fact, the emphasis on thinking in terms of moles is to encourage development of problem-solving techniques that can handle any situation.

EXAMPLE 24

Suppose you spill 40.0 ml of 0.100 M H_2SO_4. How much solid $NaHCO_3$ would you have to dump on it to neutralize the acid by the following reaction?

$$2NaHCO_3(s) + H_2SO_4 \longrightarrow Na_2SO_4 + 2H_2O + 2CO_2(g)$$

Solution

Moles of H_2SO_4 spilled $= (0.0400 \text{ liter}) \left(0.100 \frac{\text{mol}}{\text{liter}}\right) = 0.00400 \text{ mol}$

Moles of $NaHCO_3$ required $= (0.00400 \text{ mol of } H_2SO_4) \left(\frac{2 \text{ mol of NaHCO}_3}{1 \text{ mol of H}_2SO_4}\right)$

$= 0.00800 \text{ mol of NaHCO}_3$

Grams of $NaHCO_3 = (0.00800 \text{ mol}) \left(84.0 \frac{\text{g}}{\text{mol}}\right) = 0.672 \text{ g}$

Besides being useful for acid-base neutralization, solution stoichiometry is often involved in synthetic procedures where a variety of reagents have to be put together, some as weighed-out solids and some as poured-out solutions. In the following situation, for example, a complex salt $K_3Fe(C_2O_4)_3$* is to be synthesized from a solid starting material $Fe(NH_4)_2(SO_4)_2 \cdot 6H_2O$, where the dot indicates that in some unspecified way 6 molecules of H_2O are bound to 1 formula-unit of $Fe(NH_4)_2(SO_4)_2$. Also needed are some solid potassium oxalate ($K_2C_2O_4$), a solution of oxalic acid ($H_2C_2O_4$), and an excess of hydrogen peroxide (H_2O_2). We will not bother here with the detailed chemistry of the reaction, since the course of reaction is complicated. However, we can see from the following example that useful stoichiometric calculations can be carried out even without having all the detailed equations.

EXAMPLE 25

What are the minimum amounts of $Fe(NH_4)_2(SO_4)_2 \cdot 6H_2O$ and of $K_2C_2O_4$ needed for the synthesis of 10.0 g of $K_3Fe(C_2O_4)_3$?

---

*The correct systematic name for this is "tripotassium trisoxalatoferrate(III)," following the rules for nomenclature as spelled out in Appendix A2.1. However, most chemists would probably recognize it as potassium iron oxalate.

Solution

$$10.0 \text{ g of } K_3Fe(C_2O_4)_3 = \frac{10.0 \text{ g}}{437.2 \text{ g/mol}} = 0.0229 \text{ mol of } K_3Fe(C_2O_4)_3$$

1 mol of $K_3Fe(C_2O_4)_3$ requires 3 mol of K atoms, or $\frac{3}{2}$ mol of $K_2C_2O_4$, and 1 mol of Fe atoms, or 1 mol of $Fe(NH_4)_2(SO_4)_2 \cdot 6H_2O$.

$$\text{Amount of } K_2C_2O_4 \text{ required} = 0.0229 \text{ mol of } K_3Fe(C_2O_4)_3 \times \frac{\frac{3}{2} \text{ mol of } K_2C_2O_4}{1 \text{ mol of } K_3Fe(C_2O_4)_3}$$

$$= 0.0343 \text{ mol of } K_2C_2O_4$$

$$= (0.0343 \text{ mol}) \left(166.2 \frac{\text{g}}{\text{mol}}\right) = 5.70 \text{ g}$$

$$\text{Amount of } Fe(NH_4)_2(SO_4)_2 \cdot 6H_2O \text{ required} = 0.0229 \text{ mol of } K_3Fe(C_2O_4)_3 \times$$

$$\frac{1 \text{ mol of } Fe(NH_4)_2(SO_4)_2 \cdot 6H_2O}{1 \text{ mol of } K_3Fe(C_2O_4)_3}$$

$$= 0.0229 \text{ mol of } Fe(NH_4)_2(SO_4)_2 \cdot 6H_2O$$

$$= (0.0229 \text{ mol}) \left(392.1 \frac{\text{g}}{\text{mol}}\right) = 8.98 \text{ g}$$

2.9
OXIDATION-REDUCTION

Besides acid-base reactions and other reactions in which charged ions come together or go apart, there is a large class of reactions known as *oxidation-reduction* reactions. They are also known as *redox* reactions. Originally, the term "oxidation" was used to describe the addition of oxygen to an element or to a compound, and the term "reduction" was used to designate the removal of oxygen from a compound. Later, as the processes became more fully understood, the meaning of the terms was broadened. Oxidation-reduction reactions are now defined as reactions in which there is *transfer of an electron* (a unit of negative charge) from one atom to another. The *loss of electrons* is described as *oxidation;* the *gain of electrons* is called *reduction.* The substance that loses the electrons is said to be *oxidized;* the substance that gains the electrons is said to be *reduced.* Let us consider as an example the reaction

$$2Na + Cl_2 \longrightarrow 2NaCl$$

In this reaction the Na is considered to go from a neutral state on the left side of the equation to a positively charged state (Na^+) in NaCl on the right side of the equation. At the same time, the chlorine goes from a neutral state (Cl_2) on the left side of the equation to a negatively charged state (Cl^-) in NaCl. There has been a transfer of electrons from Na to Cl_2. The Na, which has lost the electrons, is oxidized; the Cl_2, which has gained the electrons, is reduced. Na does the reducing and is therefore called the *reducing agent;* Cl_2 does the oxidizing and is therefore called the *oxidizing agent.* Oxidizing and reducing agents always go together since the electron transfer always involves an electron donor (the reducing agent) and an electron acceptor (the oxidizing agent).

A less-obvious example of electron transfer occurs when hydrogen combines with oxygen to form water by the reaction

$$2H_2 + O_2 \longrightarrow 2H_2O$$

Although it is not so simple, we can visualize the H atom as going from a neutral state in H_2 to a positive state in H_2O; the O atom goes from a neutral state in O_2 to a negative state in H_2O. It appears that there is an electron transfer from H to O. The H_2 is oxidized; the O_2 is reduced. However, as we shall see later, the charge transfer is only partial and is perhaps better described as an electron shift rather than a complete loss by H and a gain by O.

In order to keep track of electron shifts in chemical reactions, it will be convenient to introduce a concept known as oxidation number. The *oxidation number,* also referred to as oxidation state, is defined as the *charge which an atom appears to have* when the net electric charge on a chemical species is apportioned according to certain rules. The basis of the rules, which will be discussed more fully in Chap. 5, is that atoms are essentially electrical in nature, and binding of atoms results either from transfer of electrons from one atom to another or sharing of electrons between adjacent atoms. The sharing depends on how different the atoms are from each other, but the net result is that the following operational rules can be used:

1   In uncombined or free elements, each atom is assigned an oxidation number of 0.   This rule holds no matter how complicated the molecule is. Hydrogen in H_2, sodium in Na, sulfur in S_8, oxygen in O_2, and phosphorus in P_4 all have oxidation numbers of 0.

2   In simple ions (i.e., charged species which contain but one atom), the oxidation number is equal to the charge on the ion.   In these cases, the apparent charge of the atom is the real charge of the ion. Thus, in tripositive aluminum ion (Al^{3+}) the oxidation number of the aluminum is +3. Iron, which can form either dipositive (Fe^{2+}) or tripositive ions (Fe^{3+}), has an oxidation number that is +2 in the former and +3 in the latter. In the dinegative oxide ion, O^{2-}, the oxidation number of oxygen is −2. It is useful to remember that the elements sodium (Na) and potassium (K) form only +1 ions; their oxidation numbers are +1 in all their compounds. The elements magnesium (Mg) and calcium (Ca) form only +2 ions. They always have an oxidation number of +2 in all their compounds.

3   In compounds containing oxygen, the oxidation number of each oxygen atom is generally −2.   There are two kinds of exceptions: One arises in the case of peroxides, compounds of oxygen in which oxygen atoms are directly linked to each other. An example is hydrogen peroxide (H_2O_2). In peroxides, each oxygen is assigned an oxidation number of −1. The second exception occurs more rarely, i.e., when oxygen is bonded to fluorine. In such compounds, e.g., oxygen difluoride (OF_2), the oxygen is assigned an oxidation number of +2.

4   In compounds containing hydrogen, the oxidation number of hydrogen is generally +1.   This rule covers practically all the hydrogen compounds. It fails in the case of the hydrides, where the hydrogen is combined with an element such as Na or Ca. An example of a hydride is NaH, in which the sodium is considered to be Na^+ (oxidation number +1) and the hydrogen is H^- (oxidation number −1). Hydrides are rather rare and will be labeled as such until familiarity with them is established.

5   All oxidation numbers must be consistent with the conservation of charge.   This is just another way of saying that if we add up all the oxidation numbers in a compound and multiply by the number of atoms of each kind, then the sum must be equal to zero if we are dealing with a neutral molecule, or be equal to the net charge if we are dealing with an ion. As an example of a neutral molecule, we can take H_2O. In this compound, the oxidation number of H is +1, and the oxidation number of O is −2. Since we have two H's and one O, the net charge will be $2(+1)$ plus $1(-2) = 0$.

The neutrality rule enables us to assign oxidation numbers to atoms with which we may not be familiar. For example, in the case of sulfuric acid (H_2SO_4), we can figure out that the oxidation number of sulfur must be +6. The reasoning goes as follows: The H in H_2SO_4 has an oxidation number of +1; there are two H's, and so the apparent charge contribution of H is +2. The O in H_2SO_4 has an oxidation number of −2; there are four O's; so the apparent charge contribution of O is $4(-2)$, or −8. What must the sulfur be? If +2 comes from the H's and −8 comes from the O's, we have −6; to counterbalance this, we need a +6. Therefore, the sulfur in H_2SO_4 must have an oxidation number of +6.

Oxidation number is a per-atom concept. If a net charge is distributed over several identical atoms, each contributes an equal share. Thus, in $Na_2Cr_2O_7$ (sodium dichromate), Na is +1 and O is −2. Since there are two Na's and seven O's, their apparent charge contribution is $2(+1)$ plus $7(-2) = -12$. The counterbalance offered by the chromium must be +12. Since there are two Cr's, each must contribute +6. Therefore, the oxidation number of Cr in $Na_2Cr_2O_7$ is +6.

Since oxidation numbers are quite arbitrary, they sometimes have values that look strange. For example, in sucrose, $C_{12}H_{22}O_{11}$, the oxidation number of carbon comes out to be 0. The total apparent charge of the hydrogen, $(22)(+1)$, is just balanced by the total apparent charge of the oxygen, $(11)(-2)$. Hence, the carbon atoms look like they contribute nothing to the net charge of the neutral molecule $C_{12}H_{22}O_{11}$. Fractional oxidation numbers are also possible, as in $Na_2S_4O_6$ (sodium tetrathionate), where the oxidation number of sulfur is $+\frac{10}{4}$.

In complex ions (charged particles that contain more than one atom), the apparent charge of all the atoms must add up to equal the charge on the ion. This is true in the hydroxide ion (OH^-), for example, where the superscript "minus" indicates that the ion has a net charge of −1. Since oxygen has an oxidation number of −2 and hydrogen has an oxidation number of +1, the total apparent charge is $(-2) + (+1) = -1$, the same as the actual charge on the OH^- ion. In the dichromate ion, $Cr_2O_7^{2-}$, a dinegative ion, the seven oxygen atoms contribute $7(-2)$, or -14, and so the chromium must contribute $+12$ in order to make the whole ion have a net charge of -2. Since there are two chromium atoms, each must have an oxidation number of $+6$.

In order to minimize confusion, the actual charge on an ion is written as a superscript, and the oxidation number of an atom, when needed, is written beneath the atom to which it applies. For example, in

$$\underset{+5\ -2}{P_2O_7^{4-}}$$

the charge on the ion is -4; the oxidation number of P is $+5$, and the oxidation number of O is -2. It must be emphasized that oxidation numbers are not real but only apparent charges on atoms. In the specific case of $P_2O_7^{4-}$, it can be shown experimentally that the complex aggregate carries a -4 charge, but it cannot be shown experimentally that the real charge of P is $+5$ and that of O is -2. The $+5$ and -2 are arbitrarily assigned numbers. We must not conclude that $P_2O_7^{4-}$ contains P^{5+} and O^{2-} ions.

Once we have the concept of oxidation number, we can redefine *oxidation as a chemical process in which an atom shows an increase in oxidation number; reduction is a process in which an atom shows a decrease in oxidation number.* For the redox reaction

$$2H_2 + O_2 \longrightarrow 2H_2O$$

the hydrogen changes oxidation number from 0 to $+1$ (is oxidized) and the oxygen changes oxidation number from 0 to -2 (is reduced). When sucrose, $C_{12}H_{22}O_{11}$, is burned in oxygen to give carbon dioxide, CO_2, the carbon goes from 0 to $+4$. The sucrose is oxidized. Figure 2.3 summarizes the terms used to describe oxidation-reduction.

Listed in Fig. 2.4 are some typical examples of oxidation-reduction reactions. The numbers below the formulas indicate the oxidation numbers of interest. It should be noted that the terms "oxidizing agent" and "reducing agent" refer to the entire substance and not to just one of the atoms contained therein. For example, in the next-to-last reaction of the table, the oxidizing agent is $KClO_3$, not $+5Cl$. Because the rules for assigning oxidation numbers

FIG. 2.3 Terms Used in Describing Oxidation-Reduction

Term	Change in oxidation number	Change in electrons
Oxidation	Increase	Loss of electrons
Reduction	Decrease	Gain of electrons
Oxidizing agent	Decrease	Accepts electrons
Reducing agent	Increase	Donates electrons
Substance oxidized	Increase	Loses electrons
Substance reduced	Decrease	Gains electrons

FIG. 2.4 Examples of Oxidation-Reduction Reactions

Oxidizing agent	+	reducing agent	$\longrightarrow$	product
O_2		H_2		H_2O
0		0		+1 -1
O_2		CH_4		$CO_2 + H_2O$
0		-4		+4 -2 -2
S		Zn		ZnS
0		0		+2 -2
Cl_2		H_2		HCl
0		0		+1 -1
H^+		Zn		$Zn^{2+} + H_2$
+1		0		+2 0
$KClO_3$		S		$KCl + SO_2$
+5		0		-1 +4
H_2O_2		H_2O_2		$H_2O + O_2$
-1		-1		-2 0

are quite arbitrary, it only *appears* that it is the +5Cl that accepts the electrons. The only thing that can be demonstrated experimentally is that $KClO_3$ accepts electrons; how the accepting is assigned is a matter of choice. Two other points in Fig. 2.4 are noteworthy: (1) Sulfur in the third reaction acts as an oxidizing agent; in the next-to-last reaction, it acts as a reducing agent. (2) H_2O_2 in the last reaction acts both as an oxidizing agent and a reducing agent. It oxidizes and reduces itself and is said to undergo *autooxidation* or *disproportionation*.

2.10
BALANCING OXIDATION-REDUCTION EQUATIONS

Given the reactants and products of a redox reaction, the chemical equation can generally be balanced by inspection, but such a procedure is likely to be a hit-or-miss affair that can end up in blind alleys of wasted effort. It is usually wiser to use a systematic method based on matching up the electron transfer. As an example, suppose we consider a reaction in which H_2 combines with O_2 to form H_2O. We write the problem as

$$?H_2 + ?O_2 \longrightarrow ?H_2O$$

The question marks indicate the coefficients we are looking for. First we assign oxidation numbers; then we note what changes occur; finally we adjust the coefficients to balance the changes. Hydrogen goes from 0 in H_2 to $+1$ in H_2O; oxygen goes from 0 in O_2 to -2 in H_2O. We can indicate the changes by arrows as follows:

$$?H_2 \quad + \quad ?O_2 \longrightarrow H_2O$$

$$\begin{array}{cccc} 0 & 0 & +1 & -2 \end{array}$$

1e^- per atom ↑ 2e^- per atom
×2   ×2

The e^- stands for a unit of transferred charge. The downward pointing arrow below H_2 indicates that 1e^- per atom has to be *released* to change one atom of H from 0 to $+1$. We multiply this by 2 because we have a subscript 2 in the formula H_2, and so, like it or not, we have two H atoms to worry about. Similarly, the upward pointing arrow under the O_2 indicates that 2e^- per oxygen atom have to be *gained* to change one atom of O from 0 to -2. Again, we have to multiply the 2e^- by two because of the two atoms in O_2. At this stage we see 2e^- coming out of H_2 and 4e^- going into O_2. To conserve charge, we need to take two of the former changes for each of the latter. This gives us

$$2H_2 + 1O_2 \longrightarrow ?H_2O$$

The final step is to adjust the right side to be consistent with the left side by putting a 2 before the H_2O. The balanced equation is then

$$2H_2 + O_2 \longrightarrow 2H_2O$$

As a more complicated example let us now consider the reaction in which a mixture of $KClO_3$ and sugar ($C_{12}H_{22}O_{11}$) is ignited to produce CO_2, H_2O, and KCl as products. The problem is to balance the equation

$$?KClO_3 + ?C_{12}H_{22}O_{11} \longrightarrow ?CO_2 + ?H_2O + ?KCl$$

So far as electron transfer is concerned, we need worry only about those atoms that change oxidation number. On applying the rules for assigning oxidation numbers, we can see that Cl changes from $+5$ in $KClO_3$ to -1 in KCl and C changes from 0 in $C_{12}H_{22}O_{11}$ to $+4$ in CO_2. In going from $+5$ to -1, the Cl appears to gain six negative charges; in going from 0 to $+4$, the C appears to lose four negative charges. We can indicate this as follows:

$$?KClO_3 + ?C_{12}H_{22}O_{11} \longrightarrow ?CO_2 + ?H_2O + ?KCl$$

$$\begin{array}{cccc} +5 & 0 & +4 & -1 \end{array}$$

↑6e^- ↓4e^- per atom
48e^- per formula-unit

Since each formula-unit of $C_{12}H_{22}O_{11}$ contains 12 carbon atoms, it will donate 12×4, or 48, electrons. These 48 electrons must be picked up by the $KClO_3$.

Since each $KClO_3$ accepts 6 electrons, we need $8KClO_3$ units for each $C_{12}H_{22}O_{11}$ unit. Hence we can write

$$8KClO_3 + 1C_{12}H_{22}O_{11} \longrightarrow ?CO_2 + ?H_2O + ?KCl$$

Eight K atoms on the left require 8 K atoms on the right, and so we next put an 8 in front of the KCl. Twelve C atoms on the left mean 12 C atoms on the right, and so we put a 12 in front of the CO_2. The final step is to balance the H_2O. This can be done by counting O on the left ($8 \times 3 + 1 \times 11 = 35$) and noting that $12CO_2$ on the right accounts for $2 \times 12 = 24$ of these, leaving 11 more to be accounted for in the H_2O, i.e., $11H_2O$. The final balanced equation is

$$8KClO_3 + C_{12}H_{22}O_{11} \longrightarrow 12CO_2 + 11H_2O + 8KCl$$

Here, in summary, for future reference are the steps followed:

1 Assign oxidation numbers for those atoms that change.

2 Decide on number of electrons to be shifted per atom.

3 Decide on number of electrons to be shifted per formula-unit.

4 Compensate electron gain and loss by writing appropriate coefficients for the oxidizing agent and the reducing agent.

5 Insert other coefficients consistent with the conservation of atoms.

Once the balanced redox equation has been obtained, it can, of course, be used for stoichiometric calculations, as indicated in Sec. 2.7. The following examples illustrate typical calculations.

EXAMPLE 26

On being heated, the white solid $KClO_3$ decomposes to form the white solid KCl and oxygen gas O_2. How many grams of $KClO_3$ must be decomposed to give 0.96 g of O_2?

Solution

First we need to balance the equation

The Cl changes from $+5$ to -1; it appears to gain 6 electrons. The O changes from -2 to 0; it appears to lose 2 electrons. There are three oxygen atoms for every chlorine atom in $KClO_3$, and so the compound itself has taken care of the electron gain and electron loss. To fix the right side of the equation, we note that one $KClO_3$ on the left produces one KCl on the right. Three oxygen atoms on the left require three oxygen atoms on the right. We can get these three oxygen atoms on the right by placing the coefficient $\frac{3}{2}$ before the formula O_2, giving

$$KClO_3(s) \longrightarrow KCl(s) + \tfrac{3}{2}O_2(g)$$

Multiplying through by 2 to get rid of the fraction gives

$$2KClO_3(s) \longrightarrow 2KCl(s) + 3O_2(g)$$

We now have the balanced equation and can proceed to solve the problem by converting the given oxygen weight to moles and calculating how much $KClO_3$ is required to produce this:

$$0.96 \text{ g of } O_2 = \frac{0.96 \text{ g}}{32 \text{ g/mol}} = 0.030 \text{ mol of } O_2$$

$$(0.030 \text{ mol of } O_2) \frac{2 \text{ mol of } KClO_3}{3 \text{ mol of } O_2} = 0.020 \text{ mol of } KClO_3$$

$$(0.020 \text{ mol of } KClO_3) (122.5 \text{ g/mol}) = 2.5 \text{ g}$$

EXAMPLE 27

On being heated, 4.90 g of $KClO_3$ shows a weight loss of 0.384 g. What percentage of the original $KClO_3$ has decomposed?

Solution

The weight loss of 0.384 g is due to the fact that a gas is driven off. The only gas formed in the reaction is oxygen.

$$2KClO_3(s) \longrightarrow 2KCl(s) + 3O_2(g)$$

$$\text{Moles of } O_2 \text{ driven off} = \frac{0.384 \text{ g of } O_2}{32.0 \text{ g/mol}} = 0.0120 \text{ mol}$$

$$\text{Moles of } KClO_3 \text{ decomposed} = (0.0120 \text{ mol of } O_2) \frac{2 \text{ mol of } KClO_3}{3 \text{ mol of } O_2}$$

$$= 0.00800 \text{ mol}$$

$$\text{Moles of } KClO_3 \text{ originally available} = \frac{4.90 \text{ g of } KClO_3}{122.55 \text{ g/mol}} = 0.0400 \text{ mol}$$

$$\text{Percent decomposed} = \frac{\text{moles decomposed}}{\text{moles available}} \times 100$$

$$= \frac{0.00800}{0.0400} \times 100 = 20.0\%$$

EXAMPLE 28

In the reaction of gaseous carbon monoxide (CO) with solid iron oxide (Fe_2O_3), the products are gaseous CO_2 and solid Fe. How many grams of Fe can be formed from 1.00 g of CO and 5.00 g of Fe_2O_3?

Solution

In solving this problem, the first thing to do is to write the balanced equation:

$$3CO(g) + Fe_2O_3(s) \longrightarrow 3CO_2(g) + 2Fe(s)$$

Next we decide which reactant limits the amount of product and which reactant is present in excess.

To do this, we convert the given data into moles. The formula weight of CO is 28.0 amu; the formula weight of Fe_2O_3 is 159.7 amu.

$$\frac{1.00 \text{ g of CO}}{28.0 \text{ g/mol}} = 0.0357 \text{ mol of CO}$$

$$\frac{5.00 \text{ g of } Fe_2O_3}{159.7 \text{ g/mol}} = 0.0313 \text{ mol of } Fe_2O_3$$

The equation tells us there should be 1 mol of Fe_2O_3 per 3 mol of CO, so that 0.0357 mol of CO would require

$$(0.0357 \text{ mol of CO}) \frac{1 \text{ mol of } Fe_2O_3}{3 \text{ mol of CO}} = 0.0119 \text{ mol of } Fe_2O_3$$

Clearly, there is more than enough Fe_2O_3 to supply this needed amount, so that the Fe_2O_3 is in excess. CO is the limiting reagent. The calculation of Fe product should be based on CO:

$$(0.0357 \text{ mol of CO}) \frac{2 \text{ mol of Fe}}{3 \text{ mol of CO}} = 0.0238 \text{ mol of Fe}$$

$$(0.0238 \text{ mol of Fe}) \left(55.85 \frac{\text{g}}{\text{mol}}\right) = 1.33 \text{ g}$$

Important Concepts

stoichiometry
atomic weight
mole
Avogadro number
simplest formula
molecular formulas

formula weight
molecular weight
chemical equations
balancing equations
calculations using equations
limiting reagent

molarity
acid-base reactions
oxidation-reduction
oxidation number
balancing oxidation-reduction equations

Exercises

*2.1 Atomic weight Given the Dalton finding that water was composed of 12.5% by weight hydrogen and 87.5% by weight oxygen, what atomic weight, referred to H = 1, would Dalton have assigned to O if he had assumed water is formulated of one H atom per two O atoms.

*2.2 Atomic weight Using the present relative values for the atomic weights, what would be the

atomic weight of uranium on Berzelius' scale, referred to O = 100?

**2.3 Atomic mass unit What is the mass of 1 amu in grams?

**2.4 Atomic weight Suppose that prior to 1960, physicists set their atomic weight scale to $O^{18} = 18.0000$ amu whereas the chemists left theirs alone. On which scale would atomic weights be greater?

*2.5 Stoichiometry (a) How many moles of NaCl are there in 117 g of NaCl? (b) How many moles of HCl can be made from 117 g of NaCl by the reaction

$$NaCl(s) + H_2SO_4 \longrightarrow NaHSO_4(s) + HCl(g)$$

(c) How many grams of HCl can be made from 117 g of NaCl?

*2.6 Stoichiometry A given bottle contains some colored compound. Analysis of the contents shows the presence of 0.300 mol of Fe and 0.400 mol of O atoms. (a) How many grams of compound are there in the bottle? (b) Calculate the percent of the compound's weight contributed by each element? (c) What is the simplest formula of the compound? (d) How many moles of the compound are there in the bottle?

*2.7 Stoichiometry How many moles are there in each of the following: (a) 128 g of CO_2; (b) 10 molecules of SO_2; (c) 0.032 g of O_2.

*2.8 Stoichiometry Given the reaction $Pb + S \longrightarrow PbS$, what is the maximum weight of PbS that can be made from 32 g of Pb and 207 g of S?

**2.9 Atomic abundance Lithium consists of isotopes lithium 6 (mass 6.01513 amu) and lithium 7 (mass 7.01601 amu). If the natural atomic weight is 6.941 amu, what is the percentage relative abundance of the two isotopes?

Ans. 7.49 and 92.51%

*2.10 Atomic abundance Natural bromine is found by mass-spectrometric analysis to consist of 50.54% of the bromine-79 isotope (mass 78.9183 amu) and 49.46% of the bromine-81 isotope (mass 80.9163 amu). What is the chemical atomic weight of natural bromine?

*2.11 Atomic weight Natural copper consists of the isotopes copper 63 and copper 65. The natural abundance is 69.09% of the former and 30.91% of the latter. If the mass of the copper-63 isotope is 62.9298 amu, what must be the mass of the copper-65 isotope?

*2.12 Mole The total population of the world is now believed to be about 4.2 billion people. How many moles of people is this? If you had one sulfur atom for each person, what would be the weight of the sulfur sample?

Ans. 6.9 fmol: 0.223 pg

**2.13 Mole How many moles of atoms are there in one atom?

**2.14 Mole How many moles of phosphorus atoms are there in each of the following: (a) 3.1 g of phosphorus; (b) 31 atoms of phosphorus; (c) 3.1×10^{23} atoms of phosphorus; (d) 3.1 g of P_4?

*2.15 Mole What is the atomic weight of an element X for which a sample containing 1.58×10^{19} atoms weighs 1.05 mg?

*2.16 Mole In a chemical reaction requiring two atoms of aluminum for three atoms of sulfur, how many grams of aluminum are required per gram of sulfur?

**2.17 Simplest formula On being heated in oxygen, 3.120 g of metal M converts to 4.560 g of oxide. If the atomic weight of M is 52.0, what is the simplest formula of the oxide?

*2.18 Simplest formula A given sample of oxalic acid shows on analysis 26.68% C, 71.08% O, and 2.24% H by weight. What is the simplest formula of the compound?

*2.19 Percent composition Which compound contains a greater percent by weight of water; $MgSO_4 \cdot 7H_2O$ or $KAl(SO_4)_2 \cdot 12H_2O$? Which of these compounds contains a greater percent by weight of oxygen?

*2.20 Simplest formula The principal active constituent of marijuana contains 71.23% by weight of carbon, 12.95% hydrogen, and 15.81% oxygen. What is its simplest formula?

Ans. $C_{21}H_{30}O_2$

**2.21 Molecular formula A compound consisting only of carbon and chlorine shows on analysis 10.15% carbon and 89.85% chlorine. If each molecule weighs 3.93×10^{-22} g, what is the molecular formula of the compound?

*2.22 Formula weights What is the formula weight of each of the following: (a) H, (b) H_2, (c) H_2O, (d) $CaSO_4$, (e) $CaSO_4 \cdot 2H_2O$.

**2.23 Molecular weight What is the molecular weight of a compound which contains in 1.04×10^{22} molecules of itself 0.8266 g of carbon and 1.04×10^{23} atoms of hydrogen?

*2.24 Percent composition When FeS_2 is burned in air, it converts to Fe_2O_3. What happens to the percent by weight of iron in the process?

Ans. From 46.55% to 69.95%

2.25 **Chemical equations** Balance each of the following equations:

a $?C_2H_6 + ?O_2 \longrightarrow ?CO_2 + ?H_2O$

b $?C_2H_6 + ?O_2 \longrightarrow ?CO + ?H_2O$

c $?C_2H_6O + ?O_2 \longrightarrow ?CO_2 + ?H_2O$

d $?C_2H_6O + ?O_2 \longrightarrow ?CO + ?H_2O$

2.26 **Chemical equations** If the following equation is balanced as it stands, what must be the values of the subscripts x, y, and z?

$$2K_xC_yO_z + O_2 \longrightarrow 2K_2CO_3 + 2CO_2$$

2.27 **Chemical equations** When strychnine, $C_{21}H_{22}N_2O_2$, is burned in oxygen, it forms CO_2, H_2O, and NO_2. Write a balanced equation for the reaction.

2.28 **Chemical equations** Assuming that fuel oil is $C_{16}H_{34}$ and that it burns in air (O_2) to form CO_2 and H_2O, how many tons of fuel oil must I burn to get a ton of H_2O?

2.29 **Chemical equations** When HCl is heated with MnO_2, the products are Cl_2, H_2O, and $MnCl_2$. Given 1.00 g of HCl and 1.00 g of MnO_2, what is the maximum weight of Cl_2 you could prepare? *Ans. 0.486 g*

2.30 **Chemical equations** An unknown mixture of C_2H_6 and C_3H_8 is burned in O_2 to give CO_2 and H_2O as products. Analysis of the products shows 63.04% by weight CO_2 and 36.96% by weight H_2O. What was the makeup of the original mixture?

2.31 **Chemical equations** Given the reaction $Al_4C_3 + 12H_2O \longrightarrow 4Al(OH)_3 + 3CH_4$, what would be the minimum weight of Al_4C_3 needed to make at least 0.500 kg of $Al(OH)_3$ and at least 0.500 kg of CH_4?

2.32 **Solutions** How many moles of HCl are there in each of the following:

a 35.0 ml of 0.225 M HCl

b 25.0 ml of 0.150 M HCl plus 2.50 g of HCl

c 25.0 ml of 0.250 M HCl plus 50.0 ml of 0.300 M HCl

2.33 **Solutions** Assuming volumes are additive, what would be the final concentration of a solution made by mixing 25.0 ml of 0.150 M HNO_3, 35.0 ml of 0.250 M HNO_3, and 45.0 ml of 0.350 M HNO_3?

2.34 **Solutions** How much water should I add to a mixture of 0.250 liter of 6.00 M HCl and 0.650 liter of 4.00 M HCl to make a final solution that is 4.00 M HCl? Assume volumes are additive. *Ans. 0.125 liter*

2.35 **Molarity** Calculate the molarity of each of the following solutions:

a 2.92 g of HCl in enough water to make 50.0 ml of solution

b 5.0×10^{-3} mol of NaOH in 0.100 liter of solution

c A mixture of 25.0 ml of 0.350 M H_2SO_4 and 35.0 ml of 0.250 M H_2SO_4 (assume volumes are additive)

d 3.71 g of $Ca(OH)_2$ plus 20.0 ml of 0.500 M $Ca(OH)_2$ in enough water to make 25.0 ml of solution

e 1.96 g of H_3PO_4 plus 2.50×10^{-3} mol of H_3PO_4 plus 30.0 ml of 0.250 M H_3PO_4 and enough water to make 40.0 ml of solution

2.36 **Molarity** For protecting a typical automobile radiator from freezing in the winter one might mix 8.50 liters of water (density, 0.997 g/ml) and 4.50 liters of ethylene glycol ($HOCH_2CH_2OH$, density 1.109 g/ml). If the density of the final solution is 1.05 g/ml, what is the molarity of the solution?

2.37 **Neutralization** How many milliliters of 0.250 M H_2SO_4 would it take to neutralize 50.0 ml of 0.300 M NaOH?

2.38 **Neutralization** What is the concentration of an unknown HCl solution of which it takes 0.0300 liter to neutralize 0.0600 liter of 0.150 M $Ca(OH)_2$ solution?

2.39 **Neutralization** What volume of 0.0150 M H_2SO_4 would it take to neutralize a mixture of 25.0 ml of 0.240 M NaOH and 45.0 ml of 0.060 M $Ca(OH)_2$ solution? *Ans. 0.380 liter*

2.40 **Neutralization** An unknown solution of $Ca(OH)_2$ is being neutralized with an unknown solution of H_3PO_4. If it requires 1.80 ml of the former to neutralize 1.00 ml of the latter, what must be the relative concentration (molarity) of each? What must be the absolute concentrations if the neutralization product from 1.80 ml of the $Ca(OH)_2$ solution contains 0.310 g of $Ca_3(PO_4)_2$?

2.41 **Oxidation-reduction** In the flame of a Bunsen burner a possible reaction is $C_2H_6 + \frac{7}{2}O_2 \longrightarrow 2CO_2 + 3H_2O$. Indicate specifically the oxidizing agent, the reducing agent, what is oxidized, and what is reduced in this reaction. Indicate why this reaction is referred to as an electron-transfer reaction.

*2.42 Oxidation numbers Assign oxidation numbers to the underlined elements in each of the following:

a NaH$\underline{S}$O$_4$ b Ca$\underline{S}$O$_3$
c H$_4\underline{P}_2$O$_7$ d KAl($\underline{S}$O$_4$)$_2 \cdot$ 12H$_2$O
e $\underline{C}_3$H$_8$ f Ca$\underline{O}_2$ (peroxide)
g Na$_2\underline{Mn}$O$_4$ h NaAl$\underline{H}_4$(hydride)

**2.43 Oxidation numbers What are the oxidation numbers of the underlined elements in each of the following: (a) $\underline{I}_3^-$, (b) $\underline{S}_4$O$_6^{2-}$, (c) $\underline{Fe}_3$O$_4$, (d) $\underline{C}_2$H$_5$OH, (e) $\underline{C}$H$_3\underline{C}$HO?

**2.44 Oxidation-reduction equations Using electron transfer, balance each of the following equations:

a ?Al + ?Cl$_2 \longrightarrow$?AlCl$_3$
b ?Al(s) + ?Zn$^{2+} \longrightarrow$?Zn(s) + ?Al^{3+}
c ?C$_8$H$_{18}$ + ?O$_2 \longrightarrow$?CO$_2$ + ?H$_2$O
d ?K$_2$O + ?NO + ?KMnO$_4 \longrightarrow$
$$+1\ -2\ +2-2\ \ +1+7\ -8$$
?KNO$_3$ + ?MnO
$$+1+7 \qquad +2$$
e ?C$_6$H$_{12}$O$_6$ + ?NaClO$_2 \longrightarrow$
$$0\ +1\ 2\ -1\ 2 \quad +1\ 3\ -4\ \quad ?CO_2 + ?NaCl + ?H_2O$$

*2.45 Oxidation-reduction If 1.00 g of KClO$_3$ is decomposed by heating to KCl and O$_2$, how many grams of O$_2$ will be formed?

Ans. 0.392 g

**2.46 Oxidation-reduction When NaNO$_3$(s) is heated, it decomposes to NaNO$_2$(s) and O$_2$(g). If a particular sample of pure NaNO$_3$(s) is heated so it loses 5.0% of its weight, what fraction of the NaNO$_3$ has been decomposed?

**2.47 Oxidation-reduction When iron rusts, the essential reaction is the oxidation of Fe(s) by O$_2$(g) to form Fe$_2$O$_3$(s). If a given sample of pure iron gains 10.0% of its weight on partial rusting, what fraction of the iron would have converted as above?

**2.48 Oxidation-reduction Given a reaction in which NH$_3$(g) is oxidized by O$_2$(g) to give NO(g) and H$_2$O(g), what is the maximum weight of NO that could be obtained starting only with 1.00 g of NH$_3$ and 2.00 g of O$_2$? *Ans. 1.50 g*

**2.49 Oxidation-reduction You are given 400.0 g of a mixture that consists only of C$_8$H$_{18}$ and C$_9$H$_{20}$. When burned in O$_2$ so as to convert completely to CO$_2$ and H$_2$O, the weight of water collected is 565.0 g. What percent by weight of the original mixture must have been C$_8$H$_{18}$?

***2.50 Oxidation-reduction You are given a 1.000-g sample consisting only of a mixture of Zn, Mg, and Al. The sample contains a total of 0.0252 mol of atoms. When exactly half the sample is treated with excess H$^+$ to convert it to Zn^{2+}, Mg^{2+}, and Al^{3+}, 0.0135 mol of H$_2$ is released. When the other half of the sample is burned in excess O$_2$ to convert it to ZnO, MgO, and Al$_2$O$_3$, the weight of the product is 0.717 g. What is the percent by weight of each element in the initial mixture?

Chapter 3

PERIODICITY IN CHEMICAL BEHAVIOR

In the early days of the nineteenth century, chemistry was nourished by numerous unsung investigators who patiently did their experiments and recorded their observations in their notebooks. They all thus contributed to the flashing insight of a genius who was inspired to see the great design in which the pieces came together. Such was the development of the great cornerstone of chemistry, the periodic law. Thousands of experiments had to be done, many with extremely impure materials, challenged, repeated, verified, until finally they could stand as reliable witnesses of how specific chemicals behaved under specific conditions. It seemed obvious very early that the mass of atoms had something to do with regularities in chemical behavior. Although it turned out to be false, the belief that mass controlled the properties of the elements led to exact atomic weight experiments which eventually overthrew the belief and led to our present systematization of chemistry.

The periodic law, as originally stated, was simplicity itself: the properties of the elements are periodic functions of their atomic weights. But it was not easy to discover. Some experimenters reported correct results on incorrectly identified elements; others reported erroneous measurements on properly labeled elements. The great credit that goes to Dmitri Mendeleev and to Lothar Meyer for recognizing the underlying true generality should be for their courage to disregard wrong results and to follow their beliefs with sure instinct by posing questions for crucial experiments that would stand or fall on the validity of their newly discovered natural law.

The periodic law and the periodic table are now accepted as key principles of chemistry. Chemists use them almost by second nature to sort known facts and to make predictions, but it is to the newcomer that the periodic law displays its greatest favor. From a few facts and the appropriate framework, we can make predictions that would have astonished the ancients.

In this chapter we seek to understand not why periodic chemical behavior occurs but what it is and how it works for us to make available a world of chemical information.

3.1
PERIODIC LAW

Although Dmitri Mendeleev of Russia and Lothar Meyer of Germany are usually given credit for independently discovering the periodic law, about 1869, it would be unjust not to recognize that they were but following an idea that had begun developing as soon as the concept of atomic weight was created. Not long after atomic weights started being assigned to the elements (around 1803), it was noted that certain regularities in behavior began to appear when the elements were arranged in order of increasing atomic weight. At first, these regularities appeared simply as mild curiosities; few elements were known, and no grand design could be discerned. For example, of the three elements, calcium, strontium, and barium, which have remarkably similar properties, it was noted that the atomic weight of the middle one, strontium (atomic weight 88), was roughly the average of the other two, that is, 40 for calcium and 137 for barium. Similarly, in the sequence chlorine (35.5), bromine (80), and iodine (127), three similar elements can be arranged so that the middle one has an atomic weight that is the mean of the other two. Clearly, there must be some regularity connecting atomic weight and similarity of properties.

As chemical research progressed and more elements were discovered, the regularities became more pronounced. By 1864, for example, the English chemist John Newlands was able to state: "If the elements are arranged in the order of atomic weights, the eighth element, starting from a given one is a kind of repetition of the first, like the eighth note in an octave of music." This "law of octaves" was at first ignored and ridiculed as a crazy notion unworthy of serious consideration, but the idea was basically correct. It was only a few years later that Mendeleev and Lothar Meyer extended the generalization and posed it, not as a mere system of classifying elements, but as a "law of nature"—a sweeping generalized observation that summarized known behavior but also could be extrapolated to predict behavior of substances that were then as yet unknown. One of the spectacular successes of Mendeleev was to call attention to three gaps in his array of elements, to predict what sort of elements would be discovered to fill those gaps, and even to foretell what properties the compounds of these elements would have. It was only a matter of time, shortened, of course, because chemists now knew what they were looking for, before the missing three elements (gallium, scandium, and germanium) were discovered.

As initially formulated, the *periodic law* stated that *the chemical elements if arranged according to their atomic weights showed a periodicity of properties.* We now know that the original statement was not entirely exact; it is not the atomic weight that decides the regular periodicity in properties but a more fundamental quantity called the atomic number. The *atomic number* tells us the positive charge in the central part of an atom, but, as we shall see in the next chapter, it was not discovered until about 50 years later. For the present we simply introduce it as a serial number, related to the atomic weight, which indicates the order in which the complexity of atoms is built up. Neither Mendeleev nor Lothar Meyer nor their colleagues had the sophisticated equipment needed to determine atomic number, but, fortunately, except for a few contrary elements, atomic weight increases roughly as the atomic number. Therefore, the ordering of the elements, which was the key to establishing the periodic table, proceeded almost as well on the basis of atomic weight as it would have on atomic number. It is a testimony to the insight of Mendeleev that in those few cases where the order disagreed, Mendeleev boldly insisted that the accepted atomic weights were wrong. In the case of cobalt and nickel, where the atomic weights really are inverted, the atomic weights later checked out to be correct, but by then the periodic table was so

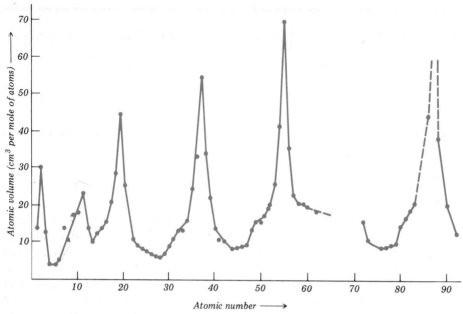

FIG. 3.1 Atomic volumes of the elements, in cubic centimeters per mole of atoms, as a function of atomic number.

firmly entrenched in chemical thinking that the response was to seek an alternative explanation for the inverted atomic weights rather than junk the periodic system.

In its present form the *periodic law* states that the *properties of the elements recur periodically if the elements are arranged in order of increasing atomic number.* Let us look first at a typical physical property to appreciate what is meant by periodicity. Figure 3.1 is a graph showing how the atomic volume, i.e., the volume occupied by 1 mol of atoms, varies for the different elements if they are arranged in order of increasing atomic number. (We would get almost exactly the same curve if the atomic volumes were plotted against atomic weight.) The atomic-volume behavior is periodic; it goes through cycles, dropping down from a sharp maximum to a trough and then sharply rising again. Each of the cycles is called a period, the portion of the curve that has to be retraced to get back to the same behavior. The location of elements on the peaks or in the troughs has an important correlation with their chemical reactivity; the elements at the peaks are the most reactive of all; the elements in the troughs are characteristically more inert. The elements at the peaks can be considered to form a group by themselves.

Atomic volume was one of the first periodic properties to be recognized. This is partly because atomic volume is relatively easy to measure. All we need is the atomic weight and the density of the element. (For elements that are normally gases at room temperature, the convention is to take the density at the temperature at which the element has just solidified.)

70

EXAMPLE 1

Sodium has atomic weight 22.99 amu and density 0.97 g/cm³. What is its atomic volume?

Solution

$$\frac{22.99 \text{ g mol}}{0.97 \text{ g/cm}^3} = 24 \text{ cm}^3/\text{mol}$$

3.2
PERIODIC TABLE

Although the periodic law is generally accepted and there is general agreement that the elements should be arranged in cycles, there is still much argument as to how the elements should be displayed to represent these cycles. Figure 3.2 shows one of the most commonly used forms of the periodic table. The basic features of the arrangement are horizontal rows called *periods* and vertical columns called *groups*. The horizontal periods represent the cycles that have to be traversed to get back to similar properties; the vertical groups collect together the elements that have similar properties. As shown in Fig. 3.2, across the top, the groups are numbered by Roman numerals. Group I contains hydrogen and the elements lithium, sodium, potassium, rubidium, cesium, and francium. Group II contains beryllium, magnesium, calcium, strontium, barium, and radium. At the extreme right is group 0, comprising helium, neon, argon, krypton, xenon, and radon. Between groups II and III, which are sometimes called *main groups,* there are 10 short groups, marked "transition elements," which are called *subgroups.* In some designations these columns are labeled IIIB, IVB, VB, VIB, VIIB, VIII (covering three successive columns), IA, and IIA; in other designations, the A and B are interchanged.

Since chemists themselves cannot agree on the A and B designation and frequently get them mixed up, it is better to omit these designations. They are relics of the original Mendeleev table, which was squashed together so that for the long periods two elements were made to appear in the same box. Instead of referring to the transition elements as A or B subgroups, we shall refer to them by the name of the head element. Thus, scandium (element 21) heads the scandium subgroup, which contains scandium, yttrium, the series of elements marked by an asterisk, and the series of elements marked by a dagger. The elements marked by an asterisk, fifteen in number, are called the *lanthanides.* They all occupy the same position in the periodic table. The elements marked by a dagger, also fifteen in number, are called the *actinides.* The idea behind putting 15 elements in the same box at the single asterisk and 15 elements at the dagger is that they all have remarkably similar chemical properties, and so for classification purposes they are lumped together.

Along the left side of the periodic table in Fig. 3.2 the period designations are given in Arabic numerals. Period 1, the top horizontal row, contains only

71

Group I

FIG. 3.2 Periodic Table (The numbers above the symbols of the elements denote the atomic numbers. The numbers below the element symbols are the atomic weights.)

Period 1 1 H 1.008																	2 He 4.003
2 3 Li 6.941	4 Be 9.012										5 B 10.81	6 C 12.011	7 N 14.007	8 O 15.999	9 F 18.998	10 Ne 20.179
3 11 Na 22.990	12 Mg 24.305			— Transition Elements —							13 Al 26.982	14 Si 28.086	15 P 30.974	16 S 32.06	17 Cl 35.453	18 Ar 39.948
4 19 K 39.098	20 Ca 40.08	21 Sc 44.956	22 Ti 47.90	23 V 50.941	24 Cr 51.996	25 Mn 54.938	26 Fe 55.847	27 Co 58.933	28 Ni 58.70	29 Cu 63.546	30 Zn 65.38	31 Ga 69.72	32 Ge 72.59	33 As 74.922	34 Se 78.96	35 Br 79.904	36 Kr 83.80
5 37 Rb 85.468	38 Sr 87.62	39 Y 88.906	40 Zr 91.22	41 Nb 92.906	42 Mo 95.94	43 Tc 98.906	44 Ru 101.07	45 Rh 102.906	46 Pd 106.4	47 Ag 107.868	48 Cd 112.41	49 In 114.82	50 Sn 118.69	51 Sb 121.75	52 Te 127.60	53 I 126.904	54 Xe 131.30
6 55 Cs 132.905	56 Ba 137.33	57–71 *	72 Hf 178.49	73 Ta 180.948	74 W 183.85	75 Re 186.207	76 Os 190.2	77 Ir 192.22	78 Pt 195.09	79 Au 196.966	80 Hg 200.59	81 Tl 204.37	82 Pb 207.2	83 Bi 208.980	84 Po (210)	85 At (210)	86 Rn (222)
7 87 Fr (223)	88 Ra 226.025	89–103 †	104 Ku (257)	105 Ha (260)	106 ?	107 ?											

sub group

*	57 La 138.906	58 Ce 140.12	59 Pr 140.908	60 Nd 144.24	61 Pm (145)	62 Sm 150.4	63 Eu 151.96	64 Gd 157.25	65 Tb 158.925	66 Dy 162.50	67 Ho 164.930	68 Er 167.26	69 Tm 168.934	70 Yb 173.04	71 Lu 174.97
†	89 Ac (227)	90 Th 232.038	91 Pa 231.036	92 U 238.029	93 Np 237.048	94 Pu (242)	95 Am (243)	96 Cm (247)	97 Bk (249)	98 Cf (251)	99 Es (254)	100 Fm (253)	101 Md (256)	102 No (254)	103 Lr (257)

two elements: hydrogen and helium. Period 2 consists of 8 elements: lithium, beryllium, boron, carbon, nitrogen, oxygen, fluorine, and neon. Period 3, the third horizontal row, also has 8 elements: sodium, magnesium, aluminum, silicon, phosphorus, sulfur, chlorine, and argon. The fourth period is a long period; it has 18 elements ranging from potassium, number 19, through krypton, number 36.

Period 5, rubidium (Rb) through xenon (Xe), also has 18 elements, but period 6, cesium (Cs) through radon (Rn), has 32 elements. The latter, of course, includes the 15 elements designated by an asterisk. Similarly, period 7 includes the 15 elements designated by a dagger. We do not know how many elements there are in period 7. It starts with francium (Fr, number 87), but no one is quite sure where it ends. New elements are synthesized and added to it every other year or so, and the last element shown, unnamed but marked 107, designates the most recent arrival.

Professional chemists find it useful to remember the positions of the elements in the periodic table. The reason for this is twofold: (1) properties vary gradually and systematically in going from left to right, and (2) they vary equally logically in going from top to bottom. If we know the position of an element in the periodic table and have a rough idea of how properties vary across a period or down a group, we can make a fair guess at the properties of any element. Equally important, if we are looking for an element with a desired set of properties, we will have a good idea of where to look.

3.3
THE NOBLE GASES

The periodic table was discovered by noting the systematic variation of properties in the elements. One of its main uses at present is to provide a framework for remembering these variations. Let us illustrate by considering the property of chemical reactivity. If we string the elements out in a line in order of increasing atomic number, we will find a regular drop in chemical reactivity at element numbers 2 (helium), 10 (neon), 18 (argon), 36 (krypton), 54 (xenon), and 86 (radon). This is represented schematically in Fig. 3.3. Until 1962, these elements were believed to be totally unreactive chemically and would not form compounds. This remains true of the first three (helium, neon, and argon), but the last three (krypton, xenon, and radon) are now known to form compounds with oxygen and fluorine under special conditions. In the old days, all these elements were called the *inert elements,* or, because they were gases at room temperature, *inert gases.* As shown in Fig. 3.2, they were all lumped together at the extreme right of the periodic table in a vertical column labeled group 0. The zero actually implied something about the state of chemical combination, more precisely the lack of chemical reactivity. The label has stuck and so has the descriptive adjective *inert.* Because the word *noble* was used for metals that were not particularly reactive (for example, the *noble metal* gold), some people prefer to call the group 0 elements *noble gases;* others call them the *rare gases.* Whatever we call them, they are now firmly fixed as terminal points of the periods in the periodic table.

As can be seen from Fig. 3.3, the intervals between the noble gases vary in length, thus corresponding to the different numbers of elements in the successive periods of the periodic table. Period 1 has but 2 elements, periods 2 and 3 each have 8 elements, periods 4 and 5 each have 18 elements. Why should there be different-length cycles between successive inert elements? This was a question that had to be resolved by any theory that proposed to explain the nature of atoms.

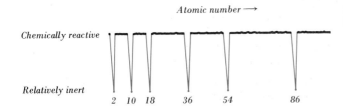

Atomic number $\longrightarrow$

Chemically reactive

Relatively inert

2   10 18   36   54   86

FIG. 3.3 Periodic occurrence of low reactivity in the elements.

FIG. 3.4   Components of Dry Air

Component	Percent by volume
Nitrogen (N_2)	78.09
Oxygen (O_2)	20.95
Argon (Ar)	0.93
Carbon dioxide (CO_2)	0.023–0.050
Neon (Ne)	0.0018
Helium (He)	0.0005
Krypton (Kr)	0.0001
Hydrogen (H_2)	0.00005
Xenon (Xe)	0.000008

To keep the record straight, it should be noted that Mendeleev knew nothing about the rare gases when he set up the periodic table. They were not discovered until 1894 when Lord Rayleigh noticed that absolutely pure nitrogen prepared from decomposition of compounds was of slightly lower density (1.2505 g/liter) than the residual gas obtained from the atmosphere by removal of what was believed to be its only other constituents, oxygen, carbon dioxide, and water (1.2572 g/liter). In conjunction with Sir William Ramsay, Rayleigh removed the nitrogen from the air residue by various reactions, such as with hot magnesium metal. After removal of the nitrogen, there was still some remaining gas which, unlike any gas known at the time, was completely unreactive. It was christened "argon" from the Greek word *argos* meaning "lazy." Later investigations showed that crude argon, and hence the atmosphere, contains the other noble-gas elements helium, neon, krypton, and xenon. Except for argon, the amounts of the noble gases in the atmosphere are relatively small. Figure 3.4 shows the average composition of the earth's atmosphere. In addition to the noble gases listed, there are traces of radon, Rn, in the atmosphere. It is a radioactive element, and its concentration is very low and variable because it is produced near deposits of radioactive minerals.

3.4
THE HALOGENS

Just preceding the noble-gas elements (group 0) in the periodic table of Fig. 3.2 is a group of elements labeled group VII. These are the halogens, and they include fluorine (atomic number 9), chlorine (17), bromine (35), iodine (53), and astatine (85). They are one of the most famous families in chemistry. The elements are quite vigorous in their chemical reactivity, they produce an enormous variety of interesting compounds, and both the elements and compounds show quite well-behaved patterns of group behavior. Actually the group behavior of the halogens is so neatly logical that one can well be misled into expecting too much regularity in other parts of the periodic table.

The word "halogen" comes from the Greek *halos*, "salt," and *genes*, "born." It describes the fact that these elements are "salt formers." They all combine readily with elements such as sodium to give the familiar white crystalline solids typified by ordinary salt. Salt is a general chemical term that covers a neutralization product of an acid and base reaction. In everyday vocabulary, however, salt is likely to mean sodium chloride. In snowy regions of the country, where salt is frequently used for road clearance, the "salt" is likely to be a mixture of sodium chloride and calcium dichloride ($CaCl_2$).

The most characteristic property of the halogens is that all of them, in the elemental state, are very good oxidizing agents. This means they tend to react so as to attract electrons to themselves and go to a more negative condition, the state in which, except for iodine, they are most likely to be found in nature. *Fluorine,* the top element of the group, is the best oxidizing agent of the lot. It is so reactive chemically that one is hard put to find a container in which to keep the element. One of the most spectacular demonstrations in chemistry is to see a jet of fluorine gas directed at a concrete wall. The fluorine, by virtue of its vigorous oxidizing reaction, cuts through the wall like a hot knife through butter. Fluorine as an element deserves special respect: not only can it disintegrate an experimenter as it does a concrete wall, but it will also leave nasty sores that take ages to heal.

Chlorine, the second element of the halogen group, is also quite reactive, but not nearly so much as is fluorine. It occurs in seawater in enormous amounts and also in salt deposits that are presumably the evaporation residues of ancient seas. In both cases, the essential form is as sodium chloride, NaCl. To get the free element from NaCl as in the following reaction requires expenditure of considerable energy.

$$2NaCl(l) \longrightarrow 2Na(l) + Cl_2(g)$$

The reaction is driven uphill energetically by pumping electric current through molten sodium chloride. The energy that goes into the chlorine is one of the reasons why chlorine is so reactive.

Bromine is a vile-smelling, evil-looking brown liquid. In fact, the name bromine comes from the Greek *bromos* for "stink." Although less reactive than the elements fluorine and chlorine, bromine has its own nasty disposition, and a drop of the liquid caught under the fingernail can cause a sore that will last for 6 months. No wonder, then, that chemists handling bromine are advised to wear rubber gloves. Even so, they should avoid breathing the volatile vapors since the stuff easily evaporates and is just as corrosive in the gaseous state as it is in the liquid state. Bromine, it might be mentioned, is produced from seawater in enormous amounts because it is used to make one of the compounds that goes into gasoline. In a Catch-22 sequence, civilized man adds tetraethyl lead [$(C_2H_5)_4Pb$] to gasoline to improve the smooth burning properties (antiknock) of the gasoline but then has to add dibromoethane ($C_2H_4Br_2$) to scavenge the lead and prevent it from accumulating in the engine. [The lead bromide comes out the exhaust and accumulates in the environment instead!] Bromine is also important as a component of silver bromide emulsions in photography.

The most gentle of the halogens is the fourth member of the group, *iodine.* Unlike fluorine and chlorine, which are normally gases, and bromine, which is normally a liquid, iodine is a bluish black solid at room temperature. However, it is very volatile, and a favorite demonstration in chemistry is to heat solid iodine so as to generate beautiful dense clouds of violet vapor. Iodine is commonly encountered as an antiseptic in medicine in the form of tincture of iodine, which is a solution of iodine in alcohol. An intriguing question is how come iodine molecules in the vapor are purple but in alcohol solution are brown? The answer tells us something about the nature of the binding of iodine atoms to each other. Far-fetched as it may seem, this problem is related to the action of a common testing indicator found in the chemical laboratory—starch–potassium iodide paper. This is a slip of paper, white at the start, which, when moist, turns blue on exposure to any oxidizing agent. The iodide (I^-) is colorless at first, but when it loses a negative charge to an oxidizing agent, it converts to I_2 which would be brown in water (as in alcohol) but turns blue (or violet) in contact with starch. Truly, the colors of molecules depend importantly on what their environment is.

Iodine is also unusual in that it does not come from the sea like the other halogens, although its principal source used to be seaweed. It now comes mainly from two other sources: (1) the brine from oil wells, some of which are surprisingly rich in iodine, and (2) as an impurity, probably calcium iodate, in the natural deposit of Chile saltpeter ($NaNO_3$).

Astatine, the fifth and last member of the halogen family, is a recent acquisition in chemistry. It is radioactive, in that the atom may spontaneously disintegrate into smaller fragments, but it can be prepared artificially. It was made for the first time in a cyclotron in 1940. The name astatine comes from the Greek *astatos,* meaning "unstable." Not much is known about the element. It is rarely encountered because it is unstable and hazardous to work with and quickly disintegrates as it is being investigated.

There are many properties of the halogens that can be used to illustrate the general trends as one goes from the top of the periodic table to the bottom. Figure 3.5 shows some of the properties that are typical. As indicated in the second column, the atomic number jumps progressively by 8, 18, 18, and 32 units, consistent with the increasing number of elements in successive periods. The atomic weight goes up in roughly proportional value. (Note that the atomic weight of astatine is given in parentheses. This number applies to the mass of the most stable atom. Since astatine is radioactive, only the least unstable mass is generally given.) The fourth column gives the melt-

FIG. 3.5 Properties of the Halogens

Element	Atomic number	Atomic weight, amu	Melting point, °C	Boiling point, °C	Atomic radius, nm
Fluorine	9	18.998	−223	−188	0.071
Chlorine	17	35.453	−103	−34.6	0.099
Bromine	35	79.904	−7.2	58.78	0.114
Iodine	53	126.904	113.9	184.35	0.133
Astatine	85	(210)	?	?	?

ing point, the temperature at which the solid normally changes to the liquid state. As can be seen, the melting point increases as one goes down the group from the lighter to the heavier halogens. This is typical of elements on the right side of the periodic table.

The fifth column of Fig. 3.5 gives the boiling point, the temperature at which liquid halogens go over to the gaseous state. The temperatures steadily increase as we go down the group, again in line with regular group trends on the right side of the periodic table. This trend of increasing boiling point is intimately connected with the structure of the atoms. A hint of the changes in this structure is shown in the last column. Values shown are for the atomic radius, a concept that we shall see is a bit fuzzier than we might like to admit. The units, nanometers ($1 \text{ nm} = 1 \times 10^{-9} \text{ cm}$), are perhaps unfamiliar, but the trend is unmistakably clear. As one goes down a group, the atomic size generally increases. This increase in size is an important feature in controlling group properties.

3.5
THE ALKALI ELEMENTS

Way over on the left of the periodic table is a column of elements, the alkali elements, for which the group similiarities are even more pronounced than for the halogens. These elements are lithium (atomic number 3), sodium (11), potassium (19), rubidium (37), cesium (55), and francium (87). They are collectively called the alkali elements. Alkali comes from the Arabic word *al-qali*, which means "ashes" and recalls that the elements sodium and potassium occur in the ashes left over from the burning of wood or plants.

The alkali elements are quite different from the halogens. They are all solids at room temperature and have many of the properties typical of *metals,* i.e., good electric conductivity, shiny luster, excellent reflectivity for all kinds of light, and good heat conduction. *Nonmetals,* as the name implies, lack these properties. In general, the elements on the left of the periodic table are metals, those on the right are nonmetals. As we go from left to right, metallic character decreases.

Chemically, the alkali elements are very reactive. When a small bit of sodium is dropped into water, the sodium fizzes vigorously and may even burst out of the container because of the vigor of the reaction. The alkali elements react with oxygen in the air to form a scaly coat of oxide which obscures their real appearance. With moderate care, one can pick up a piece of sodium in air with forceps, set it on a glass plate, and slice away the surface coat of oxide to disclose the shiny metallic surface underneath. However, especially if the air is moist, the shiny surface quickly tarnishes (loses its luster) and reverts to oxide surface. In practice, the alkali elements are usually handled in evacuated glass vessels which have been filled with helium or one of the inert gases. Lithium, the lightest element of the group, is so reactive that it even tarnishes under nitrogen, which is normally rather inert.

Unlike the halogens, which are good oxidizing agents with great tendency to pick up electrons, the alkali elements are good reducing agents. They generally give up electrons, and most of their chemical reactivity is associated with this process. For use as chemical reducing agents, the alkali elements are generally purchased as blocks of metal or lengths of wire packed in hermetically sealed cans under inert atmosphere. Sodium is by far the most abundant and cheapest of the group.

Lithium is interesting as the lightest of the alkali elements. It occurs naturally in salt deposits in small amounts as LiCl together with NaCl and KCl. When placed in a burner flame, LiCl gives a very characteristic red color, the study of which did much to unravel the mysteries of atomic structure. Currently, lithium is of special interest because of two uses. It is used as a nuclear fuel in the hydrogen bomb, where the salt lithium deuteride, LiD, furnishes the nuclear fuel hydrogen in the form of the heavy isotope, deuterium, and also furnishes a light isotope of lithium, Li 6, for setting off a chain reaction. A potentially important use of lithium, just recently suggested, is as the reducing agent in a lightweight electric cell that may ultimately replace the conventional lead storage battery in applications where saving weight is an important factor.

Sodium is the most common of the alkali elements. It ranks sixth in order of abundance of all elements in the earth's crust. The most familiar occurrence is as the salt NaCl in seawater and also in solid mineral deposits, such as near the shores of inland salt seas (e.g., Great Salt Lake and Dead Sea) but also underground (e.g., Michigan, Louisiana). World production of NaCl comes to some 60 million tons per year; its importance may be judged from the fact that Roman soldiers used to get paid in salt, from which comes our present word *salary*. The amount of NaCl in the sea is relatively large (each liter of seawater contains about 27 g of NaCl), but it is less than 10 percent of what the water could theoretically hold. The large amount of NaCl in human body fluids may reflect the fact that we originally evolved from the sea and at one time lived in a seawater environment.

Preparation of metallic sodium from NaCl is an important commercial process. Electric current is pumped through cells containing molten NaCl, and the NaCl decomposes to give metallic sodium at the negative electrode and free chlorine gas at the positive electrode. This is an energy-consuming process, and great pains must be taken to keep the elemental sodium and elemental chlorine away from each other; otherwise, they would recombine with explosive violence to re-form NaCl. A variant of the molten NaCl electrical decomposition is to use an aqueous solution, but with liquid mercury as the negative electrode. As soon as the elemental sodium forms, it dissolves in the liquid mercury (where it does not have the usual violent reactivity against water). One major environmental problem is that, when the cells are washed, some of the mercury gets swept in trace amounts into streams and lakes.

Potassium is only a bit less abundant in the earth's crust than is sodium, but potassium is relatively little used as a chemical reagent. For one thing, it is more expensive; for another, it is somewhat more reactive than sodium, so that there is a hazard working with it. As with sodium, the element potassium can be cut with a knife like a piece of cheese on a glass plate. However, it often bursts into flame when not immediately protected from chemical reaction with the air. Both sodium and potassium are usually stored under oil so as to keep them away from air attack. Both oxygen and water vigorously react with potassium and should be kept away from the metal.

Sodium and potassium are among the indispensable constituents of animal and plant tissue. Na^+ is the principal positive ion in the fluids outside the cells, whereas K^+ is the main positive ion inside the cells. Besides filling general physiological roles, such as aiding water retention, these ions have specific functions. For example, Na^+ depresses the activity of muscle enzymes and is required for contraction of all animal muscle. In plants, K^+, but not Na^+, is a primary requirement. As a result, more than 90 percent of the alkali content of ashes is due to potassium. Plants have such a high demand for potassium that, even in soils in which the sodium content predominates manyfold, the potassium is taken up preferentially. Since an average crop extracts from the soil about 20 kg of potassium per acre, the necessity for potassium fertilizers is obvious. The old idea of putting wood ashes in the garden makes sense because trees are plants that are rich in potassium and burning them produces a potassium-rich product.

Rubidium is a relatively rare element. Seawater contains about 0.2 ppm (parts per million, by weight) of rubidium, which although not very abundant, is about twice the concentration of lithium. When salts of rubidium are placed in a burner flame, they impart to it a characteristic red color with a bluish tinge. The flame color, when resolved into its component energies, is relatively simple, unlike, for example, the case of iron, where the flame color is a complex mix of very many components.

Cesium, the next to last of the alkali elements, is very reactive. It is a soft, shiny, metallic element, with the relatively low melting point of 28.5°C. Since body temperature is about 37°C, just the heat of your hands would be sufficient to melt a bar of cesium metal to a liquid puddle. Of course, you would be ill-advised to try the experiment, since the molten cesium would burn a hole through your hands; it is a powerful reducing agent. Cesium metal is used as a coating on the light-detecting surface in some photoelectric cells. It turns out that negative electrons are more easily ejected out of cesium metal than from any other metal, so that cesium can serve as a very sensitive detector for low-energy irradiation.

Francium, the bottom member of the alkali group, is relatively poorly known. It is radioactive. Experiments with large amounts of francium are extraordinarily difficult because the radioactive decomposition produces highly energetic particles that can cause severe medical damage to exposed human tissue. Studies with trace amounts indicate that the chemical properties of francium are much like those of the other alkali elements.

All the alkali elements bear strong resemblance to each other. They are all shiny metals that corrode in air; they react vigorously when placed in water; they impart characteristic colors to flames. How do they differ? Figure 3.6 lists some typical properties. As we go down the group, atomic number and atomic weight increase regularly. However, as shown in the fourth col-

FIG. 3.6 Properties of the Alkali Elements

Element	Atomic number	Atomic weight, amu	Density, g/cm³	Melting point, °C	Boiling point, °C	Atomic radius, nm
Lithium	3	6.941	0.53	186	1336	0.134
Sodium	11	22.990	0.97	97.5	880	0.154
Potassium	19	39.098	0.86	62.3	760	0.196
Rubidium	37	85.468	1.53	38.5	700	0.216
Cesium	55	132.905	1.87	28.5	670	0.235
Francium	87	(223)	?	?	?	?

umn, the density is not so well-behaved. Instead of getting a steady increase down the group, we observe a drop from sodium to potassium. For some reason, potassium is less dense than sodium. There are two possible explanations: (1) the K atom is lighter than it should be; (2) the K atom is bigger than it should be. The first explanation does not seem to hold when one scrutinizes the atomic weights in column three—the progression seems quite orderly. The second explanation is supported somewhat by the values for the atomic radius, as shown in the last column. As can be seen, the jump in size in going from Na to K is about twice as large as the change elsewhere in the group. What is the reason for this curious progression in size? We need to develop clearer ideas about size of atoms and their structures, as in the next chapters, before we can hope to answer this question.

The fifth and sixth columns of Fig. 3.6 show that the melting points of the alkali elements are relatively low, compared for example with 1535°C for iron, and they decrease regularly down the group. The melting point gives a crude measure of binding forces in a solid, and the decrease down the group suggests that lithium atoms in a solid are bound to other lithium atoms more strongly than are other elements of group I. Boiling points give similar measures of binding in the liquid. The forces here appear to be much stronger than in the solid, but we must not overinterpret our data. For one thing, in *melting* we simply go from solid to liquid: we are not really separating atoms from each other. We are just changing their arrangement. However, in *boiling,* where we go from liquid to gas, we are separating the atoms from each other to a greater degree.

3.6
THE SECOND PERIOD

In the preceding two sections, we looked briefly at how some selected properties vary as one goes down a group. Let us look now at how properties vary as one goes from left to right across a period. Since the first period has but two elements, hydrogen and helium, it is not very interesting. The second period, however, has eight elements—lithium, beryllium, boron, carbon, nitrogen, oxygen, fluorine, and neon—so it ought to be more interesting. We would expect, on the basis of what was mentioned in the preceding sections, that properties will have to change from a typical alkali metal (Li) on the left to a typical inert gas (Ne) on the right. Figure 3.7 shows the course of some characteristic properties. As we progress from left to right, the atomic number (first row) and atomic weight (second row) increase progressively. The density (third row) increases to a maximum at carbon, then abruptly drops to the value for nitrogen. Several points are worth noting here. There are two values listed under carbon: 3.52 g/cm³ for D (diamond) and 2.25 g/cm³ for G (graphite). The element carbon can be found in two forms which differ in the way the atoms are packed. In diamond the atoms are packed more tightly—about 50 percent more atoms are contained per unit volume—and therefore it has a higher density. The implication is that density depends not only on the mass of atoms but also on how they are packed together. This actually is the reason for the strange behavior of the density throughout the rest of the period: nitrogen atoms pack in pairs as diatomic molecules, quite different from extended networks as are characteristic of carbon. The abrupt drop in density from 2.25 g/cm³ at graphite carbon to 0.956 g/cm³ at nitrogen is mainly due to a drastic drop in the efficiency of occupying space.

The fourth row of Fig. 3.7 shows a rise in melting point from the relatively modest 186°C for lithium to the relatively astronomical value of 3500°C for carbon. This is followed by a spectacular drop to the −200°C level for the rest of the period. A similar trend is seen in the boiling points, as shown in the fifth row of the table. Why this steep climb and precipitous drop? Obviously, there is an abrupt drop in the forces holding together the liquid and solid when one passes from carbon to nitrogen. Yet the amazing thing is that the force holding one nitrogen atom to another is considerably greater (almost twice as great) as the force holding one carbon to another. Clearly, there must be more to the equation than just interatomic forces.

The last row of Fig. 3.7 is most satisfying. It shows a smooth regular drop in atomic radius from 0.134 nm for Li to 0.071 nm for F. No value is listed for neon, the end member, because there is disagreement as to exactly what is meant by atomic radius for an element such as neon. One way to define atomic radius is as half the distance between an atom and another atom to which it is bound. Neon does not bind to other neon atoms, and so an atomic radius cannot be deduced. There are other ways of deriving an atomic radius, which we shall consider in Chap. 6, but values so obtained are not directly compara-

FIG. 3.7 Properties of the Second-Period Elements

	Lithium	Beryllium	Boron	Carbon	Nitrogen	Oxygen	Fluorine	Neon
Atomic number	3	4	5	6	7	8	9	10
Atomic weight, amu	6.941	9.012	10.81	12.011	14.007	15.999	18.998	20.179
Density of solid, g/cm³	0.53	1.85	2.46	3.52(D) 2.25(G)	0.956	1.426	1.67	1.54
Melting point, °C	186	1280	2040	3500	−210	−219	−223	−249
Boiling point, °C	1336	2970	4100	4200	−196	−183	−188	−246
Atomic radius, nm	0.134	0.125	0.090	0.077	0.075	0.073	0.071	

ble with the values listed in Fig. 3.7. Ignoring neon, we still have an unresolved question: Why does the atomic size decrease as one goes from left to right across a period of the periodic table? Such behavior is quite typical.

The above brief resumé says something about the typical variation in some standard physical properties across the periodic table. However, it tells little of the changes in *chemical* behavior, which actually are quite remarkable. The whole wide diversity of the chemical world is typified in this row of elements. Let us first survey the chemical aspect of these elements when exposed to two quite ordinary chemical reagents, water and oxygen.

When a small piece of lithium is dropped into a beaker of water, there is a vigorous fizzing reaction, while the piece of lithium skids around in rather erratic fashion on the surface of the water. (The density of lithium is only about half that of water, and so it readily floats on top of the water.) The whole reaction is over in a matter of seconds.

Now let us try the same experiment with beryllium. It is also a nice shiny metal, but its behavior is quite different. For one thing, unlike lithium which is generally kept under mineral oil to prevent corrosion, beryllium can be left in air almost indefinitely with but a slight impairment of its beautiful shiny luster. When dropped in water, nothing happens. The beryllium (almost twice as dense as water) simply settles to the bottom. There is no obvious chemical reaction. Actually, this is all quite deceptive. Beryllium ought to react quite vigorously with water. It does not because of a peculiar quirk of its chemistry. When beryllium reacts with water, it immediately forms a thin invisible coating of beryllium oxide, which has the property of adhering very tightly to the underlying metal and *preventing further reaction*. Thus, the product BeO acts as an insulating barrier to protect deeper layers of the metal.

What happens when boron is placed in water? Again, nothing. Boron is not a very reactive element at room temperature, and again there is formation of a superficial oxide coat that protects the bulk element. Incidentally, boron (unlike the two preceding elements Li and Be) does not look metallic. It has a sort of hard-looking glassy luster, which at very low angles reflects light extraordinarily well, almost as well as a metal, but which diminishes in reflectivity efficiency when viewed head on.

Now we come to the fourth element of the period, carbon. What happens when we put it in water? Nothing, and it makes no difference whether it is diamond (glassy-looking, transparent) or graphite (dark, almost metallic-looking). Carbon in both forms is inert to water or air at room temperature. If the temperature is raised sufficiently, reaction will occur. Both diamond and graphite gradually disappear when exposed to very hot steam or very hot oxygen. They get converted to $CO(g)$ and $CO_2(g)$. However, the temperature has to be of the order of 1000°C or higher.

The fifth element of the period is nitrogen. It is, under normal circumstances, a diatomic gas at room temperature, and so the question of chemical reactivity comes down to this: What chemical reaction occurs when $N_2(g)$ is bubbled through water? The answer is, nothing. Nitrogen is a relatively inert element and does not react at room temperature. However, there are some bacteria which apparently can convert nitrogen at room temperature into nitrogen compounds. Such compounds are currently under intensive investigation, since they are connected with an important economic problem: What is the cheapest way to take N_2 from the air and convert it into useful nitrogen compounds in the soil?

The sixth element of the period, oxygen, is also inert to water. As a gas at room temperature, it can be bubbled through water without a sign of chemical reaction. A slight amount of oxygen can dissolve in water, which is highly critical for survival of fish life, but the amount is almost negligible and can almost be ignored. There is no chemical reaction.

The seventh element of the second period, fluorine, is quite reactive to water. In fact, it is a rather dangerous reaction to carry out. Gaseous fluorine bubbled into water spits and crackles in ominous fashion, quite clear evidence that a chemical reaction is going on. With dry oxygen, the reaction of fluorine is less vigorous.

The final element of the period, neon, does nothing when exposed to air or water. It is inert, apparently to all reagents, not only at room temperature but even at very high temperatures.

The net result of the above tour is not very great. Of the eight elements in the second period, only two, lithium and fluorine, show much reactivity to water. The others are relatively inert. This is actually quite typical. The most reactive elements are generally toward the ends of a period—those in the middle of a period are likely to be relatively nonreactive. Why? The answer is complicated and is a function of atomic structure. However, the main point is that elements on the far left tend to part with their electrons relatively easily; similarly, elements at the far right, excluding the inert, or group 0, elements, tend to accept electrons readily. These two tendencies make for vigorous reaction.

Another interesting way to monitor the chemical behavior of the elements across a period is to look at the kind of compounds that are formed.

Figure 3.8 shows the formulas of typical oxygen compounds of the second-period elements. Those at the left of the periodic table tend to be high melting solids with considerable stability toward thermal decomposition. Those at the right side of the table tend to be very volatile, either as gases or low-boiling liquids. Carbon dioxide, in the middle of the period, is doubly strange. It does not have a normal melting point—that is, when solid CO_2 is heated under a normal atmosphere, it does not pass into the liquid state as one would expect. Instead the solid goes directly to gas; i.e., it sublimes. In

FIG. 3.8   Oxygen Compounds of Second-Period Elements

Element	Li	Be	B	C	N	O	F	Ne
Compound	Li_2O	BeO	B_2O_3	CO_2	N_2O_5	—	F_2O	None
Melting point, °C	>1700	2530	460	−56.6 (5.2 atm)	30	—	−224	—
Boiling point, °C	Decomposes	3900	ca. 1860	−78.5 (sublimes)	Decomposes	—	−145	—

order to see the melting of solid CO_2, one has to put it under pressure. Only then would we see ordinary melting. The other strange thing about CO_2 is that when it goes from solid to gas under normal pressure, the temperature at which this happens is considerably lower than the temperature at which liquid CO_2 exists under pressure.

Another important parameter for oxides is reactivity with the environment in which the compound is placed. For example, when Li_2O is placed in water, the solution formed acts basic. The action can be described by the equation

$$Li_2O + H_2O \longrightarrow 2LiOH$$

It shows that hydration, or addition of water to Li_2O, converts it to a hydroxide. Alternatively, we can regard the Li_2O as the base LiOH minus water, that is, $Li_2O = 2LiOH - H_2O$. Such substances which are derived from other substances by subtraction of water are called *anhydrides*. Li_2O is the *base anhydride* of LiOH. It is typical of all the oxides at the left side of the periodic table that they are basic anhydrides—when water is added to them, they form compounds with basic properties (turn litmus blue, neutralize acids, etc.).

EXAMPLE 2

What would be the anhydride of $Al(OH)_3$?

Solution

The trick in solving such problems is to note that all the hydrogen has to be removed as H_2O. However, before we can remove the H_2O, we have to multiply the original formula by an appropriate number so that we can take out a whole number of H_2O molecules. In the present case, we can double $Al(OH)_3$ and then remove $3H_2O$. This can be written schematically as follows:

$$2Al(OH)_3 - 3H_2O = Al_2O_6H_6 - O_3H_6 = Al_2O_3$$

The second oxide in this period, BeO, is somewhat like Li_2O in that it, too, reacts with water to form a base. Thus, BeO is also a base anhydride. However, it is a weaker base. The solution does not quite give the basic reaction that Li_2O does. When we move over to the right to boron oxide, B_2O_3, the basicity is even less pronounced. Indeed, when B_2O_3 is placed in water, the resulting solution shows slightly acidic properties (turns litmus red, neutralizes bases, etc.). Thus, we would call B_2O_3 an *acid anhydride*. It is an acid with the water removed. The equation for the *hydration* can be written as

$$\underset{\text{Boric oxide}}{B_2O_3} \; + \; 3H_2O \longrightarrow \underset{\text{Boric acid}}{2H_3BO_3}$$

The reverse process, *dehydration,* occurs when water is removed from H_3BO_3. These two processes, hydration and dehydration, are of immense practical importance. Our world is essentially an aqueous world, so that absorption of water and liberation of water are important for maintaining the water balance.

The fourth oxide shown in the second period is CO_2, carbon dioxide. It is the end product of combustion when unlimited oxygen is available. When the supply of oxygen is limited, the product is carbon monoxide, CO. Both CO and CO_2 are colorless gases. CO is odorless (which makes it rather an insidious poison, since it has the nasty property of impeding oxygen uptake by the blood and thus possibly leading to death by suffocation even when present in relatively small quantities). CO_2 is more friendly. It is nonpoisonous and advertises itself by its rather pronounced sour odor. The sour sensation is not really an odor, but more a taste, since CO_2 reacts with water to produce an acid reaction. The physiological reaction of CO_2 is probably mainly due to acid formation on hydration in the nasal passages.

The acid-formation reaction of CO_2 is easy to observe. All we need do is drop a piece of dry ice (which is nothing but solid CO_2) into water. As the CO_2 vaporizes, because the water is warmer than the dry ice temperature of $-78°C$, the bubbles of CO_2 gurgle through the water and dissolve in it as they rise to escape. A sharp-eyed observer would note that CO_2 bubbles rising through water get smaller as they rise, unlike most other gases. This is due to solubility of the CO_2 in the water. That the reaction occurs can be easily demonstrated by noting that an indicator, say litmus, added with a drop of $NaOH$ gradually changes color from blue (basic) to red (acidic) as the CO_2 proceeds to dissolve and change the solution. There has been considerable controversy over exactly what happens when CO_2 dissolves in water. The simplest view is that CO_2 reacts with water to give carbonic acid, as per the equation

$$CO_2 + H_2O \longrightarrow H_2CO_3$$

The fifth oxide of period 2 is N_2O_5, nitrogen pentoxide. It is normally a white solid but reacts quite vigorously with water to give HNO_3, nitric acid. The reaction can be written as

$$N_2O_5 + H_2O \longrightarrow 2HNO_3$$

Thus N_2O_5 can be considered to be the *acid anhydride* of HNO_3. HNO_3 is normally purchased as a clear colorless solution, which is largely HNO_3 dissolved in water. Action of a *dehydrating agent* (a substance which attracts water to itself) not only soaks up the solvent water but can pull H and O out of HNO_3 in proportion to make water, thus leaving N_2O_5.

The last oxide shown in Fig. 3.8 for the oxides of period-2 elements is F_2O, which we might call difluorine oxide, but which would better be written as OF_2 and called oxygen difluoride. The reason for the preference of the formula OF_2 rather than F_2O is that conventionally formulas are written with the more positive component first. Normally, oxygen in a compound is the more negative component. But not with fluorine! Flourine has such a great drive to become negative that it drags charge away from oxygen and makes the oxygen look positive. Thus, OF_2 is more a *fluoride* than it is an *oxide*. This is illustrated further by the reaction with water. When OF_2 is placed in water, it spits and reacts vigorously per the following reaction:

$$OF_2 + H_2O \longrightarrow 2HF + O_2$$

The result, besides producing hydrofluoric acid HF, gives elemental oxygen. Since an acid is formed, one might be tempted to call OF_2 an acid anhydride, but this would not be strictly correct. An acid anhydride is a compound which adds to water to give acid only, no other product.

What have we learned from the above brief description of the oxide compounds of period 2? Besides the physical change from Li_2O solid to gaseous OF_2, the most striking gradation chemically is from a base-producing oxide on the left of the period to an acid-producing oxide at the right of the period. This is a general trend throughout the periodic table: *Oxides become more acidic as one goes from left to right in the periodic table.*

In the third row of the periodic table, we would list as typical oxides the following sequence:

$$Na_2O \qquad CaO \qquad Al_2O_3 \qquad SiO_2 \qquad P_2O_5 \qquad SO_2 \qquad Cl_2O_7 \qquad none$$

Again there is a pronounced change in acid-base character as one progresses from left to right. Na_2O, for example, is a basic anhydride and reacts with water to form NaOH:

$$Na_2O + H_2O \longrightarrow 2NaOH$$

On the extreme right, on the other hand, we have Cl_2O_7, chlorine heptoxide, which is an acid anhydride. It reacts as

$$Cl_2O_7 + H_2O \longrightarrow 2HClO_4$$

The product is perchloric acid, $HClO_4$, one of the strongest acids known. In this sequence we go from a very strong base-producing element, Na, to a very strong acid-producing element, Cl.

The situation toward the middle of this sequence is especially interesting. Al_2O_3, aluminum oxide, is neither a good acid-producing nor a good base-producing element. Indeed it does both, and Al_2O_3, as well as its hydrated equivalent, $Al(OH)_3$, can act both as an acid (i.e., it can neutralize bases) and as a base (i.e., it can neutralize acids). Such elements and such oxides are referred to as *amphoteric,* coming from the Greek word *amphoteros,* which means "both." As a general rule oxides toward the middle of the periodic table can be expected to be amphoteric. Experimentally, amphoteric behavior can be demonstrated by adding a sample of the suspected oxide to an acid and a sample to base. If it reacts with both, then the oxide has acted, respectively, as a base and as an acid and should be classified as being amphoteric.

At this point it might be appropriate to ask how the acid-base character of oxides changes in going down a group of the periodic table. Group III, comprising the elements B, Al, Ga, In, and Tl, offers a typical pattern. The respective oxides are B_2O_3, Al_2O_3, Ga_2O_3, In_2O_3, and Tl_2O_3. The top oxide, B_2O_3, boric oxide, dissolves in water to give an acidic solution, boric acid (H_3BO_3). The second oxide, Al_2O_3, is amphoteric (acid when in the presence of base, base when in the presence of acid). The third oxide, Ga_2O_3 (gallium oxide), leans to the basic side. Although it can function both ways, as acid and base, it is better at neutralizing acids. There is a definite gradation from acid B_2O_3, to amphoteric Al_2O_3, to base-leaning Ga_2O_3. Further down the group, the trend becomes more pronounced. In_2O_3, indium oxide, is a basic oxide with good ability to neutralize acids. The bottom oxide, Tl_2O_3 (thallium oxide), acts only as a base.

Why does the oxide character change from base at the top of a group to acid at the bottom? The answer is connected with size. In general, the bigger an atom, the more likely it is to form a basic oxide. As we go down a group of the periodic table, atoms generally get bigger, and so we expect an increase in base character. A deeper question is: Why should big atoms tend to form bases? This cannot be answered until we know more about what fundamentally distinguishes an acid from a base. Size is important, but, as we shall see in subsequent chapters, the environment in which the atom finds itself is even more important.

Important Concepts

periodic law	alkali elements
periodic table	metals vs. nonmetals
period	behavior of elements across the second period
group	oxygen compounds of the second period
subgroup	base vs. acid anhydride
noble gases	amphoteric
halogens	

Exercises

*3.1 **Atomic volume** Replot the data in Fig. 3.1 as a function of atomic weight instead of atomic number. What differences, if any, can you discern?

3.2 **Periodic law Which elements of the periodic table would you likely discard in order to make Newlands' "law of octaves" apply to the whole of the periodic table?

3.3 **Density Using data read from the curve of Fig. 3.1 and the atomic-weight table, calculate the corresponding density for each element. Make a plot of density vs. atomic number.

*3.4 **Terms** Distinguish clearly the following terms as applied to the periodic table: group, subgroup, period, halogens, alkali elements, noble gases.

***3.5 **Periodic table** One of the many alternative forms suggested for a periodic table is a helical form with different-size loops for the differing-length periods. By cutting strips of paper and holding them together with tape, construct a model of such a helical periodic table. Point out any special advantages of such a form.

*3.6 **Noble gases** On the basis of information given in this chapter, discuss the relative merits of putting the noble gases at the extreme left or the extreme right of the periodic table. What difference does it make?

*3.7 **Halogens** Without referring to any notes or other crutches, write down, in descending group order, the symbols of the halogens and give next to each element the symbol of the adjacent noble gas. Take any property you feel confident about, and tell how it changes going down the group.

3.8 **Iodine Tell briefly how starch–potassium iodide paper acts to detect presence of oxidizing agents. Suppose the potassium iodide were replaced by potassium bromide; speculate on how this change would affect the action of the paper.

3.9 **Alkali elements Give the names of the alkali elements in descending group order. From memory, write down what halogen is coupled in the same periodic row with each alkali element. What is the chief chemical reactivity difference between the alkali elements as a group and the halogens as a group?

*3.10 **Alkali elements** Indicate one outstanding feature (either occurrence, preparation, or use) that distinguishes each of the alkali elements.

3.11 **Seawater Besides hydrogen and oxygen, the two most abundant elements in seawater are chlorine [18,980 ppm (parts per million by weight)] and sodium (10,561 ppm). If all the sodium were paired up with chlorine as the compound $NaCl$, what concentration of chlorine would be left over?   *Ans. 2690 ppm*

*3.12 **Application** Explain why it makes good sense to dispose of wood ashes in a garden.

*3.13 **Alkali elements** Using the data from Fig. 3.6 make a plot of density of the alkali elements against (*a*) atomic number; (*b*) atomic weight. What information do you gather from a comparison of these plots?

3.14 **Second period Without referring to notes or text, write down in proper sequence the elements of the second period. Which of these elements has the highest density? What is unusual about the course of density vs. element across this period?

3.15 **Atomic radius Why in a periodic sequence would it not be surprising to find a value of the atomic radius omitted for the end member at the right?

3.16 **Chemical reactivity Tell briefly what would be observed when each of the elements of the second period is brought into contact with water.

3.17 **Oxygen compounds Write the formulas of the most likely oxides of the third-period elements and predict how they would behave when placed in water so far as acid-base reaction is concerned.

3.18 **Anhydride What is meant by the term "anhydride"? Distinguish between "base anhydride" and "acid anhydride." In which class would you put Al_2O_3?

3.19 **Anhydride What would be the anhydride of each of the following: $Ca(OH)_2$, H_3PO_4, H_2SO_4, HNO_2?

*3.20 **Dry ice** What is dry ice? What is strange about the melting of dry ice? How can you demonstrate that dry ice in water acts as an acid anhydride?

*3.21 **Oxygen compounds** Explain why it might not be appropriate to label OF_2 an acid anhydride.

3.22 **Stoichiometry How many grams of O_2 could you produce at the most from no more than 1.00 g of OF_2 and 1.00 g of H_2O?

3.23 **Oxides In general, how does the acid-base character of oxides change across the periodic table? Write some specific equations for period-4 elements to illustrate your point. How does the acid-base character generally change down a group?

Chapter 4

Na$_2$mnO$_4$

INTERNAL STRUCTURE OF ATOMS

The idea of an atom as an ultimate building block of matter appears to have originated with the early Greek philosophers. Two schools of thought were in contention: One believed that progressive subdivision of any sample of matter could be continued indefinitely; the other, that the process of subdivision had an ultimate limit. In the first view, even the smallest portion of a sample of matter is continuous, a sort of jelly that could be cut up into ever smaller and smaller regions, each being still jellylike in its consistency. The atomists, on the other hand, believed that any sample of matter is ultimately discontinuous. The process of chopping it up into bits would have to stop when it reached the ultimate indivisible particles. These particles were called *atoms* after the Greek word *atomos,* meaning "indivisible."

In the modern view, the atom is believed to be composed of more "fundamental" units (e.g., protons, neutrons, and electrons), but the way these are held together is not completely understood. Furthermore, as we shall see in this chapter, the atom is not a very solid object but has a spatial extent that depends to some degree on what else is in the immediate vicinity. In particular, when two or more atoms "interact" to form the more complicated unit we call a *molecule,* the distinct identity of the individual atoms is no longer retained; for many purposes, it is more useful to think in terms of the new unit formed.

In this chapter we consider the experiments that led to the unraveling of the internal structure of atoms and the conceptual framework that has been developed to account for atomic properties.

4.1
EXPERIMENTS ON THE ELECTRICAL NATURE OF ATOMS

Michael Faraday, English physicist-chemist, reported in 1832 that when electric current is passed through a molten salt or a salt solution in order to decompose it into elementary substances, "the chemical action of the current of electricity is in direct proportion to the absolute quantity of electricity which passes." For example, a steady flow of electricity through molten sodium chloride produces twice as much sodium and chlorine in 10 min as in 5 min. This is an example of what is called the *first law of electrolysis.* Faraday also discovered that the weight of elemental product generated in the decomposition is directly proportional to the atomic weight of the element divided by a small whole number. Thus, for example, if a suitable amount of electricity is used to decompose molten sodium chloride, we get 23 g of sodium plus 35.5 g of chlorine; if the same amount of electricity is used to decompose molten calcium chloride, we get 20 g of calcium plus 35.5 g of chlorine; if the same amount of electricity is used to decompose molten aluminum chloride, we get 9 g of aluminum plus 35.5 g of chlorine. These numbers illustrate the

89

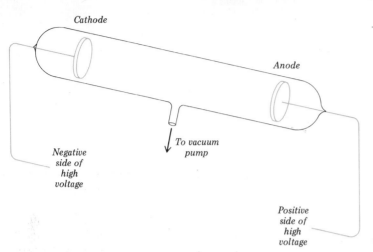

Cathode

Anode

To vacuum
pump

Negative
side of
high
voltage

Positive
side of
high
voltage

FIG. 4.1 Electric discharge tube.

second law of electrolysis; the masses of sodium, calcium, aluminum, and chlorine formed are the corresponding atomic weights (23, 40, 27, and 35.5) divided, respectively, by 1, 2, 3, and 1 [Clearly the fact that chemical change is produced by electricity indicates a connection between atoms and electricity; the appearance of whole numbers as divisors suggests that the electric structure of atoms involves discrete units of electricity.]

The nature of these discrete units was considerably clarified during the latter half of the nineteenth century by the study of electric discharges through gases. Given a glass vessel, such as is shown in Fig. 4.1, containing metal electrodes sealed in at either end, across which a high voltage can be imposed, we observe as we evacuate the vessel that the residual gas commences to glow when the gas pressure is sufficiently reduced. Subsequent study suggested that electric rays, called *cathode rays,* come out from the negatively charged electrode (the *cathode*) and go into the positive electrode (the *anode*). In traveling through the space between cathode and anode, the cathode rays were believed to gain energy because of the accelerating effect of the high voltage. If the gas pressure in the tube is low enough, there are so few gas molecules left in the vessel that the cathode rays can gain lots of energy before colliding with a gas molecule. When this energy gets big enough, a process called *ionization* occurs. "Negative electricity" is stripped off the neutral molecule, and the result is to leave a positive fragment called a positive *ion.* The positive ions are then believed to attract the "negative electricity," and the recombination process emits energy in the form of light. Clever design, notably by William Crookes of England and Eugen Goldstein of Germany, enabled them to select out a beam of charged fragments for study. Goldstein, for example, drilled a hole through the cathode so that some of the positive fragments that were attracted to the cathode could drift through the hole into the space on the other side of the cathode (see Fig. 4.2). By applying electric and magnetic forces to the beams of charged particles, it was possible to learn a great deal about the charges and masses of the particles that made up the

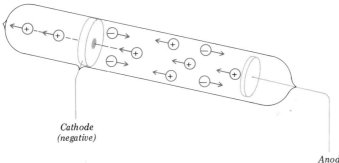

Anode
(positive)

FIG. 4.2 Electric discharge tube for obtaining a beam of positive particles.

beams. In this way the English physicist J. J. Thomson was led to the discovery of the *electron* in 1897. By careful quantitative study of beam behavior, Thomson was able to show that cathode rays consist of negative particles, or electrons, which are all alike no matter what electrode material they come from and no matter what gas is in the tube. On the other hand, the beams of positive rays differ, depending on what gas is in the tube.

4.2
CHARGE AND MASS OF THE ELECTRON

A typical cathode-ray tube, such as can be used for Thomson-type quantitative measurements, is shown in Fig. 4.3. The negatively charged cathode at the left emits electrons, which are attracted to the right by the positively charged anode. The anode has a hole bored along its axis through which some of the electrons pass. The result is to give a narrow beam continuing to the right and impinging on a zinc sulfide coating covering the inner face of the tube. Zinc sulfide is a remarkable solid that has the property of converting the kinetic energy of incident electrons into visible light. Thus, a spot of light appears on the face of the tube where the electron beam hits the surface.

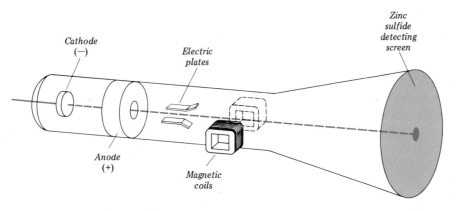

Cathode
(−)

Electric
plates

Anode
(+)

Magnetic
coils

Zinc
sulfide
detecting
screen

FIG. 4.3 Cathode-ray tube for studying deflection of electrons by electric and magnetic forces.

Electric plates in the tube, one located above the beam and the other below, can deflect the beam up and down as the negative plate repels the beam and the positive plate attracts it. The amount of deflection, which can be determined by measuring the movement of the spot, is directly proportional to the voltage between the plates.

Magnetic coils which clamp around the tube from the outside are set so there is a magnetic field at right angles to the electron beam. The nature of electric-magnetic interactions is such that the electron beam is bent into a curved path as it passes through the magnetic field. The Thomson experiment consists of two parts: In one part, a deflection is measured with zero voltage on the plates (the deflection is due to magnetic field alone). In the second part of the experiment, the voltage between the plates is adjusted until there is no net deflection of the beam (the electric and magnetic forces just cancel each other).

Figure 4.4 shows the relation of the quantities involved. An electron with charge e moves to the right with velocity v. As it moves through magnetic field H (perpendicular to the beam), it experiences a magnetic force Hev which bends the electron into a curved path of radius r. The value of r can be determined by measuring the deflection x, for once the electron leaves the magnetic field, it continues in a straight line until it hits the face of the cathode-ray tube.

A fundamental law of physics states that force = mass × acceleration. We can therefore set $Hev = m(v^2/r)$, where m is the mass of the electron and v^2/r is the acceleration of a particle in a curved path. The equality $Hev = mv^2/r$ leads directly to $e/m = v/Hr$. This relation, however, which tells us the charge-to-mass ratio, is of little use unless we know v.

We can determine v from the second part of the Thomson experiment. Putting a voltage between the electric plates (Fig. 4.3) exerts an electric field E on the electron. The magnitude of E is equal to the voltage divided by the distance between the plates. We adjust the voltage until Ee, the electric force on the electron, just cancels Hev, the magnetic force. In other words, we crank up the plate voltage until the beam spot returns to its undeflected position. The equality $Ee = Hev$ leads to $v = E/H$, which can be substituted in the earlier equation $e/m = v/Hr$ to give $e/m = E/H^2r$.

The combined electric and magnetic measurements of the Thomson experiment give us a numerical value for the charge-to-mass ratio e/m. For all

FIG. 4.4 Schematic representation showing how an electron with velocity v is deflected by a magnetic field H.

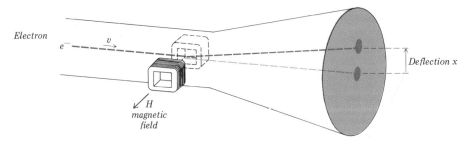

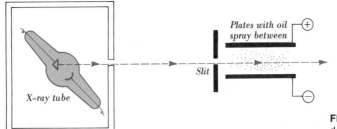

FIG. 4.5 Millikan's oil-drop experiment.

electrons, the value obtained is always the same, -1.7588×10^8 C/g. (The *coulomb,* C, a unit for measuring electric charge, is the amount of charge passing a given point in one second when the electric current is one ampere. See Appendix A5.5.)

Once the charge-to-mass ratio has been measured, another experiment has to be performed to get either the charge or the mass separately. If, for example, the charge is measured independently, the mass can be calculated from the value of e/m. The charge of the electron was first measured precisely in a classic experiment by the American physicist R. A. Millikan. Figure 4.5 shows the essential features of the experiment. A cloud of oil droplets is sprayed between two charged plates. Because of gravity the droplets tend to settle. The gravitational force is equal to Mg, where M is the mass of a drop and g is the acceleration due to gravity. By irradiating the oil-drop chamber with X rays, the oil droplets can be given a negative charge. (The X rays knock electrons off the molecules of air, and one or more of these electrons can be picked up by an oil droplet in order to make it negative.) If the droplets carry a negative charge, they will experience an upward force because of the electric field between the charged plates. The magnitude of this electric force is Eq, where E is the electric field strength and q is the total electric charge on a particular drop. Different droplets may carry different charges.

It is observed that, although the charges on different oil droplets vary, the values obtained are always small, whole-numbered multiples of -1.60×10^{-19} C. Apparently, this is the smallest possible charge that any one oil droplet can pick up, and it is assumed to be the charge of an individual electron. Combining the measured charge of the electron (-1.60×10^{-19} C) with the measured charge-to-mass ratio (-1.76×10^8 C/g) gives a measured mass for the electron of 9.1×10^{-28} g.

4.3
ATOMIC SPECTROSCOPY

At the same time as the above investigations into the electric nature of matter were being conducted, parallel studies were going on into the nature of light given off when materials are heated. This kind of investigation, called *spectroscopy,* reached a peak late in the nineteenth century when Bunsen perfected his burner for getting an almost colorless, very hot flame from combustion of natural gas. Injection by spraying of salt solutions into such a flame produces

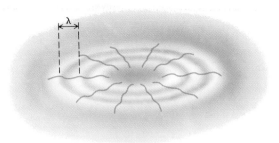

FIG. 4.6 Wave motion generated on the surface of a liquid.

various colors, which can be analyzed by passing the light through a prism or a diffraction grating. The prism separates component colors because light moves more slowly in glass than in air, depending on its color. The diffraction grating, which consists of a series of parallel grooves on a flat surface, separates light because of selective interference of light waves along particular directions.

Light is a form of electromagnetic radiation, composed of electric and magnetic pulsations transmitted through space. The pulsations, or waves, move at a speed of 2.998×10^8 m/sec. They can be characterized in terms of a wavelength and a frequency. Figure 4.6 illustrates what is meant by wavelength, λ (Greek letter lambda), for the particular case of waves spreading out from a central point on the surface of a liquid. The distance between any pair of corresponding points, such as two adjacent wave crests, is the wavelength. It is equal to the speed at which the train of pulsations is propagated divided by the frequency of the waves. The frequency, ν (Greek letter nu), can be pictured as the number of complete waves, or pulses, that pass a given point per second. The *speed* of the waves equals the product of wavelength times frequency.

EXAMPLE 1

Given that the speed of light, c, is 2.998×10^8 m/sec, calculate the frequency of light that has wavelength 400×10^{-9} m.

Solution

Speed of light = (frequency) (wavelength)

$c = (\nu) (\lambda)$

$\nu = \dfrac{c}{\lambda} = \dfrac{2.998 \times 10^8 \text{ m/sec}}{400 \times 10^{-9} \text{ m}} = 7.50 \times 10^{14} \text{ sec}^{-1}$

Note that the unit for frequency is reciprocal second.

The energy E of a light wave is proportional to its frequency; that is, $E = h\nu$, where the proportionality constant h, called the Planck constant, has the value of 6.6262×10^{-34} J sec.

EXAMPLE 2

What is the energy of the light wave in Example 1?

Solution

$E = h\nu$

$E = (6.63 \times 10^{-34} \text{ J sec}) (7.50 \times 10^{14} \text{ sec}^{-1})$

$E = 4.97 \times 10^{-19} \text{ J}$

White light is a combination of waves of different wavelengths, extending from about 400 nm for violet light to about 700 nm for red light.* The range from 400 to 700 nm is called the *visible spectrum* since it is the part which the human eye can distinguish. Ultraviolet radiation lies on the short-wavelength side of 400 nm; infrared radiation, which we sense as heat waves, lies on the long-wavelength side of 700 nm. In terms of energy, ultraviolet radiation is on the high-energy side, and infrared radiation is on the low-energy side of the visible spectrum.

If white light, as from a glowing filament, is passed through a prism, it is observed that the resulting pattern consists of a *continuous spectrum* of colors, a gradual blending from one color to the next. If, instead of white light, colored light from a flame is passed through a prism, the resulting pattern is not continuous but consists of a series of narrow lines of different color. An example of such a *line spectrum* is shown in Fig. 4.7.

The actual pattern obtained differs from element to element and is characteristic of the particular atoms present. The absorption of energy by the atoms from the flame raises the energy of the atoms in a particular way. When the atoms return to lower energy states, they emit characteristic energies. It is these characteristic energies that we see as variously colored bands of light.

If we express energy as frequency (recall that energy E is proportional to frequency ν), we find that for every line spectrum the observed frequencies can be described in a simple systematic way. For example, for hydrogen the

---

*Until recently, wavelengths were generally expressed in units of 1×10^{-8} cm or angstrom units (Å). The International Union of Pure and Applied Chemistry (IUPAC) has recommended that angstroms be abandoned as a unit. In its stead, the recommended unit is the nanometer (nm), which is 1×10^{-9} m, or 1×10^{-7} cm. Thus, we have

$$1 \text{ nm} = 1 \times 10^{-9} \text{ m} = 1 \times 10^{-7} \text{ cm}$$
$$= 10 \text{ Å}$$

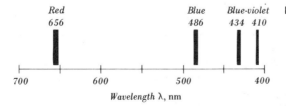

FIG. 4.7 Hydrogen line spectrum.

frequencies seen in the visible part of the spectrum can be represented by the equation

$$\nu = 3.290 \times 10^{15} \left(\frac{1}{4} - \frac{1}{b^2} \right)$$

Here b can be any whole number greater than 2, that is, 3, 4, 5, 6, etc. This gives us what is called the *Balmer series* of hydrogen. It extends from a bright red line at wavelength 656.3 nm (corresponding to $b = 3$), through 486.1 nm ($b = 4$), 434.0 nm ($b = 5$), 410.1 nm ($b = 6$), and so forth, until it converges at 364.6 nm ($b = $ infinity) in the blue region of the spectrum.

In the ultraviolet part, hydrogen shows another series of lines related by the equation

$$\nu = 3.290 \times 10^{15} \left(1 - \frac{1}{b^2} \right)$$

where b now can have values 2, 3, 4, 5, 6, etc. This is called the *Lyman series*. It falls in the wavelength region between 121.5 and 91.2 nm. A third series, called the *Paschen series,* occurs in the infrared region, with a series of lines fitting the equation

$$\nu = 3.290 \times 10^{15} \left(\frac{1}{9} - \frac{1}{b^2} \right)$$

where $b = 4, 5, 6, 7$, etc.

It should be noted that the Balmer, Lyman, and Paschen series differ only in that the first term inside parentheses is $\frac{1}{4}$, 1, and $\frac{1}{9}$, respectively. These can be expressed as $(\frac{1}{2})^2$, $(\frac{1}{1})^2$, and $(\frac{1}{3})^2$. For a long time the integral character of these numbers and also of the b values appeared to be just so much magic, but eventually they proved to be the main clue for unlocking the secret of the electron structure of atoms. Mysterious as they were, the observed regularities in the line spectra clearly suggested an underlying regularity in the energy levels of atoms.

4.4
DISCOVERY OF THE NUCLEUS

By the beginning of the twentieth century it was generally accepted that atoms can be broken up into negatively charged electrons and positively charged residual fragments. Beam-deflection experiments such as those described in Sec. 4.1 indicated that electrons are light and identical, whereas the positive fragments are massive and dependent on the particular neutral atom they are derived from. J. J. Thompson, in 1898, put the two concepts together and proposed that an atom be considered as a sphere of "positive electricity" in which the negative electrons were embedded like raisins in a pudding. Practically all the mass of the atom (better than 99.9 percent) was associated with the "positive electricity," and so the mass was assumed to be spread out uniformly over the entire volume of the atom. The electrons contributed practically nothing to the total mass.

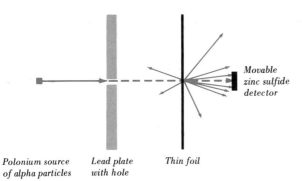

Polonium source   *Lead plate*   *Thin foil*
of alpha particles   *with hole*

FIG. 4.8 Rutherford's experiment that led to the discovery of the nucleus.

The English physicist Ernest Rutherford, while studying the effect of thin metal foils on the scattering of positive beams, discovered in 1911 that Thomson's model was incorrect. Instead, the positive charge and its associated mass must be concentrated in a tiny central core, the *nucleus*. The details of Rutherford's classic experiment are represented schematically in Fig. 4.8. A beam of *alpha particles,* dipositive helium ions, was obtained from a source containing the element polonium. This element, like all the very heavy elements, is radioactive; its atoms spontaneously disintegrate in random fashion to give simpler fragments. One of these fragments is the alpha particle.

A well-defined beam of alpha particles can be obtained by placing the polonium in a lead box and letting the alpha particles come out through a pinhole in the lead. Lead is a good absorber for particles resulting from radioactive decay. It stops the alpha particles except for the ray that comes through the pinhole. Rutherford's experiment was to place thin sheets of metal in the path of the alpha ray in order to see how various metals would affect the alpha-particle trajectory. Alpha particles are energetic enough to penetrate thin foils readily. What happens to them as they go through the foil, specifically how they are scattered from a straight-line path, can be monitored by placing a detector such as a zinc sulfide screen at various places on the exit side of the foil. On the basis of a Thomson model for the atoms in the foil, the alpha particles going through would have little reason to swerve from their original path. They should plow right through, and the detector would be expected to pick up at most only slight deflections. On the contrary, Rutherford found that some of the alpha particles were deflected at astonishingly large angles. A few were actually reflected back toward the source. To Rutherford this was absolutely unbelievable. In his own words, "It was almost as incredible as if you fired a 15-inch shell at a piece of tissue paper and it came back and hit you."

The old Thomson model could not account for such large deflections. If mass and positive charge were spread uniformly throughout the metal, a positively charged alpha particle would not encounter a large repulsion or major obstacle anywhere in its path. According to Rutherford, the only way to account for the large deflections would be to say that the "positive electricity" and mass in the metal foil are concentrated in very small regions. Although most of the alpha particles go through without any deflection, occasionally one comes very close to a high concentration of positive charge. The positive charge is essentially immovable because it contains practically all the mass of the atom. As like charges get closer together, they repel each other, and the repulsion between alpha particles and nucleus may be great enough to cause the relatively light alpha particle to swerve considerably from its original path. Hence, to explain his observations, Rutherford suggested that an atom has a *nucleus,* or center, in which its positive charge and mass are concentrated.

The quantitative results of scattering experiments such as Rutherford's indicate that the nucleus of an atom has a radius of about 10^{-13} cm, which is only about one hundred-thousandth the size generally ascribed to atoms. Later, when neutrons and protons were discovered, it was noted that the radius of a particular nucleus can be expressed roughly as r (in centimeters) $\sim 1 \times 10^{-13} A^{1/3}$, where A is the total number of neutrons and protons in that nucleus.

4.5
DISCOVERY OF THE ATOMIC NUMBER

About the same time that Rutherford was led to postulate the existence of the nucleus, H. G. J. Moseley was measuring the energies of X rays emitted by various elements. These measurements, which contributed greatly to further characterization of the nuclear atom, can be carried out with a high-voltage gas-discharge tube such as the one shown in Fig. 4.9. High-energy electrons from the curved cathode on the right are focused to impinge on the anode on the left, giving rise to the emission of X rays when the electrons hit the target material. It is found that wavelengths emitted are characteristic of the element from which the anode is made.

X rays, like light, are a form of electromagnetic radiation. They are very energetic and have wavelengths much shorter than those of visible light. The frequencies, i.e., the energies, are some 1000 times as large. As a specific example, if an X-ray tube has an anode made of copper, the X ray that it emits has $\lambda = 0.1541$ nm, corresponding to a frequency ($\nu = c/\lambda$) of 1.945×10^{18} sec^{-1} or an energy ($E = h\nu$) of 1.289×10^{-15} J. For comparison, blue light with $\lambda = 400$ nm corresponds to $\nu = 7.495 \times 10^{14}$ sec^{-1} or $E = 4.966 \times 10^{-19}$ J. Moseley's experiments, in which he successively substituted different materials for the anode, showed that there is a simple relation between the X rays

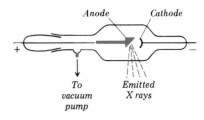

FIG. 4.9 Gas-discharge tube with interchangeable anode which can be used for production of X rays.

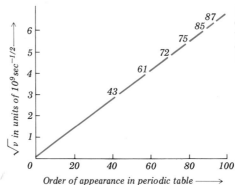

FIG. 4.10 Moseley's plot of X-ray frequencies for the elements. (Numbers on the line indicate missing elements.)

emitted and the nature of the anode material. Specifically, when he plotted the square root of the X-ray frequency against the order in which the element appeared in the periodic table, he obtained a straight line.

As shown in Fig. 4.10, to get a smooth curve Moseley found it necessary to leave gaps in a few places (corresponding to undiscovered elements). He also used an order based on trends of chemical properties rather than on strict adherence to increasing atomic weight. For example, in the case of cobalt (58.93 amu) and nickel (58.71 amu), cobalt would come first on the basis of chemical trends, but nickel would come first on the basis of increasing atomic weight. From his X-ray data Moseley was led to choose the order cobalt followed by nickel, thus overturning the notion that atomic weight was most decisive in fixing atomic properties. Instead, there is a characteristic number called the *atomic number,* generally designated as Z, which is believed to be equal to the positive charge (the number of protons) of the nucleus and which determines the ordering of the elements by chemical behavior.

4.6
ISOTOPES

Why is it that in some cases chemical properties of the elements do not follow a strict order based on atomic weight but instead follow an order based on atomic number? Put another way, why is there not a direct parallel between increasing atomic number and increasing atomic weight? The answer lies in the fact that different atoms of the same element may differ in mass. Atoms that have the same atomic number but differ in mass are called *isotopes.* Their existence was discovered in 1907 by the American chemists, H. N. McCoy and W. H. Ross. The word *isotope* comes from the Greek *isos,* meaning "same" and *topos,* meaning "place"; it refers to the fact that isotopes occur in the *same place* in the periodic table.

The existence of isotopes can be shown by use of the *mass spectrometer.* Figure 4.11 shows the essential features of this device. A positively charged filament on being heated by electric current emits electrons which ionize the gas present in the vacuum chamber near the the filament. The positive ions formed are accelerated by attraction through the negatively charged first slit to give a beam. They are bent into a circular path by the magnetic field, which is perpendicular to the plane of the diagram. Ions having different values of

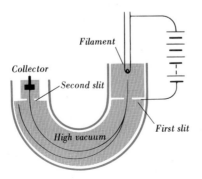

FIG. 4.11 Mass spectrometer. (Magnetic field, not shown, is perpendicular to the plane of the diagram.)

the charge-to-mass ratio follow different paths. This is shown by the curved lines in the figure. Separation of ion paths occurs for the following reasons: For an accelerating voltage V between filament and first slit, a particle with charge q is given an amount of energy equal to qV. This energy is added to the particle as kinetic energy and so equals $\frac{1}{2}mv^2$. From the equality

$$qV = \tfrac{1}{2}mv^2$$

the particle velocity can be found to be

$$v = \sqrt{\frac{2qV}{m}}$$

As shown on page 92, the charge-to-mass ratio of a particle moving in a magnetic field H is related to the radius of curvature r by $q/m = v/Hr$. Substituting v in this equation, we get

$$\frac{q}{m} = \frac{2V}{H^2r^2}$$

This equation tells us that for a fixed value of voltage V and magnetic field H, particles of larger charge-to-mass ratio follow paths of smaller radius. In other words, light particles are bent more than heavy ones. Thus, particles of different charge-to-mass ratio can be separated in a mass spectrometer.

In practice, a mass spectrometer operates with a fixed collector to catch only ions of a particular radius of curvature. By varying either the voltage or the magnetic field, ions of a particular q/m can be focused into the collector. For a particular element it is found that as V is progressively increased, several beams are collected, differing from each other in q/m value. For example, typical values for neon gas are 4.81×10^3, 9.62×10^3, and 4.33×10^3 C/g. The second of these values, which is exactly twice the first, suggests that some of the neon atoms have lost two electrons instead of one; hence they have twice as large a charge-to-mass ratio. However, the third value, 4.33×10^3 C/g, bears no simple relation to the other two and cannot be explained by assuming loss of an integral number of electrons. *The fact that the charge-to-mass ratio differs by something other than a whole-number multiple indicates that particles of different mass must be present.* In other words, the sample of element neon contains nonidentical atoms, or isotopes.

For every element two or more isotopes are known, although all do not necessarily occur in nature. The relative abundance can be determined by measuring the relative intensity of the different isotope beams in the mass spectrometer. As an illustration of isotope abundances, the element oxygen in its natural mixture consists of three isotopes, which can be designated ^{16}O, ^{17}O, and ^{18}O. The superscripts indicate the *mass number,* which is the whole number closest to the mass of the isotope. The actual masses of the isotopes are 15.9949 amu for ^{16}O, 16.9991 amu for ^{17}O, and 17.9991 amu for ^{18}O. Relative abundances in a "natural mix" are 99.759% ^{16}O, 0.037% ^{17}O, and 0.204% ^{18}O. The atomic weight used for calculations involving natural oxygen is a weighted average of the three, which turns out to be 15.9994 amu.

Practically all the mass of an atom resides in the nucleus. However, there is no simple relation between increasing mass and increasing atomic number. For example, in going from ^{12}C to ^{13}C, the mass changes from 12 to 13.0034 amu, but the atomic number stays at 6. In going from ^{12}C to ^{12}N, the mass change is only from 12 to 12.0187 amu, but the atomic number changes from 6 to 7. The observation that nuclear charge and nuclear mass do not simply go together can be accounted for by assuming that the nucleus contains *both* protons and neutrons. The *neutron,* as the name implies, is a neutral particle. It was discovered in 1932. It has a mass nearly equal to that of the proton but has no charge. Protons in a nucleus account for all the positive charge; protons and neutrons together account for the total mass. Counting the electron charge as -1, the proton has a charge of $+1$.

On the atomic-weight scale, the masses of protons, neutrons, and electrons are 1.00727, 1.00866, and 0.000549 amu, respectively.* The masses of the neutron and proton are very close to unity, and so the mass number (i.e., the whole number closest to the mass of an isotope) directly gives the total number of neutrons plus protons in the nucleus. It follows, then, that isotopes of the same element have the same number of protons per nucleus but differ in the number of neutrons. Thus, for example, ^{12}C has six protons and six neutrons; ^{13}C has six protons and seven neutrons. Similarly, ^{12}N has seven protons and five neutrons; ^{13}N has seven protons and six neutrons.

Frequently, nuclei are designated by symbols such as $^{18}_{8}$O. In such symbols the subscript indicates the atomic number Z, which gives the nuclear charge, and the superscript indicates the mass number A, which gives the total number of neutrons plus protons. $A - Z$ gives the number of neutrons. The nucleus $^{18}_{8}$O has 8 protons and 10 neutrons; its atomic number is 8, and its mass number is 18.

4.7
THE BOHR ATOM

Major credit for the first workable theory of atomic structure belongs to the Danish physicist Niels Bohr. He proposed in 1913 what was then a revolutionary model. Three important sets of observations had to be explained: systematic recurrence of chemical properties when elements are ordered as in the periodic table; systematic regularity of observed spectral-line frequencies; and an apparent contradiction between classical electrodynamics and its application to electrons in atoms. By this last point is meant the fact that electrically charged particles moving in curved paths are expected to radiate energy to the surroundings. (The bluish glow emitted from a particle accelerator such as a cyclotron is a modern manifestation of this.) Similarly, if an atom consists of negative electrons in motion around a nucleus, then the atom would be expected to radiate energy to the surroundings. Such loss of

---

*These masses apply to isolated particles at rest and are not exactly equal to the masses in the atomic nuclei. In the formation of a nucleus from isolated protons and neutrons, some of the mass disappears and is converted to energy.

energy by radiation would eventually result in collapse of the atom. In fact, collapse is not observed; an atom will not radiate unless it has previously been excited to a higher energy state.

In setting up Bohr's model of the atom, we go back to the fundamental principle of physics: force = mass × acceleration. The force is the force of electric attraction between the positively charged nucleus and the negatively charged electron. It is proportional to the charge of the nucleus times the charge of the electron divided by the square of the distance between them. For a nucleus of positive charge Ze at a distance r from an electron of negative charge e, the force of electric interaction is $K(Ze)(e)/r^2$. K is a universal constant, with the value 8.98×10^9 N m^{2}C^{-2}. As was seen on page 92, the acceleration experienced by an electron of mass m moving with velocity v along a path with radius of curvature r is v^2/r. Equating the appropriate terms gives us the following:

$$\text{Force} = \text{mass} \times \text{acceleration}$$

$$K\frac{Ze^2}{r^2} = m\frac{v^2}{r} \tag{1}$$

From this equation it would seem that all values of r would be possible. However, Bohr then introduced an additional restriction, which was hard to accept at the time. He stated that the angular momentum (mass × velocity × radius of path) of the electron can take on only certain permitted values. This requirement, which is referred to as the *quantum* condition,* can be expressed as

$$mvr = \frac{nh}{2\pi} \tag{2}$$

where mvr is the angular momentum; n is the *quantum number,* which can take on the integral values 1, 2, 3, 4, 5, . . . ; and h is Planck's constant. In words, Eq. (2) states that the angular momentum is restricted to taking on values that are whole-number multiples of $h/2\pi$. Each of these values would correspond to a permitted state of the atom. The corresponding permitted values of r can be obtained by combining Eqs. (1) and (2) to get rid of v:

$$r = \frac{n^2}{Z}a_0 \tag{3}$$

Here all the constants ($h^2/4\pi^2Kme^2$) have been lumped together and simply written as a_0. a_0 is called the Bohr radius and has the numerical value 0.05292 nm. In words, Eq. (3) states that the only permitted values for the radius of the electron path are those proportional to the square of a whole number n divided by the atomic number Z. For hydrogen, where $Z = 1$, allowed values of r are $1a_0$, $4a_0$, $9a_0$, $16a_0$, $25a_0$, ...

In principle, atoms with higher Z can also be treated by Eq. (3); however,

---

*The word *quantum* comes from the Latin *quantus,* meaning "how much." It was introduced by Max Planck in 1900 as the name for a packet of radiant energy.

since the derivation does not take into account repulsions between electrons in the same atom, Eq. (3) can strictly be used only for the combination of a nucleus and one electron.

Actually, the radius of an electron path in an atom, although interesting, is not an observable quantity. What is observable in an experiment is the energy state of an atom, or rather differences between energy states. In the simple Bohr atom described above, the electron's energy can be calculated as the sum of the kinetic energy, $\frac{1}{2}mv^2$, and the potential energy, $-KZe^2/r$. (The reason for the minus sign is that when two opposite charges attract each other, the potential energy becomes more negative as the charges come closer together.) Designating the total energy as E, we write

$$E = \text{kinetic energy} + \text{potential energy}$$

$$= \tfrac{1}{2}mv^2 - \frac{KZe^2}{r} \tag{4}$$

From Eq. (1) we see that mv^2 is KZe^2/r. Substituting this into Eq. (4) and replacing r by n^2a_0/Z from Eq. (3), we obtain finally

$$E = -\frac{KZ^2e^2}{2n^2a_0} \tag{5}$$

Equation (5), with different values 1, 2, 3, 4, etc., substituted for n, gives the values of the energy permitted to a one-electron atom. Figure 4.12 shows how these energy states are related to each other. The bottom line, marked $n = 1$, corresponds to the lowest permitted state. It is called the *ground state* and has an energy $E_1 = -KZ^2e^2/2a_0$. The next higher line, marked $n = 2$, corresponds to a higher (less negative) value of the energy $E_2 = -\frac{1}{4}KZ^2e^2/2a_0$. This represents an *excited state;* energy must be added to raise the atom from ground state E_1 to excited state E_2. As n subsequently increases, we move upward on the scale; E becomes progressively less negative and eventually approaches 0 from the negative side. In the limit, when $n = \infty$, we reach the state corresponding to $E_\infty = 0$.

As can be seen from Eq. (3), the electron radius increases as the square of n. Electrons with $n = 1$ are closest to the nucleus; they are referred to as being in the innermost *shell* or *orbit,* also called the *K shell*. Electrons with $n = 2$ are next farther from the nucleus; they are referred to as being in the L shell, or in a higher orbit. Higher orbits, or higher energy levels, are also referred to as outer energy levels. The letter designations K, L, M, N, ... were originally introduced by workers in X-ray spectroscopy. Transitions of electrons from outer shells to the K shell give rise to the K series of X rays, which were the ones used by Moseley in working out his correlation of frequency with atomic number. As we shall see, the Bohr idea of orbits, or shells, for describing electron motions is no longer acceptable, but the terms "orbit" and "shell" are still often used to refer to energy levels.

Although it is not possible to measure individual energy levels, the lines observed in spectra correspond to transitions between such levels—i.e., to energy differences between atomic states. For calculating these differences we can make use of Eq. (5). Designating E_b and E_a as the energies of two dif-

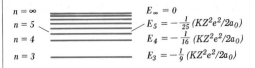

$n = \infty$ $E_\infty = 0$

$n = 5$ $E_5 = -\frac{1}{25}(KZ^2e^2/2a_0)$

$n = 4$ $E_4 = -\frac{1}{16}(KZ^2e^2/2a_0)$

$n = 3$ $E_3 = -\frac{1}{9}(KZ^2e^2/2a_0)$

$n = 2$ $E_2 = -\frac{1}{4}(KZ^2e^2/2a_0)$

Energy ⟶

FIG. 4.12 Energy-level diagram for an atom consisting of a nucleus with atomic number Z and one electron with various possible values of n.

$n = 1$ $E_1 = -(KZ^2e^2/2a_0)$

ferent states corresponding to quantum numbers n_b and n_a, we get the following:

$$E_b - E_a = \frac{KZ^2e^2}{2a_0}\left(\frac{1}{n_a^2} - \frac{1}{n_b^2}\right) \tag{6}$$

The energy difference can be converted to a frequency by using $E_b - E_a = h\nu$, where h is the Planck constant and ν is the frequency of the light wave emitted by the transition of the atom from state E_b to state E_a. This gives us

$$\nu = \frac{KZ^2e^2}{2ha_0}\left(\frac{1}{n_a^2} - \frac{1}{n_b^2}\right)$$

If we choose $n_a = 1$, that is, select only those transitions that go down to the ground state, and substitute numerical values (including $Z = 1$) for the various constants, we finally get

$$\nu = 3.290 \times 10^{15}\left(\frac{1}{1} - \frac{1}{n_b^2}\right)$$

This is precisely the same as the expression given on page 96 for the observed Lyman series of lines in the ultraviolet region of the hydrogen spectrum. In like manner, if E_a represents one of the higher states, it is possible to derive a similar equation for each of the other series of spectral lines. Thus, at least for the hydrogen atom, the Bohr theory accurately describes observed atomic spectra. With some modification to take care of effects due to electron-electron repulsion, the Bohr theory can be extended to elements containing more than one electron. However, as discussed below, there are major flaws in the Bohr model, and they have led to its replacement by a more sophisticated theory of the atom.

4.8
BUILDUP OF ATOMS ON BOHR MODEL

For a while, the Bohr model was extremely attractive; it not only explained the puzzling regularities in line spectra, but, by adding another assumption—that chemical properties depend on the number of outer electrons—it also made it possible to rationalize the periodic recurrence of chemical properties in the elements. The reasoning went as follows: Only specified energy levels (designated by $n = 1, 2, 3, 4, \ldots$) are permitted for electrons in atoms. Let us assume in addition that the maximum number of electrons allowed in any one energy level is also limited. Specifically, as Bohr discovered, this maximum number is equal to $2n^2$. Then the lowest energy level ($n = 1$) will have a maximum population of 2; the second level, $2(2)^2 = 8$; the third level, $2(3)^2 = 18$; the fourth level, $2(4)^2 = 32$; and so on. Each time a shell is completed, at 2, 8, 18, 32, . . ., respectively, addition of another electron produces an atom with but one electron in its outermost occupied energy level. Chemical properties depend mainly on the number of electrons in the outermost energy level, so that limitation of each shell's population to $2n^2$ automatically leads to periodic reappearance of the same number of electrons in the outer shell and, hence, the same properties.

Imagine the successive buildup of atoms by addition of electrons to a nucleus. To have a neutral atom, we have to add as many electrons as there are protons in the nucleus. Each electron enters the lowest energy level available. In the case of hydrogen ($Z = 1$) the lone electron goes into the K shell. The next electron, which would give an atom of helium ($Z = 2$), also enters the K shell. However, the third electron, which would give lithium ($Z = 3$), has to go into the L shell, since the K shell is full when it has two electrons.

Figure 4.13 lists the first 18 elements in order of increasing atomic number and shows the number of electrons in the various shells. Since the K

FIG. 4.13   Assignment of Electrons to Various Shells in Atoms of the First 18 Elements

Atomic no.	1	2	3	4	5	6	7	8	9	10	11	12	13	14	15	16	17	18
Element	H	He	Li	Be	B	C	N	O	F	Ne	Na	Mg	Al	Si	P	S	Cl	Ar
Electron population																		
K shell	1	2	2	2	2	2	2	2	2	2	2	2	2	2	2	2	2	2
L shell			1	2	3	4	5	6	7	8	8	8	8	8	8	8	8	8
M shell											1	2	3	4	5	6	7	8
		"inert"								"inert"								"inert"

shell can accommodate only two electrons, it becomes completely populated at the chemically unreactive element helium. Proceeding from helium, the L-shell population increases from one to eight in the sequence lithium, beryllium, boron, carbon, nitrogen, oxygen, fluorine, and neon. In neon the situation is like that in helium. With two electrons in the K shell and eight electrons in the L shell, the shells are completely filled. Neon is chemically unreactive. In other words, after a period of eight atoms, the property of low reactivity has reappeared.

In the next eight elements, electrons add to the third, or M, shell, building it up gradually from one to eight electrons. The element argon, number 18, would not be expected to be unreactive because, according to the maximum permitted population, 10 more, or a total of 18, electrons can be put into the M shell. However, the *observed fact is that argon is not very reactive*. It must be that eight electrons in the third shell behave like a full shell. This point will be considered in more detail later in this section.

In setting up the periodic table, elements with similar properties are placed under each other. This corresponds to grouping the atoms which have the same number of electrons in their outermost levels. Thus, as can be seen on page 72, the element sodium, which has one electron in its outermost level, is placed under lithium in group I; magnesium is placed under beryllium in group II; aluminum under boron in group III; silicon under carbon in group IV; phosphorus under nitrogen in group V; sulfur under oxygen in group VI; chlorine under fluorine in group VII; and argon under neon in group 0. Because they have the same number of electrons in the outermost level, the atoms of each pair just mentioned have chemical similarity.

In the periodic table the first period contains but two elements (H and He). The second period contains eight elements (Li, Be, B, C, N, O, F, and Ne). The third period (Na, Mg, Al, Si, P, S, Cl, and Ar) also contains only eight elements. How come there are only eight elements in the third period, whereas the simple energy-level picture suggests there should be 18? The reason for this apparent discrepancy is associated with the fact that after eight electrons have been added to the third shell, the next two electrons go into the fourth shell, even though the third shell is not yet fully completed.

The fourth period, K through Kr, has 18 elements, ranging from atomic number 19 to atomic number 36. Of these 18 elements, the first two (K and Ca) and the last six (Ga, Ge, As, Se, Br, and Kr) correspond to addition of electrons to the outermost (fourth) shell. The 10 intervening elements (Sc, Ti, V, Cr, Mn, Fe, Co, Ni, Cu, and Zn) involve belated filling of the next-to-outermost (third) shell.

Before going on, we need to clear up one question raised previously. In the third period we find 8 elements; from simple shell-filling, we expect 18. Such an expectation rests on the assumption that all the electrons in a given shell (i.e., with the same value of n) are of the same energy. This is strictly true only for the hydrogen atom, where there are no electron-electron repulsions to worry about. However, in multielectron atoms, electrons do have an effect on each other, and the result is to split or separate some of the energy levels into *sublevels,* or *subshells.* The separation of sublevels is illustrated schematically in Fig. 4.14.

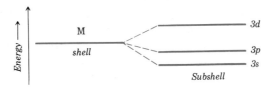

FIG. 4.14 Energy sublevels of the M shell.

The number of subshells in any main shell turns out to be equal to the quantum number *n*. Thus, the K shell (*n* = 1) consists of only one energy level. The L shell (*n* = 2) consists of two subshells, one group of electrons having an energy somewhat higher than that of the other group. In the M shell (*n* = 3), there are three possible energy levels; in the N shell, four energy levels; etc. The subshells are generally designated by a number followed by a letter, for example, 3*s*. The number specifies the main shell to which the subshell belongs; the letter specifies the subshell. The lowest subshell of a shell is labeled *s*; the next higher *p*; the next higher *d*; the one above that *f*.* Superscripts are often used to indicated the electron population of a particular subshell. Thus, $3s^2$ means two electrons in the *s* subshell of the main shell having *n* = 3. For Mg(*Z* = 12), the full electron population can be written $1s^2 2s^2 2p^6 3s^2$.

The energy-level diagram previously given in terms of main shells (Fig. 4.12) can now be redrawn to take account of subshells. Figure 4.15 shows

*The letters *s*, *p*, *d*, and *f* were originally chosen on the basis of observations of line spectra. Certain lines were observed to belong to a *sharp* series, and these were associated with energy transitions involving the *s* subshell; other spectrum lines belonged to what were called *principal*, *diffuse*, and *fundamental* series; hence, the designations *p*, *d* and *f*.

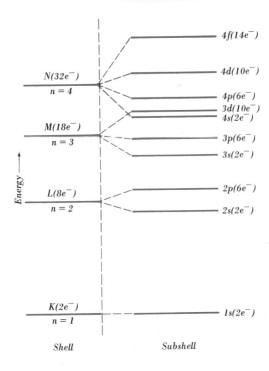

FIG. 4.15 Energy-level diagram showing component subshells. Numbers in parentheses indicate maximum electron populations.

schematically how main shells and subshells are related. Main shells are shown to the left of the vertical dashed line; subshells to the right. The relative spacing of the subshells is not the same for all elements but varies with Z. A distinctive feature of the diagram is the overlapping of the higher-energy subshells, an overlapping which gets more complicated as the fifth and sixth main shells are added to the picture.

Just as the number of electrons that can be put in any main shell is limited to $2n^2$, so the population of a subshell is similarly limited. An s subshell can hold 2 electrons; a p subshell, 6; a d subshell, 10; and an f subshell, 14.

How does the existence of subshells affect the buildup of atoms from electrons and nuclei? So far as the first 18 elements are concerned, the number of electrons per main shell is as predicted before. However, for later elements there is a difference. As shown in Fig. 4.13, element 18, argon, has electron configuration 2, 8, 8, corresponding to two electrons in the $1s$ subshell, two in the $2s$, six in the $2p$, two in the $3s$, and six in the $3p$. We can also write this as $1s^2 2s^2 2p^6 3s^2 3p^6$. The next subshell $3d$ is so much higher in energy that argon behaves as an "inert" atom. In the next element, potassium, number 19, the nineteenth electron goes into the $4s$ subshell, since $4s$ is lower in energy than $3d$ (see Fig. 4.15), *even though the third shell is not yet completely populated*. In calcium, element 20, another electron is added to the $4s$ energy level. In element 21, scandium, the twenty-first electron goes back into the third shell, in the $3d$ level which is the next available state. With minor irregularities, buildup of the third subshell proceeds in this fashion in the next eight elements. The addition of electrons to the $3d$ subshell after the $4s$ subshell is occupied has an interesting effect on the chemistry of the elements from calcium to zinc. The chemical properties of these elements do not change as drastically as might be expected with increasing atomic number. In the sixth period there is an even better example of this. The elements 57 through 71, the lanthanides, or rare-earth elements, are built up by the addition of $4f$ electrons to the third shell from the outside. Such changes deep within the atom do not affect chemical properties very much; all the lanthanides have nearly identical properties. Elements 89 through 103, the actinides, also exhibit electron buildup in a third outermost shell.

The electronic configurations of all the elements are shown in Fig. 4.16. These configurations apply to the atoms in their lowest energy states (in their ground states). The detailed assignment is based on observations of the spectra and on the magnetic properties of the individual elements. The question marks that occur in Fig. 4.16 denote cases where the assignment is in doubt. It should be noted, furthermore, that electronic configurations by themselves do not account for chemical properties of all the elements. Predictions made from these configurations are sometimes not borne out.

4.9
WAVE NATURE OF THE ELECTRON

One of the great weaknesses of the Bohr theory (Bohr himself recognized and repeatedly called attention to it) is that it offered no real explanation of why the electrons were restricted to only certain orbits. Bohr *assumed* that the angular momentum was quantized, i.e., limited to certain values; he did not pretend to know why. He simply noted that if he made that assumption, then the mathematics worked out to give good agreement between predictions and experimental observations.

In 1924, the French physicist Louis de Broglie suggested a possible explanation for the quantized nature of electrons in atoms. He proposed that every moving particle has associated with it a wave nature like that of light. The wavelength of the de Broglie wave is given by

$$\lambda = \frac{h}{mv}$$

where h is the Planck constant and mv is the momentum of the particle. For describing the motion of massive particles, the wave character is of little practical consequence since the associated wavelength is very small relative to particle dimensions. However, for describing the motion of low-mass particles, such as electrons, the wavelength is comparable to the dimensions of the atom in which the electron finds itself.

EXAMPLE 3

Given the speed of light as 3.0×10^8 m/sec and the electron mass as 9.1×10^{-28} g, calculate the de Broglie wavelength for an electron traveling at 1 percent the speed of light.

Solution

$$\lambda = \frac{h}{mv}$$

$m = 9.1 \times 10^{-28}$ g $= 9.1 \times 10^{-31}$ kg

$v = (0.01)\,(3.0 \times 10^8$ m/sec$) = 3.0 \times 10^6$ m/sec

$h = 6.6 \times 10^{-34}$ J sec

$$\lambda = \frac{6.6 \times 10^{-34} \text{ J sec}}{(9.1 \times 10^{-31} \text{ kg})\,(3.0 \times 10^6 \text{ m/sec})}$$

$$= 2.4 \times 10^{-10} \text{ J sec}^2 \text{ kg}^{-1} \text{ m}^{-1}$$

From Appendix 1 we note that J $=$ kg m^2 sec^{-2}

$$\lambda = 2.4 \times 10^{-10} \text{ m}$$

Although it represents a highly oversimplified argument, the following line of reasoning suggests why the existence of de Broglie waves leads to the quantum condition. Imagine an electron moving in a Bohr orbit. It has associated with it a wave of wavelength λ, given by the de Broglie condition. If this wavelength did not divide into the path length (the circumference of the orbit) a whole number of times, then the wave would destructively interfere with itself. A stable orbit would correspond only to one in which there is an integral multiple of wavelengths along the circumference of the orbit. Mathematically this corresponds to saying that

$$n\lambda = 2\pi r$$

FIG. 4.16 **Electron Configurations of Atoms of All the Elements**

Z	Element	1 s	2 s	2 p	3 s	3 p	3 d	4 s	4 p	4 d	4 f	5 s	5 p	5 d	5 f	6 s	6 p	6 d	6 f	7 s
1	H	1																		
2	He	2																		
3	Li	2	1																	
4	Be	2	2																	
5	B	2	2	1																
6	C	2	2	2																
7	N	2	2	3																
8	O	2	2	4																
9	F	2	2	5																
10	Ne	2	2	6																
11	Na	2	2	6	1															
12	Mg	2	2	6	2															
13	Al	2	2	6	2	1														
14	Si	2	2	6	2	2														
15	P	2	2	6	2	3														
16	S	2	2	6	2	4														
17	Cl	2	2	6	2	5														
18	Ar	2	2	6	2	6														
19	K	2	2	6	2	6		1												
20	Ca	2	2	6	2	6		2												
21	Sc	2	2	6	2	6	1	2												
22	Ti	2	2	6	2	6	2	2												
23	V	2	2	6	2	6	3	2												
24	Cr	2	2	6	2	6	5	1												
25	Mn	2	2	6	2	6	5	2												
26	Fe	2	2	6	2	6	6	2												
27	Co	2	2	6	2	6	7	2												
28	Ni	2	2	6	2	6	8	2												
29	Cu	2	2	6	2	6	10	1												
30	Zn	2	2	6	2	6	10	2												
31	Ga	2	2	6	2	6	10	2	1											
32	Ge	2	2	6	2	6	10	2	2											
33	As	2	2	6	2	6	10	2	3											
34	Se	2	2	6	2	6	10	2	4											
35	Br	2	2	6	2	6	10	2	5											
36	Kr	2	2	6	2	6	10	2	6											
37	Rb	2	2	6	2	6	10	2	6			1								
38	Sr	2	2	6	2	6	10	2	6			2								
39	Y	2	2	6	2	6	10	2	6	1		2								
40	Zr	2	2	6	2	6	10	2	6	2		2								
41	Nb	2	2	6	2	6	10	2	6	4		1								
42	Mo	2	2	6	2	6	10	2	6	5		1								
43	Tc	2	2	6	2	6	10	2	6	6		1?								
44	Ru	2	2	6	2	6	10	2	6	7		1								
45	Rh	2	2	6	2	6	10	2	6	8		1								
46	Pd	2	2	6	2	6	10	2	6	10										
47	Ag	2	2	6	2	6	10	2	6	10		1								
48	Cd	2	2	6	2	6	10	2	6	10		2								
49	In	2	2	6	2	6	10	2	6	10		2	1							
50	Sn	2	2	6	2	6	10	2	6	10		2	2							
51	Sb	2	2	6	2	6	10	2	6	10		2	3							
52	Te	2	2	6	2	6	10	2	6	10		2	4							
53	I	2	2	6	2	6	10	2	6	10		2	5							
54	Xe	2	2	6	2	6	10	2	6	10		2	6							

FIG. 4.16

Z	Element	1	2		3			4				5				6				7
		s	s	p	s	p	d	s	p	d	f	s	p	d	f	s	p	d	f	s
55	Cs	2	2	6	2	6	10	2	6	10		2	6			1				
56	Ba	2	2	6	2	6	10	2	6	10		2	6			2				
57	La	2	2	6	2	6	10	2	6	10		2	6	1		2				
58	Ce	2	2	6	2	6	10	2	6	10	2	2	6			2?				
59	Pr	2	2	6	2	6	10	2	6	10	3	2	6			2?				
60	Nd	2	2	6	2	6	10	2	6	10	4	2	6			2				
61	Pm	2	2	6	2	6	10	2	6	10	5	2	6			2?				
62	Sm	2	2	6	2	6	10	2	6	10	6	2	6			2				
63	Eu	2	2	6	2	6	10	2	6	10	7	2	6			2				
64	Gd	2	2	6	2	6	10	2	6	10	7	2	6	1		2				
65	Tb	2	2	6	2	6	10	2	6	10	9	2	6			2?				
66	Dy	2	2	6	2	6	10	2	6	10	10	2	6			2?				
67	Ho	2	2	6	2	6	10	2	6	10	11	2	6			2?				
68	Er	2	2	6	2	6	10	2	6	10	12	2	6			2?				
69	Tm	2	2	6	2	6	10	2	6	10	13	2	6			2				
70	Yb	2	2	6	2	6	10	2	6	10	14	2	6			2				
71	Lu	2	2	6	2	6	10	2	6	10	14	2	6	1		2				
72	Hf	2	2	6	2	6	10	2	6	10	14	2	6	2		2				
73	Ta	2	2	6	2	6	10	2	6	10	14	2	6	3		2				
74	W	2	2	6	2	6	10	2	6	10	14	2	6	4		2				
75	Re	2	2	6	2	6	10	2	6	10	14	2	6	5		2				
76	Os	2	2	6	2	6	10	2	6	10	14	2	6	6		2				
77	Ir	2	2	6	2	6	10	2	6	10	14	2	6	7		2				
78	Pt	2	2	6	2	6	10	2	6	10	14	2	6	9		1				
79	Au	2	2	6	2	6	10	2	6	10	14	2	6	10		1				
80	Hg	2	2	6	2	6	10	2	6	10	14	2	6	10		2				
81	Tl	2	2	6	2	6	10	2	6	10	14	2	6	10		2	1			
82	Pb	2	2	6	2	6	10	2	6	10	14	2	6	10		2	2			
83	Bi	2	2	6	2	6	10	2	6	10	14	2	6	10		2	3			
84	Po	2	2	6	2	6	10	2	6	10	14	2	6	10		2	4?			
85	At	2	2	6	2	6	10	2	6	10	14	2	6	10		2	5?			
86	Rn	2	2	6	2	6	10	2	6	10	14	2	6	10		2	6			
87	Fr	2	2	6	2	6	10	2	6	10	14	2	6	10		2	6			1?
88	Ra	2	2	6	2	6	10	2	6	10	14	2	6	10		2	6			2
89	Ac	2	2	6	2	6	10	2	6	10	14	2	6	10		2	6	1		2?
90	Th	2	2	6	2	6	10	2	6	10	14	2	6	10	2	6	2			2
91	Pa	2	2	6	2	6	10	2	6	10	14	2	6	10	2	2	6	1		2?
92	U	2	2	6	2	6	10	2	6	10	14	2	6	10	3	2	6	1		2
93	Np	2	2	6	2	6	10	2	6	10	14	2	6	10	4	2	6	1		2?
94	Pu	2	2	6	2	6	10	2	6	10	14	2	6	10	6	2	6			2?
95	Am	2	2	6	2	6	10	2	6	10	14	2	6	10	7	2	6			2?
96	Cm	2	2	6	2	6	10	2	6	10	14	2	6	10	7	2	6	1		2?
97	Bk	2	2	6	2	6	10	2	6	10	14	2	6	10	8	2	6	1		2?
98	Cf	2	2	6	2	6	10	2	6	10	14	2	6	10	10	2	6			2?
99	Es	2	2	6	2	6	10	2	6	10	14	2	6	10	11	2	6			2?
100	Fm	2	2	6	2	6	10	2	6	10	14	2	6	10	12	2	6			2?
101	Md	2	2	6	2	6	10	2	6	10	14	2	6	10	13	2	6			2?
102	No	2	2	6	2	6	10	2	6	10	14	2	6	10	14	2	6			2?
103	Lr	2	2	6	2	6	10	2	6	10	14	2	6	10	14	2	6	1		2?
104	Ku	2	2	6	2	6	10	2	6	10	14	2	6	10	14	2	6	2		2?
105	Ha	2	2	6	2	6	10	2	6	10	14	2	6	10	14	2	6	3		2?
106	?	2	2	6	2	6	10	2	6	10	14	2	6	10	14	2	6	4		2?
107	?	2	2	6	2	6	10	2	6	10	14	2	6	10	14	2	6	5		2?

where n is a whole number and r is the radius of the orbit. Substituting $\lambda = h/mv$, we get

$$n\left(\frac{h}{mv}\right) = 2\pi r$$

and rearranging we obtain

$$mvr = n\left(\frac{h}{2\pi}\right)$$

This is the same as Eq. (2) on page 102 for the quantum condition assumed by Bohr!

The concept of a de Broglie wave for a particle such as an electron means that the particle cannot be precisely localized. Instead, the electron must be thought of in the somewhat tenuous manner we use for imagining waves. The problem, a general one for particles of low mass, has been treated mathematically by Heisenberg. In a famous theorem, called the *uncertainty principle*, Heisenberg showed (in 1927) that there is an inherent indeterminancy in knowing the combination of a particle's position and momentum. This indeterminancy can be expressed by saying that the uncertainty in the position times the uncertainty in the momentum is of the order of the Planck constant h. What this means is that the more precisely we try to specify momentum or position, the more uncertain we are of the other. In the limit, if we precisely specify the position, we do not know the momentum; if we know the momentum we do not know where the particle is. For large masses the uncertainty of position or momentum is trivial, but for small masses the uncertainty principle places a real restriction upon the extent of permitted knowledge. Specifically, we cannot draw precise orbits for the electrons to follow.

Since tracks cannot be drawn for electrons in atoms, the best we can do is to speak of the probability or relative chance of finding an electron at a given location within the atom. The calculation of such a probability is quite involved mathematically. It uses equations which describe the motion of waves and applies them to the de Broglie wave associated with an electron. This procedure is a basic concern of the field of study called *wave mechanics*.

The probability of finding an electron can be specified in various ways. Three of these are shown in Fig. 4.17 for the case of a 1s electron. At the top of the figure, the probability of finding a 1s electron at a given location in space is plotted as a function of the distance of that location from the nucleus. The position of greatest probability is at the nucleus. As the distance from the nucleus increases, the probability of finding the electron decreases. Nowhere, however, does the probability go to zero. Even at points very far from the nucleus there is some chance, although it is small, of finding the electron. Figure 4.17b is another way of representing the same electronic distribution. Here the decreasing relative probability of finding the electron is represented by a decrease in the intensity of shading. Consistent with this picture, one can visualize the 1s electron as forming a rather fuzzy spherical charge cloud about a central nucleus. Sometimes it is convenient simply to indicate the shape of the charge cloud. This can be done by a boundary contour, as in Fig. 4.17c. Remembering that atoms are

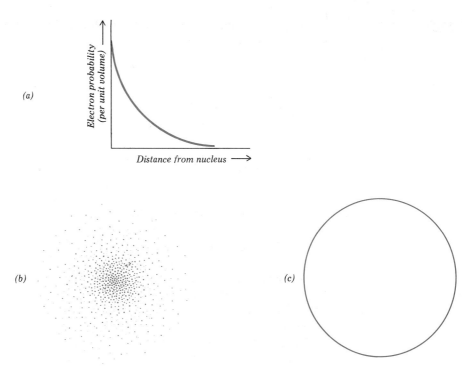

(a)

(b)

(c)

FIG. 4.17　Various ways of representing the spatial distribution of a 1s electron.

three-dimensional, we should think of Fig. 4.17*c* as a sphere within which the chance of finding the electron is great.

　The distributions shown in Fig. 4.17 are three different ways of representing the spatial distribution of an electron in a 1*s* energy level. Since these representations replace the Bohr idea of a simple orbit, they can properly be said to represent 1*s* orbits. To reduce any possible confusion between the old and new ideas, it has become customary to use the term *orbital* when referring to a *specific electronic probability distribution in space.*

　There is still another way of describing an electron in a 1*s* orbital, a way that serves to relate the old idea of electronic shells to probability concepts. First, we ask this question: Starting from the nucleus and working along a straight line from the nucleus to the outside of the atom, how does the chance of finding the 1*s* electron change? Obviously, the chance decreases, as is described in Fig. 4.17*a*. But suppose, as we move out of the atom, at each radial distance *r* from the nucleus we investigate all the possible locations in three-dimensional space at that distance *r* from the nucleus and determine the total chance of finding the 1*s* electron at that distance *r*. Then we move farther from the nucleus and investigate all the locations at a slightly bigger *r*. How does the chance of finding the 1*s* electron change? The answer is not immediately obvious, since we have to consider two factors: (1) the chance of finding the electron at a given location decreases as we move away from the nucleus, and (2) the number of locations to be investigated increases. Mathe-

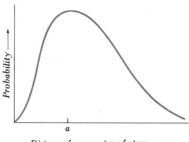

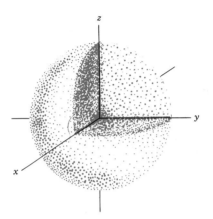

FIG. 4.18 Probability of finding a 1s electron in a spherical shell at a distance r from the nucleus. Bottom part of figure shows relation to center of atom.

matically, this is equivalent to considering the atom to be divided like an onion into concentric shells and multiplying the probability *per unit volume* in a given shell by the *volume* of that shell. The result for a 1s electron is the probability curve shown in Fig. 4.18. As can be seen, the probability of finding the electron goes up to a maximum and then decreases. The distance *a*, at which the probability reaches a maximum, can be thought of as corresponding to the radius of an electron shell.

Electrons that are in different levels differ from each other in having different probability distributions. Figure 4.19 compares the probability distributions for 1s, 2s, and 2p electrons. It may be noted that the distances of maximum probability for the 2s and 2p electrons are roughly the same; they are larger than the distance of maximum probability for the 1s electron. This is consistent with the fact that the 2s and 2p electrons are of about the same energy and that this energy is greater than that of the 1s electron. The extra little bump at small distances in the 2s distribution indicates that the 2s electron spends, on the average, more of its time close to the nucleus than does the 2p electron. This can account for the fact that the 2s electron is bound more tightly to the nucleus (is of lower energy) than the 2p electron. Furthermore, it should be noted that all three of the distributions shown in Fig. 4.19 overlap, implying that electrons can "occupy" the region already "occupied" by other electrons. Specifically, electrons that are in outer orbitals can penetrate into the region occupied by electrons that are in inner orbitals.

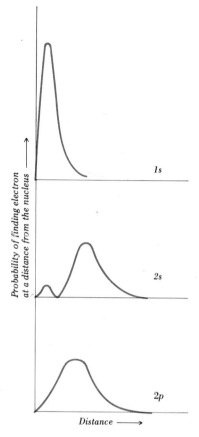

FIG. 4.19 Probability plots for various electron orbitals.

There is another essential difference between *s* and *p* electrons which is not evident from Fig. 4.19. The spatial distribution of an *s* electron is spherically symmetric; i.e., its probability of being found is identical in all directions from the nucleus. On the other hand, *p* electrons are more probably found in some directions than in others. In fact, the probability distribution of a *p* electron can be thought of as forming two diffuse blobs, one on each side of the nucleus, as shown in three different representations in Fig. 4.20. This is called a *p* orbital, and the electron in a *p* orbital has equal probability of being found on either side of the nucleus. A *p* subshell consists of three such orbitals, all perpendicular to each other. The one concentrated along the *x* axis is called the p_x orbital; the one along the *y* axis, the p_y orbital; the one along the *z* axis, the p_z orbital. The respective orientations are shown in Fig. 4.21. Although it is not obvious, the combined distribution of three electrons (one each in p_x, p_y, p_z) or of six electrons (two each in p_x, p_y, and p_z) is a sphere.

The *d* subshell, which can accommodate 10 electrons, can be resolved into five orbitals. These are somewhat more complicated than *p* or *s* orbitals. The 3*d* set is represented in Fig. 4.22. One of the 3*d* orbitals, the one labeled d_{z^2}, is symmetric about the *z* axis and can be visualized as a squashed doughnut

what should about orbital shapes

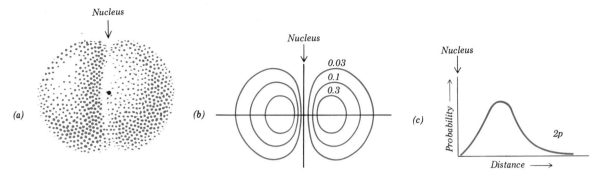

(a)

(b)

(c)

FIG. 4.20 Three representations of spatial distribution of a 2p electron. In the middle representation, curved lines are contour lines corresponding to different values of constant probability given as fractions of the maximum.

around the pinched waist of an hourglass. The other four $3d$ orbitals look like inflated four-leaf clovers. The one designated $d_{x^2-y^2}$ has its maximum electron-probability density along the x and y axes. The other three, d_{xy}, d_{yz}, and d_{xz}, have their probability maxima lying along lines that make 45° angles with the axes. As with the set of p orbitals, it turns out that having one electron in each (or two electrons in each) of the five d orbitals gives a spherical electronic distribution.

The f subshell consists of seven orbitals. The distributions are even more complex than those of the d orbitals. The geometry of the $4f$ orbitals is such that equal populations of all seven orbitals again add up to a sphere.

A relative comparison of s, p, and d orbitals shows an interesting characteristic. The $1s$ orbital has no plane of zero probability except at large distances from the nucleus. The $2p$ electron, on the other hand, has a plane (right through 'the nucleus, between the two blobs) where the probability goes to zero. This plane is called a *node, a place where the probability vanishes*. The $3d$ electron has two nodes, most easily seen as the zy and zx planes for the d_{xy} electron. The $4f$ electrons have three nodes, etc. In general, the more nodes there are in an electron distribution, the higher its energy.

FIG. 4.21 Relative orientations of p_x, p_y, and p_z orbitals.

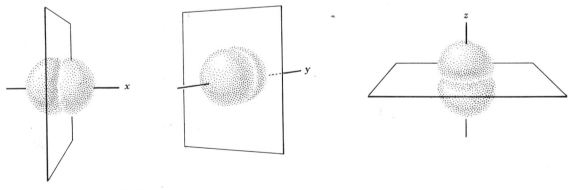

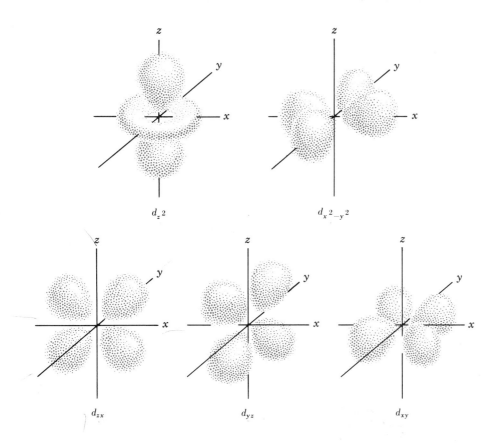

d_{z^2}

$d_{x^2-y^2}$

d_{zx}

d_{yz}

d_{xy}

4.10
ELECTRON SPIN

In the preceding section it was noted that there is one orbital in an s subshell, three orbitals in a p subshell, five in a d, and seven in an f. Since these subshells can accommodate 2, 6, 10, and 14 electrons, respectively, it follows that any orbital can hold 2 electrons. The 2 electrons in the same orbital are alike in probability distribution, but they differ in one important respect. They are said to be of opposite *spin*. The reason for talking about electron spin comes from studies on the magnetic behavior of substances.

It is a familiar observation that certain solids, such as iron, are strongly attracted to magnets. Such materials are called *ferromagnetic*. Other substances, such as oxygen gas and copper sulfate, are moderately attracted to magnets. These are called *paramagnetic*. Still other substances, such as sodium chloride, are very feebly repelled by magnets; they are called *diamagnetic*. Ferromagnetism is a special property of the solid state, but all three types of magnetic behavior just described are believed to arise from electrons in atoms.

117

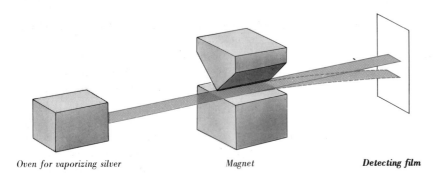

FIG. 4.23 Stern-Gerlach experiment showing magnetic splitting of a beam of silver atoms.

Oven for vaporizing silver   *Magnet*   **Detecting film**

Information about paramagnetic behavior of individual atoms can be obtained from an experiment like the one first performed by the German physicists Otto Stern and Walter Gerlach in 1921. In their experiment, shown schematically in Fig. 4.23, a beam of *neutral* silver atoms (obtained from the vaporization of silver) was passed between the poles of a specially designed magnet. The beam was found to be split into two separate beams; i.e., half the atoms were deflected in one direction, and the rest in the opposite direction.

In interpreting this experiment, the main problem is to explain why the silver atoms are behaving as if they were tiny magnets. The atoms are electrically neutral, so that there should be no deflection of the beam. Moreover, the beam is split into two. This means that half the silver atoms are acting like magnets pointing in one direction and the other half, in the opposite direction. How do these tiny magnets come about? Why are there two kinds?

One way by which electrons can give rise to paramagnetism is to circulate in a loop. This is called *orbital motion*. However, in a *filled* subshell of electrons the total electronic distribution comes out to be spherically symmetric, and so any paramagnetic effect due to orbital motion cancels out. Silver atoms ($Z = 47$) have the electron configuration $1s^2 2s^2 2p^6 3s^2 3p^6 3d^{10} 4s^2 4p^6 4d^{10} 5s^1$. All the electrons except the $5s$ are in filled subshells, and so there is no net orbital motion. Furthermore, the s electron is special; it is spherically symmetric about the nucleus and also should produce no net paramagnetic effect due to its orbital motion. The only thing left is to assume that each electron is *rotating* on its own axis. Since any rotating charge is magnetic, behavior as a tiny magnet would be consistent with saying that the electron has *spin*. Apparently, only two directions of spin are possible: clockwise or counterclockwise. In the clockwise case, the electron behaves as a magnet pointing in one direction; in the counterclockwise case, as a magnet pointing in the opposite direction. If we had two electrons of opposite spin, we might expect them to attract each other, as two oppositely directed magnets would; however, electric repulsion due to like negative charges is very much greater than the magnetic attraction. When two electrons are forced to be together, as in a completely filled subshell of an atom, each electron will pair up with another electron of opposite spin. Such an electron pair is nonmagnetic because the magnetism of one spin is canceled by the magnetism of the opposite spin.

In silver atoms, the one $5s$ electron is unpaired. Hence, its uncanceled spin gives rise to paramagnetism. The two deflections observed in the Stern-Gerlach experiment result from a separation of silver atoms of two types, which differ in the direction of spin of the unpaired electron, one having clockwise spin, and the other counterclockwise.

Any atom which, like the silver atom, contains an odd number of electrons must be paramagnetic. Furthermore, atoms which have an even number of electrons can also be paramagnetic, provided that there is an un-filled subshell of electrons. These more complex cases will be considered in Sec. 18.9. When all the electrons in an atom are paired, there is no paramagnetism. There is only diamagnetism, which occurs in all matter, even though it may be obscured in paramagnetic substances. Diamagnetism arises not from the spin of electrons but from their electric charge. A detailed discussion is beyond the scope of this book.

4.11
QUANTUM NUMBERS

From the complete mathematical treatment of the wave nature of electrons in atoms there emerges a description of each electron in terms of four index numbers, the so-called "quantum numbers." One of these has been mentioned above, the quantum number n. It is called the *principal* quantum number and gives, other things being equal, the order of increasing distance of the average electron distribution from the nucleus. Hence, it is related to the order of the electronic energies. In most cases, however, the electronic energy depends significantly on the second quantum number also.

The second quantum number, usually designated by l, is called the *orbital* quantum number. It denotes the subshell which the electron occupies and indicates the angular shape of the electron distribution. Permitted values for l are 0, 1, 2,..., up to a maximum value of $n-1$. $l=0$ corresponds to the s subshell; $l=1$, to the p subshell; $l=2$, to the d subshell; $l=3$, to the f subshell.

The third quantum number, usually designated by m_l, is called the *magnetic* quantum number. It tells something about the orbital circulation of the electric charge, which gives rise to magnetism. This magnetism causes the orbitals within a given subshell to separate out into different discrete energy levels when a magnetic field is applied. The splitting, called the *Zeeman effect,* can be observed experimentally by measuring spectral lines in a magnetic field. One line in the absence of a field may split into several on application of the field. For a given value of l, permitted values of m_l are the integers, including 0, from $+l$ to $-l$. Thus, for example, in the p subshell, where $l=1$, allowed values of m_l are $+1$, 0, and -1. In the d subshell, where $l=2$, permitted values of m_l are $+2$, $+1$, 0, -1, and -2. As seen, there are three permitted values of m_l in the p subshell, and there are five permitted values of m_l in the d subshell.*

*The mathematical operations are somewhat complicated, but it can be shown that the three permitted values of m_l in the p subshell are related to the three kinds of p orbitals shown in Fig. 4.21 (p_x, p_y, and p_z). Unfortunately, one cannot make a direct match on a one-to-one basis between the set $m_l = +1$, 0, -1 and the set p_x, p_y, and p_z. The mathematical description involving $m_l = +1$ or -1 has complex "imaginary" terms which have to be combined in special ways to get the "real" orbitals pictured as p_x, p_y, and p_z. Similarly, the five permitted values of m_l in the d subshell are related to the five kinds of d orbitals shown in Fig. 4.22 (d_{z^2}, $d_{x^2-y^2}$, d_{xz}, d_{yz}, and d_{xy}).

The fourth quantum number, usually designated m_s, is called the *spin* quantum number. It can have values of either $+\frac{1}{2}$ or $-\frac{1}{2}$, corresponding to the two possible orientations of electron spin. Instead of $+\frac{1}{2}$ and $-\frac{1}{2}$, the orientations are often designated by arrows pointing in opposite directions: ↑ and ↓.

An electron in an atom is completely described once its four quantum numbers have been specified. Furthermore, there is a fundamental principle, called the *Pauli exclusion principle,* which states that no two electrons in the same atom can be completely identical, i.e., have the same set of four quantum numbers. Because of this restriction and the restriction on permitted values of quantum numbers, the number of electrons in a shell is limited to $2n^2$, as

FIG. 4.24　Permitted Combinations of Quantum Numbers for the Various Electrons

n	l	m_l	m_s
1 (K shell)	0 (s subshell)	0	$+\frac{1}{2}, -\frac{1}{2}$
2 (L shell)	0 (s subshell)	0	$+\frac{1}{2}, -\frac{1}{2}$
	1 (p subshell)	$+1$	$+\frac{1}{2}, -\frac{1}{2}$
		0	$+\frac{1}{2}, -\frac{1}{2}$
		-1	$+\frac{1}{2}, -\frac{1}{2}$
3 (M shell)	0 (s subshell)	0	$+\frac{1}{2}, -\frac{1}{2}$
	1 (p subshell)	$+1$	$+\frac{1}{2}, -\frac{1}{2}$
		0	$+\frac{1}{2}, -\frac{1}{2}$
		-1	$+\frac{1}{2}, -\frac{1}{2}$
	2 (d subshell)	$+2$	$+\frac{1}{2}, -\frac{1}{2}$
		$+1$	$+\frac{1}{2}, -\frac{1}{2}$
		0	$+\frac{1}{2}, -\frac{1}{2}$
		-1	$+\frac{1}{2}, -\frac{1}{2}$
		-2	$+\frac{1}{2}, -\frac{1}{2}$
4 (N shell)	0 (s subshell)	0	$+\frac{1}{2}, -\frac{1}{2}$
	1 (p subshell)	$+1$	$+\frac{1}{2}, -\frac{1}{2}$
		0	$+\frac{1}{2}, -\frac{1}{2}$
		-1	$+\frac{1}{2}, -\frac{1}{2}$
	2 (d subshell)	$+2$	$+\frac{1}{2}, -\frac{1}{2}$
		$+1$	$+\frac{1}{2}, -\frac{1}{2}$
		0	$+\frac{1}{2}, -\frac{1}{2}$
		-1	$+\frac{1}{2}, -\frac{1}{2}$
		-2	$+\frac{1}{2}, -\frac{1}{2}$
	3 (f subshell)	$+3$	$+\frac{1}{2}, -\frac{1}{2}$
		$+2$	$+\frac{1}{2}, -\frac{1}{2}$
		$+1$	$+\frac{1}{2}, -\frac{1}{2}$
		0	$+\frac{1}{2}, -\frac{1}{2}$
		-1	$+\frac{1}{2}, -\frac{1}{2}$
		-2	$+\frac{1}{2}, -\frac{1}{2}$
		-3	$+\frac{1}{2}, -\frac{1}{2}$

mentioned on page 105. How this comes about is shown in Fig. 4.24. The number of combinations within each shell indicates the number of electrons that can be accommodated in that shell. For the K shell, we can have one electron ($n = 1$, $l = 0$, $m_l = 0$) with $m_s = +\frac{1}{2}$ and another with $m_s = -\frac{1}{2}$, and so the total possible number of electrons is 2; for the L shell, we can have $+\frac{1}{2}$ and $-\frac{1}{2}$ in the s orbital and $+\frac{1}{2}$ and $-\frac{1}{2}$ in each of three p orbitals, and so the total possible number is 8; for the M, 18; and for the N, 32. In terms of the principal quantum number n, the maximum possible population is $2(1)^2$, $2(2)^2$, $2(3)^2$, $2(4)^2$, ..., $2n^2$.

The assignment of electrons to orbitals for building up the periodic table is governed by several factors. These include the Pauli exclusion principle, the relative energies of subshells, and the repulsions between electrons in orbitals belonging to the same subshell. Figure 4.25 shows an orbital-filling diagram in which each circle represents schematically an orbital belonging to the subshell for which the designation is given in the left margin. There is a single circle for the s subshell since there is but one s orbital for each value of n; there are three circles for each of the p subshells, standing, respectively, for p_x, p_y, and p_z orbitals or their equivalent; there are five circles, standing for the five orbitals of a d subshell. The orbitals are filled from the bottom of the diagram working upward. At any one horizontal level, i.e., within any subshell, the electrons are spread out one to an orbital before any pairing (two electrons ⇅ in the same orbital) is allowed to occur. The reason for this is that electrons, being of the same negative electric charge, repel each other; hence, they try to spread over as much space as possible. They can do this by occupying separate orbitals, each of which corresponds to a different distribution in space. After all the orbitals belonging to a given subshell are populated by one electron, pairing generally begins to occur. Repulsion due to two electrons in the same orbital is usually less than the energy jump to the next higher subshell.

It should be especially noted that Fig. 4.25 is a *filling* diagram and not an

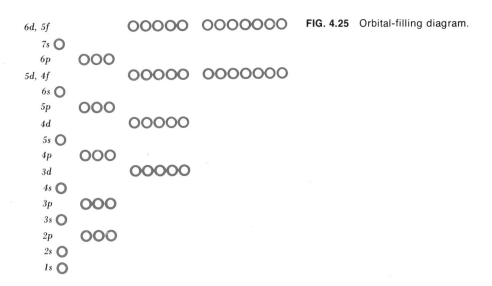

FIG. 4.25 Orbital-filling diagram.

energy-level diagram. The reason for this distinction is that there is no single energy-level diagram that would be appropriate for all the elements. The relative energies depend on the nuclear charge Z and the number of other electrons present. In building up the periodic table by adding an electron to a previously formed atom, the nuclear charge Z is also being increased by one unit. When this is done, the energy levels shift, and in some cases the relative order changes.

The order shown in Fig. 4.25 is that appropriate for stepwise buildup of the elements. It gives the relative order in which subshells become available when electron filling is occurring in a particular region. As an example, the chart shows $4s$ to lie lower than $3d$. This corresponds to the fact that $4s$ begins to fill at potassium ($Z = 19$), whereas $3d$ does not begin to fill until scandium ($Z = 21$). Apparently, in potassium and scandium the $4s$ subshell is lower than the $3d$. However, it says nothing about the relative ordering of the same subshells in a later element such as zinc ($Z = 30$), in which, in fact, the $3d$ is lower. A related point to note is that Fig. 4.25 should not be used to decide which electrons are removed when an ion is formed from a particular atom. Specifically, in forming Ti^{2+} from a neutral titanium atom ($Z = 22$, $1s^2 2s^2 2p^6 3s^2 3p^6 4s^2 3d^2$), it is not the two $3d$ electrons that are lost, but the two $4s$ electrons. In Ti^{2+} ion, the $3d$ level is lower in energy than the $4s$, contrary to what might be expected from just looking at the order-of-filling diagram.

4.12
MANY-ELECTRON ATOMS

The preceding discussion assumes that electrons in atoms can be described by superposition of hydrogen-like descriptions. It assumes that any particular orbital retains all its characteristic features no matter how many other electrons are added to the same atom. This cannot be quite true because electrons repel each other. Adding a second electron to a one-electron atom, for example, changes the problem. It is no longer a simple question of considering the attraction of a light, negative electron to a heavy, positive nucleus, but now one needs to consider that the light, negative electron is repelled by another light, negative electron, neither of which is fixed in space. The problem is a complex one because the force acting on each electron is a combination of the attraction to the nucleus and the repulsion from the other electron. At present, the problem has not yet been solved exactly, and so we have to be satisfied with approximate solutions. The gist of these approximation methods is that each electron is considered to move separately in an average field generated by the nucleus and the other electrons. The other electrons are then adjusted one at a time to allow for what was found out in each step of the calculation. The process is continued until small adjustments of one electron do not appreciably affect the others. With the advent of high-speed computers, results can be obtained to any desired degree of accuracy, but the computer expense quickly mounts up as more and more electrons are added to the problem. The interesting point to note is that the elegant results are not

123   Section 4.13
Properties
in Relation to
Electron
Structure

FIG. 4.26   Ground-State Spin Configurations for the First 10 Elements

Element	Electron configuration	Spin arrangement				
H	$1s^1$	↑	—	—	—	—
He	$1s^2$	↓↑	—	—	—	—
Li	$1s^2 2s^1$	↓↑	↑	—	—	—
Be	$1s^2 2s^2$	↓↑	↓↑	—	—	—
B	$1s^2 2s^2 2p^1$	↓↑	↓↑	↑	—	—
C	$1s^2 2s^2 2p^2$	↓↑	↓↑	↑	↑	—
N	$1s^2 2s^2 2p^3$	↓↑	↓↑	↑	↑	↑
O	$1s^2 2s^2 2p^4$	↓↑	↓↑	↓↑	↑	↑
F	$1s^2 2s^2 2p^5$	↓↑	↓↑	↓↑	↓↑	↑
Ne	$1s^2 2s^2 2p^6$	↓↑	↓↑	↓↑	↓↑	↓↑

much different from what one gets by assuming that electrons move independently of each other in hydrogen-like orbitals with repulsive interactions added on as small perturbations.

The Pauli exclusion principle is actually a way of minimizing electron-electron repulsions. For example, in the case of the lithium atom, where the ground-state configuration is $1s^2\ 2s^1$, there is lower electron-electron interaction than if all three electrons were squeezed into the $1s$ orbital. Besides the Pauli exclusion principle, there is another important generalization for helping to figure out proper electron distributions. It is called *Hund's rule* and simply states that the electron configuration for which the spin magnetism is the greatest has the lowest energy. Consider, for example, a case in which we have two p electrons in the same subshell. Using short horizontal lines to represent the three orbitals of a p subshell, we can envisage the following possibilities for a p^2 configuration: ↓↑ _ _ , ↑ ↓ _ , or ↑ ↑ _ . The last of these is the lowest energy state. It corresponds to a total spin of $(m_s = +\frac{1}{2})$ plus $(m_s = +\frac{1}{2}) = 1$, whereas the others correspond to a total spin of $(m_s = +\frac{1}{2})$ plus $(m_s = -\frac{1}{2}) = 0$. The high-spin state is favored because it is most effective at keeping the electrons apart. The state ↓↑ _ _ suffers from the fact that both electrons are in the same orbital, and so electron-electron repulsion is a maximum; the state ↑ ↓ _ suffers from the fact that since the spins are not the same, one electron can "leak" over into the other's orbital; the state ↑ ↑ _ does not permit such leak-over since the Pauli exclusion principle forbids two electrons of the same spin ever to occupy the same orbital.

Figure 4.26 shows schematically the magnetic spin situation for the ground states of the first 10 elements.

4.13
PROPERTIES IN RELATION TO ELECTRON STRUCTURE

In this section, we apply some of the knowledge gained about electronic structure to get an understanding of why some atomic properties vary as they do. The first property we want to look at is *atomic size*. It is a difficult property to determine.   For one thing, the electronic probability distribution never becomes exactly zero, even at great distances from the nucleus. Therefore, the distance designated as the boundary of the atom is an arbitrary choice. For another thing, the electronic probability distribution is affected by neighboring atoms; hence, the size of an atom may change in going from one condition to another, as, for example, in going from one compound to another. Therefore, in examining any table of atomic radii we must remember that the values listed may be meaningful only in providing a relative comparison of sizes. Figure 4.27 gives a set of atomic radii. They are deduced by equally splitting the observed distance between centers of identical adjacent atoms. Figure 4.28 shows these same values plotted as a function of atomic number. As can be seen, there are marked regularities. The alkali elements Li, Na, K,

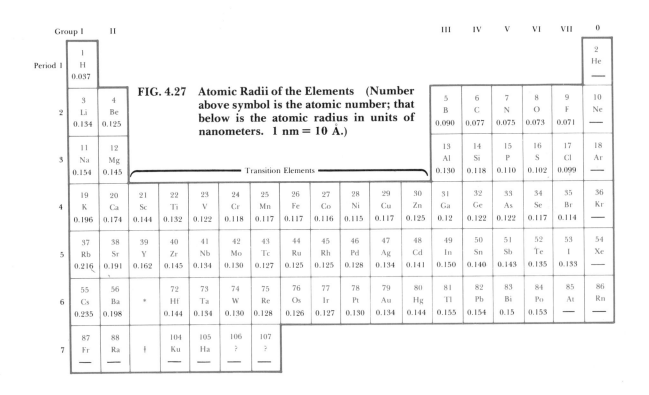

FIG. 4.27   **Atomic Radii of the Elements   (Number above symbol is the atomic number; that below is the atomic radius in units of nanometers. 1 nm = 10 Å.)**

Rb, and Cs all have outstandingly large radii; these radii increase systematically as one goes down the group. Following each alkali element, there is a progressive shrinkage in size as one moves across a period. For the second and third periods, the shrinkage is regular, but for the fourth and subsequent periods, irregularities appear. The general trends are easy to remember—size increases down a group of the table and generally decreases from left to right in a period.

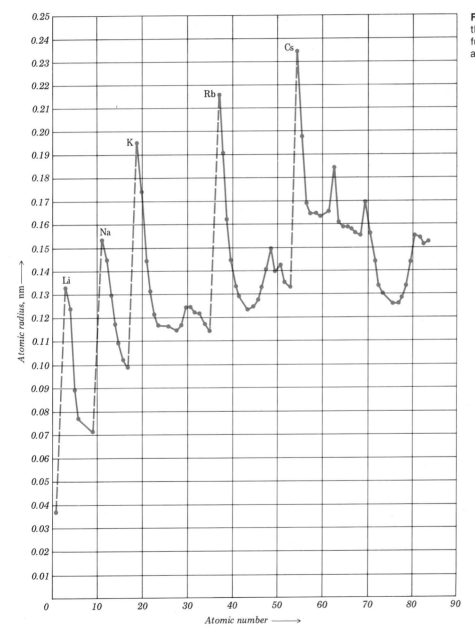

FIG. 4.28 Atomic radii of the elements plotted as a function of increasing atomic number.

FIG. 4.29   Change of Atomic Radius in Second Period

	Li	Be	B	C	N	O	F
Atomic radius, nm	0.134	0.125	0.090	0.077	0.075	0.073	0.071
Nuclear charge	+3	+4	+5	+6	+7	+8	+9
K shell	$2e^-$	$2e^-$	$2e^-$	$2e^-$	$2e^-$	$2e^-$	$2e^-$
L shell	$1e^-$	$2e^-$	$3e$	$4e^-$	$5e^-$	$6e^-$	$7e^-$

How can we explain the general trend across a period in terms of electronic structure? Figure 4.29 shows the atomic radii for the elements of the second period together with the nuclear charges and the electronic populations of the K and L shells. Within the period the nuclear charge increases from +3 to +9. What effect would this have on the K electrons? In each of these elements there are two K electrons, attracted to the nucleus by a force proportional to the nuclear charge. As nuclear charge increases, the pull on the electrons is increased. The maximum in the K probability distribution curve is moved closer to the nucleus. The K shell shrinks.

What about the L shell? Here the problem is complicated by the fact that the L electrons are screened from the nucleus by the intervening K electrons. The attractive force exerted by the nuclear positive charge is reduced. In lithium, for example, the outermost electron is attracted not by a charge of +3 but by a charge of +3 screened by two intervening negative electrons. The effective attractive charge is closer to +1 than it is to +3. In the beryllium atom, the L electrons are attracted by a +4 nucleus screened by two negative charges, or effectively a +2 charge. Even taking screening into account, in going from left to right across the period, the L electrons have a higher and higher positive charge attracting them to the center of the atom. Just as the K shell becomes smaller because of this effect, the L shell gets smaller also.

How can we explain the change of atom size within a group? Figure 4.30 shows the situation for the alkali elements. There is an increase in size from 0.134 to 0.235 nm going down the group. As shown by the shell population, the number of occupied shells increases also. As expected, the more shells used, the bigger the atom. However, the nuclear charge also increases down the sequence, and so we might expect the individual shells to get smaller. Nevertheless, adding a shell is apparently such a big effect that it dominates. Similar behavior is found for many of the other groups of the periodic table.

FIG. 4.30   Change of Atomic Radius Within the First Group

Element	Atomic radius, nm	Nuclear charge	Electron shell population
Li	0.134	+ 3	2, 1
Na	0.154	+11	2, 8, 1
K	0.196	+19	2, 8, 8, 1
Rb	0.216	+37	2, 8, 18, 8, 1
Cs	0.235	+55	2, 8, 18, 18, 8, 1

There are some places in the periodic table where size does not change much going down a group. This is particularly true in the center of the periodic table, especially when elements number 57 through 71 intervene between the atoms compared. Thus, for example, Zr ($Z = 40$) has practically the same radius (0.145 nm) as does the element just below it, Hf ($Z = 72$), which has atomic radius 0.144 nm. The reason for this near identity in size is that elements 57 through 71 intervene. They correspond to electron expansion in the $4f$ orbitals deep inside the atom where increasing nuclear charge and increasing electron population have just about canceled each other. As might be expected, the great similarity in size between Zr and Hf leads to great similarity in their chemical behavior. For instance, in nature they always occur together in minerals.

The second property we need to consider is the *ionization potential,* also called *ionization energy.* It is the energy required to pull an electron off a neutral atom and corresponds to the energy required for the reaction

$$M(g) \longrightarrow M^+(g) + e^-$$

The phase notation (g), for the gas phase, emphasizes that the atoms must be isolated from each other, as they would be in a gas. To measure the ionization potential, energy can be added (for example, as heat) to the neutral atom to raise the outermost electron to higher energy levels. When the electron falls back, it emits energy, which can be viewed as a spectral line. A whole series of lines will be obtained corresponding to re-emission of energy from various excited states. As can be seen from Fig. 4.31, for the element lithium, the energies emitted converge on a series limit corresponding to return of an electron from the furthest removed orbit to the L shell. This series limit is equal to the ionization potential—it gives a direct measure of the energy required to remove the electron from the L shell to the furthest removed orbit, i.e., completely away from the atom.

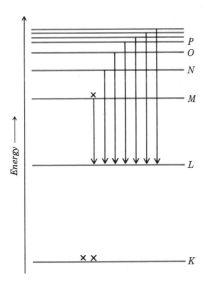

FIG. 4.31 Schematic representation of how spectral lines in the lithium series get closer together and finally converge on a series limit.

The actual experimental situation is somewhat more complicated than this because inner electrons (from the K shell) are also being excited and produce additional spectral lines that can confuse the fixing of the series limit. However, complete analysis of the spectrum can usually resolve the different processes.

A further complication is that not just one electron can be removed from an atom, corresponding to what is called the *first ionization potential,* but additional electrons can also be stripped off. These give rise to what are known as second, third, fourth, etc., ionization potentials, corresponding, respectively, to removal of a second, third, fourth, etc., electron. The energy required to remove the first electron is always the lowest, since the first electron is being pulled away from the neutral atom. Subsequent electrons are removed from ever-more-positive ions, so that the expended energy has to be greater. In the case of lithium, the respective values for the first, second, and third ionization potentials are 5.39, 75.64, and 122.45 eV per atom.*

If we restrict ourselves to consider only the first ionization potential, we see values for the various elements as displayed in the periodic format of Fig. 4.32. These same values are plotted in Fig. 4.33. As can be seen, there are some rough regularities, but the detailed changes seem to be quite complicated and anything but regular. The rough regularities are as follows:

1 There is a series of spikes corresponding to extraordinarily high values for the noble gases He, Ne, Ar, Kr, Xe, and Rn.

2 The spikes get smaller as one goes to higher atomic numbers, i.e., as one moves down the group of the periodic table.

3 Just after each noble gas, there is a sharp downward spike corresponding to an alkali element Li, Na, K, Rb, or Cs.

4 The downward spikes are more or less at the same level, but there is a slight progressive decrease as one again goes down a group of the periodic table.

5 Between each alkali element on the downside and the following noble gas at the topside, there is a rough but irregular rise in ionization potential. This corresponds to going from left to right across a period.

Let us examine first the trend in a group. Going down group I, the alkali elements Li, Na, K, Rb, and Cs show the *decreasing* set of values 5.39, 5.14, 4.34, 4.18, and 3.89 eV. The decrease is regular and matches the regular *increase in size* down the group. All these elements have similar electron configurations, consisting of filled shells and one *s* electron in the outermost shell. The

---

*The unit most frequently used to express the ionization potential is the *electronvolt* (eV). It is the amount of energy an electron gains when it is accelerated through a voltage difference of one volt. To convert from electronvolts per atom to kilojoules per mole, we multiply by 96.488. The conversion factor comes from multiplying the number of joules per electronvolt (1.6022×10^{-19}) by the Avogadro number (6.0222×10^{23}) and then converting to kilojoules.

FIG. 4.32 First Ionization Potentials of the Elements in Units of Electronvolts

	Group I	II						Transition Elements						III	IV	V	VI	VII	0
Period 1	1 H 13.598																		2 He 24.587
2	3 Li 5.392	4 Be 9.322												5 B 8.298	6 C 11.260	7 N 14.534	8 O 13.618	9 F 17.422	10 Ne 21.564
3	11 Na 5.139	12 Mg 7.646												13 Al 5.986	14 Si 8.151	15 P 10.486	16 S 10.360	17 Cl 12.967	18 Ar 15.759
4	19 K 4.341	20 Ca 6.113	21 Sc 6.54	22 Ti 6.82	23 V 6.74	24 Cr 6.766	25 Mn 7.435	26 Fe 7.870	27 Co 7.86	28 Ni 7.635	29 Cu 7.726	30 Zn 9.394		31 Ga 5.999	32 Ge 7.899	33 As 9.81	34 Se 9.752	35 Br 11.814	36 Kr 13.999
5	37 Rb 4.177	38 Sr 5.695	39 Y 6.38	40 Zr 6.84	41 Nb 6.88	42 Mo 7.099	43 Tc 7.28	44 Ru 7.37	45 Rh 7.46	46 Pd 8.34	47 Ag 7.576	48 Cd 8.993		49 In 5.786	50 Sn 7.344	51 Sb 8.641	52 Te 9.009	53 I 10.451	54 Xe 12.130
6	55 Cs 3.894	56 Ba 5.212	*	72 Hf 7.0	73 Ta 7.89	74 W 7.98	75 Re 7.88	76 Os 8.7	77 Ir 9.1	78 Pt 9.0	79 Au 9.225	80 Hg 10.437		81 Tl 6.108	82 Pb 7.416	83 Bi 7.289	84 Po 8.42	85 At ?	86 Rn 10.748
7	87 Fr ?	88 Ra 5.279	†	104 Ku ?	105 Ha ?	106 ?	107 ?												

*	57 La 5.577	58 Ce 5.47	59 Pr 5.42	60 Nd 5.49	61 Pm 5.55	62 Sm 5.63	63 Eu 5.67	64 Gd 6.14	65 Tb 5.85	66 Dy 5.93	67 Ho 6.02	68 Er 6.10	69 Tm 6.18	70 Yb 6.254	71 Lu 5.426
†	89 Ac 6.9	90 Th ?	91 Pa ?	92 U ?	93 Np ?	94 Pu 5.8	95 Am 6.0	96 Cm ?	97 Bk ?	98 Cf ?	99 Es ?	100 Fm ?	101 Md ?	102 No ?	103 Lr ?

nuclear charge increases down the group, but we can expect that this will roughly be screened out as we add additional shells of electrons between the outermost electron and the nucleus. It is the outermost electron that is being removed in the process of ionization, and it feels attraction to essentially a +1 charge. Since, however, the *s* shell in question gets bigger as we go down the group, the attraction should decrease as we go from Li (pulling off $2s^1$) to Cs (pulling off $6s^1$).

How does the ionization potential change across a period? As can be seen, for example, in the second period, there is a steady increase from 5.39 eV for lithium to 21.56 eV for neon, with a minor setback at boron and again at oxygen. Why the steady increase? At least two factors must be considered.

129

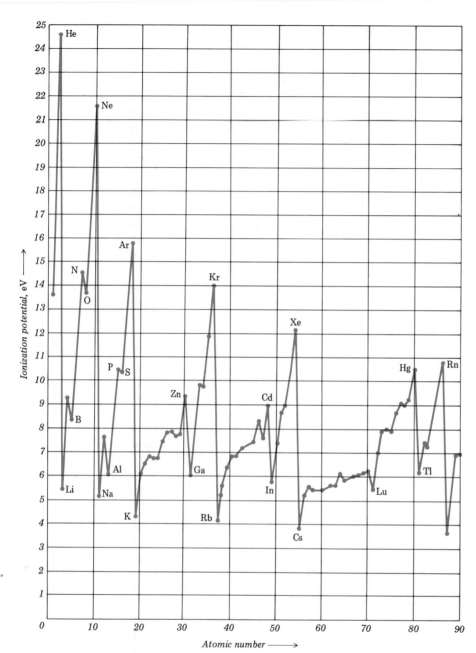

FIG. 4.33 Ionization potentials, in electronvolts, plotted as a function of increasing atomic number of the elements.

First, the nuclear charge (even taking account of shielding by inner shells) increases from left to right across a period. By itself alone, this effect would predict that the ionization potential should increase from lithium to neon. Second, the atomic radius decreases from left to right. This shrinking-size effect would also predict that ionization potential should increase, since the closer an electron is to the nucleus, the harder it is to pull off.

130

Why are there irregularities at boron and at oxygen? In going from beryllium $(Z = 4)$ to boron $(Z = 5)$, the electron configuration changes from $1s^2 2s^2$ to $1s^2 2s^2 2p^1$. The added electron is $2p$. As can be seen from Fig. 4.19, $2p$ differs from $2s$ in not having an extra probability bump close to the nucleus. This extra probability bump means two things: (1) the $2s$ will be bound tighter to the nucleus than the $2p$, and (2) the $2s$ electron will partly shield the $2p$ electron against full attraction by the nucleus. Hence, the added $2p$ electron in boron is detached more readily than one might expect.

In going from nitrogen $(Z = 7)$ to oxygen $(Z = 8)$, the electron configuration changes from $1s^2 2s^2 2p^3$ to $1s^2 2s^2 2p^4$. Showing only the p electrons, this looks like $\underline{\uparrow}\ \underline{\uparrow}\ \underline{\uparrow}$ for nitrogen going to $\underline{\uparrow\downarrow}\ \underline{\uparrow}\ \underline{\uparrow}$ for oxygen. As can be seen, the added electron is going into an orbital that is already occupied by an electron. Because of increased electron-electron repulsion between two electrons in the same orbital, the added electron in oxygen should be more readily detachable.

The third and final property that we need to tie to the electronic structure is the property called *electron affinity*. This is the energy released when an electron is added to an isolated neutral atom, corresponding to the reaction

$$X(g) + e^- \longrightarrow X^-(g)$$

Experimental values for electron affinity are hard to get. Figure 4.34 shows the few values available. Electron affinity can be measured by looking at the ionization spectra of the corresponding negative ion. In other words, the energy released in the process $X + e^- \longrightarrow X^-$ is equal in magnitude to the energy required to bring about the reverse process $X^- \longrightarrow X + e^-$. Stated another way, the electron affinity of X equals the ionization potential of X^-.

The halogen atoms (F, Cl, Br, and I) have the largest electron affinities of any elements in the periodic table. This is related to the fact that the elements are on the extreme right side of the periodic table where atomic sizes are relatively small, where the effective nuclear charge is relatively large, and where there is but one electron vacancy before the attainment of noble-gas configurations which are especially stable. In fluorine $(Z = 9)$, for example, the electron configuration is $1s^2 2s^2 2p^5$. There is room for one more electron in the $2p$ subshell, where the attraction to the nucleus is effectively to a $+7$ charge. (This is calculated as the nuclear charge, less shielding by the two electrons in the K shell. There is also some incomplete shielding by the $2s$ electrons, but this we can ignore.)

How do the electron affinities change down the group? The values for F, Cl, Br, and I are 3.45, 3.61, 3.36, and 3.07, respectively. Excluding F, the electron affinity decreases down the group. This is roughly what we would expect on the basis of increasing size. The attracted electron corresponds to $2p$ for F, $3p$ for Cl, $4p$ for Br, and $5p$ for I. The $5p$ orbital has a larger average radius than does the $2p$, and so, assuming in each case an effective attracting charge of $+7$, we expect the attraction to be least in iodine.

Why should fluorine be out of line? Chemists still argue this one, but the main reason appears to be that fluorine, being the smallest atom of the halogen group, cannot have its outer-shell electron-electron repulsions ignored. To convert F to F^-, the eighth electron is being squeezed into the $n = 2$ shell. This shell is small, and electron-electron repulsions in it are relatively large. This leads to a relatively large expansion (from 0.071 nm for F to 0.133 nm for F^-), which amounts to 87 percent. In contrast, when iodine picks up an electron, the eighth electron goes into a much larger ($n = 5$) shell. Electron-electron repulsions are less, and the radial expansion (from 0.133 nm for I to 0.220 nm for I^-) is only 65 percent.

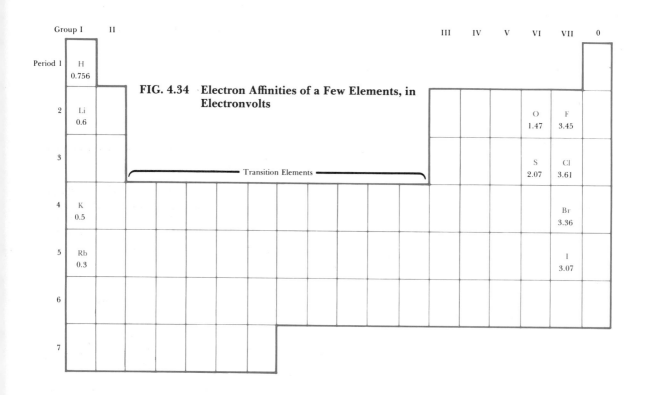

FIG. 4.34 **Electron Affinities of a Few Elements, in Electronvolts**

Important Concepts

cathode rays	mass spectrometer	uncertainty principle
electron	Bohr atom	electron probability distribution
wavelength	quantum numbers	node
frequency	shell	electron spin
spectrum	orbit	orbital filling
nucleus	electronic buildup of atoms	Pauli exclusion principle
atomic number	subshells	atomic radius
mass number	s, p electrons	ionization potential
isotope	wave nature of electron	electron affinity

Exercises

**4.1 Electrolysis Explain how Faraday's observations on electrolysis indicated that electric charge comes in discrete units.

*4.2 Discharge tubes How can one account for the glowing light that is emitted when a high voltage is imposed on a low-pressure gas in a discharge tube?

**4.3 Discharge tubes Benjamin Franklin assumed that when an electric current moves through a metal, it is the positive charge that moves and the negative charges, being massive, stand still. If this were true, how would the observations on discharge-tube experiments be changed?

*4.4 Cathode-ray tube Describe the function of each component shown in the cathode-ray tube pictured in Fig. 4.3.

**4.5 Cathode-ray tube Given the electron beam shown in Fig. 4.4, what would happen to the deflection x:

    a If the electron velocity were decreased

    b If the magnetic field were increased

    c If the electron were made less massive

    d If the electron charge were increased

**4.6 Thomson experiment Explain why the Thomson experiment by itself cannot give the mass of the electron.

*4.7 Millikan experiment What charge would you assign to an electron if in a Millikan-type experiment you observed the following charges for the oil droplets: -1.6×10^{-19}, -2.4×10^{-19}, -3.2×10^{-19}, and -4.8×10^{-18} C?

*4.8 Light Calculate the frequency and energy of red light ($\lambda = 700$ nm).

Ans. 2.84×10^{-19} J

*4.9 Line spectrum Calculate, in terms of wavelength, the positions of the first five lines of the Lyman series for hydrogen. Plot these as a spectrum in the same fashion as Fig. 4.7.

*4.10 Line spectrum What are the wavelength positions of the first and last lines of the Paschen series of hydrogen?

***4.11 Rutherford experiment If you were John Dalton and believed in hard atoms, how might you explain the observations of the Rutherford experiment?

***4.12 Rutherford experiment Suppose you were a disbeliever of atomic theory and maintained, like some of the early Greek philosophers, that matter ultimately is continuous. How might you account for the observations of the Rutherford experiment?

**4.13 X rays For the elements 26 to 30 (Fe, Co, Ni, Cu, and Zn), the observed X-ray lines are 0.1936, 0.1789, 0.1658, 0.1541, and 0.1435 nm, respectively. Show that these fit a Moseley plot of $\sqrt{\nu}$ vs. Z.

**4.14 Moseley experiment By examining the periodic table in Fig. 3.2, find where else besides cobalt and nickel Moseley would have had to invert the order of elements arranged by increasing atomic weight.

**4.15 Mass spectrometer How is a mass spectrometer related to a cathode-ray tube? Point out similarities and differences.

**4.16 Mass spectrometer Referring to the mass spectrometer pictured in Fig. 4.11, tell which of the two paths shown would correspond to (*a*) a particle of higher charge and (*b*) a particle of higher mass. Assume all other things are equal.

**4.17 Mass spectrometer If you ran natural oxygen (^{16}O, ^{17}O, and ^{18}O) through a mass spectrometer under conditions where $+1$ and $+2$ ions could form, what values of charge-to-mass ratio should you observe?

*4.18 Nuclear symbols Tell how many neutrons and how many protons there are in each of the following: ^2_1H, ^6_3Li, $^{19}_9F$, $^{56}_{26}Fe$, $^{107}_{47}Ag$. What regularity can you discover?

***4.19 Bohr atom Calculate the velocity of the electron in the innermost orbit of the hydrogen atom. *Ans. 2.18×10^8 cm/sec*

**4.20 Bohr atom Draw to scale circular orbits, showing the first five electron orbits of the hydrogen atom. How big would Z have to become so that the fifth orbit around Z would fit inside the first orbit around $Z = 1$?

***4.21 Bohr atom Suppose the electron were 10 times heavier than it really is. On a Bohr picture, what would this do to the size of the electron orbits and to the magnitude of the energy difference between ground and first excited state?

***4.22 Bohr atom Assuming that the radius of the nucleus is given by $r = 1 \times 10^{-13} A^{1/3}$ cm and noting that for heavier nuclei the number of neutrons is roughly 1.5 times the number of protons, estimate crudely at what value of Z the first Bohr orbit should shrink into the nucleus. *Ans. 3000*

***4.23 Bohr model What is the physical meaning of the topmost energy level shown in Fig. 4.12? What would be the significance of any additional levels above this line?

*4.24 Bohr model Without referring to the text or any other notes, write out the electron population by shells for each of the first 18 elements. Identify each element by name and symbol.

**4.25 Periodic table Explain on the basis of a Bohr model why there are only 8 elements instead of 18 in the third period.

*4.26 Subshells Explain how the existence of subshells accounts for the fact that the fourth shell starts filling before the third shell has its maximum number of electrons. Illustrate by reference to elements 19 through 21.

*4.27 Electron configuration Give the electron population by subshells for each of the following elements: $Z = 9, 19, 29, 39, 49$.

*4.28 Electron configuration Using the linear electron-population notation (e.g., $1s^22s^22p^3$ for N), write out the electron configuration for each of the following atoms in its ground state:

a Be ($Z = 4$)   b Mg ($Z = 12$)
c P ($Z = 15$)   d Zn ($Z = 30$)
e Sn ($Z = 50$)

*4.29 Electron configuration A neutral atom of an element in its ground state has two K electrons, eight L electrons, and five M electrons. Indicate as many of the following as possible on the basis of the given information. If not enough information is given, tell what fact is needed:

a Atomic number
b Atomic weight
c Total number of *s* electrons
d Total number of *p* electrons
e Total number of *d* electrons

*4.30 Electron configuration Without looking at a periodic table, select from each of the following lists those elements which belong to the same group or subgroup: (*a*) $Z = 11, 16, 26, 55$; (*b*) $Z = 37, 24, 4, 42$; (*c*) $Z = 4, 10, 18, 26$; (*d*) $Z = 2, 8, 16, 32$.

*4.31 de Broglie Assuming Monsieur de Broglie weighed 70 kg, what was his wavelength when he was jogging at 15 km/h?

**4.32 de Broglie What is the de Broglie wavelength for an electron in the innermost orbit of the hydrogen atom? *Ans. 3.34×10^{-8} cm*

**4.33 Electron probability distribution Draw a graph showing how the electron probability per unit volume changes as one goes along a diameter of a hydrogen atom in its ground state.

*4.34 Electron probability distribution How is the graph of electron probability per unit volume, shown in Fig. 4.17a, related to the circle shown in Fig. 4.17c?

**4.35 Electron probability distribution (a) Draw a graph showing how the volume of a spherical shell of uniform thickness changes with r of the shell. (b) On the same axes, draw a graph showing how the electron probability per unit volume changes with r for a $1s$ electron. (c) Multiply curves (a) and (b). Compare the result with Fig. 4.18.

**4.36 Electron probability distribution Assuming Fig. 4.18 is to scale for a $1s$ electron in hydrogen, estimate roughly to what distance from the nucleus one should go to encompass 99 percent of the $1s$ electron. *Ans. 0.16 nm*

*4.37 Orbital What is the difference in meaning between "orbit" and "orbital"?

**4.38 Electron probability distribution Given the p_x electron as shown in Fig. 4.20, what is the locus of points where its electron probability per unit volume is zero?

***4.39 Node What is meant by a node? Show that each of the $3d$ orbitals shown in Fig. 4.22 has two nodal surfaces.

*4.40 Electron spin What is the experimental basis for postulating that an electron has spin?

**4.41 Stern-Gerlach experiment How would you expect beams of Na atoms and Na^+ ions to differ in their behavior in a Stern-Gerlach–type of experiment? Explain.

*4.42 Electron spin Predict the total electron spin for the first five elements of the fourth period (K, Ca, Sc, Ti, V), assuming each is in its ground state.

**4.43 Quantum numbers Give all four quantum numbers for each of the six electrons in the carbon atom (ground state).

**4.44 Quantum numbers What is the atomic number of the element which, in its ground state, has its highest-energy electron characterized by the following quantum numbers: $n = 4$, $l = 0$, $m_l = 0$, $m_s = -\frac{1}{2}$?

***4.45 Pauli exclusion principle What is the Pauli exclusion principle? Show in general that

it leads to the restriction that no electron shell can hold more than $2n^2$ electrons.

*4.46 Orbital-filling diagram Use the orbital-filling diagram of Fig. 4.25 to predict the electron configuration and the number of unpaired electrons in an atom for which $Z = 50$.

*4.47 Atomic radii In general, how do atomic radii change in the periodic table (a) going down a group and (b) going across a period. Explain in terms of electron configuration.

*4.48 Atomic sizes Draw a series of circles with radii proportional to atomic sizes to represent the elements of group I and period 4. Comment on the appropriateness of treating atomic sizes as representable by hard balls.

**4.49 Periodic table A very useful generalization for using the periodic table is the so-called "diagonal relation." This states that elements on a diagonal (i.e., over one to the right and then one down) are likely to be similar in properties. Suggest an explanation.

**4.50 Atomic sizes Suggest a reason why adding the assigned radii of H and Cl atoms, as given in Fig. 4.27, does not give the observed interatomic distance in HCl. How might you improve the assigned atomic radii to make the above prediction more accurate?

**4.51 Atomic sizes Assuming the chlorine atom is a hard sphere of radius 0.099 nm, how long a line (in centimeters) of chlorine atoms would it take to accommodate the Avogadro number of atoms? Supposing that the chlorine atoms were arranged in a planar-square array of touching spheres, how big would the square be? Supposing that the chlorine atoms were arranged in a three-dimensional cubic array, what would be the edge length of the cube? Always assume the spheres to be in contact along the three cartesian directions x, y, and z.

**4.52 Ionization potential What is meant by "ionization potential"? How does the ionization potential change going down the group I elements? Make plots of ionization potential of group I elements versus (a) atomic number; (b) atomic size as shown in Fig. 4.30. Which gives a better correlation, (a) or (b)?

**4.53 Electron affinity Define the term "electron affinity." How is it related to ionization potential? How can electron affinity be determined experimentally? Suggest a reason why so few data are available for electron affinities.

4.54 Ionization You are given a sample consisting of a mixture of fluorine atoms and chlorine atoms. Removal of an electron from each atom of the sample requires a total of 284 kJ; addition of an electron to each atom of the assembly releases a total of 68.8 kJ. How many atoms of each kind are there in the original sample?

Ans. 4.6×10^{22} F and 7.5×10^{22} Cl

***4.55 Electron affinity** As can be seen from Fig. 4.34, the electron affinity of oxygen (1.47

eV) is less than that of sulfur (2.07 eV). Yet the frequently quoted generalization is that electron affinity decreases going down a group. How then might one account for the anomaly of oxygen?

*4.56 Electron affinity and ionization potential How many chlorine atoms can you ionize in the process

$$Cl \longrightarrow Cl^+ + e^-$$

by the energy liberated from the process

$$Cl + e^- \longrightarrow Cl^-$$

for 6.0×10^{23} atoms?

Chapter 5

MOLECULES AND MOLECULAR STRUCTURE

The concept of a molecule is an old one, the term itself apparently having been introduced in about 1860 by Stanislao Cannizzaro in his description of gases as consisting of tiny aggregates of atoms. The idea of atomic clusters is much older; John Dalton, for example, talked about "compound atoms" around 1800. At present, any electrically neutral aggregate of atoms held together strongly enough to be considered as a unit is called a *molecule*. The net attractive interaction between two adjacent atoms within a molecule is called a *chemical bond*.

Hydrogen gas is composed of aggregates of two hydrogen atoms. Water vapor is composed of aggregates of two hydrogen atoms and one oxygen atom. Solid sulfur consists of aggregates of eight sulfur atoms. In each of these cases the aggregate is usefully called a "molecule." On the other hand, in solid sodium chloride there are no simple aggregates consisting of a few atoms. All the sodium ions and all the chloride ions in a given crystal are bound into one giant aggregate. The term "molecule" is not useful in such a case.

The fact that there are molecules is generally agreed on. The problem is what is the best way to give an adequate description of the electron distribution in them? Why do they hold together? Why do they have the shape and properties they do? Why is the hydrogen molecule H_2 and not H_3 or H_4? Why is H_2O normally neither linear nor right-angled but bent at an angle of 104.52°? The answers to these questions are important not only for a fundamental understanding of the behavior of matter but also as bases for considering applied problems. Why do many materials of construction generally show failure to tensile or shear stress at considerably less than their theoretical strength? Why do molecules such as DNA, the carrier of the genetic code, fold into funny shapes such as a double helix? Why does advancing the spark on an automobile engine reduce hydrocarbon emissions but increase pollution by oxides of nitrogen? An intelligent route to solving such problems requires knowing something about electrons in molecules.

5.1
ELECTRONS IN MOLECULES

In an isolated atom each electron is under the influence of only one nucleus and the other electrons. When two atoms come together, the electrons of one atom come under the influence of the electrons and nucleus of the other atom. The interaction might produce an attraction between the two atoms. If that is so, an electronic rearrangement must have occurred to give a more stable state. In other words, the formation of a chemical bond suggests that the molecule represents a state of lower energy than the isolated atoms.

A detailed description of electrons in molecules is a difficult problem.

Two general approaches can be used: One is to consider the entire molecule as a new unit, with all the electrons moving under the influence of all the nuclei and all the other electrons. This approach recognizes that each electron belongs to the molecule as a whole and may move throughout the entire molecule. The spatial distributions of the electrons in the molecule are called *molecular orbitals* and can be thought of in the same way as the orbitals of electrons in isolated atoms. The other approach to describing molecules is simpler but less correct. It assumes that the atoms in a molecule are very much like isolated atoms except that one or more electrons from the outer shell of one atom are accommodated in the outer shell of another atom. This method of describing molecules is called the *valence-bond method;* it utilizes directly the orbitals of isolated atoms.

In order to point up the difference in the two ways of viewing molecules, let us consider the hydrogen molecule. It is formed from two hydrogen atoms, each with one proton and one electron. In the molecular-orbital approach, H_2 is visualized as consisting of two H nuclei at some distance apart, with two electrons in an energy level that is spread out over the whole molecule. In the valence-bond approach, H_2 is visualized as consisting of two individual hydrogen *atoms* sitting side by side, with the electron shell of each atom the same as if the hydrogen atom were all by itself except that now electrons may be exchanged between atoms and each atom may contain both electrons part of the time.

No matter which picture is used, molecular orbital or valence bond, the molecule is held together because the electrons have been spread over more space and because the attraction between two positive protons and two negative electrons exceeds the total repulsion in the system.

In the more complicated case of hydrogen fluoride, the molecule is formed from one hydrogen atom and one fluorine atom. The hydrogen atom contributes a +1 nucleus and one electron; the fluorine atom, a +9 nucleus and nine electrons. In the molecular-orbital approach, the molecule is visualized as consisting of two nuclei at some distance apart, with ten electrons placed in various energy levels of the molecule as a whole. In the valence-bond approach, the molecule is visualized as consisting of a hydrogen *atom* and a fluorine *atom* side by side. The hydrogen atom is assumed to be the same as if it were alone except that part of the time it now may contain, besides its own electron, one of the electrons borrowed from the fluorine atom. For the fluorine atom, the inner shell is assumed to be unchanged. However, part of the time the outer shell contains, besides the original seven electrons, one additional electron from the hydrogen atom. The one electron from the hydrogen and one electron from the fluorine are considered as a *pair of electrons* shared between the atoms. The pair is regarded as holding the molecule together because it is attracted simultaneously to both nuclei.

Certainly in a molecule the energy levels of many, if not all, of the electrons are changed from those of isolated atoms. Therefore, it would be desirable to discuss chemical bonding exclusively in terms of molecular orbitals. However, the valence-bond approach is so much simpler that it remains in great use among chemists and will be used in the following discussion. Detailed consideration of molecular orbitals will be postponed until we get to Chap. 16.

It will be convenient to consider only the electrons in the outermost shell as being the ones involved in the *valence*, or chemical, binding. These electrons are therefore referred to as the *valence electrons*. Sometimes they are shown as dots around the symbol for the element. Examples of electronic symbols are

$$\text{Li}\cdot \qquad \text{Be}: \qquad \cdot\dot{\text{B}}\cdot \qquad \cdot\dot{\text{C}}\cdot \qquad \cdot\dot{\ddot{\text{N}}}\cdot \qquad :\dot{\ddot{\text{O}}}\cdot \qquad :\ddot{\text{F}}\cdot$$

The letters can be considered to represent the entire *core*, or *kernel*, of the atom. The core includes not only the nucleus but also the electrons in inner shells. The placing of the dots has no significance so far as the actual positions of the electrons are concerned. The dots are simply a convenient way of counting up the outermost electrons.

5.2
IONIC BONDS

In discussing chemical bonds, we shall assume first that molecules can be described in terms of individual atoms. Second, we shall assume that bonds can be described as being *ionic bonds*, in which electrons have been completely transferred from one atom to another, or as *covalent bonds*, in which electrons are shared between atoms.

The formation of an ionic bond is favored when an atom of low ionization potential interacts with an atom of high electron affinity. As shown in Fig. 5.1, an example of such a reaction is the one between sodium atoms and chlorine atoms. A sodium atom in its ground state is $1s^2 2s^2 2p^6 3s^1$; it has a low ionization potential; i.e., not much energy is required to pull off the outer electron. A chlorine atom in its ground state is $1s^2 2s^2 2p^6 3s^2 3p^5$; it has a high electron affinity; i.e., considerable energy is released when an electron is added to its outer shell. Suppose these two atoms collide. As shown in the figure, sodium starts with one valence electron, and chlorine, seven. In the collision, an electron is believed to transfer from the sodium to the chlorine. The sodium now has a positive charge because of the loss of a negative electron. The chlorine has a negative charge because of the gain of an electron. Thus, a positive ion and a negative ion are formed. Because the ions are of opposite electric charge, they attract each other in what is known as an *ionic bond*. The ionic bond is sometimes called the *electrovalent bond*.

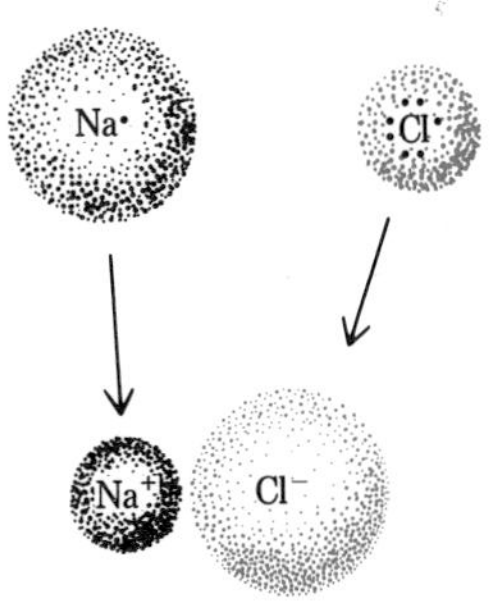

The formation of an ionic bond can be thought of in three steps:

$$\text{Na}\cdot \longrightarrow \text{Na}^+ + e^- \tag{1}$$

$$:\overset{..}{\text{Cl}}\cdot + e^- \longrightarrow \left[:\overset{..}{\underset{..}{\text{Cl}}}:\right]^- \tag{2}$$

$$\text{Na}^+ + \left[:\overset{..}{\underset{..}{\text{Cl}}}:\right]^- \longrightarrow [\text{Na}^+]\,[\text{Cl}^-] \tag{3}$$

FIG. 5.1 Formation of an ionic bond.

Step (1) requires energy equal to the ionization potential of sodium, i.e., 5.14 eV. Step (2) releases energy equal to the electron affinity of chlorine, i.e., 3.61 eV. As can be seen, there is a deficit of 1.53 eV. Step (3) releases energy, because of attraction between positive and negative ions. The ionic bond is formed only because the energy released in steps (2) plus (3) is greater than that required in step (1).

For step (3), energy is liberated when a positive ion of q_1 electron charge units attracts a negative ion of q_2 electron charge units to bring the nuclear centers r nm apart; the magnitude of the liberated energy in electronvolts is $1.44\, q_1 q_2 / r$. In the case of Na^+ and Cl^- the internuclear spacing will be about 0.28 nm, and thus the energy liberated by the ion pairing is 1.44(1)(1)/0.28, or 5.1 eV. This more than makes up the deficit of 1.53 eV in the first two steps. Using the symbol Δ (Greek "delta") to stand for change, with $+$ for increase and $-$ for decrease, we can write the energy change as follows:

$$
\begin{array}{ll}
Na \longrightarrow Na^+ + e^- & \Delta E = +5.14 \text{ eV} \\
e^- + Cl \longrightarrow Cl^- & \Delta E = -3.61 \text{ eV} \\
\underline{Na^+ + Cl^- \longrightarrow [Na^+]\,[Cl^-]} & \underline{\Delta E = -5.1 \ \ \text{eV}} \\
\text{Net } Na + Cl \longrightarrow [Na^+]\,[Cl^-] & \Delta E = -3.6 \ \ \text{eV}
\end{array}
$$

Clearly, it is step (3) that makes ionic-bond formation possible. Actually, for forming NaCl in the *solid* state, step (3) is even more favorable than calculated here for ion pairing in the gas state; in the solid each positive ion has more than one negative-ion neighbor, and vice versa. (See Fig. 7.11.)

In forming compounds by electron transfer, there must be a balance of electrons gained and lost. The reaction between sodium and chlorine requires one atom of sodium for every atom of chlorine. When barium reacts with chlorine, each barium atom loses two valence electrons ($6s^2$) to form Ba^{2+}. Hence, two chlorine atoms, each picking up one electron, are required to balance this. The compound formed, barium chloride, contains one doubly positive barium ion for every two singly negative chloride ions, as indicated by the formula $BaCl_2$. It should be noted that after reaction all these atoms are left with octets of electrons in their outermost shells. It is a general rule (except for transition elements) that when electrons are transferred the ions produced usually end up with filled electronic groupings in the outermost shell.

Since, in general, the elements on the left of the periodic table have low ionization potential and the elements on the right have high electron affinity, ionic bonds are favored in reactions between these elements. Thus, any alkali metal (group I) reacts with any halogen (group VII) to form an ionic compound. Similarly, most of the group II elements react with the halogens, or with group VI elements, to form ionic bonds. In general, these ionic compounds resemble sodium chloride in that at room temperature they are white, brittle solids which dissolve in water to give conducting solutions. They melt at relatively high temperatures.

5.3
COVALENT BONDS

Most bonds cannot be adequately represented by assuming a complete transfer of electrons from one atom to another. For example, in the hydro-

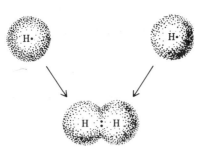

FIG. 5.2 Formation of a covalent bond.

gen molecule H_2 it would be unreasonable to say that one hydrogen atom captures an electron from the other. Rather, in such cases it is assumed that electrons are shared, and the bond is called *covalent*. In H_2, F_2, and HF, as discussed below bonds are described as *shared-electron pairs*.

The simplest case of a shared-pair bond is found in the case of hydrogen. When two hydrogen atoms come together, as shown in Fig. 5.2, a shared pair of electrons results because each atom can accommodate another electron in its shell. Neither atom is able to gain complete possession of both electrons. The conventional way of showing this, using electronic symbols, is

$$H \cdot + H \cdot \longrightarrow H : H$$

However, it must be emphasized that this is only a schematic representation of a complicated situation and tells little about the electron probability distribution. The actual charge distribution is one in which the shared-electron pair is spread over the whole molecule, spending equal time in the vicinity of each nucleus.

What happens to the energy of the system when two H atoms come together to form the diatomic H_2 molecule? Figure 5.3 shows how the potential energy changes during bond formation. When the two atoms are far apart, their potential energies are independent of each other and are arbitrarily set at zero (far right of the diagram). As the two atoms approach, there is an attraction between them; the potential energy of the system decreases. This decrease is shown by the curve which drops slowly first, then steeply to a

FIG. 5.3 Change of potential energy as two hydrogen atoms change their distance apart.

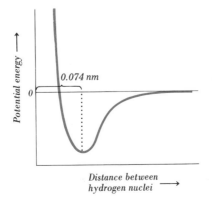

minimum value. To the left of the minimum, the potential energy shoots up steeply because of repulsion between like charges at very close distances of approach. The position of the minimum, which occurs at 0.074 nm, corresponds to the *bond length,* i.e., the average distance between the nuclei. Because the potential energy is lowest at this distance, the two H atoms tend to favor this spacing.

Why should there be a minimum in the potential-energy curve? Why are the two atoms attracted? These questions are difficult to answer. No single method gives a completely satisfactory answer. The valence-bond method assumes that when two H atoms come together, there is an increasing chance that an electron from the $1s$ orbital of one atom transfers to the $1s$ orbital of the other atom. However, because of repulsion between electrons, it is not likely that both electrons should stay on the same atom at the same time. Hence, an exchange is believed to occur. As a consequence, either electron may be found on either atom. This is equivalent to saying that both electrons occupy the same space and, according to the Pauli exclusion principle, is possible only if the two electrons exchanging have opposite spins. Consequently, *the two electrons forming the covalent bond must have their spins opposite,* or paired.

However, the question still remains: Why does the energy decrease on bond formation? Magnetic attraction between opposite spins is much too feeble to account for the energies involved. There are two major contributions to energy lowering:

A general result of wave mechanics is that spreading an electron wave over more space lowers its energy. Thus, as the electron distribution changes from confinement on one atom to being spread over two atoms, the energy of the electron diminishes.

As the two H atoms come together, each electron feels an added attraction due to attraction to the other nucleus. This attraction more than compensates for repulsion between like charges. Still, at very small internuclear distances like-charge repulsion becomes dominant. Hence, there is a steep rise in potential energy at very small distances.

The sharing of an electron pair as described above constitutes a *single bond.* It is customarily represented by a single dash; e.g., H—H. Since the sharing results from electron exchange between *two* atoms, the covalent bond is thought of as joining *two* atoms. In the case of hydrogen, having but one electron, only one bond can be formed to another atom. This is the reason why hydrogen does not normally form H_3 or H_4. To do so, at least one hydrogen atom would have to form more than one covalent bond.

The fluorine molecule F_2 is somewhat more complicated. It can be visualized as resulting from the sharing of an electron pair between the p orbitals of two fluorine atoms. A fluorine atom has in its ground state the electron configuration $1s^2 2s^2 2p^5$. All the subshells are filled except for the $2p$, which can accommodate six electrons. Of the three orbitals making up the $2p$ subshell, two are fully occupied by electron pairs and are not involved in the

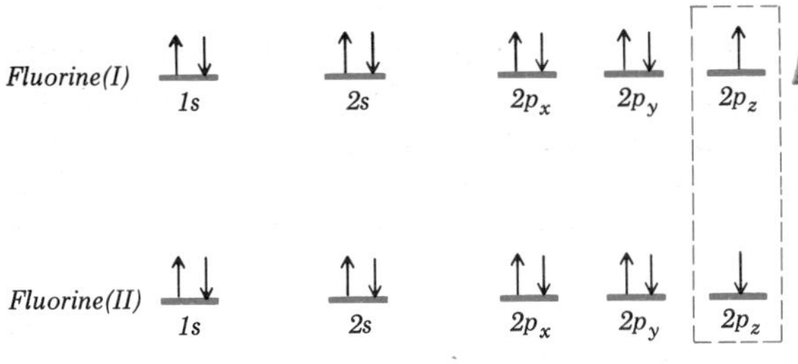

Fluorine(I) $\uparrow\downarrow$ $\uparrow\downarrow$ $\uparrow\downarrow$ $\uparrow\downarrow$ $\uparrow$
  1s   2s $2p_x$ $2p_y$ $2p_z$ *why*

Fluorine(II) $\uparrow\downarrow$ $\uparrow\downarrow$ $\uparrow\downarrow$ $\uparrow\downarrow$ $\downarrow$
  1s   2s $2p_x$ $2p_y$ $2p_z$

FIG. 5.4 Electron sharing involving exchange of a p electron on each of two fluorine atoms. (z is the axis of the molecule.)

bonding. The third orbital has a single electron which can exchange with a corresponding electron of opposite spin on a second atom. This is indicated schematically in Fig. 5.4; the dashed box indicates the shared pair of electrons. Thus, the valence-bond description in F_2 leads to a single covalent bond which, like the bond in H_2, involves a single pair of electrons. Unlike H_2, it involves p orbitals rather than s orbitals. Using electronic symbols for each fluorine atom, the equation for formation of F_2 can be written

$$:\overset{..}{\underset{..}{F}}\cdot + \cdot\overset{..}{\underset{..}{F}}: \longrightarrow :\overset{..}{\underset{..}{F}}:\overset{..}{\underset{..}{F}}:$$

The formula on the right is called an *electron-dot formula*. In such a formula, the dots between the chemical symbols are taken to represent electrons that are shared between the atoms.

When a hydrogen atom and a fluorine atom are brought together, the molecule HF is formed. The process can be written

$$H\cdot + \cdot\overset{..}{\underset{..}{F}}: \longrightarrow H:\overset{..}{\underset{..}{F}}:$$

The HF molecule may be visualized as resulting from the sharing of an electron pair between the $1s$ orbital of the hydrogen and a $2p$ orbital of the fluorine. The HF bond is also a single covalent bond, but it differs in a fundamental way from the bonds in H_2 and F_2. The reason for the difference is the unequal attraction of hydrogen and fluorine for a shared pair of electrons. In this case the electron pair spends more time on the fluorine than on the hydrogen. We shall return to this question in the following section.

In each of the covalent bonds so far described, the atom involved can be visualized as having attained a stable, i.e., noble-gas, electron configuration. In the bonding of hydrogen the additional electron can be thought of as completing the $1s$ subshell, as occurs in the noble-gas helium. In the bonding of fluorine the additional electron can be thought of as completing the $2p$ subshell, as occurs in the noble-gas neon. In the majority of chemical compounds, covalent bonding occurs so as to produce noble-gas electron configurations. Except for bonds to hydrogen, this means completing the s and p subshells of the outermost shell so that they contain eight electrons. This is the basis of the so-called "octet rule," which states that *when atoms combine, the bonds formed are such that each atom is surrounded by an octet of electrons.* In F_2, for example, each fluorine has completed its octet and does not bind additional fluorine atoms.

FIG. 5.5   Bond Lengths l and Bond Strengths D

Molecule	l, nm	D, kJ/mol
H_2	0.074	432
Li_2	0.267	105
B_2	0.159	289
C_2	0.131	628
N_2	0.109	941
O_2	0.121	494
F_2	0.143	151
Cl_2	0.199	239
Br_2	0.228	190
I_2	0.267	149

Two parameters are generally used to describe covalent bonds: One is the length of the bond, i.e., the distance between the nuclei. The other is the strength of the bond, i.e., the energy required to break the bond apart. Figure 5.5 tabulates bond lengths l and bond strengths D for some representative diatomic molecules. The bond lengths are given in nanometers; the bond strengths, in kilojoules per mole of molecules. As can be seen, the molecules of the second period Li_2, B_2, C_2, N_2, O_2, F_2 show a maximum in strength at N_2 in about the middle of the period. A question that might be raised is why such an effect occurs. Why should the bond in N_2 be almost twice as strong as the bond in O_2? Why should the bond in F_2 be so much weaker than either N_2 or O_2? A related puzzle occurs in the group VII sequence F_2, Cl_2, Br_2, I_2. Except for F_2, there is a regular decrease going down the group. Why should the F_2 be out of line? As we shall see later, an important factor is that as we go to the right in a period, the number of non-bonding electron pairs (i.e., electron pairs that do not appear to be shared between atoms) increases. Repulsion between nonbonding pairs serves to weaken bonds. This effect is especially important for fluorine, the smallest of the halogens.

Besides single bonds, *double* and *triple* bonds may be formed in order that an atom can complete its octet. These correspond, respectively, to sharing of two pairs and three pairs of electrons between the bonded atoms. An example of a double bond is that found in the molecule CO_2 for which the electron-dot formula can be worked out as follows.

EXAMPLE 1

Write the electron-dot formula for CO_2, given that C lies on the line between the O's.

Solution

Carbon has 4 valence electrons, $\cdot \overset{\displaystyle \cdot}{C} \cdot$

Oxygen has 6 valence electrons, $\overset{\displaystyle \cdot\cdot}{\underset{\displaystyle \cdot}{:O}} \cdot$

We can first write

$$:\ddot{O}\cdot\cdot\dot{C}\cdot\cdot\ddot{O}:$$

This would give us a single bond between oxygen and the central carbon. Then we move the other unpaired electrons to get pairing:

$$:\ddot{O}\,\overset{\curvearrowright}{\underset{\curvearrowleft}{\,_{\cdot}\,C\,_{\cdot}\,}}\ddot{O}:$$

This gives

$$:\ddot{O}::C::\ddot{O}:$$

In general, we try to have all the electrons paired, and, if possible, eight electrons around each atom (except hydrogen which has two).

The pairs of dots shown between the C and the O are considered to belong simultaneously to the octets of C and of O. The other pairs of dots represent unshared pairs of electrons belonging to the oxygen only. An example of a triple bond is that in N_2. In this case each N atom (ground-state configuration $1s^2 2s^2 2p^3$) completes its octet by sharing an electron in each of its p orbitals with an electron in a p orbital of the other N. This can be represented by $:N:::N:$, where each N atom is considered to be surrounded by three pairs of shared electrons and one pair of unshared $2s$ electrons. In general, for the same pair of bonded atoms, triple bonds are shorter than double bonds and double bonds are shorter than single bonds.

In all the cases discussed above, each shared pair of electrons involves one electron from each of the bonded atoms. There are also cases in which one atom in the bond contributes *both* of the electrons that are to be shared. Some examples are

$$
\begin{array}{c}
\text{H} \\
\text{H}:\ddot{\text{N}}: + \text{H}^+ \\
\text{H}
\end{array}
\longrightarrow
\left[
\begin{array}{c}
\text{H} \\
\text{H}:\ddot{\text{N}}:\text{H} \\
\text{H}
\end{array}
\right]^+
$$

$$
\begin{array}{cc}
\text{H} & :\ddot{\text{F}}: \\
\text{H}:\ddot{\text{N}}: + & \ddot{\text{B}}:\ddot{\text{F}}: \\
\text{H} & :\ddot{\text{F}}:
\end{array}
\longrightarrow
\begin{array}{c}
\text{H}:\ddot{\text{F}}: \\
\text{H}:\ddot{\text{N}}:\ddot{\text{B}}:\ddot{\text{F}}: \\
\text{H}:\ddot{\text{F}}:
\end{array}
$$

$$
\left[:\ddot{\text{S}}:\right]^{2-} + \cdot\ddot{\text{S}}: \longrightarrow \left[:\ddot{\text{S}}:\ddot{\text{S}}:\right]^{2-}
$$

Such bonds are sometimes called *coordinate covalent,* or *donor-acceptor,* bonds. The use of such names is generally unnecessary, since the final bond is independent of the way it was formed. For example, in the first equation above, the four bonds in the ammonium ion NH_4^+ turn out to be identical, although one of them seems to be set up in a different way from the others.

145

Because electrons may be shared unequally between atoms, it is necessary to have some way of describing electric-charge distribution in a bond. The usual way is to classify bonds as *polar* or *nonpolar*. As examples, the bonds in H_2 and F_2 are called nonpolar; the bond in HF, polar.

Why are the covalent bonds in H_2 and F_2 called nonpolar? In both H_2 and F_2 the "center of gravity" of the negative-charge distribution is at the center of the molecule. The shared-electron pair is just as probably found with one nucleus as with the other. Thus, the center of negative charge is the same as the center of the positive-charge distribution. The molecule is electrically neutral in two senses of the word. Not only does it contain an equal number of positive and negative charges (protons and electrons), but also the center of the positive charge coincides with the center of the negative charge. The molecule is a *nonpolar molecule;* it contains a *nonpolar bond* in which an electron pair is *shared equally* between two atomic kernels.

In the case of HF the bond is called polar because the center of the positive charge does not coincide with the center of the negative charge. The molecule as a whole is electrically neutral because it contains equal numbers of positive and negative particles. However, owing to unequal sharing of the electron pair, the F end of the molecule appears negative with respect to the H end. Polarity arises mainly because the shared pair of electrons spends more time on the F than on the H, *not just because* F *has more electrons than* H. (Note that the charge of unshared electrons is balanced by greater positive charge of the F nucleus.)

As another example of a polar covalent bond, consider the bond formed when Cl combines with F to form the molecule ClF. The formation equation can be written as

$$: \overset{..}{\underset{..}{Cl}} \cdot + \cdot \overset{..}{\underset{..}{F}} : \longrightarrow : \overset{..}{\underset{..}{Cl}} : \overset{..}{\underset{..}{F}} :$$

Cl, like F, has one vacancy in its p subshell. Exchange involving one p orbital of the Cl with one p orbital of the F produces a single covalent bond. However, in the bond between Cl and F, the shared pair of electrons is distributed unequally, spending more time on the F than on the Cl. The F end of the molecule therefore appears negative with respect to the Cl end. As shown in Fig. 5.6, this polarity can be indicated by a δ^+ and a δ^- to show where the centers of positive and negative charge are, or by a positive-tailed arrow pointing in the direction of the electron imbalance. Each molecule as a whole is electrically neutral—that is, there are just as many positive charges as there are negative charges—but there is a dissymmetry in the electric distribution.

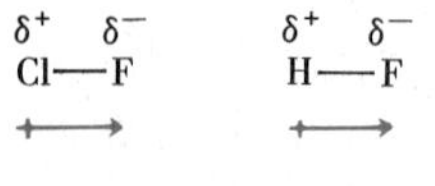

A molecule (or bond) such as ClF is called a *dipole*. It is characterized by *a positive charge and an equal negative charge separated by some distance.* The product, charge × distance, is called the *dipole moment*. The magnitude of the

$$\overset{\delta^+ \quad \delta^-}{Cl-F} \qquad \overset{\delta^+ \quad \delta^-}{H-F}$$
$$\longmapsto \qquad \qquad \longmapsto$$

FIG. 5.6 Polar molecules.

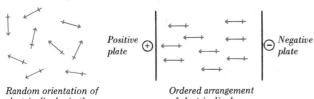

Random orientation of
electric dipoles in the
absence of a field

Ordered arrangement
of electric dipoles
in an electric field

FIG. 5.7 Behavior of dipoles in an electric field.

dipole moment* measures the tendency of the dipole to turn when placed
in an electric field. As shown in Fig. 5.7, each dipole turns because its positive end (e.g., H in HF) is attracted to the negative plate and its negative end
(F in HF) to the positive plate. Since the positive and negative centers are
part of the same molecule, the molecules can only turn; there is no net
migration toward the plates.

The behavior of dipoles in an electric field gives an experimental method
for distinguishing between polar and nonpolar molecules. The experiments
involve the determination of a property called the *dielectric constant.* It can be
measured as follows: An electric capacitor consisting of two parallel, metallic
plates, such as those shown in Fig. 5.7, has the ability to store electric charge.
The *capacitance*—that is, the amount of charge that can be put on the plates
for a given voltage—depends upon the substance between the plates. The
dielectric constant is defined as the ratio of the capacitance when the substance is between the plates to the capacitance when a vacuum is between
them.

In general, a substance which consists of polar molecules has a *high* dielectric constant; i.e., a capacitor can store much more charge when such a substance is between the plates. This high dielectric constant can be thought of as
arising in the following way: As shown in Fig. 5.7, dipoles tend to turn in a
charged capacitor so that negative ends are near the positive plate and positive ends are near the negative plate. This partially neutralizes the charge on
the plates and permits more charge to be added. Thus, measurement of the
dielectric constant gives information about the polarity of molecules. The fact
that hydrogen gas has a *low* dielectric constant (1.00026 as compared with
1.00000 for a perfect vacuum) confirms the idea that H_2 molecules are
nonpolar.

The quantitative calculation of dipole moments of individual *bonds* from
measured dielectric constants of *molecules* is complicated: (1) The relative orientation of the bonds, i.e., the bond angle, must be known. (2) Unshared

---

*The unit for measuring a dipole moment is called the *debye,* after Peter J. W. Debye, a Dutch
chemist who first described polar molecules. One debye corresponds to the dipole moment which
would be produced by a negative charge equal to 0.208 of an electron separated by 1×10^{-8} cm
from an equal but opposite charge. The 0.208 comes from the reciprocal of 4.80. The electron
charge used to be given as 4.80×10^{-10} esu (electrostatic units), and the combination of a positive
electron 1×10^{-8} cm away from a negative electron was assigned a dipole moment of 4.80×10^{-18}
esu cm, or 4.80 debyes.

147

electrons may contribute to the electric dissymmetry of the molecule. (3) The presence of charged plates can *polarize*, or temporarily distort, the charge distribution in molecules. Such *induced* dipoles can be distinguished from the *permanent* dipole by switching the plate voltage at such high frequency that the molecules cannot turn their permanent dipole moments rapidly enough to keep up. In such a case only the induced dipole contributes to the dielectric constant. The permanent dipole can then be found by taking the difference. The permanent dipole can also be found by measuring dielectric constant as a function of temperature. The contribution of the permanent dipole moment decreases as the temperature is raised because of increasing disorder at high temperatures. High temperatures hinder the lining up of dipoles.

It is easy to predict whether a given diatomic molecule will be polar or nonpolar. If the two atoms are alike, the *bond* between them must be nonpolar, and therefore the *molecule* is nonpolar. If the two atoms are different, the *bond* is polar, and the *molecule* is also polar. The degree of polarity increases as the atoms become more different in electron-pulling ability. It is not so easy to predict the polar nature of a molecule containing more than two atoms. Such a *molecule* can be nonpolar even though the individual *bonds* are polar. Carbon dioxide, CO_2, is an example. As shown in Fig. 5.8, the two oxygen atoms are bonded on opposite sides of the carbon atom. Oxygen attracts shared electrons more than carbon does, and so each carbon-oxygen bond is polar. The shared electrons spend more time near the oxygen than near the carbon, and the polarity of each bond is as shown in the figure. However, because the molecule is linear, the effect of one dipole just cancels the effect of the other. As a result, when a carbon dioxide molecule is placed in an electric field, there is no tendency for the molecule to turn. Any turning action of one bond is counteracted by the opposite turning action of the other bond. Thus, carbon dioxide has a low dielectric constant.

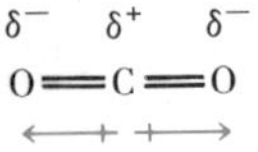

Water, H_2O, is a triatomic molecule in which two hydrogen atoms are bonded to the same oxygen atom. There are two different possibilities for its structure: The structure may be linear, with the three atoms arranged in a straight line, or the atoms may be arranged in the form of a bent chain. The two possibilities are shown in Fig. 5.9. The experimental fact is that water has a very high dielectric constant; this supports the bent structure on the right. The linear structure on the left would represent a nonpolar molecule in which two polar bonds would be placed in line so that there would be no net dipole moment. In the actual molecule of water the two bond dipoles do not cancel out but instead, owing to the bent structure, give a resultant moment as shown on the right of the figure.

Ammonia, NH_3, also acts as a polar molecule. This rules out the possibility that the molecule might be planar with three H—N bonds all pointing toward the middle of an equilateral triangle and canceling each other out. Instead, the true structure has the three H atoms lying at the corners of the base of a triangular pyramid with the N atom at the apex. The two alternative structures are shown in Fig. 5.10.

In light of the foregoing discussion of polar bonds, it is important to note

$$\overset{\delta^-}{O} = \overset{\delta^+}{C} = \overset{\delta^-}{O}$$
$$\longleftarrow + \; + \longrightarrow$$

FIG. 5.8 Nonpolar CO_2 molecule in which individual C—O bond dipoles compensate each other.

$$\overset{\delta^+}{H} - \overset{\delta^-}{O} - \overset{\delta^+}{H}$$
$$\longmapsto \quad \longleftarrow \; \longmapsto$$

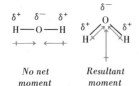

No net moment   *Resultant moment*

FIG. 5.9 Possible configurations of the H_2O molecule.

that *there is no sharp distinction between ionic and covalent bonds.* In a chemical bond between atoms A and B, all gradations of polarity are possible, depending on the specific nature of A and B. If A and B have equal ability to attract electrons, the bond is nonpolar. If the electron-pulling ability of B exceeds that of A, the shared electrons spend more time on B, and the bond becomes more polar (i.e., more ionic) the greater the difference. If the electron-pulling ability of B greatly exceeds that of A, then in the limit the electron pair will not be shared at all but will spend all its time on B. The end result will be a negative ion B⁻ and a positive ion A⁺; the bond will be 100 percent ionic.

5.5
ELECTRONEGATIVITY

In the preceding section we referred to the relative electron-pulling ability of atoms in molecules. This *relative ability to attract shared electrons* is known as the *electronegativity.* R. S. Mulliken suggested that a quantitative measure of this property could be obtained by taking the average of ionization potential and electron affinity of individual atoms. That both these quantities must be considered can be seen from the following argument: The bond in ClF consists of an electron pair shared unequally between F and Cl. The preference of the electron pair for F or Cl depends on how much energy (the ionization potential) is required to pull an electron from one atom and how much energy (the electron affinity) is released when it is added to the other atom. The net energy required to transfer an electron from Cl to F is 12.97 eV minus 3.45 eV, or 9.52 eV. This is less than the net energy required to transfer an electron in the other direction, i.e., from F to Cl ($17.42 - 3.61 = 13.81$ eV). Hence the electron pair spends more time on F than Cl. We say that F is more electronegative. In calculating the energy required for such transfers it is necessary to know both the ionization potential and the electron affinity of each atom. Unfortunately, electron affinities are available for only a very few elements; so the Mulliken method cannot be used in most cases.

The concept of electronegativity was actually first introduced by Linus Pauling in 1932. By using, as described in the next section, the various properties of molecules, such as dipole moments and energies required to break bonds, he was able to set up a useful scale of electronegativity covering most of the elements. The numerical values, which should be used with some caution, are shown in Fig. 5.11. Fluorine has been assigned the highest value (4.0). Otherwise as we go from right to left across a period (decreasing nuclear charge), there is a general decrease. The elements at the far left have low values. The elements at the right have high values. For the group VII elements, which are assigned the values F, 4.0; Cl, 3.0; Br, 2.8; and I, 2.5, the decreasing order is regular, unlike the order of electron affinities. In general, electronegativity decreases as we go down a periodic group (size increases).

Of what use are these values of electronegativity? For one thing, they can be used in predicting which bonds will be predominantly ionic and which will

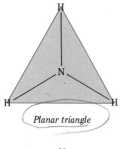

Planar triangle

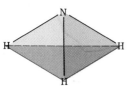

Triangular pyramid

FIG. 5.10 Two possible arrangements of atoms in NH₃ molecule.

FIG. 5.11 Pauling Scale of Electronegativities for the Various Elements

1 H 2.1																	2 He —
3 Li 1.0	4 Be 1.5											5 B 2.0	6 C 2.5	7 N 3.0	8 O 3.5	9 F 4.0	10 Ne —
11 Na 0.9	12 Mg 1.2											13 Al 1.5	14 Si 1.8	15 P 2.1	16 S 2.5	17 Cl 3.0	18 Ar —
19 K 0.8	20 Ca 1.0	21 Sc 1.3	22 Ti 1.5	23 V 1.6	24 Cr 1.6	25 Mn 1.5	26 Fe 1.8	27 Co 1.8	28 Ni 1.8	29 Cu 1.9	30 Zn 1.6	31 Ga 1.6	32 Ge 1.8	33 As 2.0	34 Se 2.4	35 Br 2.8	36 Kr 2.9
37 Rb 0.8	38 Sr 1.0	39 Y 1.2	40 Zr 1.4	41 Nb 1.6	42 Mo 1.8	43 Tc 1.9	44 Ru 2.2	45 Rh 2.2	46 Pd 2.2	47 Ag 1.9	48 Cd 1.7	49 In 1.7	50 Sn 1.8	51 Sb 1.9	52 Te 2.1	53 I 2.5	54 Xe 2.6
55 Cs 0.7	56 Ba 0.9	57–71 * 1.1–1.2	72 Hf 1.3	73 Ta 1.5	74 W 1.7	75 Re 1.9	76 Os 2.2	77 Ir 2.2	78 Pt 2.2	79 Au 2.4	80 Hg 1.9	81 Tl 1.8	82 Pb 1.8	83 Bi 1.9	84 Po 2.0	85 At 2.2	86 Rn —
87 Fr 0.7	88 Ra 0.9	89–103 † 1.1–1.3	104 Ku —	105 Ha —													

be predominantly covalent. Since electronegativity indicates relative attraction for shared electrons, two elements of very different electronegativity, e.g., Na (0.9) and Cl (3.0), are expected to form ionic bonds. Thus, electronegativities support the expectation that group I and II elements form essentially ionic bonds with elements of groups VI and VII. Two elements of about equal electronegativity, on the other hand, such as C (2.5) and H (2.1), are expected to form covalent bonds. There is no sharp crossover point from ionic to covalent bonding, but generally it takes an electronegativity difference of about 1.9 to give more than 50 percent ionic character to a bond.

Furthermore, electronegativities can be used to predict the degree of polarity of covalent bonds. The further apart in electronegativity two elements are, the more polar the bond should be. Thus, the bond between H (2.1) and N (3.0) is more polar than that between H (2.1) and C (2.5). In both cases the H end should be positive, since H has the lower electronegativity.

5.6
BOND ENERGIES AND THE SCALE OF ELECTRONEGATIVITY

One method for setting up the scale of electronegativities involves the use of bond energies. *Bond energy* is defined as the *energy required to break a bond so as to form neutral atoms.* It can be determined experimentally by measuring the heat involved in the decomposition reaction or by measuring spectroscopically the energy difference between the molecule in its lowest vibrational state and in the completely dissociated state. The relation between bond energy and electronegativity can be seen from the following example: It is found that 431 kJ of heat is required to break the Avogadro number of H_2 molecules into individual atoms. Thus, the bond energy of H_2 is 431 kJ per Avogadro number of bonds, or 7.16×10^{-22} kJ per bond. Because the sharing of the electron pair is equal between the two H atoms, it would be reasonable to assume that each bonded atom contributes half the bond energy, or 3.58×10^{-22} kJ. Similarly, from the bond energy found for Cl_2, 239 kJ per Avogadro number of bonds, we deduce that a Cl atom should contribute 1.99×10^{-22} kJ to any bond in which the sharing of an electron pair is equal.

Suppose we now consider the bond in HCl. This bond is polar, but for the moment let us imagine that it is nonpolar; i.e., that the electron pair is shared equally. This amounts to picturing H in HCl to be the same as in H_2, and Cl to be the same as in Cl_2. If H contributes 3.58×10^{-22} kJ and if Cl contributes 1.99×10^{-22} kJ, the expected bond energy of HCl should be the sum of these contributions, or 5.57×10^{-22} kJ. Actually, the bond energy of HCl found by experiment is 427 kJ per Avogadro number of bonds, or 7.09×10^{-22} kJ. This is significantly greater than the calculated value, which means that HCl is more stable than our model predicts.

The enhanced stability of HCl can be attributed to unequal sharing of the electron pair. If the electron pair spent more time on the Cl, that end of the molecule would become negative, and the H end positive. Since the positive and negative ends would attract each other, there would be additional binding energy. The amount of additional binding energy would depend on the relative electron-pulling ability of the bonded atoms, since the greater the charge difference between the ends of the molecule, the greater the additional binding energy. Thus, it should be possible to estimate relative electronegativities from the difference between experimental bond energies and those calculated by assuming equal sharing.

In Fig. 5.12 experimental values of bond energies of the hydrogen halides (HX) are compared with values calculated by assuming equal sharing of electrons. The discrepancy is greatest in HF and least in HI. This implies that the sharing of electrons between H and F is more unequal than the sharing between H and I. We would say that HF is more ionic than HI. The fact that for HI there is practically no difference between observed bond energy

FIG. 5.12 Bond Energies

| Bond | Energy, kJ per Avogadro number of bonds | | | |
	X = F	X = Cl	X = Br	X = I
H—H	431	431	431	431
X—X	151	239	190	149
H—X (calculated)	293	336	311	290
H—X (observed)	565	427	359	295
Discrepancy	272	91	48	5

and that calculated under the assumption of equal sharing suggests that the electron-pulling ability, or electronegativity, of H and I are nearly the same. Specific numerical values of electronegativity have been selected by a complex procedure to account for differences such as those listed in Fig. 5.12. The value assigned to H is 2.1. The values assigned to F (4.0), Cl (3.0), Br (2.8), and I (2.5) are consistent with the trend toward equal sharing in the sequence HF, HCl, HBr, HI.

Support for the assignment of electronegativity values comes from measurements of dipole moments. For the hydrogen halides the observed dipole moments in debye units are HF, 1.94; HCl, 1.08; HBr, 0.78; and HI, 0.38. The decreasing polarity indicates a trend toward equal sharing of electrons, consistent with decreasing electronegativity from F to I.

5.7
SATURATION OF VALENCE

In Chap. 2 we considered methods for determining formulas of compounds from experimental data. When such methods are applied to most substances, it is found that there is a limit to the combining ability of one kind of atom for another. For example, when magnesium is combined with chlorine, no more than two chlorine atoms per magnesium are found; when carbon is combined with hydrogen, no more than four hydrogen atoms per carbon are found. If we use the term *valence* to describe the *ability of atoms to bind together,* then we can summarize the above observations by saying that there is a *limit to the valence one atom shows for others;* i.e., there is a *saturation of valence.*

Consider the compounds that chlorine forms with sodium, magnesium, and aluminum. All three of these compounds are considered to be mostly ionic, because chlorine is considerably more electronegative than sodium, magnesium, or aluminum. In the combination of chlorine with sodium, there is but one chlorine atom per sodium atom. This can be accounted for by noting that each sodium atom (Na·) has one valence electron, and each chlorine atom $\left(: \overset{..}{\underset{..}{Cl}} \cdot \right)$ has seven. If a chlorine atom takes one electron away from a sodium atom, the resulting chloride ion $\left(: \overset{..}{\underset{..}{Cl}} :^- \right)$ has a complete octet of electrons. The sodium ion (Na$^+$) has no more valence electrons that it can easily lose to other chlorine atoms. Therefore, in NaCl, only one chlorine atom is combined per sodium atom.

In the combination of chlorine with magnesium, there are two chlorine atoms per magnesium. Two valence electrons of the magnesium atom (Mg :) can be lost to two chlorine atoms. However, no more than two electrons are lost, because the magnesium ion (Mg^{2+}) has a stable octet of inner electrons. When chlorine is combined with aluminum, no more than three chlorine atoms react per aluminum atom, because the aluminum atom $\left(\cdot \overset{.}{Al} \cdot \right)$ has but three valence electrons.

When covalent bonds are formed, not only is the relative number of atoms fixed, but also the actual number of atoms in the molecule may be limited. For example, in the combination of carbon with hydrogen, there are no more than four hydrogen atoms per carbon atom. Furthermore, the compound formed (methane) consists of discrete molecules, each of which has one carbon atom and four hydrogen atoms. How can we account for this saturation of valence? Each carbon atom $\left(\cdot \overset{\cdot}{\underset{\cdot}{C}} \cdot \right)$ has four valence electrons; each hydrogen $(H \cdot)$ has one. Since the electronegativities of carbon and hydrogen are similar, covalent rather than ionic bonds are expected. If a carbon atom contributes one electron to each covalent bond formed, four such bonds can be established. We can represent the formation of the compound as follows:

$$4H \cdot + \cdot \overset{\cdot}{\underset{\cdot}{C}} \cdot \longrightarrow H \overset{\overset{\displaystyle H}{\cdot\cdot}}{\underset{\underset{\displaystyle H}{\cdot\cdot}}{\vdots C \vdots}} H$$

By sharing electrons, the carbon atom gets a complete octet of electrons in its outer shell, and each hydrogen gets two electrons, all that its valence shell can accommodate. Since the molecule as a whole is electrically neutral and since all valence shells are filled with shared electrons, no other atoms can bind to the molecule. The valence is saturated.

Other examples of valence saturation in covalent bonds are provided by hydrogen compounds of fluorine, oxygen, and nitrogen. These lead, respectively, to

$$H \colon \overset{\cdot\cdot}{\underset{\cdot\cdot}{F}} \colon \qquad H \colon \overset{\cdot\cdot}{\underset{\cdot\cdot}{O}} \colon \qquad H \colon \overset{\cdot\cdot}{N} \colon H$$

$$\qquad\qquad\qquad H \qquad\quad H$$

Hydrogen fluoride   Water   Ammonia

In each case, the number of hydrogen atoms bonded is equal to the number of electrons required to complete the octet. It should be noted that in these compounds, but not in methane, there are pairs of valence electrons which do not appear to be shared between atoms. We might imagine that these electron pairs could be used to bind to other atoms, but this can occur only if the additional atom has room for two more electrons in its valence shell. The unshared electron pairs cannot bind additional hydrogen atoms $(H \cdot)$, because three electrons cannot normally be accommodated by a single H atom.

Saturation of valence is not restricted to bonding between unlike atoms. It also may occur when an atom forms covalent bonds with atoms of its own kind, as in the following examples:

$$: \overset{\cdot\cdot}{\underset{\cdot\cdot}{F}} : \overset{\cdot\cdot}{\underset{\cdot\cdot}{F}} : \qquad : \overset{\cdot\cdot}{\underset{\cdot\cdot}{O}} : \overset{\cdot\cdot}{\underset{\cdot\cdot}{O}} : \qquad H : \overset{\cdot\cdot}{N} : \overset{\cdot\cdot}{N} : H \qquad H : \overset{\overset{\displaystyle H}{\cdot\cdot}}{C} : \overset{\overset{\displaystyle H}{\cdot\cdot}}{C} : \overset{\overset{\displaystyle H}{\cdot\cdot}}{C} : H$$

$$\qquad\qquad\qquad H \quad H \qquad\qquad H \quad H \qquad\qquad H \quad H \quad H$$

Fluorine   Hydrogen peroxide   Hydrazine   Propane

In order to get complete octets, the elements fluorine, oxygen, nitrogen, and carbon must form one, two, three, and four electron-pair bonds, respectively. The number of bonds formed is the same, whether the atom is bonded only to hydrogen or to other atoms of its own kind.

In addition to "saturation of valence," the terms "saturated" and "unsaturated" are also used when describing compounds. In unsaturated compounds of carbon there are double or triple bonds between adjacent carbon atoms. These compounds are called unsaturated because they can undergo chemical reaction in which atoms are added to the molecule. For example,

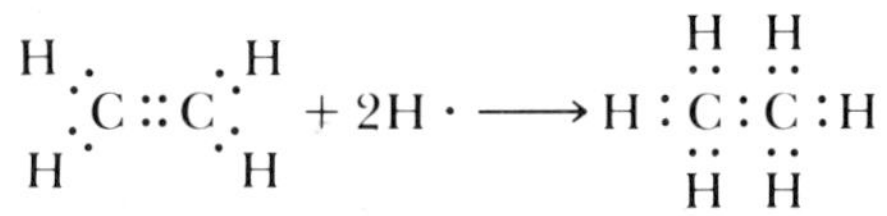

$$H_2C{=}CH_2 + 2H \longrightarrow H:C:C:H$$

The compound on the left, ethylene, is unsaturated; each carbon atom has a complete octet of electrons, made possible by sharing two pairs of electrons with the other carbon atom. The compound on the right, ethane, is saturated; the carbon atoms are joined by sharing one pair of electrons. If each electron pair is designated by a dash, the above equation is written:

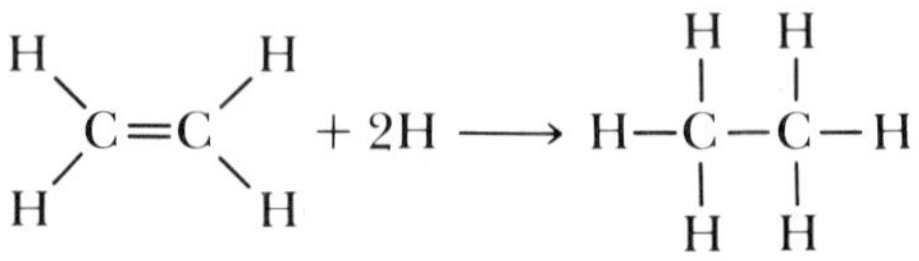

To save space, it can also be written on a single line: $H_2C{=}CH_2 + 2H \longrightarrow CH_3CH_3$. Another unsaturated compound, acetylene, consists of molecules containing two carbon atoms and two hydrogen atoms. These atoms are arranged in a straight line with the carbons in the center. To write the electronic formula, we note that each carbon makes available four valence electrons and each hydrogen one valence electron, giving a total of 10 bonding electrons. Bonds between each hydrogen and its carbon take care of four electrons. There are six electrons left. To satisfy the octet rule, these six occur as three pairs of electrons shared between the two carbon atoms. This is a triple bond; it can be written as

$$H:C:::C:H \quad \text{or} \quad H{-}C{\equiv}C{-}H$$

5.8
RESONANCE

From the two preceding sections it should be evident that there is generally no simple way to describe the electron distribution in a molecule or in a bond in order to describe completely all its properties. Thus, we are led to qualifying descriptions such as "the bond in HCl is *polar* covalent," not just "covalent." Alternatively, we can even attempt quantitative descriptions, such as "the bond in HCl has 17 percent ionic-bond character." What this means is that no single picture of a molecule will be adequate, but it will best be represented as a composite of several pictures. Such a problem is encountered in more obvious terms in the case of a molecule such as sulfur dioxide, SO_2. This molecule has a high dipole moment; hence, we conclude that it is nonlinear, with the atoms arranged in a bent chain. Sulfur has six outer-shell electrons, and oxygen also has six. There are thus a total of 18. These can be disposed of in several ways:

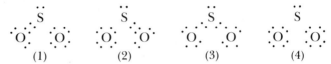

Neither formula (1) nor formula (2) is consistent with experimental fact because each formula indicates that the sulfur dioxide molecule should have one double bond (a short one) and one single bond (a longer one). Experiments show that the two bonds in SO_2 are exactly the same length. Formula (3) is excluded because it contains unpaired electrons. Molecules containing unpaired electrons are paramagnetic; sulfur dioxide is not. Formula (4) is traditionally excluded because it does not comply with the octet rule.

A situation in which *no single electronic formula conforms both to observed properties and to the octet rule* is described as *resonance*. The SO_2 molecule can be described as a combination of formulas (1) and (2), in which the actual electron distribution in the molecule is said to be a *resonance hybrid* of the contributing formulas. The choice of the word "resonance" for this situation is unfortunate because it encourages people to think that the molecule resonates from one structure to the other or that the extra electron pair jumps back and forth from one bond to the other. *Such is not the case.* The molecule has only one real electron structure. The problem is in describing it. The properties of a resonance hybrid do not oscillate from those of one contributing resonance structure to those of the other. The properties are fixed and are those of the actual hybrid structure.

Another common example of resonance is benzene, C_6H_6. This molecule is composed of six carbon atoms arranged as a hexagon with a hydrogen atom attached to each. What should we write for its electronic formula? Six carbon atoms would give 6×4, or 24, valence electrons; six more from the hydrogen atoms give a total of 30 electrons. Twelve of the 30 will be needed as six pairs to bind the six hydrogens to the six carbons. This leaves 18 electrons for nine bonds between carbon atoms to hold the hexagon together. Two electron arrangements would appear possible:

H
|
H   C   H
 C═C
| ‖
 C═C
H C H
|
H
(1)

or

H
|
H   C   H
 C═C
‖ |
 C═C
H C H
|
H
(2)

Although each of these formulations satisfies the octet rule, neither *by itself* is a satisfactory representation of the benzene molecule.   Each representation would be faulty in that it would show alternating double and single bonds. Double bonds between carbon atoms, as in ethylene, $H_2C = CH_2$, are 0.133 nm in length; single bonds, as in ethane, $CH_3 — CH_3$, are 0.154 nm.   Thus, formulation (1) or (2) would be expected to show two characteristic carbon-to-carbon distances, 0.133 and 0.154 nm.   Yet the experimental fact is that in benzene all the carbon-to-carbon bonds are alike, 0.140 nm.   The way out of the dilemma is to say that the electronic distribution in benzene is a resonance hybrid of *both* formulations (1) and (2).   For convenience, the symbol for ben-

zene is often given as ⬡ , which is to be thought of as simultaneously rep-

resenting ⬡ and ⬡ , equivalent shorthand designations of (1) and

(2) above.

Resonance represents an attempt to patch up the valence-bond description of certain molecules.   The difficulty lies in the description and not in the molecule itself.   In the molecular-orbital description the problem does not arise.   All the electrons belong to the molecule as a whole.

5.9
SHAPES OF MOLECULES AND HYBRID ORBITALS

Molecules that contain but two atoms are necessarily linear, but those containing three or more atoms usually have interesting shapes.   For example, the hydrogen compounds of period 2 elements show a variety of shapes ranging from the linear for HF to bent for H_2O to tetrahedral for CH_4.   Why do molecules have the shapes they do?   How can we rationalize the shapes observed? How can we predict what shapes might be expected?

Let us start with a relatively simple problem.   Why is the water molecule nonlinear?   To answer this question, we must consider (1) the nature of the orbitals involved in bonding the hydrogen to the oxygen and (2) the spatial distribution of the electron charge clouds about each of the nuclei.   Imagine assembling the molecule H_2O from two H atoms and one O atom.   Each H atom has initially a $1s$ electron, which is spherically symmetric about the H nucleus.   The O atom has, in its outer shell, two $2s$ electrons (spherically symmetric) and four $2p$ electrons.   Recalling the three p-type orbitals shown in Fig. 4.21, we find two electrons in one of the p orbitals and one electron in each of the other two.   The electron distribution in the O atom thus looks as follows:

$$\underset{1s}{\uparrow\downarrow} \quad \underset{2s}{\uparrow\downarrow} \quad \underset{2p}{\uparrow\downarrow \; \uparrow \; \uparrow}$$

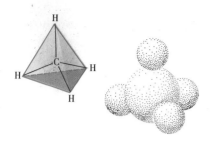

FIG. 5.13 Tetrahedral CH_4 molecule.

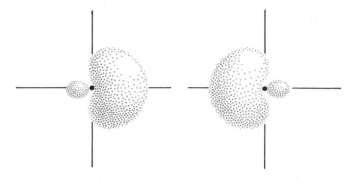

FIG. 5.14 *sp* hybrid orbitals. (The black dots represent a given nucleus. The left half of the figure is to be superimposed on the right half of the figure so that the black dots coincide into one.)

In the valence-bond description the O-to-H bond arises from sharing the $1s$ electron of H with one of the unpaired $2p$ electrons of O. Such sharing favors bonding along the direction of the $2p$ orbital used. To tie on two H atoms requires use of two $2p$ orbitals, which are at right angles to each other. Thus, in this simple picture we expect the two O — H bonds in H_2O to be perpendicular to each other. Actually, they form the somewhat greater angle of 104.52°. We shall try to explain this discrepancy later.

Methane, CH_4, has a tetrahedral shape, as shown in Fig. 5.13, with carbon at the center of a tetrahedron and four hydrogens at the corners. The angles between the C — H bonds are 109.47°. If we are to use p orbitals, why are the bond angles not right angles? Furthermore, why are there four equivalent C — H bonds formed, whereas we have but three p orbitals? To form four bonds to the central carbon atom, we need to use four orbitals of the central carbon, not just the three $2p$ orbitals, but the $2s$ orbital as well.

The reason we get into the above problems is that the concept of independent electron orbitals is an oversimplification when there is more than one electron in an atom. It is all right to talk about independent $2s$ and $2p$ orbitals in a hydrogen atom, where there is but one electron to worry about, but such is not the case when both kinds of electrons are present simultaneously. The presence of the $2s$ electron perturbs the motion of the $2p$ electron, and vice versa. Specifically, presence of a $2s$ electron makes the $2p$ electron take on some s-like character—i.e., the $2p$ orbital becomes more spherical, less elongated. Similarly, the presence of a $2p$ electron makes the $2s$ electron take on some p-like character—i.e., the $2s$ orbital becomes less spherical, more elongated. The net result is that the original hydrogen-like $2s$ and $2p$ orbitals have to be replaced by new *orbitals that contain the combined characteristics of the original orbitals*. These new orbitals are called *hybrid orbitals*.

The process of combining orbitals on a given atom is called *hybridization*. As a general rule, the number of hybrid orbitals resulting from a hybridization is equal to the number of orbitals that are being mixed together. As an example, if we mix an s orbital with a p orbital, we get as a result two hybrid orbitals, each designated as an sp orbital. These are shown in Fig. 5.14. The black dots represent the nucleus of a given atom. To appreciate the relative arrangement of the two *sp* hybrids, one must mentally superimpose the left half of the figure over the right half of the figure so that the two black dots coincide. As can be seen, one of the *sp* hybrids corresponds to having most of the electron density on one side of the atom. The other *sp* hybrid corresponds to having most of the electron density on the other side of the atom. Hence, two *sp* hybrids on an atom would be ideal for binding to two other atoms along a 180° line from the central nucleus.

Mixing one *s* orbital with two *p* orbitals on the same atom gives what are called *sp²* hybrids. (In designating hybrid orbitals, superscripts tell the number of orbitals of each kind going into the hybridization. This should not be confused with the number of electrons in an orbital, as when one gives electron configuration.) There are three hybrids of the *sp²* type. They are all in the same plane and are mainly directed to the corners of an equilateral triangle, as shown in Fig. 5.15. Each *sp²* hybrid consists of a big blob (called a *lobe*) directed toward a corner of the triangle and a small lobe directed in the opposite direction. The existence of the small lobes is usually neglected. Thus, the set of three *sp²* hybrids can be visualized as three lobes (or regions of high electron density) extending outward from the nucleus at 120° angles.

The most common situation for us will be the one where we have one *s* orbital and three *p* orbitals simultaneously involved in the bonding. In such a case we have what are called *sp³* hybrids. There are four of them, as shown in Fig. 5.16, each directed toward a corner of a tetrahedron. (A simple way to derive a tetrahedron is to start with a cube and chop off alternate corners so that only four of the eight corners are left.) The formation of a CH_4 molecule can be pictured as follows:

1   Start with a ground-state carbon atom, $1s^2 2s^2 2p^2$.

2   Promote one of the $2s$ electrons to the empty $2p$ orbital so that we have an excited carbon atom, $1s^2 2s^1 2p^3$.

3   Ignoring the inner shell ($1s^2$), mix up the one $2s$ and three $2p$ orbitals so that we now have four *sp³* hybrids.

4   Put one of the carbon's electrons in each of the *sp³* hybrids so that we have four electrons pointing in tetrahedral directions.

5   Bring up a hydrogen atom ($1s$) to each of the *sp³* hybrids and pair the electron in the $1s$ orbital ($\downarrow$) with an electron in the *sp³* hybrid ($\uparrow$).

Schematically, this corresponds to replacing

Carbon (ground state)

by

Carbon (excited, hybridized)

and then pairing with hydrogen ($1s$) electrons

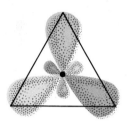

FIG. 5.15 Three *sp²* hybrid orbitals. (The black dot represents the nucleus. Each kind of shading shows the region that belongs to one of the *sp²* hybrids.)

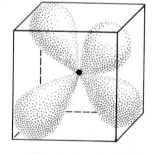

FIG. 5.16 *sp³* hybrid orbitals. (The black dot represents the nucleus. The small lobes, such as those shown in Figs. 5.14 and 5.15, have been omitted.)

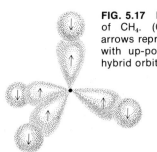

FIG. 5.17 Electron pairing in formation of CH_4. (Circles with down-pointing arrows represent $1s$ orbitals of H; lobes with up-pointing arrows represent sp^3 hybrid orbitals of C.)

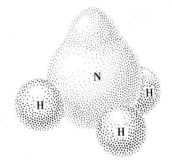

FIG. 5.18 NH_3 molecule. (The apex position is occupied by an unshared pair of electrons.)

Here, the Roman numerals indicate the four tetrahedral directions. Figure 5.17 is another way of representing the same thing.

The use of hybrid orbitals can account for the observed shapes of molecules other than methane, even when there are not four attached atoms. For example, the NH_3 molecule can be imagined as being built from a nitrogen atom with the five outer-shell electrons $2s^2 2p^3$ distributed among four equivalent tetrahedral orbitals. Two of the electrons are paired in one sp^3 orbital, and the other three electrons are singly placed in the other three sp^3 hybrids.

$$\underline{\uparrow\downarrow \quad \uparrow \quad \uparrow \quad \uparrow}$$
$$sp^3 \text{ hybrids}$$

The three unpaired electrons, one in each of the sp^3 hybrids, are exchange-shared with s electrons of the hydrogen atoms. The result, as shown in Fig. 5.18, is a pyramidal molecule in which the three hydrogens form a base, and the *lone pair* of electrons the apex. The observed angles between N — H bonds in NH_3 are 108°, not very far from what is expected for a regular tetrahedron (109.47°).

Tetrahedral orbitals can also help explain the observed bond angle in H_2O. Following the reasoning of the preceding paragraph, we expect the O atom to offer $\underline{\uparrow\downarrow \; \uparrow\downarrow \; \uparrow \; \uparrow}$ for bonding with H. Two H's can share the unpaired electrons, and the two lone pairs will be left to occupy the other two tetrahedral directions. The expected bond angle for H — O — H would be 109.47°. This is a lot closer to the observed bond angle in H_2O, 104.52°, than the 90° we predicted at the beginning of this section. Figure 5.19 attempts to show that the two lone pairs of electrons and the two bound hydrogens in water are directed approximately toward the corners of a tetrahedron.

The sp^3-hybrid orbitals are useful not only for describing simple molecules such as CH_4 but also for describing more complicated molecules such as C_2H_6, C_3H_8, and C_4H_{10}. In these chainlike hydrocarbons, each C atom can be regarded as having four sp^3 hybrids directed toward the corners of a tetrahedron. Consequently, the preferred bonding directions are toward the corners of a tetrahedron. Electron sharing with a $1s$ orbital of an H atom forms a C—H bond in one direction, whereas electron sharing with an sp^3 hybrid of another C atom forms a C — C bond in another direction. Struc-

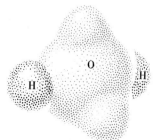

FIG. 5.19 H_2O molecule. (Two of the tetrahedral directions are occupied by lone pairs of electrons.)

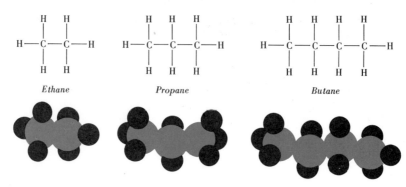

FIG. 5.20 Shapes of chain hydrocarbons.

tures that may result are shown in Fig. 5.20. The chains of C — C bonds can be extended almost indefinitely. The giant molecules that result, called *macromolecules,* are important in some commonly encountered materials (e.g., polyethylene).

How can we account for the shape of the ethylene molecule? This molecule, which can be written $H_2C = CH_2$, is planar. All six atoms lie in the same plane, as shown in the bottom part of Fig. 5.21. Putting together of the molecule can be visualized, as shown in the top part of the figure, in the following way: We mix one s orbital and two p orbitals (let us say p_x and p_y) of a carbon atom to produce three sp^2-hybrid orbitals, all lying in the same plane at angles of 120° to each other. The remaining p orbital (which would be p_z) is left unhybridized; it sits perpendicular to the xy plane. Each carbon atom, when ready for bonding, can be represented as follows:

$$\underline{\uparrow \quad \uparrow \quad \uparrow} \qquad \underline{\uparrow}$$
$$sp^2 \text{ hybrids} \qquad \text{Unhybridized } p$$

Two of the sp^2 hybrids can be used to bind hydrogen atoms; the other sp^2 hybrid can be used to join carbon to carbon. When the two CH_2 fragments come together, as shown schematically in Fig. 5.21, there results a side-to-side merging of the unhybridized p_z orbitals. Electron exchange can occur between the unhybridized p_z orbital of one carbon and the unhybridized p_z orbital of the other carbon. This produces another bond, unusual in shape since it results from side-to-side pairing of p orbitals rather than end-to-end pairing. Thus, the two carbon atoms in ethylene are held together by two bonds. One is concentrated directly between the two nuclei, and the other is composed of a double streamer lying to both sides of the internuclear line. A bond of the first type is called a σ *(sigma) bond;* a bond of the second type, a π *(pi) bond.* The π bond, like the σ bond, is only one bond, in that it involves but a single pair of electrons shared between two atoms. The combination of one σ bond and one π bond constitutes what is called a *double bond.*

The acetylene molecule, which has the linear structure $H — C \equiv C — H$, can be described by a combination of sp hybridization and π bonding. For each carbon, we imagine formation of two sp-hybrid orbitals from a carbon $2s$ and one of its $2p$ orbitals (let us call it p_x). One of the sp-hybrid orbitals is used to bind on a hydrogen atom by means of a σ bond; the other sp orbital forms a

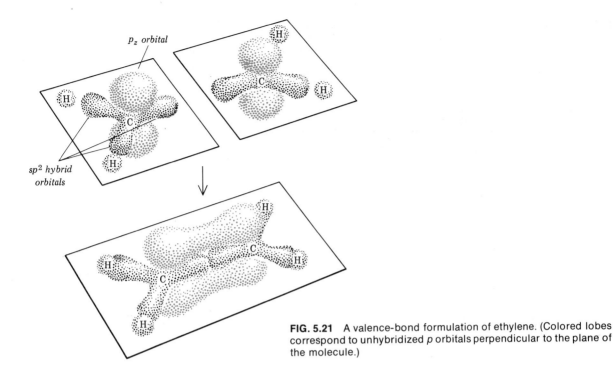

FIG. 5.21 A valence-bond formulation of ethylene. (Colored lobes correspond to unhybridized p orbitals perpendicular to the plane of the molecule.)

σ bond to the other carbon. As shown in Fig. 5.22, the unhybridized p orbitals (p_y and p_z) of the carbon are perpendicular to each other and to the line through the two carbon atoms. Electron exchange between the p_y orbital of one carbon and the p_y orbital of the other carbon gives one π bond; electron exchange between the p_z of one carbon and the p_z of the other carbon gives

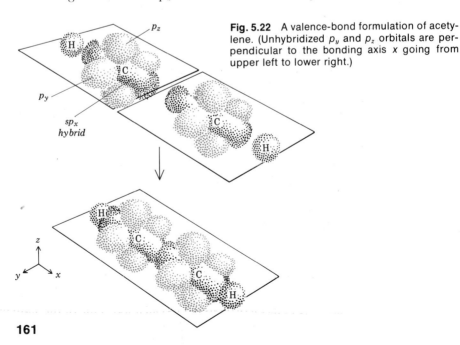

Fig. 5.22 A valence-bond formulation of acetylene. (Unhybridized p_y and p_z orbitals are perpendicular to the bonding axis x going from upper left to lower right.)

FIG. 5.23　Hybrid Orbitals

Designation	Typical combination	Bond angle, deg	Geometry
sp	$s + p_x$	180	Linear
sp^2	$s + p_x + p_y$	120	Triangular
sp^3	$s + p_x + p_y + p_z$	109.47	Tetrahedral
dsp^2	$d_{x^2-y^2} + s + p_x + p_y$	90	Square-planar
d^2sp^3	$d_{x^2-y^2} + d_{z^2} + s + p_x + p_y + p_z$	90	Octahedral

another π bond. Thus, the carbon atoms in acetylene are held together by a triple bond composed of one σ bond and two π bonds.

In the examples discussed above, hybrid orbitals that consist of various combinations of s and p orbitals are utilized. The geometry of these hybrids is summarized in Fig. 5.23. The hybrids sp, sp^2, and sp^3 are most useful for describing the bonding of the second- and third-period elements. For elements where d electrons come into the picture, other hybrids are possible. Two commonly used sets are dsp^2 (read "d-s-p-squared") and d^2sp^3 (read "d-squared-s-p-cubed"). The dsp^2 hybrids are four orbitals directed toward the corners of a square. They are best visualized as being concentrated along the x and y axes. An example of square-planar geometry is afforded by $PtCl_4^{2-}$ in K_2PtCl_4, where the four chlorine atoms are arranged in a square around the platinum. The d^2sp^3-hybrid set consists of six equivalent orbitals directed toward the corners of an octahedron. The set is constructed from the two d orbitals that are directed along the axes (d_{z^2} and $d_{x^2-y^2}$) and a full set of s, p_x, p_y, and p_z orbitals. The molecule SF_6, which has an octahedral structure, can be described as having six σ bonds about sulfur, each arising from electron exchange between a d^2sp^3-hybrid orbital on the sulfur and a p orbital on the fluorine. The use of d orbitals is discussed further in Chap. 18.

5.10
VALENCE-SHELL–ELECTRON-PAIR REPULSION

The above discussion does not really tell *why* molecules have the shapes they do; it simply describes how the language of hybrid orbitals can be used to describe each shape. For further insight as to why particular shapes result, we need to look further at the problem of electron repulsions. We can be remarkably successful in predicting molecular geometry by using a simple qualitative approach in which the shape is essentially dictated by the net effect of all the repulsions between nonbonding pairs and bonding pairs in the valence shell of an atom. This approach is called the valence-shell–electron-pair repulsion theory. It is often designated VSEPR.

The VSEPR theory says that the arrangement of bonds around an atom is determined by the number of electron pairs around an atom and the size and shape of the orbitals in which these electrons are housed. The assumption is that these electron pairs will arrange themselves so that the least repulsion occurs between them. In other words, the electron pairs tend to get as far apart as possible. For example, if there are two electron pairs, the preferred arrangement will be to have one pair on each side of the atom. This will lead to a linear bonding pattern, as in $H - Be - H$. If there are three pairs of electrons in the valence shell of an atom, they will arrange themselves as an equilateral triangle. Presumably, this is what happens in BCl_3, which has a triangular shape (B at the center, Cl at each corner). For four pairs of electrons, minimum repulsion results and the pairs are furthest from each other when the arrangement of pairs is tetrahedral, as it is in CH_4. Figure 5.24 summarizes the characteristic arrangements.

Regular shapes will result only if all the pairs of electrons are used to bond to identical atoms. If one or more of the bonded atoms are different, distorted arrangements may occur. Thus, for example, in chloromethane (CH_3Cl), where one of the H atoms of methane has been replaced by Cl, the bond angles are no longer exactly equal to 109.5°, as they would be in a perfect tetrahedron. The bond angle $H - C - H$ is 110.9°, and the bond angle $H - C - Cl$ is 108.0°.

If an electron pair in the valence shell of an atom is not used to bond an atom but resides in the valence shell as a "lone pair," then it will occupy one position of one of the arrangements shown in Fig. 5.24. Thus, for the case of ammonia, where we have four pairs of electrons (one nonbonding and three bonding), the nonbonding pair occupies one of the tetrahedral positions, forcing the three bonding pairs all to be on one side of the nitrogen. As a result, the NH_3 molecule is not flat but is shaped like a pyramid. Similarly, the two nonbonding pairs in H_2O occupy two of the tetrahedral positions, leaving the other two for the bonding pairs.

FIG. 5.24   Predicted Arrangement of Electron Pairs Around an Atom

Number of pairs	Arrangement	Example
2	Linear	$H-Be-H$
3	Equilateral triangle	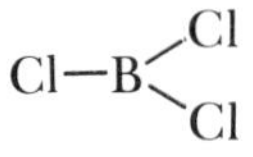 $Cl-B\begin{smallmatrix}\diagup Cl\\ \diagdown Cl\end{smallmatrix}$
4	Tetrahedron	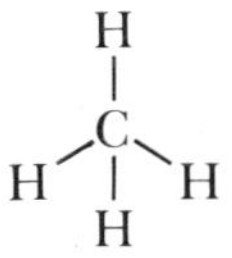 $H-\overset{\displaystyle H}{\underset{\displaystyle H}{C}}-H$
5	Trigonal bipyramid	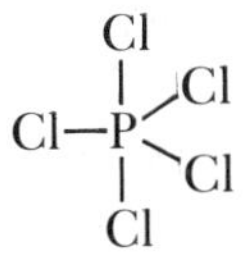 $Cl-\overset{\displaystyle Cl}{\underset{\displaystyle Cl}{P}}\begin{smallmatrix}\diagup Cl\\ \diagdown Cl\end{smallmatrix}$
6	Octahedron	 $\begin{smallmatrix}F\\ F\diagdown\mid\diagup F\\ S\\ F\diagup\mid\diagdown F\\ F\end{smallmatrix}$

**FIG. 5.25   Decrease of Bond Angle with
Increased Number of Lone Pairs**

Molecule	Number of lone pairs	Bond angle
CH_4	0	109.5°
NH_3	1	107.3°
OH_2	2	104.5°

Some of the finer details of molecular geometry can also be understood qualitatively by making additional assumptions:

1   A nonbonding pair of electrons is larger and takes up more room than does a bonding pair. As illustrated by the series CH_4, NH_3, OH_2, shown in Fig. 5.25, successive replacement of bonding pairs by lone pairs squeezes the remaining bonding pairs closer together and decreases the bond angle.

2   A bonding electron pair takes up less space as the electronegativity of the attached atom is increased. An illustration of this is the change in bond angle in going from NH_3 to NF_3. As shown in Fig. 5.26, the bond angle decreases from 107.3° in NH_3 to 102.1° in NF_3. The greater electronegativity of F compared with H makes the N — F bonding orbital smaller than the N — H. The large lone pair of electrons on N can thus push the smaller N — F bonds closer together.

A useful summary of VSEPR is that repulsion between electron pairs decreases in the following order:

Lone pair–lone pair > lone pair–bonding pair >

bonding pair–bonding pair

Thus, in NH_3 the bond angle is squeezed down to less than tetrahedral because the repulsion between the lone pair and bonding pair is greater than between the bonding pairs.

The shapes of molecules containing double or triple bonds can also be accounted for if one assumes that the double or triple bond occupies only one position in space on the surface of the atom. Furthermore, the effective size of the two electron pairs of a double bond (or of the three electron pairs of a triple bond) appears to be somewhat larger than the size of the one electron pair of a single bond. Thus, in ethylene, $H_2C = CH_2$, the bond angle H — C — H is 116°, a bit less than the 120° expected for three groups around a carbon.

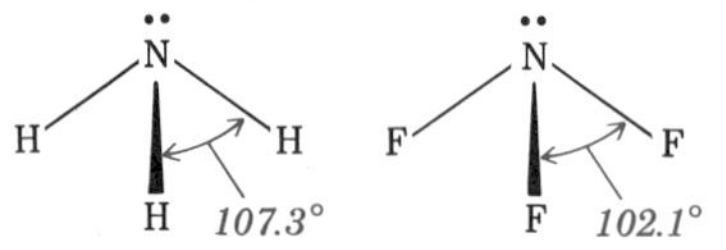

FIG. 5.26   Comparison of bond angles in NH_3 and NF_3. (The pair of dots represents a nonbonding pair of electrons. The wedge-shaped bond sticks out of the paper, with the blunt end toward the observer.)

Important Concepts

valence	triple bond	dipole	octet rule
valence electrons	σ bond	dipole moment	shapes of molecules
ionic bonds	π bond	electronegativity	hybrid orbitals
covalent bonds	donor-acceptor bonds	bond energy	VSEPR
single bond	polar vs. nonpolar	saturation of valence	
double bond	electron-dot formulas	resonance hybrid	

Exercises

*5.1 Valence-bond vs. molecular-orbital method In very broad terms sketch out the essential difference between the valence-bond and molecular-orbital descriptions of the H_2 molecule.

*5.2 Ionic vs. covalent bond What is the essential distinction between an ionic bond and a covalent bond?

**5.3 Ionic bond Using data from Figs. 4.32 and 4.34, show that the process $Cs + F \longrightarrow Cs^+ + F^-$ is energetically unfavorable. Compare with the process $Cs + F \longrightarrow [Cs^+]\,[F^-]$. Assume radii of 0.167 nm for Cs^+ and 0.133 nm for F^-. *Ans. -0.44 eV vs. $+4.4$ eV*

**5.4 Ionic bond What four requirements would you specify in order to make formation of the ionic bond M^+X^- most favorable? Which pair of elements of the periodic table come closest to your specifications?

***5.5 Ionic bond Using data from Figs. 4.32 and 4.34 and from Appendix 8, calculate the net energy liberated in the process

$$M + X \longrightarrow M^+X^-$$

where M is each of the alkali elements and X is each of the halogen elements. Display your results in tabular form. Which pair is most favorable for ionic-bond formation?

**5.6 Ionic bond The first ionization potential of barium is 5.212 eV, the second is 10.004 eV. The electron affinity of chlorine is 3.61 eV. Using 0.135 nm for the radius of Ba^{2+} and 0.181 nm for the radius of Cl^-, predict the energy change for the reaction

$$Ba + 2Cl \longrightarrow [Ba^{2+}]\,[Cl^-]_2$$

Ans. Releases 1.12 eV

**5.7 Ionic bond Try an ionic model for H_2. The ionization potential of H is 13.6 eV; the electron affinity of H is 0.76 eV. The radius of H^- in crystals is about 0.144 nm. Calculate the net energy change for the process $H + H \longrightarrow H^+[H^-]$. What is the main flaw in this calculation?

*5.8 Covalent bond Describe qualitatively in terms of electron distribution what happens in the setting up of the covalent bond in the Li_2 molecule.

*5.9 Covalent bond Explain why the curve in Fig. 5.3 rises so steeply at the left.

**5.10 Covalent bond Assuming the axes in Fig. 5.3 are linearly scaled, about how far would you have to stretch the H_2 molecule before the bond "breaks"?

*5.11 Electron-dot formulas Draw electron-dot formulas for the hydrogen compounds of the elements of the second period.

*5.12 Electron-dot formulas Draw electron-dot formulas for each of the following: Na_2O, MgO, PH_3, Ca_3N_2, ClF.

5.13 Electron-dot formulas Write electron-dot formulas for each of the following: H_2S, SO_2, SO_3, $(HO)_2SO$, $(HO)_2SO_2$.

5.14 Bond energies Going down the periodic table, the sequence F_2, Cl_2, Br_2, I_2 (bond energies 151, 239, 190, 149 kJ/mol) shows the first member out of line, whereas the sequence Li_2, Na_2, K_2, Rb_2, Cs_2 (105, 72, 49, 45, 44 kJ/mol) is perfectly regular. How might you rationalize the difference? Explain also why the alkali series might be expected to decrease more slowly.

5.15 Polar bonds If two bonded atoms are identical, why must the bond between them be nonpolar? Discuss your answer with respect to H_2N-NH_2 and $H_2N-NHCl$.

5.16 Polar bonds Suppose you have an H_2 molecule in which one atom is the light isotope ^1H and the other is the heavier isotope ^2H. Is the bond polar or nonpolar?

5.17 Dipole moment Given a fixed electronegativity for X, greater than that of H, which would you expect to have a higher dipole moment: $H:X$ or $H:\overset{..}{X}:$? Explain your answer.

5.18 Polar bonds Recognizing that Cl has a larger electron affinity than F but a lower electronegativity than F, which end of the ClF molecule would you expect to be more positive? Explain your answer.

5.19 Molecular polarity Which of the following molecules would you expect to be polar? $H_2C=CH_2$, $H_2C=CHCl$, $ClHC=CHCl$ (two forms), $H_2C=CCl_2$. Justify your answer using structural formulas.

5.20 Molecular polarity How would you expect the dipole moment of NF_3 to compare with that of NH_3? Tell how each of the following contributes to your answer: (a) electronegativity of F exceeds that of H; (b) bond angle F—N—F is less than H—N—H; (c) N—F bond length is greater than N—H.

5.21 Dipole moment The observed dipole moment of H_2O is 1.85 debyes. Given that the bond angle is 104.52°, what dipole moment would you assign to each H—O bond?

5.22 Dipole moment Assuming 100 percent ionic character, predict the value of the dipole moment for NaCl. The radii of Na^+ and Cl^- are 0.102 and 0.181 nm, respectively.

Ans. 13.6 debyes

5.23 Dipole moment The observed net dipole moment for the NH_3 molecule is 1.47 debyes. The H—N—H bond angle is 107.3°. Calculate the apparent dipole moment of each N—H bond.

Ans. 1.33 debyes

5.24 Electronegativity What is the essential difference between electron affinity and electronegativity? Tell what each is.

5.25 Electronegativity Elements X and Y have ionization potentials of 5 and 10 eV, respectively, and electron affinities of 5 and 3 eV, respectively. Which element would you expect to be the more electronegative?

5.26 Electronegativity If the bond energies are 50, 70, and 90 kJ/mol for X_2, Y_2, and Z_2, respectively, and the observed bond energies for HX, HY, and HZ are 250, 270, and 290 kJ/mol, respectively, what might you conclude as to the electronegativity of X, Y, and Z relative to hydrogen?

5.27 Saturation of valence Show how the octet rule accounts for each of the following compounds: Li_2O, MgO, Al_2O_3.

5.28 Saturation of valence By writing electronic formulas for each of the following, show how saturation of valence is attained:
a NH_3
b Mg_3N_2
c N_2
d HNO_3 [N in center, 2(O) and 1(OH) attached to it]

5.29 Resonance The three species CO_3^{2-}, NO_3^-, and SO_3 have identical planar structures in which the three oxygen atoms are at the corners of an equilateral triangle centered on the other atom. Write electronic formulas for the resonance forms appropriate to these species.

5.30 Molecular shape Would you expect the molecule H_2S to be linear or nonlinear? Explain. How would you expect the bond angle H—X—H in H_2S to compare with that in H_2O?

5.31 Molecular shape Show that the tetrahedral angle is indeed 109.47°. (*Hint:* Set up a cube with carbon in the center. Put hydrogens at alternate corners.)

5.32 Hybrid orbitals How would you expect the hybrid orbitals made from mixing s and p_x to differ from those made by mixing s and p_z?

5.33 Hybrid orbitals Using hybrid orbitals, predict the bond angle in the molecule $BeCl_2$.

5.34 Hybrid orbitals Describe the bonding in BCl_3, using hybrid-orbital language. What shape do you expect for this molecule?

5.35 Hybrid orbitals Using hybrid orbitals describe the bonding in the molecule $H_2C=CHCH_3$ and describe the spatial arrangement of neighbors around each carbon atom.

***5.36 Molecular shape** Describe the bonding and the probable spatial arrangement of atoms in propadiene, for which the formula is $H_2C=C=CH_2$.

5.37 VSEPR Show how the octet rule plus valence-shell repulsion leads to a tetrahedral geometry for CH_4. What would be the effect of replacing one H by Cl?

5.38 VSEPR In the sequence SiH_4, PH_3, and SH_2, the bond angle changes from $109.47°$ to $93.8°$ to $92.2°$. How would VSEPR theory account for this trend? Why are the PH_3 and SH_2 values so much smaller than the corresponding values for NH_3 and OH_2?

***5.39 VSEPR** In the SF_4 molecule, there are five pairs of electrons in the valence shell (one lone pair and four bonding pairs). Assuming trigonal-bipyramid arrangement (see Fig. 5.24) of the five pairs, predict what would be the probable arrangement of F atoms in SF_4.

Chapter 6

THE GASEOUS STATE

In preceding chapters, we have considered some aspects of chemical behavior in terms of atoms and the forces that hold them together in molecules. We turn now to large collections of atoms and molecules, as in a solid, liquid, or gas, where the emphasis will be not on the specific chemical identity of the material, but on its general properties as decided by its state of aggregation. Thus, for example, although gaseous oxygen is a chemical oxidizing agent, we shall not be concerned with that aspect of its behavior but with the set of properties that it shares in common with all other gases, e.g., easy compressibility, rapid diffusion, high thermal expansion. Since in many respects the gaseous state is the simplest, it will be considered first. The general approach will be first to define the terms used to describe gases, then to discuss the laws that summarize their observed behavior, and finally to consider the theories that have been proposed to account for the observations. The breakdown of these theories will ultimately lead us to liquids and solids, which we will take up in the next chapter.

6.1
VOLUME

The volume of a substance is the space occupied by that substance. If the substance is a gas, the volume of a sample is the same as the volume of the container in which the sample is held. Ordinarily, this volume is specified in units of liters, milliliters (ml), or cubic centimeters (cm³). The officially recommended SI unit for volume is the cubic meter (m³), but chemists generally ignore this recommendation. As the name implies, one *cubic centimeter* is the volume of a cube one centimeter on an edge. One *liter* is the volume of a cube that is one decimeter (dm = 0.1 m = 10 cm) on an edge. Thus, a liter is exactly 1000 times as great as a cubic centimeter. It follows that one *milliliter,* which is a thousandth of a liter, is exactly equal to one cubic centimeter. This equality was not always true; prior to 1964, the liter was defined as the volume occupied by one kilogram of water at the temperature of its maximum density. However, the difference is only 27 ppm.

The volume of liquids and solids does not change much with change of pressure or temperature. Consequently, to describe the amount being handled, i.e., the number of moles, it is usually sufficient to specify only the volume. For gases, this is not enough. As an example, 1 liter of hydrogen at 1 atm pressure and 0°C contains 0.0446 mol, whereas 1 liter at 2 atm and 25°C contains 0.0817 mol. In order to fix the number of moles in a given sample of gas, it is necessary to specify the pressure and temperature as well as the volume.

When solids or liquids are mixed together, the total volume is roughly equal to the sum of the original volumes. This is not true of gases. The final volume after mixing depends strongly on the final pressure. If the final pressure is allowed to rise sufficiently, two or more gases can occupy the same volume as each gas does alone. Since all gases *can mix in any proportion,* they are said to be *miscible.*

6.2
TEMPERATURE

It is a familiar observation that a hot substance and a cold substance placed in contact with each other change so that the hot substance gets colder and the cold substance gets hotter. This is interpreted as resulting from a flow of heat energy from the hot body to the cold body. The hot body is said to have a higher temperature; the cold body, a lower temperature. Thus, *temperature is a property that determines the direction of heat flow;* heat always flows from a region of higher temperature to one of lower temperature. (See Sec. 1.5.)

The international scale for measuring temperature is an absolute scale; it starts with *absolute zero.* Absolute zero is the lower limit of temperature, and temperatures lower than it are unattainable. The scale is sometimes called the Kelvin scale after the English thermodynamicist Lord Kelvin, who proposed it in 1848. It is defined by assigning the value 273.16°K (degrees Kelvin) to the temperature at which H_2O coexists in the liquid, gaseous, and solid states. This point, called the *triple point,* corresponds to the temperature at which liquid water, water vapor, and ice are all in equilibrium with each other. The number 273.16 was picked on the basis of observations that under ideal conditions all gases expand uniformly with rise in temperature. At 0°C the rate of expansion is always 1/273.16 of the volume observed. (It should be noted that IUPAC has recommended that the unit on the Kelvin scale be called a "kelvin," not a "degree Kelvin." In line with this recommendation, we should write 273.16 K instead of 273.16°K. Physicists are generally moving to accept this recommendation, but chemists appear to be resisting it. For clarity, we shall stick to °K in this book.)

The size of the degree Kelvin is defined as 1/273.16 of the temperature difference between absolute zero and the triple point of H_2O. The triple point of H_2O turns out to be 0.01° higher than the normal freezing point of water. The *normal freezing point* is the temperature at which water freezes under one atmosphere pressure; the *triple point* is the temperature at which water freezes under its own vapor pressure, which is only 0.006 atm. Thus, on the Kelvin scale, the normal freezing point of water is 273.15°K.

On the Celsius scale the normal freezing point of water is set as the zero point. The size of the degree, designated by °C, is taken to be the same as the degree Kelvin. The normal boiling point of H_2O, i.e., the boiling point at an atmosphere of pressure, turns out to be 100°C on the Celsius scale. Because

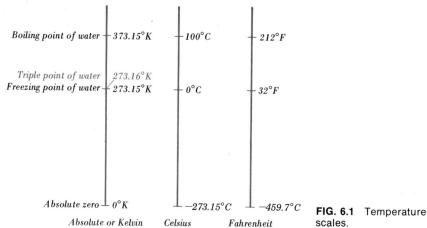

Boiling point of water	373.15°K	100°C	212°F
Triple point of water	273.16°K		
Freezing point of water	273.15°K	0°C	32°F
Absolute zero	0°K	−273.15°C	−459.7°C
	Absolute or Kelvin	Celsius	Fahrenheit

FIG. 6.1 Temperature scales.

there are 100° between the normal freezing and boiling points of H_2O, this temperature scale is sometimes called the *centigrade* scale. A comparision of the Kelvin scale with the Celsius and Fahrenheit scales is shown in Fig. 6.1.

The Fahrenheit scale, which is supposed to be on its way out of use, was originally set up by the German physicist Gabriel Fahrenheit, on the basis of 0° for the lowest temperature he could get with a mixture of snow and salt. The size of the degree on the Fahrenheit scale is only five-ninths as large as the degree on the Celsius and Kelvin scales.

Temperature on the Celsius, or centigrade, scale is converted to temperature on the Kelvin scale by addition of 273.15°:

$$°C + 273.15 = °K$$

To convert Fahrenheit temperature to Kelvin temperature, it is also necessary to correct for the difference in the size of the degree:

$$(°F - 32) \times \tfrac{5}{9} + 273.15 = °K$$

6.3
PRESSURE

Just as temperature determines the direction of heat flow, so pressure is a property which determines the direction of mass flow. Unless otherwise constrained, matter tends to move from a place of higher pressure to a place of lower pressure. Quantitatively, *pressure* is defined as *force per unit area*. Force is defined as that which tends to change the state of rest or motion of an object. The fundamental unit of force is the *newton* (N) (after Isaac Newton), which is the force required to accelerate one kilogram of matter by one meter per second in a time interval of one second. A related unit of force is the *dyne*, which is the force required to accelerate one gram of matter by one centimeter per second. As can be seen,

$$1 \text{ N} = \frac{1 \text{ kg m}}{\text{sec}^2} = \frac{(10^3 \text{ g}) (10^2 \text{ cm})}{\text{sec}^2} = 10^5 \frac{\text{g cm}}{\text{sec}^2} = 10^5 \text{ dyn}$$

Force can also be expressed in terms of pound weight. Units for expressing pressure can be derived from any of the above. The fundamental unit would be newtons per square meter, for which the name *pascal* (Pa) (after the French scientist Blaise Pascal) has recently been recommended. More traditionally, one would speak of dynes per square centimeter or pounds per square inch. The relative equivalence of the units is as follows:

$$1 \text{ Pa} = \frac{1 \text{ N}}{\text{m}^2} = \frac{10^5 \text{ dyn}}{(10^2 \text{ cm})^2} = 10 \text{ dyn/cm}^2$$

$$1 \text{ lb/in}^2 = \frac{(453.6 \text{ g}) (980.6 \text{ cm/sec}^2)}{(2.54 \text{ cm})^2} = 68,900 \text{ dyn/cm}^2$$

In high-pressure work, a frequently encountered unit is the *bar,* which is 10^6 dyn/cm². A bar is very close to 1 atm.

In *fluids,* which is a general term that includes both *liquids* and *gases,* the pressure at a given point is the same in all directions. This can be visualized by considering a swimmer under water. At a given depth, no matter how the swimmer turns, the pressure exerted on all sides by the water is always the same. However, as the swimmer's depth increases, so does the pressure. This comes about because of the pull of gravity on the water. We can picture the swimmer's body as being compressed from above by the weight of the column of water above it and supported from below by the pressure of the water beneath it. In general, for all fluids, the greater the depth of immersion, the greater the pressure.

The earth is surrounded by a blanket of air approximately 800 km thick. Thus, we live at the bottom of a fluid, the atmosphere, which exerts a pressure. The existence of this pressure can be shown by filling a long tube, closed at one end, with mercury and inverting it in a dish of mercury. (Any other liquid would do, but mercury has the advantage of not requiring too long a tube.) Some of the mercury runs out of the tube, but not all of it. This setup, called a *barometer,* is represented in Fig. 6.2. No matter how large the diameter of the tube and no matter how long the tube, the *difference* in height between the mercury levels inside and outside the tube is the same. The fact that all the mercury does not run out shows that there must be a

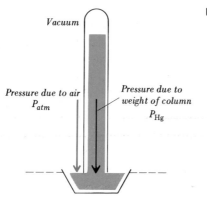

Vacuum

Pressure due to air
P_{atm}

Pressure due to weight of column
P_{Hg}

FIG. 6.2 Barometer.

pressure exerted on the surface of the mercury in the dish sufficient to support the column of mercury.

To a good approximation, the space above the mercury level is a vacuum (contains only a negligible amount of mercury vapor) and exerts no pressure on the upper mercury level. The pressure at the bottom of the mercury column, therefore, is due only to the weight of the mercury column. As noted, it is a general property of fluids that at any given level in the fluid the pressure is constant. In Fig. 6.2 the dashed line represents the level of interest. At this level, outside the tube, the force per unit area is due to the atmosphere and can be labeled as P_{atm}. The pressure inside the tube is due to the pressure of the column of mercury and can be labeled P_{Hg}. The equality $P_{atm} = P_{Hg}$ provides us with a method for measuring the pressure exerted by the atmosphere.

Atmospheric pressure changes from day to day and from one altitude to another. A *standard atmosphere,* referred to as 1 atm, is defined as the pressure which supports a column of mercury that is 760 mm high at 0°C at sea level.*

Pressure can be expressed in terms either of number of atmospheres or number of millimeters of mercury (mmHg). The pressure unit mmHg is also frequently referred to as a *torr,* in honor of Evangelista Torricelli, the inventor of the barometer. (However, IUPAC recommends that "torr" be abandoned.) We can also express pressure by the height of a water column. Since water has a density of 1 g/ml, whereas mercury has a density of 13.6 g/ml, a given pressure supports a column of water that is 13.6 times as high as one of mercury. One atmosphere of pressure supports 76 cm of mercury or 76(13.6) cm of water, the latter being roughly 34 ft. In terms of pounds per square inch, 1 atm is 14.7 lb/in². In terms of pascals, 1 atm is 1.01×10^5 pascals.

The device shown in Fig. 6.3 is a *manometer,* used to measure the pressure of a trapped sample of gas. The manometer is constructed by placing a liquid in the bottom of a U-shaped tube with the gas sample in one arm of the U. If the right-hand tube is open to the atmosphere, the pressure which is exerted on the right-hand surface is the atmospheric pressure P_{atm}. At the same liquid level in both arms of the tube, the pressures must be equal; otherwise, there would be a flow of liquid from one arm to the other. At the level indicated by the dashed line in Fig. 6.3, the pressure in the left arm is equal to the pressure of the trapped gas P_{gas} plus the pressure of the column of liquid above the dashed line P_{liq}. We can therefore write

$$P_{atm} = P_{gas} + P_{liq}$$

or

$$P_{gas} = P_{atm} - P_{liq}$$

The atmospheric pressure can be measured by a barometer, and P_{liq} can be obtained by measuring the difference in height between the liquid level in the

---

*If we think of pressure as weight per unit area, we can see why it is necessary that both 0°C and sea level need to be specified in defining the standard atmosphere. The density of liquid Hg changes with temperature; therefore, the weight of a 760-mm-high column of Hg of fixed cross section changes with temperature. Hence, the temperature must be specified. Similarly, the force of gravity changes slightly with altitude; the weight of the Hg column changes when moved away from sea level. Hence, sea level must be specified.

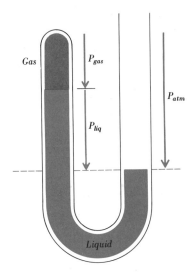

FIG. 6.3 Manometer.

right and left arms and correcting for the known density of the liquid. The same units must be used to express P_{atm} and P_{liq}. For example, if P_{atm} is in millimeters of mercury and the manometer liquid is water instead of mercury, the difference in water levels must be converted to its mercury equivalent by dividing by 13.6. If the bottom of the U tube consists of flexible rubber tubing, the right arm can be raised with respect to the left arm until the two liquid levels are at the same height, in which case $P_{liq} = 0$ and $P_{gas} = P_{atm}$.

6.4
P-V RELATION

A characteristic property of gases is their great compressibility. This behavior is summarized quantitatively in Boyle's law (1662). Boyle's law states that *at constant temperature a fixed mass of gas occupies a volume inversely proportional to the pressure exerted on it.* For example, if the pressure is doubled, the volume goes to one-half of the original. Boyle's law can be summarized by a pressure-

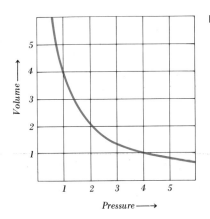

FIG. 6.4 *P-V* plot for a gas.

volume, or *P-V*, plot like that shown in Fig. 6.4. The abscissa (horizontal axis) represents the pressure of a given sample of gas, and the ordinate (vertical axis), the volume occupied by it. The curve is a hyperbola, the equation for which is

$$PV = \text{constant}$$

or $V = \text{constant}/P$. The size of the constant is fixed once the mass of the sample and its temperature are specified. The following examples show how Boyle's law is used. In both cases it is assumed that the temperature stays constant.

EXAMPLE 1

An 8-g sample of a gas occupies 12.3 liters at a pressure of 40.0 cmHg. What is the volume when the pressure is increased to 60.0 cmHg?

Solution

P changes to 60.0/40.0 of the original.
V changes inversely.
V changes to 40.0/60.0 of the original.
Final volume = (40.0/60.0) (12.3 liters) = 8.20 liters.

EXAMPLE 2

To what pressure must a gas be compressed in order to get into a 1.25-m³ tank the entire mass of gas that occupies 3.25 m³ at atmospheric pressure?

Solution

V changes to 1.25/3.25 of the original.
P changes inversely.
P changes to 3.25/1.25 of original, or 2.60 atm.

The behavior specified by Boyle's law is not always observed. For most gases the law is best followed at low pressures and at high temperatures; as the pressure is increased or as the temperature is lowered, deviations may occur. This can be seen by considering the experimental data listed in Fig. 6.5. There are two series of experiments, one at 100°C and the other at −50°C. The pressure is measured when a fixed mass of gas, 39.95 g of argon, is contained in different volumes. The *PV* products in the last column, obtained by multiplying the observed values in the second and third columns, should, according to Boyle's law, be constant at each temperature. The data indicate that, although at the high temperature Boyle's law is closely obeyed, at the low temperature the *PV* product is not constant but drops off significantly as the pressure increases. In other words, Boyle's law is not obeyed. As the temperature of argon is decreased, its behavior deviates from that specified in Boyle's law.

FIG. 6.5   Pressure-Volume Data for 39.95 g of Argon Gas

Temperature, °C	V, liters	P, atm	P × V, liter atm
100	2.000	15.28	30.56
	1.000	30.52	30.52
	0.500	60.99	30.50
	0.333	91.59	30.50
−50	2.000	8.99	17.98
	1.000	17.65	17.65
	0.500	34.10	17.05
	0.333	49.50	16.48

How High pressure Lower temp (handwritten margin note)

That deviations from Boyle's law also increase at higher pressures can be seen from some experimental data for acetylene as given in Fig. 6.6. When the pressure is doubled from 0.5 to 1.0 atm, the *PV* product is essentially unchanged; so in the low-pressure range acetylene follows Boyle's law reasonably well. However, when the pressure is doubled from 4.0 to 8.0 atm, the *PV* product does not remain constant but decreases by more than 3 percent; in this higher-pressure range, Boyle's law is not followed so well. For any gas, the lower the pressure, the closer the approach to Boyle's law behavior. When the law is obeyed, the gas is said to show *ideal* behavior.

A careful study of the deviations from ideal behavior led the Dutch physicist J. D. van der Waals to propose a modified version of the *P-V* relation. His modification is based on the observation that observed pressures generally are smaller than those predicted by Boyle's law and observed volumes are generally greater than those so predicted. The van der Waals relation is usually written

$$\left(P + \frac{n^2a}{V^2}\right)(V - nb) = \text{constant}$$

It is related to the Boyle's law expression

$$P_{ideal}V_{ideal} = \text{constant}$$

by a correction term n^2a/V^2 added to the observed pressure and a correction term nb subtracted from the observed volume, n being the number of moles in the sample; a and b are constants that are characteristic of the gas in question. As indicated in Sec. 6.12, a can be related to molecular attractions, and b to molecular volumes. Representative values of the van der Waals constants are given in Fig. 6.7. Finally, it should be noted that at very high pressures

FIG. 6.6   *PV* Products for a Sample of Acetylene at 0° C

P, atm	0.5	1.0	2.0	4.0	8.0
PV, liter atm	1.0057	1.0000	0.9891	0.9708	0.9360

FIG. 6.7   van der Waals Constants

Gas	a, liter^2 atm/mol^2	b, liter/mol
Helium	0.0341	0.0237
Argon	1.35	0.0322
Nitrogen	1.39	0.0391
Carbon dioxide	3.59	0.0427
Acetylene	4.39	0.0514
Carbon tetrachloride	20.39	0.1383

and low temperatures even the van der Waals relation fails to represent the observed data. Other special relations have been proposed to handle such cases.

6.5
V-T RELATION

Another characteristic property of gases is their high thermal expansion. Like most solids and liquids, gases increase in volume when their temperature is raised, but for gases the relative effect is much bigger. Experimentally, the dependence can be measured by confining a fixed mass of gas in a glass tube with a blob of liquid mercury, as shown in Fig. 6.8. The mercury blob, which is free to move, acts as a piston to confine the gas sample at constant pressure. As the gas is heated, the mercury moves out, and the volume of the gas sample increases.

Typical numerical data are plotted in Fig. 6.9. The points fall on a straight line; the volume varies linearly with temperature. If the temperature is lowered sufficiently, the gas liquefies, and no more experimental points can be obtained. However, if the straight line is extended, or extrapolated, to lower temperatures, as shown by the dashed line, it reaches a point of zero volume. The temperature at which the volume would go to zero is $-273.15°C$. It is significant that this value, $-273.15°C$, does not depend on the kind of gas used or on the pressure at which the experiment is performed. Designating $-273.15°C$ as absolute zero is thus reasonable, since temperatures below this would correspond to negative volume.

The volume-temperature data of Fig. 6.9 can be expressed by the relation

$$V = V_0 + \alpha t$$

FIG. 6.9 Plot of gas volume as a function of temperature.

FIG. 6.8 Simple device for measuring gas volume as a function of temperature.

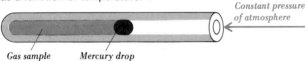

Constant pressure of atmosphere

Gas sample   *Mercury drop*

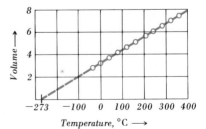

Volume →

-273 -100 0 100 200 300 400

Temperature, °C →

where V = observed volume at any temperature t (in degrees Celsius)
V_0 = volume at 0°C
α = a constant to be determined

Since V approaches 0 when t approaches -273.15°C, we can write for that special point

$$0 = V_0 + \alpha \, (-273.15)$$

Solving for α, we get

$$\alpha = \frac{V_0}{273.15}$$

In other words, the constant α, which represents the slope of the graph in Fig. 6.9, has a value equal to $1/273.15$ of the volume at 0°C. Substituting for α in the original equation, we get

$$V = V_0 + \frac{V_0}{273.15} \, t$$

which can be written

$$V = V_0 \left(1 + \frac{t}{273.15} \right) = V_0 \left(\frac{273.15 + t}{273.15} \right)$$

Since absolute temperature, generally designated by T, is equal to $273.15 + t$, we can also write

$$V = \frac{V_0}{273.15} \, T = \text{constant} \times T$$

In words, this summarization of gas behavior, called Charles' law (1787), is as follows: *At constant pressure, the volume occupied by a fixed mass of gas is directly proportional to the absolute temperature.* The value of the proportionality constant depends on pressure and on the mass of gas. The following example shows how Charles' law is used.

EXAMPLE 3

A sample of nitrogen gas weighing 9.3 g at a pressure of 0.99 atm occupies a volume of 12.4 liters when its temperature is 455°K. What do you expect will be its volume when the temperature is 305°K? Assume pressure stays constant.

Solution

T changes to $\frac{305}{455}$ of its original value.
V changes proportionally.
V changes to $\frac{305}{455}$ of its original value.
Final volume is $(\frac{305}{455})(12.4 \text{ liters}) = 8.31$ liters.

Actually, Charles' law, like Boyle's law, represents the behavior of an *ideal,* or *perfect,* gas. For any real gas, especially at high pressures and at temperatures near the liquefaction point, deviations from Charles' law are observed.

Near the liquefaction point the observed volume is less than that predicted by Charles' law.

Because of the Charles' law relation of volume to absolute temperature, calculations involving gases require that temperatures be expressed on the Kelvin scale. It is also convenient in working with gases to have a reference point. The customary reference point is 273.15°K (0°C) and 1 atm (760 mmHg) pressure. These conditions are called *standard temperature and pressure* (STP).

6.6
PARTIAL PRESSURES

The behavior observed when two or more gases are placed in the same container is summarized in *Dalton's law of partial pressures* (1801). Dalton's law states that the *total pressure exerted by a mixture of gases is equal to the sum of the partial pressures* of the various gases. The *partial pressure* is defined as *the pressure the gas would exert if it were alone in the container.* As illustration, suppose a sample of hydrogen is pumped into a 1-liter box and its pressure is found to be 0.065 atm. Suppose further that a sample of oxygen is pumped into a second 1-liter box and its pressure is found to be 0.027 atm. If both samples are now transferred to a third 1-liter box, the pressure is observed to be 0.092 atm. For the general case, Dalton's law can be written

$$P_{total} = P_1 + P_2 + P_3 + \cdots$$

where the subscripts denote the various gases occupying the same volume. Actually, Dalton's law is an idealization, but it is closely obeyed by most mixtures of nonreacting gases.

In laboratory experiments dealing with gases, the gases may be collected above water, in which case the water vapor contributes to the total pressure measured. Figure 6.10 illustrates such an experiment in which oxygen gas is collected by water displacement. If the water level is the same inside and outside the bottle, then we may write

$$P_{atm} = P_{oxygen} + P_{water\ vapor}$$
$$P_{oxygen} = P_{atm} - P_{water\ vapor}$$

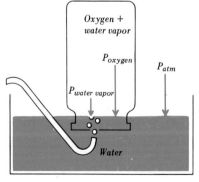

Oxygen + water vapor

P_{oxygen}

P_{atm}

$P_{water\ vapor}$

Water

FIG. 6.10 Pressure contributions when oxygen is collected over water.

Atmospheric pressure readings are obtained from a barometer. As we shall see later (Sec. 7.2), $P_{water\ vapor}$ is a function only of the temperature of the water and can be obtained from tables such as the one given in Appendix 3. Thus, the partial pressure of oxygen can be determined from a measured total pressure, a temperature, and reference to a table of vapor-pressure data. The following example shows how Dalton's law of partial pressure may enter into calculations involving gases.

EXAMPLE 4

If 40.0 liters of nitrogen are collected over water at 22°C when the atmospheric pressure is 0.957 atm, what would be the volume of the dry nitrogen at standard temperature and pressure, assuming ideal behavior?

Solution

The initial volume of the nitrogen is 40.0 liters. The final volume is unknown. The initial pressure of the nitrogen gas is the atmospheric pressure, 0.957 atm, minus the vapor pressure of water. From Appendix 3, we find that at 22°C water has a vapor pressure of 0.026 atm; this makes the initial pressure of nitrogen 0.957 − 0.026, or 0.931, atm. The final pressure is 1.000 atm. The initial temperature is 273 + 22 = 295°K, and the final temperature is 273°K. The problem is solved by considering separately how the volume is affected by a change in pressure and then by a change in temperature.

Pressure changes to 1.000/0.931 of its original value.
Volume changes inversely, or to 0.931/1.000 of its original value.
Temperature changes to $\frac{273}{295}$ of its original value.
Volume changes proportionally, or to $\frac{273}{295}$ of its original value.

$$V_{final} = V_{initial} \times \left(\begin{array}{c}\text{correction for}\\\text{pressure change}\end{array}\right) \times \left(\begin{array}{c}\text{correction for}\\\text{temperature change}\end{array}\right)$$

$$= 40.0 \text{ liters} \times \frac{0.931}{1.000} \times \frac{273}{295}$$

$$= 34.5 \text{ liters}$$

6.7
AVOGADRO PRINCIPLE

In the preceding section we assumed that when gases are mixed, they do not react with each other. However, sometimes they do react. For example, when a mixture of hydrogen and chlorine gas is exposed to ultraviolet light, reaction occurs to form the gas hydrogen chloride. In such a reaction, it is observed that the volumes of the individual gases which actually react are simple multiples of each other and bear a simple whole-number ratio to the volume of any gaseous product.

As specific examples, we have the following observations:

$$\text{Hydrogen} + \text{chlorine} \longrightarrow \text{hydrogen chloride}$$
$$\quad\quad\text{1 liter}\quad\quad\quad\text{1 liter}\quad\quad\quad\quad\quad\text{2 liters}$$

$$\text{Hydrogen} + \text{oxygen} \longrightarrow \text{gaseous water}$$
$$\quad\quad\text{2 liters}\quad\quad\quad\text{1 liter}\quad\quad\quad\quad\quad\text{2 liters}$$

$$\text{Hydrogen} + \text{nitrogen} \longrightarrow \text{ammonia}$$
$$\quad\quad\text{3 liters}\quad\quad\quad\text{1 liter}\quad\quad\quad\quad\quad\text{2 liters}$$

The occurrence of simple whole-number ratios between reacting volumes of gases suggested that there is a simple relation between the volume of a gas sample and the number of molecules it contains. The Italian physicist Amadeo Avogadro was the first to propose, in 1811, the following principle: *Equal volumes of gases at the same temperature and pressure contain equal numbers of molecules.* That this principle accounts for the observation of simple reacting volumes can be seen from the following argument: Hydrogen chloride (HCl) contains in each molecule an equal number of H atoms and Cl atoms. These come from equal volumes of hydrogen gas and chlorine gas, so each of these volumes must contain the same number of atoms. Since the volume of hydrogen used (as well as the volume of chlorine) is only one-half the volume of hydrogen chloride produced, the hydrogen atoms in the initial reactant gas must be paired up, or at least be in molecules divisible by two.

The assumption that hydrogen and chlorine molecules are diatomic rather than monatomic can be justified as follows: If hydrogen were monatomic, i.e., consisted of individual hydrogen atoms, and if chlorine were also monatomic, then 1 liter of hydrogen (x atoms) would combine with 1 liter of chlorine (x atoms) to give 1 liter of HCl gas (x molecules). This is contrary to observation; the volume of HCl formed is *twice* as great as the volume of either hydrogen or chlorine that is used up. It must be that the hydrogen and chlorine molecules are more complex than monatomic, at least to the extent of containing an even number of atoms. If, as in fact turns out to be the case, hydrogen and chlorine are *diatomic*, then 1 liter of hydrogen (x molecules, or $2x$ atoms) will combine with 1 liter of chlorine (x molecules, or $2x$ atoms) to form 2 liters of hydrogen chloride ($2x$ molecules). This agrees with experiment.*

As was first shown by the Italian chemist Stanislao Cannizzaro (1858), the Avogadro principle can be used as a basis for the determination of molecular weights. If two gases at the same temperature and pressure contain the same number of molecules in the same volume, then the relative weights of the volumes give directly the relative weights of the two kinds of molecules. For example, at STP 1 liter of acetylene is observed to weigh 1.17 g, whereas 1 liter of oxygen weighs 1.43 g. Since, according to the Avogadro principle, the

---

*This reasoning can also be used to prove that water is H_2O and not HO. (Recall John Dalton's problem on p. 31.) Since one volume of oxygen gives two volumes of water, the oxygen molecule must contain an even number of oxygen atoms. If oxygen, like hydrogen, is diatomic, the fact that two volumes of hydrogen are needed per volume of oxygen implies that the water molecule contains twice as many hydrogen atoms as oxygen atoms.

FIG. 6.11   Molar Volume at STP

Gas	Molar volume, liters
Hydrogen	22.432
Nitrogen	22.403
Oxygen	22.392
Carbon dioxide	22.263
Ideal gas	22.414

number of molecules is the same in both samples, the acetylene molecule must be 1.17/1.43, or 0.818, times as heavy as the oxygen molecule. Given that the diatomic oxygen molecule has a molecular weight of 32.0 amu, the molecular weight of acetylene must be (0.818) (32.0), or 26.2 amu.

The volume occupied at STP by 32.00 g of oxygen (1 mol) has been determined by experiment to be 22.4 liters. This is called the *molar volume*. Figure 6.11 shows observed molar volumes at STP for several gases. They are all approximately 22.4 liters. The value 22.414 shown for the ideal gas is obtained from measurements made on gases at high temperatures and low pressures (where gas behavior is more nearly ideal) and then extrapolating to STP by using Boyle's and Charles' laws. For the first three gases, hydrogen, nitrogen, and oxygen, agreement with ideality is quite satisfactory. Even for the fourth, carbon dioxide, the agreement is better than 1 percent. Consequently, we usually assume that at STP the molar volume of any gas will be 22.4 liters. The following example shows how this molar volume can be used to determine molecular weight and molecular formulas.

EXAMPLE 5

Chemical analysis shows that ethylene has a simplest formula corresponding to one atom of carbon per two atoms of hydrogen. It has a density of 1.25 g/liter at STP. What would be the corresponding molecular weight and molecular formula?

Solution

At STP 1 mol of gas (if ideal) has a volume of 22.4 liters. Each liter of ethylene weighs 1.25 g; so 1 mol of ethylene would weigh 22.4×1.25 g, or 28.0 g. One mole is equal to the molecular weight in grams. Since the simplest formula is CH_2, the molecular formula must be some multiple of that, or $(CH_2)_x$. For $(CH_2)_x$, the formula weight is equal to x times 14.0. By experiment, this is equal to 28.0; so x must be equal to 2. The molecular formula of ethylene is therefore $(CH_2)_2$, or C_2H_4.

6.8
EQUATION OF STATE

Boyle's law, Charles' law, and the Avogadro principle can be combined to give a general relation between the volume, pressure, temperature, and number of moles of a gas sample. Such a general relation is called an *equation of state*. It tells how in going from one gaseous state to another the four variables, $V, P, T,$ and n change. Boyle's law, Charles' law, and the Avogadro principle can be written, respectively, as

$$V \sim P^{-1} \quad \text{this holds at constant } T \text{ and } n$$
$$V \sim T \quad \text{this holds at constant } P \text{ and } n$$
$$V \sim n \quad \text{this holds at constant } T \text{ and } P$$

Combining all these conditions we write

$$V \sim (P^{-1})\,(T)\,(n)$$

Written as a mathematical equation, the general relation becomes

$$V = R\frac{1}{P}Tn \quad \text{or} \quad PV = nRT$$

The symbol R is inserted as the constant of proportionality; it is called the *universal gas constant.* The equation $PV = nRT$ is called the *equation of state for an ideal gas* or the *perfect-gas law.*

The numerical value of R can be found by substituting experimental quantities in the equation. At STP, $T = 273.15°K$, $P = 1$ atm, and for 1 mol of gas ($n = 1$), $V = 22.414$ liters. Consequently,

$$R = \frac{PV}{nT} = \frac{(1 \text{ atm})\,(22.414 \text{ liters})}{(1 \text{ mol})\,(273.15°K)} = 0.082057 \frac{\text{liter atm}}{\text{mol } °K}$$

In order to use this value of R in the equation of state, P must be expressed in atmospheres, V in liters, n in moles, and T in degrees Kelvin.

EXAMPLE 6

The density of an unknown gas at 98°C and 0.974 atm pressure is 2.50 g/liter. What is the molecular weight of this gas, assuming ideal behavior?

Solution

Temperature = 98 + 273 = 371°K.
Pressure = 0.974 atm.
From the equation of state $PV = nRT$, we can calculate the number of moles in 1 liter:

$$\frac{n}{V} = \frac{P}{RT} = \frac{0.974 \text{ atm}}{(0.0821 \text{ liter atm mol}^{-1} \text{ deg}^{-1})\,(371 \text{ deg})} = 0.0320 \text{ mol/liter}$$

Since 0.0320 mol weighs 2.50 g, 1 mol weighs 2.50/0.0320, or 78.1 g. Therefore, the molecular weight is 78.1 amu.

From the equation of state $PV = nRT$, it is seen that the constant for Boyle's law ($PV = $ constant) is just equal to nRT. As noted in Sec. 6.4, van der Waals showed that it is not the simple product PV that remains constant at constant temperature but the more complex expression $(P + n^2a/V^2)\,(V - nb)$.

Setting this product equal to nRT, we obtain the so-called "van der Waals equation of state":

$$\left(P + \frac{n^2a}{V^2}\right)(V - nb) = nRT$$

It describes real gases over a wider range of conditions than does the ideal equation of state.

EXAMPLE 7

By using the data in Fig. 6.7, calculate the pressure exerted by 0.250 mol of carbon dioxide in 0.275 liter at 100°C, and compare this value with the value expected for an ideal gas.

Solution

$P = ?;\ n = 0.250$ mol; $a = 3.59$ liter² atm/mol²; $V = 0.275$ liter; $b = 0.0427$ liter/mol; $R = 0.08206$ liter atm deg⁻¹mol⁻¹; $T = 373°K$.

$$\left[P + \frac{(0.250)^2\,(3.59)}{(0.275)^2}\right][0.275 - (0.250)\,(0.0427)] = (0.250)\,(0.08206)\,(373)$$

$$P = 26.0 \text{ atm}$$

From the ideal equation of state $PV = nRT$, we would have calculated $P_{ideal} = 27.8$ atm. (The actual observed value under the conditions specified is 26.1 atm.)

6.9
DIFFUSION

A gas spreads to occupy any volume accessible to it. This spontaneous spreading of a substance throughout a phase is called *diffusion*. Diffusion can readily be observed by liberating some ammonia gas in a room. Its odor soon fills the entire room. Further, it is found for a series of gases that the lightest gas diffuses most rapidly. Quantitatively, the rate of diffusion is observed to be inversely proportional to the square root of the molecular weight. This is known as Graham's law of diffusion. In mathematical form it can be written

$$R = \frac{\text{constant}}{\sqrt{m}} \quad \text{or} \quad \frac{R_1}{R_2} = \frac{\sqrt{m^2}}{\sqrt{m_1}}$$

where R_1 and R_2 are the rates of diffusion of gases 1 and 2, and m_1 and m_2 are the respective molecular weights. In the case of oxygen and hydrogen

$$\frac{R_{H_2}}{R_{O_2}} = \frac{\sqrt{m_{O_2}}}{\sqrt{m_{H_2}}} = \sqrt{\frac{32}{2}} = \sqrt{16} = 4$$

The fact that heavier gases diffuse more slowly than light gases has been applied on a mammoth scale to effect the separation of uranium isotopes. Natural uranium, consisting of 99.3% ^{238}U and 0.7% ^{235}U, is converted to the gas UF_6, and the mixture of the gases is passed at low pressure through a porous solid. The heavier $^{238}UF_6$ diffuses less rapidly than $^{235}UF_6$; hence, the gas mixture that first emerges from the solid is enriched in the ^{235}U isotope. Since the square root of the ratio of molecular weights is only 1.0043, the step must be repeated thousands of times, but eventually substantial enrichment of the desired ^{235}U isotope is obtained for use in nuclear fission.

6.10
KINETIC THEORY

An aspect of observed gas behavior that originally gave the strongest clue to the nature of gases is the phenomenon known as *brownian motion*. This motion, first observed by the Scottish botanist Robert Brown in his study of pollen (1827), is an *irregular zigzag movement of extremely minute particles when suspended in a liquid or gas*. Brownian motion can be observed by focusing a microscope on a particle of smoke which is strongly illuminated from the side. The particle observed does not settle to the bottom of its container but moves continually to and fro and shows no sign of coming to rest. The smaller the particle, the more violent is this permanent condition of irregular motion. The higher the temperature, the more vigorous is the movement.

The existence of brownian motion suggests that molecules of matter, though invisible, are constantly moving. A particle of smoke appears to be jostled by unseen neighboring molecules; its motion reflects the motion of the submicroscopic, invisible molecules of matter. Here then is powerful support for the suggestion that matter consists of extremely small particles which are ever in motion. This "moving-molecule" theory is known as the *kinetic theory of matter*. Its two basic postulates are that molecules are in motion and that heat is a manifestation of this motion.

Like any theory, the kinetic theory represents a model which is proposed to account for an observed set of facts. In order that the model be practical, certain simplifying assumptions must be made. The validity of each assumption and the reliability of the whole model can be checked by how well the facts are explained. For a perfect gas the following assumptions are made:

1 Gases consist of tiny molecules, which are so small and so far apart on the average that the actual volume of the molecules is negligible compared with the empty space between them.

2 In the perfect gas there are no attractive forces between molecules. The molecules are completely independent of each other.

3 The molecules of a gas are in rapid, random, straight-line motion, colliding with each other and with the walls of the container. In each collision it is assumed that there is no net loss of kinetic energy, although there may be a transfer of energy between the partners in the collision.

4 At a particular instant in any collection of gas molecules, different molecules have different speeds and, therefore, different kinetic energies. However, the average kinetic energy of all the molecules is assumed to be directly proportional to the absolute temperature.

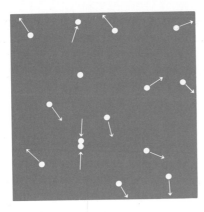

FIG. 6.12 Kinetic model of a gas.

Before discussing each of these assumptions, we ask how the model is related to the observable quantities V, P, and T. A gas is mainly empty space in which tiny points representing molecules are in violent motion. Figure 6.12 shows a schematic version of the model. The *volume* is mostly empty space, but it is *occupied* in the sense that moving particles occupy the entire region in which they move. *Pressure*, defined as force per unit area, is exerted because the molecules collide with the walls of the container. Each collision produces a tiny impulse, and the sum of all the impulses is the pressure. *Temperature* gives a quantitative measure of the average motion of the molecules.

That assumption 1 above is reasonable can be seen from the fact that the compressibility of gases is so great. In oxygen gas, for example, 99.96 percent of the total volume is empty space at any instant. Since at STP there are 2.7×10^{19} molecules per cubic centimeter, the average spacing between the molecules is about 3.7 nm, which is about 13 times the molecular diameter of oxygen. When oxygen or any other gas is compressed, the average spacing between molecules is reduced; i.e., the fraction of free space is diminished.

The validity of assumption 2 is supported by the observation that gases spontaneously expand to occupy all the volume accessible to them. This behavior occurs even with a highly compressed gas, in which the molecules are fairly close together and any intermolecular forces would be expected to be greatest. Hence, there must be no appreciable binding of one molecule of a gas to its neighbors.

Assumption 3 ties in with the observation of brownian motion, which implies that molecules of a gas move. Like any moving body, a gas molecule has kinetic energy equal to $\frac{1}{2}ms^2$, where m is the mass of the molecule and s is its speed. That the molecule moves in a straight line follows from the assumption that there are no attractive forces.

Because there are so many molecules and because they are moving so rapidly (at 0°C the average speed of oxygen molecules is about 1700 km/h), there are frequent collisions. It is necessary to assume that these collisions are elastic (like collisions between billiard balls). Otherwise, kinetic energy would be lost by conversion to potential energy. Motion of the molecules would eventually stop, and the molecules would settle to the bottom of the container. It might be noted that the distance a gas molecule has to travel before colliding with another is much greater than the average spacing between molecules. The molecules have many near misses. In oxygen at STP the average distance between successive collisions, called the *mean free path,* is approximately 300 times the molecular diameter. As already mentioned, the average spacing in oxygen at STP is only 13 times the molecular diameter.

Assumption 4 has two parts: (*a*) there is a distribution of kinetic energies, and (*b*) the average kinetic energy is proportional to the absolute temperature. The distribution comes about because molecular collisions continually change the speed of a particular molecule. A given molecule may move with a certain speed, but it soon hits another molecule, to which it may lose kinetic energy; perhaps later it may get hit by a third molecule from which it gains kinetic energy; and so on. The exchange of kinetic energy goes on constantly; it is only the total that stays constant (provided, of course, no energy is lost to the surroundings or added to the gas sample from the outside, as by heating). The situation is summarized in Fig. 6.13, which indicates the usual distribution of kinetic energies in a gas sample. Each point on the curve tells what fraction of the molecules are in a particular range of kinetic energy.

The temperature of a gas can be raised by the addition of heat. What happens to the molecules as the temperature is raised? Figure 6.14 shows with a dashed curve the situation at higher temperature. The fraction of molecules having higher kinetic energies is larger; the average kinetic energy is increased. Temperature serves to measure the average kinetic energy.

The assumption that average kinetic energy is directly proportional to the absolute temperature is supported by the fact that predictions based on the assumption agree so well with experiment. For example, it immediately follows that two different gases at the same temperature must have equal average kinetic energies. If two gases A and B have different molecular masses m_A and m_B, then the average speeds s_A and s_B must be related as the inverse ratio of the square roots of the masses.

FIG. 6.13 Energy distribution in a gas.

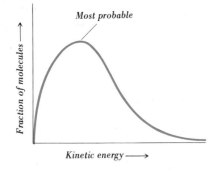

FIG. 6.14 Comparison of energy distribution in a gas sample at two different temperatures.

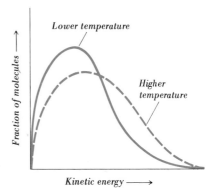

Average kinetic energy of A = average kinetic energy of B

$$\tfrac{1}{2}m_A s_A^2 = \tfrac{1}{2}m_B s_B^2$$

$$\frac{s_A}{s_B} = \sqrt{\frac{m_B}{m_A}}$$

It is reasonable to assume that the rate of diffusion is directly proportional to average speed. Thus, we can write

$$\frac{\text{Rate of diffusion of A}}{\text{Rate of diffusion of B}} = \frac{s_A}{s_B} = \sqrt{\frac{m_B}{m_A}}$$

This inverse proportionality was found experimentally as Graham's law.

6.11
KINETIC THEORY AND EQUATION OF STATE

The equation of state of an ideal gas, $PV = nRT$, can be deduced from kinetic theory by considering in detail how the pressure of a gas arises from molecular impacts. Suppose we imagine a gas confined in a cubic box, as in Fig. 6.12. The pressure on the walls of the container will be proportional to the *number of molecular impacts* per square centimeter of wall area per second times the *impulse* (see Appendix A5.3) *of each impact.*

Let N be the number of molecules in the box; m, the mass of each molecule; s, the average speed; and l, the edge length of the box. The motion of any molecule in the box can be resolved into three components along the three edge directions of the box. We thus assume that the net effect on the walls is the same as if one-third of the molecules in the box were constrained to move perpendicular to a pair of opposite faces. In other words, $\tfrac{1}{3}N$ molecules will be colliding with any chosen face of the box. How frequently will each of these molecules make a molecular impact on the wall of interest? Between successive impacts on the same wall the molecule has to travel the entire length of the box and back, i.e., the distance $2l$. Since it is traveling at a speed of s cm/sec, it will collide with the wall $s/2l$ times per second.

$$\text{Number of impacts per face per second} = \tfrac{1}{3}N\frac{s}{2l}$$

These impacts will be spread over the whole side of the box, which has an area of l^2 cm².

$$\text{Number of impacts per square centimeter per second} = \tfrac{1}{3}N\frac{s/2l}{l^2}$$

Since the volume V of the box is l^3 cm³, the expression can be written

$$\text{Number of impacts per square centimeter per second} = \frac{Ns}{6V}$$

To get the *impulse per impact,* it is necessary to note that the impulse is equal to the change of momentum. In a collision in which a molecule of mass m and speed s bounces off the wall with the same speed but in an opposite direction (denoted by $-s$), the momentum changes from an initial value ms to a final value $-ms$, i.e., by an amount equal to $2ms$.

$$\text{Pressure} = \left(\frac{\text{impacts}}{\text{area} \times \text{time}}\right)\left(\frac{\text{impulse}}{\text{impact}}\right)$$

$$= \left(\frac{Ns}{6V}\right)(2ms)$$

$$= \frac{N}{3V}ms^2$$

The proportionality between average kinetic energy $\frac{1}{2}ms^2$ and absolute temperature T can be written as

$$\tfrac{1}{2}ms^2 = \tfrac{3}{2}kT$$

where the constant k, called the Boltzmann constant, has the value 1.3806×10^{-23} J/deg, or 1.363×10^{-25} liter atm/deg. By replacing ms^2 by $3kT$ in the pressure equation given above we find

$$P = \frac{N}{V}kT$$

The number of molecules N is just equal to n, the number of moles, times 6.0222×10^{23}. Since the Boltzmann constant k is the gas constant R divided by 6.0222×10^{23}, we have

$$P = \frac{nRT}{V}$$

Thus we are led to the equation of state $PV = nRT$ from first principles of molecular motion and molecular collisions.

6.12
DEVIATIONS FROM IDEAL BEHAVIOR

The model of a gas as a collection of point masses moving independently leads to $PV = nRT$. Yet we have seen that gases are better described by other equations, such as that of van der Waals:

$$\left(P + \frac{n^2a}{V^2}\right)(V - nb) = nRT$$

What is incorrect about the kinetic molecular model? In the first place, molecules are not point masses but have finite dimensions. This means that the molecules are not free to move in all of volume V. The van der Waals quantity $(V - nb)$ can be though of as representing only the *free volume* actually accessible to the molecules. The quantity nb is an excluded volume which is not accessible because molecules are impenetrable to each other.

Figure 6.15 suggests how one can show that the magnitude of b is directly related to the size of molecules. Let r be the radius of a gas molecule. The vol-

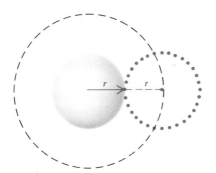

FIG. 6.15 Schematic diagram showing how finite sizes of molecules lead to excluded volume. Dashed circle represents area from which center of dotted molecule is excluded.

ume of the molecule is $\frac{4}{3}\pi r^3$, and obviously no other molecule can come into this space. But the situation is more complicated than this. Even in direct collision, another molecule cannot approach closer than shown by the dotted sphere. The centers of other molecules are excluded (as shown by the dashed line) from the volume $\frac{4}{3}\pi(2r)^3$, or $8(\frac{4}{3}\pi r^3)$. The excluded volume is eight times the volume of an individual molecule! For n mol of molecules, the excluded volume would be as follows:

$$V_{\text{excl}} = 8\left(\frac{4\pi r^3}{3}\right)(6.02 \times 10^{23}\, n)\left(\frac{1}{2}\right)$$

The final factor $\frac{1}{2}$ comes in because the excluded volume shown in Fig. 6.15 is actually generated by a pair of molecules; we have to be careful not to count the molecules twice. Since $nb = V_{\text{excl}}$, we can use b to get an idea of molecular sizes. Figure 6.16 lists values of b and the molecular radii r deduced from them.

A second reason for nonideal behavior is that, contrary to what we assumed above, there *are* attractive forces between molecules. The effect of such forces would be to draw the molecules together and hence reduce the pressure. The quantity $(P + n^2a/V^2)$ can be thought of as the ideal pressure. The actual pressure P is less than ideal by the amount n^2a/V^2. The number of moles per unit volume, n/V, is a *concentration*. When squared, it gives the probability of collision between molecules. The proportionality constant a indicates the magnitude of cohesive force between molecules in a collision.

FIG. 6.16 Values of van der Waals Constant b and Derived Molecular Radius r

	b, liters/mol	r, nm
Helium	0.0237	0.133
Neon	0.0171	0.119
Argon	0.0322	0.147
Krypton	0.0398	0.158
Hydrogen	0.0266	0.138
Nitrogen	0.0391	0.157
Water	0.0305	0.145
Benzene	0.115	0.225

Thus, *a,* which is a measure of the deviation from ideality, tells how much drop there is from ideal pressure due to attractive forces.

The attractive forces described by the van der Waals constant *a* are special in that they operate only at very short distances, i.e., during collision. For this reason, they are called *short-range forces.* In some cases it is easy to see where these forces come from. For example, with polar molecules the positive end of one molecule may attract the negative end of another molecule. It is not surprising, therefore, that polar substances such as water deviate markedly from ideal behavior.

It is not so easy to see the reason for attractive forces between nonpolar molecules. For example, why should two neon atoms attract each other? We get a hint of the answer by considering two neon atoms extremely close together, as shown in Fig. 6.17. We can imagine that *at a particular instant* the electron distribution in atom 1 might be unsymmetric, with a slight preponderance of electron charge density on one side. The atom would appear to have one end slightly negative with respect to the other end. The neighboring atom, as a result, would be distorted because the positive end of atom 1 would attract the electrons in atom 2. As shown, there would be an instantaneous dipole in both of the atoms, with consequent attraction. This situation would persist for only an extremely short time because the electrons are in rapid motion. As electrons in atom 1 move to the other side, electrons in atom 2 follow. In fact, we can think of van der Waals forces as arising because electrons in adjacent molecules are beating in time, so as to produce synchronized fluctuating dipoles, which give rise to an instantaneous attraction. The attraction is strong when the particles are close together but rapidly weakens as they move apart. Also, the more electrons there are in a molecule and the less tightly bound these electrons are, the greater are the van der Waals forces.

Attractive forces become less significant as temperature increases. A rise in temperature produces a disordering effect because of molecular motion, which increases in speed as the temperature rises. The attractive forces try to draw the molecules together, but the molecules, because of their motion, stay apart. As the temperature is raised, the molecules have more ability to overcome the attractive forces. The attractive forces are unchanged, but the motion of the molecules increases; hence, the attraction becomes relatively less important.

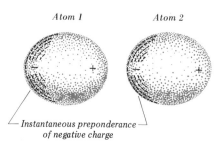

Atom 1   *Atom 2*   **FIG. 6.17** Model of van der Waals attraction.

Instantaneous preponderance
of negative charge

At sufficiently low temperature, attractive forces, no matter how weak, take over and draw the molecules together to form a liquid. This temperature is called the *liquefaction temperature*. Liquefaction is easier at high pressures, where distances between molecules are smaller and hence the influence of intermolecular forces is greater. The higher the pressure, the easier the gas is to liquefy, and the less it needs to be cooled to accomplish liquefaction. Thus, the liquefaction temperature increases with increasing pressure.

6.13
CRITICAL TEMPERATURE

There is for each gas a characteristic temperature above which attractive forces are not strong enough to produce liquefaction no matter how strong a pressure is exerted on the system. This temperature is called the *critical temperature* (T_c). It is defined as the *temperature above which a substance can exist only as a gas*. Above T_c, which is different for different substances, molecular motion is so vigorous that, no matter how high the pressure, the molecular forces are not able to coalesce the molecules into a liquid. Thus, T_c gives a measure of the magnitude of the attractive forces among the molecules.

Figure 6.18 contains values of the critical temperature for some common substances. Listed also is the *critical pressure*, the *minimum pressure that must be exerted to produce liquefaction at the critical temperature*. At temperatures higher than T_c, no amount of pressure can produce liquefaction. For example, in the case of water, above 647°K H_2O exists only as a gas. Such a critical temperature is relatively high and indicates that the attractive forces among H_2O molecules must be quite high. They are great enough that even with the violent thermal motions that exist at 647°K they can produce the liquid state. In the case of sulfur dioxide, the attractive forces between SO_2 molecules are apparently less.

FIG. 6.18   Critical Constants

Substance	Critical temperature, °K	Critical pressure, atm
Water, H_2O	647	217.7
Sulfur dioxide, SO_2	430	77.7
Ammonia, NH_3	406	112.5
Hydrogen chloride, HCl	324	81.6
Carbon dioxide, CO_2	304	73.0
Oxygen, O_2	154	49.7
Nitrogen, N_2	126	33.5
Hydrogen, H_2	33	12.8
Helium, He	5.2	2.3

In the extreme case of He the attractive forces are so weak that liquid He can exist only below 5.2°K. In Fig. 6.18 the order of decreasing critical temperature is also the order of decreasing attractive forces.

6.14
COOLING BY EXPANSION

Substances with high critical temperatures are easy to liquefy; substances with low critical temperatures are difficult to liquefy. In general, the latter must be cooled before they can be liquefied. For example, oxygen cannot be liquefied at room temperature. It must be cooled below 154°K before liquefaction can occur. This cooling would be quite difficult were it not that gases usually cool themselves on expansion.* When a gas expands against a piston, the gas does work in pushing the piston. If the energy for this work comes from the kinetic energy of the gas molecules, a decrease in the kinetic energy of the molecules is observed as a lowering of the temperature. However, a temperature drop may be observed even for a gas which expands into a vacuum and therefore does no external work.

The cause of cooling by unrestrained expansion can be seen by considering the experiment shown in Fig. 6.19. The box shown is perfectly insulated. It is divided into two compartments by a diaphragm. The left compartment contains compressed gas; the right compartment is originally empty. If a hole is punched in the diaphragm, the gas streams into the vacuum. A thermometer would show a drop in temperature. Why? As the gas expands into the empty space, molecules on the average move further apart and hence do work against the attractive forces of their neighbors. This requires energy. Since

*It should be noted that cooling does not occur at *any* temperature but only below the so-called *inversion temperature*. The inversion temperature for most gases is roughly six times the critical temperature. Above the inversion temperature, a gas warms when expanded; below the inversion temperature, it cools. Since most inversion temperatures turn out to be above room temperature, most gases at room temperature cool on expansion and heat on compression. However, the inversion temperatures of hydrogen and helium are well below room temperature (195 and 45°K, respectively); consequently, these gases heat up when allowed to expand (into a vacuum) at room temperature.

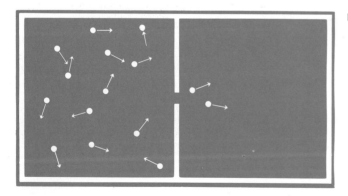

FIG. 6.19 Free expansion of a gas.

no outside energy is available, the molecules must use up some of their kinetic energy. The average kinetic energy, as measured by the temperature, drops. If there were no attractive forces, there would be no cooling effect. Indeed, the fact that cooling is observed indicates that there *are* attractive forces.*

The commercial liquefaction of gases makes use of cooling by expansion. In order to make liquid air, for example, air is first compressed to high pressure, cooled with a refrigerant to remove the heat that accompanies compression, and then allowed to expand.

Important Concepts

volume	Charles' law	ideal-gas law	van der Waals equation
temperature	partial pressure	Graham's law of diffusion	critical temperature
pressure	Avogadro principle	brownian motion	critical pressure
triple point	molar gas volume	kinetic theory	cooling by expansion
Boyle's law	equation of state	nonideal behavior	

Exercises

*6.1   Volume   A distinctive characteristic of a gas is that "it does not retain its volume." Tell what is meant by this statement. What feature of molecular gas structure accounts for it?

*6.2   Volume   IUPAC does not like liter as the unit for volume. In terms of an approved unit, decimeter, what would you call a liter?

*6.3   Volume   How many liters are there in each of the following:
 a One cubic meter (m³)
 b One cubic decimeter (dm³)
 c One cubic centimeter (cm³)
 d One cubic millimeter (mm³)
                                   Ans. 10^3, 1, 10^{-3}, 10^{-6}

*6.4   Temperature   Express each of the following in degrees Celsius and in degrees Kelvin:
 a Room temperature 68°F
 b Body temperature, 98.6°F
 c Book-burning temperature, "Fahrenheit 451"
 d Fahrenheit's snow-salt mix, 0°F

*6.5   Temperature   Suppose you invent a new absolute temperature scale where the triple point of H_2O is given a value of 100°U. What would room temperature (20°C) be on this scale?

*6.6   Temperature   At what temperature do readings on the Fahrenheit and Celsius scales become identical?   *Ans.* $-40°$

---

*The opposite effect, warming on unrestrained expansion, may also be observed above an inversion temperature. Such warming indicates that there must also be repulsive forces between gas molecules. Below the inversion temperature, repulsions are masked by the larger attractions; however, as the temperature is raised, the attractive forces become less important, and the repulsive forces, small as they are, dominate. The repulsions can be viewed as arising from the noninterpenetrability of molecules, and hence they are related to the term *b* of the van der Waals equation.

*6.7 Pressure Suppose you take the barometer shown in Fig. 6.2 and you stick it under 1 m of water. What happens to the mercury level in the tube?

*6.8 Pressure "High vacuum" means about 1×10^{-6} mmHg; "ultrahigh vacuum" means about 1×10^{-8} mmHg. Express these as pascals.

**6.9 Pressure Given the manometer shown in Fig. 6.3 with molten gallium (density 6.095 g/ml) as the liquid. The gallium in the left arm is 40.0 cm above the bottom of the U and the gallium in the right arm is 20.0 cm above the bottom of the U. If $P_{atm} = 0.975$ atm, what is the pressure of the gas?

Ans. 0.857 atm

*6.10 Gas laws Make a graph showing how the product PV (vertical axis) varies as a function of P (horizontal axis) for an ideal gas at constant temperature. How would this graph be changed if the absolute temperature were doubled?

*6.11 Gas laws A commercially delivered tank of compressed oxygen is observed to show a pressure of 2000 lb/in² at 18°C. What should happen to the pressure as the room warms up to 25°C? Assume ideal behavior.

**6.12 Gas laws Given the curve shown in Fig. 6.4 for a specified weight of gas at a fixed temperature, draw on the same axes what the curve should look like if the absolute temperature were increased by 25 percent. Assume ideal behavior.

*6.13 Gas laws A 1-liter balloon filled with helium at a pressure of 1.10 atm breaks in a room that is $4 \times 4 \times 3$ m³. Assuming ideal behavior, what will the pressure of helium become in the room?

*6.14 Gas laws Assuming ideal behavior, how much pressure would be exerted by a sample of gas that expands from 2.5 liters at 0.975 atm to 3.5 liters?

*6.15 Gas laws At constant pressure, a given sample of gas is heated from −25 to +25°C. What should happen to its volume, assuming ideal behavior?

*6.16 Gas laws Assuming ideal behavior what would be the volume at STP of a sample of dry gas occupying 1.00 liter at 100°C and 1.15 atm?

*6.17 Gas laws The allowed threshold limit for the poison gas H_2S is 10 mg/m³. What pressure

does this correspond to? Assume ideal behavior at room temperature 20°C. *Ans. 7.1×10^{-6} atm*

**6.18 van der Waals Given the data shown in Fig. 6.7, which of the gases listed will deviate most from ideal behavior? Explain.

**6.19 Nonideal gases In working with nonideal gases, the volume is usually less than expected and the pressure is usually more than expected. Give a qualitative justification for each of these.

**6.20 Gas laws Point out the flaw in the following line of reasoning: (1) According to Boyle's law, volume is inversely proportional to pressure, so $V_1/V_2 = P_2/P_1$. (2) According to Charles' law, volume is directly proportional to temperature, so $V_1/V_2 = T_1/T_2$. (3) Therefore, setting these equal to each other, we can conclude that $P_2/P_1 = T_1/T_2$, i.e., pressure is inversely proportional to temperature.

**6.21 van der Waals equation Suppose you had a gas in which all the molecules carried a negative charge. How would the van der Waals equation take care of this? Predict how the observed pressure in this case would compare with that assuming ideal behavior.

*6.22 Gas laws How would the slope of the line in Fig. 6.9 be affected by each of the following:
 a Doubling the size of the sample?
 b Doubling the pressure of the sample?
 c Doubling the amount of mercury in the setup used to get the data (Fig. 6.8)?

*6.23 Gas laws Calculate the total pressure in a 5-liter container containing 0.689 g of O_2 and 0.311 g of N_2 at 25°C. Assume ideal behavior.

Ans 0.159 atm

**6.24 Gas laws Given three boxes A, B, and C. Box A contains 1.00 g of helium at 1.25 atm and 25°C, B contains 2.00 g of neon at 2.50 atm and 50°C, and C contains 3.00 g of argon at 75°C. The contents of A and B are now added to C. With the final temperature adjusted to 75°C, the total pressure observed is 6.00 atm. What is the volume of each box? What was the initial pressure in box C?

**6.25 Partial pressures A bubble of dry nitrogen, diameter 1.00 cm, is released at the bottom of a 1-meter cylinder of water. As the bubble rises, it expands because of the decrease in pressure and because of the wetting of the nitrogen gas. Assuming ideal behavior, calculate the diameter of the bubble at the top of the column when the external pressure is 0.975 atm and the water temperature is 27°C.

*6.26 Gas laws If you wanted to collect 0.300 g of methane over water by water displacement as shown in Fig. 6.10, what volume would you need to collect when the atmospheric pressure is 0.975 atm and the water temperature is 28°C? Assume ideal behavior. *Ans. 0.494 liter*

**6.27 Gas laws A given solar cell decomposes water by the reaction $2H_2O(l) \longrightarrow 2H_2(g) + O_2(g)$. What volume of hydrogen (collected over water at a wet pressure of 1.04 atm and a temperature of 40°C) would you expect to get from transfer of 6.0×10^{21} electrons in the above reaction? Assume 100 percent efficiency and ideal behavior.

**6.28 Avogadro principle Assuming hydrogen and nitrogen are diatomic how can you use the observation

Hydrogen + nitrogen $\longrightarrow$ ammonia
  3 liters   1 liter   2 liters

together with the Avogadro principle to deduce the correct molecular formula for ammonia? Make sure your logic is sound.

*6.29 Molecular formula A particular compound X analyzes to 37.83% by weight carbon, 6.35% hydrogen, and 55.82% chlorine. It takes 0.384 g of X to fill a gas volume of 100 cm³ at 114°C and 0.960 atm. What is the molecular formula of X?

**6.30 Air A typical sample of clean, dry air will analyze as follows: 780,900 ppm (parts per million by volume) N_2, 209,500 ppm O_2, 9300 ppm argon, and 300 ppm CO_2. (a) Calculate the composition expressed as percent by weight. (b) Calculate the density at a total pressure of 1.00 atm and 25°C. *Ans. (b) 1.184 g/liter*

**6.31 Internal combustion engine Using isooctane (C_8H_{18}) as a typical fuel, the combustion reaction in an automobile engine can be written

$$C_8H_{18} + 12\tfrac{1}{2}O_2 \longrightarrow 8CO_2 + 9H_2O$$

Taking into account that the oxidant is usually air, which contains 78.09% by volume N_2, 20.95% O_2, and 0.93% Ar, calculate the ideal air-to-fuel ratio, i.e., weight of air required per unit weight of fuel.

*6.32 Stoichiometry A 5.000-g sample of impure MnO_2 (used in flashlight cells) when heated with HCl generates 0.746 liter of chlorine gas at 298°K and 0.965 atm by the reaction

$$MnO_2 + 4H^+ + 2Cl^- \longrightarrow Mn^{2+} + Cl_2(g) + 2H_2O$$

What is the purity of the sample?

*6.33 Diffusion How would you expect the rate of diffusion of $H_2O(g)$ to compare with that of $O_2(g)$? *Ans. 33% faster*

**6.34 Diffusion If the average velocity of O_2 molecules at 0°C is 1700 km/h, what would you predict for $^{235}UF_6(g)$ molecules at the same temperature? What would it be at 25°C?

**6.35 Diffusion One of the classic experiments in chemistry is to inject $NH_3(g)$ and $HCl(g)$ simultaneously into the opposite ends of a glass tube and watch where the product $NH_4Cl(s)$ forms as a ring on the wall of the tube. If, in such an experiment, using a 1-m tube and HX as the acid, the product forms 30 cm from the HX end, what is the molecular weight of HX?

**6.36 Kinetic theory Given two samples (A and B) of H_2 gas, each at the same pressure, but sample A has one-fourth the absolute temperature of sample B. What relation, if any, must exist between each of the following quantities in samples A and B? Justify each answer.
  a Average speed of the molecules
  b Number of molecules per cubic centimeter
  c Volume of the samples
  d Density of the samples

***6.37 Kinetic theory Given a sample each of CH_4 and O_2. Both samples are at the same temperature and have the same density (g/cm³). What relation, if any, must exist between each of the following quantities in the two samples? Justify each answer.
  a Average speed of the molecules
  b Impulse of each impact on the container wall
  c Number of impacts per unit wall area per second

**6.38 Kinetic theory What is meant by the term "mean free path"? Why, in a gas sample, is the mean free path greater than the average spacing between molecules?

***6.39 Kinetic theory With reference to Fig. 6.13, it turns out that the most *probable* value that one would find for the $\frac{1}{2}ms^2$ of a molecule in a gas sample is somewhat less than the *average* value calculated by summing up all the kinetic energies and dividing by the number of molecules. What features of the curve in Fig. 6.13 ensure that this will be true?

6.40   van der Waals   The van der Waals constant b for mercury is 0.01696 liter/mol. Calculate a molecular radius r for Hg and compare with the values listed in Fig. 6.16.

***6.41   van der Waals**   Using the model shown in Fig. 6.15, what would you expect would happen to the excluded volume nb at very high pressures?

6.42   Nonideal behavior   Carbon monoxide, CO, and nitrogen, N_2, are interesting because they have the same masses and the same numbers of electrons. Which of the two would you expect to be closer to ideal behavior? Where would you probably put CO in the table shown in Fig. 6.18? Predict a value for its critical temperature.

***6.43   Critical   temperature**   In   the sequence HF, HCl, HBr, HI, the critical temperatures are 461, 324, 363, and 423°K, respectively. How might you account for this trend in values?

***6.44   Cooling by expansion**   You have a tire at room temperature pumped up to a high pressure with hydrogen. If you let the hydrogen out into the air, it cools down; if you did the experiment in a vacuum, it would heat up. Explain.

Chapter 7

LIQUIDS, SOLIDS, AND CHANGES OF STATE

When a sample of gas below its critical temperature is cooled or compressed, it liquefies. In the process molecules, originally far apart on the average, are slowed down and brought close enough together that attractive forces become significant. The molecules coalesce into a cluster, which settles to the bottom of the container as liquid. What are the general properties of liquids? How can they be accounted for by the kinetic molecular theory?

Whereas the gaseous and liquid states are characterized by disorder, the solid state is ordered. Atoms in a solid are arranged in a regular pattern. The deciphering of the pattern is a challenging problem, since the observed properties are determined by it. As a concept, ordered arrangement in the solid state is very old, being originally based on the observation that crystals show planar faces with characteristic angles between them. However, the working out of the detailed atomic arrangements was made possible only after the discovery of X rays and their application to the study of crystals.

In this chapter, we first consider the liquid state. How does it differ from the gaseous state? Then we take up the solid state and compare its special properties with those of a liquid or a gas. Finally, we take a look at some of the special conditions that hold when all these states exist together.

7.1
PROPERTIES OF LIQUIDS

Liquids are practically incompressible. Unlike gases, there is little change in volume when the pressure on a liquid is changed. Theory accounts for this by assuming that the free space between molecules of a liquid is minimal. Any attempt to compress the liquid meets with resistance as the electron cloud of one molecule repels that of an adjacent molecule.

Liquids maintain their volume no matter what the shape or size of the container. A 10-ml sample of liquid water occupies a 10-ml volume whether in a small beaker or a large flask. Gaseous water, on the other hand, spreads out to fill the whole volume accessible to it. Gases do not maintain their volume because the molecules are independent of each other and move freely into any space available. In liquids, mutual attractions are strong; consequently, the molecules are clustered together.

Liquids have no characteristic shape. In the absence of gravity, as in an orbiting space vehicle, the molecules of a liquid pack into a globular cluster. The

198

Chapter 7
Liquids,
Solids, and
Changes of
State

cluster can readily be deformed, as by poking it, but left to itself it goes back to a spherical shape. When gravity comes into the picture, as on Earth, a liquid sample assumes the shape of the bottom of its container. Apparently there are no fixed positions for the molecules of a liquid. They are free to slide over each other in order to minimize the potential energy with respect to each other and with respect to gravitational attraction by the Earth.

Liquids diffuse, but slowly. When a drop of ink is released in water, there is at first a sharp boundary between ink and water. Eventually the color diffuses throughout the liquid. In gases, diffusion is more rapid. Diffusion is able to occur because molecules have kinetic energy and move. In a liquid, they do not move far before colliding with a neighbor. The mean free path, i.e., the average distance between collisions, is short and comparable to the average spacing between the molecules. Eventually each molecule of a liquid does migrate from one side of its container to the other, but it takes a long time and it has to suffer many billions of collisions in doing so. In gases there is less obstruction. Because a gas is mostly empty space, the mean free path is much longer. Hence the molecules of one gas mix rather quickly with those of another.

Liquids evaporate from open containers. Although there are attractive forces which hold molecules of a liquid together, those molecules having enough kinetic energy can escape into the gas phase. In any collection, a given molecule does not have the same kinetic energy all the time. There are perpetual exchanges due to collisions. As an example, the molecules in a collection might all start out with the same energy, but this situation would not persist. Two or more molecules may simultaneously collide against a third one. Molecule 3 will now have higher than average kinetic energy. If it happens to be near the surface of the liquid, it may be able to overcome the attractive forces of neighbors and go off into the gas phase.

Figure 7.1 shows typical energy distributions for molecules of a liquid. The curves are quite similar to those given previously for a gas (Fig. 6.13). If the value marked *E* corresponds to the minimum kinetic energy required to overcome attractive forces and escape, then all the molecules in the shaded area have enough energy to overcome the attractive forces. These are the "hot" molecules, which have the possibility of escaping provided that they are close enough to the liquid-gas surface.

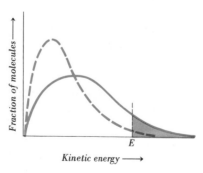

FIG. 7.1 Energy distribution in a liquid. Solid curve applies at one temperature; dashed curve, at a lower temperature.

If it is only highly energetic molecules that leave the liquid, then the average kinetic energy of those left behind must be lowered. Each molecule that escapes carries with it more than an average amount of energy. Since the remaining molecules end up with lower average kinetic energy, their temperature must be lower. Thus, evaporation is generally accompanied by cooling.

When a liquid evaporates from a noninsulated container, such as a beaker, the temperature does not fall very far before there is heat flow from the surroundings into the liquid. Provided the rate of evaporation is not too great, this inflow of heat will be sufficient to match the energy required for evaporation. As a consequence, the temperature of the liquid remains close to that of the room, even though the liquid continues to evaporate. When evaporation proceeds from an *insulated* container, the heat flow from the surroundings is less; so the temperature of the liquid may drop significantly. An example of such an insulated container is the *vacuum bottle* (also called the *Dewar flask*). The essential feature is double-walled construction with vacuum between the walls. The vacuum acts as an insulator to inhibit heat flow.

When a liquid evaporates from a Dewar flask, the flow of heat from the surroundings is generally not fast enough to compensate for the evaporation. The temperature of the liquid will drop, the average kinetic energy of the molecules will decrease, and the rate of evaporation will diminish until the heat flow into the liquid just equals the heat required for evaporation. A lower temperature will be established at which the distribution of the kinetic energies will be shifted to the left, as shown by the dashed curve in Fig. 7.1. The temperature is constant, but at a value which may be considerably lower than the original temperature. Liquid air, an extremely volatile liquid, remains in an open Dewar flask at approximately $-190°C$ for many hours.

7.2
EQUILIBRIUM VAPOR PRESSURE

When a bell jar is placed over a beaker of evaporating water, as shown in Fig. 7.2, the water level drops for a while and then becomes constant. This can be explained as follows: Molecules escape from the water into the gas, or vapor, phase. After escaping, they are confined to a limited space. As they accumulate in the space above the liquid, there is increasing chance that some of them will return to the liquid. Eventually, molecules return to the liquid just as fast as other molecules leave it. At this point the liquid level no longer drops; the number of molecules evaporating per second is equal to the number of molecules condensing per second. Such a condition in which two opposing changes are occurring at exactly the same rate is referred to as *dynamic equilibrium*. The system is not at a state of rest, but it shows no *net* change. The amount of liquid stays constant; the concentration in the vapor is also constant. A particular molecule spends part of its time in the liquid and part in the vapor phase, but as molecules pass from liquid to gas, other molecules move from gas to liquid, keeping the number of molecules in each phase constant.

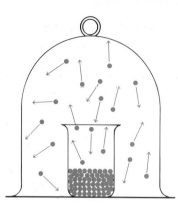

FIG. 7.2   Evaporation in a closed system.

The molecules that are in a vapor exert a pressure. At equilibrium, this pressure is known as the *equilibrium vapor pressure*. It is the *pressure exerted by a vapor when the vapor is in equilibrium with its liquid*. The magnitude depends (1) on the nature of the liquid and (2) on its temperature.

1   The nature of the liquid is involved since each liquid has characteristic attractive forces between its molecules. Molecules such as H_2O which have large mutual attractions have a small tendency to escape into the vapor phase. Such a liquid has a relatively low equilibrium vapor pressure. Liquids composed of molecules with small mutual attractions, such as CH_4, have a high escaping tendency and therefore a high equilibrium vapor pressure.

2   As the temperature of a liquid is raised, the average kinetic energy of the molecules of the liquid increases. The number of high-energy molecules capable of escaping also becomes larger; so the equilibrium vapor pressure increases.

There are various devices for demonstrating equilibrium vapor pressure, one of which is shown in Fig. 7.3. It consists of a mercury barometer set

Fig. 7.3   Measurement of equilibrium vapor pressure.

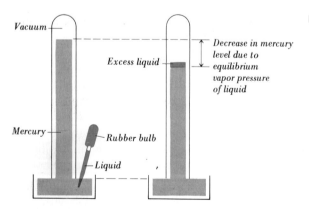

Vacuum

Excess liquid

Decrease in mercury level due to equilibrium vapor pressure of liquid

Mercury

Rubber bulb

Liquid

up in the usual manner. Above the mercury there is initially a vacuum. By squeezing the rubber bulb of the medicine dropper, a drop of the liquid to be measured can be ejected into the mercury under the mouth of the barometer tube. Since practically any liquid is less dense than mercury, the drop will float to the top of the mercury. Enough of it will evaporate to establish an equilibrium between the liquid and vapor. The corresponding vapor pressure now pushes down the mercury column. (The excess liquid also acts to push down the mercury column, but this is a negligible effect, especially when there is little excess.)

The extent to which the mercury level is depressed gives a quantitative measure of the vapor pressure of the liquid. At 20°C, water depresses the column by 17.5 mm; hence, it has an equilibrium vapor pressure of 17.5/760, or 0.023, atm. Similarly, one would find for carbon tetrachloride, CCl_4, 0.120 atm, and for chloroform, $CHCl_3$, 0.211 atm. These vapor-pressure values give an idea of the increasing escaping tendencies of the molecules from these liquids.

By repeating the above experiment at different temperatures, it is possible to determine the vapor pressure as a function of temperature. Appendix 3 is a table showing the results for water in great detail. The general behavior of water, carbon tetrachloride, and chloroform is shown by the graph in Fig. 7.4. As can be seen, when the temperature increases, the vapor pressure rises, first slowly and then more steeply. It continues only up to the critical temperature because above its critical temperature a liquid cannot exist. Steeply rising properties of this kind are better presented as plots of the logarithm of one function against the reciprocal of the other. A plot of the logarithm of the pressure against the reciprocal absolute temperature is shown in Fig. 7.5. As can be seen, the curves of Fig. 7.4 have been transformed into straight lines. The equations for these straight lines are given by

$$\log p = \frac{-\Delta H}{19.15T} + C$$

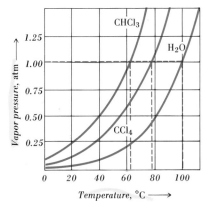

FIG. 7.4 Equilibrium vapor pressure as a function of temperature.

202

Chapter 7
Liquids,
Solids, and
Changes of
State

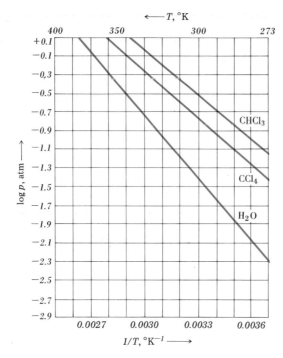

FIG. 7.5 Logarithmic plot of vapor pressure as a function of reciprocal temperature.

where ΔH (in joules per mole) is the heat required to evaporate one mole of liquid* and C is a constant which depends on the liquid and on the units used for expressing pressure.

The constant C can be eliminated by subtracting the above equation for a liquid at a given temperature from that for the same liquid at another temperature. Denoting the different temperatures as T_1 and T_2 and the corresponding vapor pressures as p_1 and p_2, we get

$$\log p_1 - \log p_2 = \frac{-\Delta H}{19.15 T_1} + \frac{\Delta H}{19.15 T_2}$$

$$\log \frac{p_1}{p_2} = \frac{\Delta H}{19.15} \left(\frac{1}{T_2} - \frac{1}{T_1} \right)$$

This last equation, called the *Clausius-Clapeyron equation*, makes it possible to evaluate ΔH by measuring p_1 and p_2 at T_1 and T_2. Conversely, if ΔH is known, the vapor pressure p_1 at one temperature T_1 can be calculated from a measured vapor pressure p_2 at some other temperature T_2.

---

*We are making two assumptions: (1) The vapor is ideal, and (2) ΔH does not change with temperature. If both these assumptions are valid, straight-line plots are obtained as in Fig. 7.5. If either assumption is a poor one, then the value of ΔH will not generally agree with the directly measured heat of vaporization.

EXAMPLE 1

The vapor pressure of carbon tetrachloride is 0.132 atm at 23°C and 0.526 atm at 58°C. Calculate the ΔH in this temperature range.

Solution

$p_1 = 0.132$ atm $\quad$ at $T_1 = 296°K$
$p_2 = 0.526$ atm $\quad$ at $T_2 = 331°K$

$$\log \frac{0.132}{0.526} = \frac{\Delta H}{19.15} \left(\frac{1}{331} - \frac{1}{296} \right)$$

$\Delta H = 32{,}000$ J/mol

EXAMPLE 2

Given that $\Delta H = 32{,}000$ J/mol for carbon tetrachloride and that the vapor pressure is 0.132 atm at 23°C, calculate the vapor pressure expected at 38°C.

Solution

$p_1 = 0.132$ atm $\quad$ at $T_1 = 296°K$

$p_2 = \;?$ atm $\quad$ at $T_2 = 311°K$

$$\log \frac{0.132}{p_2} = \frac{32{,}000}{19.15} \left(\frac{1}{311} - \frac{1}{296} \right)$$

$p_2 = 0.247$ atm

7.3
BOILING POINTS

Boiling is a special case of vaporization; it is the rapid passage of a liquid into the vapor state by means of the formation of bubbles.* A liquid boils at its *boiling point*, the temperature at which the vapor pressure of the liquid is equal to the atmospheric pressure. At the boiling point (abbreviated bp), the vapor pressure of a liquid is high enough that it can push aside the atmosphere (Fig. 7.6) to make room for bubbles of vapor in the interior of the liquid. In general, a molecule can evaporate only if it meets two requirements. It must have enough kinetic energy, and it must be close enough to a liquid-vapor boundary. At the boiling point, bubbling enormously increases the extent of the liquid-vapor boundary. Any heat added to a liquid at its boiling point is used to give molecules the extra energy needed to escape. The average kinetic energy *does not increase*. The *temperature of a pure boiling liquid remains constant while boiling continues.*

*When water is heated in an open container, it is usually observed that as the liquid is warmed, tiny bubbles gradually form at first, and then at a higher temperature, violent bubbling commences. The first bubbling should not be confused with boiling. It is due to the expulsion of air that is usually dissolved in water.

203

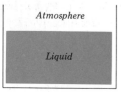

Atmosphere

Liquid

No bubbles

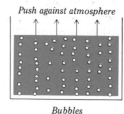

Push against atmosphere

Bubbles

FIG. 7.6 Bubble formation in boiling occurs when vapor pressure exceeds atmospheric pressure.

The boiling point of a liquid depends on atmospheric pressure. For instance, when the pressure is 0.921 atm, water boils at 97.7°C; at 1.00 atm it boils at 100°C. The higher the prevailing pressure, the higher the boiling point. To avoid ambiguity, it is necessary to define a *standard*, or *normal*, boiling point. The normal boiling point is the temperature at which the vapor pressure of a liquid is equal to one standard atmosphere. It is the one usually listed in tables of data and can be determined from a vapor-pressure curve by finding the temperature which corresponds to 1 atm pressure. Figure 7.4 shows that the normal boiling point of water is 100°C, that of carbon tetrachloride is 76.8°C, and that of chloroform is 61°C. In general, the higher the normal boiling point, the greater the attractive forces.*

The change of boiling point with pressure can be computed from the Clausius-Clapeyron equation if we know ΔH and the normal boiling point or the vapor pressures at two temperatures. The following example illustrates such a calculation.

EXAMPLE 3

Given that ΔH of water at 100°C is 40,700 J/mol, calculate the boiling point of water at an altitude of 2600 m where the pressure is 0.750 atm.

Solution

$p_1 = 1.00$ atm   at $T_1 = 373°$K

$p_2 = 0.750$ atm   at $T_2 = ?$

$$\log \frac{1.00}{0.750} = \frac{40,700}{19.15} \left(\frac{1}{T_2} - \frac{1}{373} \right)$$

$T_2 = 365°$K $= 92°$C

7.4
PROPERTIES OF SOLIDS

Solids are nearly incompressible and have low thermal coefficients of expansion. This means that quite large changes of pressure or temperature are required to make an appreciable change in the volume occupied by a given

*It is often remarked in a series of similar compounds, such as CH_4, C_2H_6, and C_3H_8, that the normal boiling point is highest for the compound of greatest molecular weight. Although it is tempting to explain this in terms of gravitational attraction between molecules, such gravitational attraction is exceedingly small. A more reasonable explanation is that heavy molecules usually contain more electrons and hence have greater van der Waals attractions.

solid. According to kinetic molecular theory, this results from strong attractive and strong repulsive forces between closely spaced molecules. Repulsive forces arise from the Pauli exclusion principle (Sec. 4.11) which keeps electron clouds from penetrating each other very much.

Solids diffuse very slowly. For example, rock layers have been in contact with each other for millions of years and still retain sharp boundaries. In some solids, however, more rapid diffusion has been demonstrated; e.g., some metals penetrate each other in a matter of hours when held just below the melting point. Based on the fact that solids consist of atoms in virtual contact, it is reasonable that solid-state diffusion should be a slow process. In fact, from this point of view one might ask why diffusion occurs at all. One principal reason is that solids are imperfect, having, for example, vacancies where atoms or molecules should be. Motion into such vacancies permits diffusion to occur at a rate proportional to the number of vacancies per unit volume.

Solids form crystals, definite geometric forms that are distinctive for the substance in question. As an example, sodium chloride generally crystallizes in the form of cubes with faces that intersect at an angle of 90°. When a crystal is broken, it splits, or shows *cleavage,* along certain preferred directions. Characteristic faces and angles result even when the material is ground to a fine powder. The same chemical substance can crystallize, under different conditions, to form different kinds of crystals. For instance, although sodium chloride almost always crystallizes as cubes, it can be made to crystallize in the shape of octahedra by placing some urea in the solution. The occurrence of different crystal forms for the same chemical substance is called *polymorphism.*

Not all solids are crystalline. Some, such as glass, have the solid-state properties of extremely slow diffusion and virtually complete maintenance of shape and volume but do not have the ordered crystalline state. These substances are called *amorphous solids.* Examples include waxes, plastics, and supercooled liquids.

7.5
DETERMINATION OF SOLID-STATE STRUCTURE

Information about the arrangement of atoms in solids can be gained from the external symmetry of crystals. However, more information is obtained from X-ray diffraction. X rays are like light but more energetic (Sec. 4.5). They can be produced by bombarding a metal target with electrons. The incoming electrons excite the atoms of the metal to a higher energy level. When the atoms go back to a low energy level, they emit energy in the form of X rays.

Figure 7.7 shows the setup for the study of crystalline materials. X rays are collimated into a beam by a lead shield with a hole in it. A crystal is mounted in the path of the X-ray beam. As the X rays penetrate the crystal, the atoms which make up the crystal scatter, or deflect, some of the X rays from their original path. These X rays can be detected by photographic film. The X-ray beam exposes the film, and when the film is developed, a spot appears at each point where part of the beam strikes it. The developed film shows not just one spot, but a pattern which is uniquely characteristic of the crystal investigated.

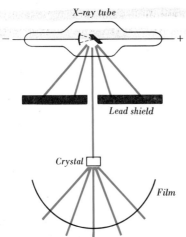

FIG. 7.7 X-ray determination of structure.

Figure 7.8 shows a typical X-ray photograph of a crystal. The center corresponds to the main unscattered beam. (Generally a hole is cut at this point so that the intense blackening that goes with the center spot is avoided.) The surrounding spots represent scattering of part of the original beam through various characteristic angles. The creation of many beams from one is like the effect observed when a light beam falls on a diffraction grating—a piece of glass on which are scratched thousands of parallel lines. The lines form slits for light to pass through. The spreading of light waves from each slit results in interference between waves from different slits, giving rise to a spatial pattern of alternate regions of light and dark. The diffraction of X rays by crystals is similar. Lines of atoms act somewhat like slits for the X rays.

The diffraction of X rays is a complex phenomenon involving the interaction between incoming X rays and the electrons that make up atoms. We can consider X rays to consist of waves of electric and magnetic pulsations. These pulsations interact with electrons in each atom so as to produce corresponding pulsations going out from that atom. Thus, each atom in the path of an X-ray

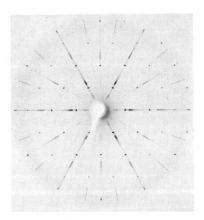

FIG. 7.8 Typical X-ray photograph of a crystal.

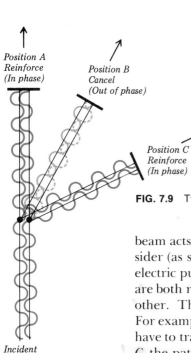

Position A
Reinforce
(In phase)

Position B
Cancel
(Out of phase)

Position C
Reinforce
(In phase)

FIG. 7.9 Two-atom model for X-ray scattering.

Incident
X-ray beam

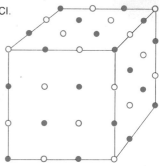

FIG. 7.10 Space lattice of NaCl.

beam acts like a radio relay station that receives and transmits a signal. Consider (as shown in Fig. 7.9) two atoms side by side in an X-ray beam set into electric pulsation at the same frequency. Viewed from position *A*, the signals are both received in phase; the signal from one atom reinforces that from the other. This in general will not be true when viewed at some arbitrary angle. For example, viewed from position *B*, the signals will be out of phase; one will have to travel a bit farther than the other. If, however, as happens at position *C*, the path difference is just equal to a whole wavelength, there will again be reinforcement of the signals.

For a real crystal consisting of many atoms regularly arranged, there will be certain angles at which there will be reinforcement in the emergent beam so as to produce spots on a photographic film. It is possible to work back from the angles at which the spots appear and determine the distance between the various planes of atoms. The relation is given by the *Bragg equation:*

$$n\lambda = 2d \sin \theta$$

Here *n* is a whole number (usually equal to 1), λ is the wavelength of the X-ray beam, and *d* is the distance between the planes of atoms that make an angle 2θ with the direction of the incident beam.

7.6
SPACE LATTICE

A careful mathematical analysis of the spot pattern resulting from X-ray diffraction enables X-ray crystallographers to calculate the positions of the particles that produce such a pattern. The pattern of points which describes the arrangement of molecules or atoms in a crystal is known as a *space lattice.* Figure 7.10 shows the space lattice of NaCl.* Each of the points corresponds to the position of the center of an ion. The open circles locate the

*In the strictest sense, a space lattice is concerned only with points, and all the points in a space lattice must be identical. In this sense, NaCl can be represented by two identical interpenetrating space lattices, one for the positions of the sodium ions and one for those of the chloride ions.

207

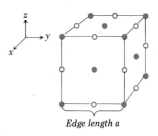

FIG. 7.12   Unit cell of NaCl.

Edge length a

FIG. 7.11   Space-filling model of NaCl. Small spheres represent Na$^+$ ions; large spheres, Cl$^-$ ions.

centers of the positive sodium ions, and the filled circles locate the positions of the negative chloride ions. It must be emphasized that the points do not represent the actual sodium ions and chloride ions; they represent only the positions occupied by the ion centers. In fact, in NaCl the ions are of quite different size and are practically touching each other, as is shown in Fig. 7.11.

The space lattice extends through the entire crystal. However, it is sufficient to consider only enough of it to represent the symmetry of the arrangement. This small fraction, which sets the pattern for the whole lattice, is called the *unit cell.* It is defined as *the smallest portion of the space lattice which when moved repeatedly a distance equal to its own dimension along each direction generates the whole space lattice.* A unit cell of NaCl is the cube shown in Fig. 7.12. If this cube is moved through its edge length *a* in *x, y,* and *z* directions many times, the whole space lattice can be reproduced.

Several kinds of symmetry occur in the unit cells of crystalline substances. The simplest, known as *simple cubic,* is shown in Fig. 7.13*a.* Each point at the corner of the cube represents a position occupied by an atom or a molecule. The dashed lines marked *a, b,* and *c* represent *axes* along which the structure must be extended to reproduce the entire space lattice. The *c* axis is generally oriented up and down; the *b* axis, left and right; the *a* axis, front and back. In the cubic system the directions of the *a, b,* and *c* axes are perpendicular to each other. Thus, with simple cubic, the points are equally spaced along the three mutually perpendicular directions of space. Closely related to simple cubic symmetry is *body-centered cubic,* the unit cell for which is shown in Fig. 7.13*b.* It has points at the corners of a cube plus an additional point in the center. In *face-centered cubic* there are points at the corners with additional points in the middle of each face, as shown in Fig. 7.13*c.*

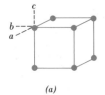

(a)

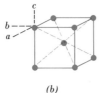

(b)

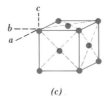

(c)

FIG. 7.13   Unit cells: (*a*) simple cubic, (*b*) body-centered cubic, (*c*) face-centered cubic.

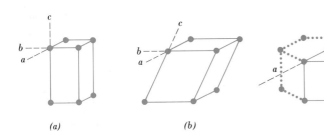

FIG. 7.14 Unit cells: (a) tetragonal, (b) monoclinic, (c) hexagonal.

Other kinds of symmetry are more complicated. To produce *tetragonal* symmetry, the cube needs to be elongated along the *c* direction. The lines of points still form right angles with each other, but the distance between points along one axis differs from that along the other two. Figure 7.14*a* shows a tetragonal unit cell. Although not shown in Fig 7.14, a *rhombic* (also called orthorhombic) unit cell retains mutually perpendicular edges but has point separation unequal in the *a, b,* and *c* directions. In *monoclinic* crystals the *c* axis is tilted with respect to the *ab* plane. An example is shown in Fig. 7.14*b*. In *triclinic* symmetry none of the three axes *a, b,* and *c* is perpendicular to any of the others. *Hexagonal* symmetry, as shown in Fig. 7.14*c*, has lattice points arranged so as to form hexagons. Each hexagon has an additional lattice point in its middle so as to create a rhombus-based prism, as indicated by the solid lines in Fig. 7.14*c*. For describing hexagonal symmetry, either the solid or dotted entity can be used as the unit cell.

7.7
PACKING OF ATOMS

(a)

(b)

FIG. 7.15 Close-packing of spheres.

The unit cells just discussed concern only points that locate atomic or molecular centers. Atoms are space-filling entities, and structures can be described as resulting from the packing of spheres. The most efficient, called *closest-packing,* can be achieved in two ways, one of which is called *hexagonal close-packing* and the other, *cubic close-packing.* Hexagonal close-packing can be built up as follows: Place a sphere on a flat surface. Surround it with six equal spheres as close as possible in the same plane. Looking down on the plane, the projection is as shown in Fig. 7.15*a*. Now form over the first layer a second layer of equally bunched spheres, offset as shown in Fig. 7.15*b* so as to nestle into the depressions formed by the first-layer spheres. The third layer is now added so each sphere is directly above a sphere of the first layer. Succeeding layers follow in alternating fashion *ababab . . . .*

The other close-packed structure, called *cubic close-packed,* results if the buildup of layers *a* and *b* is as described above but the third layer (call it *c*) is added so it is neither directly above layer *a* nor layer *b*. For this structure the sequence of layers is *abcabc . . . .*

Among the common materials that crystallize with hexagonal close-packing are many of the metals, such as magnesium and zinc. Also showing hexag-

210

Chapter 7
Liquids,
Solids, and
Changes of
State

onal close-packing is solid H_2, where apparently tumbling of the H_2 molecules makes them equivalent to spheres. Cubic close-packing is shown by other metals, such as copper, silver, and gold, as well as by simple rotating molecules such as CH_4 and HCl.

Close-packing of spheres can also be used to describe many ionic solids. For example, NaCl can be viewed as a cubic close-packed array of chloride ions in which sodium ions are fitted into "holes" between the chloride layers. Holes can be of two kinds. One (called a *tetrahedral "hole"*) has four spheres adjacent to it, and the other (called an *octahedral "hole"*) has six. The difference is illustrated in Fig. 7.16 which is a projection view of two close-packed layers. The *a* layer is represented by filled balls, and the *b* layer above it by open balls. In the left part of the figure, marked by color, is shown a grouping of four adjacent balls (three from layer *a* and one from layer *b*) that forms a hole. This is called a tetrahedral hole because a small atom inserted in the hole would have four neighboring atoms arranged at the corners of a regular tetrahedron. The grouping of four spheres and its relation to a tetrahedron is shown at the bottom left of the figure. In the right part of Fig. 7.16, also marked by color, is shown the grouping of six adjacent balls (three from layer *a* and three from layer *b*) that forms an octahedral hole. A small atom in this hole would have six neighbors at the corners of an octahedron.

In the NaCl structure the Na^+ ions can be regarded as being in the octahedral holes created between the layers of Cl^- ions. Similarly, MgO, which has the same structure as NaCl (called the *rock-salt structure*), can be regarded as an array of close-packed oxide ions with a magnesium ion in each of the octahedral holes. The rock-salt structure is found in many oxides and sulfides.

More complicated structures can be built up by using both kinds of holes simultaneously. A most important example is the *spinel* structure, named after the mineral spinel, $MgAl_2O_4$. It consists of a cubic close-packed array of oxide ions with Mg^{2+} in some of the tetrahedral holes and Al^{3+} in some of the

FIG. 7.16 Tetrahedral "holes" and octahedral "holes" between layers of close-packed spheres.

octahedral holes. Spinels in which dipositive ions (Mg^{2+}) reside in the tetrahedral holes and tripositive ions (Al^{3+}) are in the octahedral holes are referred to as *normal* spinels. There are other spinels in which the dipositive ions go into the octahedral holes and the tripositive ions are distributed half-and-half between the tetrahedral and the octahedral holes. Such compounds are called *inverse* spinels. The mineral magnetite (also called lodestone), Fe_3O_4, is an interesting example of an inverse spinel. It can be written $Fe(II)Fe(III)_2O_4$, corresponding to the presence of Fe^{2+} and Fe^{3+}. The Fe^{2+} ions occupy octahedral holes, and the Fe^{3+} ions are divided between tetrahedral and octahedral interstices. This gives rise to high electric conductivity, intense light absorption (responsible for the black color), and ferromagnetism. Information storage in computers makes use of magnetic alignment of ions in spinel-type materials.

7.8
SOLID-STATE DEFECTS

The preceding discussion implied ideal crystals. An ideal crystal is one that contains no *lattice defects*. There are several important kinds of lattice defects. One, called *lattice vacancies*, arises if some of the lattice points are unoccupied, i.e., some of the atoms are missing. Another, called *lattice interstitials*, arises if atoms are squeezed in so as to occupy extra positions between normal lattice points. All crystals are imperfect to a slight extent and contain defects. For example, in NaCl some of the sodium ions and chloride ions are always missing from the regular pattern (Fig. 7.17*a*). In silver bromide, AgBr, not only are some silver ions missing from regular positions, but also they are found squeezed in at abnormal sites between other ions (Fig. 7.17*b*). Lattice vacancies help to explain how diffusion and ionic conductivity, small as they are, do occur in the solid state. Lattice interstitials are considerably less probable; they occur when small, positive ions move between large, negative ions. The presence of interstitial Ag^+ in AgBr is believed to be important in the formation of a photographic image when AgBr crystals are exposed to light. (See Sec. 19.11.)

Another kind of defect commonly found in solids is the *dislocation*. There are two generals types: *edge dislocations* and *screw dislocations*. As shown in Fig. 7.18, an edge dislocation results from a plane of atoms inserted only part way into a crystal. In a screw dislocation, there is a line of atoms about which the crystal planes are warped, to give an effect similar to

```
+ - + - + - +        + - + - + - +
- + ( ) + - + -      - + - + - + -
+ - + - + - +        + - ( ) - + ( ) +
- + - ( ) - + -      - + - + - ( ) + -
+ - + - + - +        + - + - + - +
- + - + - + -        - + - + - + -
      (a)                  (b)
```

FIG. 7.17 Lattice defects: (*a*) vacancies in NaCl, (*b*) misplaced Ag^+ in AgBr.

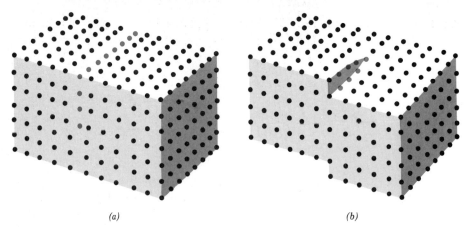

FIG. 7.18 Dislocations: (*a*) edge dislocation, (*b*) screw dislocation. (The edge dislocation can be more clearly seen by tilting the page and viewing at a small angle.)

the threads of a screw. The points where edge or screw dislocation lines come out the surface of a crystal represent points of strain and enhanced chemical reactivity. Etching of crystals occurs preferentially at such points. Current research aimed at producing materials more resistant to corrosive attack and greater mechanical strength involves elimination of dislocations and other solid-state defects.

In addition to structural imperfections, there are defects of a more chemical nature. These are associated with chemical impurities, which may be there accidentally or might have been deliberately introduced. Such impurities can change the properties of materials and hence be exploited in producing new materials with desirable combinations of properties.

Practical utilization of impurity defects is made for silicon crystals. The electric conductivity of this group IV element in the pure state is extremely low. However, on addition of trace amounts of elements from either group III or group V the conductivity is greatly enhanced. Silicon has the diamond structure in which each atom is bonded to four tetrahedral neighbors by four single covalent bonds. These bonds require all four valence electrons of the Si atom. In pure Si there are no extra electrons, so there is no conductivity (except at very high temperatures). When, however, a group V element, e.g., P or As, is substituted for Si, an extra electron above that required for forming four covalent bonds is introduced. This extra electron is easily mobile; hence, a phosphorus-doped Si crystal exhibits marked conductivity. If, on the other hand, a group III element, e.g., B or Al, is substituted for Si, an electron deficiency is introduced in the covalent bonding network. Such an electron vacancy, or "hole" as it is sometimes called, can move through the covalent structure as a nearby electron moves in to fill the "hole." The vacancy thus moves to a neighbor; and so on. In this way, the introduction of electron vacancies allows electron motion to occur where it normally would not. Conductors in which electric transport is mainly by "excess electrons" are called *n* type (*n* for negative), and those in which electric transport is mainly by "holes" are called *p* type (*p* for positive).

An interesting application of impurity conductors results from combination of n- and p-type materials to form a junction, the so-called "np" junction. This device can pass electric current more easily in one direction than in the reverse, and hence it can act as a *rectifier* for converting alternating current (ac) to direct current (dc).

Finally, we might note formation of nonstoichiometric compounds as a special kind of solid-state defect. Classic examples include $Cu_{2-x}S$, MnO_{2-x}, TiO_{1-x}, and Na_xWO_3. In these materials there are generally vacancies that can be populated by excess of one element over what is required by simple rules of stoichiometry. For example, TiO is cubic with a rock-salt arrangement of Ti^{2+} and O^{2-} except that 15 percent of each kind of site is vacant. Addition of excess Ti or O can change the Ti/O ratio by as much as 0.15. One striking result of deviation from stoichiometry is intensification of color; most nonstoichiometric materials are deep blue or black. Another result is usually enhancement of electric conductivity.

7.9
BONDING IN SOLIDS

Instead of classifying solids by symmetry or by mode of packing, it is sometimes more useful to classify them by the types of units that occupy the lattice points and their kinds of binding. There are four such common classifications: *molecular, ionic, covalent,* and *metallic.* Figure 7.19 lists the nature of the units that occupy the lattice points, the forces that bind these units together, the characteristic properties of the solids, and some typical examples.

FIG. 7.19   Classification of Solids

	Molecular	Ionic	Covalent	Metallic
Units	Molecules	Positive ions Negative ions	Atoms	Positive ions in electron gas
Binding force	van der Waals Dipole-dipole	Electrostatic attraction	Shared electrons	Electric attraction between + ions and − electrons
Properties	Very soft Low melting point Volatile Good insulators	Quite hard and brittle Fairly high melting point Good insulators	Very hard Very high melting point Nonconductors	Hard or soft Moderate to very high melting point Good conductors
Examples	H_2 H_2O CO_2	NaCl KNO_3 Na_2SO_4 $MgAl_2O_4$	Diamond, C Carborundum, SiC Quartz, SiO_2	Na Cu Fe

214

Chapter 7
Liquids,
Solids, and
Changes of
State

Molecular solids are those in which lattice points are occupied by molecules. The bonding *within* the molecules is covalent and, in general, much stronger than the bonding between the molecules. The bonding between the molecules can be of two types, dipole-dipole or van der Waals. Dipole-dipole attraction is encountered in those solids consisting of polar molecules, e.g., ice; van der Waals attractions (Sec. 6.12) are present in all solids. Because the total intermolecular attraction is generally small, molecular crystals usually have low melting temperatures. They are usually quite soft because the molecules can easily be displaced from one site to another. Finally, they are poor conductors of electricity because there is no easy way for an electron to jump from one molecule to another. Most substances which exist as gases at room temperature form molecular solids.

In an *ionic solid* the units that occupy the lattice points are positive and negative ions. The forces of attraction are high. Hence ionic solids usually have fairly high melting points, well above room temperature. Sodium sulfate, for example, melts at 884°C. Also, ionic solids tend to be brittle and fairly hard, with great tendency to fracture by cleavage. In the solid state, the ions are generally not free to move; therefore, these ionic substances are generally poor conductors of electricity. However, when melted, they become good conductors.

In a *covalent solid* the lattice points are occupied by atoms, which share electrons with their neighbors. The classic example is diamond, in which each carbon atom is joined by pairs of shared electrons to four other atoms, as shown in Fig. 7.20. Each of these carbon atoms, in turn, is bound to four carbon atoms, and so on, giving a giant three-dimensional molecule. In any solid of this type the bonds between the individual atoms are covalent and, usually, are quite strong. Substances with covalent structures generally have high melting points, are quite hard, and frequently are poor conductors of electricity.

In a *metallic solid* the points of the space lattice are occupied by positive ions. This array of positive ions is permeated by a cloud of highly mobile electrons, the *electron gas,* derived from the outer atomic shells. In solid sodium, for example, Na^+ ions are arranged in a body-centered cubic pattern with a cloud of electrons arising from the $3s$ electrons. The crystal is held together by the attraction between the positive ions and the electron gas. In some

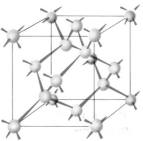

Diamond

FIG. 7.20 Diamond structure. (Black atoms represent back face of cube.)

metals such as tungsten, there is also considerable covalent binding between the positive ions superimposed on the ion–electron gas attraction. Because electrons can wander at will throughout a metal, metallic solids are characterized by high electric and thermal conductivity. Their other properties may vary widely. Sodium, for example, has a low melting point; tungsten has a very high melting point. Sodium is soft and can be cut with a knife; tungsten is very hard.

What does the term "molecule" mean when applied to these various solids? In a molecular crystal, for example, solid CO_2, it is possible to distinguish discrete molecules. Each carbon has relatively close oxygens; all other atoms are at considerably greater distances. In an ionic substance, such as NaCl, this is not true. Each Na^+ is equally bound to its neighboring six Cl^-. But each Cl^- in turn is bonded to its neighboring six Na^+, so they also must be counted as part of the same aggregate. Actually, all the ions in the crystal belong to the same giant molecule. A similar situation occurs in metallic and covalent crystals. The term "molecule" is not particularly useful for such cases.

7.10
CRYSTAL ENERGIES

The magnitude of the attractive forces in crystals can be gauged by the *crystal energy* (also called *lattice energy*). This is the amount of energy required to convert one mole of material from the solid state to a gaseous state composed of the same units as occupy the lattice points. In the case of a molecular solid, as, for example, in the process

$$H_2O(s) \longrightarrow H_2O(g)$$

the crystal energy, 43.5 kJ/mol, corresponds to the *energy of sublimation*, i.e., the energy needed to convert one mole of the solid to one mole of molecular gas. For ionic solids, the crystal energy is the energy needed to separate one mole of solid into a gaseous mixture of positive and negative ions. As example, 770 kJ/mol is needed for the process

$$NaCl(s) \longrightarrow Na^+(g) + Cl^-(g)$$

For a covalent solid, the crystal energy is that needed to form an atomic gas. In the case of diamond, the process would be

$$C(s) \longrightarrow C(g)$$

for which the energy is 710 kJ/mol.

Some typical values of the crystal energy are listed in Fig. 7.21. As can be seen, the values for molecular substances, such as CO_2, are relatively low, indicating small intermolecular attractions. The largest values are generally found for ionic solids.

FIG. 7.21   Observed Crystal Energies, kJ/mol

Ne	2.5	NaCl	770	C	710
CO_2	23.4	MgO	3920	SiO_2	1720
Cl_2	31.0	CaF_2	2610	Na	105
H_2O	43.5	$MgAl_2O_4$	19,800	W	840

7.11
HEATING CURVES

When heat is added to a solid, the solid warms up. However, when it reaches the melting point, even though heat continues to be added, the solid stays at a fixed temperature while the melting process occurs. A similar phenomenon occurs at the boiling point. Addition of heat to a boiling liquid does not raise the temperature but is used to convert liquid to gas. Why is it that at some temperatures addition of heat results in an increase in temperature; at others, not? To answer this question, we look at the general problem of adding heat to a sample of substance. We start at absolute zero with solid, melt it to give a liquid, and finally boil it to produce the gaseous state. The temperature variations that accompany these changes are represented in Fig. 7.22. The curve shown is a *heating curve*. It corresponds to the uniform addition of heat. Because heat is added at a constant rate, distance on the time axis is also a measure of the amount of heat added.

At time t_0 the temperature starts out at absolute zero. As heat is added, each particle vibrates back and forth about a lattice point, which thus represents only the center of this motion. As heat is added, the vibration becomes more vigorous. Though no change is visible, the crystal progressively becomes slightly expanded and slightly less ordered. The heat added increases the average kinetic motion of the particles. Since temperature measures average kinetic energy, temperature rises along this portion, marked 1 in Fig. 7.22.

At the melting point (mp) the vibration of particles has become so energetic that any added heat serves to loosen the binding forces that keep neigh-

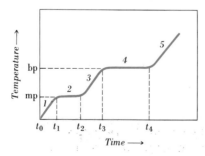

FIG. 7.22 Heating curve.

boring particles in an ordered array. Consequently, from time t_1 to t_2, along portion 2 of the curve, added heat goes not to increase the average kinetic energy but to increase the *potential energy*. Work is done against attractive forces. There is no change in average kinetic energy, so the substance stays at the same temperature. The *temperature at which solid and liquid coexist at equilibrium* is defined as the *melting point* of the substance.

The amount of heat necessary to melt one mole of solid is called the *molar heat of fusion*. It is equal to the amount of heat energy that must be added to overcome the extra attraction that exists in the solid. At the normal melting point, the heat of fusion of NaCl is 30.33 kJ/mol; that of H_2O is 6.02 kJ/mol. These values reflect the greater attractive forces in NaCl.

Eventually (at time t_2), sufficient heat has been added to tear all the particles from the solid crystal structure. Along portion 3 of the curve, added heat again increases the average kinetic energy; so the temperature rises. This continues until the boiling point (bp) is reached. At the boiling point added heat is used to overcome the attraction of one particle for its neighbors in the liquid. There is an increase in the potential energy of the particles but no change in their average kinetic energy. Along portion 4 of the curve, the liquid sample is converted to gas. Finally, after all the liquid has been boiled off, added heat again raises the average kinetic energy of the particles; this is shown by the rising temperature along portion 5.

The amount of heat necessary to vaporize one mole of liquid is called the *molar heat of vaporization*. This quantity gives a measure of the attractive forces characteristic of the liquid. At the normal boiling point, the heat of vaporization of water is 40.7 kJ/mol, and that of chloroform is 29.5 kJ/mol. The higher value for water indicates that the attractive forces between water molecules are greater than those between chloroform molecules.

7.12
COOLING CURVES

The *cooling curve* results when heat is removed at a uniform rate from a substance. The temperature as a function of time looks like the curve shown in Fig. 7.23. As heat is removed from gas, the temperature drops along the line marked *g*. During this time, the average kinetic energy decreases in order to compensate for removal of energy to the outside. Slowing down proceeds until particles are so sluggish that attractive forces become dominant.

At t_1 the particles coalesce to form a liquid. Because energy is required to take a particle from liquid to gas, the reverse process, in which a particle is taken from gas to liquid, releases energy. This decrease of potential energy compensates for the heat being removed from the system. Thus, as liquefaction proceeds, the temperature does not fall; particles on the average do not slow down. As a result, gas and liquid both stay at the same temperature, and the average kinetic energy of the particles in both phases is the same. During

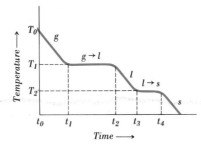

FIG. 7.23 Cooling curve.

the time interval t_1 to t_2, the temperature remains constant at T_1, the *condensation*, or *liquefaction*, temperature.

At time t_2 all the gas particles have condensed into the liquid state. Continued removal of heat causes the particles to slow down further. The average kinetic energy decreases, and the temperature drops, as is shown along the line marked l. This drop continues until t_3, when the liquid begins to crystallize.

As each succeeding particle is attracted into position to form the crystal structure, its potential energy drops. Thus, the removal of heat to the outside is compensated by energy available from the decrease in potential energy; the average kinetic energy stays constant during the solidification process. At the solidification temperature motion is not on the average slower in the solid than it is in the liquid, but it is more restricted. From time t_3 to t_4 temperature remains constant as liquid converts to solid. When all the particles have crystallized, further removal of heat drops the temperature, as shown along the final part s of the curve.

The cooling curve (Fig. 7.23) is just the reverse of the warming curve (Fig. 7.22). The temperature at which gas converts to liquid (liquefaction point) is the same as the temperature at which liquid converts to gas (boiling point). Similarly, the temperature at which liquid converts to solid (freezing point) is the same as the temperature at which solid converts to liquid (melting point).

7.13
SUPERCOOLING

Most cooling curves are not quite so simple as the one shown in Fig. 7.23. An undershoot may occur on going from gas to liquid and again from liquid to solid. Figure 7.24 shows the situation for the liquid-to-solid transition. Instead of following the dashed, flat portion, the temperature follows the dip. The liquid does not crystallize as it should at the freezing point but instead *supercools*. Supercooling arises in the following way: The particles of a liquid have little recognizable pattern and move around in a disordered manner. At the freezing point they should line up in a characteristic crystalline arrangement. Often they do not snap into the correct pattern immediately. If heat continues to be removed without crystallization, the temperature may fall

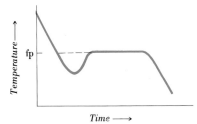

FIG. 7.24 Supercooling.

below the freezing point. Particles continue moving through various patterns until enough of them hit the right one. Once the correct pattern has been built to sufficient size, other particles rapidly crystallize on it. Many particles crystallize simultaneously, and potential energy converts to kinetic energy faster than energy is being removed to the outside; the effect is to heat up the sample. The temperature increases until it coincides with the freezing-point temperature. From there on, the behavior is normal.

Supercooling may be reduced by two simple methods: One is to stir the liquid as vigorously as possible. This increases the chance of forming the right crystal pattern. The other method involves introducing a *seed* crystal to provide a proper structure on which further crystallization can occur.

Some substances never crystallize in cooling experiments, but instead they remain permanently in the *undercooled*, or *supercooled*, state. Such substances are frequently called *glasses*, or *amorphous materials*. They owe their existence to the fact that supercooled atoms may be cooled so far that they are trapped in a disordered arrangement.

Amorphous materials are quite common. They include, besides glass, many plastics, such as polyethylene, vinyl polymers, and Teflon. They have many of the properties of solids, but their X-ray pictures are quite different from the X-ray pictures of crystalline solids. Instead of a spot pattern, the X-ray pictures show concentric rings, like those for a liquid. The existence of these rings indicates that there is a certain amount of order, but it is far from perfect. Another indication that amorphous materials are not true solids is their behavior on being broken. Instead of showing cleavage with formation of flat faces and characteristic angles between faces, they generally break to give *conchoidal fractures*. These are shell-shaped depressions such as are observed on the chips of a broken bottle.

7.14
SOLID-GAS EQUILIBRIUM

As with a liquid, particles of a solid can occasionally escape into the vapor phase to establish a vapor pressure. Those particles which at any one time are of higher than average energy and are near the surface can overcome the attractive forces of their neighbors and escape into the vapor phase. If the solid is confined in a closed container, eventually there will be enough particles in the vapor phase that the rate of return becomes equal to the rate of escape. A

220

Chapter 7
Liquids,
Solids, and,
Changes of
State

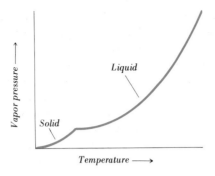

FIG. 7.25 Temperature variation of vapor pressure.

dynamic equilibrium is set up; in it there is an equilibrium vapor pressure characteristic of the solid. Since the escaping tendency depends on the intermolecular forces in the particular solid considered, the equilibrium vapor pressure differs from one substance to another. If the attractive forces are small, as with solid hydrogen, the escaping tendency is great, and vapor pressure is high. In an ionic crystal, such as NaCl, the binding forces are large; the vapor pressure is low.

The vapor pressure of solids also depends on temperature. The higher the temperature, the more energetic the particles, and the more easily they can escape. The more that escape, the higher the vapor pressure. Figure 7.25 shows how the vapor pressure of a typical substance changes with temperature. At absolute zero the particles have no escaping tendency, so the vapor pressure is zero. As the temperature is raised, the vapor pressure rises. It rarely gets to be very high before the solid melts. Above the melting point the vapor-pressure curve is just that of the liquid.

In Fig. 7.25, any point along the portion of the curve marked "Solid" corresponds to equilibrium between vapor and solid. The number of particles leaving the solid is equal to the number returning. When the temperature is raised, more particles shake loose than return. This causes a net increase in the concentration of particles in the vapor phase. This in turn causes an increased rate of condensation to the solid. Eventually, equilibrium is reestablished with a higher vapor pressure at the new temperature.

The behavior of an equilibrium system when it is upset by external action is the subject of a famous principle enunciated by Henry-Louis Le Châtelier in 1884. Le Châtelier's principle states that *if a stress is applied to a system at equilibrium, then the system readjusts, if possible, to reduce the stress*. Raising the temperature of a solid-vapor equilibrium applies a stress in the form of added heat. Since the conversion from solid to gas uses up heat

$$\text{Solid} + \text{heat} \rightleftharpoons \text{gas}$$

the stress of added heat can be absorbed by converting a portion of the solid into gas. So, on addition of heat, the system moves to a state of higher vapor pressure.

At the point where the vapor-pressure curve of a liquid intersects the vapor-pressure curve of the solid (i.e., where the vapor pressure of solid equals the vapor pressure of liquid), there is simultaneously an equilibrium between solid and gas, between liquid and gas, and between solid and liquid. This point of intersection, where *solid, liquid, and gas coexist in equilibrium with each other*, is called the *triple point.* Every substance has a characteristic triple point. For H_2O the triple-point temperature is $273.16°K$ and the triple-point pressure is 0.00603 atm. The triple-point temperature is not precisely the same as the normal melting point, $273.15°K$. The difference comes about because normal melting is under 1 atm pressure, whereas triple-point melting is under the vapor pressure of the substance only. As we shall see in the next section, the melting point of H_2O decreases under increasing pressure.

7.15
PHASE DIAGRAMS

The relation between solid, liquid, and gas can be summarized on a single graph known as a *phase diagram.* Every substance has its own particular phase diagram and this can be worked out from experimental observations at various temperatures and pressure. Figure 7.26 shows the phase diagram of

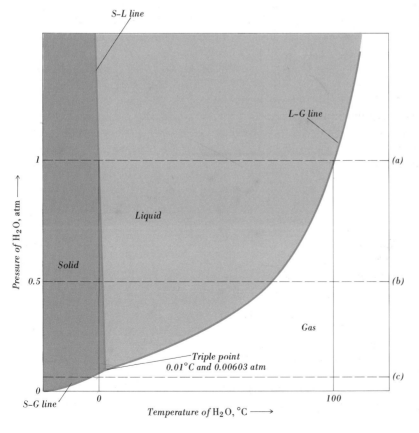

FIG. 7.26 Phase diagram of H_2O. (Scale of axes is somewhat distorted.)

H_2O. On this diagram, which has coordinates of pressure and temperature, the substance H_2O is represented as being in the solid, liquid, or gaseous state, depending on its temperature and pressure. Each of the three differently labeled regions corresponds to a one-phase system. For all values of pressure and temperature falling inside such a single-phase region, the substance is in the state specified. The heavy lines that separate one region from another are equilibrium lines, representing an equilibrium between the two phases shown. The *S-L* line represents equilibrium between solid and liquid; the *L-G* line, equilibrium between liquid and gas; and the *S-G* line, equilibrium between solid and gas. The intersection of the three lines corresponds to the triple point, where all three phases are in equilibrium with each other.

The usefulness of a phase diagram can be illustrated by considering the behavior of H_2O when heat is added at constant pressure. This corresponds to moving across the phase diagram from left to right. We distinguish three representative cases:

1 Pressure of the H_2O is kept at 1.0 atm The experiment is this: A chunk of ice is placed in a cylinder so as to fill the cylinder completely. A piston resting on the ice carries a weight which corresponds to 1 atm of pressure. The H_2O starts as a solid. As heat is added, the temperature of the H_2O moves along the dashed line *a* in Fig. 7.26. When the *S-L* line is reached, the added heat melts the ice. Solid-liquid equilibrium persists at the normal melting point of 0°C until all the solid has been converted to liquid. There is no gaseous H_2O thus far. As heating continues, liquid H_2O warms up from 0°C until the *L-G* line is reached. At 100°C the vapor pressure of the liquid has become great enough to move the piston out to make room for the vapor phase. Since the external pressure is fixed at 1 atm, the piston keeps moving out as heat is added and liquid converts completely to gas at the normal boiling temperature. From then on, the gas simply warms up.

2 Pressure of the H_2O is kept at 0.5 atm Again the H_2O starts out as the solid. The temperature is raised, moving to the right along dashed line *b*. The H_2O stays as a solid until it reaches the *S-L* line. This time, because of the slight tilt of the *S-L* line toward the left (the tilt has been somewhat exaggerated in Fig. 7.26), the temperature at which melting occurs is slightly higher than it was at 1.0 atm. At 0.5 atm ice melts not at 0°C but at 0.005°C. After all the ice has melted, further addition of heat warms the liquid up until boiling occurs when the *L-G* line is reached. Boiling occurs when the vapor pressure of the water reaches the externally set 0.5 atm. The temperature at which this would happen is 82°C, considerably lower than the normal boiling point of 100°C. Above 82°C at 0.5 atm only gaseous water exists.

3 Pressure of the H_2O is kept at 0.001 atm If the pressure exerted by the H_2O is kept at 0.001 atm, the H_2O will exist only along the dashed line marked *c*. As the temperature is raised, solid H_2O warms up until it reaches

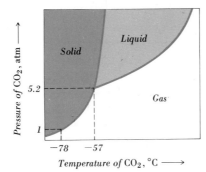

FIG. 7.27   Phase diagram of CO_2.

the *S-G* line. Solid-gas equilibrium is established; solid converts to gas. For 0.001 atm, this would happen at $-20°C$. When all the solid has evaporated at this temperature, the temperature of the gases rises. *There is no melting in this experiment and no passage through the liquid state.*

An interesting aspect of the H_2O phase diagram is that the *S-L* line, representing equilibrium between solid and liquid, tilts to the left as one goes to higher pressure. This is unusual; for most substances the *S-L* line tilts to the right. The direction of the tilt is important since it tells whether the melting point rises or falls with increased pressure. In the case of H_2O, as pressure is increased (moving upward on the diagram), the temperature at which solid and liquid coexist decreases (moving leftward on the diagram). The melting point decrease is $0.01°C/atm$.

The lowering of the ice melting point by increased pressure is consistent with Le Châtelier's principle. The argument goes as follows: The density of ice is 0.9 g/cm^3; the density of liquid water is 1.0 g/cm^3. A given mass of H_2O occupies a larger volume as ice than as liquid. An equilibrium system consisting of water and ice at 1 atm will be at the normal melting point of $0°C$. If the pressure on the H_2O is now increased, there is a stress, which the system can relieve by shrinking in volume. It can shrink in volume by converting some of the ice to water; hence, melting is favored. But melting is an endothermic process; it requires heat. If the system is insulated, the only source of heat is the kinetic energy of the molecules. Consequently, the molecules must slow down, and the temperature must drop.

Another phase diagram of interest is that of carbon dioxide, CO_2. In general appearance, as shown in Fig. 7.27, it is similar to that of H_2O. However, as is more normal, the solid-liquid equilibrium line tilts to the right instead of to the left; i.e., the melting point of CO_2 rises with increased pressure. The triple-point pressure is 5.2 atm. Since this is considerably above normal atmospheric pressure, the liquid phase of CO_2 is not generally observed. To get liquid CO_2, the pressure has to be higher than 5.2 atm. At 1 atm, only solid and gas exist. When solid CO_2 is used as a refrigerant in the form of dry ice, the conversion of solid to gas normally occurs at $-78°C$.

Important Concepts

liquid state
evaporation
equilibrium vapor pressure
normal boiling point
Clausius-Clapeyron equation
solid state
X-ray diffraction
space lattice
unit cell
close-packing
tetrahedral vs. octahedral site

lattice defects
vacancies
interstitials
dislocations
n type
p type
molecular solid
ionic solid
covalent solid
metallic solid
electron gas

crystal energy
heating curves
cooling curves
supercooling
Le Châtelier principle
phase diagram
triple point
melting point
boiling point
changes of state

Exercises

*7.1 Properties of liquids How does kinetic theory account for the fact that liquids are practically incompressible? How might you account for the fact that the compressibility of a liquid generally increases with a rise in temperature?

*7.2 Properties of liquids If molecules of a liquid tend to stick to each other, why does a liquid sample not ball up into one corner of a container?

**7.3 Properties of liquids If molecules of a liquid were tightly packed together, it would be impossible for diffusion to occur. What does this suggest about the kinetic molecular model for a liquid?

**7.4 Properties of liquids What molecular properties should a liquid have in order to show little evaporation?

**7.5 Properties of liquids If you pour ether on your hand it feels much colder than if you poured water of an equal temperature on it. Explain.

**7.6 Dewar flask When a Dewar flask for containing liquid air goes bad, in the sense that it will no longer keep liquid air, it is said to "have a leak." Why should this make the flask bad?

**7.7 Vapor pressure Suppose the gas phase shown under the bell jar in Fig. 7.2 were completely pumped out at a certain time. What would you then observe? What effect does changing the air pressure have on the equilibrium vapor pressure shown at a given temperature?

*7.8 Vapor pressure Why is it necessary to have some excess liquid on top of the mercury column for the experiment shown in Fig. 7.3?

*7.9 Vapor pressure On a given day when the barometric pressure is 0.965 atm and the temperature is 20°C, precisely how many millimeters would the mercury level fall on injection of chloroform into the setup shown in Fig. 7.3?

**7.10 Vapor pressure The vapor pressure of benzene is 40 mmHg at 7.6°C and 100 mmHg at 26.1°C. Predict its value at 20.0°C.

4.19×10^2 °K

Ans. 74 mmHg

**7.11 Vapor pressure The vapor pressure of chlorobenzene is 0.1316 atm at 70.7°C and 0.5263 atm at 110,0°C. Predict the normal boiling point of chlorobenzene.

***7.12 Vapor pressure Chlorobenzene, C_6H_5Cl, has a vapor pressure of 0.132 atm at 70.7°C and 0.526 atm at 110.0°C. Bromobenzene, C_6H_5Br, has a vapor pressure of 0.132 atm at 90.8°C and 0.526 atm at 132.3°C. Which of these substances requires more heat to evaporate 1 mol of liquid? How much more?

*7.13 **Boiling point** Explain clearly why the boiling point of a liquid depends on the prevailing atmospheric pressure.

***7.14 **Boiling point** Suppose you have a 1-m-high column of water at the normal boiling point of 100.0°C. The density of liquid H_2O at 100.0°C is 0.958 g/ml. What difference in boiling point should there be between the top and bottom of the column? *Ans. 2.5°C*

7.15 **Boiling point Airplanes frequently cruise at an altitude of 36,000 ft, which is about 11 km. At this altitude the normal outside pressure is 0.223 atm. Calculate what the boiling point of water would be under this condition. *Ans. 61.7°C*

7.16 **Properties of solids, liquids, and gases Given a sample of H_2O at its triple point so that solid, liquid, and gaseous phases coexist, compare qualitatively each of the following properties in the three phases: (*a*) average kinetic energy per molecule, (*b*) average potential energy per molecule, (*c*) number of molecules per cubic centimeter, and (*d*) mean free path.

***7.17 **Solids** How might you explain the observation that diffusion in solids rapidly increases just below the melting point?

7.18 **X-ray diffraction Why would the diffraction effect illustrated in Fig. 7.9 not be feasible using light waves instead of X rays?

***7.19 **X-ray diffraction** Given a square array of points equidistant by 0.387 nm along both the *x* and *y* directions, imagine an X ray of wavelength 0.154 nm traveling in the *xy* plane. At what angle to the incident beam would you look for the X-ray diffraction from the atomic planes that cut the *x* and *y* axes at 45°? *Ans. 32.7°*

*7.20 **Space lattice** By referring to Fig. 7.10, show that in the space lattice of NaCl, each Na^+ has six near-neighbor Cl^- ions and each Cl^- has six near-neighbor Na^+ ions.

7.21 **Unit cells Counting each corner point as $\frac{1}{8}$, each edge point as $\frac{1}{4}$, and each face point as $\frac{1}{2}$ (to allow for sharing with neighboring unit cells), calculate how many lattice points there are

per unit cell for each of the following: simple cubic, body-centered cubic, face-centered cubic.

7.22 **Unit cells For NaCl, the unit-cell edge length *a* as shown in Fig. 7.12 has a value of 0.5627 nm. Calculate in grams per cubic centimeter the theoretical density of NaCl assuming a perfect structure. *Ans. 2.178 g/cm³*

*7.23 **Unit cell** A lattice point may be occupied by an atom, a molecule, or some other convenient unit. Using a hypothetical (Na^+) (Cl^-) pair as such a unit, show that the unit cell in Fig. 7.12 is describable as face-centered cubic.

7.24 **Unit cell If each of the unit cells shown in Fig. 7.13 has an edge length *a*, what is the distance between nearest-neighbor lattice points?

***7.25 **Packing of atoms** Suppose you had a collection of glass marbles in hexagonal close-packing. If the density of glass is 2.6 g/cm³, what would be the average density of matter in the collection? What fraction of the space is actually unoccupied? *Ans. 1.93 g/cm³, 0.2595*

***7.26 **Packing of atoms** Suppose you had identical hard spheres of radius 1.00 cm packed together in each of the following arrangements: (*a*) simple cubic, (*b*) body-centered cubic, (*c*) face-centered cubic. How big an impurity sphere could you slip into the interstices of each of the above packings?

7.27 **Defects X-ray diffraction shows that the unit-cell edge length of NaCl is 0.5627 nm. The measured density is 2.164 g/cm³. Attributing the difference between theoretical and observed density to lattice vacancies, what percent of the Na^+ and Cl^- must be missing? *Ans. 0.63%*

7.28 **Defects In a given crystal of silver bromide, 1.10 percent of the Ag^+ ions and Br^- ions are completely missing. Another 1.86 percent of the Ag^+ ions have been moved from their normal lattice sites to interstitial sites. Calculate the density of the crystal. The structure is a rock-salt structure with unit-cell edge length equal to 0.5755 nm.

7.29 **Defects The density of pure silicon is 2.33 g/cm³. How many *n*-type electrons per cubic centimeter would you expect if the silicon were contaminated with 0.100 weight percent of phosphorus? How many *p*-type holes would there be from 0.100 weight percent boron? What would you predict if both B and P were present at these concentrations?

***7.30 Defects** Suppose you had a piece of silicon, the left half of which was n type and the right half was p type. On an atomic basis, why might you expect this junction to conduct electron current more readily from left to right than vice versa?

7.31 Bonding in solids Classify each of the following in terms of the units that occupy lattice points and the nature of the bonding between the units: $CaCO_3$, O_2, CO_2, CaO, Ca, Fe, Fe_3O_4, $FeSO_4$, SO_2, P-doped Si.

7.32 Bonding in solids Given an alkali element A and a halogen B. Compare A, B, and AB in the solid state with respect to probable units that occupy lattice points, nature of the bonding between the units, and probable properties of each solid.

****7.33 Types of solid** In solid CO_2, each C has two near-neighbor oxygen atoms. In solid SiO_2, each Si has four near-neighbor oxygen atoms. CO_2 is low-melting and volatile; SiO_2 is high-melting and nonvolatile. Show how this comes about. (*Hint:* Si is below C in the periodic table.)

****7.34 Bonding in solids** Both nitrogen and carbon have a great tendency to form covalent bonds. Yet elemental nitrogen forms a molecular solid whereas elemental carbon forms a covalent solid. Describe the bonding in each and suggest a reason for the difference.

7.35 Crystal energies NaCl, NaBr, and NaI all have the same rock-salt structure. How would you expect the crystal energies of these to change in the sequence? Explain.

7.36 Cooling curve Draw a temperature vs. time plot for the uniform withdrawal of heat from a sample that starts out as molten naphthalene ($C_{10}H_8$, mp 80.55°C) and ends up as solid. Account for each characteristic feature of the curve. How would the curve be changed if you started with a bigger sample?

****7.37 Cooling curve** Why should the melting point of a substance be the same as its freezing point?

****7.38 Supercooling** Glycerin, $C_3H_8O_3$, has a great tendency to supercool. A story is told about a logging camp in the cold North where glycerin was used on the hands of the loggers to protect their hands against chapping. The glycerin (mp 17.8°C) always stayed liquid, no matter how cold it got. Then, one day, when it was particularly cold, the glycerin suddenly crystallized. From then on, they could never get the supercooled glycerin again. Explain.

7.39 Supercooling Window glass is obviously "solid," yet it is more properly described as a supercooled liquid. What would be the essential distinction?

****7.40 Triple point** Given a sealed box that contains nothing but ice, liquid water, and water vapor and another box that contains all the above plus enough helium in the gas phase to bring the total pressure up to 1 atm. What is the pressure of the H_2O in each phase in each of the boxes? What is the temperature in each?

****7.41 Triple point** Criticize the following definition: The triple point is where solid, liquid, and gas coexist.

7.42 Phase diagram In a typical phase diagram, what significance is to be associated with each of the following features:
a An arbitrary point
b A line or curve
c A place where curves intersect

7.43 Phase diagram Using the phase diagram for H_2O shown in Fig. 7.26, describe what you would observe if you (a) moved from left to right across the diagram at $P_{H_2O} = 0.9$ atm, (b) moved from bottom to top at $T_{H_2O} = 90°C$.

****7.44 Le Châtelier principle** Use the Le Châtelier principle to explain the following observation: Two heavy weights joined by a fine wire are hung across a block of ice. As time goes on, the wire works its way down through the ice leaving the block intact.

****7.45 Le Châtelier principle** Use the Le Châtelier principle to account for the following observation: A tank containing liquid carbon dioxide at room temperature has its valve opened into a canvas bag. The bag fills with a solid (dry ice) and the whole system gets quite cold.

Chapter 8

SOLUTIONS

The preceding discussion of the solid, liquid, and gaseous states was limited to pure substances. In practice, we continually deal with mixtures; hence the question that arises is the effect of mixing in a second component. A mixture is classified as *heterogeneous* or *homogeneous*. A heterogeneous mixture consists of distinct phases, and the observed properties are the sum of those of the individual phases; a homogeneous mixture or solution consists of a single phase which has properties that may differ in several important ways from those of the individual components.

8.1
TYPES OF SOLUTIONS

Solutions, defined as *homogeneous mixtures of two or more components,* can be gaseous, liquid, or solid. Since all gases mix in all proportions, any mixture of gases is homogeneous and is a solution. Air is a gaseous solution of nitrogen, oxygen, carbon dioxide, argon, and other molecules. The kinetic picture of a gaseous solution is like that of a pure gas, except that the molecules are of different kinds. Ideally, the molecules move independently of each other.

Liquid solutions are made by dissolving a gas, liquid, or solid in a liquid. If the liquid is water, the solution is called an *aqueous* solution. In the kinetic picture of an aqueous sugar solution, the sugar molecules are distributed at random throughout the bulk of the solution.

Solid solutions are solids in which one component is randomly dispersed on an atomic scale throughout another component. As in a pure crystal, the packing is orderly, but there is no particular order in which the lattice points are occupied by one or another kind of atom. Solid solutions are of great practical importance since they make up a large fraction of the class of substances known as *alloys.* An alloy may be defined as a combination of two or more metallic elements which itself has metallic properties; an alloy may be a compound or a solution. *Sterling silver,* for example, is an alloy consisting of a solid solution of copper in silver. In *brass,* which is a general term for alloys of copper and zinc, some of the copper atoms of face-centered cubic copper are randomly replaced by zinc atoms. Some kinds of *steel* can be considered solid solutions in which carbon atoms are located in some of the spaces between iron atoms. Not all alloys are solid solutions. Some alloys, such as bismuth-

227

cadmium, are heterogeneous mixtures containing tiny crystals of the constituent elements. Others, such as $MgCu_2$, are compounds which contain atoms of different metals combined in definite proportions.

Two terms that are commonly used in the discussion of solutions are *solute* and *solvent*. The substance present in larger amount is generally referred to as the solvent, and the substance present in smaller amount, as the solute. However, the terms can be interchanged whenever it is convenient. The term *solvated* is often used to describe species that result from interaction of one or more molecules of solvent with the solute. If the solvent is water, the term *hydrated* may be used instead.

8.2
CONCENTRATION

The properties of solutions, e.g., the color of a dye solution or the sweetness of a sugar solution, depend on the relative amount of solute compared to solvent or to total amount of solution. This is what is called the solution concentration. There are several common methods for describing concentration. They include mole fraction, molarity, molality, normality, percent solute, and formality.

Mole Fraction

The *mole fraction* is the ratio of the number of moles of one component to the total number of moles in the solution. For example, in a solution containing 1 mol of alcohol and 3 mol of water, the mole fraction of alcohol is $\frac{1}{4}$, and that of water $\frac{3}{4}$. The mole fractions of all the components must add up to give unity.

EXAMPLE 1

Given a solution containing 10.0 g of C_2H_5OH and 30.0 g of H_2O, what are the mole fractions of the two components?

Solution

$$\frac{10.0 \text{ g of } C_2H_5OH}{46.07 \text{ g/mol}} = 0.217 \text{ mol of } C_2H_5OH$$

$$\frac{30.0 \text{ g of } H_2O}{18.015 \text{ g/mol}} = 1.67 \text{ mol of } H_2O$$

Total moles $= 0.217 + 1.67 = 1.89$ mol

$$\text{Mole fraction of } C_2H_5OH = \frac{0.217 \text{ mol}}{1.89 \text{ mol}} = 0.115$$

$$\text{Mole fraction of } H_2O = \frac{1.67 \text{ mol}}{1.89 \text{ mol}} = 0.885$$

Molarity

The *molarity,* as already discussed in Sec. 2.8, is the number of moles of solute per liter of solution and is usually designated by M. A 3.0-molar solution of sulfuric acid would be labeled 3.0 M H_2SO_4. It corresponds to adding 3.0 mol of H_2SO_4 to enough water to make a liter of solution.

EXAMPLE 2

Given a solution made by dissolving 10.0 g of H_2SO_4 in enough water to make 100.0 ml of solution, what is the molarity of H_2SO_4?

Solution

$$\frac{10.0 \text{ g of } H_2SO_4}{98.07 \text{ g/mol}} = 0.102 \text{ mol of } H_2SO_4$$

$$\frac{0.102 \text{ mole of } H_2SO_4}{0.100 \text{ liter of solution}} = 1.02 \ M \ H_2SO_4$$

Molality

The *molality* is the number of moles of solute per kilogram of solvent. It is usually designated m. The label 3.0 m H_2SO_4 is read "3.0 molal" and represents a solution made by adding 1 kg of solvent to every 3.0 mol of H_2SO_4.

EXAMPLE 3

Given a solution made by dissolving 10.0 g of H_2SO_4 in 200.0 g of H_2O, what is the molality of H_2SO_4?

Solution

$$\frac{10.0 \text{ g of } H_2SO_4}{98.07 \text{ g/mol}} = 0.102 \text{ mol of } H_2SO_4$$

$$\frac{0.102 \text{ mol of } H_2SO_4}{0.200 \text{ kg of solvent}} = 0.510 \ m \ H_2SO_4$$

Normality

The *normality,* discussed in greater detail in Chap. 9, is a special concentration unit used for acids and bases or for oxidizing and reducing agents. It tells the number of equivalents of solute per liter of solution. One equivalent of acid furnishes one mole of H^+; one equivalent of base picks up one mole of H^+. Similarly, one equivalent of reducing agent furnishes one mole of e^-; one equivalent of oxidizing agent picks up one mole of e^-. As a concentration unit, normality is usually designated as N. The label 0.25 N $KMnO_4$ is read "0.25 normal" and represents a solution which contains 0.25 equivalent of potassium permanganate per liter of solution.

EXAMPLE 4

Given a solution made by dissolving 15.0 g of $KMnO_4$ in enough water to make 250.0 ml of a solution that is to be used for a reaction where MnO_4^- goes to Mn^{2+}, what is the normality of the solution?

Solution

$$\frac{15.0 \text{ g of } KMnO_4}{158.0 \text{ g/mol}} = 0.0949 \text{ mol of } KMnO_4$$

Oxidation number (Sec. 2.9) of Mn changes from $+7$ in MnO_4^- to $+2$ in Mn^{2+}. Therefore, each MnO_4^- ion picks up five electrons. Each mole of $KMnO_4$ picks up 5 mol of electrons, and so is equal to 5 equivalents.

$$(0.0949 \text{ mol of } KMnO_4)\left(\frac{5 \text{ equiv}}{1 \text{ mol}}\right) = 0.475 \text{ equiv of } KMnO_4$$

$$\frac{0.475 \text{ equiv of } KMnO_4}{0.250 \text{ liter of solution}} = 1.90 \text{ } N \text{ } KMnO_4$$

Percent of Solute

The *percent of solute* is a frequently used designation for solutions, but it is ambiguous. *Percent by weight* is the percent of total solution weight contributed by the weight of solute. Thus, 3% H_2O_2 *by weight* means 3 g of H_2O_2 per 100 g of solution. *Percent by volume* is the percent of *final* volume represented by the volume of solute used to make the solution. For example, 12% alcohol *by volume* means 12 ml of alcohol plus enough solvent to bring the total volume up to 100 ml. It is a curious fact that, whereas percents by weight for a given solution always add up to 100%, percents by volume generally do not.

EXAMPLE 5

Given a solution made by mixing 10.00 ml of C_2H_5OH and 30.00 ml of H_2O in which the total volume is observed to be 39.09 ml. What is the percent by volume of each component?

Solution

$$\frac{10.00 \text{ ml of } C_2H_5OH}{39.09 \text{ ml of solution}} \times 100 = 25.58\% \text{ by vol}$$

$$\frac{30.00 \text{ ml of } H_2O}{39.09 \text{ ml of solution}} \times 100 = 76.75\% \text{ by vol}$$

Formality

Occasionally one finds use of still another concentration unit called *formality*. It is essentially the same as molarity and refers to the number of formula-weights of solute per liter of solution. It is used to make a distinction between what is placed in solution and what is actually there. Thus, for example, 1 F HCl stands for "one formal hydrochloric acid," made by dissolving 1 formula-weight (or 36.5 g) of HCl in enough water to make 1 liter of solution. Actually there are no HCl molecules in the solution since they are all broken up into H^+ and Cl^- ions.

8.3
PROPERTIES OF SOLUTIONS

How are the properties of a solvent affected by addition of solute? Suppose we consider the following experiment: Two beakers, one containing pure water (beaker I) and the other containing a sugar-water solution (beaker II), are set under a bell jar, as shown in Fig. 8.1. As time goes on, the level of the pure water in beaker I is observed to drop while the level of the solution in beaker II rises. There is net transfer of water from pure solvent to solution. Apparently, the escaping tendency, or vapor pressure, of H_2O from pure water is higher than that from the sugar-water solution.

Another observation which supports the idea that addition of a solute lowers the escaping tendency of solvent molecules is lowering of the freezing point. For example, when sugar is added to water, the solution does not freeze at 0°C, but it is found necessary to cool *below 0°C* in order to freeze out the ice. The tendency of H_2O to escape from liquid into solid is decreased by the presence of solute.

The *lowering of the freezing point* and the *reduction of the vapor pressure* are found, at least in dilute solutions, to be *directly proportional to the concentration of solute particles.* Why should there be such lowerings? The first impulse might be to ascribe the lowerings to greater attractive forces. The fact is, however, that the amount of lowering of the vapor pressure and the reduction of the freezing point do not seem to depend on the strength of the interactions between solute and solvent. It is not the strength of the intermolecular attractions that is most important (though there is an effect there) but the *relative number of particles.* The effect of solute is to reduce the relative concentration of H_2O molecules. In the solution, less than 100 percent of the molecules are H_2O molecules, and therefore the escape of H_2O molecules from the solution is less probable than their escape from pure water.

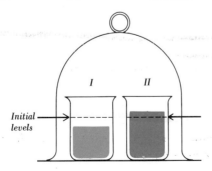

FIG. 8.1 Experimental proof that the escaping tendency of H_2O from pure water (I) is greater than that from an aqueous sugar solution (II).

Initial levels

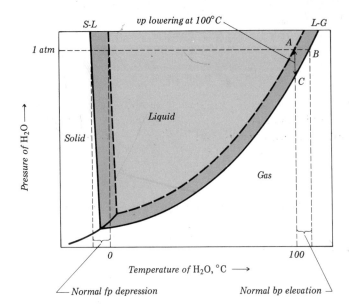

FIG. 8.2 Comparison of phase diagram of pure water (shown by dashed lines) and the phase diagram of an aqueous solution (shown by solid lines).

Figure 8.2 shows a typical effect on the solvent water of adding a particular concentration of nonvolatile solute. The dashed lines represent the phase diagram of pure H_2O; the solid lines, the aqueous solution. The solid line on the left is the new S-L line and represents the temperatures at which pure solid H_2O freezes out when the particular solution is cooled at different pressures. The solid line on the right is the new L-G line and represents the temperatures at which pure gaseous H_2O boils off when the solution is heated at various pressures.

The most striking feature of the new phase diagram is the extension of the liquid range both to higher and to lower temperatures. The liquid phase of water has been made more probable by dissolving solute in it. Associated with this is the fact that the vapor pressure of the water has been reduced. Because the vapor pressure has been lowered, the solution no longer boils at 100°C but requires some higher temperature. The new normal boiling point, which is the temperature at which the vapor pressure again becomes equal to 1 atm, will be higher because of widening of the liquid-phase area. Similarly, there is a depression of the normal freezing point.

Quantitatively, the lowering of vapor pressure is described by *Raoult's law*. It states that *the fractional lowering of the vapor pressure of a solvent is equal to the mole fraction of solute present.* Symbolically, this can be written

$$\frac{p_1^0 - p_1}{p_1^0} = x_2$$

where p_1^0 is the vapor pressure of the pure solvent, p_1 is the partial pressure of the solvent above the solution, and x_2 is the mole fraction of solute. Since the mole fraction of solvent x_1 and mole fraction of solute x_2 must add up to give unity, we can write alternatively

232

$$\frac{p_1^0 - p_1}{p_1^0} = 1 - x_1$$

$$\frac{p_1}{p_1^0} = x_1 \quad \text{or} \quad p_1 = x_1 p_1^0$$

The last relation states that there is a direct proportionality between the solvent vapor pressure p_1 and its mole fraction x_1. The proportionality constant is p_1^0. In practice, Raoult's law is an idealization that is generally best realized in dilute solutions. As the concentration of solute is increased, the vapor pressure of the solvent component deviates from ideal behavior.

EXAMPLE 6

At 25°C the vapor pressure of pure water is 0.0313 atm (23.76 mmHg). What will it become if we dissolve 10.0 g of sugar, $C_6H_{12}O_6$, in 100.0 g of water?

Solution

$$\frac{10.0 \text{ g of } C_6H_{12}O_6}{180.2 \text{ g/mol}} = 0.0555 \text{ mol of } C_6H_{12}O_6$$

$$\frac{100.0 \text{ g of } H_2O}{18.015 \text{ g/mol}} = 5.551 \text{ mol of } H_2O$$

$$\text{Mole fraction of } H_2O = \frac{5.551 \text{ mol}}{0.0555 + 5.551 \text{ mol}} = 0.990$$

$$\text{Vapor pressure of } H_2O \text{ over the solution} = \left(\begin{array}{c}\text{vapor pressure}\\\text{of pure } H_2O\end{array}\right)\left(\begin{array}{c}\text{mole fraction}\\\text{of } H_2O\end{array}\right)$$

$$= (0.0313 \text{ atm})(0.990)$$

$$= 0.0310 \text{ atm} \quad \text{or} \quad 23.6 \text{ mmHg}$$

Raoult's law explains why freezing-point lowering and boiling-point elevation are proportional to the concentration of solute molecules in solution. As the concentration of solute is increased, the mole fraction of solvent is decreased, and its vapor pressure decreases proportionately. As shown in Fig. 8.2, the extent to which the solid curve falls below the dashed curve (AC in the diagram) is the amount of vapor-pressure lowering; it controls both the boiling-point elevation (AB in the diagram) and the freezing-point lowering.

For the boiling point, because of vapor-pressure lowering, we need to increase the temperature to get boiling to occur. The required temperature increase is proportional to the vapor-pressure decrease.* Since vapor-pressure decrease is proportional to mole fraction of solute, it follows that eleva-

*To see how this comes about, look at triangle ABC in Fig. 8.2. The size of this triangle depends on how much solute is added to the solvent; no matter how big the triangle is, side AB will be proportional to AC. In words, boiling-point elevation is proportional to drop in vapor pressure. The more dilute the solution, the better the argument holds.

tion of the boiling point should be directly proportional to mole fraction of solute. These relations can be written as follows:

$$\Delta T_{bp} \sim \Delta p \sim x_2$$

where ΔT_{bp} is the elevation of the boiling point, Δp is the vapor-pressure lowering, and x_2 is the mole fraction of solute. By definition,

$$x_2 = \frac{n_{solute}}{n_{solute} + n_{solvent}}$$

where n is the number of moles. In dilute solution, n_{solute} is very small compared with $n_{solvent}$; so it can be neglected in the denominator:

$$x_2 \sim \frac{n_{solute}}{n_{solvent}}$$

Given a solution of molality m, we have m moles of solute per kilogram of solvent:

$$\frac{n_{solute}}{n_{solvent}} = \frac{m}{1000/W}$$

where W is the molecular weight of the solvent. Since $1000/W$ is a constant for a particular solvent it follows that for *dilute solutions in a given solvent*, the boiling-point elevation will be proportional to the molality of the solute. This can be written

$$\Delta T_{bp} = K_{bp}m$$

The proportionality constant K_{bp} is called the *molal boiling-point elevation constant*. Its value depends on the solvent. In water, K_{bp} equals 0.52°C per 1 m solution. This means that 1 mol of dissolved particles in 1 kg of water raises the boiling point by 0.52°C. This value compares with 2.6°C/m in benzene, 1.24°C/m in ethyl alcohol, and 5.05°C/m in carbon tetrachloride.

EXAMPLE 7

Given a solution that contains 10.0 g of sugar, $C_6H_{12}O_6$, in 250.0 g of water, what will be the boiling point?

Solution

$$\frac{10.0 \text{ g of } C_6H_{12}O_6}{180.2 \text{ g/mol}} = 0.0555 \text{ mol of } C_6H_{12}O_6$$

$$\frac{0.0555 \text{ mol of } C_6H_{12}O_6}{0.250 \text{ kg of solvent}} = 0.222 \ m$$

Boiling-point elevation = $(0.222 \ m)(0.52°C/m) = 0.12°C$

Boiling point = $100.00 + 0.12 = 100.12°C$

The quantitative description of freezing-point lowering is similar to the above. The freezing-point lowering is again proportional to the molality:

$$\Delta T_{fp} = K_{fp} m$$

where K_{fp} is a proportionality constant called the *molal freezing-point lowering constant*. As with K_{bp}, the constant K_{fp} depends on solvent, but not on solute. In water, K_{fp} equals $1.86°C/m$, which means that 1 mol of dissolved particles in 1 kg of water depresses the freezing point by $1.86°C$. This compares with $5.1°C/m$ in benzene and $6.9°C/m$ for naphthalene.

The above derivations are approximate and are justified only for dilute solutions. (In concentrated solutions K_{bp} and K_{fp} may vary somewhat with concentration.) Two other conditions must be satisfied: The first is that the solute must be dissolved *only in the liquid phase;* i.e., the solute should not be volatile, nor should it dissolve in the solid phase. The second point is that the number of solute particles should be the same as the number of molecules placed in solution. If the dissolving process breaks up molecules into smaller fragments, then it will be the total concentration of particles that determines the freezing and boiling points of the solution.

The following problem shows how we can use freezing-point lowering to find out about the molecular formula of a dissolved substance.

EXAMPLE 8

When 0.946 g of fructose (a sugar) is dissolved in 150 g of water, the resulting solution is observed to have a freezing point of $-0.0651°C$. What must be the corresponding molecular weight of fructose? If the simplest formula of fructose is CH_2O, what must be its molecular formula?

Solution

$$\frac{0.946 \text{ g of fructose}}{0.150 \text{ kg of solvent}} = 6.31 \text{ g of fructose per kilogram of } H_2O$$

The observed freezing point is $-0.0651°C$. For a 1 m solution it would have been $-1.86°C$. Therefore, the concentration of the solution is

$$\frac{0.0651°C}{1.86°C/m} = 0.0350 \ m$$

and 6.31 g of fructose must be 0.0350 mol of fructose:

$$\frac{6.31 \text{ g}}{0.0350 \text{ mol}} = 180 \text{ g per mole of fructose}$$

The molecular formula must be some multiple x of CH_2O, or $(CH_2O)_x$. The simplest-formula weight is $12 + 2 + 16 = 30$ amu. The molecular weight is 180 amu, or x times 30; so $x = 6$. The molecular formula is $(CH_2O)_6$, or $C_6H_{12}O_6$.

Osmosis is a process whereby components of a solution diffuse through a membrane. Such a membrane is said to be *semipermeable* if it allows transit of one kind of molecule but not of another. The simplest kind of semipermeability occurs when water passes through the membrane but a solute does not. Cell walls in living tissue are excellent examples of semipermeable membranes. Artificial membranes can be made by precipitating, for example, copper ferrocyanide, $Cu_2Fe(CN)_6$, in the walls of a porous clay pot. Such a membrane allows free passage of water but holds back sugar molecules.

What happens if we have a semipermeable membrane with pure water on one side and a sugar-water solution on the other? The water tends to move from the pure water side, where water concentration is higher, to the sugar-water side, where water concentration is lower. The result would be to pile up water on the sugar-water side. We can restrain the flow by imposing an extra pressure on the sugar-water side. This excess pressure required to prevent a net flow of water is called the *osmotic pressure*.

Figure 8.3 shows a device for measuring osmotic pressure. The membrane divides the cell into two compartments; one contains water plus solute and the other contains water alone. Without the piston that presses down on the solution side, the water level on the pure water side would tend to fall as water streamed through the membrane from right to left. To prevent this, pressure would need to be applied via the piston; it gives the osmotic pressure.

The osmotic pressure does not depend on the specific membrane in the apparatus but only on the *concentration of solute* in the solution and the *temperature* of the solution. In the ideal case, we find that the osmotic pressure (designated by π) is directly proportional to the molar concentration of solute and the absolute temperature. Accordingly, we can write

$$\pi = \text{constant} \left(\frac{n}{V}\right) T$$

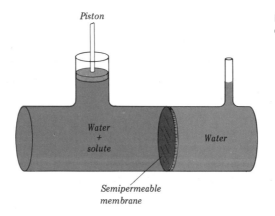

Piston

Water + solute

Water

Semipermeable membrane

FIG. 8.3 Device for demonstrating osmotic pressure.

The constant of proportionality turns out to be the same as the gas constant R; the resemblance of this equation to the ideal-gas law suggests that solute molecules can be visualized as a sort of "gas." In fact, the osmotic pressure corresponds roughly to the pressure the solute would exert if all the solvent were removed and the solute acted as a gas. For example, a sugar solution that is 0.2 m shows at 20°C an osmotic pressure of 5.06 atm. If the solute behaved as a "perfect gas" we would expect a pressure of 4.81 atm.

Of what use is osmotic pressure? One function is to maintain fluid equilibrium. For example, blood has an osmotic pressure of 8 atm. If blood cells are placed in water, they swell and tend to burst as H_2O diffuses into the cells. If they are placed in concentrated NaCl solution, they shrivel up and collapse as H_2O diffuses out of the cells. Only if the osmotic pressures are matched, as in a solution that is approximately 0.85% by weight NaCl, called *physiological saline solution,* do the cells remain intact.

Another application of osmotic pressure is to the determination of the molecular weight of high polymers and proteins. With these giant molecules (macromolecules), the number of moles present for a given weight is extremely small, perhaps too small to detect by conventional freezing-point lowering or boiling-point elevation techniques. In such cases, osmotic-pressure measurements can be useful. The following problem illustrates how osmotic pressure can be used to give information about a protein molecule.

EXAMPLE 9

Albumins are the most abundant proteins in blood. For a given albumin, 3.5 g of it in 100 ml of water at 25°C produces an osmotic pressure of 0.014 atm. What is the apparent molecular weight of this protein?

Solution

$$\pi = \frac{n}{V} RT$$

$$\frac{n}{V} = \frac{\pi}{RT} = \frac{0.014 \text{ atm}}{(0.08206 \text{ liter atm mol}^{-1} \text{ deg}^{-1})(298 \text{ deg})} = 5.7 \times 10^{-4} \text{ mol/liter}$$

$$n = (5.7 \times 10^{-4} \text{ mol/liter})(0.100 \text{ liter}) = 5.7 \times 10^{-5} \text{ mol}$$

$$\frac{3.5 \text{ g of albumin}}{X \text{ g/mol}} = 5.7 \times 10^{-5} \text{ mol of albumin}$$

$$X = 6.1 \times 10^4 \text{ g/mol}$$

In practice, solutions deviate more from ideal behavior the more concentrated they are. To get around this problem, measurements are usually carried out at a variety of concentrations. Osmotic pressure per unit concentration is calculated for each measurement, a graph is made, and the results are extrapolated back to zero concentration (infinite dilution). The extrapolated value is then used to calculate the molecular weight under conditions approaching ideal behavior. Exercise 8.17 at the end of the chapter illustrates this procedure.

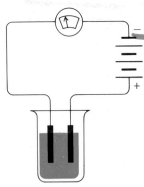

FIG. 8.4 Experiment to determine conductivity of a solution.

As first shown by the Swedish chemist Svante Arrhenius in 1887, there are many cases in which the dissolving process is accompanied by *dissociation*, or breaking apart, of molecules. The dissociated fragments are frequently electrically charged—i.e., ions—in which case electrical measurements can show whether dissociation has occurred. Figure 8.4 is a schematic diagram of an apparatus for determining whether a solute is dissociated into ions. A pair of electrodes is connected in series with an ammeter to a source of electric voltage such as a battery. As long as the two electrodes are kept separated, no electric current flows through the circuit; the ammeter reads zero. When the two electrodes are joined by an electric conductor, the circuit is complete, and the meter deflects. If the electrodes are dipped into water, the meter stays practically at zero, indicating that water does not conduct electricity very much. When sugar is added to the water, the conductivity does not change; but when sodium chloride is added, the conductivity increases enormously. By such experiments it is possible to classify substances into two groups: (1) those which produce conducting solutions and (2) those which produce nonconducting solutions. Solutes of the first class are called *electrolytes;* those of the second, *nonelectrolytes.* Figure 8.5 gives the names and formulas of several examples.

Electric conductivity requires the existence of charged particles. The greater the number of charges, the greater the conductivity. By measuring the conductivity, it is thus possible to get information about relative concentration of charges in solution. When the conductivity of 1 m HCl is compared with the conductivity of 1 m $HC_2H_3O_2$ (acetic acid), it is found that 1 m HCl conducts more. Since both solutions correspond to dissolving 1 mol of solute per kilogram of water, the HCl must break up more fully into ions than does $HC_2H_3O_2$. From observations of this kind, electrolytes may be subdivided into two groups: *strong electrolytes,* which give solutions that are good conductors of electricity, and *weak electrolytes,* which give solutions that are only slightly conducting. Figure 8.6 lists typical compounds of both classes. Strong

FIG. 8.5   Classification of Solutes as Electrolytes or Nonelectrolytes

Electrolytes		Nonelectrolytes	
HCl	Hydrochloric acid	$C_6H_{12}O_6$	Glucose
HNO_3	Nitric acid	$C_{12}H_{22}O_{11}$	Sucrose
H_2SO_4	Sulfuric acid	CH_3OH	Methyl alcohol
$HC_2H_3O_2$	Acetic acid	C_2H_5OH	Ethyl alcohol
NaOH	Sodium hydroxide	N_2	Nitrogen
$Ca(OH)_2$	Calcium hydroxide	O_2	Oxygen
NH_3	Ammonia	CH_4	Methane
NaCl	Sodium chloride	CO	Carbon monoxide
Na_2SO_4	Sodium sulfate	CH_3COCH_3	Acetone

FIG. 8.6 Classification of Solutes as Strong or Weak Electrolytes

Strong electrolytes		Weak electrolytes	
HCl	Hydrochloric acid	$HC_2H_3O_2$	Acetic acid
NaOH	Sodium hydroxide	NH_3	Ammonia
NaCl	Sodium chloride	$HgCl_2$	Mercuric chloride
KCN	Potassium cyanide	HCN	Hydrogen cyanide
$BaSO_4$	Barium sulfate	$CdSO_4$	Cadmium sulfate

electrolytes are essentially 100 percent dissociated into ions; weak electrolytes are dissociated only a few percent.

In a solution of a nonelectrolyte, where there is no dissociation, the molecules of solute retain their unbroken identity. For example, when sugar dissolves in water, as shown in Fig. 8.7, the neutral sugar molecules that were in the solid are still recognized as such in the solution. The sugar species is uncharged; so when positive and negative electrodes are inserted in the solution, there is no reason for the sugar molecules to move. Hence there is no electric conductivity.

In the case of electrolytes, the resulting solution contains ions. Where do these come from? It depends on whether the substance starts out as ionic or molecular. For ionic substances, it is not surprising that charged particles end up in the solution; the undissolved solid is already made up of charged particles. The solvent simply rips the lattice apart. Figure 8.8 shows the dissolving of sodium chloride. Positive ends (hydrogen atoms) of water molecules are attracted to the negative chloride ion. The chloride ion, weakened in its attraction to the crystal lattice, moves off into the solution with an associated

FIG. 8.7 Schematic representation of dissolving sugar.

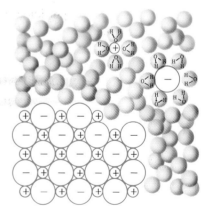

FIG. 8.8 Schematic representation of dissolving NaCl.

● *Sugar molecule*

○ *Water molecule*

cluster of water molecules. It is now a hydrated chloride ion. The entire species is negatively charged because the chloride ion itself is negatively charged. At the same time, the sodium ion undergoes similar hydration except that the negative, or oxygen, end of the water molecule faces the ion. When electrodes are inserted, the positively charged hydrated sodium ions are attracted to the negative electrode, and the negatively charged hydrated chloride ions are attracted to the positive electrode. Net transport of electric charge occurs as positive charge moves in one direction and negative charge moves in the opposite direction.

Ions may also be formed when certain *molecular substances* are dissolved in the proper solvent. For example, HCl in the pure solid, liquid, or gaseous states does not conduct electricity because it consists of neutral, distinct molecules; no ions are present. However, when HCl is placed in water, the resulting solution does conduct, indicating the formation of charged particles. The HCl molecules can be visualized to interact with the solvent as follows:

$$H\!:\!\overset{..}{\underset{H}{O}}\!: + H\!:\!\overset{..}{\underset{..}{Cl}}\!: \longrightarrow \left[H\!:\!\overset{..}{\underset{H}{O}}\!:\!H\right]^{+} + \left[:\!\overset{..}{\underset{..}{Cl}}\!:\right]^{-}$$

A proton has been transferred from HCl to H_2O to form the new species H_3O^+ and Cl^-. Thus positive and negative ions are formed, even though none is present in the original pure HCl. The positively charged H_3O^+ is referred to as a *hydronium*, or *oxonium*, ion. The negative ion is the chloride ion. Both H_3O^+ and Cl^- are hydrated, in that there are water molecules stuck on them just as on an ionic solute. The above reaction can be written

$$H_2O + HCl \longrightarrow H_3O^+ + Cl^-$$

emphasizing that the dissociation of a molecular solute by water is a chemical reaction. Because H_3O^+ is a hydrated proton, $H^+(H_2O)$, and water of hydration is usually omitted from chemical equations, the above equation is often written

$$HCl \longrightarrow H^+ + Cl^-$$

However, we should bear in mind that water is associated with each dissolved ionic species.

8.6
PERCENT DISSOCIATION

The extent of dissociation of solutes varies widely. Some substances, such as HCl, are almost completely dissociated into ions in aqueous solution. Other substances, such as $HC_2H_3O_2$, are only slightly dissociated. The difference depends on relative bond strengths. Percent dissociation can be determined by measuring any property that depends on the concentration of ions. Conductivity measurements can be used, as well as freezing-point lowering. Freezing-point lowering can be used because it is proportional to the molal concentration of particles dissolved in the solvent. As an example, consider the solution of an electrolyte AB in water. If some AB molecules are dissociated into A^+ and B^- ions, then the solution contains three kinds of species: undissociated AB, A^+, and B^-. Each particle contributes to the freezing-point depression. By determining the freezing-point lowering, the total concentration of particles and thus the percent dissociation of AB can be calculated. This is illustrated in the following example.

EXAMPLE 10

The freezing point of a 0.100 *m* HF solution is $-0.201°C$. What percent of the HF molecules have been dissociated by the water?

Solution

Let x represent the concentration of HF particles that dissociate. The dissociation equation $HF \longrightarrow H^+ + F^-$ shows that for every particle of HF that dissociates, one H^+ particle and one F^- particle are formed. This means that if x particles of HF dissociate we will get x particles of H^+ and x particles of F^-. The total number of charged particles formed will be $2x$.

At the same time, due to the dissociation, some of the HF particles will have disappeared. Specifically, to make x particles of H^+ and x particles of F^-, we must break up x particles of HF. If we start with a 0.100 *m* concentration of HF particles and let x *m* dissociate, we will be left with a $(0.100 - x)$ *m* solution of HF particles.

The total concentration of particles (HF, H^+, F^-) will be $(0.100 - x) + x + x = 0.100 + x$. This is what produces the freezing-point lowering. A 1 *m* concentration of particles will lower the freezing point by $1.86°C$, so a $(0.100 + x)$ *m* concentration will lower the freezing point by

$$[(0.100 + x) \; m] \; [1.86°C/m] = (0.100 + x) \; 1.86°C$$

Equating this to the observed freezing-point lowering, we get

$$(0.100 + x) \; 1.86 = 0.201$$
$$x = 0.081$$

$$\text{Percent of HF dissociated} = \frac{0.081}{0.100} \times 100 = 8.1\%$$

Measurements on various electrolytes indicate that the percent dissociation depends on the specific nature of the electrolyte, the nature of the solvent, the concentration of the solute, and the temperature:

1 Nature of the electrolyte When a molecule AB is dissociated into ions A^+ and B^-, the bond AB must be broken. The extent to which this process takes place depends on the specific nature of A and B. Thus, for a given temperature and total concentration, HCl in water is 100% dissociated and HF is 8% dissociated.

FIG. 8.9 Concentration Dependence of Percent Dissociation of Acetic Acid

Concentration, m	Percent dissociation
1	0.4
0.1	1.3
0.01	4.3
0.001	15
0.00001	75

2 Nature of the solvent This point is easily overlooked because we usually think only of water as the solvent. In solvents other than water the behavior of solutes may be different. As an example, under comparable conditions HCl in water is 100% dissociated, but HCl in benzene is less than 1% dissociated.

3 Temperature There is no simple rule for the effect of temperature. For some substances, the percent dissociation increases as the temperature is raised; for others, it decreases. Some substances may show a combination of both effects. For example, acetic acid is 1% dissociated in a particular solution at room temperature; if its temperature is raised, the percent dissociation decreases, but if its temperature is lowered, the percent dissociation also decreases.

4 Concentration The percent dissociation always increases as concentration of electrolyte decreases. The more dilute the solution, the higher the percent dissociation. Figure 8.9 gives numerical values for breakup of $HC_2H_3O_2$ into H^+ and $C_2H_3O_2^-$ (acetate ion). The more dilute the solutions, the higher the percentage dissociated into positive and negative ions. In infinitely dilute solutions, the percent dissociation approaches 100%. At infinite dilution, all electrolytes approach 100% dissociation.*

Definition of a strong electrolyte as one that is highly dissociated in solution and a weak electrolyte as one that is slightly dissociated has to be qualified since in very dilute solutions even weak electrolytes tend toward complete dissociation. This ambiguity can be reduced by adopting the convention that 1 M be the reference state. If a substance in 1 M solution is highly dissociated, it is generally called a strong electrolyte; if in 1 M solution it is slightly dissociated, it is generally called a weak electrolyte.

*Strictly speaking, 100% dissociation is approached only if the solvent has no dissociation fragments in common with the solute. For example, water has a trace of H^+ and OH^- which come from the self-dissociation of H_2O. The presence of this H^+ prevents complete dissociation of solutes such as HX. For the example of Fig. 8.9, the maximum attainable percent dissociation cannot exceed 99.4%, a number which is fixed by the dissociation constant of acetic acid (see Sec. 15.2).

Suppose we prepare a series of aqueous NaCl solutions at various concentrations and carefully measure their freezing points. The results will look like those shown in Fig. 8.10. The more dilute the solution, the less the freezing-point lowering. Now let us calculate the freezing-point lowering *per mole* of NaCl. What we get is shown in the last column. The more dilute the solution, the more closely we approach the ideal value 2(1.86°C), or 3.72°C. Apparently, only in very dilute solution does 1 mol of NaCl act as if totally dissociated into 2 mol of particles (Na^+ and Cl^-). The more concentrated the solution, the smaller the freezing-point lowering per mole. It almost looks as if the NaCl is not completely dissociated in more concentrated solutions. How can this be reconciled with the view that solid sodium chloride is ionic?

In 1923, Peter Debye and Ernest Hückel proposed an explanation, now known as the Debye-Hückel theory. The argument was simply this: X rays show that the salt NaCl is 100 percent composed of ions in the solid state. Thus, it is highly unlikely that NaCl is not completely composed of ions in the solution state as well. Therefore, any reduced effect of the ions Na^+ and Cl^- in solution must be due, not to association into neutral NaCl molecules, but to some other effect which reduces the activity of the ions. It was suggested that the effect was due to interionic attractions between Na^+ and Cl^- in the solution state.

The electric force between two charges, call them q_1 and q_2, separated by a distance r is proportional to the product of the charges and inversely proportional to the square of the distance between them. This is a statement of *Coulomb's law* and can be expressed symbolically as

$$F = \frac{q_1 q_2}{\epsilon r^2}$$

Here F is the force and ϵ is the dielectric constant of the medium in which the charges are immersed. (For a vacuum, $\epsilon = 1$; and for liquid water, $\epsilon \sim 80$.)

In a dilute NaCl solution, ions are far apart. On the average, r is big; so F is small and can reasonably be neglected. What happens as the concentration is increased? The average interionic distance decreases, and the attractive

---

FIG. 8.10 Concentration Dependence of Freezing-Point Lowering in NaCl Solutions

Concentration, m	Observed freezing point, °C	Freezing-point lowering, °C per unit molality
0.1	−0.347	3.47
0.01	−0.0361	3.61
0.001	−0.00366	3.66
0.0001	−0.000372	3.72

force between positive and negative ions increases. The result is that, as the concentration of NaCl is increased, Na^+ and Cl^- ions tend to draw together. However, Debye and Hückel did not propose that Na^+ and Cl^- pair up; they recognized that kinetic motion would keep particles moving; the best one could do would be to speak of an average distribution of ions. On the average, then, in an aqueous solution in the vicinity of a *positive ion* there is considerably more chance of finding *negative ions* than other ions of the same charge. Each Na^+ ion probably has more Cl^- ions than Na^+ ions in the near vicinity. The ions, being in solution, are not fixed but move in and out of any defined region; so it is a statistical preponderance of positive or negative ions that we are talking about. Each ion is in an atmosphere that is relatively rich in its *counter ion* (i.e., the ion of opposite charge).

What would be the effect on an ion of having several counter ions in the vicinity? The most obvious influence would be on the electric conductivity. If an external electric field is applied, the given ion can not move as freely as it would if the counter ions were not there. There is a drag effect, which makes the ions less mobile than expected. This, indeed, is what is found; the more concentrated the salt solution, the smaller the conductivity *per mole*. Thus, for example, if we double the concentration of NaCl from 0.01 to 0.02 m, the conductivity per mole is not doubled but raised by only 97 percent. The more concentrated the solution, the less each mole of NaCl is worth in raising the conductivity. Figure 8.11 gives some data showing how the conductivity per mole increases as the concentration of NaCl decreases.

Similar considerations apply to freezing-point lowering. Although NaCl is 100 percent dissociated to give 2 mol of particles (Na^+ and Cl^-) per mole of solute dissolved, the charged particles restrict each other's activity; their full contribution to freezing-point lowering is not exerted.

The Debye-Hückel theory accounts quantitatively for the attraction between completely dissociated ions and justifies the view that ionic solids should be considered to be completely dissociated in dilute aqueous solution. The theory can be applied to calculate the interionic attraction when the concentration of ions is very low, as in a solution of a weak electrolyte. This calculation must be made if we want to get very accurate determination of percent dissociation from freezing-point data.

For concentrated solutions, even the Debye-Hückel theory does not fully account for the observed freezing-point lowering. Concentrated solutions are not well understood at the present time.

**FIG. 8.11   Concentration Dependence of
Conductivity in NaCl Solutions**

Concentration, m	Conductivity per mole (arbitrary units)
0.1	106.74
0.01	118.51
0.001	123.74
Infinite dilution	126.45

The term "solubility" describes the qualitative idea of the dissolving process, but it is also used to describe quantitatively how much material there is in solution. The solutions considered up to now represent *unsaturated solutions*. Solute can be added successively to produce a whole series of solutions which differ slightly in concentration. However, in most cases, the process of adding solute cannot go on indefinitely. Eventually, a stage is reached beyond which addition of solute to a specified amount of solvent does not produce another solution of higher concentration. Instead, the excess solute remains undissolved. In these cases there is a limit to the amount of solute which can be dissolved in a given amount of solvent. The solution which represents this limit is called a *saturated solution*, and the concentration of the saturated solution is called the *solubility*.

The best way to ensure having a saturated solution is to maintain an excess of solute in contact with the solution. If the solution were unsaturated, solute would disappear until saturation is established. If the solution is indeed saturated, the amount of excess solute remains unchanged, as does the concentration of the solution. The system (saturated solution in contact with excess solute) is in a state of equilibrium. It is a state of dynamic equilibrium in that dissolving of solute is still occurring, but the dissolving is compensated for by precipitation of solute out of solution. The number of particles going into solution equals the number leaving. The concentration of solute in solution remains constant; the amount of solute in excess remains constant. The amount of excess solute in contact with the saturated solution does not affect the concentration of the saturated solution. In fact, it is possible to filter or separate the excess solute completely and still have a saturated solution. *A saturated solution may be defined as one which is or would be in equilibrium with excess solute.*

The concentration of the saturated solution, i.e., the solubility, depends on (1) the nature of the solvent, (2) the nature of the solute, (3) the temperature, and (4) the pressure. In considering these we should keep in mind that three important interactions operate in the dissolving process. Solute particles are separated from each other in a process that takes energy; solvent particles are pushed apart to make a hole to accommodate the solute, and this also uses energy; finally, solvent particles attract the solute particles in the process that provides energy. Three sets of interactions are involved: solute-solute, solvent-solvent, and solute-solvent. However, spontaneous dissolving is decided not just by energy balance. We also need to consider that the solution represents a more disordered and hence more probable state than the unmixed components. Thus, dissolving will frequently occur even when solute-solvent attractions are not energetically sufficient to provide the energy required to break up the solute-solute and solvent-solvent interactions.

Nature of the Solvent

A useful generalization much quoted in chemistry is "like dissolves like." What this means is that high solubility occurs when molecules of solute are similar in structure and electric properties to molecules of solvent. When there is such similarity, e.g., both solute and solvent have high dipole moments, then solute-solvent attractions are expected to be particularly strong. When solute and solvent are dissimilar, solute-solvent attractions are likely to be weak. For this reason polar substances such as H_2O are usually good solvents for polar substances such as alcohol but poor solvents for nonpolar substances such as gasoline.

In general, an ionic solid has higher solubility in a polar solvent than in a nonpolar solvent. Also, the more polar the solvent, the greater the solubility. For example, at room temperature, the solubility of NaCl in water, ethyl alcohol, and gasoline is, respectively, 311, 0.51, and 0 g/liter. The higher the polarity of the solvent, the greater the attraction for ions. Also, as was mentioned in the last section, a high dielectric constant for a medium means small attraction *between* ions. Water, being more polar than ethyl alcohol or gasoline, has a high dielectric constant. This means Na^+ and Cl^- attract each other less strongly in water than they do in ethyl alcohol or gasoline.

Nature of the Solute

Changing the solute means changing the solute-solute and solute-solvent interactions. At room temperature the amount of sucrose that can be dissolved in water is 1311 g per liter of solution. This is more than four times as great as the solubility of NaCl in water, 311 g/liter. However, these numbers are misleading. It is better to talk in terms of moles than grams. Thus, a saturated solution of NaCl is 5.3 *M*, whereas a saturated solution of sucrose is 3.8 *M*. On a molar basis NaCl has a *higher solubility* in water than sucrose has. The attractions in solid NaCl are greater than those in solid sucrose, so the higher solubility of NaCl must be due to the stronger interactions of the ions with water.

What effect does the presence of one solute have on the solubility of another solute in the same solution? As a crude generalization, unless the concentration of solute is high, there is little effect. For example, approximately the same concentration of NaCl can be dissolved in a 0.1 *M* sucrose solution as in pure water. However, the generalization does not hold when solutes have an ion in common. Thus, the solubility of NaCl is appreciably lowered by presence of KCl.

Temperature

As a general rule, the solubility of *gases in water* decreases as the temperature is raised. The tiny bubbles that form when water is first heated are due to the fact that dissolved air becomes less soluble as temperature increases. The flat taste characteristic of boiled water is due to loss of dissolved air.* In other liq-

---

*Thermal pollution of our water resources means less dissolved oxygen. For example, raising a lake's temperature from 20 to 40°C reduces dissolved oxygen from 9 to 6 ppm (by weight). This can drastically reduce the fish life that can survive in such water.

FIG. 8.12 Change of Solubility with Temperature (grams of solute per kilogram of H_2O)

Substance	0°C	20°C	40°C	60°C	80°C	100°C
$AgNO_3$	1220	2220	3760	5250	6690	9520
$Ca(OH)_2$	1.85	1.65	1.41	1.16	0.94	0.77
KCl	276	340	400	455	511	567
NaCl	357	360	366	373	384	398
Li_2CO_3	15.4	13.3	11.7	10.1	8.5	7.2
$O_2(g)$ at 1 atm	0.069	0.043	0.031	0.027	0.014	0.000
$CO_2(g)$ at 1 atm	3.34	1.69	0.97	0.058	....	
$He(g)$ at 1 atm	0.00167	0.00153	0.00152	0.00162	0.00177	

uid solvents (in fact, even in water for some gases such as helium at higher temperatures) the solubility of gases may actually increase with rising temperature. Similarly, there is no simple rule for temperature effect on solubility of *liquids* and *solids*. With increasing temperature, Li_2CO_3 in water decreases in solubility, KCl increases, and NaCl shows practically no change. Illustrative data are given in Fig. 8.12.

The change of solubility with temperature is closely related to the heat of solution. The *heat of solution* can be defined as the heat evolved when a mole of solute dissolves to give the saturated solution; it can be written as the heat that accompanies the following process:

$$\text{Solute} + \text{solvent} \longrightarrow \text{saturated solution} + \text{heat of solution}$$

The heat of solution can be positive, in which case heat is evolved to the surroundings, or it can be negative, in which case heat is absorbed from the surroundings. For example, the heat of solution of Li_2CO_3 in water is positive; that of KCl is negative. The dissolving of Li_2CO_3 evolves heat to the surroundings; the dissolving of KCl absorbs heat. We can write

$$KCl(s) + H_2O \longrightarrow \text{solution} - \text{heat}$$
or
$$\text{Heat} + KCl(s) \longrightarrow \text{solution}$$

Heat must be supplied, i.e., the process is endothermic. When a substance with negative heat of solution is dissolved, there is usually a drop in temperature. If the heat of solution is positive, the temperature rises.

Whether the heat of solution is positive or negative depends on the nature of the solute and the solvent. More precisely, the heat of solution depends on the relative magnitude of two energies: (1) the energy required to break up the solid lattice and (2) the energy liberated when the resulting particles are solvated. For KCl the two processes associated with these energies are

$$KCl(s) \longrightarrow K^+(g) + Cl^-(g)$$

$$K^+(g) + Cl^-(g) \xrightarrow{\text{water}} K^+(aq) + Cl^-(aq)$$

The first step, in which the solid is broken up, requires energy. The amount required per mole is called the *lattice energy*. The second step, in which water

247

molecules are separated from each other and attracted to the ions, liberates energy. This energy is called the *hydration energy*.* When the hydration energy is greater than the lattice energy, the overall process liberates energy; i.e., the net process is exothermic. When hydration furnishes less energy than is required to break up the lattice, the overall process is endothermic. In some cases lattice energy is approximately equal to hydration energy. For example, the lattice energy of NaCl is 770 kJ/mol and the hydration energy is 766 kJ/mol. These two energies just about balance each other; the heat of solution, which is the difference between the two, is nearly zero.

How is the heat of solution related to change of solubility with temperature? In a saturated solution at equilibrium with excess solute, the two processes, dissolving and precipitation, are occurring at the same time. If the dissolving process is endothermic, as with KCl, the precipitation process is exothermic:

$$\text{Heat} + \text{KCl}(s) + \text{H}_2\text{O} \rightleftharpoons \text{solution}$$

The upper arrow indicates the forward process, and the bottom arrow indicates the reverse process. In a beaker containing solid KCl in equilibrium with saturated solution, the two processes occur equally. If heat is now added, Le Châtelier's principle states that the system adjusts to reduce the stress of added heat. The stress can be reduced by favoring the dissolving over the precipitation: KCl goes into solution more than it comes out. The amount of dissolved solid in the solution increases. The solubility of KCl in water is therefore greater at the higher temperature. This behavior is typical of most solids. When placed in water, they dissolve by an endothermic process; hence raising the temperature increases solubility.

When, as in the case of NaCl, the heat of solution is nearly zero, there is not much change of solubility with temperature. This, however, is not the whole story, as can be seen by noting that in the case of NaCl, 770 kJ/mol is needed to break up the lattice whereas only 766 kJ/mol is liberated by the hydration energy. Why should the NaCl dissolve at all? The net process appears to be energetically *unfavorable*. We cannot answer this question now but will come back to it later when we take up the chapter on thermodynamics.

Pressure

The solubility of all gases is increased as the partial pressure of the gas above the solution is increased. For example, the concentration of CO_2 dissolved in a carbonated beverage (e.g., champagne) is dependent directly on the partial pressure of CO_2 in the gas phase. When a bottle is opened, the pressure is reduced, solubility of CO_2 is diminished, and the bubbles of CO_2 form and es-

---

*Note that hydration energy takes into account both solvent-solvent interaction (the energy required to make a hole in the water) and solvent-solute interaction (the energy released by the ion-water attraction). These are lumped together because experimentally they cannot be separated. We cannot hydrate an ion without first making room for it; nor can we make a hole in the water without putting something in it.

cape from the beverage. Quantitatively, this is expressed in *Henry's law,* which states that at constant temperature *the mole fraction of gas dissolved in a solvent is directly proportional to the partial pressure of the gas.* Henry's law is usually written

$$K = \frac{p}{x}$$

where K is the Henry's law constant, p is the partial pressure of the solute in the gas phase, and x is the mole fraction of the gas in the solution. The constant K is generally given in millimeters of mercury and changes somewhat with temperature. A typical value of K is 2.95×10^7, which is the value for oxygen in water at 20°C.

As far as liquid and solid solutes are concerned, there is essentially no change of solubility with pressure. In general, the volume change on solution is so small that pressures of the order of thousands of atmospheres are needed to change the solubility appreciably.

In closing this section we should note that it is sometimes possible to prepare solutions which have a higher concentration of solute than would correspond to a saturated solution. Such solutions are *supersaturated;* they are unstable with respect to separation of excess solute and usually have to be handled rather carefully. A supersaturated solution of sodium acetate, for example, can be made as follows: A saturated solution of $NaC_2H_3O_2$ and H_2O in contact with excess solute is heated until the increase of solubility with temperature is sufficient to dissolve all the excess solute. This unsaturated solution is filtered (to get rid of any dust) and then cooled. The system ought to return to its original equilibrium state with excess solute crystallizing out. With sodium acetate, cooling, if done carefully, can be accomplished without crystallization. The situation is reminiscent of that observed in the supercooling of a liquid below its freezing point. Supersaturation can usually be destroyed in the same manner, i.e., by seeding. When a tiny seed crystal of sodium acetate is placed in a supersaturated solution of sodium acetate, excess solute crystallizes on it until the remaining solution is just saturated. Occasionally, a mechanical disturbance such as a sudden shock may suffice to break the supersaturation. Dust particles or even scratches on the inner surface of the container may act as centers on which crystallization can start.

8.9 COLLOIDS

In introducing the topic of solutions it was implied that it is easily possible to distinguish between a homogeneous mixture and a heterogeneous one. However, this distinction is not sharp. There are systems which are not obviously homogeneous or heterogeneous. They are classed as intermediate and are known as *colloids*. Examples are cigarette smoke, fog, emulsions, foam, and egg albumin. To get an idea of what a colloid is, imagine a process in which a sample of solid is placed in a liquid and progressively subdivided. As long as distinct particles of solid are visible to the naked eye, there is no question that the system is heterogeneous. On standing, the visible particles generally separate out. They can be separated easily by filtration. However, as progressive subdivision is continued, a state is finally reached in which the dispersed particles have been broken down to individual molecules or atoms. In this limit, two phases can no longer be distinguished; a solution has been produced. Dissolved particles do not separate out on standing, nor can they be separated by filtration.

Between coarse suspensions at one extreme and true solutions at the other there is a region where heterogeneity changes to homogeneity. In this region dispersed particles are so small that they do not properly form a separate phase, but neither are they so small that they can be said to be in true solution. This intermediate state of subdivision is called the *colloidal state*. On standing, particles of a colloid do not separate out; they cannot be seen under a microscope; nor can they be separated by filtration. Usually, the definition of colloid is based on size. When particle size lies between about 10^{-7} and 10^{-4} cm, the dispersion is called a *colloid*, a *colloidal suspension*, or a *colloidal solution*.

The dispersed particle may consist of atoms, of small molecules, or even of one giant molecule. For example, colloidal gold consists of particles containing a million or more gold atoms. Colloidal sulfur can be made with particles containing a thousand or so S_8 molecules. Albumin is a giant molecule having a molecular weight of the order of 61,000 amu.

Colloids are sometimes classified on the basis of the component phases, even though the separate phases are no longer distinguishable when the colloid is formed. The more important classifications are *sols, emulsions, gels*, and *aerosols*. In *sols*, a solid is dispersed through a liquid; the liquid forms the continuous phase and bits of solid form the discontinuous phase. Milk of magnesia is a sol consisting of solid particles of magnesium hydroxide dispersed through water. Sols can be made by *dispersion* (breaking down larger particles) or *condensation* (building up small particles to colloidal dimensions). Colloidal gold, for example, can be made by striking an electric arc between two gold electrodes under water and also by chemical reduction:

$$3N_2H_5^+ + 4AuCl_4^- \longrightarrow 3N_2 + 4Au(s) + 16Cl^- + 15H^+$$

Investigation by X rays has shown that the gold is crystalline in nature.

Emulsions are colloids in which a liquid is dispersed through a liquid. A common example is milk, which consists of butterfat globules dispersed through an aqueous solution. A *gel* is an unusual type of colloid in which a liquid contains a solid arranged in a fine network extending throughout the system. Both solid and liquid phases are continuous. Examples of gels are jellies, gelatin, agar, and slimy precipitates such as aluminum hydroxide. An *aerosol* is a colloid made by dispersing either a solid or a liquid in a gas. The former is called a *smoke*, and the latter a *fog*. *Smog* is a combination of smoke and fog.

When a beam of light is passed through a colloid, an observer to one side can see the path of the beam. This is unlike the case of a true solution, as is shown in Fig. 8.13. The effect, called the Tyndall effect, can be produced by

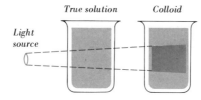

True solution   *Colloid*

*Light
source*

FIG. 8.13 Light scattering from a colloid.

shining light on an aqueous solution of sodium thiosulfate, $Na_2S_2O_3$, to which acid has been added. The ensuing chemical reaction produces elemental sulfur. The light beam is invisible until the sulfur particles aggregate to colloidal dimensions. In an ordinary solution particles of solute are much smaller than the wavelength of visible light. They are too small to reflect or scatter the wave to the side. However, when the solute particles grow to a size that is on the order of a wavelength, the light beam is scattered and becomes visible from the side. Careful studies of the scattering as a function of direction can be used to determine not only the size but also the shape of macromolecules. In this way one can deduce that albumin (molecular weight 61,000) is ellipsoidal in shape, whereas fibrinogen (a blood-clotting protein of molecular weight 400,000 amu) has an elongated, fibrillar shape.

When a microscope is focused on a Tyndall beam, colloidal particles are seen to undergo brownian motion, the rapid, random, zigzag motion previously mentioned in discussing gases (Sec. 6.10). The smaller the colloidal particle, the more violent its brownian motion.

In some colloids the particles absorb ions and thereby acquire electric charge. For example, ferric oxide sol (chemically related to rust) consists of aggregates of ferric oxide units which are positively charged due to adsorption of H^+. The resulting positive charge enhances the stability of the colloid because of repulsion between similarly charged particles. The particles stay as far away from each other as possible. In a similar way, arsenious sulfide, As_2S_3, forms a negative sol by adsorbing SH^- or OH^- ions. As one might guess, mixing a positively charged ferric oxide sol with a negatively charged arsenious sulfide sol coagulates them both.

Some colloidal particles adsorb films, which shield them from other particles. An example is found in gelatin. Gelatin is a high-molecular-weight protein (made by boiling skin, tendons, ligaments, bones, etc., in water) which has the property of tying to itself a sheath of water. The tightly bound water protects the gelatin particle from coagulating with another gelatin particle. This property of gelatin is used for stabilizing colloids of silver bromide in photographic film. When finely divided AgBr is stirred up with water, it settles out. However, if the AgBr is first mixed with gelatin, the gelatin forms a film on each silver bromide particle. The gelatin in turn adsorbs a layer of water; so essentially two protective films have formed on the silver bromide to keep it in suspension.

Colloids show high adsorptive properties for two reasons: One is extremely large surface area. For example, when 1 cm³ of sulfur is ground up into cubes which are 10^{-5} cm on the edge, the surface area becomes 6×10^5 cm².

The second reason for great adsorption is that surface atoms in general have special properties. At the surface, valence forces are not fully satisfied, since the atom is designed to bond in three dimensions. The greater the state of subdivision, the greater the fraction of atoms there are on the surface.

Important Concepts

types of solution	freezing-point lowering	saturated solution
solute	boiling-point elevation	heat of solution
solvent	Raoult's law	hydration energy
concentration	osmotic pressure	effect of temperature on solubility
mole fraction	electrolytes	effect of pressure on solubility
molarity	nonelectrolytes	supersaturation
molality	strong electrolytes	colloids
normality	weak electrolytes	sols, gels, emulsions
percent by weight	percent dissociation	Tyndall effect
percent by volume	interionic attractions	adsorption
formality	solubility	

Exercises

*8.1 Types of solution In terms of kinetic molecular theory, what is the difference between a solid solution and a pure crystalline compound?

*8.2 Mole fraction What is the mole fraction of each component in a solution made by mixing 1.5 mol of N_2, 2.5 mol of O_2, and 1.0 mol of CO_2?

*8.3 Mole fraction Calculate the mole fraction of each component in a solution made by dissolving 10.0 g of $C_6H_{12}O_6$ and 10.0 g of $C_{12}H_{22}O_{11}$ in 80.0 g of H_2O.

*8.4 Molarity How many moles of H_2SO_4 would it take to make 0.200 liter of 0.350 M H_2SO_4?

*8.5 Molarity Calculate the molarity of a solution made by dissolving 9.90 g of H_3PO_4 in enough water to make 250.0 ml of solution.

*8.6 Molality What is the molality of a solution made by mixing 0.10 mol of NaCl and 0.90 kg of water? What does the molality become on addition of 0.10 kg of water?

*8.7 Molality How many grams of $C_6H_{12}O_6$ must you dissolve in 0.100 kg of water to make a solution that is 0.150 m?

**8.8 Concentration A typical wine is 12% alcohol by volume, with the main ingredients being H_2O, C_2H_5OH, and a host of trace chemicals that add taste and bouquet. Considering only the C_2H_5OH and H_2O, calculate each of the following:
  a Mole fraction of C_2H_5OH
  b Molarity of C_2H_5OH
  c Molality of C_2H_5OH
The density of pure C_2H_5OH is 0.789 g/ml; the density of 12% by volume C_2H_5OH in H_2O is 0.983 g/ml.   *Ans. 0.040, 2.1 M, 2.3 m*

*8.9 Concentration An ethyl alcohol–water solution that is 50.0 wt % C_2H_5OH has a density of 0.9139 g/ml. Calculate the molarity and the molality of this solution.

***8.10 Concentration What is the maximum amount of 0.200 M HCl solution that you can make if you have only 20.0 ml of 6.00 M HCl and 600.0 ml of water? Assume volumes are additive.

**8.11 Vapor pressure How many grams of sugar, $C_6H_{12}O_6$, should you dissolve in 0.500 kg of H_2O at 25°C to reduce the vapor pressure of the water by 1.0%?   *Ans. 50.5 g*

**8.12 Raoult's law Show how Raoult's law leads to the prediction that the boiling-point elevation of a solvent is proportional to the molal concentration of nonvolatile solute dissolved in it. Why does the solute have to be nonvolatile? If the solute itself were volatile, how would this affect the boiling-point elevation?

*8.13 Boiling-point elevation The normal boiling point of benzene is 80.1°C. How many grams of naphthalene ($C_{10}H_8$) should I dissolve in 1 liter of benzene (density 0.8786 g/ml) to reduce its boiling point to 75.0°C?

*8.14 Freezing-point depression A given antifreeze solution, composed of water and propylene glycol, $CH_3CH(OH)CH_2OH$, is observed to freeze at −25°C. What is the probable weight percent composition of the solution?

Ans. 50.6% propylene glycol

*8.15 Molecular-weight determination From the following data calculate the molecular weight and deduce the molecular formula of phosphorus dissolved in carbon disulfide.

Normal boiling point of CS_2	46.3°C
bp elevation constant of CS_2	2.34°C/m
Weight of P in solution	70.0 g
Weight of CS_2 in solution	630.0 g
Observed bp in solution	48.4°C

*8.16 Osmotic pressure What is meant by the term "osmotic pressure"? How can the device shown in Fig. 8.3 be used to measure osmotic pressure? Predict the effect on osmotic pressure of each of the following:

a Increasing the solute concentration
b Increasing the temperature
c Increasing the size of the solute molecules

***8.17 Osmotic pressure The high polymer polyisobutylene $[H_2C=C(CH_3)_2]_x$ shows the following osmotic pressures when dissolved in the solvent cyclohexane at 25°C.

Concentration, g/ml	Osmotic pressure, atm
0.0025	0.00035
0.0050	0.00090
0.0075	0.00173
0.0100	0.0030
0.0150	0.0066
0.0200	0.0117

Use these results to calculate the molecular weight and the value of *x* at infinite dilution.

Ans. 240,000, x = 4300

*8.18 Electrolytes What is the essential distinction between an electrolyte and a nonelectrolyte? Between a strong electrolyte and a weak electrolyte?

*8.19 Electrolyte Criticize the following definition: "An electrolyte is a chemical compound that conducts electric current."

**8.20 Percent dissociation Nitrous acid, HNO_2, is a weak acid dissociating to give H^+ and NO_2^- (nitrite ion). For a given solution of 0.0100 m HNO_2, the observed freezing point is −0.0222°C. Calculate the apparent percent dissociation in this solution.

Ans. 19.1%

**8.21 Percent dissociation How might you rationalize the fact that in 0.100 m aqueous solution HCl is 100% dissociated but HF is only 8% dissociated?

**8.22 Percent dissociation If H_2SO_4 is 100% dissociated and HSO_4^- is 10% dissociated, what freezing-point lowering would you predict for 0.100 m H_2SO_4?

**8.23 Interionic attraction What evidence do you have that interionic attractions become negligible for very dilute solutions?

**8.24 Interionic attraction The dielectric constant of ethyl alcohol is 24.3 compared to 78.5 for water. How would you expect interionic attractions to compare in the two cases? What influence would this have on observed freezing-point lowering as tabulated in Fig. 8.10 for NaCl in water.

*8.25 Solubility Explain why both solute-solute and solvent-solvent interactions have to be considered as well as solute-solvent attraction in deciding whether a given solute will dissolve in a given solvent.

**8.26 Solubility H_2O and C_2H_5OH are both polar solvents. Yet NaCl is quite soluble in H_2O and almost insoluble in C_2H_5OH. Suggest an explanation for this.

***8.27 Solubility What relation is there between the heat of solution of a solid in water and the way the solubility changes with temperature? How can the Le Châtelier principle be used to account for this? When NaOH(*s*) is placed in water, there is a strong evolution of heat. Yet the solubility of sodium hydroxide in water increases with rise in temperature. How might this be explained?

**8.28 Thermal pollution Siting of power plants on fresh-water lakes is sometimes opposed on the basis of possible oxygen depletion due to temperature rise of the water. Using data from Fig. 8.12, calculate in parts per million (ppm) the solubility of O_2 in H_2O under 1 atm pressure of air at 0, 20, and 40°C. Compare with the minimum value needed for warm-water game fish, 5 ppm.

Ans. 14 ppm at 0°C

8.29 Water pollution At 0°C, the solubility of oxygen in water under 1 atm of air is 14 ppm. The required value for trout is 6 ppm. Calculate how many drops of oil, represented as $C_{20}H_{42}$, would be needed per liter of water to deplete the oxygen below the required level, assuming the oil uses up oxygen by conversion to CO_2 and H_2O. Density of oil is about 0.79 g/ml; a drop is 0.05 ml.

Ans. 0.06

8.30 Solubility As shown in Fig. 3.4, air contains 0.023 to 0.050% by volume of CO_2. Combining this with the data of Fig. 8.12, calculate the probable molarity of CO_2 in water which is in equilibrium with air at 1 atm and 20°C.

8.31 Henry's law What is Henry's law? How does it explain the formation of bubbles when a bottle of champagne is opened? Calculate the Henry's law constant for CO_2 at 20°C, using the data from Fig. 8.12.

8.32 Saturation Given three beakers containing, respectively, unsaturated, saturated, and supersaturated solutions of sodium acetate in water,

how could you tell the solutions apart, given only two crystals of sodium acetate?

8.33 Colloids Given a beaker labeled 0.1 M $FeCl_3$ which contains what appears to be a red solution, how can you tell whether you have a true solution or a colloid?

8.34 Colloids Given colloidal gold particles each containing about a million gold atoms, what would be the approximate diameter of the particles, assuming the gold atoms are in face-centered cubic arrangement with unit-cell dimension 0.406 nm?

Ans. 3.2×10^{-6} cm

8.35 Colloids Given a 3.0% by weight aqueous solution of the protein albumin having molecular weight 61,000 amu, calculate the freezing-point lowering and osmotic pressure expected at 25°C.

8.36 Colloids Charcoal is an unusual kind of colloid in which a gas phase is dispersed throughout an interconnected solid carbon network. The surface area of the carbon is enormous. It can be measured by pumping out all the gas and adsorbing on the surface a monomolecular layer of test gas. If 1 g of a particular charcoal specimen adsorbs 5.36 cm³ of H_2 gas at STP, what is the apparent adsorbing surface area? The effective diameter of a hydrogen molecule is about 0.3 nm.

Chapter 9

SOLUTION REACTIONS

One reason for the emphasis on solutions is that a large fraction of all chemical reactions are carried out in solution. In this chapter we consider various kinds of solution reactions. In the discussion we give special attention to what species are present and what changes they undergo.

9.1
ACIDS AND BASES

Two important kinds of solutions are acids and bases. How can we use the ideas of Chap. 8 to account for acid properties (sour taste, red coloration of litmus, reaction with metals to liberate hydrogen)? How can we explain bases (bitter taste, blue coloration of litmus, ability to neutralize acids)? There are several definitions of acids and bases in common use. It is necessary to be familiar with each of them.

Arrhenius, the discoverer of electrolytic dissociation, accounted for acid properties by postulating that all acids have the formula HX and dissociate in solution to give H^+ and X^-:

$$HX \longrightarrow H^+ + X^-$$

Acid properties are thus attributable to the presence of an ionizable H atom. Since X^- may also contain an ionizable H atom, it itself may be an acid. Thus, for Arrhenius, acids include HCl, HNO_3, H_2SO_4, HSO_4^-, and $HC_2H_3O_2$. Arrhenius further proposed that all bases can be written MOH and can dissociate in aqueous solution to give M^+ and OH^-:

$$MOH \longrightarrow M^+ + OH^-$$

Base properties are attributed to the presence of hydroxide ion. Examples of Arrhenius bases are $NaOH$, $Ca(OH)_2$, and $Al(OH)_3$. Since an acid produces H^+ and a base produces OH^-, a base can neutralize an acid by the reaction

$$HX + MOH \longrightarrow HOH + MX$$

to form water and the salt MX (in solution as M^+ and X^-).

The Arrhenius system has the attractive feature of simplicity. However, it suffers from lack of generality. For example, it does not account for acidic

255

solutions such as CO_2 in H_2O or basic solutions such as NH_3 in H_2O. To handle these, the Arrhenius system proposed that these substances form acids or bases on reaction with H_2O. Thus, for example, it was postulated that CO_2 reacts with H_2O to form H_2CO_3 (carbonic acid), which can then dissociate to give H^+ and HCO_3^-. Similarly, it was suggested that NH_3 reacts with H_2O to form NH_4OH (ammonium hydroxide) which can dissociate to give NH_4^+ and OH^-. In neither case, however, is the situation so simple. In aqueous CO_2 less than 1 percent of the CO_2 converts to H_2CO_3; in aqueous ammonia no discrete NH_4OH species has ever been identified.

A slight modification of the Arrhenius system gives us what is referred to as the *general-solvent system* of acids and bases. In this system it is recognized that any solvent may be somewhat dissociated into positive and negative fragments. If so, the term "acid" is applied to any substance that raises the concentration of the positive fragments; the term "base" is applied to any substance that raises the concentration of the negative fragments.

Let us consider, for example, the situation in water. Even when very highly purified, water conducts electric current to a slight extent. This can be explained by assuming that water is slightly dissociated into positive and negative ions. The dissociation can be written*

$$H_2O \longrightarrow H^+ + OH^-$$

The degree of dissociation of water is very small. On the average, only 2 out of 1 billion molecules of H_2O are dissociated. In pure water the concentration of hydrogen ion is 1.0×10^{-7} M. The concentration of hydroxide ion is, of course, the same, since each time a water molecule is split, one hydrogen ion and one hydroxide ion are formed. The balance can be upset by addition of other substances. Those *substances which increase the hydrogen-ion concentration are called acids;* those *substances which increase the hydroxide-ion concentration are called bases.* Thus, HCl is obviously an acid because when placed in water it is converted to H^+ and Cl^-, thus raising the concentration of H^+. It is a *strong acid* because it is *completely converted* to H^+ and Cl^-. On the other hand, acetic acid ($HC_2H_3O_2$) is a weak acid because it is only *slightly converted* to H^+ and $C_2H_3O_2^-$. What about CO_2? It is also an acid because it raises the H^+ concentration. The reaction can be written

$$CO_2 + H_2O \longrightarrow H^+ + HCO_3^-$$

Similarly, NH_3 is a base because its addition to water increases the hydroxide-ion concentration. The reaction can be written

$$NH_3 + H_2O \longrightarrow NH_4^+ + OH^-$$

---

*Perhaps a better way to write the dissociation would be

$$H_2O + H_2O \longrightarrow H_3O^+ + OH^-$$

corresponding to

$$H:\overset{..}{\underset{H}{O}}: + H:\overset{..}{\underset{H}{O}}: \longrightarrow \left[H:\overset{..}{\underset{H}{O}}:H\right]^+ + \left[:\overset{..}{\underset{..}{O}}:H\right]^-$$

Other more complex reactions (sometimes called *hydrolysis reactions*) can occur between salts and water so as to change the H^+ and OH^- concentrations of the water. For example, when certain aluminum salts are added to water, the solutions are acidic. This can be attributed to the net reaction

$$Al^{3+} + H_2O \longrightarrow H^+ + AlOH^{2+}$$

Similarly, when certain sulfides are added to water, the solutions are basic, which is attributed to

$$S^{2-} + H_2O \longrightarrow OH^- + SH^-$$

A third commonly used system for defining acids and bases is that proposed in 1923 by J. N. Brønsted of Denmark. In the *Brønsted system* an acid is defined as any species that acts as a *proton donor* and a base as any species that acts as a *proton acceptor*. For example, when HCl is placed in water, the reaction

$$HCl + H_2O \longrightarrow H_3O^+ + Cl^-$$

shows HCl acting as a proton donor (Brønsted acid) and H_2O acting as a proton acceptor (Brønsted base). Similarly, when H_2O undergoes the slight self-dissociation

$$H_2O + H_2O \longrightarrow H_3O^+ + OH^-$$

one of the H_2O molecules acts as a Brønsted acid and the other H_2O molecule acts as a Brønsted base.

When a proton donor HX gives up a proton it forms species X^-, which can pick up a proton to regenerate HX. Similarly, any species Y that can pick up a proton forms HY^+, which can furnish protons. Thus, every Brønsted acid is coupled with a related Brønsted base. The two together constitute a *conjugate acid-base pair*. To illustrate, HCl is the conjugate acid of Cl^-, and Cl^- is the conjugate base of HCl. Other pairs of conjugate acids and bases include the following: $HC_2H_3O_2$ and $C_2H_3O_2^-$, H_2SO_4 and HSO_4^-, HSO_4^- and SO_4^{2-}, NH_4^+ and NH_3, H_3O^+ and H_2O, and H_2O and OH^-. All these conjugate acid-base pairs differ by a single H^+. In the Brønsted system, "dissociation" of an acid HX is not simple breakup to give H^+ and X^- but rather proton transfer from HX to a molecule of solvent that functions as a base:

$$HX + H_2O \longrightarrow H_3O^+ + X^-$$
$$\text{Acid}_1 \quad \text{Base}_1 \qquad \text{Acid}_2 \quad \text{Base}_2$$

For the forward reaction, HX is the acid, and H_2O is the base; for the reverse reaction H_3O^+ is the acid, and X^- the base.

Different acids vary in their tendency to give up a proton. Picking H_2O as the common solvent, we can write an ordered list, as is done in Fig. 9.1. In the left column acids are arranged by decreasing strength; i.e., the strongest is at the top and the weakest is at the bottom. In the right column the conjugate base of each of the acids is shown. Since strong acid HX (strong tendency of HX to donate H^+ and form X^-) implies weak conjugate base X^- (weak tendency of X^- to accept H^+), the order in the right column is inverted; i.e., the weakest base is at the top, and the strongest is at the bottom.

FIG. 9.1   Relative Strengths of Brønsted Acids and Bases

	Conjugate acid	Conjugate base	
Strongest	$HClO_4$	ClO_4^-	Weakest
	H_2SO_4	HSO_4^-	
	HCl	Cl^-	
	H_3O^+	H_2O	
	HSO_4^-	SO_4^{2-}	
	HF	F^-	
	$HC_2H_3O_2$	$C_2H_3O_2^-$	
	H_2S	HS^-	
	NH_4^+	NH_3	
	HCO_3^-	CO_3^{2-}	
	H_2O	OH^-	
	HS^-	S^{2-}	
Weakest	OH^-	O^{2-}	Strongest

From the ordered list it is possible to predict the direction in which proton transfer is favored. Any acid in the list has great tendency to transfer a proton to any base *below* it; conversely, any acid has small tendency to transfer a proton to any base *above* it. For example, when HCl is placed in water, there is great tendency for the HCl to transfer its proton to H_2O (HCl on the acid side is ranked higher than H_2O on the base side). However, when HF is placed in water, there is little tendency to transfer the proton to H_2O (HF on the acid side is lower than H_2O on the base side). This only confirms the fact that HCl in water is a strong acid and HF in water is a weak acid. The order of the list can also be used to figure out that when HF in water is mixed with Cl^- and NH_3, the proton will be preferentially transferred from HF to NH_3.

Of all the acids listed in Fig. 9.1, only the so-called "strong" acids $HClO_4$, H_2SO_4, and HCl have greater tendency to give up a proton than does H_3O^+. They, therefore, transfer protons to H_2O to form H_3O^+. However, HSO_4^- lies below H_3O^+; hence, HSO_4^- has little tendency to transfer a proton to H_2O and is to be regarded as a weak acid.

Finally, we should note that even though tendencies for proton transfer may be small, some reaction will occur in all cases. Thus, HSO_4^- in H_2O does undergo limited proton transfer to form some H_3O^+ and SO_4^{2-}. Similarly, for NH_3 in H_2O there is some proton transfer

$$\underset{\text{Acid}_1}{H_2O} + \underset{\text{Base}_1}{NH_3} \xrightarrow{\text{slight}} \underset{\text{Acid}_2}{NH_4^+} + \underset{\text{Base}_2}{OH^-}$$

Likewise, in solutions of carbonate there is the limited reaction

$$H_2O + CO_3^{2-} \xrightarrow{\text{slight}} HCO_3^- + OH^-$$

These cases of limited proton transfer to give slightly basic solutions are to be contrasted with the last two bases in the table. If either S^{2-} or O^{2-} is placed in water, there is considerable production of OH^-:

$$H_2O + S^{2-} \longrightarrow HS^- + OH^-$$
$$H_2O + O^{2-} \longrightarrow OH^- + OH^-$$
$$\text{Acid}_1 \quad \text{Base}_1 \qquad \text{Acid}_2 \quad \text{Base}_2$$

In the last reaction conversion is essentially 100 percent complete.

There is yet another commonly used system for defining acids and bases which puts the emphasis on electron-pair transfer; it is called the *Lewis definition*. It applies the term "acid" to any species that acts as an *electron-pair acceptor* and the term "base" to any species that acts as an *electron-pair donor*. For instance, in the reaction

$$
H^+ + :\!\!\overset{\displaystyle H}{\underset{\displaystyle H}{N}}\!\!:H \longrightarrow \left[H:\!\!\overset{\displaystyle H}{\underset{\displaystyle H}{N}}\!\!:H \right]^+
$$

$$\text{Acid} \qquad \text{Base}$$

the Lewis acid H^+ accepts a share in the pair of electrons donated by the Lewis base NH_3. The great generality of the Lewis definition can be seen from the fact that it covers very diverse reactions such as

$$
:\!\!\overset{\displaystyle :\ddot{F}:}{\underset{\displaystyle :\ddot{F}:}{F}}\!\!:B \;+\; :\!\!\overset{\displaystyle H}{\underset{\displaystyle H}{N}}\!\!:H \longrightarrow :\!\!\overset{\displaystyle :\ddot{F}:H}{\underset{\displaystyle :\ddot{F}:H}{F}}\!\!:B\!:N\!:H
$$

and

$$
:\!\!\overset{\displaystyle :\ddot{Cl}:}{\underset{\displaystyle :\ddot{Cl}:}{Cl}}\!\!:Al \;+\; :\!\ddot{Cl}:^- \longrightarrow \left[:\!\!\overset{\displaystyle :\ddot{Cl}:}{\underset{\displaystyle :\ddot{Cl}:}{Cl}}\!\!:Al:\ddot{Cl}: \right]^-
$$

$$\text{Acid} \qquad \text{Base}$$

Since Lewis acids "look for" electrons, they are sometimes described as electron-seeking, or *electrophilic* (from the Greek *philos*, meaning "loving"). Similarly, Lewis bases, which "look for" atoms to accept electrons, are described as being *nucleophilic*. In the reaction

$$H^+ + NH_3 \longrightarrow NH_4^+$$

H^+ is said to be the *electrophile* and NH_3 is the *nucleophile*.

The Lewis definition of acids and bases is particularly useful for complex-ion formation as in the reaction

$$\underset{\substack{\text{Lewis} \\ \text{acid}}}{Cr^{3+}} + \underset{\substack{\text{Lewis} \\ \text{base}}}{6F^-} \longrightarrow CrF_6^{3-}$$

In this connection, a new variable has recently been introduced to describe such binding of acids and bases: *hardness* and *softness*. A hard acid (or base) is one that is small and difficult to polarize; a *soft* acid (or base) is one that is large and easy to polarize. An example of a hard acid is Cr^{3+}, and F^- is an example of a hard base. An example of a soft acid is Ag^+; an example of a soft base is I^-. As a general rule, it is observed that hard acids prefer to bind to hard bases and soft acids prefer to bind to soft bases.

No matter which definition of acid and base is used, it is generally agreed that acids and bases react with each other. In the Arrhenius system this reaction is called *neutralization;* as the H^+ of the acid reacts with the OH^- of the base to form water, the acidic and basic properties disappear. However, even here the term "neutralization" has to be taken with caution since the acid and base must be equally strong or equally weak to produce a strictly neutral solution. For example, when 1 mol of hydrochloric acid reacts with 1 mol of sodium hydroxide, the result is a neutral solution. However, when 1 mol of acetic acid reacts with 1 mol of sodium hydroxide, the resulting solution is not neutral but slightly basic. As for the more general definitions, e.g., that of Brønsted, the wide range of acid-base reactions makes the term "neutralization" more questionable. Thus, reaction of the acid HCl with the base H_2O can hardly be called a neutralization. Strictly speaking, then, one should define neutralization as the reaction of acids and bases of comparable strength.

In the following, we consider four possible situations involving reaction between an acid and a base in aqueous solution: strong acid–strong base, weak acid–strong base, strong acid–weak base, and weak base–weak base. As noted above, only the first and the last of these should properly be called neutralization. However, common usage often refers to all four as neutralization reactions.

1 Strong acid–strong base The neutralization of HCl with NaOH is sometimes written

$$HCl + NaOH \longrightarrow H_2O + NaCl$$

but since HCl, NaOH, and NaCl are all strong electrolytes, the species present in solution are ions. The equation is better written as

$$H^+ + Cl^- + Na^+ + OH^- \longrightarrow H_2O + Na^+ + Cl^-$$

Since Na^+ and Cl^- appear on both sides of the equation, they are "spectator ions" (Sec. 2.6) and can be omitted to give the net equation

$$H^+ + OH^- \longrightarrow H_2O$$

The net equation is general and applies to the neutralization of any strong acid by any strong base. Frequently, in recognition of the fact that H^+ is more properly represented as the hydronium ion H_3O^+, the general equation is written

$$H_3O^+ + OH^- \longrightarrow 2H_2O$$

2 Weak acid–strong base For the reaction of a weak acid with a strong base the net equation is

$$HA + OH^- \longrightarrow H_2O + A^-$$

where HA stands for any weak acid, such as acetic acid, $HC_2H_3O_2$. Since weak acids are only slightly dissociated into H^+ and A^- ions, the solution before reaction contains predominantly HA molecules. In the reaction it is the HA molecules that disappear; this must be shown in the net equation. It may be that the actual pathway involves, first, dissociation of HA into H^+ and A^-, with subsequent union of H^+ and OH^- to give H_2O, but the net equation represents only the overall reaction.

3 Strong acid–weak base For the reaction of a strong acid with a weak base, the net equation can be written

$$H^+ + MOH \longrightarrow M^+ + H_2O$$

where MOH represents a weak base. However, there are very few bases that are of the type MOH. Most weak bases are like NH_3, for which the net reaction can be written

$$H^+ + NH_3 \longrightarrow NH_4^+$$

4 Weak acid–weak base For the reaction between a weak acid and a weak base, the net equation is

$$HA + MOH \longrightarrow M^+ + A^- + H_2O$$

If the weak base is NH_3 we can write

$$HA + NH_3 \longrightarrow NH_4^+ + A^-$$

9.3
POLYPROTIC ACIDS

The term *polyprotic* acid describes those acids which can furnish more than one proton per molecule. Two examples are the diprotic acid H_2SO_4, sulfuric acid, and the triprotic acid H_3PO_4, phosphoric acid. In reaction, polyprotic acids usually dissociate one proton at a time. For example, when placed in water, H_2SO_4 dissociates

$$H_2SO_4 \longrightarrow H^+ + HSO_4^-$$

This reaction is practically 100 percent complete; in this sense, H_2SO_4 is called a strong acid. The ion HSO_4^- is an acid in its own right. Although fairly weak, it can dissociate to give H^+ and SO_4^{2-}. This reaction can be written

$$HSO_4^- \xrightarrow{\text{slight}} H^+ + SO_4^{2-}$$

It is the second step in the dissociation of the diprotic acid H_2SO_4 and occurs significantly only when there is a large demand for protons. For example, when 1 mol of H_2SO_4 is mixed in solution with only 1 mol of NaOH, evaporation of the solution produces $NaHSO_4$, sodium hydrogen sulfate; however when 1 mol of H_2SO_4 is mixed in solution with 2 mol of NaOH, evaporation of the solution produces Na_2SO_4, sodium sulfate.

The triprotic acid H_3PO_4 undergoes dissociation in three steps:

$$H_3PO_4 \xrightarrow{\text{slight}} H^+ + H_2PO_4^-$$

$$H_2PO_4^- \xrightarrow{\text{slight}} H^+ + HPO_4^{2-}$$

$$HPO_4^{2-} \xrightarrow{\text{slight}} H^+ + PO_4^{3-}$$

The extent of dissociation is again governed by the demand for protons. It is possible to get three salts from this acid, depending on the relative amount of NaOH added to the solution.

$$H_3PO_4 + OH^- \longrightarrow H_2O + H_2PO_4^-$$

$$H_3PO_4 + 2OH^- \longrightarrow 2H_2O + HPO_4^{2-}$$

$$H_3PO_4 + 3OH^- \longrightarrow 3H_2O + PO_4^{3-}$$

Evaporation of the corresponding solutions would give the salts NaH_2PO_4, monosodium dihydrogen phosphate; Na_2HPO_4, disodium monohydrogen phosphate; and Na_3PO_4, trisodium phosphate. These are sometimes referred to as the *primary, secondary,* and *tertiary* sodium phosphates, respectively.

9.4
EQUIVALENTS OF ACIDS AND BASES

For complete acid-base neutralization, it is necessary that there be an equal number of H^+ ions and OH^- ions available for the reaction. This can be expressed by writing an overall equation. For example, in the complete neutralization of $Ca(OH)_2$ by H_3PO_4 the equation is

$$3Ca(OH)_2 + 2H_3PO_4 \longrightarrow Ca_3(PO_4)_2 + 6H_2O$$

From such an equation, the usual stoichiometric calculations can be made.

It is sometimes convenient, however, to consider neutralization reactions by fixing attention only on the hydrogen and hydroxide ions. For this purpose, a new concept called the *equivalent* (equiv) is introduced. *One equivalent of acid* is defined as *the mass of acid required to furnish one mole of H^+; one equivalent of base* is the *mass of base required to furnish one mole of OH^- or accept one mole of H^+.* One equivalent of any acid just reacts with one equivalent of any base.

One of the simplest acids is HCl, 1 mol of which has a mass of 36.5 g. Since 1 mol of HCl can furnish 1 mol of H^+, 36.5 g of HCl is 1 equiv. For HCl, and for all other monoprotic acids, 1 mol is the same as 1 equiv. This means that the molarity of such solutions numerically equals the normality (Sec. 8.2). For example, a solution labeled 0.59 M HNO_3 means 0.59 mol of HNO_3 per liter of solution, which is the same as 0.59 equiv of HNO_3 per liter, so it can also be labeled 0.59 N HNO_3.

For a diprotic acid such as H_2SO_4, 1 mol of acid can furnish up to 2 mol of H^+. By definition, 2 mol of H^+ is the amount furnished by 2 equiv of acid. Therefore, for complete neutralization 1 mol of H_2SO_4 is identical with

263

Section 9.5
Oxidation-
Reduction
in Aqueous
Solutions

2 equiv. Since 1 mol $= 98$ g $= 2$ equiv, 1 equiv of H_2SO_4 has a mass of 49 g. A solution labeled 0.10 M H_2SO_4, indicating 0.10 mol of H_2SO_4 per liter of solution, can also be labeled as 0.20 N H_2SO_4.

The situation is similar for bases. For NaOH, 1 mol gives 1 mol of OH^-. Therefore, 1 mol of NaOH is 1 equiv. For all solutions of NaOH the normality is equal to the molarity. For $Ca(OH)_2$, the normality of any solution is twice the molarity.

The chief advantage of introducing equivalents in acid-base stoichiometry problems is that it becomes unnecessary to write a balanced equation for the problem, since the same information is embodied in the very definition of an equivalent. All we need to know is how many of the protons are neutralized, i.e., what the final products are. Another point to note is that the term milliequivalent (abbreviated meq) is often used when working with very dilute solutions.

EXAMPLE 1

How many milliliters of 0.0150 M $Ca(OH)_2$ would be required to neutralize completely 35.0 ml of 0.0120 M H_3PO_4?

Solution

0.0150 M $Ca(OH)_2$ means 0.0300 N $Ca(OH)_2$ since each mole of $Ca(OH)_2$ is 2 equiv.

0.0120 M H_3PO_4 means 0.0360 N H_3PO_4 since each mole of H_3PO_4 is 3 equiv.

35.0 ml of 0.0360 N H_3PO_4 contains (0.0360 equiv/liter) (0.0350 liter) $= 0.00126$ equiv of H_3PO_4.

We need 0.00126 equiv of base to neutralize 0.00126 equiv of acid. To get 0.00126 equiv of $Ca(OH)_2$ from 0.0300 N $Ca(OH)_2$ solution, we need to take

$$\frac{0.00126 \text{ equiv}}{0.0300 \text{ equiv/liter}} = 0.0420 \text{ liter} = 42.0 \text{ ml}$$

9.5
OXIDATION-REDUCTION IN AQUEOUS SOLUTIONS

In oxidation-reduction reactions (Sec. 2.9), the number of electrons furnished by the reducing agent must equal that picked up by the oxidizing agent. This is expressed by writing the balanced redox equation. Often we are not told all the species that disappear and appear in a reaction, but only the formulas of oxidizing agent, reducing agent, and their reaction products and whether the solution is acidic or basic. The insertion of H_2O, H^+, or OH^- as required is part of the process of balancing equations in aqueous solution. As an illustration, let us consider the problem of writing the equation for the oxidation of Fe^{2+} by MnO_4^- in acidic solution to form Fe^{3+} and Mn^{2+}. The proper stepwise procedure is like that summarized on page 62.

1   Assign oxidation numbers:

$$Fe^{2+} + MnO_4^- \longrightarrow Mn^{2+} + Fe^{3+}$$
$$\phantom{Fe^{2}}+2 \qquad +7 \qquad\qquad +2 \qquad 3+$$

2   Balance the electron loss and electron gain for those atoms that change oxidation number:

$$5Fe^{2+} + MnO_4^- \longrightarrow Mn^{2+} + 5Fe^{3+}$$
$$\phantom{5Fe^{2}}+2 \qquad +7 \qquad\qquad +2 \qquad +3$$

$$5 \times \boxed{1e^- = 1 \times \boxed{5e^-}}$$

After the coefficient 5 has been placed in front of Fe^{2+} on the left side, the right side is made consistent with regard to the manganese and iron atoms.

3   Count up oxygen atoms and add H_2O to the side that is deficient in oxygen. There are four oxygens on the left and none on the right, so we need $4H_2O$ on the right:

$$5Fe^{2+} + MnO_4^- \longrightarrow Mn^{2+} + 5Fe^{3+} + 4H_2O$$

4   Count up hydrogen atoms and add H^+ to the side that is deficient in hydrogen. On the left, there are no hydrogen atoms; on the right there are eight hydrogen atoms in $4H_2O$. The left side is deficient by eight hydrogens, so we add $8H^+$:

$$5Fe^{2+} + MnO_4^- + 8H^+ \longrightarrow Mn^{2+} + 5Fe^{3+} + 4H_2O$$

5   Check by comparing the net charge on left and right sides of the equation. If the balancing has been done properly, net charge on both sides is the same. Left side has $5(+2) + 1(-1) + 8(+1) = +17$. The right side has $1(+2) + 5(+3) + 4(0) = +17$.

6   If the reaction is said to be in *acidic solution,* you are finished with the balancing. On the other hand, if the reaction is said to be in *basic solution,* add enough OH^- to neutralize each H^+ to H_2O, and add an equal number of OH^- to the opposite side of the equation. Cancel any duplication of H_2O from left and right sides of the equation.

Once we have the final balanced equation, we can use it for standard stoichiometry calculations.

EXAMPLE 2

How many milliliters of 0.20 M $KMnO_4$ are required to oxidize 25.0 ml of 0.40 M $FeSO_4$ in acidic solution? The reaction which occurs is the oxidation of Fe^{2+} by MnO_4^- to give Fe^{3+} and Mn^{2+}, for which we have just developed the balanced equation.

Solution

25.0 ml of 0.40 M $FeSO_4$ supplies (0.0250 liter) (0.40 mol/liter) = 0.010 mol of Fe^{2+}.
The equation states that 1 mol of MnO_4^- is needed per 5 mol of Fe^{2+}:

$$(0.010 \text{ mol of Fe}^{2+}) \left(\frac{1 \text{ mol of MnO}_4^-}{5 \text{ mol of Fe}^{2+}} \right) = 0.0020 \text{ mol of MnO}_4^-$$

To get this quantity of MnO_4^-, the volume of $0.20\ M$ $KMnO_4$ solution is

$$\frac{0.0020 \text{ mol of MnO}_4^-}{0.20 \text{ mol/liter}} = 0.010 \text{ liter} = 10 \text{ ml}$$

An alternate way to solve the above oxidation-reduction problem is to introduce the concept of equivalent in much the same way we did for acid-base reactions. One *equivalent of oxidizing agent is the mass of substance that picks up one mole of electrons and one equivalent of reducing agent is the mass of substance that releases one mole of electrons.* The equivalents are defined in this way so that one equivalent of any oxidizing agent reacts exactly with one equivalent of any reducing agent.

For the problem shown in Example 2 the reducing agent Fe^{2+} changes to Fe^{3+}. Each Fe^{2+} loses one electron, and so each mole of Fe^{2+} loses 1 mol of electrons. Hence, for this reaction, 1 mol of $FeSO_4$ is equal to 1 equiv of $FeSO_4$. This would mean $0.40\ M$ $FeSO_4$ is $0.40\ N$. The oxidizing agent, MnO_4^-, changes to Mn^{2+} in the course of the reaction. As the manganese changes oxidation state from $+7$ to $+2$, each MnO_4^- appears to gain five electrons. Thus, 1 mol of MnO_4^- is equal to 5 equiv. Therefore, $0.20\ M$ $KMnO_4$ is $1.0\ N$. In general, the *normality of an oxidizing or reducing solution is equal to the molarity times the electron change per formula unit.*

Example 2 can now be rephrased using normality instead of molarity.

EXAMPLE 3

How many milliliters of $1.0\ N$ $KMnO_4$ are required to oxidize 25.0 ml of $0.40\ N$ $FeSO_4$ in acidic solution?

Solution

25.0 ml of $0.40\ N$ $FeSO_4$ supplies (0.0250 liter) $(0.40 \text{ equiv/liter}) = 0.010$ equiv of reducing agent. 0.010 equiv of reducing agent requires 0.010 equiv of oxidizing agent:

$$\frac{0.010 \text{ equiv}}{1.0 \text{ equiv/liter}} = 0.010 \text{ liter} = 10 \text{ ml}$$

In some cases a given oxidizing agent can be reduced to different products depending on conditions. For example, MnO_4^- can be reduced to Mn^{2+}, MnO_2, or MnO_4^{2-} when the medium is acid, neutral, or basic, respectively. In these reactions the corresponding number of electrons transferred is five, three, and one. A solution that is $1\ M$ $KMnO_4$ is $5\ N$ $KMnO_4$, $3\ N$ $KMnO_4$, or $1\ N$ $KMnO_4$, depending on which product is formed.

Labels on solution reagent bottles generally specify what the solution was made from, but usually not what the solution contains. For example, the label 0.5 M HCl might better read 0.5 M H^+ and 0.5 M Cl^-. For most quantitative considerations, however, it is not necessary to know what species are actually in the solution. It is necessary to know only what is ultimately available. The label 0.5 M $HC_2H_4O_2$ also tells that the solution was made from $HC_2H_3O_2$ and water, but in this case the solution actually contains $HC_2H_3O_2$ molecules; it is a weak electrolyte and is very slightly dissociated. There is only a trace of H^+ and $C_2H_3O_2^-$ in the solution. However, if the solution were used for a neutralization reaction, not only the trace of H^+ but also the $HC_2H_3O_2$ would be neutralized.

The use of solutions for keeping track of stoichiometry requires a clear distinction between the *number of moles* of solute and its *concentration*. To illustrate, let us suppose 15.8 g of $KMnO_4$ is dissolved to make 1 liter of 0.100 M $KMnO_4$ solution. The formula weight of $KMnO_4$ is 158 amu; hence 15.8 g is equal to 0.100 mol. To make up the solution, we place 15.8 g of the solute in a graduated container and add enough water to bring the total volume to 1 liter. The solution can now be labeled 0.100 M $KMnO_4$; it contains 0.100 mol of $KMnO_4$ per liter of solution. The concentration does not depend on how much of this solution we take. Whether we take one drop or 200 ml, the solution is still 0.100 M $KMnO_4$. However, the number of moles of $KMnO_4$ does depend on how much solution we take. The number of moles of solute in a sample is the number of moles per liter multiplied by the volume of the sample in liters. In 200 ml of 0.100 M $KMnO_4$, there is (0.200 liter) (0.100 mol/liter), or 0.0200 mol, of $KMnO_4$.

Solutions are convenient because they permit measuring amount of solute by measuring a volume of solution. For example, if a given chemical reaction calls for 0.0100 mol of $KMnO_4$, this amount can be provided by weighing out 1.58 g of $KMnO_4$ or, more easily, by measuring out 100 ml of 0.100 M $KMnO_4$ solution.

In summary we can write:

$$\text{Liters} \times \text{molarity} = \text{moles of solute in sample}$$
$$\text{Liters} \times \text{normality} = \text{equivalents of solute in sample}$$

The following examples show how these quantitative relations are applied.

EXAMPLE 4

To what volume must 50.0 ml of 3.50 M H_2SO_4 be diluted in order to make 2.00 M H_2SO_4?

Solution

50.0 ml of 3.50 M H_2SO_4 contains (0.0500 liter) (3.50 mol/liter), or 0.175 mol, of H_2SO_4. We wish the final solution to be 2.00 M.

Therefore, the final volume must be

$$\frac{0.175 \text{ mol}}{2.00 \text{ mol/liter}} = 0.0875 \text{ liter} = 87.5 \text{ ml}$$

EXAMPLE 5

If 50.0 ml of 0.50 M H_2SO_4 is added to 75.0 ml of 0.25 M H_2SO_4, what will be the concentration of the final solution assuming its volume is 125 ml?

Solution

50.0 ml of 0.50 M H_2SO_4 contains (0.0500 liter) (0.50 mol/liter).
75.0 ml of 0.25 M H_2SO_4 contains (0.0750 liter) (0.25 mol/liter).

Total moles $=$ (0.0500 liter) (0.50 mol/liter) $+$ (0.0750 liter) (0.25 mol/liter) $= 0.044$

$$\text{Final concentration} = \frac{0.044 \text{ mol}}{0.125 \text{ liter}} = 0.35 \ M \ H_2SO_4$$

EXAMPLE 6

A solution is made by mixing 10.0 ml of 0.0200 M $Ca(NO_3)_2$ and 15.0 ml of 0.0300 M $NaNO_3$. Assuming the final volume is 25.0 ml and that dissociation is complete, calculate the concentrations of Ca^{2+}, Na^+, and NO_3^- in the final solution.

Solution

Ca^{2+} (0.0100 liter) (0.0200 mol/liter) $= 0.000200$ mol

$$\text{Concentration of } Ca^{2+} = [Ca^{2+}] = \frac{0.000200 \text{ mol}}{0.0250 \text{ liter}} = 0.00800 \ M$$

Na^+ (0.0150 liter) (0.0300 mol/liter) $= 0.000450$ mol

$$\text{Concentration of } Na^+ = [Na^+] = \frac{0.000450 \text{ mol}}{0.0250 \text{ liter}} = 0.0180 \ M$$

NO_3^- (0.0100 liter) (2 $\times$ 0.0200 mol/liter) $+$ (0.0150 liter) (0.0300 mol/liter)
        $= 0.000400 + 0.000450 = 0.000850$ mol

$$\text{Concentration of } NO_3^- = [NO_3^-] = \frac{0.000850 \text{ mol}}{0.0250 \text{ liter}} = 0.0340 \ M$$

Note the use of square brackets around a formula as a shorthand way of designating the concentration of the species shown in terms of moles per liter. We shall use this designation quite frequently.

EXAMPLE 7

A solution is made by mixing 200 ml of 0.100 M $K_2Cr_2O_7$, 250 ml of 0.200 M H_2SO_3 (a weak electrolyte), and 350 ml of 1.00 M $HClO_4$ (a strong electrolyte). A reaction occurs in which H_2SO_3 and $Cr_2O_7^{2-}$ convert in acidic solution to HSO_4^- and Cr^{3+}. Assuming the final volume is 0.800 liter, calculate the final concentration of each species that was present in the initial solution.

Solution

The best way to solve such a problem is to make a table listing the species of interest, how many moles of each species are initially present, how many change, how many moles are left, and what the final concentrations are.

The balanced equation is

$$Cr_2O_7^{2-} + 3H_2SO_3 + 5H^+ \longrightarrow 2Cr^{3+} + 3HSO_4^- + 4H_2O$$

For the moles of species present initially, we have

K^+ (0.200 liter)(0.100 mol of $K_2Cr_2O_7$ per liter) $\left(\dfrac{2 \text{ mol of } K^+}{1 \text{ mol of } K_2Cr_2O_7} \right) = 0.0400$ mol

$Cr_2O_7^{2-}$ (0.200 liter)(0.100 mol of $K_2Cr_2O_7$ per liter) $= 0.0200$ mol

H_2SO_3 (0.250 liter)(0.200 mol H_2SO_3 per liter) $= 0.0500$ mol

H^+ (0.350 liter)(1.00 mol $HClO_4$ per liter) $= 0.350$ mol

ClO_4^- (0.350 liter)(1.00 mol $HClO_4$ per liter) $= 0.350$ mol

The equation states that we need 3 mol of H_2SO_3 for each mole of $Cr_2O_7^{2-}$. We have 0.0500 mol of H_2SO_3 for 0.0200 mol of $Cr_2O_7^{2-}$. Therefore, $Cr_2O_7^{2-}$ is present in excess; the extent of reaction is limited by the moles of H_2SO_3. From the equation, we see that per mole of H_2SO_3 we use up $\frac{1}{3}$ mol of $Cr_2O_7^{2-}$ and $\frac{5}{3}$ mol of H^+, or per 0.0500 mol of H_2SO_3 we use up $\frac{1}{3}(0.0500)$ mol of $Cr_2O_7^{2-}$ and $\frac{5}{3}(0.0500)$ mol of H^+. To get the values in the last column, we divide the moles present at the end of the reaction (next-to-last column) by the final volume of solution, i.e., 0.800 liter.

Species	Moles at start	Moles used up	Moles at end	Final concentration, M
K^+	0.0400	0	0.0400	0.0500
$Cr_2O_7^{2-}$	0.0200	$\frac{1}{3}(0.0500)$	0.0033	0.0041
H_2SO_3	0.0500	0.0500	0	0
H^+	0.350	$\frac{5}{3}(0.0500)$	0.267	0.334
ClO_4^-	0.350	0	0.350	0.438

9.7
HYDROLYSIS

When the salt $NaCl$ is placed in water, the resulting solution is observed to be neutral. However, when the salt $NaC_2H_3O_2$ is dissolved in water, the resulting solution is observed to be slightly basic. Other salts such as ammonium chloride, NH_4Cl, or aluminum chloride, $AlCl_3$, give slightly acid solutions. These interactions between salts and water are called *hydrolysis*.

Hydrolysis is not fundamentally different from any acid-base reaction as viewed in the Brønsted system. In the case of sodium acetate, for example, the basic nature of the resulting aqueous solution can be understood from the following equation:

$$C_2H_3O_2^- + H_2O \xrightarrow{\text{slight}} HC_2H_3O_2 + OH^-$$

Here the Brønsted base $C_2H_3O_2^-$ accepts a proton from the Brønsted acid H_2O to form the conjugate acid $HC_2H_3O_2$ and the conjugate base OH^-. The basic nature of the solution is due to formation of OH^-. This reaction goes only slightly from left to right because $C_2H_3O_2^-$ is a weak base and does not compete significantly with OH^- for a proton (see Fig. 9.1).

In principle, any negative ion can act as a base. In practice, only a few negative ions such as O^{2-} and S^{2-} compete effectively with OH^- for protons. Solutions of these ions are as basic as solutions prepared from hydroxides. Most negative ions compete only slightly with OH^- for protons; their extent of hydrolysis is small. Thus, solutions of Na_2CO_3, $NaC_2H_3O_2$, NaF, and Na_2SO_4, for example, are only slightly basic. At moderate concentration (order of 1 M) the extent of hydrolysis rarely exceeds a percent or so. The extent of hydrolysis of a few negative ions (for example, ClO_4^-, Cl^-, and NO_3^-) is so small as to be undetectable. These ions are the weakest bases, and their conjugate acids are the strongest acids.

Positive ions that hydrolyze generally produce acidic solutions. In some cases, for example, NH_4^+, the source of the proton is obvious:

$$NH_4^+ \xrightarrow{\text{slight}} NH_3 + H^+$$

In other cases, e.g., solutions of Al^{3+} salts, the source of the acidity is harder to see. It comes from the water molecules directly attached to the Al^{3+} of the hydrated aluminum ion. Although the configuration of hydrated aluminum ion is not completely known, each Al^{3+} is thought to be attached to six water molecules at the corners of an octahedron, as shown in Fig. 9.2. The dissociation of a proton from an attached water can be written

$$Al(H_2O)_6^{3+} \longrightarrow Al(H_2O)_5OH^{2+} + H^+$$

The reaction, which leaves a doubly charged complex ion, corresponds to losing one of the peripheral protons shown in Fig. 9.2 and transferring it to a molecule of solvent.

9.8
AMPHOTERISM

If a solution of sodium hydroxide is added dropwise to a solution of aluminum nitrate, a white precipitate, aluminum hydroxide, is first formed, but on

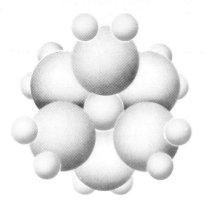

FIG. 9.2 Octahedron and Al(H₂O)$_6^{3+}$ ion. (The colored sphere buried in the middle is the Al^{3+}. The small outer spheres are H atoms attached to the O atoms.)

further addition of base or on addition of acid the precipitate dissolves. The net equations for the dissolving processes can be written

$$Al(OH)_3(s) + OH^- \longrightarrow Al(OH)_4^-$$
$$Al(OH)_3(s) + 3H^+ \longrightarrow Al^{3+} + 3H_2O$$

These two equations indicate that Al(OH)$_3$ is *amphoteric*, i.e., it is able to neutralize a base and also an acid. In other words, aluminum hydroxide itself is *able to function both as an acid and as a base.*

How can amphoterism be explained? The answer can be seen by writing the above equations in a way that emphasizes the probable species involved:

$$Al(OH)_3(H_2O)_3(s) + OH^- \longrightarrow Al(OH)_4(H_2O)_2^- + H_2O$$
$$Al(OH)_3(H_2O)_3(s) + 3H^+ \longrightarrow Al(H_2O)_6^{3+} + 3H_2O$$

Each aluminum species contains Al^{3+} at the center of an octahedron with six groups, OH$^-$ or H$_2$O, at the octahedral corners. For each complex the total number of oxygens is six. Figure 9.3 shows the relative arrangement of atoms. In the top reaction, a proton is transferred from Al(OH)$_3$(H$_2$O)$_3$ to externally added OH$^-$; in the bottom, proton transfer is from externally added H$^+$ to Al(OH)$_3$(H$_2$O)$_3$. The figure shows (in color) the two possible proton transfers. In case *A* a proton from an H$_2$O bound to the central Al^{3+} transfers to OH$^-$; in case *B* a proton from an external acid transfers to an OH$^-$ bound to the Al^{3+}. For case *A*, the aluminum hydroxide species acts as a proton donor; for case *B*, a proton acceptor.

FIG. 9.3 Two possible modes of reaction leading to amphoteric behavior of aluminum hydroxide.

Important Concepts

acids
bases
Arrhenius system
general-solvent system
Brønsted system
proton donor
proton acceptor
conjugate acid-base pair
relative strength of conjugate acids and bases
Lewis acid
Lewis base
electrophile
nucleophile

hard and soft acids
hard and soft bases
neutralization
polyprotic acids
successive dissociation
equivalent of acid
equivalent of base
equivalent of oxidizing agent
equivalent of reducing agent
stoichiometry of solutions
hydrolysis
extent of hydrolysis
amphoterism

Exercises

*9.1 Acids-bases Distinguish clearly between an acid and a base as defined in each of the following systems: Arrhenius, generalized solvent, Brønsted, Lewis.

*9.2 Acids-bases Liquid ammonia slightly conducts electricity because of the self-dissociation reaction

$$NH_3 + NH_3 \longrightarrow NH_4^+ + NH_2^-$$

What would constitute an acid and what would be a base in liquid ammonia? Give specific examples. Indicate for your examples what "neutralization" would correspond to.

*9.3 Brønsted system The reaction

$$HF + NH_3 \longrightarrow NH_4^+ + F^-$$

occurs when HF and NH_3 are placed in water. In the Brønsted system, what do the terms acid, base, and conjugate acid-base pair mean for this specific reaction? Given that this reaction goes almost to completion, what can you conclude about the relative rankings of the conjugate acid-base pairs?

**9.4 Acids-bases Given the rankings shown in Fig. 9.1, what would you predict for the predominant direction in each of the following?

$$H_2SO_4 + H_2O \rightleftharpoons H_3O^+ + HSO_4^-$$
$$HF + Cl^- \rightleftharpoons HCl + F^-$$
$$NH_4^+ + CO_3^{2-} \rightleftharpoons NH_3 + HCO_3^-$$
$$HS^- + NH_3 \rightleftharpoons NH_4^+ + S^{2-}$$
$$H_2O + C_2H_3O_2^- \rightleftharpoons HC_2H_3O_2 + OH^-$$

**9.5 Acids-bases On the basis of the rankings shown in Fig. 9.1, what would you expect to happen when H_2S in H_2O is mixed with an aqueous solution containing NH_3 and CO_3^{2-}?

**9.6 Lewis acid-base system Where in the periodic table would you look for elements most likely to form compounds that are Lewis acids? Lewis bases? Explain your reasoning.

*9.7 Neutralization Explain what neutralization means in the Arrhenius system of acids and bases. What difference in detail is there in applying the term to HCl plus NaOH vs. $HC_2H_3O_2$ plus NaOH?

*9.8 Neutralization Write net equations for each of the following "neutralization" reactions.

a The weak acid HNO_2 reacts with the strong base KOH.

b The strong acid $HClO_4$ reacts with the strong base $Ba(OH)_2$.

c The weak acid HF reacts with the weak base NH_3.

d The strong acid $HClO_4$ reacts with the weak base NH_3.

*9.9 Polyprotic acids Why would it be inappropriate to call NH_4^+ a polyprotic acid?

**9.10 Polyprotic acids By examining the data shown in Fig. 9.1, what might you conclude about the relative acid strength of the various hydrogen-containing species obtained from a polyprotic acid?

*9.11 Acid equivalent How many equivalents of acid are there in each of the following?

a 1.00 g of HCl

b 1.00 mol of HCl

c 1.00 liter of 0.10 M HCl

d 1.00 liter of 0.10 N HCl

*9.12 Acid equivalent How many equivalents of acid are there in each of the following?

a 1.00 g of H_2SO_4

b 1.00 mol of H_2SO_4

c 1.00 liter of 0.10 M H_2SO_4

d 1.00 liter of 0.10 N H_2SO_4

**9.13 Acid equivalent Given an unknown acid H_nX. It takes 1.50 g of this acid to neutralize 30.0 ml of 0.250 N NaOH solution. What is the weight of one equivalent of the acid? What other information would you need to establish the value of n? Design an experiment that would give you this information.

*9.14 Oxidation-reduction Write a complete balanced net equation for each of the following conversions:

a $Sn^{2+} + MnO_4^- \longrightarrow Sn^{4+} + Mn^{2+}$ (acidic)

b $Sn^{2+} + Cr_2O_7^{2-} \longrightarrow Sn^{4+} + Cr^{3+}$ (acidic)

c $I_2 + Cr_2O_7^{2-} \longrightarrow IO_3^- + Cr^{3+}$ (acidic)

d $SO_3^{2-} + MnO_4^- \longrightarrow SO_4^{2-} + MnO_2(s)$ (basic)

e $Sn(OH)_3^- + CrO_4^{2-} \longrightarrow SnO_2(s) + Cr(OH)_4^-$ (basic)

f $I_3^- + ClO^- \longrightarrow IO_3^- + Cl^-$ (basic)

**9.15 Oxidation-reduction A pure sample of elemental iron weighing 0.600 g is dissolved in H_2SO_4 to give Fe^{2+} and H_2. The resulting solution just reacts with 96.87 ml of $?M$ $KMnO_4$ solution by the reaction $Fe^{2+} + MnO_4^- \longrightarrow Fe^{3+} + Mn^{2+}$. What is the molarity of the $KMnO_4$ solution?

*9.16 Oxidation-reduction How many liters of $SO_2(g)$ measured at $25°C$ and 0.968 atm would it take to destroy all the MnO_4^- in 250 ml of 0.100 M $KMnO_4$ by the reaction $MnO_4^- + SO_2 \longrightarrow MnO_2 + SO_4^{2-}$ in neutral solution?

**9.17 Solution stoichiometry How many grams of $KMnO_4$ must you add to 35.0 ml of 0.100 M $KMnO_4$ solution to make it 0.600 N for a reaction where MnO_4^- goes to Mn^{2+}? Assume no change in volume. How would your answer be changed if the resulting solution were to be used for a reaction where MnO_4^- goes to MnO_2? *Ans. 0.11 g, 0.55 g*

**9.18 Solution stoichiometry How many milliliters of 0.250 M $KMnO_4$ solution must be added to 50.0 ml of 0.100 M $KMnO_4$ solution to give a solution that is 0.200 M $KMnO_4$? Assume that volumes are additive.

**9.19 Solution stoichiometry In what ratio should you mix 0.200 M $NaNO_3$ solution and 0.100 M $Ca(NO_3)_2$ so that in the resulting solution the concentration of negative ions is 50.0 percent greater than the concentration of positive ions? Assume 100 percent dissociation and additive volumes.

**9.20 Solution stoichiometry A solution is made by mixing 200 ml of 0.100 M $FeSO_4$, 200 ml of 0.100 M $KMnO_4$, and 600 ml of 1 M $HClO_4$. A reaction occurs in which Fe^{2+} and MnO_4^- convert to Fe^{3+} and Mn^{2+} in acidic solution. Calculate the concentration of each ion in the final solution. Assume volumes are additive and no other reaction occurs. *Ans. 0.568 M H^+*

***9.21 Solution stoichiometry A solution is made by mixing 200 ml of 0.100 M $Na_2S_2O_3$, 250 ml of 0.200 M Na_2CrO_4, and 350 ml of 1.00 M NaOH. A reaction occurs in which $S_2O_3^{2-}$ and CrO_4^{2-} convert to SO_4^{2-} and $Cr(OH)_4^-$ in basic solution. Assuming volumes are additive, calculate the concentration of each species in the final solution.

*9.22 Hydrolysis What is the connection between hydrolysis of the salt NaX and the action of X^- as a Brønsted base?

**9.23 Hydrolysis The hydrolysis of NaCl is much less than that of $NaC_2H_3O_2$, which in turn is much less than that of Na_2S. How can you use the ranking in Fig. 9.1 to account for this?

***9.24 Hydrolysis** In the hydrolysis of positive ions it is generally observed that the extent of hydrolysis increases in the order M^+, M^{2+}, M^{3+}. Suggest an explanation for this observation, based on the complex shown in Fig. 9.2.

**9.25 Amphoterism What is meant by the term amphoteric hydroxide? How does hydration help to account for this kind of behavior?

**9.26 Amphoterism Zinc hydroxide, $Zn(OH)_2$, is an amphoteric hydroxide. Write net equations to illustrate this action. Assume that each zinc is surrounded by four oxygen atoms.

Chapter 10

CHEMICAL THERMODYNAMICS

In Sec. 8.8, we asked why NaCl dissolves in water if it takes 770 kJ to break up the lattice whereas only 766 kJ is made available by the hydration of the ions. Clearly the overall process is energetically unfavorable; yet it occurs spontaneously. What other factor besides energy is involved?

It is generally believed that processes tend to go in the direction of lower energy. However, when ice melts at room temperature, it spontaneously goes to a state of higher energy. Why does this happen? We might argue that there is a flow of heat from room to ice. True, but why does the ice use the added heat to melt rather than simply warm up? Both processes represent an increase of energy. Why is one preferred rather than the other?

In this chapter we take a closer look at these questions. It will turn out that, besides energy, an important factor determining the direction of spontane-

ous change is the number of ways a system of given energy can be obtained. The more ways there are to have a system, the more probable it becomes. To measure the relative probability, we shall introduce a new concept called the *entropy*. We shall find that the key connection between energy and entropy is temperature. The study of heat as energy and temperature as a controlling factor in the dynamics of energy change is known as *thermodynamics*. We shall be concerned with chemical reactions, but the science of thermodynamics is extremely broad. It brings together phenomena from physics, astronomy, biology, and other fields, and has turned out to be a major unifying force in science. Because it is abstract, a full appreciation of thermodynamics can come only after accumulation of considerable background information. In this chapter, we only begin to lay a foundation for further study in this field.

10.1
SYSTEMS AND STATE FUNCTIONS

In thermodynamics the term *system* is used to designate that region of the physical world that is being considered. This might be, for example, 1 mol of CO_2 gas in a tank, 1 liter of 0.10 M $CuSO_4$ solution, or a particular crystal of NaCl. Everything else outside the system is considered to be the *surroundings*. In general, we are concerned with processes that occur in the system and the relationship of the system to its surroundings. For example, there might be an endothermic reaction in the system with enough heat flow from the surroundings to keep the system at a fixed temperature. Such a system would be an *isothermal* system. An alternative possibility would be to have the system completely insulated from its surroundings; such an isolated system would be called *adiabatic*.

To describe the state of a system, we must specify a number of variables. The most frequently used are temperature, pressure, volume, and chemical composition. The major concern of thermodynamics is with properties that depend *only* on the state of the system. ~~This means that a thermodynamic property, such as internal energy, is completely determined once the state of the system is described.~~ Such a property is called a *state function*.

In general, it is not the absolute values of the thermodynamic quantities that are of prime importance, but the changes in them. To designate these changes, we introduce the symbol Δ (Greek delta) before a property to designate a change in that property. Thus, for example, ΔE stands for the change in molar energy and represents the energy of the final state minus that of the initial state. If, by chance, the molar energy in the final state is *less* than in the initial state, then ΔE would be a negative quantity. In other words, Δ can take on either positive or negative values.

Because a thermodynamic property depends only on the state of the system, changes in thermodynamic properties are independent of how the system is taken from state 1 to state 2. For example, suppose 1 mol of O_2 gas is taken from state 1 where it occupies 100 liters at 273°K and 0.224 atm to state 2 where it occupies 100 liters at 546°K and 0.448 atm. The thermodynamic properties do not depend on whether the gas is first heated and then compressed or first compressed and then heated. Thus, if the energy of the initial state is designated E_1 and the energy of the final state E_2, the change in energy $\Delta E = E_2 - E_1$ is independent of the path.

10.2
FIRST LAW

Two laws form the basis of thermodynamics. The first of these is equivalent to the law of conservation of energy. It states that the increase in energy ΔE of a system is equal to the heat q absorbed by the system from the surroundings minus the work w performed by the system on the surroundings:

$$\Delta E = q - w$$

Since ΔE is an energy change, both q (heat) and w (work) must be in energy units. Another problem is to keep straight the sign convention used for q and w. When q is positive, heat is *absorbed by the system,* and the energy of the system increases; when q is negative, heat is *evolved by the system,* and the energy of the system decreases. When w is positive, the system does work *on its surroundings,* and the energy of the system decreases; when w is negative, work is done on the system *by the surroundings,* and the energy of the system increases.

It should also be noted that $\Delta E = 0$ for any system completely isolated from its surroundings. No heat flows into or out of the system, no work is done by the system on its surroundings, and no work is done by the surroundings on the system. Thus, $\Delta E = 0$ means that the energy of the system is constant.

Examples of heat energy q are discussed in the following sections. What are some examples of w? Suppose a gas sample trapped in a cylinder with a sliding piston expands from initial volume V_1 to final volume V_2 against a constant atmospheric pressure P. The work the gas sample does against the surrounding atmosphere w is $P(V_2 - V_1) = P\Delta V$. As a second example, suppose a lead storage battery is connected to drive an electric motor. The work w done by the battery is equal to the voltage of the battery times the current delivered times the elapsed time for which the current is delivered.

10.3
ENTHALPY AND HEAT CAPACITY

In talking about the energy of a system, we tend to focus our attention on internal forms of energy such as vibration motion of atoms, kinetic energy of molecules, and potential energy stored in chemical bonds. However, there is another, more subtle contribution to the energy. It comes from the fact that the system has a volume V and is generally under the pressure of the surrounding atmosphere P. External work of the amount PV had to be done against the surrounding atmosphere to make room for the system.* So, any system that has a finite volume V and is subject to a finite pressure P has energy associated with it of amount PV. Any changes in PV have to be accounted for in counting up the total energy. For this purpose, we introduce a new concept called the *enthalpy*, designated by H, which is the sum of the internal energy E and the external work PV. Thus, we can write, per mole of substance,

$$\underset{\text{Enthalpy}}{H} = \underset{\substack{\text{Internal} \\ \text{energy}}}{E} + \underset{\substack{\text{External} \\ \text{work}}}{PV}$$

An *increase* in enthalpy ΔH is made up of two parts: (1) increase in energy ΔE and (2) increase in the PV product. Symbolically, this is expressed as follows:

$$\Delta H = \Delta E + \Delta(PV)$$

The change in the PV product can come about because of a change in P, a change in V, or both. We shall generally restrict ourselves to situations where P is constant. In such cases, $\Delta(PV)$ is just $P\Delta V$, the work done on constant-pressure surroundings by a volume change ΔV. Thus, at constant pressure

$$\Delta H = \Delta E + P\Delta V$$

or

$$\boxed{\Delta E = \Delta H - P\Delta V}$$

Comparison of this last relation with the first law $\Delta E = q - w$ shows that if the only work w done by the system is $P\Delta V$ work, then *at constant pressure* $\Delta H = q$. In other words, the enthalpy increase ΔH is equal to the heat absorbed by the system at constant pressure. (For this reason H is sometimes referred to as the *heat content* of the system, although the term "enthalpy" is preferred.) If heat is added to a system at constant pressure, part of the heat is used to increase the internal energy of the system, and the rest is used to do work on the surroundings.

*Note that the product PV is an energy measure. Putting P in atmospheres and V in liters, we have liter atmospheres, as in the gas constant R (0.08206 liter atm mol^{-1} deg^{-1}). One liter atmosphere equals 101.32 joules.

EXAMPLE 1

For the decomposition $CaCO_3(s) \longrightarrow CaO(s) + CO_2(g)$ at 950°C and CO_2 pressure of 1 atm, ΔH is 176 kJ/mol. Assuming that the volume of the solid phase changes very little, calculate ΔE for the decomposition at 1 atm.

Solution

$$\Delta E = \Delta H - P\Delta V$$

$$\Delta V = V_{products} - V_{reactants} \sim V_{CO_2} = \frac{n_{CO_2}RT}{P}$$

where we have introduced the gas law $PV = nRT$ to calculate V_{CO_2}.

1000

$$P\Delta V = n_{CO_2}RT = (1.00 \text{ mol})(8.31 \text{ J mol}^{-1} \text{ deg}^{-1})(1223°) = 10.2 \text{ kJ}$$

$$\Delta E = 176 - 10.2 = 166 \text{ kJ}$$

In words, the preceding example states that, of the 176 kJ of heat added, 166 kJ goes into the chemical energy of decomposition, and 10.2 kJ goes into doing expansion work (*PV* work) against the atmosphere. When gases are not involved or when the number of moles of gas does not change, the $P\Delta V$ work will be negligible; in such reactions practically all the heat added goes to chemical energy. This last statement is equivalent to saying $\Delta E \sim \Delta H$, as is illustrated in the following problem.

EXAMPLE 2

At 0°C, ice has a density of 0.917 g/cm³ and liquid water has a density of 0.9998 g/cm³. For the melting process, ΔH is 6010 J/mol. Calculate ΔE and compare with ΔH.

Solution

$$\Delta E = \Delta H - P\Delta V$$

$$\Delta V = V_{liq} - V_{solid} = \frac{18.015 \text{ g/mol}}{0.9998 \text{ g/cm}^3} - \frac{18.015 \text{ g/mol}}{0.917 \text{ g/cm}^3}$$

$$\Delta V = -1.63 \text{ cm}^3/\text{mol} = -0.00163 \text{ liter/mol}$$

$$P\Delta V = (1.00 \text{ atm})(-0.00163 \text{ liter/mol}) = -0.00163 \text{ liter atm/mol}$$

$$1 \text{ liter atm} = 101.3 \text{ J}$$

$$P\Delta V = (-0.00163 \text{ liter atm/mol})(101.3 \text{ J liter}^{-1} \text{ atm}^{-1}) = -0.165 \text{ J/mol}$$

Therefore,

$$\Delta E = \Delta H - P\Delta V = 6010 + 0.165 \sim 6010 \text{ J/mol}$$

$$\Delta E \sim \Delta H$$

FIG. 10.1 Heat Capacity in Joules per Mole-Degree for Various Substances at the Temperature Indicated

$H_2O(s)$ at 239°K	33.30	$SO_2(s)$ at 198°K	69.04
$H_2O(s)$ at 271°K	37.78	$SO_2(l)$ at 270°K	86.61
$H_2O(l)$ at 273°K	75.86	$Zn(s)$ at 693°K	29.7
$H_2O(l)$ at 298°K	75.23	$Zn(l)$ at 1000°K	31.4
$H_2O(l)$ at 373°K	75.90	$Hg(l)$ at 500°K	27.6
$H_2O(g)$ at 383°K	36.28	$Xe(g)$ at 165°K	20.9

Because $P\Delta V$ is generally small compared with ΔE, chemists frequently get away with saying "energy changes" when they strictly mean "enthalpy changes."

In the above examples we considered what happens to the enthalpy of a system at a fixed temperature. What if the temperature is also allowed to change? Added heat may then be used to warm the system to a higher temperature. To describe this heating-up process, we use the *heat capacity*, defined as the amount of heat required to raise the temperature of one mole of material by one degree. If we restrict ourselves to heat capacity at constant pressure, usually designated by C_p, we find that the values of C_p depend on the chemical identity of the material, its state (whether gaseous, liquid, or solid), and its temperature. Typical values are listed in Fig. 10.1. We might also note that the heat capacity can also be given per gram instead of per mole, in which case it is sometimes called *specific heat*.*

For simplicity, let us consider 1 mol of a substance. What happens if its temperature increases from T_1 to T_2? We assume ΔT is small enough that the heat capacity C_p does not change with temperature. The amount of heat absorbed by the substance is equal to $C_p\Delta T$. This heat goes to raise the enthalpy H of the material by an amount ΔH. Hence, we can write

$$\Delta H = C_p\Delta T$$

For example, to take 1 mol of $H_2O(l)$ from 293 to 303°K requires that 75.23 J mol^{-1} deg^{-1} (from Fig. 10.1) times 10°, or 752.3 J, be added to the H_2O; that is $\Delta H = 752.3$ J.

If a phase change occurs between T_1 and T_2, it is necessary to take into account the change in enthalpy associated with the phase change as well as the difference in heat capacity between the two phases.

---

*There is a mild problem here for those who like to worry about such things. Some handbooks give specific heat in units such as calories per gram; others give no units, arguing that specific heat was originally defined as the ratio of the heat capacity per gram of substance divided by the heat capacity per gram of H_2O. Because the heat capacity per gram of H_2O is very close to 1 cal/deg, dividing by this quantity does not change the numerical value but does get rid of the units. In spite of this, most practicing chemists use the term "specific heat" as if it had the units of energy per gram.

EXAMPLE 3

By using data from Fig. 10.1, calculate the increase of enthalpy for 100 g of H_2O going from ice at $-10°C$ to liquid water at $+15°C$. The molar heat of fusion of ice is 6.01 kJ.

Solution

$$\frac{100 \text{ g}}{18.0 \text{ g/mol}} = 5.55 \text{ mol of } H_2O$$

To heat the ice, one needs (5.55 mol) (37.78 J mol^{-1} deg^{-1}) (10 deg) = 2100 J.
To melt the ice, one needs (5.55 mol) (6.01 kJ/mol) = 33,400 J.
To heat the liquid, one needs (5.55 mol) (75.86 J mol^{-1} deg^{-1}) (15 deg) = 6300 J.
The total increase in enthalpy equals $\Delta H = 2100 + 33,400 + 6300 = 41,800$ J.

10.4
ENTHALPY CHANGES IN CHEMICAL REACTIONS

In general, when a chemical reaction occurs, heat is either evolved to or absorbed from the surroundings; so the enthalpy of the system changes. As a specific example we consider the reaction between hydrogen and oxygen to form water. The initial state consists of 1 mol of $H_2(g)$ and $\frac{1}{2}$ mol of $O_2(g)$ each at 1 atm pressure and 25°C; the final state, 1 mol of $H_2O(l)$ at 1 atm pressure and 25°C. The change can be written

$$H_2(g) + \tfrac{1}{2}O_2(g) \longrightarrow H_2O(l)$$

If the pressure is to remain constant at 1 atm, then clearly the volume of the system must be allowed to shrink. If the temperature is to remain constant, then a thermostated constant-temperature bath in which the reaction chamber is immersed must be provided. Both the pressure and the temperature need to be specified, since the enthalpy of a material depends on its pressure and temperature. By convention, most data are quoted for pure materials at 1 atm, this condition being called the *standard state*. In addition, it is usual to specify the temperature. As example, when the above reaction occurs at 25°C so that H_2 and O_2 in their standard states change to H_2O in its standard state, 286 kJ of heat is liberated per mole of H_2O formed. The chemical system has decreased in enthalpy by 286 kJ. We can write

$$\Delta H = -286 \text{ kJ}$$

The minus sign indicates that the enthalpy *of the system* actually decreases, the environment having gained the 286 kJ. In general for exothermic reactions (heat liberated to the environment), ΔH of the system is negative; for endothermic reactions (heat absorbed from the surroundings), ΔH is positive.

The -286 kJ/mol of the preceding example may be referred to as the *enthalpy of formation* or *heat of formation* of liquid H_2O from the elements in their standard states at 25°C. For convenience, the enthalpy of formation of the elements is set at zero. Figure 10.2 gives representative values for enthalpies of formation of several common compounds at 25°C. The ab-

FIG. 10.2 Enthalpies of Formation from the Elements for Various Compounds at 25°C and 1 atm

Compound	ΔH, kJ/mol	Compound	ΔH, kJ/mol	Compound	ΔH, kJ/mol
$H_2O(l)$	-286	$LiF(s)$	-612	$H_2SO_4(aq)$	-908
$H_2O(g)$	-242	$NaF(s)$	-569	$HNO_3(l)$	-173
$H_2O_2(l)$	-187	$NaCl(s)$	-411	$HNO_3(aq)$	-207
$HF(g)$	-269	$NaBr(s)$	-360	$H_3PO_4(aq)$	-1290
$HCl(g)$	-92	$NaI(s)$	-288	$H_3PO_3(aq)$	-972
$HBr(g)$	-36	$CaO(s)$	-636	$CaCO_3(s)$	-1210
$HI(g)$	$+26$	$BaO(s)$	-558	$BaCO_3(s)$	-1220
$H_2S(g)$	-20	$Al_2O_3(s)$	-1670	$Na_2SO_4(s)$	-1380
$NH_3(g)$	-46	$Cr_2O_3(s)$	-1130	$NaHSO_4(s)$	-1130
$CH_4(g)$	-75	$CO(g)$	-110	$NaOH(s)$	-427
$C_2H_6(g)$	-85	$CO_2(g)$	-394	$Ca(OH)_2(s)$	-987
$C_2H_4(g)$	$+52$	$SiO_2(s)$	-859	$CaSO_4(s)$	-1430
$C_2H_2(g)$	$+227$	$SO_3(g)$	-395	$CaSO_4 \cdot 2H_2O(s)$	-2020

breviations in parentheses indicate the state of the substance. When the abbreviation in parentheses is *aq,* the reference state is an ideal 1 *m* aqueous solution.

Once the enthalpies of formation from the elements are known, it is possible to calculate the enthalpy changes for other reactions. The enthalpy change for any reaction is equal to the enthalpies of formation of the substances on the right side of the chemical equation (products) minus the enthalpies of formation of the substances on the left side of the equation (reactants). For instance, at 298°K and 1 atm the enthalpy change for

$$CaO(s) + CO_2(g) \longrightarrow CaCO_3(s)$$

can be found as follows:

$$\Delta H_{\text{reaction}} = \Delta H_{CaCO_3} - \Delta H_{CaO} - \Delta H_{CO_2}$$

where ΔH_{CaCO_3}, which represents the enthalpy of formation of $CaCO_3(s)$, gives the enthalpy of $CaCO_3$ relative to the elements Ca(s), C(s), and $O_2(g)$. Similarly, ΔH_{CO_2} and ΔH_{CaO} give the enthalpies of these compounds relative to the elements. The change in enthalpy between Ca, C, and O_2 as CaO plus CO_2 and Ca, C, and O_2 as $CaCO_3$ is the net heat of reaction. The calculation is an example of *Hess's law,* which states that *the heat of reaction is the same whether the reaction takes place in one step or by several consecutive steps that add up to the same reaction.* The reaction above can be regarded as the sum of the following reactions:

$CaO(s) \longrightarrow Ca(s) + \frac{1}{2}O_2(g)$		$\Delta H = 636 \text{ kJ}$
$CO_2(g) \longrightarrow C(s) + O_2(g)$		$\Delta H = 394 \text{ kJ}$
$Ca(s) + C(s) + \frac{3}{2}O_2(g) \longrightarrow CaCO_3(s)$		$\Delta H = -1210 \text{ kJ}$
$CaO(s) + CO_2(g) \longrightarrow CaCO_3(s)$		$\Delta H = -180 \text{ kJ}$

The first reaction is the reverse of CaO formation; the second is the reverse of CO_2 formation. When we reverse a formation reaction, we get a decomposition reaction. Correspondingly, what was an exothermic reaction, e.g., $Ca(s) + \frac{1}{2}O_2(g) \longrightarrow CaO(s)$ with $\Delta H = -636$ kJ, becomes an endothermic reaction, i.e., $CaO(s) \longrightarrow Ca(s) + \frac{1}{2}O_2(g)$ with $\Delta H = +636$ kJ. Reversal of any chemical equation causes reversal of its sign of ΔH. Finally, it might be noted that when we add the equations above, the elements Ca, C, and O_2 cancel out.

EXAMPLE 4

Calculate the enthalpy change on combustion of 1 mol of $C_2H_4(g)$ to form $CO_2(g)$ and $H_2O(g)$ at 298°K and 1 atm.

Solution

$$C_2H_4(g) + 3O_2(g) \longrightarrow 2CO_2(g) + 2H_2O(g)$$

$$\begin{aligned}
\Delta H_{combustion} &= 2\Delta H_{CO_2} + 2\Delta H_{H_2O} - \Delta H_{C_2H_4} - 3\Delta H_{O_2} \\
&= 2(-394) + 2(-242) - 52 - 3(0) \\
&= -1324 \text{ kJ}
\end{aligned}$$

Note that the enthalpy change for formation of the element oxygen in its standard state O_2 is zero.

10.5
ENTROPY

The first law of thermodynamics concerns itself with the conservation of energy, $\Delta E = q - w$. For an isolated system the heat absorbed by the system q and the work done by the system w are both equal to zero; so $\Delta E = 0$. In other words, the energy of an isolated system is constant. Still, within this stipulation that energy be constant, changes may occur. So far as the first law is concerned, all such changes are equally possible, yet experience tells us that some changes do not spontaneously occur; for example, H_2O does not freeze above its melting point; heat does not flow from low to high temperature; gases do not contract spontaneously. Evidently, there are restrictions on what direction spontaneous change will take.

One familiar criterion for spontaneous change is that energy should decrease. However, this will not cover all cases, e.g., the melting of ice, where energy spontaneously increases ($\Delta H = +6.01$ kJ/mol). The additional factor that must be considered is that systems also tend to go to the *most probable* state, the one with the most disordered molecular arrangement. The reason for this tendency is that there are more ways in which a system can be disordered than there are in which it is ordered. The disorder is described quantitatively in terms of a property called *entropy*. A disordered state has a higher entropy than an ordered state. As a specific example, H_2O in the form of liquid water has a higher entropy than H_2O in the form of solid ice.

Why is a state of high entropy (i.e., a disordered state) more probable than one of low entropy (i.e., an ordered state)? In ice there is but one recognizable repeating pattern that is unique for ice. In liquid water, there are millions of arrangements all of which are equally probable and all of which together constitute the state of liquid water. The point is that there are many arrangements we call "liquid" compared with the one we call "ice." Hence, an arrangement characteristic of liquid water is more probable than the arrangement characteristic of ice.

Ludwig Boltzmann, the great Austrian physicist, suggested that the above ideas can be expressed quantitatively by writing

$$S = k \ln \Omega$$

where S is the entropy, k is Boltzmann's constant, and Ω (omega) is the number of possible configurations that a system can have for a given energy. (The designation ln stands for "natural logarithm," i.e., to the base $e = 2.71828 \ldots$.)

In the same sense that natural processes resulting in a *decrease of energy* are favored, those resulting in an *increase of entropy* are also favored. In certain cases, as in melting ice, the two tendencies oppose each other; so there is a question of which one wins out. Temperature turns out to be the decisive factor. Above the melting point the entropy increase is dominant, and spontaneous change is favored in the direction of melting; below the melting point the energy decrease is dominant, and spontaneous change is favored in the direction of freezing.

10.6
FREE ENERGY

Quantitatively, the interplay of energy and entropy can be described by using a concept called the *free energy*. The "free" emphasizes the fact that for conversion into usable work, such as that needed to drive an engine, the total energy needs to be distinguished into two parts: (1) a part that is freely available (i.e., the *free* energy) and (2) a part that is not (i.e., the *unavailable* energy). The unavailable energy is represented by the product of the entropy S and a proportionality constant which turns out to be just the absolute temperature T. The higher the temperature, the more the disorder, and the less available the energy becomes. The free energy per mole is generally symbolized by G,* and so we can write the above relation

$$\text{Total energy} = \underset{\substack{\text{Free} \\ \text{energy}}}{G} + \underset{\substack{\text{Unavailable} \\ \text{energy}}}{TS}$$

Since the total energy is $H \, (= E + PV)$, we can write for free energy the following expression:

*The symbol G is in honor of J. Willard Gibbs, a Yale professor of mathematics who founded the science of chemical thermodynamics.

$$G \quad = \quad H \quad - \quad TS$$

Free   Enthalpy   Unavailable
energy   energy

It states that, due to the disordering influence of temperature, not all the energy H in a system is available for extraction. A certain amount, given by the product TS, cannot be used. The higher the temperature, the greater becomes the TS product, and the smaller the amount of the freely available energy.

For *changes* in free energy we can write

$$\Delta G = \Delta H - \Delta(TS)$$

For an isothermal change, i.e., one that occurs at a fixed temperature T, the free-energy change is given by

$$\Delta G = \Delta H - T\Delta S$$

This is an important equation because at constant temperature and pressure a chemical reaction or some physical change can occur spontaneously or of its own accord *only if it is accompanied by a decrease in free energy*. In other words, ΔG should be negative. A negative ΔG can be produced by a decrease in enthalpy (i.e., a negative ΔH) or an increase in entropy (i.e., a positive ΔS). The former ($\Delta H < 0$) would correspond to an energetically favorable process, i.e., one in which energy decreases; the latter ($\Delta S > 0$) corresponds to one in which disorder increases and so is favored by random thermal motion.

For many processes ΔH and ΔS have the same sign; for example, for the melting of ice both are positive. In such cases T will be all-important in deciding whether ΔH or $T\Delta S$ prevails. At low T, ΔH will predominate, ΔG will be positive, and melting will not occur. At very high temperature, however, the $T\Delta S$ term predominates. Since ΔS for melting is positive, $-T\Delta S$ will be negative, and ΔG will also be negative. Hence, melting should occur. Only at one temperature does $T\Delta S$ just equal ΔH. At this temperature ΔG is equal to zero, $\Delta G = \Delta H - T\Delta S = 0$, and solid and liquid coexist in equilibrium. The temperature of this coexistence is, of course, the melting, or freezing, point.

EXAMPLE 5

For the melting of sodium chloride the heat required is 30.3 kJ/mol. The entropy increase is 28.2 J mol^{-1} deg^{-1}. Calculate the melting point from these data.

Solution

At the melting point $\Delta G = \Delta H - T\Delta S = 0$.

$$\Delta H = T_{mp}\Delta S$$

$$T_{mp} = \frac{\Delta H}{\Delta S} = \frac{30,300 \text{ J/mol}}{28.2 \text{ J mol}^{-1} \text{ deg}^{-1}} = 1070°K$$

EXAMPLE 6
At 0°C the entropies of $H_2O(s)$ and $H_2O(l)$ are 37.95 and 59.94 J mol^{-1} deg^{-1}, respectively. Calculate ΔS and ΔH for conversion of 1 mol of ice to liquid water at 0°C.

Solution

$$\Delta S = S_{liq} - S_{solid} = 59.94 - 37.95 = 21.99 \text{ J mol}^{-1} \text{ deg}^{-1}$$

$$\Delta H = T_{mp}\Delta S = (273.15°) (21.99 \text{ J mol}^{-1} \text{ deg}^{-1}) = 6010 \text{ J/mol}$$

10.7
SECOND LAW

For our purposes the second law of thermodynamics can be formulated as follows: *Any system when left to itself will tend to change toward a condition of greater probability.* As was noted in Sec. 10.5, a condition of greater probability is one of greater randomness, or disorder. Entropy S is the thermodynamic property that measures this increased probability of disordered systems. Hence, the second law of thermodynamics can be stated in another way: *For spontaneous change to occur in an isolated system, the entropy must increase; that is, ΔS must be greater than zero.*

If a system is not isolated then it follows that the total entropy (system plus surroundings) must increase for spontaneous change to occur:

$$\Delta S_{total} = \Delta S_{system} + \Delta S_{surroundings} > 0 \tag{1}$$

The $\Delta S_{surroundings}$ corresponds to the amount of heat transferred to the surroundings divided by the temperature at which the transfer occurs. For the situations most usually encountered, where temperature and pressure are kept constant, we can write

$$\Delta S_{surroundings} = \frac{\Delta H_{surroundings}}{T} = -\frac{\Delta H_{system}}{T}$$

We note here that $\Delta H_{surroundings}$, due to heat transfer from the system, is equal and opposite to ΔH_{system}. Since we usually consider only what happens to the system, we can substitute in Eq. (1) to get

$$\Delta S_{total} = \Delta S_{system} - \frac{\Delta H_{system}}{T} > 0 \tag{2}$$

Viewed from the system, then, the second law states that the quantity which must increase in a spontaneous change at constant T and P is $\Delta S - \Delta H/T$. As we see in the following, this is just the negative of the free-energy change divided by the absolute temperature:

The definition of free energy, as given in Sec. 10.6, is

$$G = H - TS$$

The change in free energy is $\Delta G = \Delta H - T\Delta S - S\Delta T$. At constant temperature and pressure, where $\Delta T = 0$,

$$\Delta G = \Delta H - T\Delta S$$

Dividing this equation by T gives

$$\frac{\Delta G}{T} = \frac{\Delta H}{T} - \Delta S$$

or, by changing sign and rearranging,

$$-\frac{\Delta G}{T} = \Delta S - \frac{\Delta H}{T} \qquad (3)$$

According to Eq. (2), spontaneous change requires that the right-hand side of Eq. (3) must be greater than zero; so the left-hand side must also be greater than zero:

$$-\frac{\Delta G}{T} > 0$$

Since T is always positive, the only way $-\Delta G/T$ can be greater than zero is for ΔG to be negative. In other words, *for spontaneous change to occur at constant temperature and pressure the free energy of the system must decrease.*

Consider, for example, the conversion $H_2O(s) \longrightarrow H_2O(l)$. The ΔH for this process is 6.01 kJ/mol; the ΔS is 21.99 J mol^{-1} deg^{-1}. Let us compare ΔG at three different temperatures:

At 253°K: $\quad \Delta G = \Delta H - T\Delta S = 6010 - 253(21.99) = +450$ J/mol

At 273.15°K: $\quad \Delta G = \Delta H - T\Delta S = 6010 - 273.15(21.99) = 0$

At 293°K: $\quad \Delta G = \Delta H - T\Delta S = 6010 - 293(21.99) = -430$ J/mol

As can be seen, the free energy would increase at 253°K; therefore, melting is not favored. At 273.15°K there is no change in free energy in going from solid to liquid, so neither state is favored; both solid and liquid continue to coexist. At 293°K, the free energy would decrease, so the process of melting would tend to go spontaneously.

As another example of the foregoing principles, let us consider the reaction

$$Br_2(l) + Cl_2(g) \longrightarrow 2BrCl(g)$$

At 25°C and 1 atm pressure, ΔH for the reaction is +29 kJ. This means that when 1 mol of liquid bromine combines with 1 mol of gaseous chlorine to form 2 mol of gaseous bromine chloride, 29 kJ of heat is absorbed from the surroundings. The change is endothermic. Since most spontaneous reactions are exothermic ($\Delta H < 0$), it might seem that this particular reaction ($\Delta H > 0$) would not occur spontaneously. However, the second law requires that we consider also the ΔS for the reaction. At 25°C, ΔS is equal to +105 J mol^{-1} deg^{-1}. Hence, at 25°C, or 298°K, we can write for the free-energy change

$$\Delta G = \Delta H - T\Delta S$$
$$= 29 \text{ kJ} - (298°)(0.105 \text{ kJ/deg})$$
$$= -2 \text{ kJ}$$

Because ΔG is negative, the free energy would decrease; the reaction $Br_2(l)$ plus $Cl_2(g)$ to form $BrCl(g)$ should occur spontaneously. We should note, however, that thermodynamics does not tell how fast the change will occur but merely in what direction it should go.

10.8
STANDARD FREE-ENERGY CHANGES

The free energy of a substance (as well as its enthalpy and its entropy) depends on the state of the substance, i.e., whether it is solid, liquid, or gas. Furthermore, the free energy depends on temperature and pressure and, in the case of solutions, on concentration. For convenience, standard states have been chosen for reference. These correspond to 1 atm pressure and, if the substance is pure, the state, solid, liquid, or gas, in which the substance exists at 1 atm and the specified temperature. For *gaseous* solutions the standard state is generally a partial pressure of 1 atm. For *liquid* solutions the standard state can be taken as unit mole fraction or, more generally, ideal 1 *m* concentration. For *solid* solutions, the standard state corresponds to unit mole fraction.

As a specific example of the significance of standard states we consider the reaction

$$Zn(s) + 2H^+(aq) \longrightarrow H_2(g) + Zn^{2+}(aq)$$

For this reaction at 25°C, the standard free-energy change $\Delta G°$ is -148 kJ. The superscript ° indicates that the number applies to substances in their standard states only. A subscript is frequently used to show the absolute temperature at which the data apply. $\Delta G°_{298} = -148$ kJ means that the free energy of the system at 25°C and 1 atm decreases by 148 kJ when 1 mol of solid zinc reacts completely with 2 mol of hydrogen ion, from an aqueous solution in which the concentration of H^+ is an ideal 1 *m*, to produce 1 mol of gaseous H_2 at a pressure of 1 atm and 1 mol of zinc ion in aqueous solution at a concentration that is ideal 1 *m*. "Ideal one molal" means that the concentration is one mole per kilogram of solvent except that corrections have been applied for nonideal behavior, such as interionic attraction.

Handbooks usually contain tabulations of standard free-energy data for formation of substances from their elements. Figure 10.3 shows representative data at 25°C. By combining these data, one can calculate the standard free-energy change of any given reaction and tell whether it should occur. For example, the $\Delta G°$ of formation of $H_2O(l)$ is given as -237 kJ/mol. It applies to the reaction

$$H_2(g) + \tfrac{1}{2}O_2(g) \longrightarrow H_2O(l) \tag{1}$$

At the same temperature $\Delta G°$ of formation of $CaO(s)$ is given as -604 kJ/mol.

FIG. 10.3   Standard Free Energies of Formation from the Elements at 25°C and 1 atm

Compound	$\Delta G°$, kJ/mol	Compound	$\Delta G°$, kJ/mol	Compound	$\Delta G°$, kJ/mol
$H_2O(l)$	−237.2	$LiF(s)$	−584	$H_2SO_4(aq)$	−742
$H_2O(g)$	−228.6	$NaF(s)$	−541	$HNO_3(l)$	−80
$H_2O_2(l)$	−118	$NaCl(s)$	−384	$HNO_3(aq)$	−111
$HF(g)$	−271	$NaBr(s)$	−333	$H_3PO_4(aq)$	−1150
$HCl(g)$	−95	$NaI(s)$	−261	$H_3PO_3(aq)$	−860
$HBr(g)$	−53	$CaO(s)$	−604	$CaCO_3(s)$	−1129
$HI(g)$	+1	$BaO(s)$	−528	$BaCO_3(s)$	−1139
$H_2S(g)$	−3.3	$Al_2O_3(s)$	−1576	$Na_2SO_4(s)$	−1267
$NH_3(g)$	−17	$Cr_2O_3(s)$	−1047	$NaHSO_4(s)$	−1030
$CH_4(g)$	−51	$CO(g)$	−137	$NaOH(s)$	−377
$C_2H_6(g)$	−33	$CO_2(g)$	−394	$Ca(OH)_2(s)$	−897
$C_2H_4(g)$	+68	$SiO_2(s)$	−805	$CaSO_4(s)$	−1320
$C_2H_2(g)$	+209	$SO_3(g)$	−370	$CaSO_4 \cdot 2H_2O(s)$	−1796

It applies to the reaction

$$Ca(s) + \tfrac{1}{2}O_2(g) \longrightarrow CaO(s) \tag{2}$$

Similarly, the $\Delta G°$ of formation of $Ca(OH)_2(s)$ is −897 kJ/mol. It applies to the reaction

$$Ca(s) + O_2(g) + H_2(g) \longrightarrow Ca(OH)_2(s) \tag{3}$$

Given these three pieces of data we can immediately calculate the $\Delta G°$ for the reaction

$$CaO(s) + H_2O(l) \longrightarrow Ca(OH)_2(s) \tag{4}$$

By subtracting reactions (1) and (2) from reaction (3), we get

$$\Delta G° = \Delta G°_{Ca(OH)_2} - \Delta G°_{CaO} - \Delta G°_{H_2O}$$
$$= -897 - (-604) - (-237) = -56 \text{ kJ}$$

The fact that $\Delta G°$ is less than zero for reaction (4) tells us that the hydration reaction at 25°C should go spontaneously as written.

Important Concepts

system	enthalpy	free energy
surroundings	heat capacity	free-energy change
isothermal system	enthalpy change	spontaneous process
adiabatic system	standard state	second law
thermodynamic property	enthalpy of formation	standard free-energy change
state function	Hess's law	
first law	entropy	

Exercises

*10.1 Terms What is the difference between an isothermal system and an adiabatic system?

*10.2 Terms If 1 mol of oxygen gas is taken from a state where it occupies 100 liters at 273°K and 0.224 atm to a state where it occupies 50 liters at 546°K and 0.896 atm, what are ΔV, ΔT, and ΔP for the process?

Final − Initial

*10.3 First law What is the first law of thermodynamics? Justify the statement that it is equivalent to the law of conservation of energy.

gas in bags volume

10.4 Given a system that consists of a burning candle, what is the sign of q and of w for the burning process?

**10.5 First law Given a system composed of a large insulated box which is completely evacuated except that it contains a much smaller box containing helium gas at some pressure. Suppose a hole is now made in the smaller box so that the helium expands into the bigger box. Tell what happens to each of the variables in the first-law statement $\Delta E = q - w$.

**10.6 ΔE and ΔH A gas-filled balloon absorbs 1 kJ of heat and expands from 20.0 liters to 22.0 liters against 1 atm of pressure. What are ΔE and ΔH for the process?   *Ans. $\Delta E = 800$ J*

*10.7 Enthalpy Which of the following reactions has the smallest difference between ΔE and ΔH? Assume ideal behavior under identical conditions.

a $CaCO_3(s) \longrightarrow CaO(s) + CO_2(g)$
b $H_2(g) + \frac{1}{2}O_2(g) \longrightarrow H_2O(g)$
c $2H_2(g) + O_2(g) \longrightarrow 2H_2O(g)$
d $H_2(g) + I_2(g) \longrightarrow 2HI(g)$

*10.8 Enthalpy What is the definition of enthalpy? From this definition and the first law, show that at constant pressure the increase in enthalpy of a system is equal to the heat absorbed by the system.

**10.9 Enthalpy For the reaction $2KClO_3(s) \longrightarrow 2KCl(s) + 3O_2(g)$ at 400°C and 1 atm pressure, ΔH is −80.0 kJ per 2 mol of $KClO_3$ decomposed. Assuming that the volume of the solid phase changes by little compared with the volume of gas generated, calculate ΔE for the decomposition. Is the reaction exothermic or endothermic?

**10.10 Enthalpy (a) At its normal melting point, 5.53°C, benzene shows a density of 0.8786 g/ml in the liquid state and 0.9934 g/ml in the solid state. If the heat of fusion is 127.4 J/g, calculate the values of ΔE and ΔH per mole for the fusion process. (b) At its normal boiling point, 80.1°C, benzene shows a density of 0.8145 g/ml in the liquid state and 0.002732 g/ml in the gaseous state. If the heat of vaporization is 395.4 J/g, calculate the values of ΔE and ΔH per mole for the vaporization process.
   Ans. (a) $\Delta E = 9951$ J/mol, $\Delta H = 9952$ J/mol

**10.11 Heat capacity What is the essential distinction between heat content per mole and heat capacity per mole?

**10.12 Heat capacity What is the heat capacity of an ice-water mixture at 1 atm pressure and 273.15°K? $\Delta H = nCp\Delta T$

*10.13 Enthalpy Which requires more heat: to take 1.00 g of $H_2O(s)$ from 239 to 271°K, or to take 1.00 g of $H_2O(l)$ from 273 to 298°K? How much more? (See Fig. 10.1 for data.)

**10.14 Heat capacity Using data from Fig. 10.1, calculate how many 30-g ice cubes at 0°C you need to add to 1000 g of liquid water at 25°C to bring the temperature down to 0°C. It takes 6.01 kJ to fuse 1 mol of ice.

**10.15 Heat capacity What will be the final temperature if you mix 125 g of ice at 239°K with 250 g of liquid water at 373°K? The molar heat of fusion of ice is 6.01 kJ.   *Ans. 308°K*

*10.16 Enthalpy of reaction How many kilojoules of heat are liberated to the surroundings when 1.00 g of Zn reacts by the following reaction?

$$Zn(s) + \tfrac{1}{2}O_2(g) \longrightarrow ZnO(s) \qquad \Delta H = -485 \text{ kJ/mol}$$

*10.17 Enthalpy of reaction How many kilojoules of heat are *absorbed from the surroundings* when 1.00 g of S reacts by the following reaction?

$$S(s) + O_2(g) \longrightarrow SO_2(g) \qquad \Delta H = -297 \text{ kJ/mol}$$

*10.18 Enthalpy of reaction Using data from Fig. 10.2, calculate the enthalpy change for the reaction $CaO(s) + SO_3(g) \longrightarrow CaSO_4(s)$.

*10.19 Enthalpy of reaction For the reaction $NH_3(g) + HCl(g) \longrightarrow NH_4Cl(s)$, ΔH is −177 kJ mole. Given the data of Fig. 10.2, what must be the enthalpy of formation from its elements of $NH_4Cl(s)$?

*10.20 Hess's law Using the data from Fig. 10.2, calculate ΔH for the reaction $2CO(g) + O_2(g) \longrightarrow 2CO_2(g)$.

*10.21 Hess's law On the basis of data given in Fig. 10.2, predict whether the reaction $2Al(s) + Cr_2O_3(s) \longrightarrow Al_2O_3(s) + 2Cr(s)$ is exothermic or endothermic. By how much? What does this suggest about the relative stability of these oxides?

**10.22 Hess's law Given the following data:

$$Na(s) \longrightarrow Na(g) \qquad \Delta H = 108.7 \text{ kJ}$$
$$Na(g) \longrightarrow Na^+(g) + e^- \qquad \Delta H = 502.1 \text{ kJ}$$
$$Cl_2(g) \longrightarrow 2Cl(g) \qquad \Delta H = 242.8 \text{ kJ}$$
$$Cl^- \longrightarrow Cl(g) + e^- \qquad \Delta H = 365.3 \text{ kJ}$$
$$Na(s) + \tfrac{1}{2}Cl_2(g) \longrightarrow NaCl(s) \qquad \Delta H = -411.1 \text{ kJ}$$

Calculate how much heat is liberated when 1 mol of gaseous Na^+ combines with 1 mol of gaseous Cl^- to form solid NaCl. *Ans. −778.0 kJ*

**10.23 Enthalpy of reaction Given the data of Fig. 10.2, calculate which liberates more heat: the burning of 1.00 g of CH_4 or the burning of 1.00 g of C_2H_6. Assume 25°C, 1 atm pressure, and products $CO_2(g)$ and $H_2O(g)$.

**10.24 Enthalpy of reaction Calculate the enthalpy change for the reaction $H_2SO_4(aq) + 2NaOH(s) \longrightarrow Na_2SO_4(s) + 2H_2O(l)$ at 1 atm pressure and 25°C. What difference would it make if the H_2O product were gaseous instead of liquid?

*10.25 Entropy Tell whether the entropy increases, decreases, or stays constant in each of the following changes: (*a*) 1 mol of gas expands into an evacuated space, (*b*) 1 mol of H_2O goes from liquid to solid state at −10°C, (*c*) 1 liter of H_2 at STP mixes with 1 liter of O_2 at STP to give 2 liters of H_2–O_2 mixture at STP, (*d*) 1 mol of H_2O (*g*) at 373°K condenses to 1 mol $H_2O(l)$.

**10.26 Entropy Given state 1 consisting of 1 mol of $NaCl(s)$ plus 1 liter of $H_2O(l)$ and state 2 consisting of 1 *m* NaCl solution, which state has the higher entropy? Justify your answer.

***10.27 Entropy Figure 8.2 shows how the phase diagram of H_2O is broadened in its liquid

range when solute is dissolved in the liquid phase. Show how the concept of entropy helps to account for this.

***10.28 Entropy For 1 mol of gas in volume V, Boltzmann's entropy equation $S = k \ln \Omega$ goes over to $S = R \ln V$ where R is the universal gas constant. Show that this can be used to explain why 1 mol of gas placed in a corner of an insulated box tends to expand to fill the whole box.

***10.29 Entropy In rolling a pair of dice, the point total "lucky seven" is more probable than any other possible point total. In this sense, 7 can be said to have the maximum entropy. Figure out the relative entropy for all the point totals from 2 ("snake eyes") to 12 ("boxcars").

*10.30 Free energy What is the difference between free energy G and internal energy E?

**10.31 Free energy Give the signs of ΔG, ΔH, and ΔS for the process $H_2O(l) \longrightarrow H_2O(s)$ at −10, 0, and +10°C.

**10.32 Free energy Benzene has a normal melting point of 5.53°C and a heat of fusion of 127.4 J/g. Calculate ΔG at 0°C and at +10°C for the process $C_6H_6(s) \longrightarrow C_6H_6(l)$.

**10.33 Second law For a spontaneous change in an isothermal, constant-pressure system, what relation must exist between ΔS and $\Delta H/T$ of the system?

***10.34 Second law Why is it necessary to specify *constant temperature* when stating that $\Delta G < 0$ for spontaneous change?

***10.35 First and second laws A succinct statement of the first and second laws of thermodynamics is: "The energy of the universe is constant; the entropy strives to reach a maximum." Show that this statement is equivalent to the statements given in this chapter.

**10.36 Standard free-energy change By looking at the data in Fig. 10.3, what can you conclude about the relative stability of hydrogen halides with respect to decomposition to the elements at 25°C?

***10.37 Standard free-energy change From the data given in Fig. 10.3 and other information in this chapter, predict a value for the standard free-energy change of formation of $H_2O(s)$ at 25°C. Indicate any assumptions you have had to make. *Ans. −236.7 kJ/mol*

Chapter 11

ELECTROCHEMISTRY

The energy changes associated with chemical reactions generally show up as heat absorbed from or emitted to the surroundings. When there is work done by the chemical system on the environment, it is usually of the mechanical, or $P\Delta V$, type. Occasionally, however, it is possible to arrange the chemical reaction so that it does electric work. In this chapter we consider the transport of electric energy through matter, the conversion of electric energy into chemical energy, and the conversion of chemical energy into electric energy. These topics constitute the field of *electrochemistry*.

11.1
ELECTRIC CONDUCTIVITY

Electric energy may be transported through matter in the form of an *electric current* (see Appendix, Secs. A5.5 to A5.7 for a discussion of electrical terms). In order that the electric current exist, there must be carriers of electric charge and there must be a force that makes the carriers move. The charge carriers can be electrons, as is the case in metals, or positive and negative ions, as in solutions and molten salts. The one kind of conduction is called *electronic* or *metallic;* the other, *ionic* or *electrolytic*. The electric force that makes charges move is usually supplied by a battery, generator, or some similar source of electric energy. The region of space in which there is an electric force is called an *electric field*.

As pointed out in Sec. 7.9, a solid metal consists of an ordered array of positive ions immersed in a sea of electrons. For example, silver metal consists of Ag^+ ions permeated by a cloud of electrons. The Ag^+ ions are fixed in positions from which they do not move except under great stress. The electrons of the cloud, on the contrary, are free to roam throughout the crystal. When an electric field is impressed on the metal, the electrons migrate and carry negative charge through the metal. In principle, it should be possible to force all the loose electrons toward one end of a metal sample. In practice, it is extraordinarily difficult to violate electrical neutrality. The only way to keep a sustained flow of charge is to add electrons to one end and drain electrons off the other as fast as they accumulate. The metal conductor remains everywhere electrically neutral all the time since just as many electrons move into a region as move out per unit time.

290

EXAMPLE 1

When a current of 1.00 ampere (A) flows in a metallic conductor, how many electrons pass a given point in 1 sec?

Solution

$1 A = 1$ coulomb/sec (C/sec)

Charge of electron $= 1.60 \times 10^{-19}$ C

$$\frac{1 \text{ C/sec}}{1.60 \times 10^{-19} \text{ C/electron}} = 6.25 \times 10^{18} \text{ electrons/sec}$$

Most of the electrons that make up the electron cloud of a metal are of very high kinetic energy. Metallic conductivity would therefore be extremely high were it not for a *resistance* effect. Electric resistance comes from the fact that lattice ions vibrate about their lattice points and interfere with free electron motion. At higher temperatures the thermal vibrations of the lattice increase, and therefore it is not surprising to find that as the temperature of a metal is raised, its conductivity diminishes.

In solutions the positive carriers are also free to move. When an electric field is applied, as shown in Fig. 11.1, the positive ions experience a force in one direction, while the negative ions experience a force in the opposite direction. The simultaneous motion of positive and negative ions in opposite directions constitutes the *electrolytic current*. The current would stop if positive ions accumulated at the negative electrode and negative ions at the positive electrode. In order that the electrolytic current continue, appropriate chemical reactions must occur at the electrodes to maintain electrical neutrality throughout all regions.

That ions migrate when electrolytic solutions conduct electricity can be seen from the experiment diagramed in Fig. 11.2. A U-shaped tube is half filled with a purple aqueous solution of copper permanganate $Cu(MnO_4)_2$.

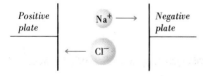

FIG. 11.1 Electric forces on ions in solution.

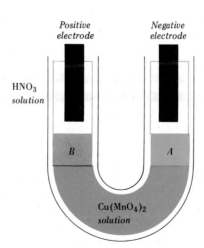

FIG. 11.2 Experiment to demonstrate migration of ions in electrolytic conductivity.

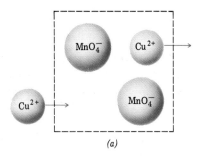

(a)

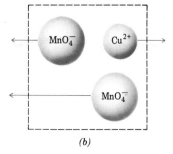

(b)

FIG. 11.3 Two ways that migrating ions could maintain electrical neutrality in a region of solution.

The blue color of hydrated Cu^{2+} ions is effectively masked by the purple of MnO_4^- ions. A colorless aqueous solution of nitric acid, HNO_3, is then floated above the $Cu(MnO_4)_2$ solution in each arm of the U. If an electric field is exerted across the solution by the two electrodes, after a while it is observed that the blue color characteristic of hydrated Cu^{2+} ions has moved toward the negative electrode into the region marked *A*. At the same time, the purple color characteristic of MnO_4^- has moved toward the positive electrode into the region marked *B*.

As in metallic conduction, electrical neutrality must be preserved at all times. Otherwise, the current would cease. Figure 11.3 shows two possible ways by which electrical neutrality can be preserved. In (*a*), one Cu^{2+} ion enters a region to compensate for the charge of a departing Cu^{2+} ion. In (*b*), as one Cu^{2+} ion leaves the region, two MnO_4^- ions depart in the opposite direction. *Both* effects (*a*) and (*b*) occur simultaneously, their relative importance depending on the relative mobilities of the positive and negative ions.

Unlike metallic conduction, electrolytic conduction is usually increased when the temperature of a solution is raised.* In metals the conducting electrons are already of such high energy that a rise in temperature does not appreciably affect their kinetic energy. In solutions, not only is the average kinetic energy of the ions increased with a rise in temperature, but also the viscosity of the solvent is diminished; so the ions can migrate faster, and the solution becomes a better conductor of electricity.

---

*There are exceptions to this generalization. For example, with some weak electrolytes, such as acetic acid above 25°C, the percent dissociation (Sec. 8.6) decreases with rising temperature. The decrease in the concentration of ions may be big enough to cause a *decrease* in conductivity with rising temperature.

In order to maintain an electric current, it is necessary to have a complete circuit. If the complete circuit includes an electrolytic conductor, then chemical reaction must occur at the electrode interface between the metallic conductors and the electrolytic solution. Electric energy is thus used to produce chemical change; this process is called *electrolysis*.

A typical electrolysis circuit is shown in Fig. 11.4. The two vertical lines at the top of the diagram represent a battery; the curved lines represent strips of connecting wire, usually copper, that join the battery to the electrodes. The electrodes may be any inert, conducting material, such as graphite or platinum. They dip into the electrolytic conductor, which contains the ions M^+ and X^- that are free to move. When operating, the battery creates an electric field which pushes the electrons in the external wires in the directions shown by the arrows.

The circuit is not complete, and current does not flow unless there is some way by which electrons can be used up at the left electrode and supplied at the right electrode. Chemical changes must occur. At the left electrode a *reduction* process must occur; some ion or molecule must accept electrons. Such an electrode at which reduction occurs is called a *cathode*. At the right-hand electrode electrons must be donated by some ion or molecule; in other words, an *oxidation* process must occur. Such an electrode is called an *anode*. In order for reduction to keep going at the cathode, ions will have to keep moving toward it. These ions are the positive ions; they are called *cations*. Simultaneously, negative ions move toward the anode; they are called *anions*.

11.3
ELECTROLYSIS OF MOLTEN NaCl

For a more specific example we consider the electrolysis of *molten* NaCl. Figure 11.5 shows the electrolysis cell. As in later cases, we assume the electrodes are inert and do not react chemically. Of the two ions present, Na^+ and Cl^-,

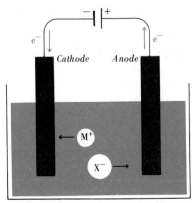

FIG. 11.4 Typical electrolysis circuit.

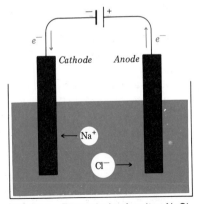

FIG. 11.5 Electrolysis of molten NaCl.

only the Na^+ can be reduced. At the cathode, where reduction must occur, the following reaction takes place:

$$Na^+(l) + e^- \longrightarrow Na(l)$$

The designations (l) emphasize that the Na^+ comes from NaCl in the liquid state and the Na formed is also in the liquid state. This is called the *cathode half-reaction*. At the anode, oxidation occurs. Of the two species in the cell, only Cl^- can be oxidized. The exact mechanism is not known but we can imagine Cl^- releases an electron to the anode to form a neutral Cl atom, which then combines with another Cl to produce the gas molecule Cl_2. The *net anode half-reaction* can be written

$$2Cl^-(l) \longrightarrow Cl_2(g) + 2e^-$$

At the cathode, electric energy converts Na^+ into liquid Na metal; at the anode, Cl^- into gaseous Cl_2. By addition, the two half-reactions can be combined into a single overall *cell reaction*. To ensure electron balance, the half-reactions are multiplied by appropriate coefficients so that when the half-reactions are added, the electrons cancel out of the final equation. Thus, for the electrolysis of molten NaCl we have the following:

Cathode half-reaction: $2Na^+(l) + 2e^- \longrightarrow 2Na(l)$

Anode half-reaction: $2Cl^-(l) \longrightarrow Cl_2(g) + 2e^-$

Overall reaction: $2Na^+(l) + 2Cl^-(l) \xrightarrow{\text{electrolysis}} 2Na(l) + Cl_2(g)$

This reaction if left to itself would go spontaneously in the reverse direction. To emphasize that it goes the other way on consumption of electric energy, the word "electrolysis" has been written over the arrow.

11.4
ELECTROLYSIS OF AQUEOUS NaCl

When *aqueous* NaCl is electrolyzed, it is observed that hydrogen gas is liberated at the cathode and chlorine gas is liberated at the anode. How can these observations be accounted for? Figure 11.6 shows the electrolysis cell. It contains, besides Na^+ and Cl^- ions, H_2O molecules and traces of H^+ and OH^- from the dissociation of H_2O. Molecules of H_2O can be either oxidized to O_2 by removal of electrons or reduced to H_2 by addition of electrons. The H_2O must thus be considered as a possible reactant at either electrode. At the cathode, reduction must occur. Three different reactions are possible:

$$Na^+ + e^- \longrightarrow Na(s) \tag{1}$$
$$2H_2O + 2e^- \longrightarrow H_2(g) + 2OH^- \tag{2}$$
$$2H^+ + 2e^- \longrightarrow H_2(g) \tag{3}$$

When there are several possible reactions at a cathode, it is necessary to consider not only which reactant is the best oxidizing agent but also which one

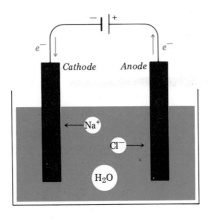

FIG. 11.6   Electrolysis of aqueous NaCl.

is reduced most *rapidly*. Complications generally arise when currents become very large and when concentrations of reactants become very small. The fact that hydrogen gas, not metallic sodium, is actually observed in the electrolysis suggests that reaction (1) does not occur. Also, in NaCl solution, the concentration of H^+ (which is only $1 \times 10^{-7} M$) is not large enough to make reaction (3) reasonable as a *net change*. Therefore, reaction (2) is usually written as the best description for the cathode reaction.

Reaction (2) shows that OH^- accumulates in the region around the cathode; so positive ions (Na^+) must move toward the cathode to preserve electrical neutrality. In addition, some of the OH^- migrates away from the cathode. Both migrations are consistent with the requirement that cations migrate toward the cathode and anions toward the anode.

At the anode, oxidation must occur. Two reactions are possible:

$$2Cl^- \longrightarrow Cl_2(g) + 2e^-$$
$$2H_2O \longrightarrow O_2(g) + 4H^+ + 4e^-$$

It was stated that chlorine gas is liberated at the anode, so experiment shows that the first of these predominates. As the chloride-ion concentration around the anode is depleted, fresh Cl^- moves in, and Na^+ moves out.

In summary, the equations for the electrolysis of aqueous NaCl can be written as follows:

Cathode: $2e^- + 2H_2O \longrightarrow H_2(g) + 2OH^-$

Anode: $2Cl^- \longrightarrow Cl_2(g) + 2e^-$

Overall reaction: $2Cl^- + 2H_2O \xrightarrow{\text{electrolysis}} H_2(g) + Cl_2(g) + 2OH^-$

As indicated, the concentration of Cl^- diminishes, and the concentration of OH^- increases. Since there is always Na^+ in the solution, the solution is gradually converted from aqueous NaCl to aqueous NaOH. Indeed, commercial production of chlorine by electrolysis of aqueous NaCl yields NaOH as a by-product.

When aqueous Na_2SO_4 is electrolyzed with inert electrodes, H_2 gas is formed at the cathode and O_2 is formed at the anode. Changes at the electrodes can be demonstrated by running the electrolysis with litmus in a two-compartment cell such as is shown in Fig. 11.7. Before electrolysis, the solution contains Na^+, SO_4^{2-}, and H_2O. It is essentially neutral; therefore litmus takes the usual violet coloration. After electrolysis has proceeded for a while, the cathode compartment becomes blue, indicating the solution is basic; the anode compartment becomes red, indicating the solution is acidic. Consistent with these observations are the following electrode reactions:

Cathode: $2e^- + 2H_2O \longrightarrow H_2(g) + 2OH^-$

Anode: $2H_2O \longrightarrow O_2(g) + 4H^+ + 4e^-$

The overall reaction, obtained by doubling the cathode reaction and adding the result to the anode reaction, is

$$6H_2O \xrightarrow{\text{electrolysis}} 2H_2(g) + O_2(g) + 4H^+ + 4OH^-$$

In this equation, both H^+ and OH^- appear as products, which would seem to be a self-contradiction, except that the H^+ and OH^- are formed in different regions. If the H^+ and the OH^- are allowed to mix, neutralization occurs, and the above net reaction becomes

$$2H_2O \xrightarrow{\text{electrolysis}} 2H_2(g) + O_2(g)$$

In this electrolysis only the H_2O disappears. The Na^+ and SO_4^{2-} are unchanged. Is the Na_2SO_4 necessary? Because of the need for electrical neutrality, some kind of electrolytic solute must be present. Positive ions must move into the cathode region to counterbalance the charge of the OH^- produced. Negative ions must be available to move to the anode to counterbalance the H^+ produced.

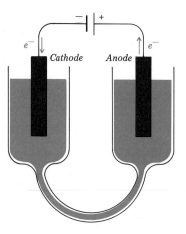

FIG. 11.7 Two-compartment electrolysis cell.

Almost any ionic solute makes possible the electrolysis of water. The only requirement is that the ions of the solute not be oxidized or reduced. This would happen, for example, when aqueous $CuSO_4$ is electrolyzed because Cu^{2+} is more easily reduced than H_2O. During electrolysis, copper plating forms on the cathode. The reactions are

Cathode: $Cu^{2+} + 2e^- \longrightarrow Cu(s)$

Anode: $2H_2O \longrightarrow O_2(g) + 4H^+ + 4e^-$

Overall reaction: $2Cu^{2+} + 2H_2O \xrightarrow{\text{electrolysis}} 2Cu(s) + O_2(g) + 4H^+$

In the above cells the electrodes were assumed to be inert. This would almost always be the case if the electrodes were made of graphite or the inert metal platinum. If, however, the electrode material were reactive, it would also have to be considered as a possible reactant. For example, copper or silver anodes are oxidized when no other species is present that is more readily oxidized.

11.6
QUANTITATIVE ASPECTS OF ELECTROLYSIS

As was indicated in Sec. 4.1, Michael Faraday empirically established the quantitative laws of electrolysis: The weight of product formed at an electrode is proportional to the amount of electricity transferred at the electrode and to the equivalent weight of the material. These observations can be accounted for by considering the electrode reactions. For example, in the electrolysis of molten NaCl, the cathode reaction

$$Na^+(l) + e^- \longrightarrow Na(l)$$

tells us that one Na atom is produced when one Na^+ disappears and one electron is transferred. If the Avogadro number of electrons were transferred, 1 mol of Na^+ would disappear, and 1 mol of Na atoms would be formed. For this reaction, 1 equiv of Na is 22.99 g; hence, transfer of the Avogadro number of electrons liberates 22.99 g of Na. Increasing the amount of electricity transferred increases proportionately the mass of Na produced.

The Avogadro number of electrons is such a convenient measure that it is given a special name, the *faraday*. In electrical units it is equal to 96,500 coulombs. As described in Appendix A5.5, a *coulomb* (C) is the amount of electricity that is transferred when a current of one ampere flows for one second. Relations important for electrolysis calculations are the following:

1 faraday = 96,500 coulombs

1 ampere = 1 coulomb/sec

1 coulomb = 1 ampere sec

EXAMPLE 2

Given an electric current of 1 mA that flows for 1 h, how many faradays of charge are transferred?

Solution

$1 \text{ mA} = 1.00 \times 10^{-3} \text{ C/sec}$

$1 \text{ h} = 3600 \text{ sec}$

$$\frac{(1.00 \times 10^{-3} \text{ C/sec}) (3600 \text{ sec})}{96,500 \text{ C/faraday}} = 3.73 \times 10^{-5} \text{ faraday}$$

Electrode half-reactions expressed in atoms and electrons can be read in terms of moles and faradays. Thus,

$$Na^+(l) + e^- \longrightarrow Na(l)$$

can be read "one mole of sodium ions reacts with one faraday of electricity to form one mole of sodium atoms."

EXAMPLE 3

How many grams of chlorine can be produced by the electrolysis of molten NaCl with a current of 1.00 A for 5.00 min?

Solution

$(1.00 \text{ C/sec}) (5.00 \text{ min}) (60 \text{ sec/min}) = 300 \text{ C}$

$$\frac{300 \text{ C}}{96,500 \text{ C/faraday}} = 0.00311 \text{ faraday}$$

The half-reaction $2Cl^-(l) \longrightarrow Cl_2(g) + 2e^-$ tells us that we produce 1 mol of chlorine per 2 faradays of electricity.

$$(0.00311 \text{ faraday}) \left(\frac{1 \text{ mol Cl}_2}{2 \text{ faradays}}\right) \left(\frac{70.9 \text{ g Cl}_2}{1 \text{ mol Cl}_2}\right) = 0.111 \text{ g Cl}_2$$

EXAMPLE 4

A current of 0.0965 A is passed for 1000 sec through 50.0 ml of 0.100 M NaCl. What will be the average concentration of OH^- in the final solution?

Solution

$$\frac{(0.0965 \text{ C/sec}) (1000 \text{ sec})}{96,500 \text{ C/faraday}} = 0.00100 \text{ faraday}$$

The cathode reaction $2e^- + 2H_2O \longrightarrow H_2(g) + 2OH^-$ shows that 2 faradays liberate 2 mol of OH^-; so 0.00100 faraday would liberate 0.00100 mol of OH^-. Assuming the final volume of solution

298

remains at 50.0 ml, the final concentration of OH^- is

$$\frac{0.00100 \text{ mol}}{0.0500 \text{ liter}} = 0.0200 \ M$$

11.7
PHOTOELECTROLYSIS OF WATER

In each of the above examples of electrolysis, a battery or similar device was needed as a source of energy to bring about the decomposition. Recently, much interest has focused on the possibility of using the sun as a source of energy for electrolysis of H_2O:

$$H_2O \xrightarrow{\text{light}} H_2(g) + \tfrac{1}{2}O_2(g) \qquad \Delta H = +286 \text{ kJ/mol}$$

This reaction uses 286 kJ/mol. The energy could be retrieved by burning the hydrogen so as to let the reverse reaction proceed. Only the hydrogen fuel would need to be stored, since we have abundant oxygen in the atmosphere.

Unfortunately, water does not absorb the visible and ultraviolet light that accounts for approximately 50 percent of the sun's energy. One way to get around this is to use TiO_2 to absorb the light. In its pure stoichiometric state, this material is a colorless nonconductor, so it is not much good for light absorption. However, by trace removal of oxygen (less than 1 percent) it can be made nonstoichiometric, in which case it becomes a black semiconductor and absorbs very well. The idea then is to shine sunlight on TiO_{2-x} in a cell and arrange the circuit so that the absorbed energy goes to decompose water.

Figure 11.8 shows a schematic diagram of such a cell. Light strikes the photoelectrode and generates an electron which travels through the external circuit to the counterelectrode. There, cathode reaction occurs to liberate hydrogen. At the same time, the photoelectrode (which is now deficient in electrons) acts as an anode and liberates oxygen from the solution. The hydrogen and oxygen can be kept separate by using a two-compartment cell such as is shown in Fig. 11.9.

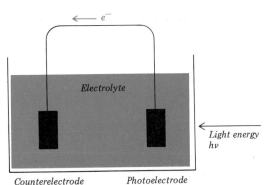

FIG. 11.8 Schematic representation of cell for photoelectrolysis of water.

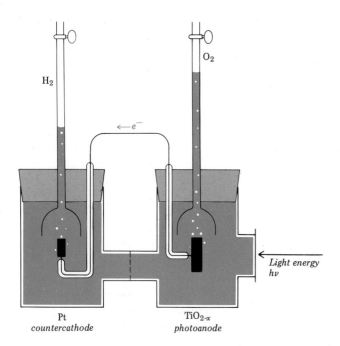

H₂

O₂

$\leftarrow e^-$

Light energy
hν

Pt
countercathode

TiO₂₋ₓ
photoanode

FIG. 11.9 Two-compartment cell for photoelectrolysis. Typical electrolyte would be 5 *M* NaOH.

At the present time devices such as these are relatively inefficient. TiO_{2-x} suffers from the fact that a small voltage (~ 0.2 V) must be imposed from the outside to get any appreciable evolution of gas. Also it utilizes only a small fraction of the solar spectrum. The search is on, in earnest, for better photoelectrodes. The ideal requirements are light-absorption properties that match the solar spectrum, relatively low cost, and stability for long periods of time in aqueous solution. For basic solutions, where evolution of gas is generally easier than from acidic solutions, the electrode reactions can be written as follows:

Cathode: $2H_2O + 2e^- \longrightarrow H_2 + 2OH^-$
Anode: $4OH^- \longrightarrow 2H_2O + O_2 + 4e^-$

Direct use of solar energy is attractive since there is an increasing demand for energy and the end of petroleum supplies looms in sight.

11.8 GALVANIC CELLS

In the ordinary electrolysis cell, electric energy in the form of current is used to bring about oxidation-reduction. It is also possible to do the reverse, i.e., use oxidation-reduction to produce electric current. The main requirement is that the oxidizing and reducing agents be kept separate from each other so that electron transfer is forced to occur through a wire. Any device which accomplishes this is called a *galvanic*, or *voltaic*, cell. Luigi Galvani (1780) and Alessandro Volta (1800) made the basic discoveries.

When a bar of zinc is dipped directly into a solution of copper sulfate, copper metal quickly deposits over the zinc surface. The reaction which occurs,

$$Zn(s) + Cu^{2+} \longrightarrow Zn^{2+} + Cu(s)$$

can be split into two half-reactions:

$$Zn(s) \longrightarrow Zn^{2+} + 2e^- \quad \text{(oxidation)}$$
$$Cu^{2+} + 2e^- \longrightarrow Cu(s) \quad \text{(reduction)}$$

A galvanic cell operates on the principle that two separated half-reactions can be made to take place simultaneously in separate compartments with the electron transfer occurring through a wire. Figure 11.10 shows such a galvanic cell. The dashed line down the middle represents a porous partition which divides the container but still permits limited diffusion of ions from one compartment to the other. In the left compartment is a solution of $ZnSO_4$ with a zinc bar; in the right compartment, a solution of $CuSO_4$ with a copper bar. When the zinc and copper are connected by an external wire, electric current flows, as shown by an ammeter in the circuit. As time progresses, zinc is "eaten" away and the copper bar becomes plated with a fresh deposit of copper. The net reaction is the same as that shown above.

The cell operates as follows: At the zinc bar, oxidation occurs. By definition, this makes zinc the anode. The half-reaction

$$Zn(s) \longrightarrow Zn^{2+} + 2e^-$$

produces Zn^{2+} ions and electrons. The zinc ions migrate into the solution, and the electrons move through the wire, as indicated by arrows in the figure. At the copper bar, reduction occurs. By definition, this makes copper the cathode. Electrons which have come through the wire accumulate at the cathode surface, where they are picked up and used in the reaction

$$Cu^{2+} + 2e^- \longrightarrow Cu(s)$$

Copper ions in the solution are thus depleted, and new copper ions have to move into the vicinity of the cathode surface. The electric circuit is now complete. The cell runs until either the zinc bar or the Cu^{2+} is depleted.

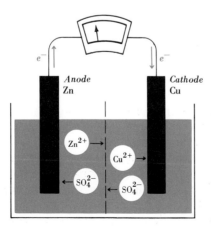

FIG. 11.10 Example of a galvanic cell.

EXAMPLE 5

Setting up a galvanic cell from a bar of zinc weighing 100 g and 5.00 liters of 0.500 M CuSO$_4$ solution, how long would the cell run assuming it delivers a steady current of 1.00 A?

Solution

$$\frac{100 \text{ g of Zn}}{65.38 \text{ g/mol}} = 1.53 \text{ mol of Zn}$$

$$(5.00 \text{ liters}) \left(\frac{0.500 \text{ mol of Cu}^{2+}}{\text{liter}} \right) = 2.50 \text{ mol of Cu}^{2+}$$

The equation $Zn(s) + Cu^{2+} \longrightarrow Zn^{2+} + Cu(s)$ shows that Cu^{2+} is in excess. For the reaction $Zn(s) \longrightarrow Zn^{2+} + 2e^-$:

$$(1.53 \text{ mol of Zn}) \left(\frac{2 \text{ faradays}}{1 \text{ mol of Zn}} \right) = 3.06 \text{ faradays}$$

$$(3.06 \text{ faradays}) (96{,}500 \text{ C/faraday}) = 2.95 \times 10^5 \text{ C}$$

$$\frac{2.95 \times 10^5 \text{ C}}{1.00 \text{ C/sec}} = 2.95 \times 10^5 \text{ sec, or } 82.0 \text{ h}$$

In describing the operation of a galvanic cell it is not necessary to specify the charge of the electrode. In fact, a simple assignment of charge cannot be made. To account for the *electron current* (which goes in the external wire from anode to cathode), the anode must be labeled *negative* with respect to the cathode. To account for the *ion current* (negative ions move toward the anode and positive ions toward the cathode), the anode must be labeled *positive* with respect to the cathode. How can the anode be positive and negative at the same time? The discrepancy can be resolved by considering the electrode in detail. The Zn^{2+} ions released into the solution form a layer which makes the anode appear positive as *viewed from the solution*. The electrons produced in the zinc bar because of Zn^{2+} formation make the anode appear negative as *viewed from the wire*.

Strictly speaking, to get current from the above cell, Zn^{2+} ions and the copper bar need not be initially present. Any metal support would do in place of the copper bar. Any positive ion that does not react with zinc metal would do in place of zinc. However, as cell reaction proceeds, Zn^{2+} would necessarily be produced.

Is the porous partition necessary? It serves only to hinder Cu^{2+} from getting over to the zinc metal, where *direct* electron transfer would *short-circuit* the whole cell. It must be porous in order to allow limited diffusion of ions from one compartment to the other. Otherwise, the solution in the anode compartment would become positively charged (owing to accumulation of Zn^{2+}), and that in the cathode compartment would become negatively charged (owing to depletion of Cu^{2+}). Such charge imbalance would cause the current to stop.

There are numerous oxidation-reduction reactions that have been made into useful sources of electric current. Probably the most famous example is

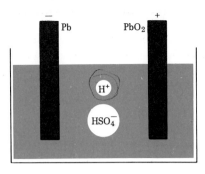

FIG. 11.11   Cell of a lead storage battery.

the *lead storage battery*. As shown in Fig. 11.11, its basic features are electrodes of lead (Pb) and lead dioxide (PbO_2) dipping into a strong aqueous solution of H_2SO_4. When the cell *discharges*, the reactions can be written as follows:

Anode: $Pb(s) + HSO_4^- \longrightarrow PbSO_4(s) + 2e^- + H^+$

Cathode: $PbO_2(s) + HSO_4^- + 3H^+ + 2e^- \longrightarrow PbSO_4(s) + 2H_2O$

Overall cell reaction: $Pb(s) + PbO_2(s) + 2HSO_4^- + 2H^+ \longrightarrow$
$$2PbSO_4(s) + 2H_2O$$

The electrode reactions can be reversed so as to restore or charge the cell to its original condition. In *discharge*, as shown by the overall cell reaction, Pb and PbO_2 are depleted, and the concentration of H_2SO_4 is diminished. Since the density of the solution is dependent on the concentration of H_2SO_4 measurement of the density can be used as a simple way to tell how far the cell has discharged.

Another familiar galvanic cell is the *Leclanché cell,* also known as the *dry cell*, which is frequently used as a power source in flashlights and other porta-ble devices. The acid version, as shown in Fig. 11.12*a*, consists of a Zn can con-taining a centered graphite stick surrounded by a moist paste of manganese dioxide (MnO_2), zinc chloride ($ZnCl_2$), and ammonium chloride (NH_4Cl). The Zn serves as the anode, and the graphite rod as the cathode. At the anode Zn is oxidized; at the cathode MnO_2 is reduced. The electrode reac-tions are extremely complex and vary depending on how much current is being drawn from the cell. For the delivery of very small currents, the follow-ing reactions occur:

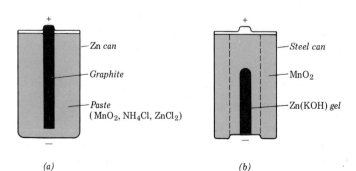

FIG. 11.12   Dry cell: (*a*) acid version, (*b*) alka-line version.

(a)   *(b)*

Anode: $\mathrm{Zn}(s) \longrightarrow \mathrm{Zn}^{2+} + 2e^-$

Cathode: $2\mathrm{MnO}_2(s) + \mathrm{Zn}^{2+} + 2e^- \longrightarrow \mathrm{ZnMn}_2\mathrm{O}_4(s)$

Overall cell reaction: $\mathrm{Zn}(s) + 2\mathrm{MnO}_2(s) \longrightarrow \mathrm{ZnMn}_2\mathrm{O}_4(s)$

In the alkaline version, as shown in Fig. 11.12*b*, $\mathrm{NH}_4\mathrm{Cl}$ is replaced by KOH. Although the alkaline dry cell costs more, it lasts longer because there is not the problem of acidic $\mathrm{NH}_4\mathrm{Cl}$ corroding the zinc.

Another technologically important galvanic cell is the *Edison* cell, which uses the oxidation of Fe by $\mathrm{Ni}_2\mathrm{O}_3$ in basic medium. The overall reaction can be written

$$\mathrm{Fe}(s) + \mathrm{Ni}_2\mathrm{O}_3(s) + 3\mathrm{H}_2\mathrm{O} \longrightarrow \mathrm{Fe(OH)}_2(s) + 2\mathrm{Ni(OH)}_2(s)$$

11.9
FUEL CELLS

In principle, any oxidation-reduction reaction can be separated into half-reactions and used to drive a galvanic cell. In particular, the reaction for oxidation of a fuel gas such as CH_4 should be so separable. Major interest was attracted to this possibility because *direct* conversion of chemical energy to electric energy can be made considerably more efficient (i.e., 60 percent) than burning fuel and using the heat in a steam-driven turbine (40 percent efficient).

For the oxidation of natural gas, we can write

$$\mathrm{CH}_4(g) + 2\mathrm{O}_2(g) \longrightarrow \mathrm{CO}_2(g) + 2\mathrm{H}_2\mathrm{O}$$

which, by the method to be outlined in Sec. 11.13 can be separated into two half-reactions. For acidic solution, these can be written as follows:

Anode: $\mathrm{CH}_4(g) + 2\mathrm{H}_2\mathrm{O} \longrightarrow \mathrm{CO}_2(g) + 8\mathrm{H}^+ + 8e^-$

Cathode: $\mathrm{O}_2(g) + 4\mathrm{H}^+ + 4e^- \longrightarrow 2\mathrm{H}_2\mathrm{O}$

In practice, the reaction is better utilized in basic medium, where the product CO_2 exists as carbonate ion, CO_3^{2-}:

Anode: $\mathrm{CH}_4(g) + 10\mathrm{OH}^- \longrightarrow \mathrm{CO}_3^{2-} + 7\mathrm{H}_2\mathrm{O} + 8e^-$

Cathode: $\mathrm{O}_2(g) + 2\mathrm{H}_2\mathrm{O} + 4e^- \longrightarrow 4\mathrm{OH}^-$

The detailed construction of a workable cell has proved to be a challenging engineering task. The basic design principles are the same as for any galvanic cell, viz., two electrode compartments each containing the reactants shown in the respective half-reactions. In the present case two of the reactants are gases and must be bubbled into the cell from the exterior. This is shown schematically in Fig. 11.13. To make electric contact with the reactant gases, conducting, but otherwise inert, electrodes are suspended in the bubble streams.

The first really workable fuel cell uses the reaction

$$2\mathrm{H}_2(g) + \mathrm{O}_2(g) \longrightarrow 2\mathrm{H}_2\mathrm{O}$$

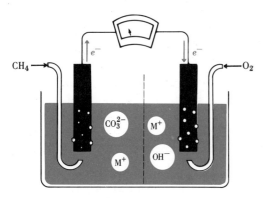

FIG. 11.13 Schematic representation of a fuel cell.

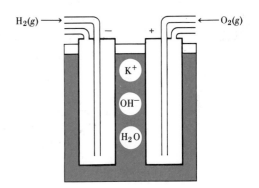

FIG. 11.14 Hydrogen-oxygen fuel cell.

Half-reactions in basic solution are:

Anode: $H_2(g) + 2OH^- \longrightarrow 2H_2O + 2e^-$

Cathode: $O_2(g) + 2H_2O + 4e^- \longrightarrow 4OH^-$

Figure 11.14 shows one design which has been used successfully. Two chambers constructed of porous carbon dip into an aqueous solution of KOH. Hydrogen gas is pumped into one chamber, while oxygen gas is introduced into the other. Because H_2 and O_2 react very slowly at room temperature, catalysts are mixed in and pressed with the carbon. At the anode, suitable catalysts are finely divided platinum or palladium; at the cathode, cobaltous oxide (CoO), platinum, or silver.

11.10
ELECTRODE POTENTIALS

A voltmeter connected between the two electrodes of a galvanic cell shows a characteristic voltage, which depends on what reactants take part in the electrode reactions and on what their concentrations are. For example, in the Zn–Cu cell of Fig. 11.10, if Zn^{2+} and Cu^{2+} are at 1 m concentrations and the temperature is 25°C, the voltage measured would be 1.10 V, no matter how big the cell or how big the electrodes. This voltage is characteristic of the reaction

$$Zn(s) + Cu^{2+} \longrightarrow Zn^{2+} + Cu(s)$$

It measures the work done in moving unit charge from one electrode to the other. Hence, it gives a quantitative measure of the relative tendency of the oxidation-reduction reaction to occur.

Figure 11.15 shows on the left a galvanic cell set up to study the reaction

$$Zn(s) + 2H^+ \longrightarrow H_2(g) + Zn^{2+}$$

In the anode compartment a Zn bar dips into an aqueous solution of a Zn

305

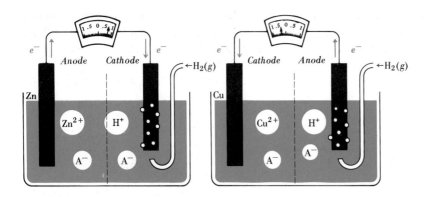

FIG. 11.15 Comparison of zinc and of copper electrodes against the hydrogen electrode.

salt. In the cathode compartment H_2 gas bubbles over an inert Pt electrode dipped into an acidic solution. At 25°C, when the concentrations of H^+ and Zn^{2+} are each at 1 m and when the pressure of the H_2 gas is 1 atm, the voltmeter reads 0.76 V. The deflection is in such direction as to indicate that Zn has greater tendency than H_2 to give off electrons. In other words, the half-reaction $Zn(s) \longrightarrow Zn^{2+} + 2e^-$ has greater tendency to occur than $H_2(g) \longrightarrow 2H^+ + 2e^-$, to the extent of 0.76 V.

The galvanic cell on the right in Fig. 11.15 makes use of the reaction

$$H_2(g) + Cu^{2+} \longrightarrow 2H^+ + Cu(s)$$

At 25°C, when the concentrations of H^+ and Cu^{2+} are 1 m and when the pressure of H_2 is 1 atm, the voltmeter reads 0.34 V; the deflection direction indicates that Cu has a smaller tendency than H_2 to give off electrons. In other words, the half-reaction $Cu(s) \longrightarrow Cu^{2+} + 2e^-$ has a smaller tendency to occur than $H_2(g) \longrightarrow 2H^+ + 2e^-$, by 0.34 V.

In all cells the voltage observed between two electrodes arises from two sources: a contribution at the anode and a contribution at the cathode. If either of these were known, the other could be obtained by difference. However, it is impossible to measure the electrode potential of a separate electrode; any circuit necessarily contains two electrodes. Conventional procedure is to select one electrode as a standard reference, assign it a zero value, and then refer all other electrode potentials to this arbitrarily designated zero. As reference, international convention has agreed on the *standard hydrogen electrode* (sometimes designated as SHE) with the understanding that a zero value of the voltage will be assigned to it at 25°C, 1 atm H_2 pressure, and 1 m H^+ concentration. Thus, in a cell that contains the standard hydrogen electrode, the entire measured voltage is attributed to the half-reaction at the other electrode. Voltages thus assigned are called *oxidation-reduction potentials*, or *redox potentials*. If the half-reaction is written with the electrons on the left, the associated voltage is called a *reduction potential;* if the half-reaction is written with the electrons on the right, the associated voltage is called an *oxidation potential*. There is, unfortunately, considerable confusion as to whether electrode potential means reduction potential or oxidation potential. The convention that has been recommended internationally is that electrode potential means reduction potential.

Figure 11.16 lists some common half-reactions together with their reduction potentials. A more extensive listing is given in Appendix 7. The values given apply for the half-reaction read from left to right. For the reverse direction, the signs must be changed.

The forward reaction is a reduction process in which an oxidizing agent, shown just to the left of each arrow, is reduced to give the reducing agent shown to the right. The table is arranged so that the oxidizing agents are listed in order of decreasing strength (or tendency to be reduced). In other words, as we go from top to bottom, there is decreasing tendency of the forward half-reaction to occur. For example, of the list given, fluorine (F_2), at the top, is the best oxidizing agent; it has the greatest tendency to pick up electrons. Lithium ion, Li^+, at the bottom of the list, is the poorest oxidizing agent and has the least tendency to pick up electrons.

The numerical values given in Fig. 11.16 apply to aqueous solutions at $25°C$ in which the concentration of each dissolved species is $1\ m$. A positive value indicates that the oxidizing agent is stronger than H^+ (set at zero); a negative value indicates that the oxidizing agent is weaker than H^+.

Each oxidizing agent in Fig. 11.16 is coupled in its half-reaction with a reduced form. For example, $F_2(g)$ is coupled with F^-. The reduced form is capable of acting as a reducing agent when the half-reaction is reversed. Thus, the reduction potentials in Fig. 11.16 also give information about the relative tendency of reducing agents to give off electrons. If a half-reaction, such as the one at the top of the table, has great tendency to go to the right, it has small tendency to go to the left; the reducing agent shown is a poor one. Of the reducing agents listed (always to the right of the arrow), F^- is the poorest, and $Li(s)$ is the best. The half-reaction

$$Li(s) \longrightarrow Li^+ + e^- \qquad +3.05 \text{ V}$$

FIG. 11.16   Some Half-Reactions and Their Standard Reduction Potentials

Half-reaction	Standard reduction potential, V
$2e^- + F_2(g) \longrightarrow 2F^-$	+2.87
$2e^- + Cl_2(g) \longrightarrow 2Cl^-$	+1.36
$4e^- + 4H^+ + O_2(g) \longrightarrow 2H_2O$	+1.23
$2e^- + Br_2 \longrightarrow 2Br^-$	+1.09
$e^- + Ag^+ \longrightarrow Ag(s)$	+0.80
$2e^- + I_2 \longrightarrow 2I^-$	+0.54
$2e^- + Cu^{2+} \longrightarrow Cu(s)$	+0.34
$2e^- + 2H^+ \longrightarrow H_2(g)$	0
$2e^- + Fe^{2+} \longrightarrow Fe(s)$	−0.44
$2e^- + Zn^{2+} \longrightarrow Zn(s)$	−0.76
$3e^- + Al^{3+} \longrightarrow Al(s)$	−1.66
$2e^- + Mg^{2+} \longrightarrow Mg(s)$	−2.37
$e^- + Na^+ \longrightarrow Na(s)$	−2.71
$e^- + Li^+ \longrightarrow Li(s)$	−3.05

Increasing strength as oxidizing agents ↑

Increasing strength as reducing agents ↓

which is the reverse of

$$e^- + Li^+ \longrightarrow Li(s) \qquad -3.05 \text{ V}$$

has a much greater tendency to occur than

$$2F^- \longrightarrow F_2(g) + 2e^- \qquad -2.87 \text{ V}$$

Figure 11.16 lists, to the right of the arrows, reducing agents in order of increasing strength. Such a list is sometimes called the *electromotive force*, or *emf, series.*

The potential of a half-reaction is a measure of the tendency of the half-reaction to occur, no matter what the other half of the complete reaction is. Thus, the potential of any complete reaction can be obtained by adding potentials of its two half-reactions. The potential so obtained is a measure of the tendency of the complete reaction to occur and gives the voltage measured for a galvanic cell which uses the overall reaction. For example, in the Zn–Cu cell we have

Cathode: $2e^- + Cu^{2+} \longrightarrow Cu(s)$		$+ 0.34$ V
Anode: $Zn(s) \longrightarrow Zn^{2+} + 2e^-$		$+ .76$ V
Complete cell: $Zn(s) + Cu^{2+} \longrightarrow Zn^{2+} + Cu(s)$		$+ 1.10$ V

The voltage $+1.10$ V is positive, which means that the reaction tends to go spontaneously as written. It should be noted that the value 1.10 V holds only when the concentrations of the ions are 1 m, since redox potentials are defined for concentrations of 1 m.

Any oxidation-reduction reaction for which the overall potential is positive tends to take place spontaneously as written. This can be determined just from the relative positions of the two half-reactions. For example, in Fig. 11.16 any oxidizing agent tends to react spontaneously with any reducing agent below it. Thus, I_2 oxidizes Cu, H_2, Fe, and so on, but does not oxidize Br^-, H_2O, Cl^-, and so on. Similarly, Zn reduces Fe^{2+}, H^+, Cu^{2+}, and so on, but does not reduce Al^{3+}, Mg^{2+}, Na^+, and so on.

EXAMPLE 6

I_2 and Br_2 are added to a solution containing I^- and Br^-. What reaction should occur when the concentration of each species is 1 m?

Solution

Method 1: From the relative positions of Fig. 11.16, Br_2 can oxidize I^-, but I_2 cannot oxidize Br^-. Therefore, the reaction is predicted to be

$$2I^- + Br_2 \longrightarrow I_2 + 2Br^-$$

Method 2: The half-reactions to be considered are

$$2e^- + I_2 \longrightarrow 2I^- \qquad +0.54 \text{ V}$$

$$2e^- + Br_2 \longrightarrow 2Br^- \qquad +1.09 \text{ V}$$

We want the overall voltage to be positive; so we pick the bigger voltage to be positive and invert the other half-reaction:

$$
\begin{array}{ll}
2e^- + Br_2 \longrightarrow 2Br^- & +1.09 \text{ V} \\
\underline{2I^- \longrightarrow I_2 + 2e^-} & \underline{-0.54 \text{ V}} \\
2I^- + Br_2 \longrightarrow I_2 + 2Br^- & +0.55 \text{ V}
\end{array}
$$

This reaction should occur as written since its voltage is positive. If we had made the other choice, we would have written:

$$
\begin{array}{ll}
2Br^- \longrightarrow Br_2 + 2e^- & -1.09 \text{ V} \\
\underline{2e^- + I_2 \longrightarrow 2I^-} & \underline{+0.54 \text{ V}} \\
2Br^- + I_2 \longrightarrow Br_2 + 2I^- & -0.55 \text{ V}
\end{array}
$$

However, this overall voltage is negative, so the reaction should not occur spontaneously as written.

11.11
NERNST EQUATION

The electrode potentials just discussed are *standard potentials* and, as such, describe tendency to react at fixed unit concentrations.* What happens to reaction tendencies when concentrations are changed? Qualitatively, the principle of Le Châtelier predicts that increasing the concentration of a reactant favors its tendency to react. Similarly, decreasing the concentration of a product favors the tendency toward formation of that product.

Quantitatively, the change of reaction tendency with concentration is given by the *Nernst equation*, which relates E, the potential for a reaction at nonstandard conditions, to $E°$, the standard potential for that reaction at unit concentrations. The Nernst equation is

$$E = E° - \frac{RT}{n\mathcal{F}} \ln Q$$

where R is the gas constant, T is the temperature, n is the number of electrons

---

*Strictly speaking, the tendency of a reaction to occur is influenced not only by the concentrations of species but also by interionic attractions and other complicating factors such as the change of hydration with concentration. All of these factors contribute to the *chemical activity* of the species. For precise work, it is the *chemical activity* rather than the *concentration* of the species that must be specified. *Chemical activity* has no units; it is a *dimensionless* quantity defined as the ratio of "effective" concentration of a species divided by the concentration in a standard reference state. For our purpose we shall find that activity can well be represented for dissolved species by molality, or in dilute solutions by molarity, and by atmospheres of pressure for gaseous species.

transferred in the reaction, and $\mathfrak{F}$ is the faraday. Putting in the appropriate constants and going over to base-10 logarithms, the Nernst equation at 25°C is

$$E = E° - \frac{0.0591}{n} \log Q$$

where Q, called the mass-action expression, is formed by multiplying together the concentration of each species on the right side of the chemical equation taken to a power equal to the coefficient of the species in the balanced chemical equation, and dividing by the concentration of each species on the left side of the equation, taken to the appropriate power.

For the half-reaction

$$2e^- + 2H^+ \longrightarrow H_2(g)$$

the Nernst equation at 25°C gives

$$E = E° - \frac{0.0591}{2} \log \frac{p_{H_2}}{[H^+]^2}$$

where p_{H_2} represents the concentration or pressure of hydrogen gas and $[H^+]$ represents the concentration of hydrogen ion. (Square brackets around a formula generally represent concentration of a dissolved species in moles per liter.)

For the hydrogen half-reaction, the standard potential $E°$ is equal to zero, so we can write

$$E = 0.00 - 0.0296 \log \frac{p_{H_2}}{[H^+]^2}$$

At standard conditions, where hydrogen pressure is 1 atm and hydrogen-ion concentration is 1 M, the term $p_{H_2}/[H^+]^2$ equals 1, its logarithm is 0, and $E = 0.00$ V.

As a further illustration, in pure water the hydrogen-ion concentration is 1.0×10^{-7} M. Substituting $p_{H_2} = 1$ and $[H^+] = 1.0 \times 10^{-7}$, we get

$$E = 0.00 - 0.0296 \log \frac{1.0}{(1.0 \times 10^{-7})^2} = -0.41 \text{ V}$$

The fact that this potential is less than zero tells us that the half-reaction $2e^- + 2H^+ \longrightarrow H_2$ has *less* tendency to go to the right in pure water than in 1 M acid.

As a more complex example, we can consider the following complete reaction:

$$2I^- + H_3AsO_4 + 2H^+ \longrightarrow I_2(s) + H_3AsO_3 + H_2O$$

The Nernst equation at 25°C gives

$$E = E° - \frac{0.0591}{n} \log \frac{[I_2(s)] \, [H_3AsO_3] \, [H_2O]}{[I^-]^2 \, [H_3AsO_4] \, [H^+]^2}$$

We can simplify this by using the fact that the activity of pure solids is unity; hence $[I_2(s)]$ can be omitted from the expression. Likewise, the activity of water in dilute solutions is also taken to be unity, so $[H_2O]$ can be omitted

311 Section 11.12
Free-
Energy
Change
and Cell
Voltage

from the expression. What is $E°$? By looking in Appendix 7, we find $E° = +0.54$ V for $2e^- + I_2(s) \longrightarrow 2I^-$ and $E° = +0.56$ V for $2e^- + 2H^+ + H_3AsO_4 \longrightarrow H_3AsO_3 + H_2O$. Subtracting the first from the second, we get $E° = 0.02$ for the overall reaction. For the net reaction, two electrons are transferred, so $n = 2$. Hence we get

$$E = 0.02 - 0.0296 \log \frac{[H_3AsO_3]}{[I^-]^2 \, [H_3AsO_4] \, [H^+]^2}$$

Under standard conditions, where each concentration is unity, the log term is 0. The E for the reaction is slightly positive, thus being favorable for the reaction in the direction written. However, if the hydrogen-ion concentration is depressed, as by addition of base, the lowered value of $[H^+]$ in the denominator can make E change sign. For example, in neutral solution with $[H^+] = 1.0 \times 10^{-7}$ and all other species at unit concentration, we get

$$E = 0.02 - 0.0296 \log \frac{1}{(1.0 \times 10^{-7})^2} = -0.39 \text{ V}$$

Now E for the reaction is negative and the reaction as written should not occur; instead the reverse reaction should take place.

11.12
FREE-ENERGY CHANGE AND CELL VOLTAGE

For oxidation-reduction reactions, the net free-energy change is directly related to the voltage that would be obtained if the reaction were set up as a galvanic cell. The relationship between ΔG and E is

$$\Delta G = -n\widetilde{\mathfrak{F}}E$$

where n is the number of faradays transferred and $\widetilde{\mathfrak{F}}$ is the value of the faraday, 96.49 kJ V^{-1} equiv^{-1}. As noted in Sec. 11.11, the Nernst equation tells how the voltage of a cell depends on the concentrations of the species involved. Hence, we can substitute the general form of the Nernst equation

$$E = E° - \frac{RT}{n\widetilde{\mathfrak{F}}} \ln Q$$

into $\Delta G = -n\widetilde{\mathfrak{F}}E$ to get

$$\Delta G = -n\widetilde{\mathfrak{F}} \left(E° - \frac{RT}{n\widetilde{\mathfrak{F}}} \ln Q\right)$$

$$= -n\widetilde{\mathfrak{F}}E° + RT \ln Q$$

The term $-n\widetilde{\mathfrak{F}}E°$ is just equal to $\Delta G°$, the standard free-energy change when all species are in standard states. Substituting, we get

$$\Delta G = \Delta G° + RT \ln Q$$

When all the species are in standard states, $Q = 1$, $\log Q = 0$, and $\Delta G = \Delta G°$. When the species are in nonstandard states we can use the Nernst equation and the above relation to calculate E and ΔG.

EXAMPLE 7

We are given the reaction

$$H_2(g) + 2AgCl(s) \longrightarrow 2Ag(s) + 2H^+(aq) + 2Cl^-(aq)$$

At 25°C the standard free energy of formation of $AgCl(s)$ is -109.7 kJ/mol and that of $(H^+ + Cl^-)$ (aq) is -131.2 kJ/mol. Calculate the cell voltage and ΔG if this reaction is run at 25°C in a cell in which the $H_2(g)$ pressure is unity and the $H^+(aq)$ and $Cl^-(aq)$ concentrations are each at 0.0100 M. Recall that the free energy of formation of any element in its standard state is zero.

Solution

$$\Delta G° = 2\Delta G°_{H^+,Cl^-} - 2\Delta G°_{AgCl}$$

$$= 2(-131.2) - 2(-109.7) = -43.0 \text{ kJ}$$

$$E° = \frac{\Delta G°}{-n\mathcal{F}} = \frac{-43.0 \text{ kJ}}{-2(96.49 \text{ kJ/V})} = 0.223 \text{ V}$$

$$E = E° - \frac{0.0591}{n} \log \frac{[H^+]^2 [Cl^-]^2}{p_{H_2}}$$

Note that *solid* Ag and *solid* AgCl can be omitted.

$$E = 0.223 - \frac{0.0591}{2} \log \frac{(0.0100)^2 (0.0100)^2}{1.00}$$

$$= 0.223 + 0.236 = 0.459 \text{ V}$$

$$\Delta G = -n\mathcal{F}E = -(2 \text{ equiv}) (96.49 \text{ kJ V}^{-1} \text{ equiv}^{-1}) (0.459 \text{ V})$$

$$= -88.6 \text{ kJ}$$

11.13
BALANCING EQUATIONS BY HALF-REACTIONS

An oxidation half-reaction must always be paired with a reduction half-reaction. The requirement of electron balance makes possible a general method of balancing redox equations in which the artificially devised oxidation number (Sec. 2.9) is no longer necessary.

First we consider a simple example. The balanced equation for the reaction $Zn(s) + Ag^+ \longrightarrow Zn^{2+} + Ag(s)$ can be written by noting that Zn must release two electrons to form Zn^{2+} and it takes two Ag^+ ions to pick up these two electrons. The principle of the method is to write the two half-reactions and then multiply each by an appropriate factor so as to match electron loss and gain when the results are added up. Thus,

$$Zn(s) \longrightarrow Zn^{2+} + 2e^-$$

$$+ 2[e^- + Ag^+ \longrightarrow Ag(s)]$$

$$\overline{Zn(s) + 2e^- + 2Ag^+ \longrightarrow Zn^{2+} + 2e^- + 2Ag(s)}$$

Cancellation of the $2e^-$ from each side of the equation gives

$$Zn(s) + 2Ag^+ \longrightarrow Zn^{2+} + 2Ag(s)$$

The balanced equation for the change $Fe^{2+} + MnO_4^- \longrightarrow Fe^{3+} + Mn^{2+}$ (acidic solution) can be written from the two half-reactions

$$Fe^{2+} \longrightarrow Fe^{3+} + e^-$$

$$5e^- + 8H^+ + MnO_4^- \longrightarrow Mn^{2+} + 4H_2O$$

Multiplying the first by 5 and adding to the second gives

$$5Fe^{2+} + 8H^+ + MnO_4^- \longrightarrow 5Fe^{3+} + Mn^{2+} + 4H_2O$$

When given an equation to balance *in acidic solution* the steps to follow are:

1 Separate the change into half-reactions.

2 Balance each half-reaction separately:

a Adjust coefficients to balance all atoms except H and O.

b Add H_2O to side deficient in O.

c Add H^+ to side deficient in H.

d Add e^- to side deficient in negative charge.

3 Multiply each half-reaction by the appropriate number to balance electron gain and loss. Add.

4 Subtract any duplications on left and right.

The following example illustrates the above stepwise procedure.

EXAMPLE 8

Write a balanced equation for the oxidation of H_2SO_3 by $Cr_2O_7^{2-}$ in acidic solution to form HSO_4^- and Cr^{3+}.

Solution

Step 1

$$H_2SO_3 \longrightarrow HSO_4^- \qquad\qquad Cr_2O_7^{2-} \longrightarrow Cr^{3+}$$

Step 2a

$$H_2SO_3 \longrightarrow HSO_4^- \qquad\qquad Cr_2O_7^{2-} \longrightarrow 2Cr^{3+}$$

Step 2b

$$H_2SO_3 + H_2O \longrightarrow HSO_4^- \qquad\qquad Cr_2O_7^{2-} \longrightarrow 2Cr^{3+} + 7H_2O$$

Step 2c

$$H_2SO_3 + H_2O \longrightarrow HSO_4^- + 3H^+ \qquad\qquad Cr_2O_7^{2-} + 14H^+ \longrightarrow 2Cr^{3+} + 7H_2O$$

Step 2d

$$H_2SO_3 + H_2O \longrightarrow HSO_4^- + 3H^+ + 2e$$
$$Cr_2O_7^{2-} + 14H^+ + 6e^- \longrightarrow 2Cr^{3+} + 7H_2O$$

[Two electrons have been added to the right side since, in step (2c), the left side has a net charge of 0 and the right side has a net charge of +2. The right side of (2c) is deficient in negative charge by two units.]

[Six electrons have been added to the left since, in step (2c), the left side has a net charge of +12 and the right side has +6.]

Step 3

$$3(H_2SO_3 + H_2O \longrightarrow HSO_4^- + 3H^+ + 2e^-)$$
$$\underline{Cr_2O_7^{2-} + 14H^+ + 6e^- \longrightarrow 2Cr^{3+} + 7H_2O}$$
$$3H_2SO_3 + 3H_2O + Cr_2O_7^{2-} + 14H^+ + 6e^- \longrightarrow 3HSO_4^- + 9H^+ + 6e^- + 2Cr^{3+} + 7H_2O$$

(The top half-reaction has been multiplied by 3 to get six electrons in each half of the reaction and then the two half-reactions have been added.)

Step 4

Since $3H_2O$, $9H^+$, and $6e^-$ are duplicated on left and right sides, these can be subtracted to give

$$3H_2SO_3 + Cr_2O_7^{2-} + 5H^+ \longrightarrow 3HSO_4^- + 2Cr^{3+} + 4H_2O$$

---

if the reaction occurs in *basic solution,* the equation should not show H^+. In order to add H atoms in step (2c), add H_2O molecules equal in number to the deficiency of H atoms and an equal number of OH^- ions to the opposite side. The rest of the method is the same. An example of reaction in basic solution is the change

$$Cr(OH)_3(s) + IO_3^- \longrightarrow I^- + CrO_4^{2-}$$

The half-reactions are

$$Cr(OH)_3(s) + 5OH^- \longrightarrow CrO_4^{2-} + 4H_2O + 3e^-$$
$$IO_3^- + 3H_2O + 6e^- \longrightarrow I^- + 6OH^-$$

and the net equation is

$$2Cr(OH)_3(s) + IO_3^- + 4OH^- \longrightarrow 2CrO_4^{2-} + I^- + 5H_2O$$

This method of balancing redox equations is called the *method of half-reactions.* It is also sometimes called the *ion-electron method.*

Important Concepts

electric current	faraday	standard hydrogen electrode
electric field	photoelectrolysis	oxidation potential
electrolytic current	galvanic cell	reduction potential
electrolysis	lead storage battery	emf series
anode	dry cell	Nernst equation
cathode	fuel cell	free-energy change and cell voltage
half-reaction	electrode potential	balancing redox equations

Exercises

*11.1 **Conductivity** How do metallic and electrolytic conductivity differ in terms of what moves?

*11.2 **Conductivity** When a metal is heated, its conductivity generally decreases; when a solution is heated, its conductivity generally increases. Account for the difference.

11.3 **Conductivity When an electrolytic solution carries current, part of the current is carried by positive ions moving in one direction and part by negative ions moving in the opposite direction. How might you devise an experiment to measure what fraction of the total current is carried by each? How would this fraction probably change as the size of the negative ion was increased? Explain.

***11.4 **Conductivity** When LiCl solution conducts electric current, 33 percent of the current is carried by Li^+ migration and 67 percent by Cl^- migration. When KCl conducts, 40 percent of the current is carried by the K^+. How might you explain this anomaly?

*11.5 **Electrolysis** Define each of the following terms as applied to electrolysis: cathode, cation, anode, anion.

11.6 **Electrolysis Suppose the beaker shown in Fig. 11.4 were replaced by two beakers, one around each of the electrodes. The electrolysis would not work. Why not?

11.7 **Electrolysis As electrolysis proceeds, what happens to the concentration of M^+ ions to the *right of the anode* as diagrammed in Fig. 11.4? Explain.

*11.8 **Electrolysis** Diagram a cell suitable for the electrolysis of molten KBr. Write electrode reactions, label the anode and cathode, and show by arrows the directions of motion of charged species in the internal and external circuits.

11.9 **Electrolysis What would be the probable electrode reactions if a molten mixture of NaCl and KBr were electrolyzed? Label anode and cathode. Explain your reasoning.

*11.10 **Electrolysis** In the electrolysis of aqueous NaCl, how long would you have to pass a 1.00-A current through such a cell to convert 1.00 liter of 1.00 M NaCl into 1.00 M NaOH? How much chlorine (STP) would be evolved?   *Ans. 26.8 h*

*11.11 **Electrolysis** When an aqueous solution of $CaCl_2$ is electrolyzed, the products are oxygen at the anode and hydrogen at the cathode. Write appropriate electrode reactions.

11.12 **Electrolysis Suppose you carry out the electrolysis of aqueous NaCl in a double-compartment cell such as the one shown in Fig. 11.7. As electrolysis proceeds, what happens to the relative concentration of Na^+ in the two compartments? Explain.

11.13 **Electrolysis Electrolytic hydrogen is generally made by electrolyzing an aqueous solution of NaOH using an iron cathode. Write an appropriate electrode reaction. Using an energy cost of 1 cent per kilowatthour, calculate how much it would cost to make enough hydrogen to fill a compressed-gas tank with a volume of 50.0 liters to a pressure of 136 atm at 25°C? One kilowatthour is about 33,000 coulombs.

11.14 Electrolysis Explain why the electrolysis of aqueous Na_2SO_4 containing litmus develops a blue color at the cathode and a red color at the anode. How would the result differ if you started with $NaHSO_4$ instead of Na_2SO_4?

11.15 Electrolysis If aqueous Na_2SO_4 with litmus were electrolyzed in the two-compartment cell of Fig. 11.7 using copper electrodes instead of inert platinum electrodes, what would probably be observed? Write the electrode reactions and label each.

*11.16 **Electrolysis** A convenient way to generate a precisely controlled amount of H^+ for use in a chemical reaction is to use an anode at which H_2O is electrochemically oxidized to O_2. How long would you need to pass a 50.0-mA current to generate 50.00 millimol of H^+?

*11.17 **Electrolysis** Aluminum is made commercially by electrolyzing a solution of Al_2O_3 dissolved in molten cryolite, Na_3AlF_6. How long would you have to pass a 10.0-A current through the solution to get 1 kg of aluminum?

11.18 Electrolysis Three electrolytic cells are set up in series so the same current passes through each: Cell I contains an aqueous solution of Ag_2SO_4, cell II contains an aqueous solution of $CuSO_4$, and cell III contains an aqueous solution of H_2SO_4. Assuming oxygen is given off at each anode and $Ag(s)$, $Cu(s)$, and $H_2(g)$ are formed at the respective cathodes, which *solution* will lose the most weight? If the loss is exactly 1.000 g, how much total oxygen (STP) has been evolved?

11.19 Photoelectrolysis Given the schematic setup shown in Fig. 11.8, explain how it can be used to convert solar energy into hydrogen. Assuming the electrolyte is aqueous NaOH, write probable half-reactions for each of the electrodes.

***11.20 Photoelectrolysis** Using the cell shown in Fig. 11.9, calculate the probable rate of hydrogen evolution on exposure to sunlight. Peak solar radiation is about 755 W/m². Assume that about 3 percent of this is actually effective and that the area of the photoanode is 5 cm².

*11.21 **Galvanic cell** What is the driving force that makes a galvanic cell work? Which of the components shown in Fig. 11.10 are absolutely necessary to get a current out of the cell?

*11.22 **Galvanic cell** Diagram a galvanic cell that makes use of the reaction $Zn(s) + Cl_2(g) \longrightarrow Zn^{2+} + 2Cl^-$. Label the anode and cathode, write half-reactions for each, and show with arrows the directions of motion of each of the charged species.

11.23 Galvanic cell Explain why the galvanic cell shown in Fig. 11.10 needs a porous partition but the cell shown in Fig. 11.11 does not. What general formulation would you make for when a partition is not needed?

11.24 Lead storage battery Each cell of a lead storage battery, such as the one shown in Fig. 11.11, generates a voltage difference of about 2 V. Six of these in series constitute a regular 12-V car battery. How would its voltage be affected by each of the following?
  a Adding more H_2SO_4
  b Adding more H_2O
  c Making the Pb plates bigger
  d Raising the temperature

*11.25 **Lead storage battery** When a lead storage battery is being recharged, what are the corresponding half-reactions? Label the anode and cathode. Explain why a rise in density of the electrolyte solution signifies that the battery is being charged.

11.26 Galvanic cell Diagram the Edison cell which uses the oxidation of Fe by Ni_2O_3 in basic medium. Write electrode reactions and explain why this cell should show no fall-off in voltage as it is being run.

*11.27 **Galvanic cell** Explain why the casing of an ordinary flashlight cell is usually labeled negative.

11.28 Fuel cells Write the two half-reactions for the oxidation of $C_2H_6(g)$ by $O_2(g)$ to give $CO_2(g)$ and H_2O in acidic solution. Calculate how many liters of $C_2H_6(g)$ at STP would be needed at the anode to generate a current of 1.00 A for 1 h. How many liters of $O_2(g)$ at STP would be needed at the cathode?

11.29 Fuel cells Suppose you had a fuel cell that used the oxidation of $CH_4(g)$ by $O_2(g)$ to give $CO_2(g)$ and H_2O in acidic solution. What effect would there be on the voltage and the current from the cell if the CH_4 went to $CO(g)$ rather than $CO_2(g)$? Explain.

*11.30 **Fuel cell** Given the fuel cell shown in Fig. 11.14, which way do the ions migrate? Justify your answer.

11.31 **Fuel cell Given a 50-liter tank of gaseous hydrogen and a 50-liter tank of gaseous oxygen, each at 100 atm pressure and 25°C, and using the fuel cell of Fig. 11.14, how long could you generate a 10.0-A current from this supply?

Ans. 45.7 days

*11.32 **Electrode potentials** What is the difference between electrode potential, oxidation potential, and reduction potential?

11.33 **Electrode potential Suppose that the standard hydrogen electrode was arbitrarily assigned a value of 1.00 V for $2e^- + 2H^+ \longrightarrow H_2(g)$. What would this do to the observed voltage under standard conditions for each of the following: $Zn–H_2$ cell, $Cu–H_2$ cell, $Zn–Cu$ cell? Explain.

*11.34 **Electrode potentials** Given the electrode-potential data shown in Fig. 11.16, pick out five oxidizing agents that $Fe(s)$ would react with spontaneously and five with which it would not react.

11.35 **Electrode potentials Given the data shown in Fig. 11.16, predict the voltage that would be obtained from a cell using the oxidation of $Zn(s)$ by $O_2(g)$ under standard conditions. Why would such a cell not work very well?

11.36 **Nernst equation What is the Nernst equation, what does it tell you, and what is the significance of each parameter entering into it? Show how the Nernst equation is correlated with the Le Châtelier principle.

11.37 **Nernst equation Assuming everything else at unit activity, at what concentration of hydrogen ion would you expect to get zero voltage at 25°C for the reaction

$$2I^- + H_3AsO_4 + 2H^+ \longrightarrow I_2(s) + H_3AsO_3 + H_2O$$

11.38 **Cell voltage and free energy Given the reaction $Zn(s) + 2Ag^+ \longrightarrow Zn^{2+} + 2Ag(s)$, calculate the expected cell voltage and free-energy change under standard conditions. What would happen to each if the concentration of both ions were doubled?

11.39 **Balancing equations Using the method of half-reactions, balance each of the following under the conditions noted. Give the balanced half-reactions in each case.

a $MnO_4^- + H_2C_2O_4 \longrightarrow Mn^{2+} + CO_2$ (acidic)
b $Cr_2O_7^{2-} + N_2H_4^+ \longrightarrow N_2 + Cr^{3+}$ (acidic)
c $C_8H_{18} + O_2 \longrightarrow CO_2 + H_2O$ (acidic)
d $C_8H_{18} + O_2 \longrightarrow CO_3^{2-} + H_2O$ (basic)
e $NCS^- + O_2 \longrightarrow NO_2 + CO_2 + SO_2 + H_2O$
(acidic)
f $S_2^{2-} + ClO_3^- \longrightarrow SO_4^{2-} + Cl^-$ (basic)
g $P_4 + IO_3^- \longrightarrow HPO_4^{2-} + I^-$ (basic)
h $H_2O_2 + N_2O_4 \longrightarrow HNO_2 + O_2$ (acidic)
i $S_2O_3^{2-} + NO_3^- \longrightarrow NO_2^- + SO_3^{2-}$ (basic)
j $HO_2^- + NH_3 \longrightarrow N_2H_4 + OH^-$ (basic)

***11.40 **Balancing equations** Given the three equations

$$Cr_2O_7^{2-} + 8H^+ + 3H_2O_2 \longrightarrow 2Cr^{3+} + 7H_2O + 3O_2$$
$$Cr_2O_7^{2-} + 8H^+ + 5H_2O_2 \longrightarrow 2Cr^{3+} + 9H_2O + 4O_2$$
$$Cr_2O_7^{2-} + 8H^+ + 7H_2O_2 \longrightarrow 2Cr^{3+} + 11H_2O + 5O_2$$

prove that only one of these is the correct equation for the oxidation of H_2O_2 by $Cr_2O_7^{2-}$ in acidic solution. Explain why the other two also appear to be properly balanced equations.

Chapter 12

CHEMICAL KINETICS

The term *chemical kinetics* is applied to the branch of chemistry concerned with the velocity of chemical reactions and the mechanism by which chemical reactions occur. *Reaction velocity,* or *reaction rate,* tells us how fast the concentrations of chemical reactants change with time. *Reaction mechanism* tells us the sequence of steps by which the overall change is accomplished. In many reactions the net change may actually consist of several consecutive reactions, each of which constitutes a step in the formation of final products. In discussing chemical reactions it is important to keep clear the distinction between a net reaction and one step in that reaction.

When a reaction occurs in steps, intermediate species are probably formed, and they may not be detectable because they may be promptly used up in a subsequent step. However, by investigating the influence that various factors have on the rate at which the net change occurs, it is sometimes possible to elucidate what the intermediates are and how they are involved in the mechanism of the reaction.

What factors influence the rate of a chemical reaction? The question is of practical importance because, as indicated in Sec. 10.7, to determine that a reaction *should* go does not guarantee that it *will* go fast enough to be observed. Experiments show that four important factors generally influence reaction rates: (1) nature of the reactants, (2) concentration of the reactants, (3) temperature, and (4) catalysis.

12.1
NATURE OF REACTANTS

In a chemical reaction, bonds are formed and bonds are broken. The rate should therefore depend on the specific bonds involved. Experimentally, the reaction velocity depends on the specific substances brought together in reaction. For example, reduction of MnO_4^- ion in acidic solution by Fe^{2+} is practically instantaneous. MnO_4^- disappears as fast as Fe^{2+} is added. On the other hand, reduction of MnO_4^- in acidic solution by oxalic acid, $H_2C_2O_4$, is not instantaneous. The violet color characteristic of MnO_4^- persists long after the solutions are mixed. In these two reactions everything is identical except the nature of the reducing agent, but still the reaction rates are quite different.

The rates observed for different reactants vary widely. There are reactions, such as acid-base neutralization, which may be over in a nanosecond. There are also very slow reactions, such as those in geologic processes, which may not reach completion in a million years. Changes in a lifetime may be too small to be detected. Most information has been accumulated about reactions between these extremes.

318

EXAMPLE 1

If it takes 1 nsec for a hydrogen-ion concentration of 0.010 M to disappear, what is the average rate of reaction?

Solution

$$\text{Rate} = \frac{\text{concentration change}}{\text{time}} = \frac{0.010 \text{ mol/liter}}{1.00 \times 10^{-9} \text{ sec}}$$

$$= 1.00 \times 10^7 \text{ } M/\text{sec}$$

EXAMPLE 2

In a given reaction at 25°C it is observed that the hydrogen-gas pressure falls from 0.200 to 0.150 atm in 45 min. Calculate the average rate of reaction in (a) atmospheres per minute and (b) molarity per second.

Solution

$$\text{Rate} = \frac{\text{pressure change}}{\text{time}} = \frac{0.050 \text{ atm}}{45 \text{ min}} = 0.0011 \text{ atm/min}$$

For an ideal gas, $P = \dfrac{n}{V} RT$, so that concentration in moles per liter (n/V) is P (in atmospheres) divided by RT. R is 0.08206 liter atm mol^{-1} deg^{-1}. T is 298°.

$$\text{Rate} = \frac{\left(0.0011 \dfrac{\text{atm}}{\text{min}}\right)\left(\dfrac{1 \text{ min}}{60 \text{ sec}}\right)}{\left(0.08206 \dfrac{\text{liter atm}}{\text{mol deg}}\right)(298 \text{ deg})} = 7.5 \times 10^{-7} \text{ } M/\text{sec}$$

12.2
CONCENTRATION OF REACTANTS

It is found by experiment that the rate of a *heterogeneous* reaction, i.e., one that involves more than one phase, is proportional to the area of contact between the phases. An example is the rusting of iron. Rusting involves a solid phase, iron, and a gas phase, oxygen. Rusting is slow when the surface of contact is small, as with an iron nail, but is rapid when the area of contact is great, as with steel wool.

The rate of a *homogeneous* reaction, i.e., one that occurs in a single phase, depends on the concentration (amount per volume) of reactants in that phase. For gaseous phases, concentrations can be changed by altering the pressure. For liquid phases, the concentration of an individual reactant can be changed by addition of a reactant or by changing the volume of the system, as by addition of a solvent. The specific effect on the reaction rate has to be determined by experiment. Thus, in the reaction of A with B, the addition of A may

319

cause an increase, a decrease, or no change in rate, depending on the particular reaction. Quantitatively, the rate may double, triple, become half as great, etc. It is not possible to look at the net equation and tell how the rate will be affected. The quantitative influence can be found only by experiment.

The experimental determination of how the rate of a reaction depends on concentration of reactants is not easy. The usual procedure is to do a series of experiments in which everything is kept constant except the concentration of one reactant. As its concentration is systematically changed, the reaction rate is measured, e.g., by noting the rate of disappearance of a reactant or the rate of formation of a product. Experimental difficulties usually arise in determining the instantaneous concentration of a component as it changes with time.

The reaction between hydrogen and nitric oxide

$$2H_2(g) + 2NO(g) \longrightarrow 2H_2O(g) + N_2(g)$$

is a homogeneous reaction which can be investigated kinetically by following the change in pressure. The pressure gradually drops because 4 moles of gas reactants are being replaced by 3 moles of gas products. Typical data for several experiments at 800°C are given in Fig. 12.1. Since reactants are being used up, concentrations and rates of reaction are constantly changing. The concentrations and rates listed are those at the very beginning, when little change has occurred.

The data for experiments I and II show that when the initial concentration of NO is kept constant at 0.006 M, doubling the concentration of H_2 from 0.001 to 0.002 M doubles the rate from 0.025 to 0.050 atm/min. Experiments I and III show that tripling the concentration of H_2 triples the rate. The rate of reaction is therefore found to be proportional to the first power of the concentration of H_2.

The data for experiments IV and V show that when the concentration of H_2 is kept constant, doubling the concentration of NO quadruples the rate; experiments IV and VI show that tripling the concentration of NO triply triples the rate. The rate of reaction is therefore found to be proportional to the square, or second power, of the concentration of NO. Quantitatively, the data can be summarized by stating that the reaction rate is proportional to the concentration of H_2 × the concentration of NO squared. Using square brackets to represent concentration in moles per liter, this can

FIG. 12.1   Reaction Rate Data for NO and H_2 at 800°C

Experiment	Initial molar concentration NO	H_2	Initial rate, atm/min
I	0.006	0.001	0.025
II	0.006	0.002	0.050
III	0.006	0.003	0.075
IV	0.001	0.009	0.0063
V	0.002	0.009	0.025
VI	0.003	0.009	0.056

be written mathematically as

$$\text{Rate} = k[H_2][NO]^2$$

The equation is known as the *rate law* for the reaction. The proportionality constant k is called the *specific rate constant* and is characteristic of the given reaction, although it varies with temperature.

The general form of any rate law is

$$\text{Rate} = k[A]^x[B]^y \cdots$$

where x is the appropriate power to which the concentration of A must be raised and y is the appropriate power to which the concentration of B must be raised in order to summarize the observed data. The three dots represent other reactants which may be involved in the rate law. The exponents x and y are generally integers, but they may be fractions, they may be zero, or they may even be negative. For gaseous reactions, partial pressures are often used instead of concentrations in moles per liter in the rate law.

The values of x and y indicate what is called the *order of the reaction*. If $x = 1$, the reaction is said to be first order in A; if $x = 2$, the reaction is said to be second order in A, etc. The most important thing to note is that the exponents in the rate law are determined by experiment. A common misconception is that they are the coefficients in the balanced net equation. This, *in general*, is not *true*. For example, in the above reaction the exponents in the rate law are 1 and 2, whereas the coefficients in the balanced equation are 2 and 2.

EXAMPLE 3

For the reaction $2NO(g) + H_2(g) \longrightarrow N_2O(g) + H_2O(g)$ at $1100°K$, the following data are obtained:

Initial pressure of NO, atm	Initial pressure of H_2, atm	Initial rate of pressure decrease, atm/min
0.150	0.400	0.020
0.075	0.400	0.005
0.150	0.200	0.010

Find the rate law and the numerical value of the specific rate constant.

Solution

When the P_{NO} is halved (from 0.150 to 0.075), the rate is quartered (from 0.020 to 0.005). So the rate is proportional to the square of the NO pressure.

When the P_{H_2} is halved (from 0.400 to 0.200), the rate is halved (from 0.020 to 0.010). Hence the rate is directly proportional to the H_2 pressure.

$$\text{Rate} = kP_{NO}^2 P_{H_2}$$

To find k, substitute the data from any run. From the first run

$$\left(0.020 \frac{\text{atm}}{\text{min}}\right) = k(0.150 \text{ atm})^2 (0.400 \text{ atm})$$

$$k = 2.2 \text{ atm}^{-2} \text{ min}^{-1}$$

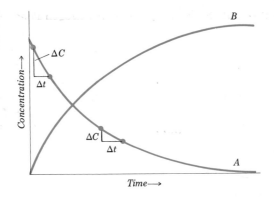

FIG. 12.2 Changes of concentration with time as 1 mol of A gradually converts to 1 mol of B.

Is it always true the ΔC is larger near the start?

In our discussion so far we have been limited to the *initial* rate, i.e., the reaction rate at the very beginning of reaction. What happens as time goes on? Figure 12.2 shows how the situation changes for a simple reaction in which one molecule of A converts to one molecule of B. (Since A and B are stoichiometrically related, the rate of reaction can be expressed either as the rate of disappearance of A or as the rate of formation of B.) Referring to curve A, the rate of disappearance of A is represented by the slope of the curve. From the two triangles shown we can see that the rate is not constant with time. For an equal time interval Δt, the concentration of A drops by an amount ΔC larger near the start of the reaction than later. This can also be expressed by saying that the slope (i.e., the tangent) of the curve decreases as time goes on. Mathematically, the slope, or rate, is given by the ratio $\Delta C / \Delta t$. For the reaction A $\longrightarrow$ B, the rate is proportional to the first power of the A concentration, and so the rate law can be written

$$\text{Rate} = \frac{\Delta C_A}{\Delta t} = -k C_A$$

The minus sign comes from the fact that the A concentration is decreasing with time—in other words, $\Delta C_A / \Delta t$ is a negative number.

In order to get at the specific rate constant k, the above equation for $\Delta C_A / \Delta t$ can be rearranged to give

$$\frac{\Delta C_A}{C_A} = -k \Delta t$$

The ratio $\Delta C_A / C_A$, which represents the *fractional change* in C_A, can also be represented by $\Delta \ln C_A$, where ln stands for the natural logarithm (see Appendix A4.2). Therefore, we can write

$$\Delta \ln C_A = -k \Delta t$$

or, again rearranging,

$$\frac{\Delta \ln C_A}{\Delta t} = -k$$

FIG. 12.3   Rate Data for Decomposition of Nitrogen Pentoxide at 45°C

Time, sec	$[N_2O_5]$	$\ln[N_2O_5]$	Time, sec	$[N_2O_5]$	$\ln[N_2O_5]$
0	0.0176	−4.04	3600	0.0029	−5.84
600	0.0124	−4.39	4200	0.0022	−6.12
1200	0.0093	−4.68	4800	0.0017	−6.38
1800	0.0071	−4.95	5400	0.0012	−6.73
2400	0.0053	−5.24	6000	0.0009	−7.01
3000	0.0039	−5.55	7200	0.0005	−7.60

This last equation says that if we plot $\ln C_A$ versus t, the slope of the curve is a constant and is equal to $-k$. Figure 12.3 shows some data for the decomposition of nitrogen pentoxide:

$$N_2O_5(g) \longrightarrow 2NO_2(g) + \tfrac{1}{2}O_2(g)$$

A plot of the data, Fig. 12.4, shows that concentration versus time does not give a straight line but that logarithm of concentration versus time does. The specific rate constant k can be deduced from the latter.

EXAMPLE 4

Deduce the value of the specific rate constant k from the graph shown in Fig. 12.4.

Solution

$$t = 0 \qquad \ln[N_2O_5] = -4$$
$$\underline{t = 8000 \text{ sec} \qquad \ln[N_2O_5] = -8}$$
$$\Delta t = 8000 \text{ sec} \qquad \Delta\ln[N_2O_5] = \text{final} - \text{initial} = -8 - (-4) = -4$$

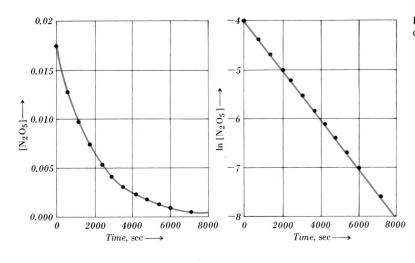

FIG. 12.4 Plots of rate data for decomposition of N_2O_5.

Slope of the line $= \dfrac{\Delta \ln [N_2O_5]}{\Delta t} = \dfrac{-4}{8000 \text{ sec}} = -k$

$k = 5.0 \times 10^{-4} \text{ sec}^{-1}$

? In general, for any first-order reaction the plot of logarithm of concentration vs. time will be a straight line. The plot indicates that the *fractional* decrease of concentration is constant with time. No matter when during the reaction we choose to examine it, the time required to use up half of the remaining reactants will stay the same. Thus, first-order reactions can be characterized either by giving the value for the specific rate constant k (for example, $5.0 \times 10^{-4} \text{ sec}^{-1}$ for the reaction above) or by specifying the half-life of the reaction. The *half-life*, the *time required for half of a given number of molecules to disappear*, is generally indicated by $t_{1/2}$. It is related to k as

$$t_{1/2} = \dfrac{0.693}{k}$$

The factor 0.693 is just the natural logarithm of 2. For the reaction given above, the half-life is 1400 sec.

EXAMPLE 5

When heated to 600°C, acetone (CH_3COCH_3) decomposes to give CO and various hydrocarbons. The reaction is found to be first order in acetone concentration with a half-life of 81 sec. Given at 600°C a 1-liter container into which acetone is injected at 0.48 atm, approximately how long would it take for the acetone pressure to drop to 0.45 atm?

Solution

$CH_3COCH_3 \longrightarrow CO + \cdots$

$\text{Rate} = k[CH_3COCH_3]$

$k = \dfrac{0.693}{t_{1/2}} = \dfrac{0.693}{81 \text{ sec}} = 0.0086 \text{ sec}^{-1}$

$\dfrac{\Delta \ln C}{\Delta t} = -k = -0.0086 \text{ sec}^{-1}$

$\dfrac{\ln 0.45 - \ln 0.48}{\Delta t} = -0.0086 \text{ sec}^{-1}$

$\Delta t = \dfrac{2.3 \, [\log 0.45 - \log 0.48]}{-0.0086 \text{ sec}^{-1}} = 7.5 \text{ sec}$

Note that we have used here the fact that $\ln x = 2.3 \log x$ where ln stands for the natural logarithm (i.e., to base e) and log stands for the logarithm to the base 10.

A *second-order reaction* is one in which the rate is proportional to the square of a reactant concentration or to the product of two reactant concentrations each taken to the first power. An example of a second-order reaction is

$$2HI \longrightarrow H_2 + I_2$$

For this reaction a plot of logarithm of concentration vs. time will *not* give a straight line. Instead, to get a straight line, it is necessary to plot the reciprocal of the concentration, $[HI]^{-1}$, against time. The slope of the line is just k.

Another example of a second-order reaction is the reverse of the above change; namely, $H_2 + I_2 \longrightarrow 2HI$. In this case a straight line will be obtained by plotting the reciprocal concentration of either H_2 or I_2 against time. However, this works *only* if the two concentrations are equal. If $[H_2]$ is not equal to $[I_2]$, this cannot be done. To appreciate the difference, consider the situation where H_2 is present in overwhelming excess so that $[H_2]$ can be considered to be a constant. The reaction would then be *pseudo first order;* i.e., a plot of $\ln [I_2]$ versus t would give a straight line.

In general, for rate analysis involving more than one reactant, excess concentrations are used for all but one reactant and the order of reaction is determined with respect to that reactant. Successive experiments with other reagents in excess can tell about the other reactants.

In summary, to close this section, the dependence of reaction rate on concentration can be obtained from graphic analysis of experimental data. The concentration of a reactant is successively plotted against time to see which plot gives the best straight line. If $\ln C$ versus t is linear, the reaction is first order; if C^{-1} versus t is linear, the reaction is second order. For higher orders, the plot of $1/C^n$ would be linear if the reaction is of the order $n + 1$.

12.3
TEMPERATURE

How does temperature affect the reaction rate? Observations indicate that a rise in temperature almost invariably increases the rate of any reaction.* The change of rate with temperature is expressed by a change in the specific rate constant k. For every reaction, k increases with increasing temperature. As to magnitude, no generalization can be made. The magnitude varies from one reaction to another and from one temperature range to another. A commonly accepted rule, which must be used with great caution, is that a 10°C rise in temperature approximately doubles the reaction rate. For each specific reaction it is necessary to determine from experiment the actual effect of a rise in temperature.

*There are a few extraordinary cases where reaction rate apparently decreases with an increase in temperature. Such odd behavior can arise when there is a sequence of forward and backward steps in which the rate of a backward step increases more rapidly than the rate of a subsequent forward step.

The quantitative relation between specific rate constant k and temperature T is usually expressed by the *Arrhenius equation:*

$$k = Ae^{-E/RT}$$

where A = a numerical constant characteristic of the reaction
e = base of natural logarithms
E = another constant called the activation energy of the reaction (Sec. 12.5)
R = gas constant in energy units, 8.314 J mol^{-1} deg^{-1}

Small changes in T may produce relatively large changes in k. However, the connection is not a simple one. It can be seen by taking the natural logarithm of both sides of the Arrhenius equation. This gives

$$\ln k = \ln (Ae^{-E/RT}) = \ln A + \ln (e^{-E/RT}) = \ln A - \frac{E}{RT}$$

Since $\ln x = 2.303 \log x$, the last equation can be rewritten

$$\log k = \log A - \frac{E}{2.303RT}$$

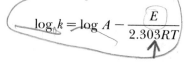

EXAMPLE 6

For a typical first-order reaction, A might be 1.0×10^{14} sec^{-1}, and E might be 80 kJ/mol. Calculate k at room temperature (300°K) and at a temperature that is 10° higher.

Solution

At 300°K:

$$\log k = \log (1.0 \times 10^{14}) - \frac{80,000}{(2.303)(8.314)(300)} = 0.07$$

$$k = 1.2 \text{ sec}^{-1}$$

At 310°K:

$$\log k = \log (1.0 \times 10^{14}) - \frac{80,000}{(2.303)(8.314)(310)} = 0.52$$

$$k = 3.3 \text{ sec}^{-1}$$

As can be seen from this example, for E = 80 kJ, a 10° rise near room temperature would roughly triple the rate. What if the activation energy were twice as large?

EXAMPLE 7

Given a first-order reaction for which A is 1.0×10^{14} sec^{-1} and E is 160 kJ/mol, compare the values of k at room temperature and 10° higher.

Solution

At 300°K:

$$\log k = \log(1.0 \times 10^{14}) - \frac{160,000}{(2.303)(8.314)(300)} = -13.85$$

$$k = 1.4 \times 10^{-14} \text{ sec}^{-1}$$

large E   small k

At 310°K:

$$\log k = \log(1.0 \times 10^{14}) - \frac{160,000}{(2.303)(8.314)(310)} = -12.95$$

$$k = 1.1 \times 10^{-13} \text{ sec}^{-1}$$

As can be seen, a large E means a much smaller rate constant, but also a 10° change in temperature has a much bigger effect. In this case a 10° rise would increase k roughly 8 times.

The value of activation energy E can be determined from experimental data by plotting the logarithm of the observed values of k against $1/T$ for several temperatures. From the equation

$$\log k = \log A - \frac{E}{2.303R}\frac{1}{T}$$

we can see that a plot of $\log k$ versus $1/T$ should give a straight line for which the slope is $-E/2.303R$. Figure 12.5 shows some typical data for the decomposition of N_2O_5 to NO_2 and O_2. The slope of the line, which turns out to be -5000, corresponds to an activation energy of 100 kJ/mol.

EXAMPLE 8

Given the graph shown in Fig. 12.5, determine the energy of activation for the reaction $N_2O_5 \longrightarrow 2NO_2 + \frac{1}{2}O_2$.

Solution

$\log k = -2.5 \quad$ when $\dfrac{1000}{T} = 3.0$

$\log k = -5.7 \quad$ when $\dfrac{1000}{T} = 3.6$

$\Delta \log k = -5.7 - (-2.5) = -3.2$

$\Delta\left(\dfrac{1}{T}\right) = 0.0036 - 0.0030 = 0.0006$

$\dfrac{\Delta \log k}{\Delta\left(\dfrac{1}{T}\right)} = \dfrac{-3.2}{0.0006} = -5000$

$\dfrac{-E}{2.303R} = -5000$

$E = (2.303)(8.314)(5000) = 100 \text{ kJ}$

327

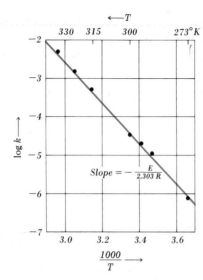

FIG 12.5 Plot of log k versus $1/T$ to show how energy of activation can be deduced from experimental data.

12.4
CATALYSIS

It is found by experiment that some reactions can be speeded up by the presence of substances which themselves remain unchanged after the reaction has ended. Such substances are known as *catalysts,* and their effect is known as *catalysis.* Often only a trace of catalyst is sufficient to accelerate reaction. However, the rate of reaction may also be proportional to some power of the concentration of catalyst. If experiments show there is such a dependence, then the catalyst concentration becomes part of the rate law the same way as the concentration of reactants.

There are numerous examples of catalysis. For instance, when $KClO_3$ is heated so that it decomposes into KCl and oxygen, it is observed that a pinch of manganese dioxide, MnO_2, considerably accelerates the reaction. At the end of the reaction the $KClO_3$ is gone, but all the MnO_2 remains. Although recovered, the catalyst must take part in the reaction; otherwise, it could not change the rate.

When hydrogen gas escapes from a cylinder into the air, no change is visible. If, however, the hydrogen is directed at finely divided platinum, the platinum glows and eventually makes the hydrogen burst into flame. In the absence of platinum, the rate of reaction between hydrogen and oxygen in air is too slow to observe. In contact with platinum, the reaction occurs, energy is liberated, the platinum gets hotter, and the rate of reaction increases until ignition eventually occurs.

Enzymes are complex substances in biological systems which act as catalysts for biochemical processes. Pepsin in the gastric juice and ptyalin in the saliva are examples. Ptyalin is the catalyst which accelerates the conversion of starch to sugar. Although starch will react with water to form sugar, it takes weeks for the conversion to occur. A trace of ptyalin is enough to make the reaction proceed at a biologically useful rate.

A special type of catalysis occasionally encountered is *autocatalysis,* or *self-catalysis.* As the name implies, this is catalysis in which one of the products is a catalyst for the reaction. For example, in the reaction of permanganate ion with oxalic acid

$$2MnO_4^- + 5H_2C_2O_4 + 6H^+ \longrightarrow 2Mn^{2+} + 10CO_2 + 8H_2O$$

the product Mn^{2+} catalyzes reaction. This can readily be observed by mixing solutions of potassium permanganate, sulfuric acid, and oxalic acid. No appreciable decolorization is observed until a tiny crystal of manganous sulfate, $MnSO_4$, is dropped into the reaction mixture, whereupon fast decolorization occurs.

Interesting catalysis is observed in the case of hydrogen peroxide. Hydrogen peroxide is thermodynamically unstable and decomposes to give water and oxygen. The reaction is rapid enough that solutions of hydrogen peroxide are ordinarily difficult to keep for any length of time without decomposition. It has been observed that certain substances, such as phosphates, can be added in trace amounts to slow down the rate of decomposition. This looks like the reverse of catalysis, and, in fact, substances which slow down rates used to be called *negative catalysts.* The name is misleading, since the function of phosphate is probably to destroy the action of catalysts already present in the hydrogen peroxide. For example, decomposition of hydrogen peroxide is catalyzed by trace amounts of Fe^{3+} ion. When phosphate is added, it ties up the Fe^{3+} and prevents the Fe^{3+} from functioning as a catalyst.

12.5
COLLISION THEORY

Many of the observed facts of chemical kinetics recounted above can be interpreted in terms of *collision theory.* For substance A to react with substance B, it is necessary that particles A, be they molecules, ions, or atoms, collide with particles B. In the collision, atoms and electrons are rearranged. There is a reshuffling of chemical bonds, and that leads to production of other species.

According to collision theory, the rate of any step in a reaction is directly proportional to (1) the *number of collisions per unit time* between the reacting particles involved in that step and (2) the *fraction of these collisions that are effective.* That the rate should depend on the number of collisions per unit time seems obvious. For instance, in a box that contains A molecules and B molecules, there is a certain frequency of collision between A and B molecules. If more A molecules are placed in the box, the collision frequency is increased. With more collisions, the reaction between A and B should go faster. However, this cannot be the full story. Calculation of the number of collisions between particles indicates that the collision frequency is extraordinarily high. In a mixture containing 1 mol of A molecules and 1 mol of B molecules as gases at STP, the number of collisions between A and B is about 10^{30} sec^{-1}. If every one of these collisions led to reaction, the reaction would be over like a shot. By observation, this is not true. It must be that only a rather small fraction of collisions lead to reaction.

Why are some collisions effective and others not? Collisions may be so gentle that there is no change upon collision. The colliding particles simply separate to resume their original identity. However, if A or B or both A and B have much kinetic energy before collision, they can easily use their kinetic energy to do work against the repulsive forces between adjacent electron clouds. If the kinetic energy available is big enough, repulsive forces can be overcome, and the molecules can penetrate into each other far enough that significant, large-scale electron and atom rearrangement results. One or more new species may be formed.

The extra amount of energy above the average level required in a collision to produce chemical reaction is the *energy of activation*. Its magnitude, which can be determined as in Sec. 12.3, depends on the nature of the reaction. Some reactions have a large energy of activation. Such reactions are slow, since only a relatively small fraction of reactant particles have enough kinetic energy to furnish the required energy of activation. Other reactions have a small energy of activation. Such reactions are fast, since a greater fraction of the collisions are effective. More of the particles have sufficient kinetic energy to furnish the required energy of activation.

Qualitatively, collision theory quite satisfactorily accounts for the four factors listed at the beginning of this chapter as being observed to influence reaction rates:

1 The rate of chemical reaction depends on the *nature of the chemical reactants,* because the energy of activation differs from one reaction to another.

2 The rate of reaction depends on the *concentration of reactants,* because the number of collisions increases as the concentration is increased.

3 The rate of reaction depends on the *temperature,* because an increase in temperature makes molecules move faster. They collide more frequently, and, what is more important, the collisions are more violent and more likely to result in reaction. Any collection of molecules has a distribution of energies (Fig. 6.14). As the temperature is raised, the whole distribution curve shifts to higher energies, and a larger fraction of the molecules becomes highly energetic. More of the collisions are therefore effective at a high temperature than at a low temperature.

4 The rate of reaction would be accelerated by the presence of *catalysts* if, somehow, a catalyst made collisions more effective. This could be done, for example, by a preliminary step in which one or more of the reactants reacted with the catalyst to give a species having lower activation energy. New reactants could be produced which would react more rapidly than the original reactants.

Another useful way to look at chemical kinetics is to focus on the change in potential energy when reactants pass over to products. To illustrate, we consider the one-step reaction in which an atom of H collides with a molecule of HBr to form a molecule of H_2 and an atom of Br:

$$H + H—Br \longrightarrow H—H + Br$$

For simplicity, the collision is assumed to occur along a straight line through all the nuclei. The H atom approaches H—Br from the left to form some kind of transient complex particle H—H—Br (called the *activated complex*), and then the Br atom breaks away to move off toward the right. Figure 12.6 shows schematically how the potential energy of the system changes as the H $\cdots$ H distance first decreases and then the H $\cdots$ Br distance increases. The horizontal scale represents the *reaction coordinate,* which tells us how far we have gone from the reactants H + HBr to the products HH + Br. As can be seen, the energy of the system passes over a hump. In the initial state, H and HBr particles are far enough apart not to affect each other. The potential energy is just the sum of the potential energy of H by itself plus that of HBr by itself. As H and HBr come together, the forces of repulsion between the electron clouds become appreciable. Work must be done on the system to squash the particles together. This means the total potential energy must increase. It increases until it reaches a maximum that corresponds to the activated complex. The activated complex then splits, and the potential energy drops as HH and Br go apart.

The difference (shown by the double-headed arrow) between the potential energy of the initial state H plus HBr and that of the activated complex is a measure of the energy which must be added to the particles in order to get them to react. This is the *activation energy of the reaction.* It usually is supplied by converting some of the kinetic energy of the particles into potential energy. If H and HBr particles do not have enough kinetic energy, all the kinetic

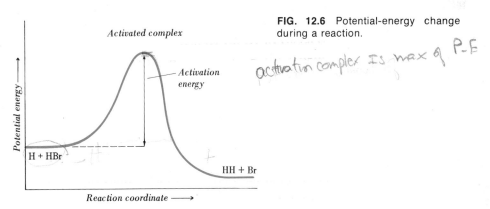

FIG. 12.6 Potential-energy change during a reaction.

activation complex is max of P-E

331

energy may be converted into potential energy without getting the system pushed up to the activated complex. In such a case H and HBr slide back down the hump and fly apart unchanged. The situation is similar to that of a ball rolled up the side of a hill. If the ball is delivered slowly, it goes part way up, stops, and rolls back. If the ball is delivered rapidly enough, it goes completely to the top and down the other side. Similarly, if H and HBr particles have enough kinetic energy, they can attain the activated complex and get over the hump from H and HBr to HH and Br. In a reaction at high temperature more molecules get over the potential-energy hump per unit time; so the reaction occurs faster.

Two other aspects of Fig. 12.6 are of interest. For the case represented, the ending state HH and Br has lower potential energy than the starting state H and HBr. There is a net decrease of 67 kJ in the potential energy as the reaction proceeds. This energy usually shows up as heat; so the net change

$$H + HBr \longrightarrow HH + Br \qquad \Delta H = -67 \text{ kJ}$$

is exothermic.

Figure 12.6 can also be read from right to left as a diagram for the reverse reaction; i.e., HH plus Br produces H and HBr. As can be seen from the diagram, the reverse reaction

$$H + HBr \longleftarrow HH + Br \qquad \Delta H = +67 \text{ kJ}$$

is endothermic. It also has a higher activation energy than the forward reaction.

When a reaction is catalyzed, there is generally a change of path or mechanism. Since the rate is now faster, the activation energy for the new path must be lower than for the old path. Figure 12.7 indicates this schematically. The dashed curve shows what the potential-energy curve might look like for a new path. Since the barrier is lower, more particles per unit time get over the hump; hence the reaction goes faster. An example of catalytic change of path would be the reaction between H_2 and O_2 in the presence of platinum. If the effect of platinum is to change H_2 molecules into H atoms, the O_2 molecules would then collide with H atoms instead of with H_2 molecules.

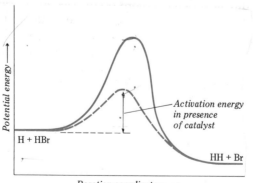

FIG. 12.7 Effect of catalysis on energy path of a reaction.

One of the trickiest aspects of chemical kinetics is to account properly for the observed power dependence of rate on concentration. Why in some cases does rate depend on the first power of concentration and in other cases on the square? For simplicity, let us consider one step of a reaction. Suppose that in this step one molecule of A reacts with one molecule of B to form a molecule AB. The balanced equation for *this step* is

$$A + B \longrightarrow AB$$

According to collision theory, the rate of formation of AB is proportional to the rate at which A and B collide. Let us imagine that we have a box that contains A molecules and one B molecule. The rate at which A molecules collide with B is directly proportional to the number of A molecules in the box. If we should double the number of A molecules, we would have twice as many A-B collisions per unit time. Suppose now we place a second B molecule in the box. We now have twice as many A-B collisions per unit time. In other words, the rate at which A and B molecules collide is directly proportional to the concentration of A and directly proportional to the concentration of B. The rate of formation of AB should therefore be

$$Rate = k[A][B]$$

It should be noted that the exponents of $[A]$ and of $[B]$ in the rate law for this step are unity, just as are the two coefficients in the balanced equation *for the step*.

What is the situation if the balanced equation for a step involves coefficients larger than 1? Consider the reaction

$$2A \longrightarrow A_2$$

In this step an A molecule must collide with another A molecule to form A_2. The rate at which A_2 forms is thus proportional to the rate at which two A molecules collide. Again we imagine a box, this time containing only molecules of type A. The rate at which *any one* A molecule collides with another A molecule is proportional to the number of other A molecules in the box. If we should double the number of other A molecules in the box, then we would double the rate at which collisions occur *with the one molecule under observation*. Now suppose we extend our observation to all the molecules in the box. The number of A-A collisions per unit time is proportional to the number of molecules hitting, multiplied by the number of molecules being hit. In effect, the rate at which two A molecules collide is proportional to the concentration of A times the concentration of A, i.e., to the square of the concentration of A. Consequently, for the step

$$2A \longrightarrow A_2$$

we can write the rate law

$$Rate = k[A]^2$$

The exponent is 2, just as is the coefficient of A in the balanced equation for the step.

For the general case of a single step for which the balanced chemical equation shows disappearance of n molecules of A and m molecules of B to form a product, we can write the rate law

$$\text{Rate} = k[A]^n[B]^m$$

This indicates that the rate of the step is proportional to the concentration of species A taken to the nth power times the concentration of species B taken to the mth power.

We note, however, that the overall chemical change generally consists of several consecutive steps. A knowledge of only the *overall* balanced chemical equation *does not* permit us to predict what the experimentally observed rate law will be. For example, in the reaction between NO and H_2, discussed in Sec. 12.2, the balanced equation for the reaction is

$$2H_2(g) + 2NO(g) \longrightarrow N_2(g) + 2H_2O(g)$$

In a one-step collision theory, it would appear that reaction could occur by collision between two H_2 molecules and two NO molecules. The number of such collisions per second would be proportional to the molar concentration of H_2 squared times the molar concentration of NO squared. This does not agree with experiment, however, which tells us the rate is proportional to the first power of H_2, not to the second. It must be that the collisions that determine the rate are not between two H_2 molecules and two NO molecules, but are something simpler. In other words, a one-step explanation does not suffice.

To account for the actual observed rate law, we have to assume that this reaction, like many others, occurs in steps. In stepwise reactions, the slow step determines the rate. It is the bottleneck. For example, let us consider a two-step reaction:

Step (1):	$A + B \longrightarrow AB$	slow
Step (2):	$AB + B \longrightarrow AB_2$	fast
Net reaction:	$A + 2B \longrightarrow AB_2$	

In step (1) a molecule of A collides with a molecule of B to form intermediate AB. In step (2) the intermediate immediately reacts with another molecule of B to form the final product AB_2. If the first step is slow and the second step is fast, the rate at which the final product AB_2 forms will depend only on the rate of the first step. The rate of the first step is determined by collision of one A and one B. Since it is the rate-determining step, the rate law for the overall change is

$$\text{Rate of production of } AB_2 = k[A][B]$$

The chemical equation for the overall change, however, is still determined by adding steps (1) and (2). We get

$$A + 2B \longrightarrow AB_2$$

The coefficients which appear in the net equation are different from the exponents in the rate law.

For the specific reaction of Sec. 12.2

$$2H_2(g) + 2NO(g) \longrightarrow N_2(g) + 2H_2O(g)$$

the rate law was determined by experiment to be

$$\text{Rate} = k[H_2][NO]^2$$

The reaction must occur in steps because the exponents differ from the coefficients in the overall equation. One set of steps, in which the first step would be slower and therefore rate-determining, might be as follows:

$$H_2(g) + NO(g) + NO(g) \xrightarrow{\text{slow}} N_2O(g) + H_2O(g)$$

$$H_2(g) + N_2O(g) \xrightarrow{\text{fast}} N_2(g) + H_2O(g)$$

However, many kinetics people object to such a mechanism because it requires a simultaneous collision of three molecules in the first step. As any billiard player knows, three-body collisions are quite improbable. An alternative set of steps without a three-body collision might be the following:

$$2NO(g) \xrightleftharpoons{\text{fast}} N_2O_2(g)$$

$$N_2O_2(g) + H_2(g) \xrightarrow{\text{slow}} N_2O(g) + H_2O(g)$$

$$N_2O(g) + H_2(g) \longrightarrow N_2(g) + H_2O(g)$$

The middle step would be rate-determining. Since the N_2O_2 would be rapidly formed by collision of two NO molecules, the concentration of N_2O_2 would be proportional to the square of the concentration of NO. The rate of the second step would thus depend on the square of the concentration of NO times the concentration of H_2, in agreement with the observed rate law. On the basis of the observed rate law, it is not possible to distinguish between these two mechanisms. In fact, it may be that neither is right, and the actual mechanism may be more complicated.

Frequently it happens that several alternative paths exist by which an overall change is accomplished. In such cases all the paths must be considered in the rate law. As an example, let us consider the situation where A reacts to give products by three alternative paths:

$$A \xrightarrow{k_1} \text{products}$$

$$A + B \xrightarrow{k_2} \text{products}$$

$$A + A \xrightarrow{k_3} \text{products}$$

If k_1, k_2, and k_3 are the respective constants for the slow step in each of the three paths, the total rate of disappearance of A by all three paths would be given by

$$\text{Rate} = k_1[A] + k_2[A][B] + 2k_3[A]^2$$

Unraveling such a rate law can be quite a chore, but often judicious choice of experimental conditions can help. If the concentration of A were made very small, for example, by diluting the system, the term in $[A]^2$ could be neglected with respect to the others. Alternatively, the concentration of B could be made much larger than that of A so that only the middle term would need to be considered.

12.8
CHAIN REACTIONS

One of the most interesting complications that may arise in chemical kinetics is the occurrence of a *chain reaction*. This comes about when intermediate species that are consumed in one step are regenerated in a later step. The result is to set up a sequence of steps that endlessly repeat themselves, like the links of a chain, until the chain is terminated or the starting materials are exhausted. Examples of chain reactions are found in flames, gas explosions, and generation of smog.

One of the first reactions which was recognized as involving a chain mechanism was the deceptively simple combination of hydrogen and bromine to form hydrogen bromide:

$$H_2(g) + Br_2(g) \longrightarrow 2HBr(g)$$

The reaction is very complicated and is believed to go as follows: First, there is a *chain-starting step* which involves dissociation of diatomic bromine to give individual atoms:

$$Br_2 \longrightarrow 2Br$$

Then there is a series of *chain-propagating steps* in which the bromine atom is supposed to react with H_2 to form HBr and H. The H, being very reactive, combines with Br_2 to form HBr and another Br which can repeat the cycle. Alternatively, some of the H can combine with HBr to form H_2 and liberate Br for carrying out the chain. The sequence can be written as

$$Br + H_2 \longrightarrow HBr + H$$
$$H + Br_2 \longrightarrow HBr + Br$$
$$H + HBr \longrightarrow H_2 + Br$$

As can be seen, the Br needed in the first step is re-formed in the second and third steps. Thus, one Br atom can carry out its job of making HBr but then be regenerated to do the whole thing over and over again. Sometimes this happens thousands of times before a Br gets sopped up by some extraneous reaction, as with a scavenger or impurity on the wall of the vessel. In any case the chain eventually is terminated; the number of cycles achieved before termination occurs is called the *chain length*. Even if there are no scavengers or impurities, propagation of the chain may come to a halt by a reaction such as

$$2Br \longrightarrow Br_2$$

which is called the *chain-terminating step*. Usually chain termination also involves a third species to carry off excess bond energy.

Chain reactions can lead to spectacular explosions if the chain-propagating steps produce heat faster than it can be conducted away. In such cases rising temperature drives the reactions to ever-increasing speed, resulting in a runaway situation.

12.9
PHOTOCHEMICAL SMOG

An important practical illustration of chemical kinetics, one that involves chain reactions, is the generation of smog. Details of smog formation and its relation to the internal-combustion engine are considered in this and the following section.

There are two kinds of smog. The oldest, called *London smog*, is a mixture of coal smoke and fog. Particles of smoke from coal combustion act as condensation nuclei on which fog droplets condense. The fog part is largely SO_2 ($+ SO_3$) and humidity; it is generally worst in the early morning hours and appears to worsen shortly after sunrise (perhaps because of light-induced oxidation of $SO_2 + \frac{1}{2}O_2 \longrightarrow SO_3$ followed by reaction with humidity to give H_2SO_4 aerosol).

The other kind of smog is called *photochemical or Los Angeles smog*. It has no relation to smoke or fog; it is worst in the sunshine, and, unlike London smog, it peaks in the afternoon. Also, whereas London smog is characterized by bronchial irritation, Los Angeles smog tends to produce eye irritation and plant damage (e.g., metallic sheen on leaves).

The first clue to the cause of photochemical smog was that articles made of rubber showed severe cracking in periods of high smog. Laboratory tests suggested the culprit was ozone. The second clue was that ozone plus double-bonded hydrocarbons give products which yield the same kind of damage to plants as does Los Angeles smog. About the same time it was discovered that double-bonded hydrocarbons plus NO_2 give products that crack rubber. Subsequently, these products were identified with ozone, O_3, an irritating air pollutant. The fourth clue came from statistical studies which suggested a correlation with the automobile, specifically that automobile exhaust is a main source of double-bonded hydrocarbons and oxides of nitrogen.

The chemical reactions that are involved in the creation of photochemical smog are complex and still only incompletely understood. They start with the combination of N_2 and O_2 in the air. At ordinary temperatures these do not react with each other, but when temperature rises, as in the combustion flame in an automobile engine, they react to form nitric oxide:

$$N_2(g) + O_2(g) \longrightarrow 2NO(g)$$

Some of this NO reacts further to form nitrogen dioxide:

$$2NO(g) + O_2(g) \longrightarrow 2NO_2(g)$$

The mixture of NO and NO$_2$ is referred to as NO$_x$. It is a serious air pollutant but is not all synthetic; it also comes from natural bacterial action. Typical background levels in the United States are 0.002 ppm NO and 0.004 ppm NO$_2$; in urban areas these often reach 100 times these values.

Nitrogen dioxide is part of the *photolytic* NO$_2$ *cycle*, which can be written as follows:

Step (1):	$NO_2 \xrightarrow{h\nu} NO + O$
Step (2):	$O + O_2 + M \longrightarrow O_3 + M$
Step (3):	$O_3 + NO \longrightarrow NO_2 + O_2$

In step (1), NO$_2$ absorbs sunlight and cleaves to give NO and atomic oxygen. In step (2), the highly reactive monatomic O combines with ordinary diatomic O$_2$ to form ozone. (M stands for a third body such as N$_2$, O$_2$, Ar, or CO$_2$, which can carry off excess energy.) In step (3), O$_3$ combines with the NO formed in step (1) to regenerate NO$_2$ and O$_2$. The cycle is a closed one and would be balanced except for competing reactions with hydrocarbons. Hydrocarbons, emitted in automobile exhausts, are believed to react either with O or O$_3$ to form secondary pollutants. In the process, NO is oxidized to NO$_2$, and lack of the NO scavenger in the atmosphere makes the O$_3$ content rise.

The components of a typical smog are CO (2 ppm), NO (0.15 ppm), NO$_2$ (0.20 ppm), unsaturated hydrocarbons (0.05 ppm), O$_3$ (0.10 ppm), aldehydes (0.20 ppm), and organic peroxides (0.03 ppm). How some of these change with time is shown in Fig. 12.8. As can be seen, they correlate differently with the time of day, because of changes in automobile traffic and sunlight. Before 6 A.M., the NO and NO$_2$ concentrations are steady. From about 6 to 8 A.M., there is a sharp increase in NO and hydrocarbons, which correlates with rise in automobile traffic. As the sun comes up, the NO content goes down but the NO$_2$ rises, presumably because of photochemically induced oxidation of

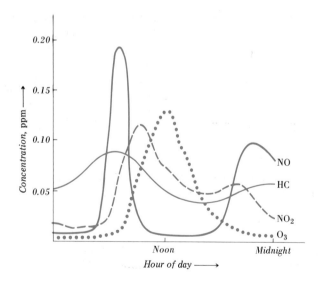

FIG. 12.8 Time change of air pollutants NO, NO$_2$, O$_3$, and hydrocarbons (HC) during a smog day. (Scale for hydrocarbon concentrations should be multiplied by 100.)

339 Section 12.10
Internal-
Combustion
Engine and
Control of
Air Pollution

NO to NO_2. As the NO drops below about 0.1 ppm, the O_3 content starts to build up, reaching a peak in the early afternoon. Again in the period 5 to 8 P.M. the NO rises as the sun goes down and traffic picks up. A belated rise in NO_2 may be attributed to late oxidation of the NO by the O_3 built up during the day.

The dynamics of the above conversions are only imperfectly understood. Some of the critical rate constants have not yet been evaluated. Attempts to simulate the problem in artificial chambers have proved to be frustrating because of large dependence on the nature, pretreatment, and area of the enclosing surface. This is typical of chain reactions.

How hydrocarbons from automobile exhaust upset the photolytic NO_2 cycle is subject to vigorous debate. Often a chain of 20 or more consecutive steps is needed to account for all the products. Even then a proposed scheme may be unacceptable because it may disagree with observed rates.

The photochemical-smog problem is by no means unique to Los Angeles. Most big cities, especially if they have lots of sunshine, are getting periods of smog alert. Los Angeles has the problem magnified because of urban sprawl, which produces an enormous number of motor-vehicle sources of pollution, and because of abundant sunshine and a geographic situation that is ideal for generating and trapping photochemical pollutants.

12.10
INTERNAL-COMBUSTION ENGINE AND CONTROL OF AIR POLLUTION

One of the prime villains in air pollution is the automobile. To understand pollution control, it is necessary to know something about the workings of the internal-combustion engine. The dilemma that we encounter is that increasing the air-fuel ratio from 13 to 16 dramatically reduces hydrocarbon and CO emissions but raises NO_x emission. To make the situation harder to control, the maximum power of a conventional internal-combustion engine comes at an air-fuel ratio of 13.

Jean Joseph Lenoir invented the idea of spark ignition of hydrocarbon-air mixtures in 1860, but the idea was a failure until 1876 when Nikolaus Otto suggested that the mixture be compressed before spark ignition. Such compression raises the temperature and increases the work per stroke. Figure 12.9 gives a schematic diagram of the so-called "Otto" or "four-stroke" spark-ignition engine. The main components are a cylinder, a piston, a connecting rod, and a crank to turn the drive shaft. There are four strokes for each power cycle, corresponding to two full rotations of the crank, clockwise about the rotation axis of the drive shaft:

1   **Intake stroke**   With the intake valve open and exhaust valve closed, a mixture of hydrocarbon and air is sucked into a cylinder as the piston moves on a downstroke.

2   **Compression stroke**   With both valves closed, the piston moves on an upstroke, compressing the fuel-air mixture by a factor ranging from 7 to 12. Spark ignition near the end of this stroke initiates combustion.

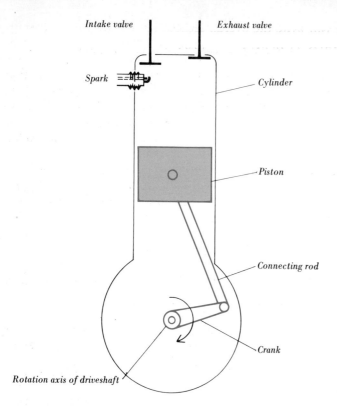

Intake valve
Exhaust valve
Spark
Cylinder
Piston
Connecting rod
Crank
Rotation axis of driveshaft

FIG. 12.9 Four-stroke spark-ignition engine.

3 **Power, or work, stroke** With both valves still closed, combustion occurs and pressure and temperature increase so as to drive the piston down. The exhaust valve opens at the end of this stroke.

4 **Exhaust stroke** With the exhaust valve open and the intake valve closed, the upstroke of the piston pushes the exhaust products out of the cylinder.

The carburetor is a device located in front of the intake valve for providing a homogeneous mixture of fuel and air. On the intake stroke, drop in pressure draws liquid fuel through a nozzle, which sprays the fuel into an air-mixing chamber. The hydrocarbon-air mix is then sucked into the piston chamber.

Liquid fuels are a complex mix of hydrocarbons, averaging 85% by weight C and 15% H (empirical formula: C_8H_{17}). An ideal fuel would be isooctane, C_8H_{18}, otherwise known as 2,2,4-trimethylpentane. Its structural formula is

$$CH_3 - \underset{\underset{CH_3}{|}}{\overset{\overset{CH_3}{|}}{C}} - CH_2 - \underset{}{\overset{\overset{CH_3}{|}}{CH}} - CH_3$$

340

341 Section 12.10
Internal-
Combustion
Engine and
Control of
Air Pollution

Using isooctane, we can write the overall combustion reaction as

$$C_8H_{18} + 12\tfrac{1}{2}O_2\,(+46.6N_2) \longrightarrow 8CO_2 + 9H_2O\,(+46.6N_2)$$

The 46.6 mol of N_2 in parentheses is a reminder that the oxidant is generally not pure oxygen but air. On a mass basis, 1 mol of C_8H_{18} corresponds to 114 g; $12\tfrac{1}{2}$ mol of O_2 and the 46.6 mol of N_2 (plus associated argon) come to 1725 g. Thus, on a mass basis, the theoretical air-fuel ratio is 1725:114, or 15:1. Ratios greater than this are called *lean;* ratios less than this are called *rich.*

It appears that the above reaction like *all* combustion reactions goes by a chain mechanism. Spark ignition produces reactive intermediates, and these propagate the chemical reaction into the unburned mixture ahead of the flame front.* Among the products of the chain reaction are very reactive hydrocarbon fragments known as *free radicals.* These are unusual molecules which do not have a normal complement of covalent bonds but have an unpaired, nonbonding electron available for quickly forming another bond. An example of a free radical is $CH_3\cdot$, the methyl radical, with three hydrogens bound to a central carbon by regular electron-pair bonds and an additional unpaired, unsatisfied lone electron (indicated by the dot).

As chain reactions, combustion reactions are very complex. For hydrocarbon combustion, the following sequence of steps seems to be involved:

1 Radicals are produced by the spark ignition in some unknown fashion to initiate the chain. One possible mechanism is *cracking,* or breaking, of a hydrocarbon, as in the following reaction:

$$C_8H_{18} \longrightarrow C_7H_{15}\cdot + CH_3$$

2 The radicals generated then proceed to strip H atoms from hydrocarbon molecules, forming new radicals in the process; for example,

$$CH_3\cdot + C_8H_{18} \longrightarrow CH_4 + C_8H_{17}\cdot$$

Some of these large radicals may break up into smaller radicals, leaving unsaturated hydrocarbons as products:

$$C_8H_{17}\cdot \longrightarrow CH_3\cdot + C_7H_{14}$$

3 The radicals may react with O_2 to form peroxide radicals:

$$R\cdot + O_2 \longrightarrow R-O-O\cdot$$

---

*It is sometimes possible to have self-ignition of the fuel-air mix ahead of a flame front. This might occur, for example, when the rise in pressure that normally accompanies combustion compresses the "end gas" so that its temperature rises above self-ignition temperature. Spontaneous ignition may spread from several point sources, unlike a smooth flame front, and the result is *knock.* It may range from a thud to a small ping. A little bit of pinging is considered desirable because one wants to hasten the final stages of combustion before the piston gets far into its expansion stroke, but too much is bad because it wastes power and produces more air pollutants.

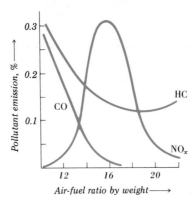

FIG. 12.10 Effect of air-fuel ratio on pollutant emissions. (Vertical scale needs to be multiplied by 40 for CO, by 0.4 for HC.)

4 Several paths now arise. The peroxide radical might decompose, or it might strip H from another hydrocarbon to set up a radical-peroxide chain:

$$R \cdot \xrightarrow{O_2} ROO \cdot \xrightarrow{RH} ROOH + R \cdot$$

Depending on how the peroxide decomposes, whether by breaking the R—O bond or the O—O bond, a variety of products is possible. Only in relatively cool flames are the final products CO_2, CO, and H_2O. Hot flames, i.e., self-ignition flames, are excessively rich in aldehydes and ketones, which are organic molecules containing $-\overset{\overset{\textstyle O}{\|}}{C}-$ groups. Engine misfiring generally leads to aggravated air pollution because of the highly reactive species created.

The auto pollutants that are the easiest to control are hydrocarbons and carbon monoxide. As shown in Fig. 12.10, increasing the air-fuel ratio (i.e., making the mixture more lean) dramatically reduces hydrocarbon and carbon monoxide emission.

However, as Fig. 12.10 makes abundantly clear, the result is to aggravate the NO_x problem. Running engines at higher air-fuel ratios means higher combustion temperatures, and that means more generation of NO and NO_2. Two solutions have been tried: One is to keep the air-fuel ratio low (this pleases the "hot-rodders," since maximum power corresponds to "rich" mixtures with air-fuel ratio $\sim$ 13) but add an afterburner in which exhaust gases are more completely burned after addition of more air. The other method is to add a catalytic reactor in which exhaust gases are passed through a bed of catalysts to convert to less noxious products. The problem is that most catalysts are deactivated (i.e., "poisoned") by the lead that is often added to gasoline to improve its antiknock qualities. High-compression engines tend to knock rather badly because of self-ignition brought about by the compression. Tetraethyllead, $Pb(C_2H_5)_4$, added to the extent of about 1 cm^3/liter, helps to cut down knocking by generating a smooth supply of free radicals for maintaining an orderly flame front.

Important Concepts

reaction velocity
reaction mechanism
homogeneous reaction
heterogeneous reaction
rate law
specific rate constant
order of a reaction
half-life of a reaction
Arrhenius equation
activation energy
catalysis
autocatalysis
collision theory

potential-energy diagrams for chemical reaction
activated complex
stepwise reactions
chain reactions
chain-starting
chain propagation
chain termination
photochemical smog
photolytic cycle
air-pollution control
internal-combustion engine
air-fuel ratio
free radical

Exercises

*12.1 Terms Distinguish clearly between each of the following terms: step, mechanism, collision, catalyst.

*12.2 Rates Explain why rates of chemical reactions are generally given in moles per liter per second rather than in moles per second.

**12.3 Activation energy What activation energy should a reaction have so that at room temperature a 10°C rise would exactly double the rate?

**12.4 Rates A drop of solution (volume 0.05 ml) contains 3.0×10^{-6} mol H^+. If the rate of disappearance of the H^+ is 1.00×10^7 M/sec, how long would it take for the H^+ in the drop to disappear? *Ans.* 6.0×10^{-9} sec

*12.5 Rates In a given reaction at 100°C, it is observed that the concentration of hydrogen gas changes from 0.0360 to 0.0340 M in 10.0 min. Calculate the rate of reaction in atmospheres per minute.

**12.6 Rate law In a kinetic study of the reaction

$$5Br^- + BrO_3^- + 6H^+ \longrightarrow 3Br_2 + 3H_2O$$

the following data were obtained at 25°C:

Initial concentration, M			Time required for BrO_3^- to disappear, sec
Br^-	BrO_3^-	H^+	
0.001	0.005	0.10	20
0.002	0.005	0.10	10
0.001	0.010	0.10	20
0.001	0.005	0.20	5

Determine the order of the reaction for each reactant. Write the rate law and calculate a value for the specific rate constant. *Ans.* $k = 5000$ M^{-3} sec^{-1}

*12.7 Mechanism The reaction of iodide ion with hydrogen peroxide

$$2I^- + H_2O_2 + 2H^+ \longrightarrow 2H_2O + I_2$$

follows a rate law of the form

$$\text{Rate} = k_1[I^-][H_2O_2] + k_2[H^+][I^-][H_2O_2]$$

What does this tell you about the mechanism of the reaction?

**12.8 Rate law For the first-order reaction A $\longrightarrow$ B, as shown in Fig. 12.2, what is the significance of the point at which the two curves cross? How must the slopes of the two curves be related at this point? If Fig. 12.2 had used logarithms of concentration instead of concentration, what would the crossing point look like?

**12.9 Concentration vs. time On the basis of the first two data points of Fig. 12.3, what would be the concentration of N_2O_5 at elapsed times of 100, 200, 300, 400, and 500 sec, respectively?

**12.10 Rate law The conversion of tertiary butyl bromide to tertiary butyl alcohol by H_2O:

$$(CH_3)_3CBr + H_2O \longrightarrow (CH_3)_3COH + HBr$$

generates the following data at 25°C:

Time, h	Concentration of $(CH_3)_3CBr$, M
0	0.1039
3.15	0.0896
4.10	0.0859
6.20	0.0776
8.20	0.0701
10.0	0.0639
13.5	0.0529
18.3	0.0380
26.0	0.0270
30.8	0.0207

Plot these data as (a) concentration vs. time and (b) log of concentration vs. time. Deduce the rate constant and figure out the half-life of the reaction.

Ans. $k = 1.4 \times 10^{-5}$ sec^{-1}

**12.11 Temperature Explain qualitatively why a 10°C rise in temperature will have a bigger effect on reaction rate (a) if the activation energy is higher; (b) if the temperature is lower.

**12.12 Temperature For the second-order reaction

$$H_2 + I_2 \longrightarrow 2HI$$

the factor A in the Arrhenius equation has a value of 3.4×10^{10} liters mol^{-1} sec^{-1} at 25°C. If the activation energy is 178 kJ/mol, what temperature rise would you need to double the rate of reaction?

**12.13 Energy of activation For the reaction $CH_3I + C_2H_5ONa \longrightarrow CH_3OC_2H_5 + NaI$ in ethyl alcohol, the following data have been obtained:

Temp., °C	k, liter mol^{-1} sec^{-1}
0	5.60×10^{-5}
6	11.8×10^{-5}
12	24.5×10^{-5}
18	48.8×10^{-5}
24	100×10^{-5}

What is the activation energy for this reaction?

Ans. 81.1 kJ

*12.14 Catalysis The rate of decomposition of acetaldehyde, CH_3CHO, into CH_4 and CO in the presence of I_2 at 800°K follows the rate law

$$Rate = k[CH_3CHO][I_2]$$

The decomposition is believed to go by a two-step mechanism:

$$CH_3CHO + I_2 \longrightarrow CH_3I + HI + CO$$
$$CH_3I + HI \longrightarrow CH_4 + I_2$$

What is the catalyst for the reaction? Which of the two steps is the slow one? Justify each of your answers.

*12.15 Collision theory What are the specific assumptions of the collision theory for the rates of chemical reactions? Show how these assumptions satisfactorily account for the observed behavior.

*12.16 Collision theory How does a rise in temperature affect the following aspects of collision theory: frequency of collision, fraction of collisions effective, activation energy, path of the reaction?

**12.17 Activated complex How is the activated complex for the reaction

$$H + HBr \longrightarrow H_2 + Br$$

related to that for the reaction

$$H_2 + Br \longrightarrow H + HBr?$$

Trace the electronic change that would have to occur in passing through the activated complex for each of these reactions.

***12.18 Activation energy With reference to the energy diagram of Fig. 12.6, tell how the activation energy is related to the difference in energy between reactants and products.

***12.19 **Energy diagram** Explain why the peaks in the two potential-energy curves shown in Fig. 12.7 do not come at the same value of the reaction coordinate.

*12.20 **Stepwise reactions** It is rare that a chemical reaction goes by a single step. What does this mean for the exponents in the rate law?

***12.21 **Stepwise reactions** For the two-step reaction mechanism

$$A + B \longrightarrow AB$$
$$AB + B \longrightarrow AB_2$$

what difference does it make in the rate law whether the first or the second step is the slow one? If it is observed that the rate of generation of AB_2 is proportional to $[B]^2$, how would you have to change the mechanism?

12.22 **Radioactive decay The spontaneous decomposition of radioactive nuclei is a first-order rate process. Uranium 238 disintegrates with the emission of an alpha particle with a half-life of 4.5×10^9 yr. (a) Write the rate law for this process. (b) Calculate the specific rate constant k. (c) Given 1 mol of uranium-238 nuclei at time $t = 0$, calculate how many nuclei will be left 1 billion years later. *Ans. (c) 5.16 × 10²³*

12.23 **Chain reactions What distinguishes a chain reaction from an ordinary reaction? How might one tell experimentally when one is dealing with a chain reaction?

12.24 **Chain reaction In nuclear reactors and also in A bombs, there is a chain reaction of the type

$$U^{234} + neutron \longrightarrow products + 3 \ neutrons$$

Suggest why in one case there is controlled release of energy; in the other, an explosion.

*12.25 **Smog** What is the essential difference between London smog and Los Angeles smog? Why is the latter called photochemical smog?

*12.26 **Smog** In the generation of photochemical smog, where is the chain-reaction part of the chemistry?

12.27 **Smog In the generation of photochemical smog, why is it necessary to have third-body particles M involved in the generation of ozone? What effect would a decrease in the concentration of M have on the concentration of ozone?

12.28 **Smog Explain qualitatively why photochemical smog generally shows the characteristic ozone maximum around noon, as illustrated in Fig. 12.8.

*12.29 **Internal-combustion engine** By referring to Fig. 12.9, tell what happens at each stroke of the four-stroke internal-combustion engine. At what stage does hydrocarbon oxidation occur?

*12.30 **Internal-combustion engine** What would be the ideal air-fuel ratio if there were no nitrogen in the air? *Ans. 3.7*

***12.31 **Internal-combustion engine** How are free radicals involved in the combustion reaction? Where do the free radicals originate? Where might free radicals come from in a diesel engine where ignition comes from high compression only?

12.32 **Pollution control Explain why it was relatively easy to reduce CO and hydrocarbon emissions from automobile engines but the nitrogen oxides pose a more difficult problem. Why was the NO_x problem aggravated when CO and hydrocarbon emissions were reduced?

***12.33 **Combustion** How is the reaction that occurs in a Bunsen burner related to the reaction that occurs in an internal-combustion engine? Suggest a mechanism by which the Bunsen flame front comes about.

Chapter 13

CHEMICAL EQUILIBRIUM

When reacting species undergo chemical reaction, conversion of reactants to products is often incomplete, no matter how long the reaction is allowed to continue. When a reaction starts, reactants are present at some definite concentration. As reaction proceeds, these concentrations decrease. Sooner or later, however, they level off and become constant. A state in which the concentrations no longer change with time becomes established. This state, which persists as long as the system is free of external perturbations, is known as the state of *chemical equilibrium*. In this chapter we take up a systematic description of the chemical-equilibrium state.

13.1
THE EQUILIBRIUM STATE

As an example of the attainment of equilibrium, we consider the reaction

$$H_2(g) + CO_2(g) \longrightarrow H_2O(g) + CO(g)$$

At the start of the experiment, H_2 and CO_2 (not necessarily in equal amounts) are mixed in a box. The concentration of H_2 and the concentration of CO_2 are measured as time passes. Results of the measurements are plotted in Fig. 13.1, where concentration is the vertical axis and time is the horizontal axis. The initial concentration of H_2 is some definite number, depending on the number of moles of H_2 and the volume of the box. As time goes on, the concentration of H_2 diminishes, at first quite rapidly, but then less rapidly. Eventually, it levels off and becomes constant. The concentration of CO_2 changes in similar fashion, though it may not start off at the same value as H_2. The initial concentrations of H_2O and CO are zero. As time goes on, H_2O and CO are produced. Their concentrations, which must be equal to each other, increase quite rapidly at first but then level off. At time t_e each of the concentrations $[H_2]$, $[CO_2]$, $[H_2O]$, and $[CO]$ becomes constant. Once this state has been established, it persists indefinitely and, if undisturbed, will last forever.

The constant state that characterized equilibrium vapor pressure (Sec. 7.2) was attributed to equality of opposing reactions, evaporation and condensation. Similarly, chemical equilibrium is due to equality of opposing reactions. H_2 and CO_2 molecules continue to react to form H_2O and CO as long as H_2 and CO_2 are present. The reaction does not stop at time t_e. However, as

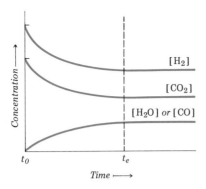

FIG. 13.1 Approach to equilibrium for the reaction $H_2(g) + CO_2(g) \longrightarrow H_2O(g) + CO(g)$.

soon as appreciable H_2O and CO have formed, they react with each other to regenerate H_2 and CO_2. After time t_e, the forward and reverse changes just balance each other out. The equality of opposing reactions is indicated by writing

$$H_2(g) + CO_2(g) \rightleftharpoons H_2O(g) + CO(g)$$

As can be seen, reversibility and chemical equilibrium are inextricably tied up with each other.

13.2 MASS ACTION

Every particular reaction has its own specific equilibrium state, characterized by a definite relation between the concentrations of the materials. To illustrate this relation, we consider the equilibrium involving $PCl_3(g)$, $Cl_2(g)$, and $PCl_5(g)$. PCl_5 is thermally unstable; so it decomposes to PCl_3 and Cl_2. When PCl_3 and Cl_2 are mixed, they react to form PCl_5. Let us consider a series of three experiments, all done at the same temperature, but differing in the starting concentration of materials:

Experiment 1 Suppose we inject 1 mol of PCl_5 into a 1-liter box at 546°K and wait for equilibrium. At equilibrium, when we examine the box, we find 0.764 M PCl_5, 0.236 M PCl_3, and 0.236 M Cl_2. Some of the PCl_5, which started out at 1 M, has decomposed into PCl_3 and Cl_2. Figure 13.2a summarizes what has happened.

Experiment 2 We proceed next to inject 1 mol of PCl_3 plus 1 mol of Cl_2 into another 1-liter box at 546°K and wait for equilibrium. We again find 0.764 M PCl_5, 0.236 M PCl_3, and 0.236 M Cl_2. Some of the PCl_3 and Cl_2, which started out at 1 M, have combined to form PCl_5. Figure 13.2b shows the change with time. We note that the final state we obtain here is exactly the same as that observed in experiment 1. It makes no difference whether we start with 1 mol of PCl_5 and let it partially decompose or with 1 mol of PCl_3

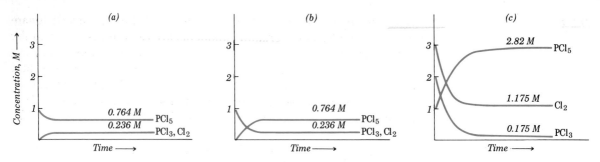

FIG. 13.2 Equilibrium $PCl_5(g) \rightleftharpoons PCl_3(g) + Cl_2(g)$ as approached from different amounts of starting materials.

and 1 mol of Cl_2 and let them partially react. Stoichiometrically, the two starting systems are equivalent; they lead to the same final equilibrium state.

Experiment 3   In our third experiment, we start with a random mixture, for example, 1 mol of PCl_5, 2 mol of PCl_3, and 3 mol of Cl_2. On examining the concentrations at equilibrium, we find 2.82 M PCl_5, 1.175 M Cl_2, and 0.175 M PCl_3. Figure 13.2c summarizes the results.

What do all three of these experiments have in common? If, at equilibrium, we form the ratio $[PCl_5]/[PCl_3][Cl_2]$, then in each case we get the same numerical value, 13.7.

$$\frac{[PCl_5]}{[PCl_3][Cl_2]} = \frac{0.764}{(0.236)(0.236)} = \frac{2.82}{(0.175)(1.175)} = 13.7$$

Also, if we had formed the inverse ratio, $[PCl_3][Cl_2]/[PCl_5]$, then we would have had 0.0730. Either of these numbers, 13.7 or 0.0730, can be used to characterize the system composed of PCl_5, PCl_3, and Cl_2. It is called the *equilibrium constant* of the system. Whenever PCl_5, PCl_3, and Cl_2 are present together in an equilibrium situation at 546°K, no matter what the individual concentrations are, the expression

$$\frac{[PCl_5]}{[PCl_3][Cl_2]} = 13.7$$

must be satisfied.

For the general case we can write the balanced equation

$$nA(g) + mB(g) + \cdots \rightleftharpoons pC(g) + qD(g) + \cdots$$

where n molecules of A plus m molecules of B, etc., react to form p molecules of C and q molecules of D. If the balanced equation is written this way, then the relationship that is constant at equilibrium is

$$\frac{[C]^p[D]^q \cdots}{[A]^n[B]^m \cdots}$$

In this ratio, which is called the *mass-action expression,*[*] the brackets designate concentrations in moles per liter, and exponents are the powers to which the concentrations must be raised. We follow the usual convention of putting materials from the right side of the chemical equation in the numerator and materials from the left side in the denominator. At equilibrium the mass-action expression is numerically equal to some number, dependent only on the temperature, which is the equilibrium constant K for the particular reaction:

$$\frac{[C]^p[D]^q \cdots}{[A]^n[B]^m \cdots} = K$$

This equilibrium condition is called the *law of chemical equilibrium.* It states that at equilibrium the concentrations of materials participating in the reaction must satisfy constancy of the mass-action expression.

EXAMPLE 1

You are given a box containing NH_3, N_2, and H_2 at equilibrium at 1000°K. Analysis of the contents shows that the concentration of NH_3 is 0.102 mol/liter, N_2 is 1.03 mol/liter, and H_2 is 1.62 mol/liter. What is the equilibrium condition for the reaction $N_2(g) + 3H_2(g) \rightleftharpoons 2NH_3(g)$ at 1000°K?

Solution

$$K = \frac{[NH_3]^2}{[N_2][H_2]^3}$$

$$K = \frac{(0.102)^2}{(1.03)(1.62)^3} = 0.00238$$

The law of chemical equilibrium is an experimental fact. It can, however, be justified by using the principles of chemical kinetics and by requiring, at equilibrium, equality of the rates of forward and reverse reactions. For example, in the equilibrium

$$A(g) + B(g) \rightleftharpoons C(g) + D(g)$$

the condition at equilibrium is $K = [C][D]/[A][B]$. This is true regardless of whether the reaction occurs in a single step or a series of steps.

---

[*]The term "mass action" derives from the original work of Cato Maximilian Guldberg and Peter Waage, Norwegian chemists, who in 1864 proposed that the reaction $A + B = C + D$ could be treated as follows: The *action force* between A and B is proportional to the *active mass* of A times the active mass of B. This is called the *law of mass action.* Similarly, the action force between C and D is proportional to the active masses of C and D. At equilibrium the action force between A and B equals the action force between C and D. Guldberg and Waage were not clear in what was meant by action force and active mass, but their work was influential in developing the ideas of chemical equilibrium.

1 Suppose the reaction proceeds through a single reversible step:

$$A(g) + B(g) \rightleftharpoons C(g) + D(g)$$

If k is the specific rate constant for the forward reaction, we can write

$$\text{Rate of forward reaction} = k[A][B]$$

Similarly, if k' is the specific rate constant of the reverse reaction, we can write

$$\text{Rate of reverse reaction} = k'[C][D]$$

At equilibrium, the rate of the forward reaction must be equal to the rate of the reverse reaction. Consequently,

$$k[A][B] = k'[C][D]$$

or

$$\frac{[C][D]}{[A][B]} = \frac{k}{k'} = \text{a constant}$$

This proves that the mass-action expression is equal to a constant.

2 Suppose the reaction above proceeds through an intermediate Q by steps such as

$$A(g) + A(g) \rightleftharpoons C(g) + Q(g) \tag{1}$$
$$Q(g) + B(g) \rightleftharpoons A(g) + D(g) \tag{2}$$

which add up to give

$$A(g) + B(g) \rightleftharpoons C(g) + D(g)$$

which is the same reaction as for a single reversible step. If we let k_1 and k_1' be rate constants for the forward and reverse directions of step (1) and k_2 and k_2' be rate constants for the forward and reverse directions of step (2), then at equilibrium we can set forward and reverse rates to be equal as follows:

$$k_1[A][A] = k_1'[C][Q] \quad \text{for the first step}$$
$$k_2[Q][B] = k_2'[A][D] \quad \text{for the second step}$$

These simultaneous equations can be combined by solving for and eliminating the chemical intermediate Q. The result is

$$\frac{k_1[A][A]}{k_1'[C]} = \frac{k_2'[A][D]}{k_2[B]}$$

which, on rearranging and simplifying, gives

$$\frac{[C][D]}{[A][B]} = \frac{k_1 k_2}{k_1' k_2'}$$

Again, the mass-action expression is shown to be equal to a constant.

The numbers observed for equilibrium constants range from large to small. If $K < 1$, the numerator of the mass-action expression is smaller than the denominator. This means that, at equilibrium, the concentration of at least one of the materials on the right side of the chemical equation is small. Thus, a small equilibrium constant implies that the reaction does not proceed far from left to right. For example, $K = 2.38 \times 10^{-3}$ for $N_2(g) + 3H_2(g) \rightleftharpoons 2NH_3(g)$ means that mixing N_2 and H_2 does not result in production of much NH_3 at equilibrium. If $K > 1$, the denominator of the mass-action expression is smaller than the numerator. This means that, at equilibrium, the concentration of at least one of the materials on the left of the chemical equation is small. Thus, a large equilibrium constant implies that the reaction proceeds from left to right essentially to completion. For example, $K = 9.39 \times 10^5$ for $2H_2(g) + S_2(g) \rightleftharpoons 2H_2S(g)$ means that mixing of H_2 and S_2 results in practically complete conversion to H_2S.

The numerical value of the equilibrium constant is determined by experiment. For example, for the equilibrium

$$H_2(g) + I_2(g) \rightleftharpoons 2HI(g)$$

the equilibrium condition is

$$K = \frac{[HI]^2}{[H_2][I_2]}$$

Measurement of a particular equilibrium state at 490°C gave the following results:

Concentration of $H_2 = 0.000862$ mol/liter

Concentration of $I_2 = 0.00263$ mol/liter

Concentration of $HI = 0.0102$ mol/liter

Since these are equilibrium concentrations, they must satisfy the equilibrium condition

$$K = \frac{[HI]^2}{[H_2][I_2]} = \frac{(0.0102)^2}{(0.000862)(0.00263)} = 45.9$$

Once the number 45.9 has been obtained, it can be used to describe any equilibrium system at 490°C containing H_2, I_2, and HI. If the mass-action expression is not equal to 45.9, the mixture is not at equilibrium, and changes occur until equilibrium is established.

In the above example the number of moles of gas on the two sides of the equation is equal. Concentration units in the numerator and denominator cancel; so the final equilibrium constant is dimensionless. In some cases, for example, $PCl_5(g) \rightleftharpoons PCl_3(g) + Cl_2(g)$, there are different numbers of moles

on the two sides of the chemical equation; so K, expressed in terms of molar concentrations, has units. Thus

$$K = \frac{[PCl_3][Cl_2]}{[PCl_5]} = 0.073 \text{ mol/liter}$$

We will generally not put any units for K since these can easily be figured out from the mass-action expression. If there may be ambiguity in what is meant by K, equilibrium constants using concentrations expressed in moles per liter are sometimes designated as K_C.

13.4
EQUILIBRIUM CALCULATIONS

The equilibrium constant for a reaction can be used to describe any system containing the chemical components of that reaction at equilibrium. Thus, 45.9 can be used to describe any system containing H_2, I_2, and HI at equilibrium at 490°C.

EXAMPLE 2

One mol of H_2 and 1 mol of I_2 are introduced into a 1-liter box at 490°C. What will be the equilibrium concentrations in the box?

Solution

Initially, there is no HI in the box. The system is not at equilibrium. In order to establish equilibrium, HI must come from the reaction

$$H_2(g) + I_2(g) \longrightarrow 2HI(g)$$

Let n equal the number of moles of H_2 that disappear in order to establish equilibrium. Every time 1 mol of H_2 disappears, 1 mol of I_2 also disappears. So, n also represents the number of moles of I_2 that disappear. According to the balanced equation, if 1 mol of H_2 and 1 mol of I_2 disappear, 2 mol of HI must be formed. If n mol of H_2 disappear, $2n$ mol of HI must appear. Since the volume of the box is 1 liter, the concentration of each component is the same as the number of moles. The situation is summarized as follows:

Initial concentration, mol/liter	Equilibrium concentration, mol/liter
$[H_2] = 1.000$	$[H_2] = 1.000 - n$
$[I_2] = 1.000$	$[I_2] = 1.000 - n$
$[HI] = 0$	$[HI] = 2n$

The concentrations at equilibrium must satisfy the condition

$$\frac{[HI]^2}{[H_2][I_2]} = 45.9$$

~~don't~~

$$n = 0.772^*$$

Substitution gives

$$\frac{(2n)^2}{(1.000 - n)(1.000 - n)} = 45.9$$

for which

$$n = 0.772*$$

Therefore, at equilibrium

$$[H_2] = 1.000 - n = 0.228 \text{ mol/liter}$$
$$[I_2] = 1.000 - n = 0.228 \text{ mol/liter}$$
$$[HI] = 2n = 1.544 \text{ mol/liter}$$

That these values represent equilibrium concentrations can be checked by calculating the value of the mass-action expression

$$\frac{[HI]^2}{[H_2][I_2]} = \frac{(1.544)^2}{(0.228)(0.228)} = 45.9$$

To emphasize the fact that it makes no difference from which side of the chemical equation equilibrium is approached, we consider what happens when only HI is placed in the box at 490°C. Since initially there is no hydrogen or iodine in the system, decomposition of HI must occur in order to establish equilibrium.

EXAMPLE 3

Two mol of HI are injected into a box of 1-liter volume at 490°C. What will be the equilibrium concentration of each species?

Solution

The equilibrium can again be written

$$H_2(g) + I_2(g) \rightleftharpoons 2HI(g)$$
$$K = \frac{[HI]^2}{[H_2][I_2]} = 45.9$$

Let x equal the number of moles of HI that must decompose in order to establish equilibrium. The back reaction shows that for each 2 mol of HI that disappear, 1 mol of H_2 and 1 mol of I_2 must be formed. If x mol of HI disappear, $x/2$ mol of I_2 and $x/2$ mol of H_2 appear. The concentrations are summarized as follows:

*The equation can be solved by taking the square root of both sides. For a more general case we can use the quadratic formula, Appendix A4.3. Of the two roots obtained here, one can be discarded as physically impossible. The root $n = 1.42$ would be thrown away because it corresponds to more than 100 percent reaction.

353

Initial concentration, mol/liter	Equilibrium concentration, mol/liter
$[HI] = 2.000$	$[HI] = 2.000 - x$
$[H_2] = 0$	$[H_2] = x/2$
$[I_2] = 0$	$[I_2] = x/2$

At equilibrium

$$\frac{[HI]^2}{[H_2][I_2]} = 45.9 = \frac{(2.000 - x)^2}{(x/2)(x/2)}$$

for which $x = 0.456$. Therefore, at equilibrium

$$[H_2] = \frac{x}{2} = 0.228 \text{ mol/liter}$$

$$[I_2] = \frac{x}{2} = 0.228 \text{ mol/liter}$$

$$[HI] = 2.000 - x = 1.544 \text{ mol/liter}$$

The above examples show that it makes no difference whether an equilibrium state is approached from the left or right side of the chemical equation. Change occurs to produce the material that is missing in sufficient concentration to establish equilibrium. Sometimes the initial system contains all components, in which case the change necessary to establish equilibrium may not be obvious.

EXAMPLE 4

One mol of H_2, 2 mol of I_2, and 3 mol of HI are injected into a 1-liter box. What will be the concentration of each species at equilibrium at 490°C?

Solution

The equilibrium is

$$H_2(g) + I_2 \rightleftharpoons 2HI(g)$$

Let x be the number of moles of H_2 that must be *used up* in order to establish equilibrium. (If we have guessed wrong, then x would turn out to be a negative number.) According to the stoichiometry, x is also the number of moles of I_2 that must be used up, and $2x$ is the number of moles of HI that must be formed. To reach equilibrium, the initial concentration of H_2 is reduced by the amount x, the concentration of I_2 is reduced by the amount x, and the concentration of HI is increased by $2x$.

Initial concentration, mol/liter	Equilibrium concentration, mol/liter
$[H_2] = 1.000$	$[H_2] = 1.000 - x$
$[I_2] = 2.000$	$[I_2] = 2.000 - x$
$[HI] = 3.000$	$[HI] = 3.000 + 2x$

At equilibrium

$$\frac{[HI]}{[H_2][I_2]} = 45.9 = \frac{(3.000 + 2x)^2}{(1.000 - x)(2.000 - x)}$$

~3,102

for which $x = 0.684$. Therefore, at equilibrium

$[H_2] = 1.000 - x = 0.316$ mol/liter

$[I_2] = 2.000 - x = 1.316$ mol/liter

$[HI] = 3.000 + 2x = 4.368$ mol/liter

13.5
HETEROGENEOUS EQUILIBRIUM

Heterogeneous equilibria are those which involve two or more phases. For example, the "water gas" equilibrium

$$C(s) + H_2O(g) \rightleftharpoons CO(g) + H_2(g)$$

which is used to make the important industrial fuel, $CO + H_2$, involves both gaseous and solid phases. The solid phase consists of pure C, and the gas phase consists of a mixture of H_2O, CO, and H_2. In mass-action expressions, the concentrations that are shown must correspond to the particular phase specified by the chemical equation. For example, for the above equilibrium, the equilibrium condition is

$$K = \frac{[CO(g)][H_2(g)]}{[C(s)][H_2O(g)]}$$

where $[C(s)]$ refers to the concentration of carbon *in the solid phase* and the other concentrations refer to the gas phase.

A simplification of the above equilibrium condition is possible because the concentration of carbon in the solid phase is not a variable. The concentration of CO, H_2O, or H_2 in the gas phase can be changed, e.g., by addition of that component at higher pressure in the same volume. For solid carbon this is not possible. If more solid carbon is added, the concentration is not changed—as the number of moles increases, the volume also increases. The number of *moles per unit volume of solid carbon* is always the same number, no matter how many moles of carbon are present.

In the general case, the concentration of a pure solid or a pure liquid can be combined with the original equilibrium constant to give a new equilibrium constant for which the mass-action expression does not include the pure condensed phase. Thus, for the equilibrium

$$C(s) + H_2O(g) \rightleftharpoons CO(g) + H_2(g)$$

$$K[C(s)] = \frac{[CO][H_2]}{[H_2O]}$$

where $[C(s)]$ is a constant. Therefore,

$$K[C(s)] = K'$$

$$K' = \frac{[CO][H_2]}{[H_2O]}$$

The last equation expresses the requirement that a system containing $CO(g)$, $H_2(g)$, $H_2O(g)$, and $C(s)$ is in equilibrium no matter how much $C(s)$ is present, provided that $[CO][H_2]/[H_2O]$ has the proper value. The simple rule is that for heterogeneous equilibria, pure solids and pure liquids are omitted from the mass-action expression.

13.6
EQUILIBRIUM CHANGES

When a system at equilibrium is disturbed, a chemical reaction occurs so as to reestablish equilibrium. As an example, we consider the equilibrium system consisting of H_2, I_2, and HI in a sealed box:

$$H_2(g) + I_2(g) \rightleftharpoons 2HI(g) \qquad K = \frac{[HI]^2}{[H_2][I_2]}$$

At 490°C, K is 45.9. The concentrations of HI, H_2, and I_2 do not vary until conditions are changed. Several kinds of change are possible: (1) H_2, I_2, or HI can be injected into the box; (2) H_2, I_2, or HI can be removed; (3) the volume of the box can be changed; (4) the temperature of the system can be changed; or (5) a catalyst can be added. How is the equilibrium state affected by each of these changes? Figure 13.3 summarizes the final results. Let us see how each of the results is arrived at.

1 Concentration of One Component Is Increased

Suppose the concentration of one component (for example, H_2) is changed by addition of that component. How would such a concentration increase affect the other components? The problem can be explored three ways:

a The equilibrium constant The equilibrium condition is of the form

$$\frac{[HI]^2}{[H_2][I_2]} = 45.9$$

FIG. 13.3 Effect on Equilibrium Concentrations in the System
$$H_2(g) + I_2(g) \rightleftharpoons 2HI(g)$$

	$[H_2]$	$[I_2]$	$[HI]$
(1) Add H_2	Increases	Decreases	Increases
(2) Remove H_2	Decreases	Increases	Decreases
(3) Decrease volume	Increases	Increases	Increases
(4) Raise temperature	Increases	Increases	Decreases
(5) Add catalyst	No effect	No effect	No effect

By increasing the concentration of H_2, the denominator is made bigger. The system would no longer be at equilibrium. To reestablish equilibrium, two things could happen: There could be a decrease in the concentration of I_2, so that the denominator would be restored to its original value. Or there could be an increase in the concentration of HI; the numerator would increase to compensate for the increased denominator. Since I atoms must exist either as I_2 or as HI molecules, a decrease of I_2 and an increase of HI occur simultaneously.

why →

   b Le Châtelier's principle predicts the effect of reduced volume as follows: 7.14), any equilibrium system subjected to a stress tends to change so as to relieve the stress. For a system in chemical equilibrium, changing the concentration of one of the components constitutes a stress. If in the present case H_2 is added to the box, the equilibrium system

$$H_2(g) + I_2(g) \rightleftharpoons 2HI(g)$$

can absorb the stress if some H_2 molecules combine with I_2 to form HI. This means that the concentration of HI would increase and the concentration of I_2 would decrease. Le Châtelier's principle thus leads to the same prediction as did use of the equilibrium constant in part (*a*) above.

   c Kinetics   The effect of added H_2 can be predicted from a consideration of reaction rates. The argument here is relatively simple if the reaction proceeds by one step. In the equilibrium state, collisions between H_2 and I_2 molecules form HI, and, simultaneously, collisions between HI molecules form H_2 and I_2. These two rates are equal. By addition of H_2 to the box, the chance for collision between H_2 and I_2 is increased and the rate of HI formation is increased. However, there is no instantaneous effect on the decomposition rate of HI. For a time, HI will be forming faster than it is decomposing. Its concentration will increase. Eventually, HI will increase to the point where there are more collisions between HI molecules; so the reverse reaction, the decomposition of HI, will begin to speed up. It will continue to speed up until it equals the increased rate of HI formation. Equilibrium will be reestablished with a net increase in HI and H_2 and a decrease in I_2.

2   Concentration of One Component Is Decreased

Suppose the concentration of one of the components is decreased. For example, suppose some H_2 is removed from the box.

   a   *The equilibrium condition*

$$\frac{[HI]^2}{[H_2][I_2]} = 45.9$$

predicts that a decrease in $[H_2]$ would be compensated for by a decrease in $[HI]$ and an increase in $[I_2]$.

   b Le Châtelier's principle predicts that the system would adjust to relieve the stress caused by the removal of H_2. Some HI would decompose to form H_2 to replace some of that removed. The effect would be to reduce the concentration of HI and increase the concentration of I_2.

c Kinetics predict that the removal of H_2 from the container would reduce the rate at which H_2 and I_2 combine to form HI. This would mean that, instantaneously, HI would be forming from H_2 and I_2 more slowly than it would be decomposing to H_2 and I_2. The result would be a net decrease in HI concentration and a net increase in I_2 concentration.

3 Volume of the System Is Decreased

Suppose the volume of the box is decreased. In cases 1 and 2 the volume of the box was kept constant. If it is decreased, the *concentrations* of all species would be increased. However, to determine how the number of moles changes, a detailed analysis is required.

a The equilibrium condition

$$\frac{[HI]^2}{[H_2][I_2]} = 45.9$$

is expressed in terms of concentration. For each component, the concentration is equal to the number of moles n divided by the volume V. By substitution of n/V for concentration, the equilibrium condition can be rewritten:

$$\frac{[HI]^2}{[H_2][I_2]} = \frac{(n_{HI}/V)^2}{(n_{H_2}/V)(n_{I_2}/V)} = \frac{n_{HI}^2}{(n_{H_2})(n_{I_2})} = 45.9$$

As can be seen, V cancels out. No matter what the volume of the box, the moles of HI squared divided by the moles of H_2 times the moles of I_2 must equal 45.9. Changing the volume of the box *in this case* does not change the relative number of moles of each species. The reason is that for

$$H_2(g) + I_2(g) \rightleftharpoons 2HI(g)$$

the number of gas molecules is the same on both sides of the chemical equation. If this were not true, volume would not cancel out of the mass-action expression. An example of a case that is volume-dependent is the equilibrium between nitrogen, hydrogen, and ammonia:

$$N_2(g) + 3H_2(g) \rightleftharpoons 2NH_3(g)$$

The equilibrium expression is

$$K = \frac{[NH_3]^2}{[N_2][H_2]^3}$$

Substituting n/V for concentration gives

$$K = \frac{(n_{NH_3}/V)^2}{(n_{N_2}/V)(n_{H_2}/V)^3} = \frac{n_{NH_3}^2}{(n_{N_2})(n_{H_2})^3} V^2$$

The volume does not cancel out, and a change in volume must be compensated for by a change in the molar ratio. Specifically, if V is decreased, $n_{NH_3}^2/n_{N_2}n_{H_2}^3$ must increase in order to maintain K constant. The number of moles of ammonia would have to increase at the expense of the moles of nitrogen and hydrogen. However, the *concentrations* of ammonia, nitrogen, and hydrogen (as distinct from *number of moles*) all increase. In the commercial production of ammonia from nitrogen and hydrogen, the reaction is carried out in as small a volume as possible in order to maximize conversion to ammonia.

b Le Châtelier's principle predicts the effect of reduced volume as follows: When the volume of the box is reduced, a stress is applied in that the molecules are crowded closer together. The stress can be relieved if the molecules could be reduced in number. In the case of

$$H_2(g) + I_2(g) \rightleftharpoons 2HI(g)$$

there is no device by which this can be accomplished since the number of gas molecules is the same on both sides of the equation. Neither the forward nor reverse reaction can absorb the stress. The number of moles of H_2, I_2, and HI stays constant. Of course, since the volume is diminished, the concentration of each component is increased. In the NH_3 equilibrium, the situation would be different. When one N_2 reacts with three H_2, only two NH_3 are formed. A decrease of the volume of the box could be compensated for by forming fewer molecules, i.e., by favoring the formation of NH_3. It is a general principle that when there is a change in the number of gas molecules, decrease in volume favors the reaction direction that produces fewer molecules.

c Kinetics predict the effect of decrease in volume by considering the effects on the rates of the forward and reverse reactions. In the system containing H_2, I_2, and HI, decrease in volume forces H_2 and I_2 molecules closer together; so they collide more frequently. There is consequently an increase in the rate of the forward reaction. At the same time, HI molecules are also brought closer together; so they too collide more frequently. The back reaction, therefore, is also increased. If the number of gas molecules is the same on both sides of the equation, the rate of the forward reaction is increased just as much as that of the back reaction. If the number of gas molecules changes on reaction, the situation is more complicated. For example, in the case

$$N_2(g) + 3H_2(g) \rightleftharpoons 2NH_3(g)$$

a decrease in volume increases the forward rate to a greater extent than the back rate. The net effect is to increase the number of NH_3 molecules present at equilibrium.

4   Temperature of the System Is Changed

Suppose the temperature of the system is changed.

a If temperature is changed, K generally changes value. For reactions which are endothermic, a rise in temperature causes K to increase; for reactions which are exothermic, a rise in temperature causes K to decrease (how this comes about is discussed in Sec. 13.9). The reaction

$$H_2(g) + I_2(g) \rightleftharpoons 2HI(g) + 13 \text{ kJ}$$

is exothermic as written. K decreases when T is increased. With increase of T, the concentration of HI diminishes, while the concentrations of H_2 and I_2

increase. This is another way of saying that HI is less stable at higher temperatures.

b Le Châtelier's principle predicts that a rise in temperature favors the change that uses up heat. When 1 mol of H_2 and 1 mol of I_2 disappear, 2 mol of HI and 13 kJ of heat are liberated. The reverse process absorbs heat. If the temperature is increased, the system tries to relieve the stress by absorbing the added heat. Since the back reaction uses heat, it is the one that is favored. Favoring the back reaction causes a net decrease in the concentration of HI and a net increase in the concentration of I_2 and H_2.

c Kinetics It is a general principle that the rate of reaction is increased by an increase in temperature (Sec. 12.3). Furthermore, for a given equilibrium the endothermic reaction (the one with larger activation energy) is increased relatively more than the exothermic reaction. For

$$H_2(g) + I_2(g) \rightleftharpoons 2HI(g) + 13 \text{ kJ}$$

the rate of decomposition (endothermic) is increased more than is the rate of HI formation (exothermic). The result, on rise of temperature, is a net decrease in the concentration of HI.

5 Catalyst Is Added to the System

What effect does a catalyst have on an equilibrium system?

a The equilibrium constant is concerned only with the materials shown in the net equation. A catalyst does not appear in the net equation or in the equilibrium-constant expression. Hence, insertion of a catalyst into an equilibrium system should have no effect on equilibrium concentrations.

b Le Châtelier's principle has nothing to say about the presence of a catalyst.

c Kinetics give the best argument that a catalyst cannot affect the composition of an equilibrium system. According to reaction-rate theory, the rate of chemical reaction depends on how fast particles get over the potential-energy barrier between the initial and final states. Figure 13.4 shows the potential-energy barrier for the reaction

$$H_2(g) + I_2(g) \rightleftharpoons 2HI(g)$$

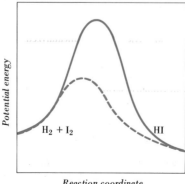

FIG. 13.4 Potential-energy diagram showing effect of catalysis. (Dashed line applies to catalyzed path.)

Potential energy

$H_2 + I_2$ HI

Reaction coordinate

The dashed line represents the path in the presence of a catalyst. As can be seen, the catalyst reduces the height of the barrier, and so it speeds up the rate. However, if the barrier is lowered for the forward change, it is likewise lowered for the reverse change. The increase in forward and reverse reaction rates must be the same; hence, the equilibrium concentrations are unchanged.*

13.7
CHEMICAL EQUILIBRIUM AND FREE ENERGY

In the above picture of chemical equilibrium, emphasis has been on a molecular picture with equality of two opposing reactions. There is another powerful way of looking at equilibrium, which makes use of the concepts of free energy and entropy developed in Chap. 10. As was indicated in Sec. 10.7, for spontaneous change to occur (at constant temperature and pressure), the free energy of the system must decrease. Since a system at equilibrium has no tendency to change, it must represent a state where the free energy has reached a minimum.

Let us see how this comes about. For concreteness, we consider the reaction

$$Br_2(l) + Cl_2(g) \longrightarrow 2BrCl(g)$$

At 25°C and 1 atm pressure, ΔH for the reaction is $+29$ kJ/mol; ΔS is $+105$ J mol^{-1} deg^{-1}. Putting these two together we get

$$\Delta G = \Delta H - T\Delta S$$
$$= 29 \text{ kJ} - (298 \text{ deg})(0.105 \text{ kJ/deg})$$
$$= -2 \text{ kJ}$$

---

*Quantitatively, we can see this last point by considering the Arrhenius equation (Sec. 12.3) for the specific rate constant:

$$k = Ae^{-E/RT}$$

where E is the activation energy. For the case shown in Fig. 13.4, the activation energy of the forward process is 167 kJ/mol whereas the activation energy of the reverse process is 180 kJ/mol. Suppose the catalyst acts to lower the activation barrier by 80 kJ/mol. For the forward process we can write

$$k_f = A_f e^{-167/RT} \qquad \text{uncatalyzed}$$
$$k'_f = A_f e^{-(167-80)/RT} \qquad \text{catalyzed}$$
$$= A_f(e^{-167/RT})(e^{80/RT})$$

For the reverse process we can similarly write

$$k_r = A_r e^{-180/RT} \qquad \text{uncatalyzed}$$
$$k'_r = A_r e^{-(180-80)/RT} \qquad \text{catalyzed}$$
$$= A_r(e^{-180/RT})(e^{80/RT})$$

It should be noted that the catalyst effect of dropping the activation energy by 80 kJ/mol is to multiply *each rate* by the same factor $e^{80/RT}$. Clearly, the equilibrium constant, which is the ratio of k_f to k_r, is left unchanged.

Because ΔG is negative, the free energy would decrease in the course of converting $Br_2(l)$ plus $Cl_2(g)$ into $2BrCl(g)$. The reaction should occur spontaneously in the direction written. But how far does it go? Does it go to completion? All that we have said so far is that the state "2 mol of $BrCl(g)$" is lower in free energy than the state "1 mol $Br_2(l)$ plus 1 mol $Cl_2(g)$." We also need to look at the free energy when only part of the Br_2 and Cl_2 gets converted. It turns out that a state short of complete reaction has a lower free energy than if total conversion to $2BrCl(g)$ occurs.

To see how this comes about, we look separately at how enthalpy and entropy change during the conversion. The statement that

$$Br_2(l) + Cl_2(g) \longrightarrow 2BrCl(g) \qquad \Delta H = +29 \text{ kJ}$$

means that the enthalpy of the system progressively increases as the Br_2 and Cl_2 change into BrCl. Figure 13.5a shows this schematically as a function of increasing fraction of conversion from reactants to products. The starting enthalpy of the elemental reactants has been arbitrarily set at zero. The point to note is that the enthalpy of the system rises linearly with fraction of conversion.

What happens to the entropy during this conversion? The statement that

$$Br_2(l) + Cl_2(g) \longrightarrow 2BrCl(g) \qquad \Delta S = +105 \text{ J/deg}$$

tells us that the entropy increases by 105 J/deg when there is complete conversion of Br_2 and Cl_2 into 2BrCl. But what about the intermediate degrees of conversion? Unlike the case for enthalpy, where the change is linear with increasing fraction of conversion, the entropy shows a more complicated dependence. The entropy of the intermediate states will be *enhanced* because mixtures of unlike species have many more configurations than do pure substances. Figure 13.5b shows how the TS product changes on progressive conversion of $Br_2(l)$ and $Cl_2(g)$ into $BrCl(g)$. (The beginning and end points of the curve are fixed at 112 and 143 kJ/mol, respectively. Intervening points are calculated by adding up the entropies of all the components in the mixture, making due allowance for the number of moles of each present and the volume of the system as it expands to keep constant pressure.*)

As can be seen, the TS product rises along a curved line as the fraction of conversion changes from zero to one. Our prime concern, however, is the free energy G, which is equal to $H - TS$. Once the TS curve is obtained, it can be subtracted from the enthalpy curve. The result is Fig. 13.5c. It shows that the free energy decreases progressively until it reaches a minimum at 64 percent conversion. This is the state of chemical equilibrium. One hundred percent conversion to BrCl would give free energy somewhat lower than the ini-

---

*The starting entropy is 152 J mol^{-1} deg^{-1} for $Br_2(l)$ plus 223 J mol^{-1} deg^{-1} for $Cl_2(g)$, or a total of 0.375 kJ. At 298°K, the corresponding TS product would be 112 kJ/mol. When conversion is complete, we have the entropy of 2 mol of $BrCl(g)$, that is, (2)(240) J mol^{-1} deg^{-1}. This gives a TS product of 143 kJ/mol. For intervening points we need to know that when a gas expands from V_1 to V_2, the entropy change is given by $\Delta S = R \ln (V_2/V_1)$. For ideal gases, V is inversely proportional to P, and so we can write $\Delta S = R \ln (P_1/P_2)$.

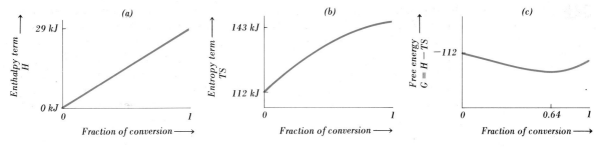

FIG. 13.5 Change of thermodynamic parameters H, TS, and G during course of reaction $Br_2\,(l) + Cl_2(g) \longrightarrow 2BrCl(g)$. (Scale is somewhat distorted.)

tial state $(Br_2 + Cl_2)$, but the lowest free energy occurs for an intermediate system where some unchanged Br_2 and Cl_2 are left along with the BrCl.

At the minimum in the free-energy curve, the curve is flat. Small additional conversion of $Br_2 + Cl_2$ into BrCl or a small amount of decomposition of BrCl into $Br_2 + Cl_2$ does not materially affect the free energy. Therefore, ΔG is equal to zero at this minimum. We shall return to this point in the following section.

13.8
CHEMICAL ACTIVITIES AND CHEMICAL EQUILIBRIUM

Chemical reactions involve mixtures of substances. Since the substances are not in pure states, we need to have a way to describe their tendency to react under nonstandard conditions. This relative tendency is often expressed in terms of *chemical activity,* the ratio of the effective concentration of a species to its concentration in a standard reference state. The *effective concentration* takes care of deviation from ideal behavior as well as dilution. For gases, concentration can be expressed as moles per liter or as atmospheres of pressure. The latter is the more usual procedure when making thermodynamic calculations. In terms of pressure, activity of a gas is set as the real pressure of the species divided by its ideal standard-state pressure. It might seem that dividing by the standard-state pressure is unnecessary since the numerical value in the standard state is unity. However, we must not forget units. By dividing as indicated, we have activity as a *dimensionless* quantity embodying a comparison with a standard state. For liquid solutions the activity of a solute is generally given as the actual molal concentration of the solute divided by its molality (i.e., 1 m) in the standard state.

To see better what is meant by activity, let us consider a mixture of gaseous hydrogen and gaseous iodine. The total free energy of the mixture G_{total} can be written as the sum of a contribution from the hydrogen plus a contribution from the iodine. Letting n be the number of moles of each component and G the free-energy contribution per mole ascribable to each component in this particular mixture, we get

$$G_{total} = n_{H_2}G_{H_2} + n_{I_2}G_{I_2}$$

How does G_{H_2} in the mixture differ from what it would be in the standard state of H_2? If we represent the molar free-energy contribution in the standard state by $G_{H_2}^\circ$, then it turns out that

$$G_{H_2} = G_{H_2}^\circ + RT \ln a_{H_2}$$

Thus, a_{H_2}, the activity of the hydrogen, gives a measure of the departure in free-energy contribution from the standard-state value. In the standard state of hydrogen, $a_{H_2} = 1$, and hence $G_{H_2} = G_{H_2}^\circ$. For activities less than unity (that is, H_2 pressure less than 1 atm), G_{H_2} will be less than $G_{H_2}^\circ$; for activities greater than unity (that is, H_2 pressure greater than 1 atm), G_{H_2} will be greater than $G_{H_2}^\circ$.

A mixture of only H_2 and I_2 is not at equilibrium since the total free energy can decrease by conversion of some H_2 and I_2 into HI. Suppose we ask what the free-energy change would be for the process

$$H_2(g) + I_2(g) \longrightarrow 2HI(g)$$

in a randomly selected mixture of H_2, I_2, and HI. ΔG equals the free energy of 2 mol of HI in the mixture minus the free energies of 1 mol each of H_2 and I_2 in the mixture:

$$\Delta G^\circ =$$
$$\Delta G = 2G_{HI} - G_{H_2} - G_{I_2}$$

Substituting

$$G_{HI} = G_{HI}^\circ + RT \ln a_{HI}$$
$$G_{H_2} = G_{H_2}^\circ + RT \ln a_{H_2}$$
$$G_{I_2} = G_{I_2}^\circ + RT \ln a_{I_2}$$

we get

$$\begin{aligned}
\Delta G &= 2G_{HI} - G_{H_2} - G_{I_2}\\
&= 2(G_{HI}^\circ + RT \ln a_{HI}) - (G_{H_2}^\circ + RT \ln a_{H_2}) - (G_{I_2}^\circ + RT \ln a_{I_2})\\
&= 2G_{HI}^\circ - G_{H_2}^\circ - G_{I_2}^\circ + RT(2 \ln a_{HI} - \ln a_{H_2} - \ln a_{I_2})
\end{aligned}$$

In place of the first three terms, $2G_{HI}^\circ - G_{H_2}^\circ - G_{I_2}^\circ$, we can write ΔG°, which stands for the molar free-energy change *if every component of the mixture were in its standard state*. Noting also that $2 \ln a_{HI} = \ln a_{HI}^2$, we can write

$$\Delta G = \Delta G^\circ + RT \ln \frac{a_{HI}^2}{a_{H_2} a_{I_2}}$$

At equilibrium, the total free energy is at a minimum—that is, the free energy curve (see Fig. 13.5c) is flat, and $\Delta G = 0$. Therefore, for equilibrium we can write

$$0 = \Delta G^\circ + RT \ln \frac{a_{HI}^2}{a_{H_2} a_{I_2}} \qquad \text{or} \qquad \Delta G^\circ = -RT \ln \frac{a_{HI}^2}{a_{H_2} a_{I_2}}$$

The activity ratio has the same form as the mass-action expression for the

365

Section 13.8
Chemical
Activities
and Chemical
Equilibrium

reaction $H_2 + I_2 \rightleftharpoons 2HI$. Hence, we can replace the activity ratio by the equilibrium constant

$$\frac{a_{HI}^2}{a_{H_2} a_{I_2}} = K$$

This leads to

$$\boxed{\Delta G^\circ = -RT \ln K}$$

or, replacing the natural logarithm by 2.303 times the base-10 logarithm,

$$\Delta G^\circ = -2.303 \, RT \log K$$

The final relation is noteworthy. It states that the final *equilibrium state*, characterized by K, is related to the free-energy change from reactants to products *in the standard state*, ΔG°. Thus, we learn about the *equilibrium* state by considering free-energy properties in the *non*equilibrium standard state.

For a chemical equation in general, ΔG° may be zero, positive, or negative. If $\Delta G^\circ = 0$, then $K = 1$; this means that the final equilibrium state will be one in which reactants and products contribute equally to the activity ratio. If $\Delta G^\circ > 0$, then $K < 1$; the products contribute less than reactants to the activity ratio in the equilibrium state. Chemical reaction would not proceed far from left to right. If $\Delta G^\circ < 0$, then $K > 1$; reaction to the right is favored, and product activities predominate in the equilibrium state.

The numerical value for K depends on the choice of standard states for the reactants and products. When the standard states are the same as those used by the National Bureau of Standards, gases turn out to be measured in atmospheres, and the equilibrium constant deduced from $\Delta G^\circ = -RT \ln K$ is K_p expressed in atmospheres. In general K_p will not have the same value as K_c, where concentrations are expressed in moles per liter. However, if the number of gas moles is the same on both sides of an equation, then K_c, expressed in concentrations, is numerically equal to the K_p, expressed in pressure.

EXAMPLE 5

The standard free energy of formation of HI from H_2 and I_2 at 490°C is -12.1 kJ/mol of HI. Calculate the equilibrium constant for $H_2 + I_2 \rightleftharpoons 2HI$. Recall that the free energy of formation of any element in its standard state is zero.

Solution

$\Delta G^\circ = (2 \text{ mol of HI})(-12.1 \text{ kJ/mol}) = -24.2 \text{ kJ}$

$\Delta G^\circ = -2.303 \, RT \log K$

$\log K = \dfrac{-\Delta G^\circ}{2.303 \, RT} = \dfrac{-(-24,200 \text{ J/mol})}{(2.303)(8.31 \text{ J mol}^{-1} \text{ deg}^{-1})(763 \text{ deg})}$

$\log K = 1.66$

$K = 46$

EXAMPLE 6

For the reaction $2NO(g) + O_2(g) \rightleftharpoons 2NO_2(g)$ at 298°K, the equilibrium constant K_p equals 1.6×10^{12}. Given that the standard free energy of formation of $NO(g)$ is 86.6 kJ/mol at 298°K, calculate the standard free energy of formation of $NO_2(g)$ at 298°K.

Solution

For $2NO(g) + O_2(g) \rightleftharpoons 2NO_2(g)$

$$\Delta G° = -RT \ln K_p$$
$$= -(8.31 \text{ J mol}^{-1} \text{ deg}^{-1})(298 \text{ deg})(2.303) \log (1.6 \times 10^{12})$$
$$= -69.6 \text{ kJ/mole of reaction*}$$

$$\Delta G° = 2\Delta G°_{NO_2} - 2\Delta G°_{NO} - \Delta G°_{O_2}$$

The $\Delta G°$ of formation of an element in its standard state is taken to be zero; so $\Delta G°_{O_2} = 0$.

$$\Delta G°_{NO_2} = \tfrac{1}{2}(\Delta G° + 2\Delta G°_{NO})$$
$$= \tfrac{1}{2}(-69.6 + 2 \times 86.6)$$
$$= 51.8 \text{ kJ/mol}$$

13.9
TEMPERATURE AND CHEMICAL EQUILIBRIUM

In Sec. 13.6, it was observed that the equilibrium constant K generally changes value when the temperature is changed. How does this come about? How is it related to ΔG, ΔH, and ΔS?

We saw in the preceding section that standard free-energy change for a reaction is directly related to the equilibrium constant for the reaction:

$$\Delta G° = -RT \ln K$$

Since the definition of free energy gives us

$$\Delta G° = \Delta H° - T\Delta S°$$

we can set these equal to each other and write

$$-RT \ln K = \Delta H° - T\Delta S°$$

from which

$$\ln K = \frac{-\Delta H°}{RT} + \frac{\Delta S°}{R}$$

This equation tells us that the equilibrium constant K depends on the heat of reaction $\Delta H°$, the entropy of reaction $\Delta S°$, and the reciprocal of the absolute temperature. Figure 13.6 shows graphically what the relation looks like when

*The term "mole of reaction" refers to the chemical equation read in moles.

366

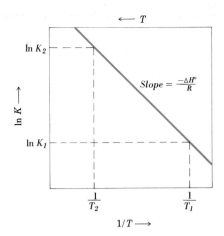

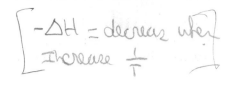

$\left[\begin{array}{l} -\Delta H = \text{decreas wher} \\ \text{Increase } \frac{1}{T} \end{array} \right]$

FIG. 13.6 Graph showing relation between equilibrium constant and reciprocal temperature.

$\Delta H°$ and $\Delta S°$ do not change with temperature. A plot of $\ln K$ vs. $1/T$ gives a straight line, the slope of which is $-\Delta H°/R$. If $\Delta H°$ is positive, the slope of the line is negative; $\ln K$ decreases as one goes to higher values of $1/T$ (i.e., lower temperature). If $\Delta H°$ is negative, the slope of the line would be positive; $\ln K$ would increase at higher values of $1/T$. If T_1 and T_2 represent two different temperatures, K_1 and K_2 will be given as follows:

$$\ln K_1 = \frac{-\Delta H°}{RT_1} + \frac{\Delta S°}{R}$$

$$\ln K_2 = \frac{-\Delta H°}{RT_2} + \frac{\Delta S°}{R}$$

Subtracting the second equation from the first, we get

$$\ln K_1 - \ln K_2 = \frac{-\Delta H°}{RT_1} + \frac{\Delta H°}{RT_2}$$

or

$$\ln \frac{K_2}{K_1} = \frac{\Delta H°}{R}\left(\frac{1}{T_1} - \frac{1}{T_2}\right)$$

This relation, called the *van't Hoff equation*, not only provides a method for calculating K at one temperature when $\Delta H°$ and K at another temperature are known, but also allows determination of $\Delta H°$ from measurements of the equilibrium constant at different temperatures.

When computing with logarithms to the base 10, the above equation becomes

$$\log \frac{K_2}{K_1} = \frac{\Delta H°}{2.303R}\left(\frac{1}{T_1} - \frac{1}{T_2}\right) = \frac{\Delta H°}{19.15}\left(\frac{1}{T_1} - \frac{1}{T_2}\right)$$

where $\Delta H°$ is in joules for the reaction as written in moles. The following examples show how this equation can be applied.

EXAMPLE 7

For the reaction

$$2NO(g) + O_2(g) \rightleftharpoons 2NO_2(g) \qquad \Delta H° = -113 \text{ kJ}$$

the equilibrium constant K_p is 1.6×10^{12} at $298°K$. Calculate K_p for the reaction at $373°K$, assuming $\Delta H°$ stays constant.

Solution

$$\log \frac{K_2}{K_1} = \frac{\Delta H°}{19.15} \left(\frac{1}{T_1} - \frac{1}{T_2} \right)$$

Let $K_1 = 1.6 \times 10^{12}$ at $T_1 = 298°K$; then

$$\log \frac{K_2}{1.6 \times 10^{12}} = \frac{-113,000}{19.15} \left(\frac{1}{298} - \frac{1}{373} \right)$$

$$K_2 = 1.7 \times 10^8 \text{ at } T_2 = 373°K.$$

The decrease of K_p from 1.6×10^{12} to 1.7×10^8 is consistent with the fact that the enthalpy change for the reaction is negative, i.e., the reaction is exothermic. Hence, a rise in temperature makes the reaction less favored.

EXAMPLE 8

The molecule NO_2 can dimerize to form N_2O_4. Calculate $\Delta H°$ for the reaction

$$2NO_2(g) \rightleftharpoons N_2O_4(g)$$

given that $K_p = 8.85$ at $298°K$ and $K_p = 0.0792$ at $373°K$.

Solution

$$\log \frac{K_2}{K_1} = \frac{\Delta H°}{19.15} \left(\frac{1}{T_1} - \frac{1}{T_2} \right)$$

$$\log \frac{0.0792}{8.85} = \frac{\Delta H°}{19.15} \left(\frac{1}{298} - \frac{1}{373} \right)$$

$$\Delta H° = -58,200 \text{ J/mol}$$

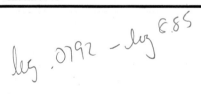

$$\log .0792 - \log 8.85$$

$$39.2$$

368

Important Concepts

equilibrium state

mass action

equilibrium constant

mass-action expression

law of chemical equilibrium

equilibrium condition

relation of rates to equilibrium

equilibrium calculations

heterogeneous equilibrium

disturbance of a system at equilibrium

relation of chemical equilibrium to free-energy change

chemical activity

dependence of equilibrium constant on temperature

van't Hoff equation

Exercises

*13.1 **Approach to equilibrium** In Fig. 13.1, tell what is happening to each of the concentrations in the time interval between $t = 0$ and the time equilibrium is reached. Explain why only one curve suffices to describe both $[H_2O]$ and $[CO]$. How would the graph be changed if the experiment were done at a higher temperature?

***13.2 **Approach to equilibrium** Suppose you wanted to reproduce the same final equilibrium state shown in Fig. 13.1 but starting from the other side of the chemical equation. How would you proceed, and what would the graph look like?

13.3 **Approach to equilibrium How might you prove that a particular equilibrium state is really the equilibrium state for a given chemical reaction?

*13.4 **Mass action** Write the mass-action expression for each of the following:

a $2NH_3(g) \rightleftharpoons N_2(g) + 3H_2(g)$

b $N_2(g) + 3H_2(g) \rightleftharpoons 2NH_3(g)$

c $NH_3(g) \rightleftharpoons \frac{1}{2}N_2(g) + \frac{3}{2}H_2(g)$

d $\frac{1}{2}N_2(g) + \frac{3}{2}H_2(g) \rightleftharpoons NH_3(g)$

If the value of K is 4.20×10^2 for (a), what will it be for (b), (c), and (d)? Assume the temperature stays the same. *Ans. (d) 4.88 × 10⁻²*

13.5 **Equilibrium Why is the final equilibrium state shown in Fig. 13.2c different from that shown in (a) and (b)? What does the equilibrium state in Fig. 13.2c have in common with that shown in Fig. 13.2a and 13.2b?

*13.6 **Law of chemical equilibrium** State the law of chemical equilibrium, with specific reference to the reaction

$$H_2(g) + CO_2(g) \rightleftharpoons H_2O(g) + CO(g)$$

*13.7 **Rates and equilibrium** Assume that the reaction $CO(g) + Cl_2(g) \rightleftharpoons COCl_2(g)$ goes by a single-step mechanism and is reversible. What must be true about the opposing rates at equilibrium? How does this lead to the equilibrium condition?

*13.8 **Equilibrium constant** You are given an 18.0-liter box at 750°C which contains, at equilibrium, 1.68 mol of H_2S, 1.37 mol of H_2, and 2.88×10^{-5} mol of S_2. Calculate K for $H_2S(g) \rightleftharpoons H_2(g) + \frac{1}{2}S_2(g)$. *Ans. 1.03 × 10⁻³*

13.9 **Equilibrium constant At high temperatures diatomic iodine I_2 is unstable to decomposition into atomic iodine. In a typical experiment at 1000°K, 2.74% of the I_2 decomposes when 0.00305 mol of I_2 is placed in a volume of 0.250 liter. Calculate K for $I_2(g) \rightleftharpoons 2I(g)$.

13.10 **Equilibrium constant Gaseous nitrogen tetroxide, $N_2O_4(g)$, is unstable with respect to decomposition into gaseous nitrogen dioxide, $NO_2(g)$, even at rather low temperatures. In a given experiment at 35°C, when 0.00628 mol of $N_2O_4(g)$ is injected into a 1-liter flask, the total pressure developed is 0.238 atm. Calculate the percent decomposition of the N_2O_4 and also the equilibrium constant in terms of concentration for $N_2O_4(g) \rightleftharpoons 2NO_2(g)$.

Ans. K = 0.0126

*13.11 Equilibrium constant A mixture consisting only of 1.000 mol of H_2 and 2.000 mol of CO_2 is placed in a 10.0-liter volume at 2000°K. When equilibrium is established, it is found that only 14.5% of the hydrogen is left. Calculate the equilibrium constant for the reaction $H_2(g) + CO_2(g) \rightleftharpoons H_2O(g) + CO(g)$ at 2000°K.

*13.12 Equilibrium calculation At 1100°K, the equilibrium constant for $H_2(g) + CO_2(g) \rightleftharpoons H_2O(g) + CO(g)$ is equal to 1.00. Starting with 1 mol of H_2 and 1 mol of CO_2 in a 1-liter box at 1100°K, what will be the concentrations at equilibrium?

**13.13 Equilibrium calculation At 1000°K the equilibrium constant K_c for the reaction $CO(g) + Cl_2(g) \rightleftharpoons COCl_2(g)$ is equal to 3.04. If 1 mol of CO and 1 mol of Cl_2 are introduced into a 1-liter box at 1000°K, what will be the final concentrations at equilibrium? *Ans. $[COCl_2] = 0.568$ M*

*13.14 Equilibrium calculation Iodine and chlorine combine to give iodine monochloride in a reaction $I_2(g) + Cl_2(g) \longrightarrow 2ICl(g)$ for which the equilibrium constant is 640 at 464°C. Starting with 1 mol of I_2 and 1 mol of Cl_2 in a 1-liter box at 464°C, what fraction of the I_2 converts to ICl?

**13.15 Equilibrium calculation Given $K_c = 0.0224$ at 500°K for $PCl_5(g) \rightleftharpoons PCl_3(g) + Cl_2(g)$, what fraction of the PCl_5 will be decomposed at equilibrium (*a*) when 1.00 mol of PCl_5 is injected into an empty 1-liter box at 500°K; (*b*) when 1.00 mol of PCl_5 is injected into a 1-liter box that initially contains 1.00 mol of Cl_2 at 500°K?

**13.16 Equilibrium calculation Given that $K_c = 13.7$ at 546°K for $PCl_5(g) \rightleftharpoons PCl_3(g) + Cl_2(g)$, calculate what pressure will develop in a 10-liter box at equilibrium at 546°K when 1.00 mol of PCl_5 is injected into the empty box. *Ans. 8.93 atm*

***13.17 Equilibrium calculation At 21.5°C and a total pressure of 0.0787 atm, N_2O_4 is 48.3% dissociated into NO_2. Calculate K_c for the reaction $N_2O_4(g) \rightleftharpoons 2NO_2(g)$. At what total pressure will the percent dissociation be 10.0%? *Ans. 2.36 atm*

*13.18 Equilibrium constant Write expres-

sions for the equilibrium condition for each of the following:

a $CaCO_3(s) \rightleftharpoons CaO(s) + CO_2(g)$
b $CuO(s) + H_2(g) \rightleftharpoons Cu(s) + H_2O(g)$
c $3Cl_2(g) + 2VCl_2(s) \rightleftharpoons 2VCl_5(s)$
d $3Fe(s) + 4H_2O(g) \rightleftharpoons Fe_3O_4(s) + 8H_2(g)$
e $Fe_2O_3(s) + 3CO(g) \rightleftharpoons 2Fe(s) + 3CO_2(g)$

*13.19 Equilibrium changes A box contains CO, Cl_2, and $COCl_2$ in equilibrium at 1000°K. Indicate qualitatively what effect there would be on (1) the concentration and (2) the number of moles of each component if the following changes were made:

a Add CO
b Remove CO
c Compress the box
d Add helium to the box

*13.20 Equilibrium changes Suppose you want to make nitrosyl chloride from the exothermic reaction $2NO(g) + Cl_2(g) \longrightarrow 2NOCl(g)$. Starting with a given amount of NO, what four things might you do to increase the yield of NOCl?

*13.21 Equilibrium changes The industrial fuel "water gas," which consists of a mixture of H_2 and CO, can be made by passing steam over hot carbon. Given the equilibrium

$$C(s) + H_2O(g) \rightleftharpoons CO(g) + H_2(g)$$

where $\Delta H° = +131$ kJ, tell how the yield of CO and H_2 at equilibrium would be affected by each of the following:

a Raising the relative pressure of the steam
b Adding hot carbon
c Raising the temperature
d Reducing the volume of the system

***13.22 Catalysis If a catalyst drops the activation energy of a reaction at room temperature from 80 to 40 kJ, what is the magnitude of the effect on the forward rate? The reverse rate?

***13.23 Thermodynamics For the reaction $H_2(g)$ (1 atm, 667°K) + $I_2(g)$ (1 atm, 667°K) $\longrightarrow 2HI(g)$ (2 atm, 667°K), $\Delta H°$ is -12 kJ and $\Delta S°$ is $+3.68$ J/deg. Make a rough graph showing how H, S, and G change during the course of this reaction as a function of the fraction of conversion.

***13.24 Thermodynamics Given the same data as in Exercise 13.23, make a rough graph showing how H, S, and G change during the reverse conversion $2HI(g) \longrightarrow H_2(g) + I_2(g)$ as a function of extent of reaction.

***13.25 Thermodynamics** Given a reaction in which 1 mol of $P_2(g)$ (1 atm, 1000°K) reacts with 1 mol of $Q_2(g)$ (1 atm, 1000°K) to form 2 mol of $PQ(g)$ (2 atm, 1000°K) for which $\Delta H° = +10,000$ J and $\Delta S° = 0$ J/deg. Assume the enthalpy and the entropy of the starting elements are zero. Show roughly on graphs of G, H, and S versus percent conversion what happens to the system as P_2 and Q_2 gradually convert to PQ. What is the equilibrium constant for the reaction?

13.26 Thermodynamics Using the expression $G = G° + RT \ln a$ for free energy as a function of composition and the condition that $\Delta G = 0$ at equilibrium, prove for the reaction $CO(g) + Cl_2(g) \rightleftharpoons COCl_2(g)$ that $\Delta G° = -RT \ln K$.

*13.27 Thermodynamics** Given that $K = 4.40$ at 2000°K for $H_2(g) + CO_2(g) \rightleftharpoons H_2O(g) + CO(g)$, calculate $\Delta G°$ for the reaction as written. What would be the $\Delta G°$ for the reverse reaction?

13.28 Thermodynamics At 298°K, the standard free energy of formation of $SO_2(g)$ is

-300 kJ/mol, that of $SO_3(g)$ is -370 kJ/mol. Calculate the equilibrium constant K_p and K_c for $2SO_2(g) + O_2(g) \rightleftharpoons 2SO_3(g)$ at 298°K. (Note that K_p can be related to K_c by using $PV = nRT$ for each gas.)

*13.29 Thermodynamics** Given an equilibrium $A(g) + B(g) \rightleftharpoons C(g) + D(g)$, what can you deduce about how K changes with temperature from each of the following:
  a $\Delta G°$ is positive.
  b $\Delta H°$ is positive.
  c $\Delta S°$ is positive.
  d $\Delta G°$ and $\Delta S°$ are both negative.

13.30 Thermodynamics For the reaction $H_2(g) + CO_2(g) \rightleftharpoons H_2O(g) + CO(g)$, K is 0.412 at 900°K and 1.37 at 1200°K. What is the value of $\Delta H°$?

***13.31 Thermodynamics** For the reaction $2NO_2(g) \longrightarrow N_2O_4(g)$, $\Delta H°$ is -58.2 kJ per mole of N_2O_4 formed. If the equilibrium constant (expressed in terms of pressure) is 0.0792 at 373°K, what will it become at 500°K? What is $\Delta G°$, $\Delta H°$, and $\Delta S°$ for this reaction at 500°K?

Ans. K = 0.00067

Chapter 14

HYDROGEN, OXYGEN, AND WATER

The most important of all oxides, possibly of all compounds, is water. It constitutes a large fraction of our viable environment, and it is *the* solvent in which our metabolic adventures occur. In this chapter we consider first the component elements, hydrogen and oxygen, and then take up the compound water. In the next chapter we consider the equilibria that are involved in aqueous solutions.

14.1
ELEMENTAL HYDROGEN

Hydrogen is unique; it is the only reactive element of period 1. Because period 1 has only two elements, there is no easy correlation with later periods. In particular, hydrogen cannot be exclusively assigned to any of the vertical groups of the periodic table; it has to be considered on its own.

Natural hydrogen consists of three isotopes: protium ($_1^1$H), deuterium ($_1^2$H, or D), and tritium ($_1^3$H, or T). The protium nucleus consists of a lone proton; deuterium, also called heavy hydrogen, consists of a proton plus a neutron; tritium consists of a proton plus two neutrons. Protium is by far the most abundant of the three. In nature, there are 7000 times as many protium atoms as deuterium and only 7×10^{-14} times as many tritium atoms. The scarcity of tritium is due to instability and consequent radioactivity of its nucleus.

In general, properties of isotopes are qualitatively rather similar. However, there may be *quantitative* differences, so-called isotope effects, especially when the percentage difference in mass is appreciable. Figure 14.1 com-

---

FIG. 14.1 Properties of Hydrogen Isotopes

Property	Protium	Deuterium
Mass of atom (H), amu	1.0078	2.0141
Freezing point (H_2), °K	14.0	18.7
Boiling point (H_2), °K	20.4	23.5
Freezing point (H_2O), °C	0	3.8
Boiling point (H_2O), °C	100	101.4
Density at 20°C (H_2O), g/ml	0.998	1.106

pares some properties of protium and deuterium. In addition, there is a quantitative difference both in equilibrium and rate properties. For example, the extent of dissociation of protium water by the reaction

$$H_2O \longrightarrow H^+ + OH^-$$

is about twice as great as is the corresponding dissociation of heavy water

$$D_2O \longrightarrow D^+ + OD^-$$

The isotope effect on the rate of reaction is even more marked. Thus, a bond to a protium can be broken as much as 18 times faster than the same bond to deuterium. For elements heavier than hydrogen, where the percentage difference in mass between isotopes is small, the isotope effect is much smaller.

The isotope effect is used as a basis for separation of protium and deuterium. Since protium bonds are broken faster than deuterium bonds, electrolysis of water releases protium faster than deuterium. This means that there will be an enrichment of the heavy hydrogen in the residual water. By continuing the electrolysis until the residual volume is very small, practically pure deuterium oxide can be obtained. In a typical experiment, 2400 liters of ordinary water produce 83 ml of D_2O that is 99 percent pure.

In the universe, hydrogen is apparently the most abundant of all elements. Spectral analysis of the light emitted by stars indicates that most of them are predominantly hydrogen. On the Earth, hydrogen is much less abundant. The Earth's gravitational attraction is too small to hold very light molecules; so there is practically no hydrogen left in the atmosphere. There may have been some at the creation, 4.6×10^9 years ago, but even if there were, most of it would have long since escaped. If we consider only the Earth's crust (atmosphere, hydrosphere, and lithosphere), hydrogen is third in abundance on an atom basis. On a mass basis, hydrogen is ninth in order. Figure 14.2 shows some abundance data. The specific values do not always agree from one compilation to another.

FIG. 14.2   Abundance of the Most Common Elements in the Earth's Crust

Element	Weight %	Atom %
Oxygen	49.4	55.1
Silicon	25.8	16.3
Aluminum	7.5	5.0
Iron	4.7	1.5
Calcium	3.4	1.5
Sodium	2.6	2.0
Potassium	2.4	1.1
Magnesium	1.9	1.4
Hydrogen	0.9	15.4
Titanium	0.6	0.2

On the Earth, free hydrogen is rare. It is found occasionally in volcanic gases. Also, as shown by studying the aurora borealis, it occurs in the upper atmosphere. On the other hand, combined hydrogen is quite common. In water, hydrogen is bound to oxygen and makes up 11.2 percent of the total weight. The human body, two-thirds of which is water, is approximately 10 percent hydrogen by weight. In fossil fuels, such as coal and petroleum, hydrogen is combined with carbon in a great variety of hydrocarbons. Clay and a few other minerals contain appreciable amounts of hydrogen, usually combined with oxygen. Finally, the proteins and carbohydrates of all plant and animal matter are composed of compounds of hydrogen with oxygen, carbon, nitrogen, sulfur, etc.

14.2
PREPARATION OF HYDROGEN

Hydrogen can be made inexpensively by passing steam over hot carbon:

$$C(s) + H_2O(g) \xrightarrow{1000°C} CO(g) + H_2(g)$$

However, H_2 from this source is not pure because CO is difficult to separate. The mixture of H_2 and CO is the important industrial fuel, *water gas;* it has a very high heat of combustion.

Purer but still relatively inexpensive hydrogen can be made by passing steam over hot iron:

$$3Fe(s) + 4H_2O(g) \longrightarrow Fe_3O_4(s) + 4H_2(g)$$

The iron can be recovered by reducing the Fe_3O_4 with water gas.

The purest (99.9 percent) but most expensive hydrogen available commercially is *electrolytic hydrogen,* made from the electrolysis of water:

$$2H_2O \xrightarrow{\text{electrolysis}} 2H_2(g) + O_2(g) \qquad \Delta H° = +565 \text{ kJ}$$

The reaction requires energy. It is the power consumption that makes electrolytic hydrogen expensive. In practice, solutions of NaOH or KOH are electrolyzed in cells with iron cathodes and nickel anodes; the cells are designed to keep the anode and cathode products separate. The electrode reactions are

Anode:	$4OH^- \longrightarrow O_2(g) + 2H_2O + 4e^-$
Cathode:	$2e^- + 2H_2O \longrightarrow H_2(g) + 2OH^-$
Net:	$2H_2O \longrightarrow 2H_2(g) + O_2(g)$

Considerable hydrogen is also formed as a by-product in the *chlor-alkali* industry, where Cl_2 and NaOH are produced by electrolysis of aqueous NaCl.

In petroleum refineries, where gasoline is made by cracking of hydrocarbons, hydrogen is a valuable by-product. Gaseous hydrocarbons are passed over hot catalyst, so that decomposition occurs to hydrogen and other hydrocarbons. The lighter hydrocarbons, such as methane, can be partially oxidized by passing them with steam over a nickel catalyst to produce hydrogen:

$$CH_4(g) + H_2O(g) \xrightarrow[900\,°C]{Ni} CO(g) + 3H_2(g)$$

In the laboratory, pure hydrogen can be made by reduction of hydrogen ion with zinc metal:

$$2H^+ + Zn(s) \longrightarrow H_2(g) + Zn^{2+}$$

In principle, such reduction should occur with any metal having an electrode potential less than zero (Sec. 11.10). For some metals, such as iron, the reaction is quite slow, even though the potential is favorable. In order to liberate H_2 from water, where the concentration of H^+ is only $1.0 \times 10^{-7}\ M$, a metal must have an electrode potential more negative than -0.41 V. Thus, the element sodium reacts with water to liberate H_2 by the reaction

$$2H_2O + 2Na(s) \longrightarrow H_2(g) + 2Na^+ + 2OH^-$$

Laboratory hydrogen can also be made conveniently from the reaction of aluminum metal with base, $2Al(s) + 2OH^- + 6H_2O \longrightarrow 2Al(OH)_4^- + 3H_2(g)$, or from the reaction of CaH_2 with water, $CaH_2(s) + 2H_2O \longrightarrow Ca^{2+} + 2OH^- + 2H_2(g)$.

14.3
PROPERTIES AND USES OF HYDROGEN

Hydrogen is a colorless, odorless, tasteless gas. It is diatomic and consists of nonpolar molecules containing two hydrogen atoms held together by a covalent bond. In order to rupture the bond, 431.8 kJ of heat must be supplied per mole. When H_2 reacts, one of the steps is usually the breaking of the H—H bond. Because of the high energy required for this step, the activation energy is high, and H_2 reactions are generally slow.

Molecular hydrogen is the lightest of all gases. A balloon filled with hydrogen rises in accord with Archimedes' principle that the buoyant force is equal to the weight of fluid displaced. Meteorologists frequently send aloft weather balloons inflated with hydrogen.

The melting point (14.0°K) and the boiling point (20.4°K) of hydrogen are very low, indicating that the intermolecular attractions are quite small. Because of the low boiling point, liquid hydrogen is used as a cryogenic fluid (to produce low temperatures).

Chemically, H_2 is able, under appropriate conditions, to combine directly with most elements. With oxygen, H_2 reacts to release large amounts of energy. The change

$$2H_2(g) + O_2(g) \longrightarrow 2H_2O(g) \qquad \Delta H° = -485\ \text{kJ}$$

occurs in the oxyhydrogen torch to produce temperatures of about 2800°C, where the reaction is self-sustaining. Mixtures of H_2 and O_2 are explosive, especially when the ratio of H_2 to O_2 is approximately 2:1.

With metals the reaction of H_2 is not nearly so violent and often requires elevated temperatures. For example, sodium hydride, NaH, is formed by bubbling H_2 through molten sodium.

Hydrogen also reacts with certain compounds. In some cases it simply adds on to the other molecule, as, for instance, in forming methyl alcohol, CH_3OH, from CO:

$$CO(g) + 2H_2(g) \xrightarrow{\text{catalyst}} CH_3OH(g)$$

Such addition reactions, called *hydrogenation* reactions, account for much of the industrial consumption of hydrogen. In other cases hydrogen removes atoms from other molecules, as in the reduction of tungsten trioxide, WO_3, to W.

Because of its high heat of combustion (120 kJ/g), liquid hydrogen is a valuable rocket fuel. Its value is enhanced because the light molecular weight of the combustion products means high velocities and therefore high thrust. In terms of a long-range solution to the energy problem, hydrogen offers considerable attraction. Raw material for its production (e.g., H_2O) is almost inexhaustible; nuclear power could be used to electrolyze water. A major unresolved problem, however, is how to transport and distribute the hydrogen safely. An intriguing recent development is the discovery that intermetallic compounds of the type $RECo_5$ and $RENi_5$ (where RE stands for a rare-earth element) can absorb large quantities of hydrogen and give them up on demand.

14.4
COMPOUNDS OF HYDROGEN

In its compounds H is found in three oxidation states: $+1$, -1, and 0.

Oxidation State $+1$

This is the most important state since it includes most of the H compounds. In these compounds, H is combined with a more electronegative element, such as any element from the right side of the periodic table. In period 2, for example, the elements more electronegative than H are C, N, O, and F. With these elements H forms compounds such as CH_4, NH_3, H_2O, and HF. In all these compounds the binding of H is covalent, and none of these compounds contains simple H^+ ion.

The compounds can be formed by direct union of the elements. The reactions are often slow, catalysts and high temperatures may be required. For example, the reaction between N_2 and H_2 to form NH_3, important industrially as the *Haber process,* is carried out under a pressure of 200 to 300 atm at about 500°C in the presence of a catalyst such as Fe.

Oxidation State −1

When H is combined with an atom less electronegative than itself, the compound is said to be a *hydride*. Hydrides may be predominantly ionic or covalent. In the hydrides of elements of groups I and II (e.g., NaH, CaH_2) the H occurs as the negative hydride ion, H^-. The compounds are saltlike and when melted can be electrolyzed to form H_2 *at the anode.*

$$2H^- \longrightarrow H_2(g) + 2e^-$$

The hydride ion is unstable in H_2O and is oxidized to H_2. Thus, for example, calcium hydride, CaH_2, in H_2O reacts as follows:

$$CaH_2(s) + 2H_2O \longrightarrow Ca^{2+} + 2OH^- + 2H_2(g)$$

The covalent hydrides, such as diborane (B_2H_6), silane (SiH_4), and arsine (AsH_3), are generally volatile liquids or gases. They do not contain H^- ion. The term "hydride" is also applied to compounds such as lithium aluminum hydride, $LiAlH_4$, where the cation is Li^+ and the anion is the complex AlH_4^-. These complex hydrides are generally solids, react with H_2O to liberate H_2, and are of great use as reducing agents.

Oxidation State 0

Hydrogen reacts with transition metals such as uranium and palladium to form hard, brittle substances that conduct electricity and have typical metallic luster. In some cases, as with uranium hydride, UH_3, the compound is stoichiometric. In others, as with palladium hydride, PdH_n, the compound is nonstoichiometric; i.e., the number of H atoms per metal atom is variable and can even be less than one. The state of the hydrogen in these metallic hydrides is in dispute. Some say the hydrogen exists as H atoms; others say the hydrogen is dissociated into an interstitial proton and a delocalized electron.

The dissolution of hydrogen in metals is important because metals which dissolve hydrogen are catalysts for hydrogenation reactions. The catalyst is thought to act by dissolving the hydrogen as H atoms, which react more rapidly than H_2 molecules. The catalysis by finely divided nickel of the hydrogenation of oils to give fats is explainable in this way.

When hydrogen dissolves in a metal, the metal may lose some of its desirable properties. This phenomenon, called *hydrogen embrittlement,* occurs even with small amounts of dissolved hydrogen, amounts that may be unavoidable in the preparation of pure metals. The large-scale use of titanium, as in supersonic aircraft, was made possible only after preparation methods were developed that avoided hydrogen entrapment.

14.5
HYDROGEN BOND

In some compounds a hydrogen atom is apparently bonded simultaneously to two other atoms. For example, in potassium hydrogen fluoride, KHF_2, the anion HF_2^- has H acting as a bridge between two F atoms. The bridge, called a *hydrogen bond*, is unusual because hydrogen is usually limited to one covalent bond at a time. Hydrogen bonds seem to be formed only between small electronegative atoms such as fluorine, oxygen, and nitrogen.

Evidence in support of hydrogen bonds comes from comparing properties of hydrogen-containing substances. For example, Fig. 14.3 shows the normal boiling points for the hydrogen halides (lower curve) and for the hydrogen compounds of group VI elements (upper curve). It is evident that the members on the extreme left, HF and H_2O, are abnormally high compared with other members of each series. In the series HF, HCl, HBr, and HI there is increasing number of electrons per molecule; so rising boiling points would be expected because of increased van der Waals attractions (Sec. 6.12). The unexpectedly high boiling point of HF is attributed to hydrogen bonds between fluorine atoms. Similarly, the unexpectedly high boiling point of H_2O might also be due to hydrogen bonding.

$$\underset{\cdots F}{\overset{F}{\diagdown}}H\underset{\cdots F}{\diagup}H\overset{F}{\diagdown}H\underset{\cdots F}{\diagup}H\overset{F}{\diagdown}H \qquad \underset{H}{\overset{|}{O}}-H\cdots\underset{H}{\overset{|}{O}}-H\cdots\underset{H}{\overset{|}{O}}-H$$

What is responsible for the hydrogen bond? The simplest view is that a proton (+) is attracted simultaneously to the charge clouds (−) of two different atoms. When hydrogen is bound to a very electronegative atom, the hydrogen has such a small share of the electron pair that it is almost like a bare proton. As such, it can be attracted to another electronegative atom. Because of its tiny size, a proton has room for only two atoms around it.

Although relatively weak (~20 kJ) compared with most other chemical bonds (~200 kJ), hydrogen bonds are extremely important in living systems. Proteins, for example, which contain both —CO and —NH groups, owe some of their structural features to hydrogen bonds between —CO groups in one part of the molecule and —NH groups in another. Hydrogen bonds also play an important role in determining the structure of liquid water and of ice. They are discussed further in Sec. 14.9.

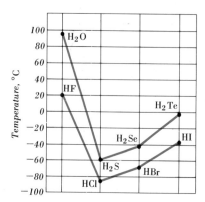

FIG. 14.3 Boiling points of some hydrogen compounds.

As was noted in Fig. 14.2, oxygen is the most abundant element in the Earth's crust both on the basis of mass and number of atoms. Almost half the mass of the Earth's crust is due to oxygen. On a number basis, O atoms are more numerous than the total of all other kinds of atoms.

In the free state, oxygen occurs in the atmosphere as O_2 molecules. In the combined state, it occurs in many rocks, plants and animals, and water. Of the oxygen-containing rocks, the most abundant are ones which contain silicon. The simplest of these is silica, SiO_2, the main constituent of sea sand. The most abundant rock that does not contain silicon is limestone, $CaCO_3$. In plant and animal materials, oxygen is combined with carbon, sulfur, nitrogen, phosphorus, or hydrogen.

From air (21% O_2 and 78% N_2) oxygen is made commercially by liquefaction and fractional distillation. The air is compressed, cooled, and expanded until liquefaction gives liquid air. On partial evaporation, the nitrogen, being lower boiling, boils away first, leaving the residue richer in oxygen. Repeated cycles give oxygen that is 99.5 percent pure.

Figure 14.4 shows how the separation works. The lower curve represents the liquid; the upper curve, the gas that is in equilibrium with it. Dashed horizontal lines are constant-temperature lines joining the liquid (at the left end of the line) with the gas (at the right end). As can be seen, the gas phase at any temperature is richer in nitrogen than is the liquid in equilibrium with it. The vertical lines represent total evaporation when a liquid of given composition is completely evaporated into the gas phase. The staircase progression from right to left represents repeated liquefaction followed by evaporation. Starting with gas 1 (80% N_2 and 20% O_2), we partially liquefy it to establish equilib-

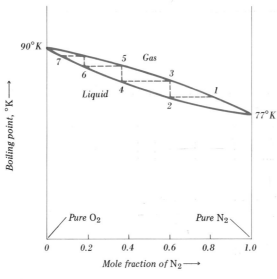

FIG. 14.4 Boiling point–composition curve for various mixtures of O_2 and N_2.

rium liquid 2 (60% N_2 and 40% O_2). We then take away this liquid and totally evaporate it to vapor (point 3). Vapor 3 is then partially liquefied to give, at equilibrium, liquid 4 (40% N_2 and 60% O_2). Again this liquid is taken off and evaporated completely to give gas 5. Partial liquefaction of gas 5 gives liquid 6 (20% N_2 and 80% O_2); and so on. Eventually, we can end up with pure liquid oxygen. In the meantime the rejected portions are relatively rich in nitrogen and can be used as a source of nitrogen. To get pure O_2, we go up the staircase from right to left; the process is called *fractional liquefaction*. To get pure N_2, we go down the staircase left to right; the process is called *fractional distillation*.

From water, very pure oxygen can be made by electrolysis as a by-product of hydrogen manufacture.

In the laboratory, oxygen is usually made by thermal decomposition of potassium chlorate, $KClO_3$. The reaction

$$2KClO_3(s) \longrightarrow 2KCl(s) + 3O_2(g)$$

is catalyzed by solids such as manganese dioxide, MnO_2, ferric oxide, Fe_2O_3, fine silica sand, and powdered glass. It appears that the main function of the catalyst is to provide a surface on which evolution of oxygen gas can occur.

14.7
PROPERTIES AND USES OF OXYGEN

At room temperature, oxygen is a colorless, odorless gas. The molecule is diatomic and paramagnetic to the extent of two unpaired electrons per molecule. Its electronic formula can be given as $:\ddot{O}:\ddot{O}:$, but this clearly violates the octet rule. Also, it is not consistent with the bond energy. The bond energy of O_2, 494 kJ, lies between that of triply bonded N_2 (941 kJ) and singly bonded F_2 (151 kJ). The bonding in O_2 presents a special problem, and we shall have more to say about it in Chap. 16.

Oxygen exhibits *allotropy;* i.e., it can exist as the element in more than one form. When energy is added to diatomic oxygen, the triatomic molecule ozone, O_3, is formed by the reaction

$$3O_2(g) \longrightarrow 2O_3(g) \qquad \Delta H° = +285 \text{ kJ}$$

Once ozone is obtained, it only slowly reverts to diatomic oxygen. In the laboratory, ozone can be made by passing oxygen between tinfoil conductors that are connected to the terminals of an electric induction coil. About 5 percent of the oxygen is converted to ozone. Ozone is also formed in appreciable amounts by lightning bolts, by ultraviolet light, and by sparking electric motors.

Traces of ozone are formed in the stratosphere, probably by absorption of ultraviolet sunlight. Ozone is very important as a screen against short-wavelength radiation ($\lambda < 310$ nm) coming from the sun. Some of the opposition to supersonic transport (SST) comes from the fact that supersonic aircraft generally operate in the ozone layer, where exhaust products (water vapor, oxides of nitrogen, and unburned hydrocarbons) may deplete the ozone concentration. Organic molecules and living organisms are particularly sensitive to light of short wavelength so that a decrease in the ozone shield may cause appreciable degradation of the human environment. Also, as was discussed in Sec. 12.9, ozone is implicated in the generation of photochemical smog.

The ozone molecule is bent with an O—O—O angle of 116.8° and O—O bonds that are 0.128 nm (slightly longer than the 0.121 nm in oxygen). Since the ozone molecule is not paramagnetic, all its electrons must be paired. If the octet rule is followed, it is necessary to write at least two contributing resonance forms for the structure:

Ozone gas has a sharp, penetrating odor. When cooled to 162°K, ozone forms a deep blue liquid that is explosive because of the spontaneous tendency of ozone to decompose to oxygen.

Some of the properties of oxygen and ozone are given in Fig. 14.5. Both oxygen and ozone are good oxidizing agents, as shown by their high reduction potentials:

$$O_2(g) + 4H^+ + 4e^- \longrightarrow 2H_2O \qquad E° = +1.23 \text{ V}$$
$$O_3(g) + 2H^+ + 2e^- \longrightarrow H_2O + O_2(g) \qquad E° = +2.07 \text{ V}$$

Of the common oxidizing agents, ozone is second only to fluorine. In most reactions, at least at room temperature, oxygen is a slow oxidizing agent, whereas ozone is rapid.

Because of its cheapness and ready availability, oxygen is one of the most widely used industrial oxidizing agents. For example, in the manufacture of steel it is used to convert carbon to carbon monoxide for reducing iron oxides to iron. In the oxyacetylene torch, used for cutting and welding metals, temperatures in excess of 3000°C can be obtained by the reaction

$$2C_2H_2(g) + 5O_2(g) \longrightarrow 4CO_2(g) + 2H_2O(g) \qquad \Delta H° = -2510 \text{ kJ}$$

When mixed with alcohol, charcoal, gasoline, or powdered aluminum, liquid oxygen gives powerful explosives.

FIG. 14.5   Properties of Allotropic Forms of Oxygen

Property	O_2	O_3
Molecular weight, amu	31.999	47.998
Bond length, nm	0.1207	0.1278
Normal melting point, °K	54.3	80.5
Normal boiling point, °K	90.2	161.7
Critical temperature, °K	154	268
Density of liquid (90°K), g/ml	1.14	1.71

The use of oxygen in respiration of plants and animals is well known. In humans, O_2, inhaled from the atmosphere, is picked up in the lungs by hemoglobin of the blood and distributed to various cells, which use it for respiration. In respiration, carbohydrates are oxidized to provide energy for cellular activities. Since oxygen itself is a slow oxidizing agent, catalysts (enzymes) must be present in order that reactions may proceed at body temperature. In the treatment of heart trouble, pneumonia, and shock, air oxygen is supplemented with additional oxygen.

The uses of ozone depend on its strong oxidizing properties. For example, it is used as a germicide, presumably because of its oxidation of bacteria. Inasmuch as oxidation of colored compounds often results in colorless ones, ozone is also used as a bleaching agent for wax, starch, fats, and varnishes. When added to the air in small amounts, ozone destroys odors; but it can be used safely only in low concentration because it irritates the lungs.

14.8
COMPOUNDS OF OXYGEN

Except for the very rare OF_2 and O_2F_2, the oxidation state of oxygen in compounds is normally negative. The oxidation numbers $-\frac{1}{2}, -1$, and -2 are observed.

Oxidation State $-\frac{1}{2}$

The heavier elements of group I (K, Rb, and Cs) react with oxygen to form compounds of the type MO_2, called *superoxides*. These are ionic solids containing the cation M^+ and the anion O_2^-. The solids are colored and paramagnetic. When they are placed in water, O_2 and H_2O_2 are formed by the reaction

$$2MO_2(s) + 2H_2O \longrightarrow O_2(g) + H_2O_2 + 2M^+ + 2OH^-$$

Oxidation State -1

Compounds that contain oxygen with oxidation number -1 are called *peroxides*. They are characterized by a direct O—O bond, which usually breaks fairly easily at elevated temperatures. Metals such as Na, Sr, and Ba form solid peroxides which contain the peroxide ion, O_2^{2-}. The ion contains no unpaired electrons.

When solid peroxides are added to acidic solutions, hydrogen peroxide, H_2O_2, is formed. For example

$$BaO_2(s) + 2H^+ \longrightarrow Ba^{2+} + H_2O_2$$

If H_2SO_4 is used, the barium ion precipitates as insoluble $BaSO_4$, leaving a dilute solution of pure H_2O_2. Commercially, most H_2O_2 is prepared by electrolysis of cold H_2SO_4 followed by distillation under reduced pressure. Because H_2O_2 is unstable, owing to the reaction

$$2H_2O_2 \longrightarrow 2H_2O + O_2(g)$$

it is difficult to keep. The decomposition is slow but is catalyzed by impurities such as transition-metal compounds. It is also accelerated in the presence of light. For these reasons, solutions of H_2O_2 are usually stored in dark bottles with additives such as diphosphate ion, $P_2O_7^{4-}$, introduced to tie up the catalysts.

Pure H_2O_2, obtained by distillation under reduced pressure, is a colorless liquid, freezing at $-0.9°C$ and boiling at $151.4°C$. The molecule corresponds to a nonplanar arrangement of atoms, as is shown in Fig. 14.6.

Because oxygen can also show oxidation states 0 and -2, compounds containing peroxide oxygen (-1) can gain or lose electrons; hence, they can act both as oxidizing agents and as reducing agents. In the decomposition

$$2H_2O_2 \longrightarrow 2H_2O + O_2(g)$$

hydrogen peroxide oxidizes and reduces itself. In the reaction

$$5H_2O_2 + 2MnO_4^- + 6H^+ \longrightarrow 5O_2(g) + 2Mn^{2+} + 8H_2O$$

hydrogen peroxide is a reducing agent (goes to O_2). In the reaction

$$H_2O_2 + 2I^- + 2H^+ \longrightarrow I_2 + 2H_2O$$

hydrogen peroxide is an oxidizing agent (goes to H_2O).

Oxidation State −2

Minus two is the most common oxidation state for oxygen in compounds. These compounds include the *oxides,* such as BaO, and the *oxy compounds,* such as $BaSO_4$.

All the elements except the lighter noble gases form oxides. Some of these oxides are ionic; others are covalent. In general, the more ionic ones (e.g., BaO) are formed with the elements from the extreme left of the periodic

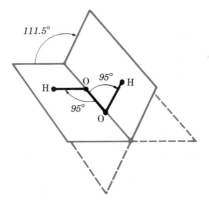

FIG. 14.6 Structure of hydrogen peroxide molecule.

table; the more covalent ones (e.g., SO_2), from elements on the right. When placed in water, O^{2-} ion reacts to give basic solutions:

$$O^{2-} + H_2O \longrightarrow 2OH^-$$

so ionic oxides are *basic oxides*. They neutralize acids, as, for example, in the reaction

$$CaO(s) + 2H^+ \longrightarrow Ca^{2+} + H_2O$$

Covalent oxides generally give acidic solutions:

$$SO_2 + H_2O \rightleftharpoons H^+ + HSO_3^-$$

so they are *acidic oxides*. They neutralize bases, as, for example, when carbon dioxide is bubbled through aqueous sodium hydroxide

$$CO_2(g) + OH^- \longrightarrow HCO_3^-$$

It is not possible to classify all oxides sharply as either acidic or basic. Some oxides, especially those toward the center of the periodic table, are able to *neutralize both acids and bases*. Such oxides are *amphoteric*. An example of an amphoteric oxide is ZnO, which undergoes both of the following reactions:

$$ZnO(s) + 2H^+ \longrightarrow Zn^{2+} + H_2O$$
$$ZnO(s) + 2OH^- + H_2O \longrightarrow Zn(OH)_4^{2-}$$

In the first reaction ZnO acts as a base; in the second, as an acid.

When any oxide, ionic or covalent, reacts with water, the resulting compound contains OH, or *hydroxyl*, groups. If the hydroxyl group exists as OH^- ion, the compound is called a *hydroxide*. Hydroxides are formed by reaction of ionic oxides with water; for example,

$$BaO(s) + H_2O \longrightarrow Ba(OH)_2(s)$$

Most hydroxides, except for those of group I elements, revert to the oxide when heated. Many, for example, $Al(OH)_3$, are insoluble in water.

Some compounds contain the OH group, not as an ion, but covalently bound to another atom. For example, in H_2SO_4 two OH groups and two O atoms are joined to a central S atom. When placed in water, such compounds give acid solutions by rupture of an O—H bond. For this reason, they are called *oxyacids*. Most oxyacids can be dehydrated by heat to give oxides. They can also be neutralized to give oxysalts such as sodium sulfate, Na_2SO_4.

14.9
WATER

The most important oxide is H_2O. The molecule is nonlinear, with an H—O—H angle equal to 104.52°. Because each bond is polar covalent, with the H end positive relative to the O end, the molecule has a considerable dipole moment. Attraction between the H atom of one molecule and the O

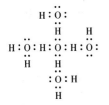

FIG. 14.7 Association of water molecules. (The H locations shown are average positions.)

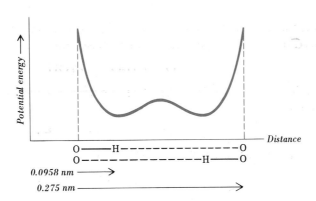

0.0958 nm $\longrightarrow$

0.275 nm $\longrightarrow$

FIG. 14.8 Potential-energy curve showing two possible H-atom positions in the O $\cdots$ O hydrogen bond.

of another leads to association of H_2O molecules into a giant cluster in both liquid water and solid ice. A two-dimensional representation is shown in Fig. 14.7. The cluster is held together by hydrogen bonds (Sec. 14.5). Each H atom is placed between two O atoms and is considered bonded equally to both. As shown in Fig. 14.8, there are actually two favored positions for the H, separated by a low potential-energy barrier. The H jumps back and forth from a position near one O atom to a position near the other; it is only the "average" positions that are shown in Fig. 14.7. The result of hydrogen bonding is to form an extended network in which each O atom is surrounded by four H atoms. That there are four H atoms can be inferred from X-ray studies. These studies do not directly see the H atoms, but they do show that there are four O atoms placed about each O. If the O atoms are joined to each other by hydrogen bonds, then there must be four H atoms about each O.

X-ray studies of ice indicate that the O atoms of neighboring H_2O molecules are located at the corners of a regular tetrahedron, as shown in Fig. 14.9. Because the arrangement is tetrahedral, the ice structure extends in three dimensions. Figure 14.10a shows part of the structure. The large spheres represent O atoms; the small spheres, H atoms. Every other O has its fourth

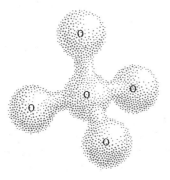

FIG. 14.9 Tetrahedral arrangement of O atoms about a given O atom in the ice structure. (Connecting region is occupied by H.)

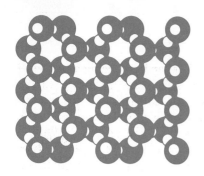
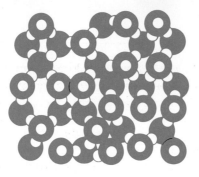

FIG. 14.10 (a) Ice structure; (b) liquid water structure.

(a)   (b)

H hidden beneath it. This hidden H joins to another O below, and so the structure continues in three dimensions. A notable feature of the structure is that it is honeycombed with hexagonal channels. Because of these holes, ice has a relatively low density.

When ice melts, the structure becomes less orderly, but it is not completely destroyed. The O atoms are still tetrahedrally surrounded by four H atoms, as in ice, but the overall arrangement of tetrahedra is more random and is constantly changing. An instantaneous view might be like that shown in Fig. 14.10b. Some of the hexagonal channels have collapsed to give a denser structure.

Data in Figs. 14.11 and 14.12 show that liquid water at 0°C is denser than ice and also that liquid water has a maximum density at 3.98°C. The maximum can be interpreted as follows: When ice is melted, collapse of the structure leads to an increase in density. As the temperature is raised, the collapse should continue further. However, there is an opposing effect. The higher the temperature, the greater the kinetic motion of the molecules. Hydrogen bonds are broken, and the H_2O molecules move farther apart on the average. This effect becomes dominant at temperatures above 3.98°C, so the density decreases. Below this temperature, collapse of structure is more important.

FIG. 14.11 Density of H_2O at Various Temperatures

Temperature, °C	State	Density, g/cm³
0	Solid	0.917
0	Liquid	0.9998
3.98	Liquid	1.0000
10	Liquid	0.9997
25	Liquid	0.9971
100	Liquid	0.9584

FIG. 14.12 Graph of density relations for H_2O. (Scales are distorted.)

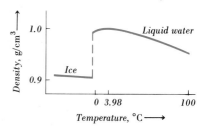

Water is the most common solvent. However, it is far from being a universal solvent, as can be seen from the large number of substances that are insoluble in water. What are the factors that influence aqueous solubility? The situation for water is more complex than with normal solvents because hydrogen bonding tends to keep water molecules associated to each other.

In general, water is a poor solvent for nonionic solutes. Hydrocarbons in particular, such as CH_4, are practically insoluble in water. In these cases, water interacts so weakly with the molecular solute that not enough energy is liberated to break down the water structure. Still there are some molecular solutes which are highly soluble in water. Examples are NH_3 and C_2H_5OH. In the case of NH_3, hydrogen bonds are established between the N of NH_3 and the O of H_2O, in the case of ethyl alcohol, hydrogen bonds are between the O of C_2H_5OH and the O of H_2O. Sugars such as sucrose, $C_{12}H_{22}O_{11}$, owe their high solubility in large measure to hydrogen bonding since they, like C_2H_5OH, have OH groups.

For ionic solutes, solubility in water depends on a delicate balance between lattice energy, hydration energy, and entropy of ions. The scheme

$$\begin{array}{ccc}
& \xrightarrow{\;(2)\;} & M^+(g) \;+\; X^-(g) \\
& & \downarrow{\scriptstyle(3)} \qquad \downarrow{\scriptstyle(4)} \\
MX(s) & \xrightarrow{\;(1)\;} & M^+(aq) + X^-(aq)
\end{array}$$

suggests that the overall process of dissolving (1) can be resolved into a lattice breakup step (2), followed by hydration of the gaseous ions (3 and 4). The energy required for the lattice breakup increases with the charge on the ions and decreases with increasing size of the ions. Figure 14.13 shows some representative values. As can be seen for the alkali halides, with increasing anion size, the lattice energy decreases. Similarly, in going from Li to Na to K salts, the lattice energy again decreases. Ag salts would be expected to be like Na salts because the ionic radius of Ag^+ (0.115 nm) is closer to Na^+ (0.102 nm) than it is to K^+ (0.138 nm). However, the lattice energies of the Ag salts turn out to be surprisingly high, probably because of larger van der Waals attractions. Figure 14.13 also shows that oxides and sulfides of +2 ions have lattice energies enormously greater than those of +1 halides. A fourfold increase in electric attraction is expected when we double charge on both positive and negative ions.

If a salt is to be appreciably soluble in water, the energy needed for lattice breakup has to be offset by hydration. Figure 14.14 lists values for $\Delta H°$ and $\Delta S°$ of hydration of some common ions. The $\Delta H°$ values refer to the process $M^+(g) \longrightarrow M^+(aq)$. The fact that the values are negative indicates that heat is liberated to the surroundings. $\Delta H°$ is smallest for the singly charged ions (about -400 kJ), roughly quadruples for doubly charged ions (about -1600 kJ), and becomes about 10 times as great for triply charged ions (about

FIG. 14.13 **Lattice Energies, kJ/mol (To get ΔH°_{298} from these values, we need to add about 5 kJ/mol)**

LiF	1008	AgF	954	TiO	3880
LiCl	828	AgCl	904	VO	3920
LiBr	791	AgBr	895	MnO	3810
LiI	732	AgI	883	FeO	3920
				CoO	3990
NaF	904	BeO	4530	NiO	4080
NaCl	770	MgO	3920	ZnO	4030
NaBr	736	CaO	3520		
NaI	690	SrO	3310		
		BaO	3120		
KF	803				
KCl	699	BeS	3740		
KBr	674	MgS	3300		
KI	632	CaS	3040		
		SrS	2900		
		BaS	2760		

FIG. 14.14 **Hydration Parameters of Some Common Ions**

	ΔH°, kJ/mol	ΔS°, J deg^{-1} mol^{-1}
Li$^+$	-506	-118.8
Na$^+$	-397	-87.4
K$^+$	-314	-51.9
Rb$^+$	-289	-40.2
Cs$^+$	-255	-36.8
Mg^{2+}	-1910	-266.5
Ca^{2+}	-1580	-210.0
Sr^{2+}	-1430	-203.8
Ba^{2+}	-1290	-157.7
Al^{3+}	-4640	-463
Ag$^+$	-469	-93.7
Zn^{2+}	-2030	-267.8
F$^-$	-506	-155.2
Cl$^-$	-377	-96.7
Br$^-$	-343	-83.3
I$^-$	-297	-60.7
OH$^-$	-502	-108.8

-4000 kJ). The larger the radius of the ion, the smaller the ΔH° of hydration.

The ΔS° of hydration also applies to the process $M^+(g) \longrightarrow M^+(aq)$. The values are all negative. In the gas phase, there are a large number of possible configurations, and the entropy is very high. When the ions go into solution, they lose much of their freedom; the entropy is considerably lower. In the process $M^+(g) \longrightarrow M^+(aq)$, as well as in $X^-(g) \longrightarrow X^-(aq)$, we lose practically all the entropy associated with a mole of gas, about 50 J deg^{-1} mol^{-1}. In addition, there is a further reduction of entropy due to ordering of the H_2O molecules in the solution.

Once we have the numerical values for ΔH° and ΔS°, we can use them to determine ΔG° for the overall reaction

$$MX(s) \longrightarrow M^+(aq) + X^-(aq)$$

Let us compare AgF and AgCl. For AgF, which is a soluble salt, we have

	ΔH°, kJ/mol	ΔS°, kJ mol^{-1} deg^{-1}
AgF(s) $\longrightarrow$ Ag$^+$(g) + F$^-$(g)	$+959$	0.200
Ag$^+$(g) $\longrightarrow$ Ag$^+$(aq)	-469	-0.094
F$^-$(g) $\longrightarrow$ F$^-$(aq)	-506	-0.155
AgF(s) $\longrightarrow$ Ag$^+$(aq) + F$^-$(aq)	-16	-0.049

$$\Delta G^\circ = \Delta H^\circ - T\Delta S^\circ = -16 - 298(-0.049)$$
$$= -16 + 15 = -1 \text{ kJ/mol}$$

As can be seen, $\Delta H°$ for the overall process is negative, favorable for dissolving. However, $-T\Delta S°$ is positive; the entropy change is unfavorable. Since the enthalpy term is bigger, it wins out, and the result is that AgF is soluble. For AgCl, in contrast, we have the following:

	$\Delta H°$, kJ/mol	$\Delta S°$, kJ mol^{-1} deg^{-1}
$AgCl(s) \longrightarrow Ag^+(g) + Cl^-(g)$	$+909$	0.200
$Ag^+(g) \longrightarrow Ag^+(aq)$	-469	-0.094
$Cl^-(g) \longrightarrow Cl^-(aq)$	-377	-0.097
$AgCl(s) \longrightarrow Ag^+(aq) + Cl^-(aq)$	$+63$	$+0.009$

$$\Delta G° = \Delta H° - T\Delta S° = +63 - 298(0.009)$$
$$= +63 - 3 = +60 \text{ kJ/mol}$$

Here $\Delta H°$ is positive, unfavorable for dissolving. The entropy change is favorable, but the enthalpy term is much bigger and so wins out. The result is that AgCl is not a very soluble salt.

In analyzing the difference between AgF and AgCl, the most important contribution is that the hydration enthalpy of Cl^- is much smaller than that of F^-. The lattice of AgCl requires less energy to break up, but this is not enough. There is insufficient hydration energy from Cl^- to make it worthwhile.

For most solids, not enough data are available to make the above type of analysis. In such cases, rough guidelines are useful. For example, if the charges on both anion and cation are simultaneously increased, insolubility is generally favored. Thus, for example, $BaSO_4$ (both ions doubly charged) and $AlPO_4$ (both ions triply charged) are much less soluble than NaCl (both ions singly charged). On the other hand, if the charge of only one ion is increased, the solubility is not much changed. So, for example, NaCl, $BaCl_2$, and $AlCl_3$ are all appreciably soluble. Another general rule is that the more dissimilar in size the anion and cation are, the more soluble a salt is likely to be. Thus, for example, $MgCrO_4$ is very soluble, whereas $BaCrO_4$ is quite insoluble. Finally, it should be noted that specific interactions may aid solubility. Barium sulfide (BaS), for example, has a specific reaction between S^{2-} and H_2O which helps make BaS more soluble than $BaSO_4$, an insoluble salt that lacks such specific interaction.

14.11
HYDRATES

Analysis shows that many solids contain H_2O molecules. These solids, called *hydrates,* are generally represented by dot formulations, e.g., $NiSO_4 \cdot 7H_2O$, which do not specify how the H_2O is bound in the crystal. For example, in $NiSO_4 \cdot 7H_2O$ all seven H_2O molecules are not equivalent. Six are bound to the Ni^{2+} ion to give $Ni(H_2O)_6^{2+}$, and the seventh is shared between $Ni(H_2O)_6^{2+}$ and SO_4^{2-}. The solid might better be represented as $Ni(H_2O)_6SO_4 \cdot H_2O$. In some hydrates, such as $Na_2CO_3 \cdot 10H_2O$, the H_2O molecules are not bound directly to the ions but serve mainly to improve the packing of the ions in the crystal.

Water of hydration can be driven off by heating to give *anhydrous material.* Such loss of water is usually accompanied by a change in crystal structure. However, some substances, such as proteins and the silicate minerals called *zeolites,* lose water on heating without much change in crystal structure. On reexposure to water, they, like sponges, take up water and swell. Apparently, water taken up this way occupies semirigid tunnels within the solid.

There are many hydrous compounds whose overall composition is known but whose structure is in doubt. For example, the substance AlO_3H_3, obtained from the reaction of base with a solution of aluminum salt, might be the hydroxide, $Al(OH)_3$, or the hydrated oxide, $Al_2O_3 \cdot 3H_2O$. Both have the same simplest formula. In order to distinguish the two possibilities, structure studies are needed, but they are difficult and in many cases have not been done.

14.12
THE AQUATIC ENVIRONMENT

Until recently little attention was given to the problem of water pollution because water is a globally abundant, renewable resource, which is constantly recycled through natural distillation (solar evaporation, cloud condensation, rain). When there were few neighbors, one could watch pollution problems flow away, but with increasing population and increasing technological waste, indifference becomes more difficult. Also, some of the dumped chemicals that were "safely" flushed away (e.g., mercury and PCB, polychlorinated biphenyls) are now coming back to haunt us.

Water quality is a relative property dependent on the use to which the water is put. Generally, it is a function of dissolved oxygen, dissolved solids, biochemical oxygen demand (BOD), suspended sediments, acidity, and temperature.

Dissolved oxygen is required by all aquatic plant and animal life. Fish require the highest levels, vertebrates next, and bacteria the least. Figure 14.15 shows how the solubility of oxygen varies with temperature; it drops

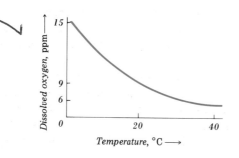

FIG. 14.15 Dissolved oxygen as a function of temperature.

from a high of 15 ppm at 0°C to about 6 ppm at 40°C. Levels below saturation arise from decay of oxygen-demanding wastes. Most of these are organic. If we represent them as carbon, we can write

$$C \; + \; O_2 \longrightarrow CO_2$$
$$12 \text{ g} \quad 32 \text{ g}$$

which indicates that a 9-ppm level of dissolved oxygen would be totally exhausted by $(\frac{12}{32})(9)$, or about 3, ppm of carbon waste. This is equivalent to about a drop of oil in 10 liters of water.

What happens as dissolved oxygen gets depleted? Plant and animal life disappear. Bacterial decomposition shifts from aerobic (O_2-requiring) to anaerobic (not requiring O_2). The products of metabolism change. Under aerobic conditions, C goes to CO_2, N to $NH_3 + HNO_3$, S to H_2SO_4, and P to H_3PO_4; however, under anaerobic conditions, C goes to CH_4, C_2H_4, etc.; N to NH_3 + amines; S to H_2S; and P to lower-valent phosphorus compounds. The point to note is that under anaerobic conditions, the decomposition products tend to stink and are more likely to be toxic.

Organic water pollutants include protein (domestic sewage, waste from creameries, slaughterhouses), fat (sewage, soap production, food processing), carbohydrates (sewage, paper mills), resin, coal, and oil. Inorganic pollutants might be acids, alkalies, heavy-metal cations, and certain anions. Acid mine drainage is a primary source of stream pollution, especially in coal-producing regions. The actual pollutants are H_2SO_4 and soluble iron salts formed by reaction of air and water on pyrites present in the coal seams. Certain types of bacteria also appear to be involved, but their role is not understood. It is estimated that about 4×10^9 kg of H_2SO_4 per year goes into United States streams, 60 percent of which originates in abandoned mines. Acid stream pollution is one of the primary causes of fish kill in the United States.

Figure 14.16 indicates some pollutant limits as recommended by the U.S. Public Health Service and as observed on a national average in public water supplies. The limits are quoted in milligrams per liter, which is essentially the same as parts per million by weight. For radium and strontium, units are picocuries per liter, where one curie is the radiation equivalent of one gram of radium (that is, 3.7×10^{10} disintegrations per second). Phosphate, which is not included in the listing, has not been considered a water pollutant in the same way as the toxic materials mentioned, but increasing runoff from fertilizer and detergent use greatly affects biological activity in streams and lakes. The problem is that phosphates are important nutrients for growth, and their excessive presence in domestic waste water can nourish biological processes beyond desirable rates. This phenomenon, known as *eutrophication* (from the Greek word *eutrophos,* meaning "well nourished"), can quickly choke an aquatic environment.

Physical pollution of water generally comes from turbidity, elevated temperature (i.e., thermal pollution), and suspended matter. Turbidity, which arises from soil erosion and colloidal wastes, can be corrected by addition of coagulants such as $FeCl_3$, alum, or $Fe_2(SO_4)_3$. Colloid particles (e.g., clay in natural waters, and proteins, fats, and carbohydrates in waste waters) are

FIG. 14.16 Water Pollutant Limits, mg/liter

Substance	Recommended USPHS	Observed U.S. average	Remarks
Ag	0.05	0.008	Limit set for cosmetic reasons, leads to discoloration of tissue
As	0.01	0.0001	Serious systemic poison, cumulative
Ba	1.0	0.034	Not common, serious toxic effect on heart
Cd	0.01	0.003	Seepage from electroplating, 15 ppm in food causes illness
Cr	0.05	0.0023	Not natural, suggests plating or tannery pollution
Cu	1	0.13	Essential and beneficial, adult needs 1 mg/day, detectable taste at 1–5 ppm, large doses may cause liver damage, used for algae control
Pb	0.05	0.013	Serious, cumulative body poison
^{226}Ra	3*	2.2*	Bone-seeking α emitter, destroys bone marrow
^{90}Sr	10*	<1.0*	Bone-seeking β emitter
Zn	5	0.19	Essential and beneficial, milky at 30 ppm, metallic taste at 40 ppm
Cl	250	27.6	Limit set for taste reasons, salty if too much
CN	0.01	0.00009	Rapid fatal poison, safety factor 100
F	1.2	0.32	Prevents dental caries in small amounts, mottling of enamel above 1.2 ppm
NO$_3$	45	6.3	Fertilizer runoff, can cause methemoglobinemia in infants
SO$_4$	250	46	Laxative effect above 750 mg/liter, often the cause of traveler's diarrhea

*Picocuries per liter.

usually stabilized by having negative charges at their surfaces, and these can be neutralized by addition of ions. Thermal pollution usually arises when manufacturing and power plants use streams for cooling. The result is decreased dissolved oxygen and increased rate of biochemical activity.

Physiological pollution of water comes from bad taste and objectionable odor. These usually go together, and the most frequent contaminants are sulfur compounds and nitrogen compounds.

Biological pollution of water may include bacteria, viruses, protozoa, parasites, and plant toxins. Infections of the intestinal tract (e.g., cholera, typhoid, and dysentery), polio, and infectious hepatitis have frequently been traced to contaminated water supplies. Generally, no check is made for these pathogenic contaminants because it is a 24-h problem to detect them, and that is usually too late. Instead, one looks for a benign indicator such as *coliform* bacteria, the presence of which alerts to fecal contamination.

Waste-water treatment can be classified into three successive stages: primary, secondary, and tertiary. In primary treatment the waste water is passed (1) through screens, to take out the large solids, (2) successively into grit and sedimentation tanks, where the smaller sediments are allowed to settle, and then (3) through a chlorine treatment, to destroy the bacteria. Most of the solids, about one-third of the BOD, and a few percent of the persistent organic compounds are removed in this way. In secondary treatment, further pollutant reduction is achieved by adding one of two possible processes: *trickling filter* or *activated-sludge treatment*. For trickling filter, a bed of gravel and rocks is provided through which the sewage is passed slowly enough that bacteria multiply on the stones and consume most of the organic matter. The process is about 75 percent effective. In the activated-sludge method, incoming sewage is inoculated with activated sludge (from recirculation), passed into an aeration tank, then into a sedimentation tank, and finally on to chlorine treatment. The process takes several hours but is 90 percent effective at removing organic wastes.

Tertiary treatments, because of their expense, are used only when drinking-quality water needs to be produced in a completely recycled system or from naturally contaminated sources. They are also used when it is necessary to remove organic compounds that do not yield to secondary treatment. One such method is to treat the nearly purified water with activated charcoal, filter off the charcoal after it has adsorbed the impurities, and then regenerate it with steam distillation. To remove phosphate, precipitation of the highly insoluble phosphates can be achieved by adding CaO, $Fe(OH)_3$, or $Al(OH)_3$. Other inorganic salts, such as nitrates, are very difficult to remove.

One promising technique for waste-water recovery is *reverse osmosis*. Instead of taking the waste out of the water, the water is squeezed out of the waste. Figure 14.17 shows a schematic representation of the setup. Salt water is fed into the top of the cell, the bottom part of which is blocked off by a semipermeable membrane.

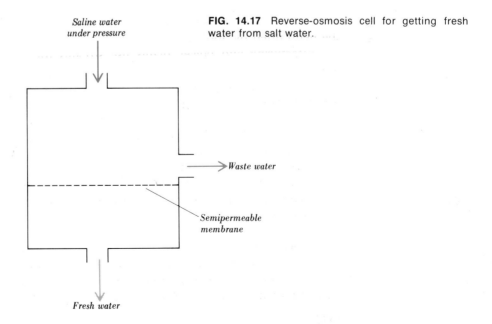

Saline water under pressure

FIG. 14.17 Reverse-osmosis cell for getting fresh water from salt water.

Waste water

Semipermeable membrane

Fresh water

Normally, as discussed in Sec. 8.4, fresh water tends to move toward the salty side, but by putting a sufficiently large pressure on the inflowing saline water, the normal osmotic flow can be reversed, and fresh water literally squeezed through the membrane so as to come out the bottom pipe.

The above process is not very cheap, but in certain installations it may be of value. One hopeful application for the future might be removal of nitrate ion. Recent generous use of nitrate fertilizers has raised NO_3^- levels in some ground waters to dangerous levels. The danger appears to be particularly great for very young infants who are particularly susceptible to meth-emoglobinemia (*blue-baby syndrome*). This comes from oxidation by nitrite, NO_2^-, of the iron in hemoglobin so it can no longer carry oxygen. Water intake by infants is disproportionately large, and the infant's digestive equipment is likely to harbor the wrong kind of bacteria, which reduce nitrate to nitrite. Nitrate removal remains an unsolved problem.

Important Concepts

isotopes of hydrogen
isotope effects
occurrence of hydrogen
relative abundance
hydrogen preparation
water gas
electrolytic hydrogen
properties of hydrogen
uses of hydrogen
compounds of hydrogen
Haber process

hydrides
hydrogen embrittlement
hydrogen bond
separation of oxygen
fractional liquefaction
allotropy
properties and uses of oxygen
properties and uses of ozone
compounds of oxygen
superoxides
peroxides

oxides
properties and structure of water
water as a solvent
free-energy change on dissolving
hydrates
water pollution
eutrophication
waste-water recovery
reverse osmosis

Exercises

*14.1 Hydrogen In some periodic classifications hydrogen is linked with group I and also with group VII. What justification can be given for this double assignment?

*14.2 Isotopes What are the three isotopes of hydrogen, and how do they differ from each other? If the respective masses are 1.0078, 2.0140, and 3.0161 amu, what would be the appropriate value to use for the atomic weight of "natural" hydrogen?

**14.3 Isotopes If you have a mole of water, how many of the molecules are likely to be HOD and how many are likely to be D_2O?

*14.4 **Abundance** One compilation of elemental abundance in the Earth's crust lists hydrogen as ninth in order; another lists it as third. Both are correct. Explain the discrepancy.

*14.5 **Hydrogen** Write equations for each of the following: (a) oxidation of hot carbon by steam, (b) oxidation of hot iron by steam, and (c) regeneration of the iron from (b) by using the product of (a).

*14.6 **Hydrogen** Diagram an electrolytic cell for preparing hydrogen from aqueous NaOH. Label anode and cathode, write electrode half-reactions, and indicate the directions of motion of ions in the system. What happens to the NaOH concentration as electrolysis proceeds? Explain briefly.

*14.7 **Hydrogen** Tell how you could make hydrogen by use of each of the following: Zn, CaH_2, Al, CH_4, H_2O.

14.8 **Hydrogen What would be the net lifting force of a balloon, 10.0 meters in diameter, filled with hydrogen at 0.95 atm pressure and 25°C in air under the same conditions? Ignore the mass of the balloon fabric. *Ans. 549 kg*

***14.9 **Hydrogen** Very low temperatures can be obtained by pumping on liquid hydrogen. If the normal boiling point of liquid hydrogen is 20.4°K and its heat of vaporization is 903 J/mol, to what pressure would you have to pump to produce a temperature of 15.0°K?

14.10 **Hydrogen The alloy $LaNi_5$ has a density of 8.2 g/cm³. It picks up enough hydrogen to form $LaNi_5H_7$. Calculate the number of H atoms stored per cubic centimeter. Compare with the number of H atoms per cubic centimeter in liquid hydrogen (density 0.070 g/cm³) and in solid hydrogen (density 0.090 g/cm³).

*14.11 **Hydrogen compounds** How can you prove that NaH contains hydrogen in the −1 oxidation state?

*14.12 **Hydrogen compounds** When added to water, which compound generates more hydrogen per gram of compound, NaH or CaH_2? How much NaH would you need to get the equivalent of a 50-liter tank of compressed hydrogen gas at 125 atm and 25°C?

***14.13 **Haber process** The equilibrium constant K_c for $3H_2 + N_2 \rightleftharpoons 2NH_3$ is equal to 6.41 at 600°K and 2.37×10^{-3} at 1000°K.

a Calculate the $\Delta H°$ of the reaction.

b Calculate the K at 500°C.

c Calculate the theoretical yield of NH_3 from 3 mol of H_2 and 1 mol of N_2 in a volume of 1.00 liter at 500°C. What will be the final total pressure?
Ans. 212 atm

14.14 **Hydrogen bond Draw a picture of two adjacent water molecules, showing population of electrons in the sp^3 hybrids. Show that the hydrogen bond can be visualized as involving migration of a proton from a normal covalent bond to a lone pair.

*14.15 **Hydrogen bond** In the HF_2^- ion the proton is located midway between two fluorine centers that are 0.226 nm apart. Draw a possible electron picture of this ion. How is the hydrogen bond in this ion different from the hydrogen bond in ice?

14.16 **Hydrogen bond How can you reconcile the fact that the strength of a hydrogen bond is about 20 kJ/mol and yet it only takes 41 kJ/mol to evaporate water from the liquid to the gaseous state? Note that the structure of water calls for bonding to about four neighbors.

14.17 **Hydrogen bond How would you expect the degree of hydrogen bonding in liquid ammonia to compare with that in liquid methane? Justify your answer.

14.18 **Hydrogen bond The molar heats of vaporization from the liquid to the gaseous state are 30.1, 16.3, 16.7, and 19.7 kJ/mol for HF, HCl, HBr, and HI, respectively. How can this be taken for evidence of hydrogen bonding?

14.19 **Oxygen How might you explain the fact that oxygen is so much more abundant in the Earth's crust than it is in the interior?

14.20 **Oxygen Describe briefly how oxygen is obtained from the air. Using Fig. 14.4, explain how a 20% O_2–80% N_2 gas mixture could be converted to a 60% O_2–40% N_2 gas mixture.

*14.21 **Oxygen** Liquid oxygen has a density of 1.149 g/cm³. Solid $KClO_3$ has a density of 2.32 g/cm³. Which would be the better source of oxygen on a per-unit-volume basis? What relative volumes would you have to take to get equal numbers of oxygen atoms?

*14.22 **Oxygen** Draw electronic formulas for N_2, O_2, and F_2. Point out what the problem is with O_2.

***14.23 Oxygen** Given that $\Delta H° = 285$ kJ for the reaction $3O_2(g) \longrightarrow 2O_3(g)$ and that the bond energy of O_2 is 494 kJ, what would you predict for each bond energy in the O_3 molecule? On this basis, how would you expect the bond length in O_3 to compare with that in O_2?

*14.24 Ozone** How is ozone related to oxygen? How can it be made from oxygen? What does ozone have to do with the opposition to the supersonic transport?

14.25 Oxygen Explain the following observations: A stream of oxygen gas is passed through a tinfoil-wrapped tube into a solution of potassium iodide and starch. Nothing happens, until a voltage is put on the tinfoil, at which point the starch–potassium iodide solution turns blue.

*14.26 Oxygen** How many liters of air at 1 atm and 25°C are required for complete oxidation of an 80-liter tank of gasoline? Assume C_8H_{18} at a density of 0.70 g/ml.   *Ans. 715,000 liters*

*14.27 Oxygen compounds** What special feature characterizes the class of oxygen compounds in which oxygen has each of the following oxidation states: $+2, -2, -1, -\frac{1}{2}$?

*14.28 Hydrogen peroxide** Hydrogen peroxide is unusual in that it is an acid, an oxidizing agent, and a reducing agent. Show how each of these functions comes about.

14.29 Hydrogen peroxide Make a model of the hydrogen peroxide molecule, starting with a tetrahedral array of sp^3 hybrid orbitals for each oxygen. Show that valence-shell–electron-pair repulsion can be used to rationalize the bond angles shown in Fig. 14.6.

14.30 Hydrogen peroxide Write balanced net equations for each of the following:

a H_2O_2 reduces $Cr_2O_7^{2-}$ to Cr^{3+} in acidic solution.

b H_2O_2 oxidizes H_2SO_3 to HSO_4^- in acidic solution.

c HO_2^- oxidizes $Cr(OH)_4^-$ to CrO_4^{2-} in basic solution.

d HO_2^- reduces IO_3^- to I^- in basic solution.
Write balanced half-reactions for the action of the peroxide in each case.

14.31 Water Describe the bonding in H_2O so as to account for the observed H—O—H angle of 104.52°. How would you expect this angle to change on substitution of deuterium for protium? Justify your answer.

14.32 Water In the ice structure there are two possible locations for each H atom. How many possibilities in all are there for the arrangement of H atoms about a given O? How many of these would correspond to a recognizable H_2O molecule?

14.33 Water Make a plot showing how the density of H_2O changes versus temperature from -10 to $+10°C$. Account for each characteristic feature of the curve.

14.34 Water Using the data of Fig. 14.11 and Archimedes' principle, calculate the fraction of an iceberg that is underwater. Assume pure water.   *Ans. 91.7%*

*14.35 Water** Explain qualitatively why C_2H_5OH is soluble in water but C_2H_6 is not.

14.36 Water The lattice energy of MgO, 3920 kJ/mol, is much greater than the lattice energy of AgCl, 904 kJ/mol. Yet, MgO is much more soluble in water than is AgCl. Suggest two reasons for the difference.

14.37 Water Using data from Figs. 14.13 and 14.14 compare NaCl and AgCl as to estimated values of $\Delta H°$, $\Delta S°$, and $\Delta G°$ for the reaction $MCl(s) \longrightarrow M^+(aq) + Cl^-(aq)$. Assume lattice vaporization increases the entropy by 0.200 kJ mol^{-1} deg^{-1}. What is the chief difference that makes one salt soluble and the other insoluble?

*14.38 Hydrates** Calculate the percent by weight of water of hydration in $NiSO_4 \cdot 7H_2O$.

*14.39 Hydrate** Epsom salt has the composition $MgSO_4 \cdot xH_2O$. If the water of hydration is driven off, the percent weight loss is 51.2%. What is the value of x?

*14.40 Hydrates** Reformulate each of the following as a hydrated oxide: MnO(OH), $SnO(OH)_2$, $VO_2(OH)$, $CrO(OH)$, $SO_2(OH)_2$, $Si(OH)_4$.

14.41 Water pollution What possible harm is there in flushing water pollutants into the sea?

14.42 Water pollution What is meant by BOD? How might it be determined? What is the simplest way to get rid of BOD in a public water supply?

14.43 Water pollution In desert areas of the southwestern United States, ground water supplies may contain appreciable amounts of nitrate. Why is this potentially dangerous? How might the nitrate be removed?

Chapter 15

CHEMICAL EQUILIBRIUM IN AQUEOUS SOLUTIONS

The key to understanding reactions in aqueous solutions is to recognize that reactions are often reversible and tend toward the equilibrium state. In this chapter we consider equilibria of two fundamental types: *dissociation,* i.e., between a dissolved, undissociated species and its component parts; and *solubility,* i.e., between a pure phase, usually a solid, and its characteristic species in solution. In addition, we shall consider simultaneous establishment of two or more equilibria in the same solution.

15.1
DISSOCIATION EQUILIBRIA IN AQUEOUS SOLUTION

When the weak acid HX is placed in water, some of it breaks up to give H^+ and X^-. When equilibrium is established, the back reaction of H^+ with X^- to form HX occurs at a rate just sufficient to balance the forward reaction. The equilibrium can be represented by the reversible equation*

$$HX \rightleftharpoons H^+ + X^-$$

for which the equilibrium condition is

$$\frac{[H^+][X^-]}{[HX]} = K$$

K, which is called the *dissociation constant* of HX, can also be designated as K_{diss}, K_{HX}, or K_a (where a stands for acid). Figure 15.1 shows representative values

---

*Strictly speaking, the dissociation of an acid in water is better written $HX + H_2O \rightleftharpoons H_3O^+ + X^-$, for which the equilibrium condition is

$$\frac{[H_3O^+][X^-]}{[HX][H_2O]} = K'$$

In dilute solutions where K' applies without correction for ionic attractions, the concentration of H_2O is essentially constant; so $[H_2O]$ is a constant number that can be combined with K'. Also, ignoring the water of hydration, we can replace $[H_3O^+]$ by $[H^+]$, which brings us to the same equilibrium condition as given above.

398

Chapter 15
Chemical
Equilibrium
in Aqueous
Solutions

FIG. 15.1   Dissociation Constants of Weak Acids (25°C)

Acid	Reaction	K_{diss}
Acetic	$HC_2H_3O_2 \rightleftharpoons H^+ + C_2H_3O_2^-$	1.8×10^{-5}
Nitrous	$HNO_2 \rightleftharpoons H^+ + NO_2^-$	4.5×10^{-4}
Hydrofluoric	$HF \rightleftharpoons H^+ + F^-$	6.7×10^{-4}
Hydrocyanic	$HCN \rightleftharpoons H^+ + CN^-$	4.0×10^{-10}
Sulfurous	$H_2SO_3 \rightleftharpoons H^+ + HSO_3^-$	1.3×10^{-2}

quoted for 25°C. The smaller the K_{diss}, the weaker the acid. Thus, HCN, with $K_{diss} = 4.0 \times 10^{-10}$, is a weaker acid than HF ($K_{diss} = 6.7 \times 10^{-4}$). When K_{diss} is 10 or greater, the acid is essentially 100 percent dissociated. For example, perchloric acid, $HClO_4$, is one of the strongest acids and has K_{diss} greater than 10. Similarly, HNO_3, HCl, and H_2SO_4 are common acids with high dissociation constants.

The dissociation constant also describes ions that dissociate as acids. For example, HSO_4^- can dissociate into H^+ and SO_4^{2-} and must be in equilibrium with these ions:

$$HSO_4^- \rightleftharpoons H^+ + SO_4^{2-}$$

$$K_{HSO_4^-} = \frac{[H^+][SO_4^{2-}]}{[HSO_4^-]} = 1.26 \times 10^{-2}$$

This is actually the second dissociation constant of H_2SO_4, since it applies to the second step of dissociation.

$$H_2SO_4 \rightleftharpoons H^+ + HSO_4^- \qquad K_I > 10$$
$$HSO_4^- \rightleftharpoons H^+ + SO_4^{2-} \qquad K_{II} = 1.26 \times 10^{-2}$$

K_I and K_{II} are, respectively, the first and second dissociation constants of sulfuric acid. The large value of K_I means that H_2SO_4 is practically 100 percent dissociated into H^+ and HSO_4^-. The moderate value of K_{II} means that a modest amount of HSO_4^- (about 10 percent in 0.1 M H_2SO_4) is dissociated into H^+ and SO_4^{2-}. Both equilibria exist simultaneously, and both constants must be satisfied by whatever is in solution, that is, H_2SO_4, H^+, HSO_4^-, and SO_4^{2-}.

For weak bases which can be written as MOH, dissociation can be represented as

$$MOH. \rightleftharpoons M^+ + OH^-$$

The equilibrium condition for this would be

$$\frac{[M^+][OH^-]}{[MOH]} = K_{diss}$$

However, most weak bases are more complicated than this. For example, aqueous ammonia is better described by the equilibrium

$$NH_3 + H_2O \rightleftharpoons NH_4^+ + OH^-$$

399

Section 15.1
Dissociation
Equilibria
in Aqueous
Solution

for which the equilibrium condition is

$$\frac{[NH_4^+][OH^-]}{[NH_3]} = K = 1.8 \times 10^{-5}$$

In the mass-action expression, the concentration of H_2O does not appear because it is essentially invariant. The constant K is referred to as the dissociation constant for aqueous ammonia.

Besides acids and bases, there are some salts that are weak electrolytes. An example is mercuric chloride, $HgCl_2$, which dissociates

$$HgCl_2 \rightleftharpoons HgCl^+ + Cl^-$$

for which

$$\frac{[HgCl^+][Cl^-]}{[HgCl_2]} = K_{diss} = 3.3 \times 10^{-7}$$

Mercuric chloride is an exception to the usual rule that salts are 100 percent dissociated in solution. However, $HgCl_2$ is not unique. For instance, cadmium sulfate, $CdSO_4$, is a weak electrolyte with a dissociation constant of 5×10^{-3}.

As was pointed out in Sec. 13.4, it makes no difference whether equilibrium is approached by starting with material on the left side of a chemical equation or material on the right side. Change occurs to form the missing material in sufficient concentration to establish equilibrium. For weak electrolytes this means that the same equilibrium state is produced by having the electrolyte dissociate as is produced by having the component ions associate. Specifically, the same final solution would result whether we placed 1 mol of $HC_2H_3O_2$ in a liter of water or 1 mol of H^+ plus 1 mol of $C_2H_3O_2^-$. In either case, the condition for equilibrium would be the same:

$$\frac{[H^+][C_2H_3O_2^-]}{[HC_2H_3O_2]} = 1.8 \times 10^{-5}$$

When ions are mixed and association occurs, the chemical equation can be written to stress the direction of the net reaction. For example, when solutions of HCl and $NaC_2H_3O_2$ are mixed, the equation can be written

$$H^+ + C_2H_3O_2^- \rightleftharpoons HC_2H_3O_2$$

for which

$$\frac{[HC_2H_3O_2]}{[H^+][C_2H_3O_2^-]} = K_{assoc}$$

The numerical value of K_{assoc} is 5.6×10^4, which is the reciprocal of $K_{diss} = 1.8 \times 10^{-5}$ for acetic acid.

Association occurs whenever the constituent parts of a weak electrolyte are mixed. Thus, when solutions of NH_4Cl and $NaOH$ are mixed, NH_4^+ ions associate with OH^- ions to form NH_3 and H_2O.

Like any equilibrium constant, K_{diss} must be experimentally determined. Once its value is known at a given temperature, it can be used for all calculations involving that equilibrium at the given temperature.

EXAMPLE 1

What is the concentration of all solute species in a solution labeled 1.00 M $HC_2H_3O_2$? What percent of the acid is dissociated?

$$HC_2H_3O_2 \rightleftharpoons H^+ + C_2H_3O_2^- \qquad K_{\text{diss}} = 1.8 \times 10^{-5}$$

Solution

Let x equal the moles per liter of $HC_2H_3O_2$ that dissociate to establish equilibrium. If x moles of $HC_2H_3O_2$ dissociate, then x mol of H^+ and x mol of $C_2H_3O_2^-$ must be formed. The situation is summarized as follows:

Initial concentration, mol/liter	Equilibrium concentration, mol/liter
$[HC_2H_3O_2] = 1.00$	$[HC_2H_3O_2] = 1.00 - x$
$[H^+] = 0$	$[H^+] = x$
$[C_2H_3O_2^-] = 0$	$[C_2H_3O_2^-] = x$

At equilibrium

$$\frac{[H^+][C_2H_3O_2^-]}{[HC_2H_3O_2]} = 1.8 \times 10^{-5} = \frac{(x)(x)}{1.00 - x}$$

Solving this equation by the quadratic formula (Appendix A4.3) gives $x = 0.0042$. Therefore, at equilibrium (with due regard for significant figures)

$$[HC_2H_3O_2] = 1.00 - x = 1.00 \; M$$
$$[H^+] = x = 0.0042 \; M$$
$$[C_2H_3O_2^-] = x = 0.0042 \; M$$

The percent dissociation is given by the number of moles of $HC_2H_3O_2$ dissociated, divided by the number of moles of $HC_2H_3O_2$ originally available:

$$\text{Percent dissociation} = \frac{100 \times 0.0042}{1.00} = 0.42\%$$

It should be noted that much of the algebraic work involved in solving equilibrium problems can be avoided by paying attention to chemical facts which may suggest laborsaving approximations. Thus, in Example 1, $HC_2H_3O_2$ is a weak acid, and so it cannot be much dissociated. In other words, x must be small compared with 1.00 and may be neglected when subtracted from 1.00. Thus, instead of solving the exact equation

$$1.8 \times 10^{-5} = \frac{x^2}{1.00 - x}$$

we can solve the approximate equation

$$1.8 \times 10^{-5} \sim \frac{x^2}{1.00}$$

We quickly get

$$x \sim \sqrt{1.8 \times 10^{-5}} = \sqrt{18 \times 10^{-6}} = 4.2 \times 10^{-3}$$

Checking the approximation and paying due attention to significant figures, we find that $1.00 - x = 1.00 - 4.2 \times 10^{-3} = 1.00$, as assumed.

EXAMPLE 2

Suppose that 1.00 mol of HCl and 1.00 mol of $NaC_2H_3O_2$ are mixed in enough water to make a liter of solution. What will be the concentration of the various species in the final solution?

Solution

Since HCl and $NaC_2H_3O_2$ are strong electrolytes, they are 100 percent dissociated in solution. The Na^+ and Cl^- do not associate and so can be ignored. The problem is thus one of associating H^+ and $C_2H_3O_2^-$ to form $HC_2H_3O_2$. If we let y represent the moles of H^+ and of $C_2H_3O_2^-$ that associate per liter, the situation is summarized as follows:

Initial concentration, M	Equilibrium concentration, M
$[H^+] = 1.00$	$[H^+] = 1.00 - y$
$[C_2H_3O_2^-] = 1.00$	$[C_2H_3O_2^-] = 1.00 - y$
$[HC_2H_3O_2] = 0$	$[HC_2H_3O_2] = y$

why

$$\frac{y^2}{1-y} = 1.8 \times 10^{-5}$$

At equilibrium

$$\frac{[H^+][C_2H_3O_2^-]}{[HC_2H_3O_2]} = 1.8 \times 10^{-5} = \frac{(1.00 - y)(1.00 - y)}{y}$$

This equation can be solved by applying the quadratic formula which gives $y = 0.996$. Since y is not small compared with 1.00, the approximation made in Example 1 that $1.00 - y \sim 1.00$ cannot be used.

A simpler way to solve the above problem would be to use the fact that the final equilibrium state does not depend on what route we take to get to it. We simply note that when 1.00 mol of H^+ and 1.00 mol of $C_2H_3O_2^-$ are mixed, the resulting solution is the same as if H^+ and $C_2H_3O_2^-$ completely reacted to form 1.00 mol of $HC_2H_3O_2$, which then dissociated to establish equilibrium. The problem thus becomes identical with Example 1. We can write directly the equilibrium concentrations as determined by the simple calculation on page 400.

$$[H^+] = 0.0042 \ M \qquad\qquad [C_2H_3O_2^-] = 0.0042 \ M$$

$$[HC_2H_3O_2] = 1.00 \ M \qquad [Na^+] = 1.00 \ M \qquad [Cl^-] = 1.00 \ M$$

EXAMPLE 3

What are the concentrations of species and the percent dissociation in 0.10 M $HC_2H_3O_2$?

Solution

Let x = moles of $HC_2H_3O_2$ that dissociate per liter.

Then x = final concentration of H^+ and $C_2H_3O_2^-$ formed.

$0.10 - x$ = final concentration of $HC_2H_3O_2$ left undissociated.

At equilibrium

$$\frac{[H^+][C_2H_3O_2^-]}{[HC_2H_3O_2]} = 1.8 \times 10^{-5} = \frac{(x)(x)}{0.10 - x}$$

Assuming that x is small compared with 0.10,

$$\frac{x^2}{0.10} \sim 1.8 \times 10^{-5}$$

$$x^2 \sim 1.8 \times 10^{-6}$$

$$x \sim 1.3 \times 10^{-3}$$

Therefore, at equilibrium

$$[H^+] = x = 0.0013 \ M$$

$$[C_2H_3O_2^-] = x = 0.0013 \ M$$

$$[HC_2H_3O_2] = 0.10 - x = 0.10 \ M$$

$$\text{Percent dissociation} = \frac{100 \times 0.0013}{0.10} = 1.3\%$$

Examples 1 and 3 illustrate a general fact: When a weak electrolyte is diluted, the percent dissociation *increases* although the concentration of each species *decreases*. There is tenfold dilution in going from 1.00 M $HC_2H_3O_2$ to 0.10 M $HC_2H_3O_2$, but the concentration of H^+ does not decrease tenfold. It decreases only from 0.0042 to 0.0013 M. This is consistent with the fact that in the more dilute solution a greater percentage of the acid is dissociated.

15.3
DISSOCIATION OF WATER; pH

In the preceding section we ignored the fact that water itself is slightly dissociated:

$$H_2O \rightleftharpoons H^+ + OH^-$$

Therefore, in pure water and in all aqueous solutions, we must satisfy the condition

$$\frac{[H^+][OH^-]}{[H_2O]} = K$$

In dilute solutions the concentration of H_2O is a constant big number which can be combined with K as follows:

$$K[H_2O] = K_w = [H^+][OH^-]$$

K_w is usually called the dissociation constant, or *ion product*, of water. It equals 1.0×10^{-14} at 25°C but increases slightly with temperature.

In pure water all the H^+ and the OH^- must come from the dissociation of water molecules. If x mol of H^+ are produced per liter, x mol of OH^- must be simultaneously produced.

$$[H^+][OH^-] = 1.0 \times 10^{-14}$$
$$(x)(x) = 1.0 \times 10^{-14}$$
$$x = 1.0 \times 10^{-7}$$

Thus, in pure water the concentrations of H^+ and OH^- are each 1.0×10^{-7} M. This very small concentration is to be compared with the water concentration of approximately 55.4 mol/liter.

If an acid is added to water, the hydrogen-ion concentration increases above 1.0×10^{-7} M. The product $[H^+][OH^-]$ must remain equal to 1.0×10^{-14}; consequently, the hydroxide-ion concentration decreases below 1.0×10^{-7} M. Similarly, when a base is added to water, the concentration of H^+ decreases below 1.0×10^{-7} M. As a convenience for working with small concentrations, the *pH scale* has been devised to express the concentration of H^+. By definition,

$$pH = -\log[H^+]$$

For example, in pure water at 25°C, where the concentration of H^+ is 1.0×10^{-7} M, the pH is 7. All neutral solutions (at 25°C) have a pH of 7; acid solutions, less than 7; basic solutions, greater than 7. (For a review of logarithms as applied to pH, see Appendix A4.2.)

EXAMPLE 4

What is the pH of 0.20 M HCl?

Solution

Practically all the H^+ comes from the 100 percent dissociation of the strong electrolyte HCl.

$$[H^+] = 0.20\ M = 2.0 \times 10^{-1}\ M$$
$$pH = -\log(2.0 \times 10^{-1}) = 1 - 0.30 = 0.70$$

EXAMPLE 5

What is the pH of 0.10 M NaOH?

Solution

NaOH is a strong electrolyte and accounts for essentially all the OH^- in the solution.

$$[OH^-] = 0.10\ M$$

$$[H^+] = \frac{K_w}{[OH^-]} = \frac{1.0 \times 10^{-14}}{0.10} = 1.0 \times 10^{-13} \, M$$

$$pH = -\log (1.0 \times 10^{-13}) = 13.00$$

In Examples 4 and 5 the H_2O dissociation contributes negligibly to $[H^+]$ in the acidic solution and to $[OH^-]$ in the basic solution. The reason for this is that acids and bases repress the dissociation of H_2O. To illustrate, in Example 5, added OH^- represses dissociation of H_2O so that only 1.0×10^{-13} mol of H^+ per liter is produced. This means that only 1.0×10^{-13} mol of OH^- per liter comes from the dissociation of H_2O, an amount that is indeed negligible compared with the 0.10 mol that comes from 0.10 M NaOH.

15.4
TITRATION AND INDICATORS

The same equilibrium constant that describes dissociation of H_2O also describes association of H^+ and OH^- to form H_2O. Such association occurs in neutralization reactions (Sec. 9.2) and is the basis of the process of *titration*, the progressive addition of an acid to a base, or vice versa.

At each step in titration the expression $[H^+][OH^-] = 1.0 \times 10^{-14}$ must be satisfied. Figure 15.2 shows what happens to H^+ and OH^- as solid NaOH is added stepwise to 0.010 mol of HCl in a liter of H_2O. (We add solid NaOH instead of aqueous NaOH in order to avoid volume expansion of the system.) As NaOH is progressively added, the original solution changes from acidic (pH less than 7) to basic (pH greater than 7). The titration could be represented graphically by plotting the concentration of H^+ against moles of added NaOH. However, since the H^+ concentration changes by a factor of 10 billion during the experiment, it would be hard to get all the values on the same scale. Not so with the pH. It changes only by a factor of 6 and is a convenient representation of what happens to the solution during the titration.

Figure 15.3 shows the change of pH as solid NaOH is added to a liter of 0.010 M HCl. Such a pH curve is typical of the titration of any strong acid with any strong base. The important thing to note is that as the neutral point is approached, there is a sharp rise in pH. At this point even a trace of NaOH increases the pH greatly. Any method that locates the point at which the pH changes rapidly can be used to detect the *equivalence point* of a titration, i.e., the point at which equivalent amounts of base and acid have been mixed.

One method for determining the equivalence point uses the fact that many dyes have colors that are sensitive to hydrogen-ion concentration. Such dyes can be used as *indicators* to give information about the pH of a solution. Indicators can be represented as weak acids, HIn, which dissociate to give H^+ and In^-. As a weak acid, HIn must satisfy the condition

$$\frac{[H^+][In^-]}{[HIn]} = K \quad \text{or} \quad \frac{[In^-]}{[HIn]} = \frac{K}{[H^+]}$$

**FIG. 15.2   Change of Concentration of H⁺ and OH⁻
on Progressive Addition of Solid NaOH
to 1 Liter of 0.010 M HCl**

Moles of NaOH added	$[H^+]$	$[OH^-]$	pH
0.000	0.010	1.0×10^{-12}	2.00
0.001	0.009	1.1×10^{-12}	2.04
0.002	0.008	1.3×10^{-12}	2.10
0.003	0.007	1.4×10^{-12}	2.15
0.004	0.006	1.7×10^{-12}	2.23
0.005	0.005	2.0×10^{-12}	2.30
0.006	0.004	2.5×10^{-12}	2.40
0.007	0.003	3.3×10^{-12}	2.52
0.008	0.002	5.0×10^{-12}	2.70
0.009	0.001	1.0×10^{-11}	3.00
0.010	1.0×10^{-7}	1.0×10^{-7}	7.00
0.011	1.0×10^{-11}	0.001	11.00
0.012	5.0×10^{-12}	0.002	11.30
0.013	3.3×10^{-12}	0.003	11.48
0.014	2.5×10^{-12}	0.004	11.60
0.015	2.0×10^{-12}	0.005	11.70
0.016	1.7×10^{-12}	0.006	11.77
0.017	1.4×10^{-12}	0.007	11.85
0.018	1.3×10^{-12}	0.008	11.90
0.019	1.1×10^{-12}	0.009	11.96
0.020	1.0×10^{-12}	0.010	12.00

from which it is evident that the ratio $[In^-]/[HIn]$ is inversely proportional to the hydrogen-ion concentration. If In^- and HIn have different colors, the color of the solution may change depending on which species is predominant. For phenolphthalein, HIn is colorless, but In^- is red. At high hydrogen-ion concentration, the ratio $[In^-]/[HIn]$ is small, and the colorless species HIn is dominant. Conversely, when $[H^+]$ is small, the red species In^- is dominant. Figure 15.4 lists the characteristic colors of some common indicators.

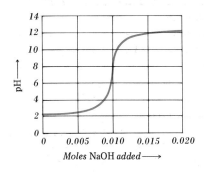

FIG. 15.3 Titration curve for addition of solid NaOH to 1 liter of 0.010 M HCl.

FIG. 15.4   Typical Indicators and Their Color Ranges

Indicator	pH at which color changes	Color at lower pH	Color at higher pH
Methyl violet	1	Yellow	Blue
Erythrosin	3	Orange	Red
Bromophenol blue	4	Yellow	Blue
Methyl orange	4	Red	Yellow
Methyl red	5	Red	Yellow
p-Nitrophenol	6	Colorless	Yellow
Bromothymol blue	7	Yellow	Blue
Phenolphthalein	9	Colorless	Red
Thymolphthalein	10	Colorless	Blue
Alizarin yellow	11	Yellow	Red

In Fig. 15.3 the pH rises so sharply at the equivalence point that any one of the indicators of Fig. 15.4 except methyl violet and alizarin yellow could be used to tell when enough NaOH had been added to neutralize 1 liter of 0.010 M HCl.

15.5
BUFFER SOLUTIONS

For practically all biological processes, as well as in many other chemical changes, it is important that the pH not deviate very much from a fixed value. For example, the proper functioning of human blood in carrying oxygen to the cells is dependent on maintaining a pH very near to 7.4. Indeed, for a particular individual, there is a difference of but 0.02 pH unit between venous and arterial blood in spite of numerous acid- and base-producing reactions in the cells.

The near constancy of pH is due to what is called *buffering* action of an acid-base equilibrium. A buffer contains both acid and base and can respond to addition of either. Let us consider, for example, a solution that contains acetic acid molecules and acetate ions. The principal equilibrium is written

$$HC_2H_3O_2 \rightleftharpoons H^+ + C_2H_3O_2^-$$

for which

$$\frac{[H^+][C_2H_3O_2^-]}{[HC_2H_3O_2]} = K$$

Solving this expression for $[H^+]$, we get

$$[H^+] = K\frac{[HC_2H_3O_2]}{[C_2H_3O_2^-]}$$

which indicates that the hydrogen-ion concentration depends on K and on the ratio of the concentrations of undissociated acetic acid to acetate ion. Taking the negative logarithm of both sides, we get

$$pH = -\log K - \log \frac{[HC_2H_3O_2]}{[C_2H_3O_2^-]}$$

Introducing the symbol pK to represent $-\log K$ of the acid, we can write

$$pH = pK - \log \frac{[HC_2H_3O_2]}{[C_2H_3O_2^-]}$$

In a particular solution containing equal moles of $HC_2H_3O_2$ and $NaC_2H_3O_2$ the ratio $[HC_2H_3O_2]/[C_2H_3O_2^-]$ will be unity, log 1 equals zero, and so pH = pK. Where the ratio $[HC_2H_3O_2]/[C_2H_3O_2^-]$ is not far from unity, the pH will not differ much from pK. Thus, a mixture of acetic acid and acetate ion is said to be a buffer for pH of $-\log (1.8 \times 10^{-5})$, or 4.74. If a small amount of strong acid is added to such a solution, some of the acetate ion is converted to acetic acid; if base is added, some of the acetic acid is converted to acetate ion. In either case the ratio $[HC_2H_3O_2]/[C_2H_3O_2^-]$ changes slightly from unity, and the pH changes even less—not nearly so much as in the absence of the buffer.

EXAMPLE 6

Calculate the pH of a solution made by adding 0.0010 mol of NaOH to 100 ml of a solution that is 0.50 M $HC_2H_3O_2$ and 0.50 M $NaC_2H_3O_2$.

Solution

In the original 100 ml of solution there are 0.050 mol of $HC_2H_3O_2$ and 0.050 mol of $C_2H_3O_2^-$. We assume that all the added 0.0010 mol of OH^- reacts to convert an equivalent amount of $HC_2H_3O_2$ into $C_2H_3O_2^-$. This gives 0.049 mol of $HC_2H_3O_2$ and 0.051 mol of $C_2H_3O_2^-$ in the final solution. Since the volume stays at 0.100 liter, the respective concentrations would be 0.49 M and 0.51 M. Hence

$$pH = -\log (1.8 \times 10^{-5}) - \log \frac{0.49}{0.51} = +4.74 + 0.017 = +4.76$$

In contrast, when 0.0010 mol of NaOH is added to 100 ml of *water*, the pH goes from 7.0 to 12.0.

In general, any solution of weak acid plus the salt of that acid can function as a buffer. The buffer region—i.e., the region in which the pH changes most slowly—is located around the pK of the acid. Similarly, a solution of weak base plus the salt of that base can function as a buffer to keep OH^- concentration approximately equal to K_{diss} of the base.

As is evident, there are as many possible buffers as there are weak acids and weak bases. In human blood there are a number of buffers acting simultaneously. These include (1) dissolved CO_2 and HCO_3^-, (2) $H_2PO_4^-$ and HPO_4^{2-}, and (3) the various proteins which can accept hydrogen ions.

The term *complex ion* refers to a charged particle that contains more than one atom. Certain complex ions, for example, SO_4^{2-}, act like simple ions because for all practical purposes they do not dissociate into smaller fragments. Others, however, may dissociate to establish an equilibrium between the complex ion and its component pieces. Thus, for example, in a solution containing the silver-ammonia complex ion, $Ag(NH_3)_2^+$, there is an equilibrium between the complex ion, the silver ion, and ammonia molecules. Although the dissociation probably occurs in steps, we can write the overall equilibrium as

$$Ag(NH_3)_2^+ \rightleftharpoons Ag^+ + 2NH_3$$

for which the equilibrium condition is

$$\frac{[Ag^+][NH_3]^2}{[Ag(NH_3)_2^+]} = 6 \times 10^{-8}$$

The small value of this constant indicates that the complex is very stable with respect to dissociation. When $AgNO_3$ and aqueous NH_3 are mixed, enough complex must form to satisfy the above equilibrium condition. Furthermore, if the concentration of NH_3 is increased, the concentration of Ag^+ must decrease, as required by constancy of the mass-action expression.

Figure 15.5 lists some common complex ions and their overall equilibrium constants. The smaller the K, the less dissociated is the respective complex. Of the three complexes shown for silver, the cyanide complex is the most stable. In a solution containing Ag^+, CN^-, $S_2O_3^{2-}$, and NH_3, the silver-cyanide complex would be preferentially formed.

15.7

AQUEOUS EQUILIBRIA INVOLVING SOLUBILITY OF IONIC SOLIDS

When excess ionic solid is placed in water, equilibrium is established between the ions in the saturated solution and the excess solid phase. For example,

FIG. 15.5   Overall Dissociation Constants for Some Complex Ions

Complex ion	Reaction	K
Copper-ammonia	$Cu(NH_3)_4^{2+} \rightleftharpoons Cu^{2+} + 4NH_3$	1.0×10^{-12}
Cobaltous-ammonia	$Co(NH_3)_6^{2+} \rightleftharpoons Co^{2+} + 6NH_3$	4.0×10^{-5}
Cobaltic-ammonia	$Co(NH_3)_6^{3+} \rightleftharpoons Co^{3+} + 6NH_3$	6.3×10^{-36}
Silver-ammonia	$Ag(NH_3)_2^+ \rightleftharpoons Ag^+ + 2NH_3$	6×10^{-8}
Silver-thiosulfate	$Ag(S_2O_3)_2^{3-} \rightleftharpoons Ag^+ + 2S_2O_3^{2-}$	6×10^{-14}
Silver-cyanide	$Ag(CN)_2^- \rightleftharpoons Ag^+ + 2CN^-$	1.8×10^{-19}
Ferric-thiocyanate	$FeNSC^{2+} \rightleftharpoons Fe^{3+} + NSC^-$	1×10^{-3}
Mercuric-cyanide	$Hg(CN)_4^{2-} \rightleftharpoons Hg^{2+} + 4CN^-$	4×10^{-42}

409

Section 15.7
Aqueous
Equilibria
Involving
Solubility of
Ionic Solids

with silver chloride the equilibrium is

$$AgCl(s) \rightleftharpoons Ag^+ + Cl^-$$

for which

$$\frac{[Ag^+][Cl^-]}{[AgCl(s)]} = K$$

The *concentration* of silver chloride in the *solid phase* is a constant and does not change, no matter how much solid there is in contact with the solution. We can therefore write

$$[Ag^+][Cl^-] = K[AgCl(s)] = K_{sp}$$

The constant K_{sp} is called the *solubility product,* and the expression $[Ag^+][Cl^-]$, the *ion product*. The equation states that the ion product equals K_{sp} when saturated solution is in equilibrium with excess solid. There is no separate restriction on what the concentrations of Ag^+ and Cl^- must be. The concentration of Ag^+ can have any value so long as the Ag^+ concentration times the Cl^- concentration is equal to K_{sp}.

The numerical value of K_{sp}, as of any equilibrium constant, must be determined by experiment. Once determined, it can be tabulated for future use. (Appendix 6 contains some typical values.) The kind of experiment that might be done is as follows: Solid barium sulfate is ground up and thoroughly agitated with a liter of H_2O at 25°C until a saturated solution is formed. The solution is then filtered and evaporated to get rid of solvent; deposited $BaSO_4$ is dried and analyzed. The solubility of $BaSO_4$ thus determined would be 3.9×10^{-5} mol/liter of H_2O at 25°C.

Like practically all salts, $BaSO_4$ is a strong electrolyte and so is 100 percent dissociated into ions. Therefore, when 3.9×10^{-5} mol/liter of $BaSO_4$ dissolves, it forms 3.9×10^{-5} M Ba^{2+} and 3.9×10^{-5} M SO_4^{2-}. Therefore, for the equilibrium

$$BaSO_4(s) \rightleftharpoons Ba^{2+} + SO_4^{2-}$$

we can write

$$K_{sp} = [Ba^{2+}][SO_4^{2-}] = (3.9 \times 10^{-5})(3.9 \times 10^{-5}) = 1.5 \times 10^{-9}$$

This means that for any solution in equilibrium with solid $BaSO_4$, the concentration of Ba^{2+} times that of SO_4^{2-} should be equal to 1.5×10^{-9}. If $[Ba^{2+}]$ multiplied by $[SO_4^{2-}]$ is less than 1.5×10^{-9}, the solution is unsaturated, and $BaSO_4$ should dissolve to increase the concentrations of Ba^{2+} and SO_4^{2-}. If the product of $[Ba^{2+}]$ and $[SO_4^{2-}]$ is greater than 1.5×10^{-9}, $BaSO_4$ should precipitate in order to decrease the concentrations of Ba^{2+} and SO_4^{2-}.

When only $BaSO_4$ is placed in pure water, the concentrations of Ba^{2+} and SO_4^{2-} are necessarily equal. On the other hand, it is possible to prepare solutions where unequal concentrations of Ba^{2+} and SO_4^{2-} are in equilibrium with solid $BaSO_4$. As an illustration, suppose unequal amounts of $BaCl_2$ and Na_2SO_4 solutions are mixed. A precipitate of $BaSO_4$ forms if K_{sp} of $BaSO_4$ is exceeded. However, there is no requirement that $[Ba^{2+}] = [SO_4^{2-}]$, since the two ions come from different sources. Alternatively, $BaSO_4$ solid might be shaken up with an Na_2SO_4 solution. Some $BaSO_4$ solid would dissolve, but in the final solution the concentration of SO_4^{2-} would be considerably greater than the concentration of Ba^{2+}.

EXAMPLE 7

Given that the K_{sp} for $RaSO_4(s) \rightleftharpoons Ra^{2+} + SO_4^{2-}$ is 4×10^{-11}, calculate the solubility of radium sulfate (*a*) in pure water and (*b*) in 0.10 *M* Na_2SO_4.

Solution

a Let $x =$ moles of $RaSO_4$ that dissolve per liter of water. Then in the saturated solution

$$[Ra^{2+}] = x \text{ mol/liter}$$
$$[SO_4^{2-}] = x \text{ mol/liter}$$
$$[Ra^{2+}][SO_4^{2-}] = K_{sp} = 4 \times 10^{-11}$$
$$(x)(x) = 4 \times 10^{-11}$$
$$x = \sqrt{40 \times 10^{-12}} = 6 \times 10^{-6} \text{ mol/liter}$$

b Let $y =$ moles of $RaSO_4$ that dissolve per liter of 0.10 *M* Na_2SO_4. This dissolving produces y moles of Ra^{2+} and y moles of SO_4^{2-}. The solution already contains 0.10 *M* SO_4^{2-}. Thus, in the final saturated solution

$$[Ra^{2+}] = y \text{ mol/liter}$$
$$[SO_4^{2-}] = y + 0.10 \text{ mol/liter}$$
$$[Ra^{2+}][SO_4^{2-}] = y(y + 0.10) = K_{sp} = 4 \times 10^{-11}$$

Since K_{sp} is very small, y is negligible compared with 0.10. Therefore

$$y + 0.10 \approx 0.10$$
$$[Ra^{2+}][SO_4^{2-}] \approx (y)(0.10) \approx 4 \times 10^{-11}$$
$$y \approx \frac{4 \times 10^{-11}}{0.10} = 4 \times 10^{-10} \text{ mol/liter}$$

It is interesting to note from the preceding example that $RaSO_4$ is less soluble in an Na_2SO_4 solution than it is in pure water. This is an example of the *common-ion effect;* i.e., the solubility of an ionic salt is decreased by presence of another solute that furnishes an ion in common. Thus, radium sulfate is less soluble in solutions containing either radium ion or sulfate ion than it is in water. The greater the concentration of the common ion, the less radium sulfate can dissolve. However, if the common ion is present in negligible concentration, it has no appreciable effect on the solubility. This is illustrated in Example 8.

As noted in Sec. 13.2 the mass-action expression contains concentrations raised to powers that correspond to coefficients in the chemical equation. The ion product is also a mass-action expression, and so it must be formed by rais-

411

Section 15.7
Aqueous
Equilibria
Involving
Solubility of
Ionic Solids

ing each concentration to the power that corresponds to the coefficient of that ion in the dissolving equation. Thus, for $Mg(OH)_2(s) \rightleftharpoons Mg^{2+} + 2OH^-$, we have $K_{sp} = [Mg^{2+}][OH^-]^2$.

EXAMPLE 8

Given that magnesium hydroxide, $Mg(OH)_2$, is a strong electrolyte and has a solubility product of 8.9×10^{-12}, calculate the solubility of $Mg(OH)_2$ in water.

Solution

$(x)\ [2x]$

Let $x =$ moles of $Mg(OH)_2$ that dissolve per liter. According to the equation $Mg(OH)_2(s) \rightleftharpoons Mg^{2+} + 2OH^-$, x mol of $Mg(OH)_2$ dissolve to give x mol of Mg^{2+} and $2x$ mol of OH^-. A trace of OH^- is also furnished by dissociation of H_2O, but H_2O is a very weak electrolyte, and so we can assume this will be only a negligible amount. At equilibrium we then have

$[Mg^{2+}] = x$ mol/liter

$[OH^-] \sim 2x$ mol/liter

For the saturated solution the equilibrium is $Mg(OH)_2(s) \rightleftharpoons Mg^{2+} + 2OH^-$, and $K_{sp} = 8.9 \times 10^{-12} = [Mg^{2+}][OH]^2$.

$[Mg][OH]^2 = 8.9 \times 10^{-2}$ x $2x$

Substituting, we get

$(4x^3) =$

$(x)(2x)^2 = 8.9 \times 10^{-12}$

$4x^3 = 8.9 \times 10^{-12}$

$x = \sqrt[3]{2.2 \times 10^{-12}} = 1.3 \times 10^{-4}$ mol/liter

Each exponent of the ion product applies to the concentration of the ion specified inside the brackets, no matter where that ion comes from. For instance, in the following example practically all the OH^- comes from NaOH, but its concentration still must be squared.

EXAMPLE 9

Calculate the solubility of $Mg(OH)_2$ in 0.050 M NaOH.

Solution

Let $x =$ moles of $Mg(OH)_2$ that dissolve per liter. This forms x mol of Mg^{2+} and $2x$ mol of OH^-. The solution already contains 0.050 mol of OH^-, and so equilibrium concentrations would be as follows:

$[Mg^{2+}] = x$ mol/liter

$[OH^-] = 2x + 0.050$ mol/liter

$[Mg^{2+}][OH^-]^2 = (x)(2x + 0.050)^2 = K_{sp}$

$(x)(2x + 0.050)^2 = 8.9 \times 10^{-12}$

On the basis of the small value of K_{sp} we can guess that x is probably a very small number; so $2x$ can be neglected where it is added to 0.050. We would then have the approximate relation

$(x)(0.050)^2 \sim 8.9 \times 10^{-12}$

$x = 3.6 \times 10^{-9}$ mol/liter

15.8
PRECIPITATION FROM AQUEOUS SOLUTIONS

One of the most useful applications of solubility product is to predict whether or not precipitation should occur when two solutions are mixed. If, after mixing, an ion product would exceed K_{sp}, then precipitation should occur.

EXAMPLE 10

Should precipitation occur when 50 ml of 5.0×10^{-4} M $Ca(NO_3)_2$ is mixed with 50 ml of 2.0×10^{-4} M NaF? The K_{sp} of CaF_2 is 1.7×10^{-10}. Assume final volume is 100 ml.

Solution

In order to solve such a problem, we first calculate the concentration of ions in the mixture, assuming no precipitation occurs. Thus, the Ca^{2+} from the 5.0×10^{-4} M $Ca(NO_3)_2$ solution would be 2.5×10^{-4} M in the final mixture because of twofold dilution when the solutions are mixed. Likewise, F^- would be diluted to 1.0×10^{-4} M. Therefore, if no precipitation occurred, the final solution would have

$[Ca^{2+}] = 2.5 \times 10^{-4}$ M and $[F^-] = 1.0 \times 10^{-4}$ M

To determine whether precipitation should occur, it is necessary to see whether the ion product exceeds the solubility product 1.7×10^{-10}. For the equilibrium

$CaF_2(s) \rightleftharpoons Ca^{2+} + 2F^-$

the ion product is $[Ca^{2+}][F^-]^2$ and, in the present mixture, has the numerical value

$[Ca^{2+}][F^-]^2 = (2.5 \times 10^{-4})(1.0 \times 10^{-4})^2 = 2.5 \times 10^{-12}$

Since this number does not exceed 1.7×10^{-10}, which is the K_{sp} of CaF_2, precipitation should not occur. The solution obtained as the final mixture is unsaturated with respect to precipitation of CaF_2.

In order to precipitate a salt, the ion product must be made to exceed the K_{sp} of that salt. This gives a method for driving ions out of solution. For example, if we are given a solution of $RaCl_2$, the Ra^{2+} can be made to precipitate as $RaSO_4$ by addition of Na_2SO_4. Practically all the valuable Ra^{2+} can be recovered by adding a large excess of SO_4^{2-}.

412

In the preceding discussions only one equilibrium was considered at a time. This is an idealized situation since usually aqueous solutions have two or more equilibria which must be satisfied simultaneously. For example, in a solution containing the weak acid $HC_2H_3O_2$, there are two dissociation equilibria:

$$HC_2H_3O_2 \rightleftharpoons H^+ + C_2H_3O_2^- \qquad \frac{[H^+][C_2H_3O_2^-]}{[HC_2H_3O_2]} = K_{diss}$$

$$H_2O \rightleftharpoons H^+ + OH^- \qquad [H^+][OH^-] = K_w$$

The solution has one characteristic concentration of H^+ which simultaneously satisfies K_{diss} and K_w. Strictly speaking, this H^+ comes partly from the dissociation of $HC_2H_3O_2$ and partly from the dissociation of H_2O. However, H_2O is so slightly dissociated that we can consider the H^+ as coming entirely from the $HC_2H_3O_2$. This assumption was implicitly made in the calculations of Sec. 15.2. To see how this comes about quantitatively, we do the following calculation for a related problem.

EXAMPLE 11

Given for HF that $K_{diss} = 6.7 \times 10^{-4}$, calculate the H^+ and OH^- concentration in 0.10 M HF solution.

Solution

First we set up the two equilibria that are involved:

$$HF \rightleftharpoons H^+ + F^-$$
$$H_2O \rightleftharpoons H^+ + OH^-$$

why consider it this time and not on the next probably

Then we define two unknowns x and y for the concentration of H^+ that is contributed by each of these equilibria. For the equilibrium state, we can write

$[H^+] = x + y$ mol/liter

$[F^-] = x$ mol/liter

$[OH^-] = y$ mol/liter

$[HF] = 0.10 - x$ mol/liter

These concentrations, if they are equilibrium values, must satisfy the two equilibrium conditions

$$K_{diss} = \frac{[H^+][F^-]}{[HF]} = 6.7 \times 10^{-4} = \frac{(x+y)(x)}{0.10 - x}$$

$$K_w = [H^+][OH^-] = 1.0 \times 10^{-14} = (x+y)(y)$$

The problem is to solve these two simultaneous equations for the two unknowns x and y. The exact solution is not easy, especially when x and y are about equal in value. Fortunately, one of the contributions usually is the dominant one. We can find out whether x or y is dominant by noting what each equilibrium would produce if not affected by the presence of the other one. Specifically, 0.10 M HF would lead to $[H^+] = 7.7 \times 10^{-3} M$; pure water would lead to $[H^+] = 1.0 \times 10^{-7} M$. Evidently, the H^+ from the HF will be dominant. A fairly good general rule is that the larger the dissociation constant, the more it dominates the final equilibrium state.

Once it has been decided that H^+ from HF is more important than H^+ from H_2O, the problem becomes simple. We need only calculate the HF dissociation problem, ignoring the H_2O dissociation, and then go back to calculate the other equilibrium.

$$\begin{array}{ccc} HF & \rightleftharpoons & H^+ + F^- \\ 0.10 - x & & x + y \quad\ x \end{array}$$

$$K_{diss} = \frac{[H^+][F^-]}{[HF]} = 6.7 \times 10^{-4} = \frac{(x+y)(x)}{0.10 - x}$$

Neglecting y with respect to x and solving for x gives us

$$x = 7.7 \times 10^{-3} M = [H^+]$$

Then we consider the water equilibrium:

$$K_w = [H^+][OH^-] = 1.0 \times 10^{-14} = (x + y)(y)$$

As we have seen, y is negligible when added to x, and so we can write

$$xy \sim 1.0 \times 10^{-14}$$

Since we have already found $x = 7.7 \times 10^{-3} M$, we can substitute for x and get

$$(7.7 \times 10^{-3})(y) = 1.0 \times 10^{-14}$$

from which it follows that

$$y = \frac{1.0 \times 10^{-14}}{7.7 \times 10^{-3}} = 1.3 \times 10^{-12} M = [OH^-]$$

As can be seen, y is truly negligible compared with x; so we were quite justified in approximating $x + y \sim x$.

EXAMPLE 12

Calculate the concentrations of H^+ and OH^- in a solution made by mixing 0.50 mol of $HC_2H_3O_2$ and 0.50 mol of HCN with enough water to make a liter of solution.

Solution

There are three simultaneous equilibria in the final solution:

$$\begin{array}{ll} HC_2H_3O_2 \rightleftharpoons H^+ + C_2H_3O_2^- & K_{HC_2H_3O_2} = 1.8 \times 10^{-5} \\ HCN \rightleftharpoons H^+ + CN^- & K_{HCN} = 4 \times 10^{-10} \\ H_2O \rightleftharpoons H^+ + OH^- & K_w = 1.0 \times 10^{-14} \end{array}$$

Only acetic acid contributes appreciable H^+; it has the largest dissociation constant. Ignoring the other dissociations, let x = moles of $HC_2H_3O_2$ that dissociate per liter. Then, at equilibrium

$$[HC_2H_3O_2] = 0.50 - x \text{ mol/liter}$$

$$[H^+] = x \text{ mol/liter}$$

$$[C_2H_3O_2^-] = x \text{ mol/liter}$$

$$\frac{[H^+][C_2H_3O_2^-]}{[HC_2H_3O_2]} = \frac{(x)(x)}{0.50 - x} = 1.8 \times 10^{-5}$$

$$x = 3.0 \times 10^{-3} \, M = [H^+]$$

Substituting this value in $K_w = [H^+][OH^-] = 1.0 \times 10^{-14}$ gives $(3.0 \times 10^{-3})[OH^-] = 1.0 \times 10^{-14}$:

$$[OH^-] = 3.3 \times 10^{-12} \, M$$

Another common example of simultaneous equilibria occurs in solutions of polyprotic acids (Sec. 9.3). For example, suppose we are asked to calculate the pH of 0.10 M H_2SO_4. It would be wrong to say that, because H_2SO_4 is a strong acid, it is completely dissociated into 0.20 M H^+. Instead, the reaction goes by two successive steps:

$$H_2SO_4 \rightleftharpoons H^+ + HSO_4^- \qquad K_I \gg 1$$
$$HSO_4^- \rightleftharpoons H^+ + SO_4^{2-} \qquad K_{II} = 1.26 \times 10^{-2}$$

The H^+ produced in the first step impedes formation of H^+ by the second step.

Quantitatively, the calculation goes like this: In 0.10 M H_2SO_4, the first reaction

$$H_2SO_4 \longrightarrow H^+ + HSO_4^-$$

goes essentially 100 percent to the right. This converts 0.10 M H_2SO_4 into 0.10 M H^+ and 0.10 M HSO_4^-. Next we consider the second step:

$$HSO_4^- \rightleftharpoons H^+ + SO_4^{2-}$$

Let x = moles per liter of H^+ generated by the second step. Then, at equilibrium, we will have

$$[H^+] = 0.10 + x \text{ mol/liter}$$
$$[HSO_4^-] = 0.10 - x \text{ mol/liter}$$
$$[SO_4^{2-}] = x \text{ mol/liter}$$

Substituting into K_{II} gives

$$K_{II} = \frac{[H^+][SO_4^{2-}]}{[HSO_4^-]} = \frac{(0.10 + x)(x)}{0.10 - x} = 1.26 \times 10^{-2}$$

This is relatively simple to solve even though x is not negligible compared with 0.10. Either the quadratic formula (Appendix A4.3) or the method of successive approximations (Appendix A4.4) leads to $x = 0.011$. Consequently, $[H^+] = 0.10 + x = 0.11$ M, from which pH $= -\log 0.11 = 0.96$.

416

Chapter 15
Chemical
Equilibrium
in Aqueous
Solutions

One of the most useful applications of simultaneous equilibria occurs in solutions of hydrogen sulfide, H_2S. There are two equilibria that correspond to stepwise dissociation of H_2S:

$$H_2S \rightleftharpoons H^+ + HS^- \qquad K_I = 1.1 \times 10^{-7}$$
$$HS^- \rightleftharpoons H^+ + S^{2-} \qquad K_{II} = 1 \times 10^{-14}$$

The first is the more important for producing H^+. Using only K_I, we can calculate that the concentration of H^+ in 0.10 M H_2S is approximately $1 \times 10^{-4}\,M$, and the HS^- concentration is $1 \times 10^{-4}\,M$. Because of the second dissociation, there is a small trace of sulfide ion, S^{2-}, in the solution. Its numerical magnitude can be calculated by using K_{II}:

$$K_{II} = \frac{[H^+][S^{2-}]}{[HS^-]} = 1 \times 10^{-14}$$

Since the concentrations of H^+ and HS^- are both $1 \times 10^{-4}\,M$, they cancel each other out of the expression, and $[S^{2-}] = 1 \times 10^{-14}\,M$.

In any solution of H_2S, both K_I and K_{II} must be simultaneously satisfied. This gives rise to two simultaneous equations:

$$\frac{[H^+][HS^-]}{[H_2S]} = 1.1 \times 10^{-7} \qquad \text{and} \qquad \frac{[H^+][S^{2-}]}{[HS^-]} = 1 \times 10^{-14}$$

Solving these for $[HS^-]$, we get

$$[HS^-] = 1.1 \times 10^{-7}\frac{[H_2S]}{[H^+]} \qquad \text{and} \qquad [HS^-] = \frac{[H^+][S^{2-}]}{1 \times 10^{-14}}$$

Equating these gives

$$\frac{1.1 \times 10^{-7}[H_2S]}{[H^+]} = \frac{[H^+][S^{2-}]}{1 \times 10^{-14}}$$

Rearranging the terms in the last equation, we get

$$(1.1 \times 10^{-7})(1 \times 10^{-14}) = \frac{[H^+]^2[S^{2-}]}{[H_2S]}$$

For a saturated solution of H_2S at atmospheric pressure and room temperature, the concentration of H_2S in solution will be constant at 0.10 M. Putting this into the denominator, we get

$$[H^+]^2[S^{2-}] = (1.1 \times 10^{-7})(1 \times 10^{-14})(0.10)$$
$$[H^+]^2[S^{2-}] = 1 \times 10^{-22}$$

The final equation is useful because it states that the sulfide-ion concentration in any saturated H_2S solution can be manipulated by changing the concentration of H^+. This possibility of changing the S^{2-} concentration by juggling the concentration of H^+ is the basis of the classic method of ion separation in qualitative analysis by sulfide precipitation.

EXAMPLE 13

A solution contains Zn^{2+} and Cu^{2+}, each at 0.02 M. The K_{sp} of ZnS is 1×10^{-22}, that of CuS, 8×10^{-37}. If the solution is made 1 M in H^+ and H_2S gas is bubbled in until the solution is saturated, should a precipitate form?

Solution

In saturated H_2S solution, $[H^+]^2[S^{2-}] = 1 \times 10^{-22}$.
If $[H^+] = 1$ M, $[S^{2-}] = 1 \times 10^{-22}$ M.
For ZnS, the ion product would be

$$[Zn^{2+}][S^{2-}] = (0.02)(1 \times 10^{-22}) = 2 \times 10^{-24}$$

This does not exceed the K_{sp} of ZnS (1×10^{-22}), and so ZnS does not precipitate.
For CuS, the ion product would be

$$[Cu^{2+}][S^{2-}] = (0.02)(1 \times 10^{-22}) = 2 \times 10^{-24}$$

This does exceed the K_{sp} of CuS (8×10^{-37}), and so CuS does precipitate.

The principles of simultaneous equilibrium can be applied to the problem of dissolving a solid. For example, ZnS is essentially insoluble in water, but it can be made to dissolve by addition of acid. The argument goes as follows: If solid ZnS is added to pure water, the equilibrium is

$$ZnS(s) \rightleftharpoons Zn^{2+} + S^{2-}$$

When acid is added, the additional equilibria

$$H^+ + S^{2-} \rightleftharpoons HS^-$$

$$H^+ + HS^- \rightleftharpoons H_2S$$

become important. Added H^+ reacts with S^{2-} to form HS^- and H_2S. As the concentration of S^{2-} is reduced, more ZnS can dissolve. The net reaction for the dissolving is the sum of these three equilibria:

$$ZnS(s) + 2H^+ \rightleftharpoons Zn^{2+} + H_2S$$

15.10
HYDROLYSIS

One of the most important applications of simultaneous equilibria is the quantitative description of hydrolysis. The subject was discussed briefly in Sec. 9.7. Specifically, a solution of sodium acetate, $NaC_2H_3O_2$, is slightly basic owing to reaction between $C_2H_3O_2^-$ and water:

$$C_2H_3O_2^- + H_2O \rightleftharpoons HC_2H_3O_2 + OH^-$$

In the forward reaction a proton is transferred from water to acetate; in the reverse reaction a proton is transferred from acetic acid to hydroxide.

418

Chapter 15
Chemical
Equilibrium
in Aqueous
Solutions

Clearly, what is involved is the relative proton affinity of $C_2H_3O_2^-$ and OH^-. The former can be described by K_{diss} of $HC_2H_3O_2$; the latter, by K_w. The relation can be seen by writing the equilibrium condition for the net hydrolysis reaction as given above:

$$\frac{[HC_2H_3O_2][OH^-]}{[C_2H_3O_2^-]} = K_{hyd}$$

Multiplying the numerator and denominator by $[H^+]$ gives

$$\frac{[HC_2H_3O_2][OH^-][H^+]}{[C_2H_3O_2^-][H^+]} = K_{hyd}$$

or, rearranging,

$$\frac{[OH^-][H^+]}{[C_2H_3O_2^-][H^+]/[HC_2H_3O_2]} = K_{hyd}$$

In the last step the terms have been collected so as to emphasize that the numerator is K_w and the denominator is K_{diss}:

$$\frac{K_w}{K_{diss}} = K_{hyd}$$

In other words, the hydrolysis constant K_{hyd} is just the ratio of the water dissociation constant to the weak-acid dissociation constant.

Once the numerical value of K_{hyd} has been obtained, it can be used for equilibrium calculations in the usual way. The following examples illustrate specific cases.

EXAMPLE 14

Calculate the pH of 0.10 M $NaC_2H_3O_2$ and the percent hydrolysis.

Solution

The net hydrolysis reaction is

$$C_2H_3O_2^- + H_2O \rightleftharpoons HC_2H_3O_2 + OH^-$$

for which

$$\frac{[HC_2H_3O_2][OH^-]}{[C_2H_3O_2^-]} = \frac{K_w}{K_{diss}} = \frac{1.0 \times 10^{-14}}{1.8 \times 10^{-5}} = 5.6 \times 10^{-10}$$

Let $x =$ moles of $C_2H_3O_2^-$ that hydrolyze per liter. This forms x mol of $HC_2H_3O_2$ and x mol of OH^- and leaves $0.10 - x$ mol of $C_2H_3O_2^-$. At equilibrium

$$[HC_2H_3O_2] = x \text{ mol/liter}$$
$$[OH^-] = x \text{ mol/liter}$$
$$[C_2H_3O_2^-] = 0.10 - x \text{ mol/liter}$$

Substituting in the mass-action expression gives

$$\frac{(x)(x)}{0.10 - x} = 5.6 \times 10^{-10}$$

Assuming that x is small compared with 0.10 gives

$$\frac{x^2}{0.10} \sim 5.6 \times 10^{-10}$$

$$x = 7.5 \times 10^{-6} \ M = [OH^-]$$

$$[H^+] = \frac{K_w}{[OH^-]} = \frac{1.0 \times 10^{-14}}{7.5 \times 10^{-6}} = 1.3 \times 10^{-9}$$

$$pH = -\log [H^+] = -\log (1.3 \times 10^{-9}) = 8.89$$

$$\frac{\text{Moles } C_2H_3O_2^- \text{ hydrolyzed} \times 100}{\text{Moles } C_2H_3O_2 \text{ available}} = \frac{7.5 \times 10^{-6} \times 100}{0.10} = 0.0075\%$$

EXAMPLE 15

What is the concentration of H^+ in 0.10 M $AlCl_3$? The hydrolysis constant of Al^{3+} is 1.4×10^{-5}.

Solution

$$Al^{3+} + H_2O \rightleftharpoons AlOH^{2+} + H^+$$

$$\frac{[AlOH^{2+}][H^+]}{[Al^{3+}]} = 1.4 \times 10^{-5}$$

Let x = moles of Al^{3+} that hydrolyze. At equilibrium

$$[Al^{3+}] = 0.10 - x \text{ mol/liter}$$

$$[AlOH^{2+}] = x \text{ mol/liter}$$

$$[H^+] = x \text{ mol/liter}$$

$$\frac{(x)(x)}{0.10 - x} = 1.4 \times 10^{-5}$$

$$x = 1.2 \times 10^{-3} \ M = [H^+]$$

Important Concepts

dissociation constant
association constant
calculations with K_{diss}
dissociation of water
ion product of water
pH
titration

equivalence point
indicator
buffer solution
complex-ion dissociation
solubility product
calculations with K_{sp}
common-ion effect

precipitation
simultaneous equilibria
dissociation of polyprotic acid
equilibria in H_2S solutions
hydrolysis
calculations with K_{hyd}

Exercises

*15.1 Dissociation How can you show that at least in dilute solution the equilibrium condition for $HX + H_2O \rightleftharpoons H_3O^+ + X^-$ is the same as that for $HX \rightleftharpoons H^+ + X^-$.

***15.2 Dissociation If an acid, such as HCl, has a dissociation constant equal to 10, what percent of the HCl will be dissociated in a 1.00-M solution?

***15.3 Dissociation Why should K_{II} for a diprotic acid H_2X be considerably smaller than K_I? In what kind of a situation might you expect to find K_{II} about the same as K_I?

*15.4 Dissociation The weak acid HOCl dissociates according to the reaction $HOCl \rightleftharpoons H^+ + OCl^-$. Write an expression for the equilibrium condition of this dissociation. If a solution that is 0.010 M HOCl shows 0.18 percent dissociation, what is the numerical value of the dissociation constant? What would be the numerical value for the association constant of $ClO^- + H^+ \rightleftharpoons HOCl$? *Ans. $K_{assoc} = 3.1 \times 10^7$*

*15.5 Dissociation What is the concentration of H^+, $C_2H_3O_2^-$, and $HC_2H_3O_2$ in a solution that is labeled 0.90 M $HC_2H_3O_2$?

**15.6 Dissociation Given the weak acid HCN for which the dissociation constant is 4.0×10^{-10}, what concentration would you have to pick in order to have 0.1 percent of the acid dissociated? How much water would you need to add to 1 liter of the original solution to double the percent dissociation?

*15.7 Dissociation Given 10.0 g of $HC_2H_3O_2$, in how much water should it be dissolved to give a hydrogen-ion concentration of 1.0×10^{-3} M? *Ans. 3.0 liters*

*15.8 Dissociation A solution is made by dissolving 0.100 mol of HCl and 0.100 mol of $NaC_2H_3O_2$ in enough water to give 0.500 liter of solution. What will be the concentration of hydrogen ion in the final solution?

*15.9 Dissociation A given solution contains 0.0010 M H^+, 0.0090 M Na^+, 0.0100 M X^-, and 4.00 M HX at equilibrium. What should be the concentration of hydrogen ion in a solution that is 1.00 M HX?

*15.10 Dissociation Given a weak acid HX in aqueous solution, what three things might you do to increase the percent dissociation of the acid?

**15.11 Dissociation The K_{diss} of HCN is 4.0×10^{-10}; that of $HC_2H_3O_2$ is 1.8×10^{-5}. What would be the hydrogen-ion concentration in a solution made by mixing 0.100 mol HCl, 0.050 mol NaCN, and 0.050 mol $NaC_2H_3O_2$ in enough water to make 0.500 liter of solution? What would be the concentration of cyanide ion? *Ans. $[CN^-] = 3.1 \times 10^{-8}$ M*

**15.12 Dissociation The percent dissociation of $HC_2H_3O_2$ in 1.00 M $HC_2H_3O_2$ is 0.42 percent. How many milliliters of 1.00 M HCl would you have to add to 1.00 liter of this solution to depress the percent dissociation to 0.10 percent?

*15.13 **Dissociation** For heavy water, D_2O, the dissociation constant is 0.2×10^{-14}. What would be the concentration of D^+ and of OD^- in pure D_2O?

*15.14 **pH** What would be the pH of each of the following solutions: (a) $1.0\ M$ HCl, (b) $0.10\ M$ HCl, (c) $10.0\ M$ HCl, (d) $2.0 \times 10^{-3}\ M$ HCl, (e) $2.0 \times 10^{-13}\ M$ HCl?

*15.15 **pH** What would be the pH of each of the following solutions: (a) $1.0\ M$ NaOH, (b) $0.10\ M$ NaOH, (c) $10.0\ M$ NaOH, (d) $2.0 \times 10^{-3}\ M$ NaOH, (e) $2.0 \times 10^{-13}\ M$ NaOH?

*15.16 **pH** Given that the K_{diss} of $HC_2H_3O_2$ is 1.8×10^{-5}, what would be the pH of a solution that is $0.35\ M$ $HC_2H_3O_2$? *Ans. 2.60*

15.17 **pH Given the information shown in Exercise 15.16, what would you have to do to make the pH = 2.80?

***15.18 **pH** What would be the pH of each of the following solutions: (a) $0.200\ M$ H_2SO_4, (b) $0.200\ M$ Na_2SO_4, (c) a solution made by mixing equal volumes of (a) and (b)? *Ans. 0.68, 7.60, 1.35*

15.19 **pH What would be the pH of each of the following solutions: (a) $0.010\ M$ NaOH, (b) $0.010\ M$ $Ba(OH)_2$ (assume both hydroxides completely dissociated), (c) a solution made by mixing equal volumes of (a) and (b)?

15.20 **Indicators Explain why an indicator changes color when the pH passes through the pK value of the indicator.

15.21 **Indicators Make a graph of the data in Fig. 15.2 showing $[H^+]$ plotted on the vertical axis vs. moles of NaOH on the horizontal axis. At which point would the indicator phenolphthalein change color? Compare with the pH plot of Fig. 15.3. What is the advantage of the pH plot? What is the disadvantage?

*15.22 **Indicators** Why would neither methyl violet nor alizarin yellow be as good as phenolphthalein for the titration shown in Fig. 15.3?

*15.23 **Buffer solution** What is a buffer solution? What determines the pH of a buffer?

15.24 **Buffer solution Given that $K_{diss} = 1.8 \times 10^{-5}$ for $NH_3 + H_2O \rightleftharpoons NH_4^+ + OH^-$, calculate the pH of a buffer made by mixing 200 ml of $1\ M$ NH_3 with 200 ml of $1\ M$ NH_4Cl. What

happens to the pH of the buffer on addition of (a) 0.010 mol of NaOH; (b) 0.010 mol of HCl? *Ans. 9.26, 9.30, 9.21*

15.25 **Buffer solution One of the buffer systems in the blood involves the equilibrium $CO_2 + H_2O \rightleftharpoons H^+ + HCO_3^-$ for which $K_{diss} = 4.2 \times 10^{-7}$. Given a pH of 7.4, what ratio of CO_2 to HCO_3^- does this correspond to? How much do you need to change this ratio to raise the pH by 0.02 unit?

***15.26 **Buffer solution** A very important buffer solution in analytical chemistry is the one prepared by mixing equal volumes of $2.0\ M$ $NaHSO_4$ and $1.6\ M$ Na_2SO_4. One of its uses is to keep hydrogen-ion concentration steady during precipitation of sulfides by the reaction $M^{2+} + H_2S(g) \longrightarrow MS(s) + 2H^+$. (a) Given that $K_{II} = 1.26 \times 10^{-2}$ for $HSO_4^- \rightleftharpoons H^+ + SO_4^{2-}$, what is the pH of the above buffer? (b) What would it become after precipitation of all the Co^{2+} out of 10.0 ml of $0.050\ M$ $CoCl_2$ in the above buffer?

*15.27 **Buffer solution** The capacity of a buffer to absorb H^+ or OH^- depends on (a) the volume of buffer solution taken and (b) the concentration of weak acid and of salt in the buffer. Show why each of these factors is involved in determining the capacity of a given buffer solution.

15.28 **Complex-ion equilibria Given the data of Fig. 15.5, which complex ion would be formed in greatest concentration when Ag^+, NH_3, $S_2O_3^{2-}$, and CN^- are mixed together in equal amounts? Justify your answer.

15.29 **Complex-ion equilibria You are given a solution containing $0.01\ M$ Ag^+ and $0.01\ M$ Hg^{2+}. Using the data shown in Fig. 15.5, calculate the concentration of each of these ions after the addition of 0.100 mol NaCN to a liter of $0.01\ M$ Ag^+ and $0.01\ M$ Hg^{2+}. *Ans. $[Ag^+] = 1.1 \times 10^{-18}\ M$*

*15.30 **Solubility product** Explain why the concentration of a dissolved solid does not depend on how many grams of extra solid are in contact with the saturated solution.

*15.31 **Solubility product** From the observed fact that one can dissolve 6.7 mg of $CaSO_4$ in 10 ml of water, calculate the solubility product for $CaSO_4$.

*15.32 **Solubility product** Given that the solubility product of silver nitrite, $AgNO_2$, is 1.2×10^{-4}, calculate the concentration of Ag^+ and of NO_2^- in a solution that is made by saturating water with $AgNO_2$. How many grams of $AgNO_2$ can be dissolved in 0.125 liter of water?

Ans. 0.21 g

**15.33 Solubility product When water is saturated with $Ca(OH)_2$, the observed pH of the solution is 12.137. Calculate the solubility product for $Ca(OH)_2 (s) \rightleftharpoons Ca^{2+} + 2OH^-$.

Ans. 1.29 × 10⁻⁶

**15.34 Solubility product Given that the K_{sp} of $Ca(OH)_2$ is 1.3×10^{-6}, calculate how many grams of $Ca(OH)_2$ can be dissolved in 0.100 liter of (a) water, (b) 0.10 M NaOH, (c) 0.10 M $CaCl_2$.

**15.35 Solubility The K_{sp} of $Mg(OH)_2$ is 8.9×10^{-12}; that of $Ca(OH)_2$ is 1.3×10^{-6}. Calculate the concentrations of Ca^{2+} and Mg^{2+} in water that is simultaneously saturated with respect to $Ca(OH)_2$ and $Mg(OH)_2$. What is the pH of the solution?

*15.36 Precipitation Given that the K_{sp} of CaF_2 is 1.7×10^{-10}, how many moles of NaF must you add to a liter of 2.0×10^{-5} M Ca^{2+} before you could start to get precipitation?

**15.37 Precipitation Given that the K_{sp} of CaF_2 is 1.7×10^{-10}, how many milliliters of 1.0 M NaF must you add to a liter of 2.0×10^{-5} M Ca^{2+} before you could start to get precipitation?

Ans. 2.9 ml

**15.38 Precipitation The K_{diss} for $NH_3 + H_2O \rightleftharpoons NH_4^+ + OH^-$ is 1.8×10^{-5}. The K_{sp} of $Ca(OH)_2$ is 1.3×10^{-6}. Calculate whether $Ca(OH)_2$ should precipitate on mixing 1 liter of 0.5 M NH_3 and 1 liter of 0.5 M $CaCl_2$.

**15.39 Dissociation Given that $K_{diss} = 1.8 \times 10^{-5}$ for $HC_2H_3O_2$ and 3.2×10^{-8} for HOCl, calculate the concentrations of H^+, OH^-, $C_2H_3O_2^-$, and OCl^- in a solution that is simultaneously 1.00 M $HC_2H_3O_2$ and 1.00 M HOCl.

**15.40 Dissociation Given a solution that is 0.200 M H_2SO_4, what will be its pH?

***15.41 Dissociation Given that K_{II} of H_2SO_4 is 1.26×10^{-2}, what concentration of H_2SO_4 should I take to have a solution with pH = 1.50?

Ans. 0.0246 M

***15.42 Dissociation Given that K_{II} of H_2SO_4 is 1.26×10^{-2}, calculate the pH of each of the following solutions: (a) 0.250 M H_2SO_4, (b) 0.250 M $NaHSO_4$, (c) a mixture made up of equal volumes of (a) and (b).

***15.43 Dissociation Given that K_{II} of H_2SO_4 is 1.26×10^{-2}, calculate the hydrogen-ion concentration in each of the following solutions: (a) 0.50 M $NaHSO_4$, (b) 0.50 M Na_2SO_4, (c) a solution that is simultaneously 0.50 M $NaHSO_4$ and 0.50 M Na_2SO_4.

***15.44 Dissociation For the triprotic acid H_3PO_4 the successive dissociation constants are: $K_I = 7.5 \times 10^{-3}$; $K_{II} = 6.2 \times 10^{-8}$; $K_{III} = 1.6 \times 10^{-12}$. Given a solution that is 0.10 M H_3PO_4, calculate the concentration of each phosphorus-containing species in the solution.

***15.45 Dissociation H_2Z is a weak diprotic acid with the following dissociation constants: $K_I = 1.0 \times 10^{-6}$; $K_{II} = 1.0 \times 10^{-10}$. (a) What is the approximate pH of 1.0 M H_2Z solution? (b) What is the approximate pH of a solution made by mixing 1.0 liter of 1.0 M H_2Z solution with 0.50 liter of 1.0 M NaOH solution? (c) What is the approximate pH of a solution made by mixing 1.0 liter of 1.0 M H_2Z solution with 1.0 liter of 1.0 M NaOH solution?

**15.46 H_2S equilibrium Given a saturated solution of H_2S at atmospheric pressure and room temperature, to what hydrogen-ion concentration should the solution be adjusted in order to precipitate ZnS ($K_{sp} = 1 \times 10^{-22}$) but not FeS ($K_{sp} = 4 \times 10^{-19}$) out of a solution containing Zn^{2+} and Fe^{2+} at 0.10 M?

Ans. 0.005 M < [H⁺] < 0.3 M

*15.47 Hydrolysis How does the extent of hydrolysis of a negative ion X^- depend on each of the following: (a) concentration of X^-; (b) dissociation constant of HX? Justify your answers.

*15.48 Hydrolysis Arrange the following solutions in the order of increasing percent hydrolysis of the negative ion: 1 M $NaC_2H_3O_2$, 1 M $NaNO_2$, 1 M NaF, 1 M NaCN, 0.1 M NaCN.

**15.49 Hydrolysis Calculate the pH and percent hydrolysis in 1.0 M $NaC_2H_3O_2$.

**15.50 Hydrolysis Given that K_{II} of H_2S is 1×10^{-14}, what would be the percent hydrolysis of S^{2-} in 0.10 M Na_2S?

Ans. 92%

**15.51 Hydrolysis Solutions of Na_2CO_3 are basic because of hydrolysis of the carbonate ion. Given that $K_{II} = 4.8 \times 10^{-11}$ for $HCO_3^- \rightleftharpoons H^+ + CO_3^{2-}$, calculate what concentration of Na_2CO_3 would be needed to give a solution of pH = 12.00.

Chapter 16

A CLOSER LOOK AT ELECTRON DESCRIPTIONS

As can be seen from preceding chapters, we have been able to go rather deeply into chemistry with only the most elementary notions about the internal structure of atoms and how atoms are held together in molecules. We now need to take a closer look at the problem of describing electrons. In the modern view, the wave nature of electrons leads to a rather complicated picture of the atom. How are these pictures to be put together to describe a molecule? How are they related to observed properties? In Sec. 4.9 we introduced the idea of an orbital as a specific electronic probability distribution in space. In this chapter we take a deeper look at this concept of orbital. We also look at how *atomic orbitals,* which describe electrons in atoms, can be put together to give *molecular orbitals,* which describe electrons in molecules. Some of the ideas in this chapter are difficult and rather abstract; for this reason we have postponed introducing them until now.

16.1 ELECTRON WAVES

In preceding chapters, electron probability distributions were described by using visual analogies as much as possible. Thus, a $1s$ electron was described as a spherical charge cloud; a $2p$ electron, as two blobs of electron density on either side of a nucleus. These visualizations, although helpful in fitting observed facts together, should not be taken too literally. The wave nature of the electron and the uncertainty principle put definite limitations on how much information we can specify about an electron. Pictures such as Fig. 4.21 frequently suggest more definite information than we actually possess.

de Broglie's idea that an electron behaves like a wave suggests that the mathematical equations used to describe wave motion can be applied to the description of the electron. There are two important aspects to wave motion: How steeply does the height of a wave build up? How fast do crests and troughs follow each other through space? Because energy and mass are conserved, there must be a connection between crests and troughs and how rapidly one changes into the other.

To get a better feeling for how waves are described, let us consider the wave shown in Fig. 16.1. The wiggly line that traces out the wave can be described by the equation

$$\psi = A \sin bx \tag{1}$$

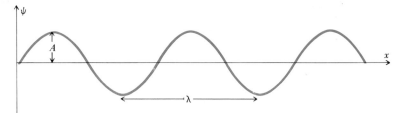

FIG. 16.1 Standing wave.

The Greek letter ψ, which is pronounced "psī" or "psē," is called the *wave function;* it tells how far any point of the wavy line is from the horizontal axis. A represents the amplitude of the wave and tells the maximum value that ψ can have anywhere along the wave. The symbol *sin* stands for the sine function of trigonometry. b is a constant characteristic of the wave and is equal to $2\pi/\lambda$, where λ is the wavelength of the wave. Trigonometric functions go through cycles, and ψ must come back to the same value after each cycle of 2π.* If we write $\pi = A \sin 2\pi(x/\lambda)$, we see that x/λ represents the fraction of a full wavelength that has been traversed. As shown in Fig. 16.1, the wave is a *stationary,* or *standing,* wave; it does not change with time.

EXAMPLE 1

How far along the above wave do we have to go to have ψ increase from 0 to its maximum value?

Solution

ψ increases from 0 to A in $\frac{1}{4}\lambda$.
It decreases from A to 0 in the next $\frac{1}{4}\lambda$, and so forth.

Let us imagine, now, that the wave moves progressively to the left as time passes. We then have a *moving,* or *running,* wave. Progressive snapshots of it would look like the sequence shown in Fig. 16.2. As noted by the marker arrow, the crests steadily advance to the left, even though the distance between crests (i.e., the wavelength) stays the same. Another feature to note is that the value of the wave function at any specific value of x now oscillates with time. In Fig. 16.2, we have illustrated this by indicating with a large black dot what happens to the ψ value at a particular x. As time goes on, the point bobs up and down very much like a cork that is agitated by passing water waves in a pond. How can we express the mathematical description of the running wave? Each of the snapshots in Fig. 16.2 is like the picture of Fig. 16.1 except for the successive displacement of the waveform. To allow for the change

*The variable bx can be expressed either in degrees or in radians. As given here, bx is expressed in radians. There are 2π radians in $360°$; one radian equals $57.30°$. The following equivalents are useful to remember: $\pi/6 = 30°$; $\pi/4 = 45°$; $\pi/3 = 60°$; $\pi/2 = 90°$; $\pi = 180°$; and $2\pi = 360°$. The corresponding values of the sine are $\frac{1}{2}, \frac{1}{2}\sqrt{2}, \frac{1}{2}\sqrt{3}, 1, 0,$ and 0.

424

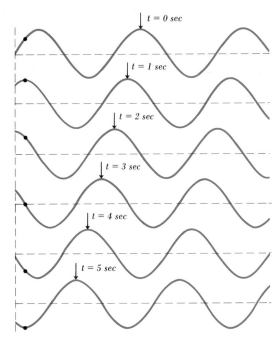

$t = 0$ sec

$t = 1$ sec

$t = 2$ sec

$t = 3$ sec

$t = 4$ sec

$t = 5$ sec

FIG. 16.2 Running wave moving to the left, shown at successive time intervals. (Position of arrow shows displacement of crest as time goes on.)

with time, we need to modify Eq. (1) by adding a term that depends on time. Specifically, we can write

$$\psi = A \sin (bx + kt) \tag{2}$$

Equation (2) is the equation for a *running wave.* It has a constant b related to the periodicity in space and a constant k related to the periodicity in time. Indeed, k is just 2π times the frequency of the wave. For example, if the frequency ν is 3 cycles/sec, it will take $\frac{1}{3}$ sec for each wavelength λ to go past a given point. Substituting $b = 2\pi/\lambda$ and $k = 2\pi\nu$ gives us the following expression for a running wave:

$$\psi = A \sin 2\pi \left(\frac{x}{\lambda} + \nu t\right) \tag{3}$$

When this equation is used to describe sound waves, ψ tells us about the pressure and how it alternates between compression and rarefaction; for electromagnetic waves, such as light, ψ is related to the intensity of the electric and magnetic fields. Figure 16.3 shows typical wavelengths and frequencies for the waves of the electromagnetic spectrum.

EXAMPLE 2

If we sit at a particular point in space and let a wave of frequency 3 cycles/sec go past, how long will it take for ψ to build up from 0 to A?

426

Chapter 16
A Closer
Look at
Electron
Descriptions

FIG. 16.3   Electromagnetic Spectrum

Kind of wave	Typical wavelength, cm	Typical frequency, cycles/sec
Cosmic rays	0.004×10^{-8}	7.5×10^{20}
X rays	1×10^{-8}	3×10^{18}
Ultraviolet light	2000×10^{-8}	1.5×10^{15}
Visible light	5000×10^{-8}	0.6×10^{15}
Infrared	1×10^{-3}	3×10^{13}
Radar	1	3×10^{10}
Radio	3×10^{5}	1×10^{5}

Solution

Frequency is 3 cycles/sec.

It takes $\frac{1}{3}$ sec for each cycle to pass.

It takes one-fourth of a cycle to build up from 0 to A.

Time required $= \frac{1}{3} \times \frac{1}{4} = \frac{1}{12}$ sec.

Equation (3) can also be used to describe the motion of particles.   All we need to do is put in the de Broglie relation (Sec. 4.9) $\lambda = h/mv$, relating the wavelength associated with a particle to its momentum mv.   There is now a problem, however.   What do we mean by the wave function ψ when we talk about particle waves?   We cannot attach any easily visualized significance to ψ itself, but ψ^2 turns out to be proportional to the probability of locating the particle.   As the wave moves through space and time, the square of the wave function tells how the probability of finding the particle changes.   Thus, Eq. (3) can be used to describe, for example, how electrons are transferred through metal or how X rays pass through space.

16.2
ELECTRON WAVES IN A BOX

What happens when an electron is confined in a box as in a piece of metal? The problem is an application of a famous kind of problem called the *particle-in-a-box* problem.   Because it is related to the problem of describing an electron in an atom (which, after all, is a kind of box), it is instructive to consider it more deeply.   In its simplest form, the problem is posed as follows: Suppose we have a one-dimensional box of length *L,* as indicated in Fig. 16.4.   We have an electron trapped in this box.   To make sure the electron stays in the box, we assume that the potential energy sharply rises at the edges of the box.   In other words, there are huge forces keeping the electron inside the box.   Between $x = 0$ and $x = L$, the potential energy is assumed constant.   We can set it at a value of zero.   What can we say about the kinetic energy of the electron in the box?

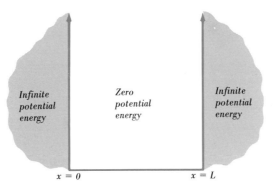

FIG. 16.4 Energy box for trapping a particle.

Infinite potential energy

Zero potential energy

Infinite potential energy

$x = 0$ $x = L$

To describe the electron, we use the standing-wave representation given in Eq. (1). (It may be that when we first put the electron in the box, a running-wave description might be more appropriate, but after the electron rattles around a bit, it should reach a steady state which does not change with time.) So, the wave function can be written as

$$\psi = A \sin 2\pi \left(\frac{x}{\lambda}\right) \tag{4}$$

where ψ^2 tells us about the probability of finding the electron. At the two edges of the box where $x = 0$ and $x = L$, the probability must go to zero. Substituting $\psi = 0$ at $x = 0$ does not tell us anything; but substituting $\psi = 0$ at $x = L$ gives us the interesting result that $\psi = A \sin 2\pi (L/\lambda) = 0$. The nature of sine functions is such that this condition will be satisfied only when $2\pi L/\lambda$ is a multiple of π. If we let n represent this multiple, then we end up with the condition that $2\pi L/\lambda = n\pi$. Canceling π and rearranging gives $L/\frac{1}{2}\lambda = n$, which says that a whole number of half wavelengths must fit into the box. The only permitted values of λ are those given by $2L/n$, where n can take on values 1, 2, 3, 4, etc.

EXAMPLE 3

For an electron in a box that has $L = 8$ nm, what are the permitted values of λ for the first three states?

Solution

$$\lambda = \frac{2L}{n} \qquad \text{where } n = 1, 2, 3$$

$$\lambda = \frac{(2)(8)}{1} = 16 \text{ nm} \qquad \lambda = \frac{(2)(8)}{2} = 8 \text{ nm} \qquad \lambda = \frac{(2)(8)}{3} = 5.3 \text{ nm}$$

428

Chapter 16
A Closer
Look at
Electron
Descriptions

The restriction to only certain permitted values of λ is reminiscent of Bohr's quantum condition (Sec. 4.7). The difference is that, whereas Bohr introduced the quantum condition as an assumption to explain spectral lines, it comes out quite naturally here just from considering the wave nature of an electron.

Figure 16.5 shows schematically what the permitted wave functions ψ look like. Shown also are the corresponding plots of ψ^2. We can see that for the lowest state $n = 1$, the electron has the highest probability of being found in the center of the box. For the second state $n = 2$, there are two regions of high probability, one in each half of the box; for this state, the probability of finding the electron in the exact center of the box is zero.

What can we say about the corresponding energy? The only permitted values of λ are given by $2L/n$. The de Broglie condition says that $\lambda = h/mv$. Equating these two, we find that $h/mv = 2L/n$. This means that the only permitted values of the momentum mv are given by $nh/2L$. Recalling that kinetic energy is $\frac{1}{2}mv^2$, which can be rewritten $(mv)^2/2m$, we can write for the energy

$$E = \frac{(mv)^2}{2m} = \frac{(nh/2L)^2}{2m} = \frac{n^2h^2}{8mL^2} \tag{5}$$

The energy values corresponding to the four lowest states are shown on the right side of Fig. 16.5. There are several points to note: (1) Since the *potential* energy in the box is zero, the total energy E, kinetic plus potential, is just equal to the kinetic energy. (2) The total energy goes up as the square of the quan-

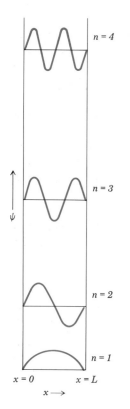

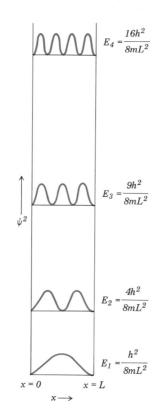

FIG. 16.5 Permitted wave functions ψ and probability distributions ψ^2 for electron in a box.

tum number n. (3) The total energy decreases when the box is made bigger (that is, when L is increased). This last point is well worth noting. It reflects an important principle of quantum mechanics: The bigger the space you provide for an electron, the lower its kinetic energy. This has significance for formation of molecules. Allowing an electron that was confined to a single atom to spread out over several atoms lowers the kinetic energy and contributes to stabilizing the system.

EXAMPLE 4

If we double the size of the box, what happens to the energy levels shown in Fig. 16.5?

Solution

E goes inversely as $1/L^2$.
If we double L, each E gets one-fourth as big.
All the levels in Fig. 16.5 would be pulled down.

16.3
ELECTRON WAVES IN AN ATOM

Actually the electron-in-a-box problem is a fairly straightforward calculation. The problem gets significantly harder when we tackle our central question: How can we describe an electron in an atom? Superficially, it is like the electron-in-a-box problem, except that the box, i.e., the atom, is no longer a simple one-dimensional box but is spherical. There is also a more serious complication. Whereas in the above box we had a uniform potential energy that could be set everywhere equal to zero, the potential energy of an electron in an atom decreases as the electron gets closer to the nucleus. In other words, there is a strong force pulling the electron to the center. How can we take this force into account?

The problem of giving a wave description of an electron in an atom was resolved brilliantly in 1926 by Erwin Schrödinger. He took the equation of motion for waves, put in the de Broglie condition, and set the total energy as the sum of kinetic and potential terms. The potential term was $-Ze^2/r$, where $+Ze$ is the charge of the nucleus and $-e$ is the charge of the electron at the distance r from the nucleus. It would take us too far afield to trace through the mathematical arguments, some of which get to be rather abstruse, but we can look at the final statement of the problem described in mathematical terms. This is expressed by the famous Schrödinger wave equation,* which can be written as shown at the top of the next page:

---

*This formidable-looking equation actually has a relatively simple significance. It relates the shape, i.e., curvature, of a wave function to a permitted value of the energy. Recalling that Δ is the general symbol for a change, we can see that $\Delta\psi/\Delta r$ represents a small change in ψ divided by a corresponding small change in r. In words, if we have a graph of ψ plotted against r, the slope of the graph is given by $\Delta\psi/\Delta r$. Now if we are interested in the curvature of this graph, i.e., how fast its slope changes, we need to take a small change in $\Delta\psi/\Delta r$ and divide it by a corresponding change in r. This gives us $[\Delta(\Delta\psi/\Delta r)]/\Delta r$ for the curvature. In general, the more curvature there is in a wave function, the more wiggles it will have, and the higher will be its kinetic energy.

430

Chapter 16
A Closer
Look at
Electron
Descriptions

$$\frac{\Delta(\Delta\psi/\Delta r)}{\Delta r} + \frac{8\pi^2 m}{h^2}\left(E + \frac{Ze^2}{r}\right)\psi = 0 \qquad (6)$$

Most of the symbols in Eq. (6) have been previously encountered. ψ is the wave function; E is the total energy; m is the mass of the electron; h is Planck's constant. It might be noted that the parenthetical term $(E + Ze^2/r)$ is just the kinetic energy, since E is the total energy and $-Ze^2/r$ is the potential energy. What the Schrödinger wave equation does is to relate the shape of the waveform to the kinetic energy of the electron in the atom.

The Schrödinger wave equation tells us how the wave function must behave but does not specify what ψ must be. ψ has a different form for each value of E. The problem for an electron in an atom is to solve the Schrödinger equation, find out what values of E are allowed, and find out what ψ looks like as a function of r for each value of E. The permitted values of E turn out to be exactly the same as previously deduced from the Bohr model of the atom (Sec. 4.7). The corresponding permitted expressions for ψ, which are generally fairly complex, are considered in the next section. Each expression for ψ, when squared, gives an electron probability distribution. It is these expressions for ψ^2 which we have been calling *atomic orbitals* in our previous discussions. In the following section, we look at them in some detail.

16.4
ATOMIC ORBITALS

The simplest orbital in an atom is the $1s$ orbital. If we confine ourselves to a one-electron atom—i.e., an atom that consists of but one electron and a nucleus of atomic number Z—then we find that the electron in the lowest energy state has a wave function of the form

$$\psi_{1s} = Ce^{-Zr}$$

where C is a constant which is proportional to Z. The constant C takes care of the fact that as Z increases, the electron is pulled closer to the nucleus. The term e^{-Zr}, the so-called "exponential term," takes care of the fact that the chance of finding the electron drops off extremely rapidly as one goes to larger values of r. In this final term, e is the base for natural logarithms.

Figure 16.6 shows what a graph of the wave function ψ_{1s} looks like plotted as a function of r. As can be seen, ψ has its maximum value when r equals zero and dies off to zero as r goes to infinity. Actually, it is not so much ψ as ψ^2 that interests us, since it is ψ^2 that is proportional to the probability of finding the electron. The course of ψ^2 as a function of r is shown in Fig. 16.7; it should be compared with Fig. 4.17.

Again, as was done on page 113, we need to make a distinction between probability per unit volume and total probability of finding the electron. The surface area of a spherical shell is $4\pi r^2$. If we multiply this area by a small thickness dr of the shell, we get $4\pi r^2 dr$ as the volume of the shell. Figure 16.8

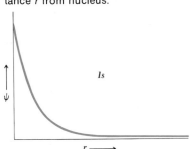

FIG. 16.6 Graph of wave function ψ for 1s electron as a function of distance r from nucleus.

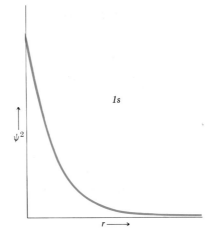

FIG. 16.7 Graph of square of wave function ψ^2 for 1s electron. Note that ψ^2 gives probability per unit volume.

shows how this quantity $4\pi r^2 dr$ increases with increasing r. We also show on the same graph how ψ^2 decreases as we go to larger r. Multiplying these two contributions together gives $4\pi r^2\psi^2 dr$, as shown by the solid curve. $4\pi r^2\psi^2 dr$ has a maximum at an r value equal to a, which we likened before (page 114) to the Bohr radius. In fact, except for language there really is not much difference in describing the innermost shell of an atom by the Bohr picture, the qualitative probability picture, or the wave-function approach.

There is a real difference that shows up when we consider the next-higher energy state, the one that corresponds to a 2s orbital. The wave function for a 2s orbital has the form

$$\psi_{2s} = C'(2 - Zr)e^{-zr/2}$$

Again C' is a constant proportional to Z, and there is an exponential term $e^{-Zr/2}$ which makes the probability die off at large values of r. In addition, we now have the parenthetic term $(2 - Zr)$ which goes to zero at some value of r. This place where ψ, and hence ψ^2, passes through zero is called a *node*. Figure 16.9 shows plots of ψ, ψ^2, and $4\pi r^2\psi^2 dr$. The node occurs when ψ switches from positive to negative. Since the probability is proportional to

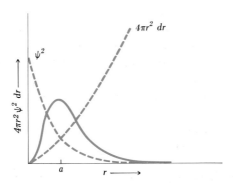

FIG. 16.8 Graph of total probability of finding 1s electron in a spherical shell of radius r and thickness dr.

431

432

Chapter 16
A Closer
Look at
Electron
Descriptions

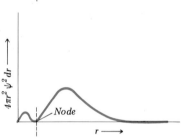

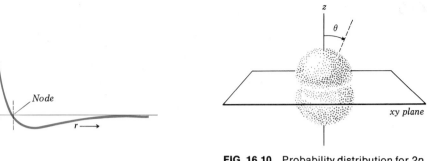

FIG. 16.10 Probability distribution for 2p electron corresponding to p_z or to $m_l = 0$.

FIG. 16.9 Graphs of ψ, ψ^2, and $4\pi r^2 \psi^2 dr$ for a 2s electron.

ψ^2, it does not matter whether ψ is positive or negative. For the 2s electron, there are two regions of high probability, consisting of an inner sphere and an outer sphere, separated by a zone of zero probability.

Wave functions for 2p introduce still another feature—a variation with angle. Unlike 1s and 2s electrons, which are spherically symmetric, 2p electrons show different orientations in space. As was indicated on page 115, there are three kinds of p electrons, designated p_x, p_y, or p_z or, alternatively, as corresponding to three possible values of the magnetic quantum number $m_l = +1$, 0, or -1. How are these related to each other, and how are they described by wave functions?

Let us consider first the p_z orbital. It turns out to be the p orbital for which $m_l = 0$. The wave function for this orbital is given by

$$\psi_{2p(m_l=0)} = C'' e^{-Zr/2} r \cos \theta$$

As before, the wave function has an exponential drop-off term $e^{-Zr/2}$; in addition, there is the multiplicative term $r \cos \theta$. The effect of r is to make the wave function go to zero as r goes to zero. The effect of $\cos \theta$ is to concentrate the probability distribution along the z axis. As shown in Fig. 16.10, θ measures the angle of tilt away from the z axis. When θ is zero, we are looking directly

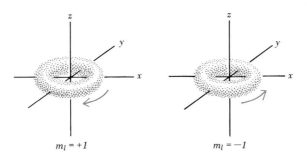

FIG. 16.11 Probability distributions for 2p electrons corresponding to $m_l = +1$ and $m_l = -1$. On the left, the running wave goes counterclockwise; on the right, clockwise.

along the z axis; cos θ is a maximum, and the chance of finding the electron is greatest. As θ increases, we look away from the z axis, and the chance of finding the electron diminishes. When θ reaches 90°, we are looking in the xy plane, cos θ has gone to zero, and we have *no* chance of finding this particular electron.

When it comes to p_x and p_y, we have a special problem. The wave functions cannot be equated directly with either $m_l = +1$ or $m_l = -1$, but to a combination of the two. Figure 16.11 shows what the probability distributions* look like for 2p electrons having $m_l = +1$ and $m_l = -1$. They can be visualized as electron running waves circulating around the z axis. One goes clockwise, and the other goes counterclockwise. For most chemical purposes, running waves are not very useful ways of representing electron probability distributions. Instead, we combine the two waves running in opposite directions into a single standing wave. If we add the two waves represented in Fig. 16.11, we get a standing wave which piles up charge along the x axis. We get a p_x electron, as is shown in Fig. 16.12. If we subtract the two, we get p_y. The p_x, p_y, p_z representation that was given in Fig. 4.21 is fully equivalent to the representation given by $m_l = +1, 0, -1$. The p_x, p_y, p_z description is more useful when talking about establishing chemical bonds in particular directions.

16.5
MOLECULAR ORBITALS

When we were discussing bonding in Chap. 5, it was pointed out that there are two general ways of describing electrons in molecules. One method, called the *molecular-orbital method,* treats each electron as belonging to the molecule as a whole and moving throughout the entire molecule. The other approach, called the *valence-bond method,* assumes the atoms in a molecule are like isolated atoms except that one or more electrons from the outer shell of

*The wave functions corresponding to these are given by

$$\psi = Ce^{-Zr/2}r \sin \theta e^{\pm i\phi}$$

where the angle ϕ measures the sweep around the z axis. The letter i stands for the square root of minus one and is not a real number. It is this $\sqrt{-1}$ which makes the wave into a running wave and circulates it around the z axis. The coefficient of $i\phi$ in the exponent is the magnetic quantum number m_l. Here it can be +1 or -1.

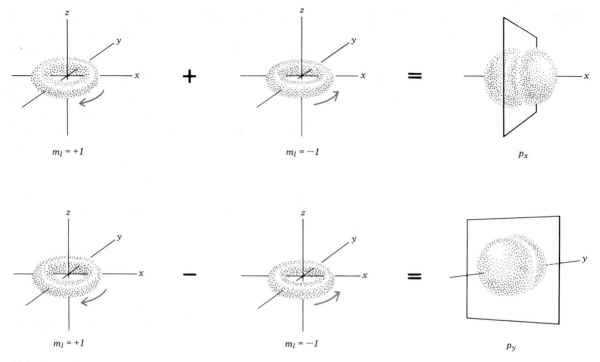

FIG. 16.12 Schematic representation of how running waves $m_l = +1$ and $m_l = -1$ can be combined to give p_x or p_y orbitals.

one atom are accommodated in the outer shell of another atom. What we will do now is look more closely at these two approaches.

One of the surprising aspects of valence-bond descriptions is that such a simple picture adequately deals with such a large variety of cases. However, from the nature of the assumptions, it must be incorrect. To assume that an atomic orbital characteristic of an electron in an isolated atom is not changed by the presence of a second atom is only an approximation. Molecular-orbital theory tries to solve the problem more exactly. It places the nuclei at the positions they occupy in the final molecule and then allows each electron to distribute itself in the electric field arising from all the nuclei and all the other electrons in the molecule.

Such molecular orbitals are exceedingly difficult to calculate exactly. Relatively few have been done. However, a molecular orbital can be approximated by realizing that the part close to one nucleus will greatly resemble an atomic orbital centered on that nucleus. Likewise, the part near a second nucleus will resemble an atomic orbital centered on the second nucleus. For the region where the electron is about equally far from both nuclei, the molecular orbital must take account of both attractions. The most common way to approximate a molecular orbital is as a sum of atomic orbitals of the bonded atoms. The addition of orbitals is not simple, however, since *two* atomic orbitals when combined must produce *two* molecular orbitals. This comes about because of the requirement in quantum mechanics that the total number of energy states must be conserved.

Let us construct the molecular orbitals for the H_2 molecule. We start by fixing the two nuclei at the observed internuclear distance, 0.074 nm. Molecular orbitals can be set up by combining the $1s$ orbital of one H atom with the $1s$ orbital of the other H. Figure 16.13 shows the two ways in which this can be done. On the left are shown the $1s$ orbitals of two isolated H atoms. The plus signs denote the positions of the positively charged nuclei; the shaded areas represent regions of high electronic probability. On the right are shown the two molecular orbitals that can be formed. The lower one results from simple addition of the two $1s$ distributions. In the region between the two nuclei, where the two individual $1s$ orbitals overlap, there is intensification of the electron density, producing a net bonding effect. For this reason the lower orbital is called a *bonding orbital*.

The upper molecular orbital represents the other possible way of combining two $1s$ orbitals. It results from addition of two waves of opposite phase (i.e., crest superimposes on trough), so that destructive interference occurs where they overlap. Electron density in the region of overlap is diminished. The two nuclei are not so well shielded from each other. Consequently, there is repulsion, which tends to push them apart. For this reason, this type of molecular orbital is called an *antibonding orbital*.

Antibonding orbitals are generally marked with asterisks. The designation σ (sigma) is a general one used for any molecular orbital in which electron density is cylindrically symmetric around the line that passes through the two nuclei. The subscript $1s$ in the designations σ_{1s} and σ_{1s}^* denotes that the molecular orbitals were formed from atomic orbitals of the $1s$ type. Finally, it should be noted that the molecular orbital σ_{1s} is lower in energy than the other molecular orbital σ_{1s}^* or the two isolated atomic orbitals. Low-

FIG. 16.13 Molecular-orbital formation in H_2 by combination of $1s$ atomic orbitals.

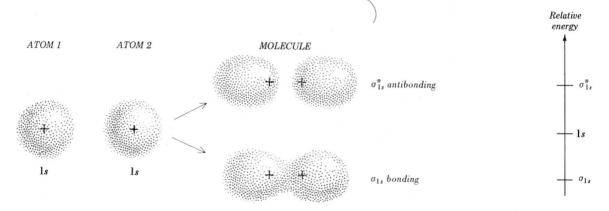

436

Chapter 16
A Closer
Look at
Electron
Descriptions

ering of total energy by transferring two electrons from two isolated $1s$ H orbitals to the σ_{1s} molecular orbital of H_2 is what corresponds to the bond energy of the H_2 molecule.

The molecular-orbital diagram of Fig. 16.13 can also be used to describe what happens when two helium atoms come together. Helium has two electrons; so there would be four electrons to be accommodated. Molecular orbitals, just like atomic orbitals, can hold only two electrons of opposite spin. If there were such a molecule as He_2, one pair of electrons would have to be in the σ_{1s}^* orbital and the other pair would be in the σ_{1s} orbital. The bonding effect due to the pair in the σ_{1s} orbital would be canceled by the antibonding effect (repulsion due to insufficiently shielded nuclei) due to the pair in the σ_{1s}^* orbital. The net bonding effect would be zero. Actually, it is generally true that antibonding orbitals are a bit more antibonding than bonding orbitals are bonding. In Fig. 16.13 the energy of the σ_{1s}^* molecular orbital is a bit further above the $1s$ energy line than the energy of σ_{1s} is below it. As a result, He_2 is energetically unstable with respect to two separated helium atoms.

For elements in the second row, both $2s$ and $2p$ orbitals become available. We can assume that the inner-shell electrons ($1s^2$) are so much lower in energy that they are not appreciably affected when two atoms come together; they remain in atomic orbitals. However, the $2s$ orbitals give rise to two molecular orbitals σ_{2s} and σ_{2s}^*. When two lithium atoms ($1s^2 2s^1$) come together, the outer $2s$ electron of each lithium atom must be accommodated in a molecular orbital. Both electrons (one from each atom) go into the σ_{2s}, which, being a bonding orbital, lowers the energy and allows formation of the molecule Li_2. This molecule has actually been detected in gaseous lithium at the boiling point.

For the next element, beryllium ($Z = 4$, $1s^2 2s^2$), formation of Be_2 would require one pair of electrons in the σ_{2s}^* molecular orbital as well as one pair in the σ_{2s}. Just as in He_2, the antibonding effect of σ_{2s}^* is somewhat greater than the bonding effect of σ_{2s}, and so no stable Be_2 molecule is formed.

For the element boron ($Z = 5$, $1s^2\,2s^2\,2p^1$) and subsequent elements, p orbitals must be considered. From the three kinds of p orbitals, p_x, p_y, and p_z, two kinds of molecular orbitals will result: σ orbitals that are cylindrically symmetric about the internuclear line and π orbitals that come from side-to-side overlap. Figure 16.14 shows how σ orbitals arise from combining p_x orbitals. (We designate as the x axis the left-right line joining the two nuclei.) Combination of p_x of atom 1 with p_x of atom 2 produces two molecular orbitals, σ_{p_x} and $\sigma_{p_x}^*$, depending on whether we add them in phase or out of phase. They correspond, respectively, to bonding and antibonding possibilities. σ_{p_x} has enhanced density between the nuclei, whereas $\sigma_{p_x}^*$ has it depleted.

Figure 16.15 represents the case in which p orbitals overlap, not end to end, but side to side. With side-to-side overlap, as shown in Fig. 16.15, the p_z of atom 1 combines with the p_z of atom 2 to produce the two molecular orbitals π_{p_z} and $\pi_{p_z}^*$. These are not cylindrically symmetric around the bond line but have the electron density to *either side of the bond line*. The bonding orbital π_{p_z} looks like two sausage-shaped clouds parallel to the bond line; the anti-

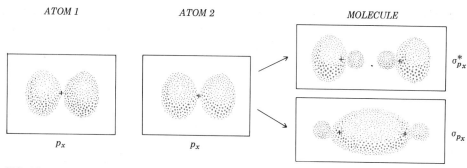

FIG. 16.14 Formation of molecular orbitals by combination of p_x atomic orbitals.

bonding orbital $\pi^*_{p_z}$ has four toplike lobes extending outward from the two nuclei. Side-to-side combination of p_y, as shown in Fig. 16.16, forms π_{p_y} and $\pi^*_{p_y}$ in the same way except that the plane of the orbitals is rotated by 90°. Except for the rotation, π_{p_y} and π_{p_z} orbitals are identical.

Figure 16.17 shows the relative placing on an energy scale of molecular orbitals derived from p. (The exact ordering, specifically whether σ_{p_x} is higher or lower than π_{p_y} and π_{p_z}, has been in dispute. The arrangement shown in Fig. 16.17 is the one that corresponds to the order in which the orbitals become available for electron filling.) In adding electrons to molecular orbitals, the principles followed are the same as those for adding electrons to atomic orbitals:

1 No more than one pair of electrons may occupy a particular molecular orbital.

FIG. 16.15 Formation of π orbitals by side-to-side combination of p_z orbitals.

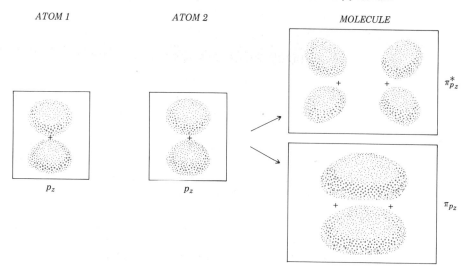

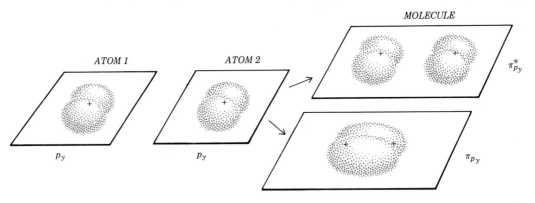

FIG. 16.16 Formation of π orbitals by side-to-side combination of p_y orbitals. (The *y* axis is to be visualized perpendicular to the plane of the paper.)

2 The lowest-energy molecular orbital that is available will fill first.

3 If there is more than one molecular orbital at the same level of energy, electrons spread out insofar as possible into separate orbitals. To ensure this, electron spins are as unpaired as possible—i.e., the total spin is at its maximum.

Figure 16.18 shows how the *p* electrons are distributed in B_2, C_2, N_2, O_2, and F_2. In boron ($Z = 5$, $1s^2\,2s^2\,2p^1$), there are two *p* electrons (one from each atom) that need to be accommodated. There are two lowest-lying orbitals π_{p_y} and π_{p_z}; so the two electrons distribute themselves one to each. Consequently, B_2 is expected to be stable (the orbitals are bonding ones) and paramagnetic (there are two unpaired electrons). In carbon ($Z = 6$, $1s^2 2s^2 2p^2$), there are four *p* electrons (two from each atom) to be accommodated. The two lowest orbitals π_{p_y} and π_{p_z} accommodate a pair each and are filled. Consequently, C_2 is diamagnetic. Also, because C_2 has twice as many electrons in bonding orbitals as B_2 has, the C_2 molecule is expected to be more stable.

FIG. 16.17 Relative energies by order of filling of some molecular orbitals.

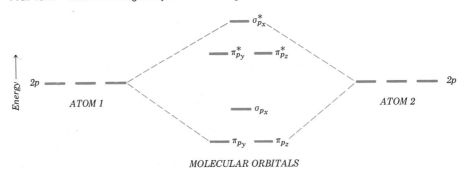

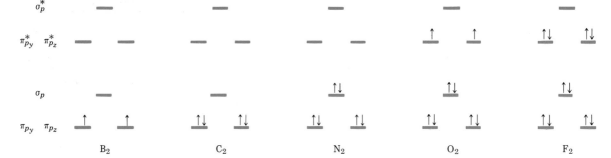

FIG. 16.18 Molecular-orbital occupancy in B_2, C_2, N_2, O_2, and F_2.

Indeed, the bond energy for C_2 (628 kJ/mol) is about twice as great as that for B_2 (290 kJ).

The next three elements give N_2, O_2, and F_2. Their respective bond energies are 941, 494, and 151 kJ/mol. How can these be accounted for? In N_2 each N ($1s^2 2s^2 2p^3$) contributes three p electrons; there are six in all. They fill the three lowest orbitals. All the electrons are paired, and so N_2 is diamagnetic; all three of the occupied orbitals are bonding, and so N_2 is more stable than C_2.

In O_2 there are two additional electrons. Since all the bonding orbitals are filled, the additional two electrons must be placed in antibonding orbitals. The next-lowest lying are $\pi_{p_y}^*$ and $\pi_{p_z}^*$. To minimize electric repulsion between like charges, one electron goes into each orbital. The spins are unpaired; hence, the O_2 molecule is paramagnetic.

The presence of the two antibonding electrons in O_2 weakens the bond relative to N_2. Whereas N_2 can be considered to have a triple bond (three bonding pairs), O_2 can be thought of as having a double bond (three bonding pairs minus two antibonding electrons is about equivalent to two bonding pairs). The *bond order,* which is defined as the number of bonding electrons minus the number of antibonding electrons all divided by 2, drops from three for N_2 to two for O_2. The bond energy drops from 941 to 494 kJ.

Finally, for F_2 there are two more electrons. These go to complete the $\pi_{p_y}^*$ and $\pi_{p_z}^*$ orbitals. The F_2 bond is thus even weaker than O_2 and is equivalent to, at most, a single bond. Indeed, the bond in F_2 is among the weakest of single covalent bonds. Apparently, the effect of two antibonding pairs is greater than that of two bonding pairs. The dominance of antibonding over bonding also shows itself in the nonexistence of Ne_2. Neon ($Z = 10$, $1s^2 2s^2 2p^6$) has six p electrons; if there were such a thing as Ne_2, there would have to be three pairs of bonding and three pairs of antibonding electrons, i.e., no net attraction.

For the above diatomic molecules the molecular-orbital approach is an improvement over the valence-bond approach. Using it, we have been able to account for both the paramagnetism and the bond strength of O_2; simple valence-bond theory cannot do this. For more complex molecules, molecular-orbital descriptions are generally at least as satisfactory as valence-bond pic-

439

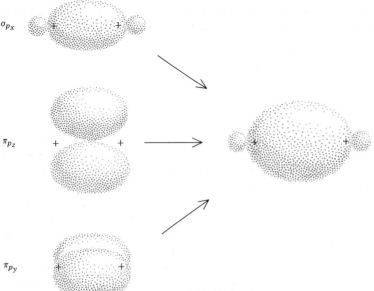

σ_{p_x}

π_{p_z}

π_{p_y}

FIG. 16.19 Superposition of molecular orbitals in N_2.

tures. They have the advantage of not requiring resonance (Sec. 5.8) and of giving more reasonable representations of electron sharing.

The final representation of electric-charge distribution is difficult when electron density from one molecular orbital overlaps that from another. As an illustration, we consider N_2. If we disregard s orbitals, the picture that emerges is somewhat like that shown in Fig. 16.19. Superposition of σ_{p_x}, π_{p_y}, and π_{p_z} leads to a total charge cloud that looks somewhat like a football. π and σ orbitals blend into each other so as to produce a uniform electron density all around the axis of the molecule.

For a more complex case, such as ethane (C_2H_6), the molecular-orbital picture leads to a charge distribution like that shown in Fig. 16.20 (which is to be compared with the valence-bond picture of Fig. 5.20). In contrast to the valence-bond description, where atomic orbitals sit side by side, the molecular-orbital picture shows electron orbitals that extend over several atoms. The whole distribution fuzzes out and is even more symmetric than that

FIG. 16.20 Electron distribution in the C_2H_6 molecule.

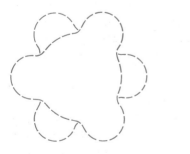

End view

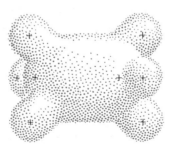

Side view

shown in the figure, which has been simplified in order to show perspective. A better representation of the charge distribution in a C_2H_6 molecule would be a cylinder-shaped charge cloud with slight bulges at the two ends.

16.6
ELECTRON WAVES IN A MOLECULE

Because of its mathematical complexity, a detailed description of electron waves in a molecule is not possible in this text. However, the wave-function description of molecular binding may help clarify the difference between valence-bond and molecular-orbital methods.

First we recall that the wave function represents a mathematical statement about electron probability distribution. ψ itself has no physical significance, but the square of the wave function ψ^2 can be identified with the probability of finding the electron in a particular volume of space. The problem for molecules is to find the wave function that describes the electron in each of its permitted energy states. As with atoms, the permitted energy states are determined from the Schrödinger equation. The potential-energy term is no longer Ze^2/r (corresponding to attraction to one nucleus) but a more complicated expression that describes simultaneous attraction to two nuclei plus repulsion by other electrons. The whole problem is very complicated, even for simple molecules. High-speed computers can handle such problems, but they are expensive, and so we have to be satisfied with approximate solutions. The valence-bond and molecular-orbital methods represent two different approaches to getting approximate solutions. The simplest molecule is H_2; in the ensuing discussion we consider it as the specific example. The test will be not how good a wave function we can get (the wave function by itself is not observable), but how well we can calculate the energy of H_2 relative to two widely separated H atoms. We start first with the valence-bond approach.

The essence of valence bond is that H atoms retain their identity when brought together in the final molecule. This means that the atomic orbital used to describe a $1s$ electron in an isolated H atom can also be used to describe a $1s$ electron in the molecule. The new mathematical idea we need to introduce is that if we have two atoms separately described by wave functions ψ_A and ψ_B, then the wave function for the two together is equal to the product $\psi_A \psi_B$. In other words, for two H atoms the wave function is ψ_{1s} on one atom times ψ_{1s} on the other atom. What happens as the two atoms are brought closer together? If we assume that the individual wave functions ψ_{1s} do not change (which is the essence of strict valence-bond theory), there will be a slight attraction as the $1s$ electron on the first atom begins to feel the pull of the second nucleus and the $1s$ electron on the second atom begins to feel the pull of the first nucleus. The attraction is only very slight; if we calculate the energy, we find that the H_2 molecule is only about 4×10^{-20} J lower in energy than two isolated H atoms. This does not look too good; experimentally, H_2 is found to be 72×10^{-20} J lower than two H. So, even though we calculate a feeble attraction, we do not account for the observed bond strength of H_2.

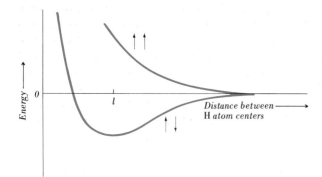

FIG. 16.21 Energy of H_2 molecule as a function of distance between H centers. (Upper curve holds when both electrons have the same spin; lower curve, when spins are opposite to each other.)

What have we forgotten? This is where electron exchange comes in. When two H atoms get close together, electron 1 on atom A may exchange places with electron 2 on atom B. Mathematically, this is expressed by writing the total wave function as the sum of two products:

$$\psi_A^{(1)} \, \psi_B^{(2)} \pm \psi_A^{(2)} \, \psi_B^{(1)}$$

The first term describes the situation where electron 1 is on atom A and electron 2 is on atom B; the second term describes the situation where electron 2 is on atom A and electron 1 is on atom B. Both terms must be used together to describe the final molecule because the electrons are indistinguishable from each other and we have no way of telling electron 1 from electron 2.

There are, furthermore, two possibilities for the total wave function—i.e., two possible states for the final molecule—depending on whether we take the plus sign or the minus sign in forming the above sum. The plus sign holds when electrons 1 and 2 are opposite in spin; the minus sign holds when electrons 1 and 2 have the same spin. Figure 16.21 shows how the energy* of the two states differs as the distance between the two H atoms is changed. The state in which both electrons have the same spin is greater in energy than the far-apart atoms; the molecule in that state tends to fly apart. On the other hand, the state in which the electrons have opposite spin shows a clear energy minimum. Calculations show that the minimum should occur at about 0.08 nm, which is not far from the experimentally observed H_2 bond length, 0.074 nm. More important, the depth of the energy minimum indicates that the H_2 molecule should be 50×10^{-20} J lower than two isolated H atoms. We are still far from the experimental value of 72×10^{-20} J, but the simple process of putting in electron exchange has significantly improved the valence-bond calculation. Further refinements can be used to improve the valence-bond calculation, but none turns out to be so dramatically effective in stabilizing H_2 as is exchange.

It is worth looking at this exchange stabilization more critically. As done above, it is a mathematical device for allowing electron 1, originally on atom A, to leak over onto atom B. In other words, exchange spreads the electron

*This is the total energy, composed of potential and kinetic energies of the electrons plus the potential energy arising from nuclear-nuclear repulsion.

442

wave function so that it is no longer localized on one atom but is now spread over the whole molecule. As noted in Sec. 16.2, spreading out a wave function lowers the energy. Still, it should not be overlooked that we would not have needed to spread out the wave function if we had not chosen to start out with the assumption that the wave function of each electron is confined to its own atom. Stated another way, we would not need such a big correction term (exchange) if we did not make such a big mistake in our first assumption. As will be seen in the following discussion of molecular-orbital theory, exchange does not come into molecular-orbital theory; exchange stabilization is only a mathematical quirk of valence-bond theory.

How do we get a wave function for H_2 in molecular-orbital theory? Conceptually, the procedure is simple. We set the two nuclei some distance apart (0.074 nm), allow the electrons to be attracted to both, and calculate the wave function. The nuclei, because they are massive, can be considered to stay put. The electrons, however, are light, so that they are constantly moving around. To avoid the complication of calculating how one moving electron affects another, we restrict ourselves to the *hydrogen-molecule* ion H_2^+, which consists of but one electron plus two nuclei.

Unlike an atomic orbital, which has one attracting center, a molecular orbital has more than one positive center of attraction. For H_2^+, there are two attracting centers. What is the best way to set up a molecular-orbital description that extends into the neighborhood of two nuclei? Clearly, when the electron is near one nucleus, the effect of the other nucleus becomes negligible. A simple procedure is to assume that when the electron is near nucleus *a*, it is describable by an atomic orbital centered on *a*, whereas when it is near nucleus *b*, it is describable by an atomic orbital centered on *b*.

Near nucleus *a* the wave function has the form $\psi_a = Ce^{-r_a}$, where r_a is the distance from nucleus *a*; near nucleus *b* the wave function has the form $\psi_b = Ce^{-r_b}$, where r_b measures the distance from nucleus *b*. The total description of the electron is obtained by superposing these two separately centered atomic orbitals. We thus can write for the molecular orbital

$$\psi = \psi_{1s \text{ on } a} \pm \psi_{1s \text{ on } b}$$

Such a method of arriving at the total wave function is known as the *linear combination of atomic orbitals;* it is often abbreviated as LCAO. The $\pm$ indicates there are two possible ways of making the combination, depending on whether we combine the separate functions in phase or out of phase. Figure 16.22 illustrates what the resulting functions look like. In one case (top part of the figure) the total wave function is positive everywhere, has two big maxima at *a* and *b*, and has a finite value midway between the two nuclei. In the other case (bottom part of the figure), the total wave function changes sign in going from the region around nucleus *a* to the region around nucleus *b*. Actually, as we have repeatedly noted, it is not the wave function itself that has physical significance; it is the square of the wave function, which is identified with the probability of finding the electron. Figure 16.23 shows what ψ^2 looks like. In the one case, ψ^2 is great between the two nuclei; in the other, it goes to zero. The top part of Fig. 16.23 corresponds to a bonding state; it is what we

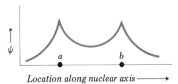

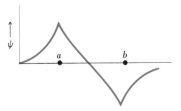

FIG. 16.22 Total wave function for electron in H_2^+. For the top curve, the 1s orbital centered on a has the same phase as the 1s orbital centered on b; for the bottom curve, the two are opposite in phase.

previously called σ_{1s}. The bottom part corresponds to an antibonding state; it is what we previously called σ_{1s}^*.

Each of the wave functions $\psi_{\sigma_{1s}}$ and $\psi_{\sigma_{1s}^*}$ corresponds to a different energy state. The energy values, as shown in Fig. 16.24, depend on how far apart the nuclei are. At very large internuclear spacing, the energies of the bonding and antibonding states become identical. As the distance decreases, the two states separate—the bonding state decreases in energy, and the antibonding state increases.

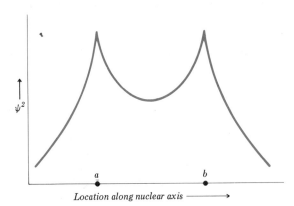

FIG. 16.23 Square of the wave functions for H_2^+ in the bonding case (top) and the antibonding case (bottom).

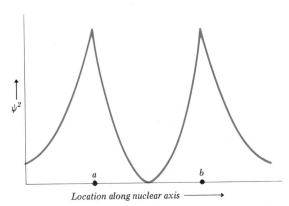

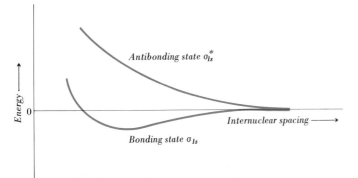

Once we have the two energy states of H_2^+, we can use them to discuss the behavior and buildup of diatomic molecules. For H_2^+ in its ground state, the electron is in the σ_{1s} molecular orbital. If we add a second electron, we get $H_2^+ + e^- \longrightarrow H_2$. Since a molecular orbital can accommodate two electrons, provided they are of opposite spin, both electrons in H_2 would be described by the $\psi_{\sigma_{1s}}$ wave function.

What do we do in molecular-orbital theory to get a wave function for a diatomic molecule such as AB, where A is different from B? The procedure is essentially the same except that we have to put in a weighting factor to take care of the difference in electronegativity. This can be expressed as follows:

$$\psi_{MO} = \psi_A + \lambda \psi_B$$

where ψ_{MO} = total wave function for the molecular orbital

ψ_A = atomic orbital on atom A

ψ_B = atomic orbital on atom B

λ = constant that describes polarity of the orbital

When A equals B, λ has values only of ± 1, but when A is not the same as B, λ can have a whole range of values. If, for example, $\lambda > 1$, it would signify that ψ_B contributes more to the molecular orbital than ψ_A does. This would correspond to having the electron spend more time on B than on A. If $\lambda < 1$, the molecular orbital looks more like ψ_A than like ψ_B. Values of λ are generally deduced from observations on dipole moments.

16.7
MOLECULAR VIBRATIONS

In Fig. 16.21 of the preceding section, a curve was presented showing how the energy of a diatomic molecule changes with distance between the nuclei. The energy that is shown there is the kinetic plus potential energy of the electrons. What is not shown is the vibration of the nuclei. Even at absolute zero there is a certain vibration.

It is sometimes useful to visualize a diatomic molecule as two weights connected by a spring (Fig. 16.25). The spring represents the chemical bond working to hold the atoms together; it also represents the repulsion force due

445

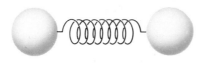

FIG. 16.25 Oscillator model for a diatomic molecule.

to repulsion between electrons and between positively charged nuclei when two atoms are forced together. Such an arrangement, in which the restoring force is proportional to the amount of elongation or compression, is called a *harmonic oscillator*. Many ideas about a harmonic oscillator can be carried over directly to describing molecular vibrations.

What happens to a harmonic oscillator when an external force is applied to stretch it? In producing the elongation, the external force does work on the system and increases its potential energy. Similarly, when a molecule is stretched, its potential energy is raised with respect to that of the undistorted molecule. This is shown in Fig. 16.26. *A* is the undistorted molecule, and *B* shows where it is after it has been stretched. The curve is a parabola, and the energy at *B* is greater than at *A*. If the external stretching force is now released, the molecule tends to snap back. The chemical bond pulls the atoms together. The potential energy drops as the molecule goes from *B* to *A*. In the process, however, the atoms have been set into motion. Kinetic energy now appears as the two nuclei move. The total energy of the molecule stays constant, but what is lost in the gradual decrease of potential energy (from *B* to *A*) shows up as a gradual increase in kinetic energy. By the time the molecule reaches *A*, the kinetic energy has reached a maximum. The kinetic energy then acts as a sort of driving force to overshoot the undistorted arrangement *A* and compress the molecule so that it rides up the curve to *C*. At *C* all the kinetic energy has been used up and converted again into potential energy. The situation is much like that of a pendulum, where potential energy is greatest at the extremes of the swing (kinetic energy is zero) and smallest at the bottom of the swing (kinetic energy is a maximum).

Figure 16.27 summarizes how the total energy of a vibrating diatomic molecule is apportioned between potential and kinetic energy. The dashed curve represents the potential energy; the dotted curve, the kinetic. As the internuclear distance changes between the two extremes l_1 and l_2, the total energy remains constant at a value E. E is the total—potential plus kine-

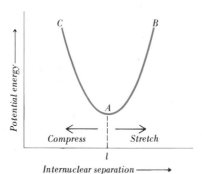

FIG.16.26 Curve showing how potential energy of a diatomic molecule changes with distortion.

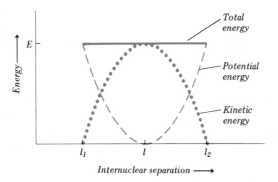

FIG. 16.27 Energy of a molecule during vibration. The solid line shows the total energy; dashed curve, potential energy; dotted curve, kinetic energy. l_1 and l_2 represent, respectively, the most compressed and the most stretched lengths of the molecular bond.

tic—energy. As the potential energy slides back and forth along the dashed line through a minimum, the kinetic energy moves back and forth along the dotted line through a maximum.

In the old days it was believed that harmonic oscillators could have any value of the total energy E. However, when quantum mechanics was applied to the problem, it was found that only certain values of E are allowed. The permitted values are given by the formula

$$E = \left(v + \frac{1}{2} \right) h\nu$$

where $v =$ the so-called "vibrational quantum number" which can take on any of the values 0, 1, 2, 3, etc. (i.e., zero or any positive integer)

$h =$ Planck constant

$\nu =$ frequency

The value of ν represents the vibration frequency that is characteristic of a particular bond. It varies with the "stiffness" of the bond and the masses of the vibrating atoms. Specifically,

$$\nu = \left(\frac{1}{2\pi} \right) \sqrt{\frac{k}{m_{\text{eff}}}}$$

where k is the *force constant* of the bond (i.e., force required to deform the bond by a given amount) and m_{eff} is the *effective mass* of the vibrating atoms. For molecule AB, the effective mass is given by $m_A m_B / (m_A + m_B)$.

Figure 16.28 shows some of the permitted values of vibrational energy states superimposed on the potential-energy curve for a typical diatomic molecule. Only a few of the vibrational states are shown. These are designated by $v = 0, 1, 2, 3,$ or 4, corresponding to total energy $E_0, E_1, E_2, E_3,$ or E_4, respectively. For each value of the total energy, the potential energy would track along the colored curve through the minimum. In the lowest energy state, $v = 0$, the length of the molecule vibrates between l_1 and l_2. In a higher state, such as $v = 4$, the vibration is considerably more vigorous; the bond length goes all the way from l_3 to l_4. The more energetic the vibrational state, the wider the swings of the vibration. In the limit, where v equals infinity, the elongation would be so great that the molecule would fly apart.

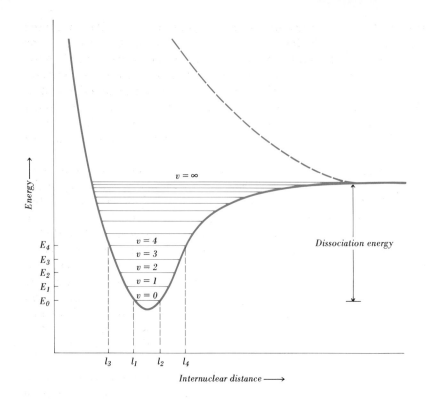

FIG. 16.28 Vibrational states of a diatomic molecule. (Dashed color curve at the top represents the antibonding state σ_{1s}^{*}.)

As indicated in Fig. 16.28, the energy difference between the molecule in its lowest vibrational state ($v = 0$) and the $v = \infty$ state is called the *dissociation energy* of the molecule.

It is useful to have in mind how big are the numbers involved in molecular vibrations. A typical molecule, such as HCl, has a characteristic vibrational frequency of about 10^{14} vibrations per second. That means it takes only about 10^{-14} sec to complete one vibration. At room temperature most of the molecules would be in the lowest vibrational state ($v = 0$); only about one in a million would be in the first excited state ($v = 1$). Infrared radiation, which has wavelength in the range 0.01 to 0.0001 cm, corresponds roughly to the energy difference between molecular vibrational states. Hence, infrared radiation (i.e., heat waves) can be absorbed by molecules in lifting them from a low vibrational state to a higher one. In general, v changes by only one unit at a time. Conversely, "hot" molecules (i.e., those in excited vibrational states) can give off infrared radiation as they make transitions to lower vibrational states. Thus the study of infrared absorption and emission gives information about vibrational states.

In general, the stronger a chemical bond, the greater will be its characteristic vibration frequency. Triple bonds generally vibrate at higher frequency than double bonds; double bonds, higher than single bonds. The heavier the atomic masses at the ends of a bond, the lower the characteristic frequency. A low vibration frequency suggests that the bond is weak and/or that the bonded atoms are heavy.

Figure 16.29 gives typical data for some common bonds. Shown are the equilibrium bond length, the force constant, and the dissociation energy. The bond length corresponds to the average distance between the nuclei when the bond is vibrating in its lowest vibrational state. The force constant measures the "stiffness" of the bond; it is given as the force in newtons needed to distort the bond per centimeter. The dissociation energy measures the energy required to break the bond from its lowest vibrational state to infinitely separated atoms; it is given as the number of kilojoules per Avogadro number of bonds broken.

As can be seen from Fig. 16.29, the force constant generally increases as the dissociation energy increases. A clear exception is the case of F_2 in the sequence F_2, Cl_2, Br_2, I_2. The force constant for F_2 is bigger than for Cl_2, but the dissociation energy is less. The apparent anomaly probably comes from the fact that the F—F bond is so short that repulsions between unshared electrons are more important than usual in reducing the dissociation energy.

Although it is not shown in Fig. 16.29, the vibration of a bond also depends on what other atoms are joined to the pair in question. Thus, the C—C bond has a force constant ranging from 4.5 to 5.6 N/cm, depending on what other atoms are attached to the carbons. Each force constant leads to a characteristic pattern of energy levels. The result is that the pattern of infrared absorption can be used as a fingerprint for identifying complex molecules.

Besides the vibrational motion discussed above, molecules can rotate in space. In general, rotations have lower frequency than vibrations. A typical rotation frequency, exemplified by HCl, is 6.3×10^{11} sec^{-1}, about 100 times slower than the vibrations. During the time it takes the molecule to complete one rotation, several hundred vibrations occur; hence, the vibration and rotation can be treated independently of each other. As with vibration, the energy

FIG. 16.29   Vibration Parameters for Some Common Bonds

Bond	Length, nm	Force constant, N/cm	Dissociation energy, kJ/mol
H—H	0.074	5.1	432
H—F	0.092	8.8	561
H—Cl	0.128	4.8	428
H—Br	0.141	3.8	362
H—I	0.160	2.9	295
H—CH_3	0.109	5.0	423
H—NH_2	0.101	6.4	427
H—OH	0.096	7.7	492
N≡N	0.109	22.4	941
O=O	0.121	11.4	494
F—F	0.143	4.5	151
Cl—Cl	0.199	3.2	239
Br—Br	0.228	2.4	190
I—I	0.267	1.7	149

450

Chapter 16
A Closer
Look at
Electron
Descriptions

of rotation can have only certain permitted values. The spacing between levels is considerably smaller than between vibration levels. Transitions between rotation levels generally occur toward the low-energy end of the spectrum, in what is called the *microwave* region. Typical wavelengths are on the order of 0.1 to 10 cm. Rotational spectra can give considerable information about molecules. Whereas vibrational spectra are most useful for getting at force constants and bond energies, rotational spectra are most informative about bond lengths and bond angles.

16.8 SYMMETRY

One of the most powerful ways to describe any object, whether it is a crystal, a molecule, or a molecular orbital, is to say something about the symmetry of the object. Thus, we say that a crystal of NaCl has cubic symmetry, that a σ orbital is cylindrically symmetric about the bond axis. We should look more closely at this idea of symmetry, since the trend in modern chemistry is to make more use of symmetry in describing molecules and related properties. Molecular vibrations, for example, and even the mechanism of chemical reactions need to be consistent with symmetry requirements.

To describe the symmetry of a molecule, we need to specify the symmetry elements that the molecule possesses. A *symmetry element* is defined as a geometric entity (point, line, or plane) with respect to which various parts of the molecule can be related. A *symmetry operation* is the act of bringing one superimposable part of the molecule into coincidence with another, either physically, by manipulating a model, or mentally. Symmetry operations are always associated with symmetry elements.

There are three important kinds of symmetry elements: axis of rotation, plane of reflection, and inversion center. The meaning of these is as follows:

1 A *rotation axis* is a line through the molecule around which the molecule can be rotated so as to bring it into an identical or equivalent configuration. (If someone carried out the rotation while your back was turned, you would not be able to distinguish the rotated arrangement from the initial arrangement.) Rotation in general is by $360°/n$, where n is the *order of the axis*. n can be 1, 2, 3, 4, etc., corresponding to rotation by 360°, 180°, 120°, 90°. These are designated, respectively, as C_1, C_2, C_3, and C_4 axes. C_1 is called a onefold axis; C_2 is a twofold axis, etc. H_2O has a C_2 axis; as shown in Fig. 16.30, it goes through the O atom and bisects the line between the two H atoms. Rotating an H_2O molecule 180° about its C_2 axis brings one of the H atoms into coincidence with the other; rotating it twice successively through 180° brings the H's back to their original positions.

2 A *mirror plane*, or a plane of reflection, implies that half the molecule is a reflection of the other half. The usual symbol for a mirror plane is *m*. As shown in Fig. 16.30, the H_2O molecule has two mirror planes, perpendicular to each other. One is the plane containing the H and O atoms.

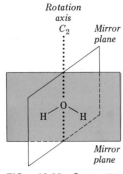

Rotation axis
C_2 *Mirror plane*

H—O—H

Mirror plane

FIG. 16.30 Symmetry elements of H_2O molecule.

The other is perpendicular to this plane, passing through the O atom and perpendicularly bisecting the line between the two H atoms. The C_2 axis of H_2O lies along the intersection of the two mirror planes.

3 An *inversion center* is the same as a center of symmetry. Its presence in a molecule implies that every point on the molecule can be reflected through the center of the molecule to match an identical point on the opposite side of the molecule. A center of inversion is usually symbolized by i. The H_2O molecule does not possess a center of inversion; a CO_2 molecule (linear O=C=O) does.

In general, a molecule has several symmetry elements. They have to be consistent with each other in the sense that an operation about one symmetry element cannot be in contradiction with another symmetry operation about some other symmetry element. How the self-consistency comes about can be seen in Fig. 16.31. Here we show a typical possible arrangement of four equivalent points on a molecule. The points could be four identical atoms, as, for example, the four Cl atoms in the square-planar complex $PtCl_4^{2-}$; we designate them as a, b, c, d so as to be able to refer to them individually.

Point O (where the Pt would be located) is an inversion center. Point a is related by inversion through the center O to point c—that is, a and c are on opposite sides of the center and equidistant from it—and b is similarly related by inversion to d. As can be seen, inversion is equivalent to changing all positive coordinates to negative ones, and vice versa.

The line AB is a twofold axis of rotation. Rotation by 180° around the line AB sends point a into b, and d into c. Simultaneously, b is sent into a, and c into d.

The plane $PQRS$ is a reflection plane. Point a is said to be related by reflection to point d—that is, a and d are on opposite sides of and equidistant from the reflection, or mirror, plane $PQRS$. Similarly, $PQRS$ is a reflection plane relating b to c.

What happens if we carry out successive operations? Suppose, for example, we first rotate the molecule about line AB and then reflect it in the mirror plane. The net result would be the same as if we had gone directly

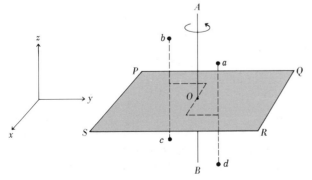

FIG. 16.31 Relation of symmetry elements relating four equivalent points on a molecule.

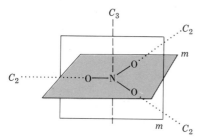

FIG. 16.32 Symmetry characteristics of nitrate ion.

through the inversion center. Specifically, rotation of b about the line AB sends b into a; subsequent reflection of a through $PQRS$ sends a into d. The net result of $b \longrightarrow a$ followed by $a \longrightarrow d$ is the same as inverting b directly through O into d. This illustrates a general principle: *Two successive symmetry operations on a molecule are equivalent to some other symmetry operation.*

As a further illustration of symmetry descriptions, let us consider the nitrate ion, NO_3^-, shown in Fig. 16.32. The three O atoms lie at the corners of an equilateral triangle, the center of which is occupied by N. The plane of the NO_3^- ion is a mirror plane and can be designated as m. There is a threefold rotation axis C_3 perpendicular to the plane; it is shown by the dashed line in Fig. 16.32.

Besides the threefold axis and mirror plane, NO_3^- is characterized by three other axes of symmetry. These are called *subsidiary* axes to distinguish them from the *main*, or *principal*, symmetry axis. As indicated by dotted lines in Fig. 16.32, there are three C_2 axes which are perpendicular to the C_3 axis. They pass through the O—N bonds; so they form angles of 120° with respect to each other.

Finally, to complete the description of NO_3^-, there are three more mirror planes defined, respectively, by the C_3 axis and each of the C_2 axes. One of them is shown in Fig. 16.32.

As a contrast to the NO_3^- case, Fig. 16.33 shows the ammonia molecule, NH_3. The number of atoms is the same, but they are disposed quite differently. The H atoms lie at the corners of an equilateral triangle, but the N

FIG. 16.33 Symmetry characteristics of ammonia molecule. The mirror plane shown passes through the left N—H bond and bisects the H—N—H angle at the right.

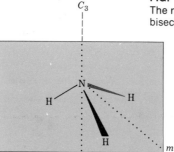

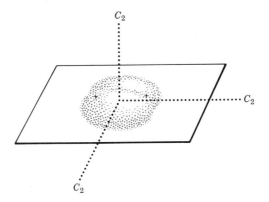

FIG. 16.34 Relative orientation of the 3 twofold axes used to describe a π_{p_y} orbital.

atom does not lie in this plane. The result is that there is no horizontal mirror plane. There is still a threefold axis, passing (as shown in Fig. 16.33) through the N, but the horizontal reflection plane and the three subsidiary C_2 axes are gone. Only the three vertical mirror planes (defined, respectively, by the C_3 axis and each of the N—H bonds) remain.

To illustrate the great conciseness of symmetry descriptions, the π_{p_y} orbital can be described as having three mutually perpendicular C_2 axes. The relative orientation of these is shown in Fig. 16.34. There are also three mutually perpendicular mirror planes passing through pairs of C_2 axes.

Important Concepts

electron waves	bonding orbital	dissociation energy
wave function	antibonding orbital	vibration frequency
standing wave	bond order	rotation frequency
running wave	valence-bond vs. molecular-orbital	symmetry
particle in a box	exchange	symmetry element
wave equation	hydrogen-molecule ion	symmetry operation
atomic orbital	molecular vibrations	rotation axis
node	harmonic oscillator	mirror plane
molecular orbital	force constant	inversion center

Exercises

*16.1 Wave motion When you throw a stone into a pool of water, a surface wave appears and spreads out from the point of impact. Tell specifi-cally what each symbol in $\psi = A \sin bx$ means for such a wave.

*16.2 Wave motion For the standing wave

shown in Fig. 16.1 what would be the value of ψ at x values of $0, \frac{1}{4}\lambda, \frac{1}{2}\lambda, \frac{3}{4}\lambda, \lambda$?

*16.3 Wave motion How would the picture in Fig. 16.2 be changed if the wave were running to the right instead of to the left?

**16.4 Wave motion Suppose you have a running wave described by $A = A \sin 2\pi[(x/\lambda) + \nu t]$. What is the special significance of x values that are multiples of λ or t values that are multiples of $1/\nu$?

**16.5 Wave motion For each of the waves shown in Fig. 16.3, how long does it take for the wave to move through one wavelength? What is the speed of each wave?

**16.6 Particle in a box Explain why the wave function $\psi = A \sin 2\pi(x/\lambda)$ must go to zero at the edges of the box.

**16.7 Particle in a box The waves that describe a particle in a box in its various energy states have been likened to the various vibration waves possible in a violin string. For sound waves, the velocity in air is 331 m/sec. Calculate the four lowest frequencies for a violin string 80 cm long.

Ans. 207, 414, 621, 828 sec^{-1}

**16.8 Particle in a box Given the energies for a particle in a box as shown in Fig. 16.5, calculate the energy of the lowest state for an electron in a one-dimensional box that is 1×10^{-8} cm in length.

**16.9 Particle in a box Explain why the energy levels allowed for a particle in a box (as shown in Fig. 16.5) are not equally spaced. What happens to the levels as the box is made smaller?

**16.10 Particle in a box Referring to Fig. 16.5, where in the box is the probability greatest of finding the electron when the electron is in the fourth energy state? Where is the probability least?

**16.11 Particle in a box What happens to the energy levels of the particle-in-a-box problem shown in Fig. 16.5 if the electron is replaced by a proton?

**16.12 Wave equation Tell what is meant by each of the symbols that appear in the Schrödinger wave equation.

**16.13 Wave equation Under what circumstance would the Schrödinger wave equation go over to the particle-in-a-box problem?

**16.14 Wave function How would you need to change the picture shown in Fig. 16.6 if Z were increased?

**16.15 Atomic orbital What is meant by the term "atomic orbital"? Suggest four different ways of representing the same atomic orbital, and show how these are related.

**16.16 Wave function What happens to the value of the $1s$ wave function at $r = 0$? What does this mean?

**16.17 Wave function The wave function for the $2s$ orbital switches from positive to negative at some value of r. How does this value of r change when Z is increased?

***16.18 Wave function What is the main difference between the wave function for the $1s$ orbital and the $2s$ orbital? $2s$ and $2p$?

**16.19 p orbitals At a given value of r, what happens to the wave function of a $2p_z$ electron as θ is increased from 0 to $\pi/2$ and then to π?

**16.20 Nodes How does the nodal surface of a $2p$ electron differ from that of a $2s$ electron?

***16.21 p orbitals Tell how the three $2p$ orbitals corresponding to $m_l = +1, 0$, and -1 differ from each other How are these related to p_x, p_y, and p_z?

*16.22 Molecular orbitals What is the essential distinction between a molecular orbital and an atomic orbital?

*16.23 Molecular orbitals By reference to Fig. 16.13, explain why the σ_{1s}^* molecular orbital is called an antibonding orbital. In other words, tell why this is a good description.

**16.24 Molecular orbitals Given the energy diagram shown in Fig. 16.13, suppose you put enough energy into an H_2 molecule so that one of the electrons is excited from the σ_{1s} orbital to the σ_{1s}^* orbital. What would this do to the relative stability of the molecule? What should happen? Explain.

*16.25 Molecular orbitals Tell what is the main difference between each of the following:

a σ_{1s} vs. σ_{1s}^* *b* σ_{2p_x} vs. σ_{2s}
c σ_{p_x} vs. π_{p_y} *d* π_{p_y} vs. $\pi_{p_y}^*$

**16.26 Molecular orbitals Which of the molecular orbitals shown in Figs. 16.14, 16.15, and 16.16 would you expect to be lowest in energy? Justify your answer.

*16.27 Molecular orbitals Using the orbital-filling diagram shown in Fig. 16.17, predict the number of unpaired electrons in each of the following: $B_2^+, B_2, C_2^+, C_2, N_2^+, N_2, O_2^+, O_2, F_2^+, F_2$.

16.28 Molecular orbitals Using the energy diagram shown in Fig. 16.17, account for the following observation: when an e^- is removed from O_2, the internuclear distance shrinks, but when an e^- is removed from N_2, the internuclear distance expands. How would you expect the bond distance in O_2 to compare with that in N_2? in O_2^+ compared with N_2^+?

16.29 Molecular orbitals Using Fig. 16.17, compare CO and N_2 with respect to which molecular orbitals are occupied and what the internuclear distances are likely to be relative to each other.

16.30 Symmetry elements Given the electron-cloud picture shown in Fig. 16.20, what symmetry elements does the cloud possess?

16.31 Wave functions of molecules How is the upper picture shown in Fig. 16.22 related to the picture shown in Fig. 16.6?

16.32 Exchange How does exchange come into the valence-bond picture of a molecule? **Why is it effective** in lowering the energy?

16.33 Molecule ion What makes the **ecule-ion problem** of H_2^+ so much easier to solve than the H_2 problem?

16.34 Electron density With reference to the top part of Fig. 16.23, explain why the electron density in the internuclear region turns out to be greater than one would get by just adding up electron density from two partially overlapped atoms. (*Hint:* $\psi = \psi_A + \psi_B$. Calculate ψ^2.)

16.35 Molecular orbitals In working out the wave functions for the molecular orbitals in a molecule such as HCl, how could one handle the problem that the electronegativity of Cl is much higher than that of H? What would that do to the total wave function?

16.36 Molecular vibrations What is meant by the term "harmonic oscillator"? How is it related to the problem of molecular vibrations?

16.37 Molecular vibrations With reference to Fig. 16.26, explain why a diatomic molecule, compressed to position C and then let go, has to swing back and forth along the curve CAB.

16.38 Molecular vibrations Suppose you had a diatomic molecule of total mass 20 amu. Assuming the force constant stays the same, how would the characteristic frequency be changed by going from 10 amu at each end to 19 amu at one end and 1 amu at the other? *Ans. 2.3 times as great*

16.39 Molecular vibrations How would each of the following affect the characteristic frequency of vibration of a diatomic molecule:
a Making the force constant twice as strong
b Doubling the mass at both ends of the molecule
c Doubling the mass at one end of the molecule only

16.40 Molecular vibrations With reference to Fig. 16.28, suggest a reason why the spacing between the vibrational states is constant in the low states but gets progressively smaller as v approaches infinity.

16.41 Molecular vibrations Given the data in Fig. 16.29, calculate the characteristic frequency for the H—O bond in water.
Ans. $1.12 \times 10^{14} \ sec^{-1}$

16.42 Molecular vibrations Given that HCl has a characteristic vibration frequency of $8.7 \times 10^{13} \ sec^{-1}$, calculate the energy spacing between the lowest and first vibrational states. Express this energy in electronvolts and tell what wavelength of radiation corresponds to this transition.

16.43 Symmetry What is the main difference between a symmetry element and a symmetry operation? Give examples of each.

16.44 Symmetry Prove that a molecule which has an inversion center and a C_2 axis must also have a mirror plane.

16.45 Symmetry Tell what symmetry elements are possessed by the basic shape of each of the digits 1, 2, 3, 4, 5, 6, 7, 8, 9, 0.

16.46 Symmetry What elements of symmetry does a cube have? Tell specifically where the rotation axes are.

16.47 Symmetry If one of the oxygen atoms of the nitrate ion were replaced by sulfur, which symmetry elements would be lost and which would be retained?

16.48 Symmetry Suppose you have a molecule XY_4 which can exist in two possible structures. Structure I has the X atom in the center of a square array of Y atoms. Structure II has the X atom in the center of a tetrahedral array of Y atoms. Tabulate the symmetry elements of each of these structures.

16.49 Symmetry elements What symmetry element, if any, is there in the σ_{p_x} orbital (Fig. 16.14) which is not in the π_{p_z} orbital (Fig. 16.15)? What symmetry element, if any, is in π_{p_z} which is not in σ_{p_x}?

Chapter 17

CHEMISTRY OF SOME TYPICAL METALS

Most chemical elements (approximately 80 percent) are metals. Unlike the nonmetals, they have physical properties associated with highly mobile electrons. These so-called "metallic" properties include shiny luster, good electric conductivity, and, most important of all, electric resistance that decreases as the temperature is lowered. Most metals are fairly good reducing agents, but some, such as platinum and gold, are almost inert. It would tax us too much to study each element in detail in order to learn the great variety of behavior that is observed. Instead, we shall look only at the most common, important, and representative elements. In this chapter, we consider the group I, or alkali, elements, the group II, or alkaline-earth, elements (especially magnesium and calcium), and aluminum, tin, and lead. In the next chapter, we consider transition elements.

17.1
ALKALI ELEMENTS

The alkali elements are lithium, sodium, potassium, rubidium, cesium, and francium; they make up group I at the extreme left of the periodic table. All crystallize with a body-centered cubic lattice in which the lattice points appear to be occupied by +1 ions. As shown in Fig. 17.1, the valence electrons (one from each atom) make up a sea, or cloud, of free electrons which permeates the whole lattice. Since the electrons are not fixed in position, they spread throughout the metal and thus produce high electric conductivity. Furthermore, high conductivity of electricity is always accompanied by high conductivity of heat. This is not surprising because thermal energy, which is nor-

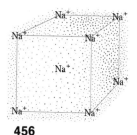

FIG. 17.1 Body-centered cubic array of Na^+ ions immersed in its electron cloud.

456

mally transported by atomic vibrations, can also be transported by conduction electrons.

The silvery luster observed in freshly cut alkali metals can be explained by the highly mobile electrons of the metallic lattice. When a light beam strikes the surface of a metal, electric fields associated with the light wave set the electrons of the metal into back-and-forth oscillation. This is easy to do because the electrons are not tightly bound to any specific atom. However, like any moving charge, oscillating electrons give off electromagnetic energy, as light. The net effect is that the beam of light is "reflected." In this respect electrons in metals act like relay stations which receive an electromagnetic signal and send it out again.

Softness, malleability, and ductility also characterize the alkali metals. These properties can be accounted for by the nature of the forces holding the lattice together. For example, metallic sodium is held together by the attraction between Na^+ ions and the valence-electron cloud. Since this attraction is uniform in all directions, there are no strongly preferred positions for the ions. The result is that Na^+ ions can easily be moved from one lattice site to another. Under pounding, the crystal easily flattens; it can be cut with a knife like soft cheese. All this is in contrast to iron or tungsten, for example, where there are strong, directed forces between adjacent positive ions because of covalent bonding.

Why do the alkali elements form crystals consisting of +1 ions and electrons? The question can be partly answered by considering the isolated gaseous atoms. Figure 17.2 shows some of their properties. The column headed "electronic configuration" indicates the population in the undisturbed neutral atom. There is but one electron in the outermost energy level. The energy required to pull off this electron is given in the column of first ionization potentials. As ionization potentials go, these are relatively small values. However, the second ionization potential, the energy required to pull off a second electron, is many times higher than the first. This means that, although it is relatively easy to form M^+, it is very difficult (practically impossible under ordinary conditions) to form M^{2+}. This is consistent with the notion that a closed shell of electrons is difficult to break into. The result is that when alkali atoms come together to form liquid or solid, M^+ ions are formed.

FIG. 17.2   Atomic Properties of Isolated Alkali Atoms

Element	Atomic number	Electronic configuration (core e^-'s in parentheses)	Ionization potential, eV* First	Second	Ionic radius, nm (M^+)
Lithium	3	(2) $2s^1$ ·	5.39	75.6	0.076
Sodium	11	(10) $3s^1$	5.14	47.3	0.102
Potassium	19	(18) $4s^1$	4.34	31.8	0.138
Rubidium	37	(36) $5s^1$	4.18	27.5	0.152
Cesium	55	(54) $6s^1$	3.89	25.1	0.167
Francium	87	(86) $7s^1$	....	....	(0.175)

*Multiply by 96.5 to get kJ/mole.

FIG. 17.3   Properties of Alkali Metals in the Condensed State

Element	Electrode potential, V	Density, g/cm^3	Melting point, °C	Boiling point, °C
Lithium	−3.05	0.53	186	1336
Sodium	−2.71	0.97	97.5	880
Potassium	−2.93	0.86	62.3	760
Rubidium	−2.93	1.53	38.5	700
Cesium	−2.92	1.87	28.5	670

The properties shown in Fig. 17.2 are well illustrative of the general changes expected in going down a group of the periodic table. For example, the radius* of the +1 ion increases progressively from lithium down. This is expected because of the increasing number of electron shells populated. Similarly, the ionization potential shows progressive decrease in going down the group. This is consistent with increased size and resulting smaller attraction for the valence electron.

The alkali metals are strong reducing agents, the most reactive metals known. Even water, which is not a very good oxidizing agent, vigorously chews them up, sometimes with explosive violence. Quantitatively, reducing strength is described by electrode potentials. As can be seen in Fig. 17.3, the electrode potentials are all large and negative. This means that the half-reaction

$$e^- + M^+(aq) \longrightarrow M(s)$$

has much less tendency to go to the right than does the reference hydrogen-electrode half-reaction

$$e^- + H^+(aq) \longrightarrow \tfrac{1}{2}H_2(g)$$

As was discussed in Sec. 11.10, a small tendency of a half-reaction to go to the right means a large tendency for that half-reaction to go to the left. Thus, the large negative numbers for the alkali electrode potentials mean a large tendency for the half-reaction

$$M(s) \longrightarrow M^+(aq) + e^-$$

to go to the right. Lithium (3.05 V) has the biggest tendency; sodium (2.71 V), the least. At first sight this may be surprising since the ionization potentials of Fig. 17.2 indicate that it is most difficult to pull an electron off lithium, not sodium. The apparent contradiction points up the great difference between gas-phase behavior and that involving solids and solutions. Whereas the ionization potential is concerned only with the *isolated gaseous* atom, the elec-

---

*From X-ray studies of ionic crystals it is possible to determine the radius of an ion. However, X-ray investigations give only the distance between centers of atoms. How should this distance be apportioned? Usual procedure is to adopt one ion as standard and assume it has a definite radius in all its compounds. Other radii are then assigned so that the sum of radii equals the observed spacing. A standard may be obtained from a salt such as LiI, where Li$^+$ is so small that the spacing can be assumed to be due to large I$^-$ ions in contact.

trode potential is concerned with the *metal* related to its ionic species in *solution*. Removal of an electron is only part of what goes on in the above half-reaction. We can see this by breaking the reaction up into its component steps:

		Li	Na	K
Step (1):	$M(s) \longrightarrow M(g)$	+122 kJ	+ 78 kJ	+ 61 kJ
Step (2):	$M(g) \longrightarrow M^+(g) + e^-$	+520 kJ	+496 kJ	+419 kJ
Step (3):	$M^+(g) \longrightarrow M^+(aq)$	−480 kJ	−371 kJ	−299 kJ
		162	203	181

In step (1) the metal is evaporated. The energy required to do this (called the *sublimation energy*) is approximately the same for all the metals of group I, except the value for Li is about twice as great. In step (2) an electron is pulled off the neutral atom to give a gaseous ion. The energy required to do this (the ionization potential) is largest for lithium. In step (3) the gaseous ion is placed in water, i.e., hydrated. Energy (hydration energy) is liberated. The tendency of the overall change to occur depends on the net effect of all three steps. The fact that for lithium the overall tendency is greatest suggests that the relatively greater difficulty of steps (1) and (2) has been more than compensated for by step (3). Hydration energy of tiny Li^+ is so great that it more than makes up for the higher energy required to pull the electron off.

Because alkali elements have such great tendency to form +1, they occur in nature only as +1 ions. Sodium and potassium are the most abundant, ranking sixth and seventh in the Earth's crust. Francium is essentially nonexistent since it has an unstable nucleus and is radioactive. Trace amounts of it have been prepared artificially by nuclear reactions.

Since most of the compounds of the alkali metals are water-soluble, they are generally found in seawater and in salt wells. However, there are many clays which are insoluble complex compounds of alkali metals with Si, O, and Al. Also, as the result of evaporation of ancient seas, there are large salt deposits which serve as convenient sources of alkali-metal compounds. To prepare the elements, it is necessary to reduce the +1 ion. This can be done chemically or electrolytically. Purely chemical methods are difficult since they require a reducing agent stronger than the alkali metals. Chemical reduction can be carried out in special cases, as in the reaction

$$Ca(s) + 2RbCl(s) \longrightarrow CaCl_2(s) + 2Rb(g)$$

The reaction occurs at high temperature in the direction indicated because rubidium is more volatile and escapes out of the reacting mixture. In the equilibrium state the concentration of rubidium would be very small.

In practice, the alkali metals are generally prepared by electrolysis of molten halides or hydroxides. For example, sodium is made commercially in ton quantities by electrolysis of fused NaOH (mp 318°C). Sodium is formed at the cathode, and oxygen at the anode. To prevent recombination, the electrode compartments are separated. Figure 17.4 shows a schematic representation of the arrangement. A circular array of nickel anodes surrounds a capped central iron cathode; a ring of fire keeps the NaOH molten as the

460

Chapter 17
Chemistry
of Some
Typical
Metals

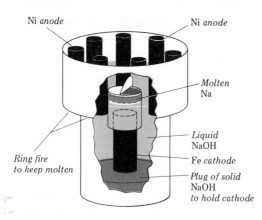

FIG. 17.4 Commercial electrolysis cell for producing sodium.

electrolysis proceeds. The cathode reaction is $Na^+(l) + e^- \longrightarrow Na(l)$, and the anode reaction is $4OH^-(l) \longrightarrow O_2(g) + 2H_2O + 4e^-$. The overall reaction is $4Na^+(l) + 4OH^-(l) \longrightarrow 4Na(l) + O_2(g) + 2H_2O$.

The alkali elements have excellent metallic properties, but they are too reactive to be generally used for this purpose. Sodium in polyethylene encased cables is used in some underground, high-voltage transmission applications. Liquid sodium is used in conducting heat energy from the center of a nuclear reactor. In both uses, expense and difficulty involved in working with sodium are partially compensated for by its excellence as an electric and heat conductor.

Cesium has the distinction of being the metal from which electrons are ejected most easily by light; such light-induced emission is termed the *photoelectric effect.* It is used in the *photocell,* a device for converting a light signal to an electric signal. As shown schematically in Fig. 17.5, it consists of an evacuated bulb containing two electrodes, one of which is coated with cesium metal, cesium oxide, or an alloy of cesium, antimony, and silver. In the absence of light the device does not conduct. When struck by light, the cesium-coated electrode emits electrons which are attracted to the positive electrode, and thus the circuit is completed. Television pickup devices use the photocell principle. Color effects are possible because cesium metal has a high response to red whereas cesium oxide is most sensitive to blue.

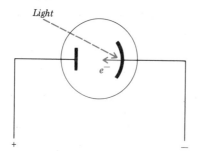

FIG. 17.5 Photocell showing how light energy can be used to eject electrons from cesium.

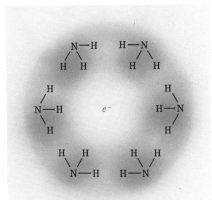

FIG. 17.6 Possible arrangement of trapped electron in sodium-ammonia solution.

Though all the alkali metals are very good reducing agents, only sodium finds extensive use for this purpose. It is used to make other metals by reducing their chlorides, and it is also used in the production of various organic compounds. For this latter purpose, sodium is frequently used in the form of its solution in liquid ammonia. It is a remarkable fact that sodium and the other alkali metals dissolve in liquid ammonia to give colored solutions. Dilute solutions are blue and are believed to contain the alkali metal dissociated into +1 ions and electrons. The blue color seems to be due to the electrons which are trapped in cages of properly oriented ammonia molecules (see Fig. 17.6). More concentrated solutions have a metallic, bronzelike appearance and have very high electric conductivity, indicating that the electrons are extremely mobile. Reducing properties in all these solutions are somewhat toned down compared with the pure alkali metals.

The great reactivity of alkali metals poses a special problem in handling them. Sodium, for example, rapidly corrodes in air. To avoid such problems, alkali metals are usually stored under kerosene or other inert hydrocarbon compounds. Great care has to be taken so that alkali metals do not accidentally come in contact with water. The prudent investigator generally keeps a bucket of sand ready to douse any fires that may occur.

17.2
COMPOUNDS OF THE ALKALI ELEMENTS

The alkali metals readily form compounds by reacting with other substances. For example, sodium metal on standing in air becomes covered with a white scale that is mostly sodium peroxide, Na_2O_2. Furthermore, water vigorously attacks sodium to liberate hydrogen:

$$2Na(s) + 2H_2O \longrightarrow 2Na^+ + 2OH^- + H_2(g)$$

Thus, the problem with the alkali metals is not to get them to form compounds but to keep them from doing so.

All the compounds, even the hydrides, are generally considered to be ionic, with the alkali metal present as a +1 ion. Most of the compounds are quite soluble in water; hence, a convenient way to get a desired anion in solution is to use its sodium salt. The alkali-metal ions do not hydrolyze appreciably. They are colorless.

The hydrides, such as NaH, can be prepared by heating alkali metal in hydrogen. The simple oxides, M_2O, are not so readily formed. Only lithium reacts directly with oxygen to form Li_2O. When sodium reacts with oxygen, the peroxide Na_2O_2 is formed instead. Potassium, rubidium, and cesium under similar conditions form superoxides of the type MO_2. In order to get simple oxides, it is necessary to reduce some alkali-metal compound such as the nitrate. For example,

$$2KNO_3(s) + 10K(s) \longrightarrow 6K_2O(s) + N_2(g)$$

All the oxides are basic and react with water to form hydroxides. Commercially, however, the hydroxides are made by electrolysis of aqueous alkali-chloride solutions. For example, as was discussed in Sec. 11.4, sodium hydroxide, or *caustic soda,* is made by the electrolysis of aqueous sodium chloride.

17.3
ALKALINE-EARTH ELEMENTS

The elements of group II are beryllium, magnesium, calcium, strontium, barium, and radium. They are called *alkaline-earth* elements because the alchemists referred to any nonmetallic substance, insoluble in water and unchanged by fire, as an "earth." The "earths" of this group, e.g., lime (CaO) and magnesia (MgO), give decidedly alkaline reactions.

In general, alkaline-earth elements exist as M^{2+} ions. Why should group II elements form M^{2+} ions whereas group I elements form M^+ ions? Figure 17.7 shows pertinent information. The point to note is that the electron configuration is now s^2 instead of s^1. This is reflected in the ionization potentials. Not only is the first ionization potential relatively modest, but so is the second.

FIG. 17.7   Atomic Properties of Isolated Alkaline-Earth Atoms

Element	Atomic number	Electronic configuration (core e^-'s in parentheses)	Ionization potential, eV* First	Second	Third	Ionic radius, nm (M^{2+})
Beryllium	4	(2) $2s^2$	9.32	18.2	153.8	0.045
Magnesium	12	(10) $3s^2$	7.64	15.0	80.1	0.072
Calcium	20	(18) $4s^2$	6.11	11.9	51.2	0.100
Strontium	38	(36) $5s^2$	5.69	11.0	(43)	0.118
Barium	56	(54) $6s^2$	5.21	10.0	(36)	0.135
Radium	88	(86) $7s^2$	5.28	10.1	....	0.126

*Multiply by 96.5 to get kJ/mole.

Unlike group I, where we have to break into an inner shell, the second elec-tron still comes from the outermost shell. The energy expenditure remains relatively reasonable. It is not until we get to the third ionization potential that we face up to a very large increase.

The above argument explains why group II elements form M^{2+} rather than M^{3+}, but we still have not explained why we do not form M^+. Why, for example, does calcium prefer to form Ca^{2+} rather than Ca^+? The data in Fig. 17.7 show that the second ionization potentials are almost twice as great as the first. Surely it would be energetically more favorable to pull two electrons, one each from two atoms, than to pull both electrons from the same atom. If only ionization potentials were involved, such would indeed be the case. For example, 6.11 eV is required to pull off the first electron from calcium, and 11.9 eV to pull off the second. This implies that return of one electron to Ca^{2+} liberates 11.9 eV of energy. [As a consequence, $Ca(g)$ and $Ca^{2+}(g)$ should be unstable with respect to conversion to $2Ca^+(g)$.]

$$Ca(g) \longrightarrow Ca^+(g) + e^- \qquad \text{requires 6.11 eV}$$
$$\underline{e^- + Ca^{2+}(g) \longrightarrow Ca^+(g) \qquad \text{liberates 11.9 eV}}$$
$$Ca(g) + Ca^{2+}(g) \longrightarrow 2Ca^+(g) \qquad \text{liberates 5.8 eV}$$

The net energy decrease is appreciable (5.8 eV, or 560 kJ); hence, Ca^+ should be the stable species *in the gas phase*.

What about the solid state or aqueous solution? Here the situation is more complicated because we now have to consider the stabilizing effect of the environment. To see the relative effect on Ca^+ and Ca^{2+}, we consider two hy-pothetical reactions:

$$Ca(s) + H^+(aq) \longrightarrow Ca^+(aq) + \tfrac{1}{2}H_2(g)$$
$$Ca(s) + 2H^+(aq) \longrightarrow Ca^{2+}(aq) + H_2(g)$$

The first ends up with $Ca^+(aq)$; the second, with $Ca^{2+}(aq)$. Which of these is more favored?

Let us break up the overall reaction into its component steps and look at the change in free energy. In the first case, we would have the following:

$$Ca(s) \longrightarrow Ca(g) \qquad \Delta G° = +159 \text{ kJ}$$
$$Ca(g) \longrightarrow Ca^+(g) + e^- \qquad \Delta G° = +590 \text{ kJ}$$
$$Ca^+(g) \longrightarrow Ca^+(aq) \qquad \Delta G° = -300 \text{ kJ}$$
$$\underline{e^- + H^+(aq) \longrightarrow \tfrac{1}{2}H_2(g) \qquad \Delta G° = -465 \text{ kJ}}$$
$$\text{Net } Ca(s) + H^+(aq) \longrightarrow Ca^+(aq) + \tfrac{1}{2}H_2(g) \qquad \Delta G° = -16 \text{ kJ}$$

As the free-energy change for the overall reaction is negative, the reaction should go spontaneously as written. In other words, the +1 aqueous ion of calcium is stable compared with solid metal, just as was found for the alkali elements!

However, the fact that formation of $Ca^+(aq)$ is favorable does not mean that it will stop at that point. Let us now look at the following sequence:

464

Chapter 17
Chemistry
of Some
Typical
Metals

$$\text{Ca}(s) \longrightarrow \text{Ca}(g) \qquad \Delta G° = +159 \text{ kJ}$$
$$\text{Ca}(g) \longrightarrow \text{Ca}^+(g) + e^- \qquad \Delta G° = +590 \text{ kJ}$$
$$\text{Ca}^+(g) \longrightarrow \text{Ca}^{2+}(g) + e^- \qquad \Delta G° = +1145 \text{ kJ}$$
$$\text{Ca}^{2+}(g) \longrightarrow \text{Ca}^{2+}(aq) \qquad \Delta G° = -1518 \text{ kJ}$$
$$2e^- + 2\text{H}^+(aq) \longrightarrow \text{H}_2(g) \qquad \Delta G° = -930 \text{ kJ}$$
$$\overline{\text{Net } \text{Ca}(s) + 2\text{H}^+(aq) \longrightarrow \text{Ca}^{2+}(aq) + \text{H}_2(g) \qquad \Delta G° = -554 \text{ kJ}}$$

As we can see, the $\Delta G°$ for forming $\text{Ca}^{2+}(aq)$ is even more favorable.

It is interesting to compare other properties of group I and group II. Figures 17.2 and 17.7 show that the ionic radius of any group II element is smaller than that of the group I element of the same period. For example, Mg^{2+} has radius 0.072 nm, compared with 0.102 nm for Na^+. Why the difference in size? Sodium ion has nuclear charge +11, with two electrons in the K shell and eight electrons in the L shell; magnesium ion has nuclear charge +12, also with two electrons in the K shell and eight electrons in the L shell. The two ions are *isoelectronic;* i.e., they have identical electronic configurations. The only difference is that Mg^{2+} has a higher nuclear charge. Increased nuclear charge means increased attraction for electrons, which in turn means a smaller K shell and smaller L shell. In any isoelectronic sequence, ionic size decreases with increased nuclear charge.

Another interesting comparison between groups I and II is the difference in electrode potentials. Values for group I were given in Fig. 17.3; values for group II are shown in Fig. 17.8 (which also gives other properties). For the elements at the top of the groups, there is a distinct difference. Group I elements have more negative potentials; therefore, they are better reducing agents. For the elements at the bottom of the group, there is little difference. Barium, for instance, has electrode potential -2.90 V, whereas cesium, of the same period, has -2.92 V.

Actually, it is not surprising that group I elements are stronger reducing agents than corresponding group II elements. The ionization potentials of group I are lower. The surprising thing is that group II elements are as good reducing agents as they are. The explanation lies in hydration energy. Although it takes extra energy to pull two electrons off a group II atom, the net

FIG. 17.8   Properties of Alkaline-Earth Metals in the Condensed State

Element	Electrode potential (from M^{2+}), V	Density, g/cm^3	Melting point, °C	Boiling point, °C
Beryllium	-1.85	1.86	1280	2970
Magnesium	-2.37	1.74	650	1100
Calcium	-2.87	1.55	850	1490
Strontium	-2.89	2.6	770	1380
Barium	-2.90	3.6	710	1140
Radium	-2.92	5(?)	700	<1700

process $M(s) \longrightarrow M^{2+}(aq) + 2e^-$ nevertheless has great tendency to occur because the doubly charged ion interacts strongly with water in forming the hydrated ion.

In compounds, the alkaline-earth elements occur as $+2$ ions. Many of these compounds are insoluble and are found as deposits in the Earth's crust. The most important of these are the silicates, carbonates, sulfates, and phosphates. Beryllium is not very abundant but is very widespread. The only important beryllium mineral found in any quantity is a silicate, beryl, for which the formula is $Be_3Al_2Si_6O_{18}$. The gemstone emerald is beryl colored deep green by trace amounts of chromium. Magnesium, the eighth most abundant element in the Earth's crust, is also widely distributed. Principal sources are magnesite ($MgCO_3$) and dolomite ($MgCO_3 \cdot CaCO_3$) as well as seawater and deep salt wells. Calcium, the most abundant of group I and group II elements, occurs as silicates, carbonate, sulfate, phosphate, and fluoride. Calcium carbonate ($CaCO_3$) as the mineral calcite is the most abundant of all nonsilicate minerals; it appears in such diverse rocks as limestone, marble, and chalk. The mineral gypsum ($CaSO_4 \cdot 2H_2O$) is also very common. It apparently owes its origin to limestone beds which have been acted on by sulfuric acid produced from oxidation of sulfide minerals. Phosphate rock is essentially $Ca_3(PO_4)_2$, an important ingredient of bones, teeth, and seashells.

The extraction of magnesium from seawater accounts for the bulk of current United States production. In the process magnesium ion in seawater is precipitated as $Mg(OH)_2$ by the addition of lime (CaO) which can be made by burning seashells, largely $CaCO_3$. The hydroxide is filtered off and converted to $MgCl_2$ by reaction with HCl. The dried $MgCl_2$ is mixed with other salts to lower the melting point and then electrolyzed.

The alkaline-earth metals are good conductors of heat and electricity, but only magnesium finds any considerable use. Surprisingly, this use is based on the structural qualities of magnesium rather than on its electrical properties. It is the lightest of all commercially important structural metals. However, it has relatively low structural strength, and so this has to be increased by alloying it with other elements.

Too rare and costly for large-scale use, beryllium is important as a trace additive for hardening other metals, such as copper. It is also used as a moderator or reflector in nuclear reactors, where it is used to coat uranium or plutonium fuel rods. In the finely powdered form, beryllium (and its compounds) must be handled carefully since it is extremely toxic.

Finely divided magnesium burns rather vigorously to emit very intense light which is particularly rich in ultraviolet radiation. For this reason, magnesium is used as one of the important light sources for photography. Flashbulbs contain wire or foil of magnesium (or aluminum) packed in oxygen atmosphere. When the bulb is fired, an electric current heats the metal and initiates the oxidation reaction.

The flame spectra of strontium salts are characteristically red, and those of barium are yellowish green. Strontium and barium salts are frequently used for color effects in pyrotechnics.

At ordinary temperatures alkaline-earth elements form compounds only in the +2 state. With the exception of beryllium, all such compounds are essentially ionic. The ions are colorless and, except for Be^{2+}, do not hydrolyze appreciably in aqueous solution. Beryllium salts hydrolyze to give acid solutions. Unlike the compounds of group I, many group II compounds are not soluble in water.

Hydrides

When heated in hydrogen gas, Ca, Sr, and Ba form hydrides. These are white powders which react with water to liberate H_2. Calcium hydride (CaH_2) is used as a convenient portable hydrogen supply:

$$CaH_2(s) + 2H_2O \longrightarrow Ca^{2+} + 2OH^- + 2H_2(g)$$

Oxides

The oxides of these elements are characteristically very high melting (refractory). They can be made by thermally decomposing the carbonates or hydroxides. For example, lime (CaO) is made from limestone ($CaCO_3$) by the reaction

$$CaCO_3(s) \longrightarrow CaO(s) + CO_2(g) \qquad \Delta H = +178 \text{ kJ}$$

Except for BeO, which is amphoteric, the oxides are basic. Both lime and magnesia (MgO) are used as brick linings in furnaces, specifically to counteract acid impurities in steel production.

Hydroxides

$CaO + H_2O \Rightarrow CaOH_2$

The hydroxides are made by adding water to the oxides in a process called *slaking*. For example, slaking of lime produces $Ca(OH)_2$. The reaction

$$CaO(s) + H_2O \longrightarrow Ca(OH)_2(s) \qquad \Delta H = -67 \text{ kJ}$$

is exothermic and is accompanied by a threefold expansion in volume, sometimes to the consternation of amateur builders whose lime supplies accidentally get wet. Lime is an important constituent of cement and is also used as an important industrial base, since it is cheaper than NaOH.

The hydroxides of group II are only slightly soluble; however, the solubility increases with increasing ionic size. The solubility products are given in Fig. 17.9. With the exception of $Be(OH)_2$, which is amphoteric, the other hydroxides are basic. They are assumed to be 100 percent dissociated in aqueous solution.

466

467   Section 17.4
Compounds
of the
Alkaline-
Earth
Elements

FIG. 17.9 K_{sp} **for Alkaline-Earth Hydroxides**

$Be(OH)_2$	Less than 10^{-19}
$Mg(OH)_2$	8.9×10^{-12}
$Ca(OH)_2$	1.3×10^{-6}
$Sr(OH)_2$	3.2×10^{-4}
$Ba(OH)_2$	5.0×10^{-3}

Sulfates

The sulfates of group II range from very soluble $BeSO_4$ to practically insoluble $RaSO_4$. Going down the group, the solubilities decrease in regular order; for $BeSO_4$, K_{sp} is very large; $MgSO_4$, about 10; $CaSO_4$, 2.4×10^{-5}; $SrSO_4$, 7.6×10^{-7}; $BaSO_4$, 1.5×10^{-9}; and $RaSO_4$, 4×10^{-11}. This decreasing order is opposite to that observed for the hydroxides.

Magnesium sulfate is well known as the heptahydrate, $MgSO_4 \cdot 7H_2O$, also known as epsom salts. In medicine it is useful as a purgative. Apparently, magnesium ions in the alimentary tract favor passage of water from other body fluids into the bowel to dilute the salt.

Calcium sulfate is well known as the mineral gypsum ($CaSO_4 \cdot 2H_2O$). When it is lightly heated, it undergoes the reaction

$$CaSO_4 \cdot 2H_2O(s) \rightleftharpoons CaSO_4 \cdot \tfrac{1}{2}H_2O(s) + \tfrac{3}{2}H_2O(g)$$

The product, plaster of paris, is often used in making casts and molds because of the reversibility of the above reaction. On water uptake, plaster of paris sets to gypsum. Expansion of volume results in remarkably faithful reproductions of the mold.

Barium sulfate and its insolubility have been repeatedly mentioned. Although Ba^{2+}, like most heavy metals, is very poisonous, the solubility of $BaSO_4$ is so low that it can safely be ingested into the stomach and intestines. The use of $BaSO_4$ in taking X-ray pictures of the digestive tract depends on this insolubility and the great scattering of X rays by Ba^{2+} ion. $BaSO_4$ is also important as a white pigment.

Chlorides and Fluorides

Beryllium chloride ($BeCl_2$) and beryllium fluoride (BeF_2) are unusual in that they do not conduct electricity in the molten state. For this reason, they are usually considered to be molecular rather than ionic salts. All the other chlorides and fluorides of group II are typical ionic solids. Calcium fluoride (CaF_2) is rather insoluble and occurs as the mineral fluorspar. $CaCl_2$ is very soluble and, in fact, has such great affinity for water that it is used as a dehydrating agent. It is also often the "salt" that is put on roads in winter to help melt ice and snow.

Carbonates

All the carbonates of group II are quite insoluble and therefore occur in nature. Calcium carbonate ($CaCO_3$), or limestone, the most common nonsilicate rock, poses a special problem for water supplies because $CaCO_3$, though almost insoluble in water, is appreciably soluble in water containing carbon dioxide. Since our atmosphere contains an average of 0.04% carbon dioxide at all times, practically all groundwaters are solutions of carbon dioxide in water. These groundwaters dissolve limestone by the reaction

$$CaCO_3(s) + CO_2 + H_2O \rightleftharpoons Ca^{2+} + 2HCO_3^-$$

The result is to give a weathering action on limestone deposits and produce contamination of most groundwaters with calcium ion and bicarbonate ion (HCO_3^-).

The dissolving action of carbon dioxide–containing water explains the many caves found in limestone regions. These caves abound in weird formations because of dissolving and reprecipitation of $CaCO_3$. Groundwater containing Ca^{2+} and HCO_3^- may seep into a limestone cave and hang as a drop from the ceiling. As the water evaporates along with the carbon dioxide, the above reaction reverses to deposit a bit of limestone. Later, another drop of groundwater seeps onto the limestone speck, and the process repeats. In time, a long shaft reaching down from the roof (stalactite) may be formed. Occasionally, drops of groundwater may drip off the stalactite to the cave floor, where they evaporate to form a spire, or stalagmite, of $CaCO_3$. The whole process of dissolving and reprecipitation is very slow and may take hundreds of years.

17.5
HARD WATER

Because limestone is so widespread, most groundwater contains appreciable concentrations of calcium ion. The presence of this Ca^{2+} (likewise Mg^{2+} or Fe^{2+}) is objectionable because it leads to formation of precipitates when such water is boiled or when soap is added. Water that behaves in this way is called *hard water*.

Hardness in water is always due to Ca^{2+}, Mg^{2+}, or Fe^{2+}. However, the hardness may be of two types: (1) *temporary*, or *carbonate*, hardness, in which case HCO_3^- is also present with the above ions; or (2) *permanent*, or *noncarbonate*, hardness, in which case no HCO_3^- ions are in the water. In either case, hardness manifests itself by reaction with soap (but not with detergents) to produce a precipitate. The usual soap is sodium stearate ($C_{17}H_{35}COONa$) and consists of Na^+ ions and negative stearate ions. When stearate ions are added to Ca^{2+}, insoluble calcium stearate forms:

$$Ca^{2+} + 2C_{17}H_{35}COO^- \longrightarrow Ca(C_{17}H_{35}COO)_2(s)$$

This precipitate is the familiar scum that forms on soapy water.

Hardness in water is also objectionable because boiling a solution containing Ca^{2+} and HCO_3^- deposits $CaCO_3$, just as in cave formation. In industrial boilers formation of $CaCO_3$ is an economic headache since, like most salts, $CaCO_3$ is a poor heat conductor. Fuel efficiency is diminished, and boilers have been put completely out of action by local overheating due to boiler scale.

How can we soften hard water effectively and economically? One way is to add excess quantities of soap. Eventually, enough stearate ion is added to precipitate all the Ca^{2+}, leaving the excess soap to carry on the cleansing action.

Another way (this works only for temporary hardness) is to boil the water. The reaction

$$Ca^{2+} + 2HCO_3^- \rightleftharpoons CaCO_3(s) + H_2O + CO_2(g)$$

is reversible, but the forward reaction can be made dominant by boiling off the CO_2. Boiling, however, is not practical for large-scale softening.

The third way is to precipitate Ca^{2+} by adding washing soda (Na_2CO_3). Added carbonate ion (CO_3^{2-}), reacts with Ca^{2+} to give insoluble $CaCO_3$. If bicarbonate ion is present, the water may be softened by adding a base such as ammonia. The base deprotonates HCO_3^- to produce CO_3^{2-}, which then precipitates Ca^{2+}. In industry, temporary hardness is often removed by adding limewater [$Ca(OH)_2$].

$$Ca^{2+} + HCO_3^- + OH^- \longrightarrow CaCO_3(s) + H_2O$$

It might seem odd that limewater, which itself contains Ca^{2+}, can be added to hard water to remove Ca^{2+}. The point is limewater adds 2 mol of OH^- per mole of Ca^{2+}. Two mol of OH^- neutralizes 2 mol of HCO_3^- and liberates 2 mol of CO_3^{2-}, thus precipitating 2 mol of Ca^{2+}—one that was added and one that was originally in the hard water.

A fourth way to soften water is to tie up the Ca^{2+} so that it becomes harmless. Certain phosphates, such as $(NaPO_3)_n$, sodium polyphosphate, act as *sequestering agents* to form complexes in which the Ca^{2+} is trapped by the phosphate.

The fifth and most clever way to soften water is to replace the offending calcium ion by a harmless ion such as Na^+. This is done by the process called *ion exchange*.

17.6
ION EXCHANGE

An *ion exchanger* is a special type of giant molecule consisting of a porous, negatively charged network filled with water molecules and enough positive ions to provide electrical neutrality. The identity of the positive ions is not very important. One cation such as Ca^{2+} can take the place of another such as Na^+ without much change in structure. In water softening, hard water containing Ca^{2+} is placed in contact with an ion exchanger whose mobile ion is

Na$^+$. Exchange occurs as represented by the equilibrium

$$Ca^{2+} + 2Na^+ \ominus \rightleftharpoons 2Na^+ + \ominus Ca^{2+} \ominus$$

where the circled minus sign represents a negative site on the exchanger. The equilibrium constant for this reaction is usually of the order of 10 or less. Therefore, to remove all the Ca^{2+}, it is necessary to run the hard water through a large amount of exchanger. Once the exchanger has given up its supply of Na$^+$, it has to be regenerated. Treatment with a concentrated solution of sodium chloride reverses the above reaction.

The early ion exchangers were naturally occurring silicate minerals called zeolites. These are composed of covalently bound silicon, oxygen, and aluminum with mobile ions (e.g., Na$^+$) in the pores. Zeolites are very closely related in structure to the clays, which also show ion exchange. Such ion exchange is important for plant nutrition since many plants receive essential minerals from the soil in this fashion.

With the advent of high-polymer techniques, chemists have been able to synthesize ion exchangers superior to the zeolites. The most common synthetic exchanger consists of a giant hydrocarbon framework called a *resin*, having a negative charge due to covalently bound SO$_3^-$ groups. These are *cation exchangers*. It is also possible to prepare ion exchangers in which the resin network is positively charged due to covalently bound groups of the type N(CH$_3$)$_3^+$. Such networks can function as *anion exchangers;* i.e., they have mobile negative ions which can be displaced by other anions.

Consecutive passage through a cation exchanger followed by an anion exchanger makes possible removal of all ions from a salt solution. If, as shown in Fig. 17.10, a solution containing M$^+$ and A$^-$ is first run through a cation exchanger whose exchangeable ions are H$^+$, the salt solution is converted from M$^+$ and A$^-$ to H$^+$ and A$^-$. If now the acid solution is run through an anion exchanger whose exchangeable ions are OH$^-$, the A$^-$ is replaced by OH$^-$. In the original solution the number of negative charges is exactly equal to the number of positive charges, so that equal amounts of H$^+$ and OH$^-$ are exchanged into the solution. Exact neutralization occurs, and pure water results. Water thus treated is said to be *deionized* and contains fewer ions than the most carefully distilled water.

Ion-exchange resins have recently become important as membranes for removing soluble salts from waste water. One of the great problems of water pollution is that, although solid and organic wastes can be removed (Sec. 14.12), there is no simple way to get out the soluble inorganic salts. Ordinary use of a water supply by an average city raises the salt content of the water by about 50 percent. This makes for problems when we attempt to recycle the water.

One promising method to desalt water is by *electrodialysis*. Positive and negative ions are separated out by being made to pass through ion-exchange membranes under the influence of an electric field. Figure 17.11 shows a typical electrodialysis cell. Many such cells are arranged in tandem so that a small effect can be multiplied. Salt water is fed into the three pipes at the top of the cell. An electric field across the flow tends to make the positive and negative

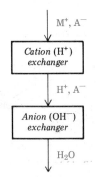

FIG. 17.10 Consecutive ion exchange to produce deionized water.

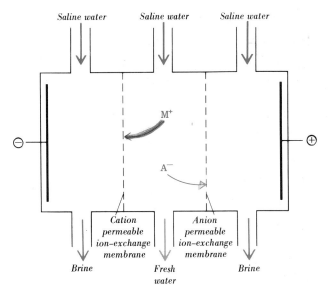

Saline water   Saline water   Saline water

M⁺

⊖   ⊕

A⁻

Cation   Anion
permeable   permeable
ion-exchange   ion-exchange
membrane   membrane

Brine   Fresh   Brine
               water

FIG. 17.11 Electrodialysis cell for getting fresh water from salt water.

ions drift off in opposite directions. By proper choice of ion-exchange resins for the membranes (shown by the dashed lines), we can make the left one permeable to cations and the right one permeable to anions. Specifically, the left membrane could be a resin containing SO_3^- groups, and the right membrane could be a resin containing $N(CH_3)_3^+$ groups. Under proper flow conditions, the water coming out of the middle effluent pipe will be considerably less salty than that coming out of the two outside pipes.

17.7
QUALITATIVE ANALYSIS FOR ALKALI AND ALKALINE-EARTH IONS

The term "qualitative analysis" means the detection of chemical elements in an unknown sample. It is to be distinguished from quantitative analysis, which means determination of relative amounts of elements.

Because alkali elements do not form many insoluble compounds and the ions are colorless, it is difficult to detect their presence by chemical methods. Instead, their presence is usually shown by flame tests. A piece of platinum wire is shaped into a loop; the loop is dipped in HCl, heated to remove volatile impurities, and then used to heat the unknown sample in a burner flame. The sodium flame is yellow and extremely intense; even traces of it can mask other colors. The main reason for cleaning the platinum loop with HCl is to expel sodium impurity as relatively volatile chloride. (In general, chlorides are more volatile than most other solids.) The potassium flame is colored a delicate violet and can be observed in many cases only through cobalt glass, which filters out interfering colors such as sodium yellow. The flames of potassium, rubidium, and cesium all look bluish violet and are so similar that definite identification requires examination of the line spectra with a spec-

471

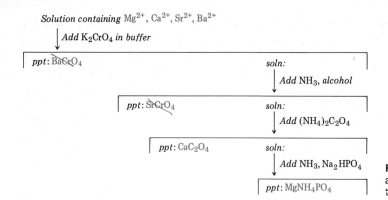

Solution containing Mg^{2+}, Ca^{2+}, Sr^{2+}, Ba^{2+}

Add K_2CrO_4 in buffer

ppt: $BaCrO_4$

soln:

Add NH_3, alcohol

ppt: $SrCrO_4$

soln:

Add $(NH_4)_2C_2O_4$

ppt: CaC_2O_4

soln:

Add NH_3, Na_2HPO_4

ppt: $MgNH_4PO_4$

FIG. 17.12 Qualitative-analysis scheme for alkaline-earth ions. (ppt stands for "precipitate"; soln stands for "solution.")

troscope. The strongest spectral lines are lithium, 670.8 nm; sodium, 589.0 and 589.6 nm; potassium, 766.5 and 769.9 nm; rubidium, 420.2 and 421.5 nm; and cesium, 455.6 and 459.3 nm.

As a group, alkaline-earth cations (excluding beryllium) can be distinguished from other common cations by taking advantage of the fact they form soluble chlorides and sulfides but insoluble carbonates. Figure 17.12 gives a schematic representation of how the alkaline-earth ions can be distinguished from each other. The barium can be precipitated first as yellow $BaCrO_4$ by addition of K_2CrO_4 in the presence of an acetic acid buffer. From the residual solution (containing Sr^{2+}, Ca^{2+}, and Mg^{2+}), light yellow $SrCrO_4$ can be precipitated by subsequent addition of NH_3 and alcohol. $BaCrO_4$ precipitates in the first step, and $SrCrO_4$ in the second, because $BaCrO_4$ ($K_{sp} = 8.5 \times 10^{-11}$) is less soluble than $SrCrO_4$ ($K_{sp} = 3.6 \times 10^{-5}$). The point of using an acetic acid buffer (Sec. 15.5) is to keep the H^+ concentration around 10^{-5}, where the CrO_4^{2-} concentration governed by the equilibrium

$$2CrO_4^{2-} + 2H^+ \rightleftharpoons Cr_2O_7^{2-} + H_2O$$

is too low to precipitate Sr^{2+} but high enough to precipitate Ba^{2+}. Subsequent addition of NH_3 reduces the H^+ concentration and raises the CrO_4^{2-} concentration sufficiently to precipitate $SrCrO_4$, especially in the presence of alcohol, which lowers salt solubility.

Calcium ion can be separated from magnesium ion by addition of ammonium oxalate to form white, insoluble calcium oxalate, CaC_2O_4. (The K_{sp} of CaC_2O_4 is 1.3×10^{-9}, compared with 8.6×10^{-5} for MgC_2O_4.) Finally, the presence of Mg^{2+} can be shown by adding more NH_3 and Na_2HPO_4, which precipitates white magnesium ammonium phosphate, $MgNH_4PO_4$.

17.8 ALUMINUM

In going left to right across the periodic table, group I and group II elements are immediately followed by 10 short columns of transition elements. These are metals, but we shall postpone discussion of them until the next chapter.

Next comes group III, which includes boron, aluminum, gallium, indium, and thallium. Boron is a nonmetal, and so we put off its discussion to a later chapter. This leaves us with aluminum, the only important metal of group III. Its ground-state electron configuration is $(Z = 13)$ $1s^2 2s^2 2p^6 3s^2 3p^1$.

Aluminum is the most abundant metal in the Earth's crust. It occurs primarily as complex aluminum silicates, such as feldspar ($KAlSi_3O_8$), from which it is economically unfeasible to separate pure aluminum. Fortunately, there are concentrated natural deposits of aluminum oxide in the form of bauxite ($Al_2O_3 \cdot xH_2O$) from which aluminum can be obtained rather easily by electrolytic reduction. However, it is necessary first to remove iron and silicon since these spoil the properties of the product aluminum.

Purification is accomplished by the *Bayer process,* which makes use of the amphoterism of aluminum. Crude oxide is treated with hot NaOH solution. The Al_2O_3 dissolves because of formation of aluminate ion $[Al(OH)_4^-]$. SiO_2 also dissolves (to form silicate ions), but Fe_2O_3 stays undissolved. The solution is filtered and cooled. On agitation with air and addition of crystalline aluminum hydroxide as a seed, aluminum hydroxide precipitates, leaving the silicate in solution.

The production of metallic aluminum is usually carried out by the *Hall-Héroult process.* Bauxite is dissolved in a molten mixture of fluorides, e.g., cryolite* (Na_3AlF_6), CaF_2, and NaF, and is electrolyzed at about 1000°C. A typical cell is represented schematically in Fig. 17.13. Graphite rods dipping into the molten mix act as the anode; a graphite lining supported by an iron box is the cathode. The electrode reactions are very complicated and only imperfectly understood. At the cathode, oxyfluoro complex ions are reduced to liquid aluminum (mp 659.7°C). At the anode, various products are formed; they include oxygen, fluorine, and carbon compounds of these elements. The carbon anodes gradually erode and must be replaced periodically. Continual addition of bauxite and recurrent draining of the liquid aluminum allow uninterrupted operation. Energy consumption is high. The process is economically feasible only near cheap sources of electric current.

*The mineral cryolite occurs in nature almost exclusively as an enormous geologic dike in Greenland. The mineral looks like glacial ice and was thought by the Eskimos to be a special kind of ice. The name comes from Greek *krios,* "frost," and *lithos,* "stone." Most cryolite now used in aluminum extraction is made synthetically from NaF and AlF_3.

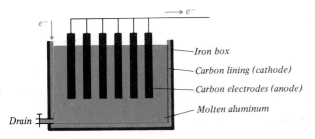

FIG. 17.13 Cell for electrolytic preparation of aluminum.

474

Chapter 17
Chemistry
of Some
Typical
Metals

Aluminum is soft and weak when pure but becomes quite strong when alloyed with other metals. Because it is so light (density 2.7 g/cm³), it finds extensive use as a structural material. Although chemically active, it resists corrosion because of formation of a self-protecting oxide coat. It is also a good conductor of heat and electricity and so is used in cooking utensils and electric equipment.

Although not so active as group I and II metals, elemental aluminum is an excellent reducing agent, as shown by the electrode potential:

$$Al^{3+} + 3e^- \longrightarrow Al(s) \qquad E° = -1.66 \text{ V}$$

In view of the high ionization potentials (first, 5.98 eV; second, 18.82 eV; and third, 28.44 eV), the large electrode potential of aluminum is somewhat surprising. As with the alkaline-earth elements, it is hydration of the ion that stabilizes the ion state; 4640 kJ of heat is evolved per mole of Al^{3+} ions hydrated. The reasons for this hydration energy are the high charge of Al^{3+} and its small size (0.054-nm radius).

The large electrode potential indicates aluminum should reduce H_2O. The reaction is too slow to detect, probably because of an oxide coat. However, the oxide (being amphoteric) is soluble in acid and in base. Consequently, aluminum liberates hydrogen from both acidic and basic solutions. The net reactions may be written

$$2Al(s) + 6H^+ \longrightarrow 2Al^{3+} + 3H_2(g)$$
$$2Al(s) + 2OH^- + 6H_2O \longrightarrow 2Al(OH)_4^- + 3H_2(g)$$

The first equation suggests that aluminum dissolves in any acid. However, this is not the case. Aluminum dissolves rapidly in HCl, but in HNO_3 no visible reaction occurs. The inertness is attributed to formation of a microscopic oxide coat. A coating of Al_2O_3 should be quite stable because of the great strength of the Al—O bond.

Further indication of the great affinity of aluminum for oxygen comes from the high heat of formation of Al_2O_3:

$$2Al(s) + \tfrac{3}{2}O_2(g) \longrightarrow Al_2O_3(s) \qquad \Delta H° = -1670 \text{ kJ}$$

This heat can be used effectively in the reduction of less stable oxides. For example, since only 824 kJ is required to decompose Fe_2O_3, aluminum can reduce Fe_2O_3 with energy left over. The overall reaction can be considered to be the sum of two separate reactions:

$$
\begin{array}{ll}
2Al(s) + \tfrac{3}{2}O_2(g) \longrightarrow Al_2O_3(s) & \Delta H° = -1670 \text{ kJ} \\
\underline{Fe_2O_3(s) \longrightarrow 2Fe(s) + \tfrac{3}{2}O_2(g)} & \underline{\Delta H° = +\ 824 \text{ kJ}} \\
2Al(s) + Fe_2O_3(s) \longrightarrow 2Fe(s) + Al_2O_3(s) & \Delta H° = -\ 846 \text{ kJ}
\end{array}
$$

The heat evolved is sufficient to produce Fe and Al_2O_3 in the molten state. The reaction, frequently called the *thermite reaction*, has been used for welding operations.

Aqueous solutions of most aluminum salts are acidic because of hydrolysis of Al^{3+}. The net reaction can be written as

$$Al^{3+} + H_2O \longrightarrow AlOH^{2+} + H^+$$

Alternatively, the acidity can be viewed as coming from dissociation of hydrated aluminum ion:

$$Al(H_2O)_6^{3+} \longrightarrow H^+ + Al(H_2O)_5OH^{2+}$$

When base is progressively added to aqueous aluminum solutions, a white, gelatinous precipitate is formed. This precipitate, variously formulated as $Al(OH)_3$ or $Al_2O_3 \cdot xH_2O$, is readily soluble in acid or excess base, but only if freshly precipitated. If it is allowed to stand, it progressively becomes more difficult to dissolve. The explanation suggested for this "aging" is that oxygen bridges are formed between neighboring aluminum atoms. In basic solutions, aluminum forms aluminate ion, $Al(OH)_4^-$, also written AlO_2^-.

Because of its small size and high charge, Al^{3+} forms quite stable complex ions. For example, progressive addition of fluoride produces AlF^{2+}, AlF_2^+, AlF_3, AlF_4^-, AlF_5^{2-}, and AlF_6^{3-}. The anion AlF_6^{3-} is found in the solid cryolite.

Like most +3 ions, Al^{3+} may be crystallized from aqueous solutions to give compounds known as *alums*. These alums are double salts, having the general formula $MM'(SO_4)_2 \cdot 12H_2O$. M is a singly charged cation, such as K^+, Na^+, or NH_4^+; M' is a triply charged cation, such as Al^{3+}, Fe^{3+}, or Cr^{3+}. Ordinary alum is $KAl(SO_4)_2 \cdot 12H_2O$. The crystals are usually large octahedra and have great chemical purity. Because of this purity, $KAl(SO_4)_2 \cdot 12H_2O$ is useful in dyeing. Al^{3+}, uncontaminated by Fe^{3+}, is precipitated on cloth as aluminum hydroxide, which acts as a binding agent (mordant) for dyes. The absence of Fe^{3+} is imperative for producing clear colors.

When aluminum hydroxide is heated to high temperature, it loses water and eventually forms Al_2O_3. This very inert material, called *alumina*, has a high melting point (about 2000°C) and finds use as a refractory in making containers for high-temperature reactions. Ordinarily, alumina is white, but it can be colored by addition of oxides such as Cr_2O_3 or Fe_3O_4. Synthetic rubies, for example, can be made by mixing Al_2O_3 and Cr_2O_3 powder and dropping it through the flame of an oxyhydrogen torch. Because of the great hardness of Al_2O_3, such synthetic jewels are used as bearing points in watches and other precision instruments.

For qualitative analysis, aluminum is usually precipitated as the hydroxide when ammonia is added to the solution from which H_2S has removed acid-insoluble sulfides. It can be separated from other cations that precipitate at the same point by exploiting the fact that of these cations only Al^{3+}, Cr^{3+}, and Zn^{2+} are amphoteric.

Zinc can be differentiated by using the fact that ZnS but not $Al(OH)_3$ precipitates when $(NH_4)_2S$ is added in the presence of $SO_4^{2-}-HSO_4^-$ buffer or by using the fact that $Zn(OH)_2$ but not $Al(OH)_3$ is soluble in excess ammonia. Chromium can be differentiated by oxidizing it in basic solution with H_2O_2 to CrO_4^{2-}, which can be precipitated as yellow $BaCrO_4$, and by precipitating

$Al(OH)_3$ from the basic solution by adding NH_4Cl. A confirmatory test for aluminum is formation of a red precipitate from $Al(OH)_3$ and the dye aluminon.

17.9
TIN

Moving to the right in the periodic table from group III, we meet the diverse and important group IV. The elements in this group, carbon, silicon, germanium, tin, and lead, range from nonmetallic at the top to metallic at the bottom. Also, there is a pronounced change from acid behavior at the top to more basic behavior at the bottom. We shall say more about the other elements of this group in later chapters. At this point, we confine ourselves to the two metals at the bottom of the group. Tin is discussed in this section; lead, in the following.

Tin has an electron configuration corresponding to $5s^25p^2$ in the outermost shell. Oxidation states +2 and +4 are common. The melting point of tin metal is relatively low (232°C); the boiling point is relatively high (2260°C). This gives tin a very long liquid range, which is typical of many metals.

The principal source of tin is cassiterite (SnO_2), from which the element is prepared by carbon reduction. Although it is usually considered a metal, the element also exists in a nonmetallic form (α, or gray, tin) which is stable below 13°C.* Ordinary tin (β, or white, tin) is a rather inert metal which resists corrosion because of a relatively low electrode potential and because of self-protection by an oxide coat. Because of its inertness, tin is widely used as a protective plating for steel especially in making "tin cans." The steel is coated, either by dipping in molten tin or by electrolytic reduction of dissolved tin salts.

Two series of tin compounds, stannous (+2) and stannic (+4), are known. The +2 state is formed when metallic tin is dissolved in acid solution; however, the rate of reaction is rather slow. In solution, the Sn^{2+} ion is colorless and hydrolyzes according to the reaction

$$Sn^{2+} + H_2O \rightleftharpoons SnOH^+ + H^+$$

for which the equilibrium constant is about 0.01. Thus, Sn^{2+} is about as strong an acid as HSO_4^-. Gradual addition of base precipitates a white solid usually described as stannous hydroxide, $Sn(OH)_2$. Further addition of base dissolves the precipitate to form stannite ion, which can be written as either $Sn(OH)_3^-$ or $HSnO_2^-$. Stannite ion is a powerful reducing agent. On standing, it disproportionates to give 0 and +4 oxidation states:

$$2Sn(OH)_3^- \longrightarrow Sn(s) + Sn(OH)_6^{2-}$$

---

*The conversion of metallic tin to gray tin was first remarked on organ pipes in early European cathedrals. At the low temperatures prevalent in these buildings, the metallic pipes slowly developed grotesque, cancerous "growths." The phenomenon, called *tin plague* or *tin disease,* was first blamed on the devil, then on microorganisms, and finally on conversion from one kind of tin to another.

Stannous chloride, $SnCl_2$, is frequently used as a mild reducing agent. The electrode potential is

$$Sn^{4+} + 2e^- \longrightarrow Sn^{2+} \qquad E^\circ = +0.15 \text{ V}$$

In the stannic state, tin is often represented as the simple Sn^{4+} ion. However, because of its high charge, Sn^{4+} probably does not exist as such in aqueous solutions. When base is added, a white precipitate forms; it may be $Sn(OH)_4$ or, more probably, a hydrated oxide $SnO_2 \cdot xH_2O$. The precipitate is amphoteric and is soluble in excess base to give stannate ion, usually written $Sn(OH)_6^{2-}$ or SnO_3^{2-}.

Both stannous and stannic sulfides are insoluble in water and can be precipitated for identification by H_2S in acid solution. Stannic sulfide, SnS_2, is a yellow solid which is soluble in excess sulfide ion. The reaction

$$SnS_2(s) + S^{2-} \longrightarrow SnS_3^{2-} \qquad \text{HN}_3$$

forms thiostannate ion, which is the sulfur analog of the stannate ion, SnO_3^{2-}. The dissolving of SnS_2 in excess S^{2-} distinguishes SnS_2 from another yellow sulfide, CdS. Owing to the great stability of SnS_3^{2-}, SnS is oxidized by polysulfide ion, S_2^{2-}, a relatively poor oxidizing agent:

$$SnS(s) + S_2^{2-} \longrightarrow SnS_3^{2-}$$

When solutions of thiostannate are acidified, SnS_2 is precipitated.

In qualitative analysis, lead and tin precipitate together as sulfides in acid solution. The lead sulfide (black) can be separated from the tin sulfide, either SnS (brown-black) or SnS_2 (yellow), by treatment with ammonium polysulfide, which converts SnS and SnS_2 to SnS_3^{2-} but leaves the PbS undissolved. If, to the solution containing SnS_3^{2-}, HCl is added in excess, the tin stays in solution, probably as a chloride complex. Presence of tin can be confirmed by evaporation (to drive off H_2S) in the presence of iron (to reduce the tin to Sn^{2+}). $HgCl_2$ addition confirms tin if a precipitate of white Hg_2Cl_2 or black Hg is observed.

17.10
LEAD

In its ground state lead has electron configuration $6s^2 6p^2$. Conforming to the usual trend of increasing metallic character down a group, it is the most metallic of the group IV elements. Like tin, it shows oxidation states of +2 and +4, but the +4 state is more highly oxidizing.

Lead occurs principally as the mineral galena, PbS. The element can be made by roasting the sulfide in air until it is completely converted to oxide and then reducing the oxide with carbon in a small blast furnace:

$$2PbS(s) + 3O_2(g) \longrightarrow 2PbO(s) + 2SO_2(g)$$
$$2PbO(s) + C(s) \longrightarrow 2Pb(l) + CO_2(g)$$

The crude lead may contain impurities such as antimony, copper, and silver. The silver, a valuable by-product, is generally recovered by extracting it with molten zinc. If lead of high purity is required, it can be refined by an electrolytic process analogous to that used for copper. Pure lead is a soft, low-melting metal which, when freshly cut, has a silvery luster that turns blue-gray on exposure to air. The tarnishing is due to formation of a surface coat of oxides and carbonates. Major uses of lead are in lead storage batteries, alloys such as type metal and solder, and "white-lead" paint (hydrated lead hydroxy-carbonate). In spite of its toxicity, "white lead" has excellent adhering and covering ability; so it continues in wide use.

Practically all the common lead compounds correspond to lead in the +2 state. This state is called *plumbous,* from the Latin name for the element, *plumbum.* When the halide concentration of plumbous solutions is increased, insoluble plumbous halides such as $PbCl_2$ form. In the presence of excess halide ions the precipitates redissolve, presumably because of formation of complex ions of the type $PbCl_3^-$. Unlike two other common insoluble chlorides, $AgCl$ and Hg_2Cl_2, $PbCl_2$ can also be dissolved by raising the temperature.

Plumbous ion hydrolyzes somewhat less than stannous ion. When base is added, white $Pb(OH)_2$ is precipitated. This is amphoteric and dissolves in excess base to form plumbite ion $Pb(OH)_3^-$, or $HPbO_2^-$. With most -2 anions, Pb^{2+} forms insoluble salts, for example, $PbSO_4$, $PbCO_3$, PbS, $PbCrO_4$. Lead sulfide ($K_{sp} = 7 \times 10^{-29}$) is the least soluble of these, and the others convert to it in the presence of sulfide ion.

The principal compound of lead in the +4, or plumbic, state is PbO_2, lead dioxide. This compound, used as the cathode of lead storage batteries (Sec. 11.8), can be made by oxidation of plumbite with hypochlorite ion in basic solution. The reaction can be written

$$Pb(OH)_3^- + ClO^- \longrightarrow Cl^- + PbO_2(s) + OH^- + H_2O$$

With acid solutions PbO_2 is a potent oxidizing agent:

$$PbO_2(s) + 4H^+ + 2e^- \longrightarrow Pb^{2+} + 2H_2O \qquad E° = +1.46 \text{ V}$$

It is made even more potent in the presence of concentrated acid and anions such as sulfate which precipitate Pb^{2+}. In very concentrated solutions of base, PbO_2 dissolves to form plumbates, variously written as PbO_4^{4-}, PbO_3^{2-}, or $Pb(OH)_6^{2-}$. Red lead, Pb_3O_4, much used as an undercoat for painting structural steel, can be considered to be plumbous plumbate, Pb_2PbO_4. Its use depends on the fact that, as a strong oxidizing agent, it renders iron passive.

Like most heavy metals, lead and its compounds are poisonous. Indeed, the decline of the Roman Empire has been attributed in part to chronic lead poisoning brought about by use of lead pipes in the water system. Since the water system was reserved for the wealthy, there was selective decimation of the aristocracy. More recently, lead has come under attack as an environmental pollutant because of its emission in automobile exhausts. Fairly large doses of lead are required for toxicity, but the danger is amplified because lead tends to accumulate in the body (central nervous system). The toxicity may be due to the fact that lead and other heavy metals are powerful inhibitors of enzyme reactions.

In qualitative analysis, lead and tin precipitate together as sulfides in acid solution. Much of the lead, however, generally precipitates in a preceding operation when HCl is added to an original unknown. $PbCl_2$ can be separated from AgCl and Hg_2Cl_2 by leaching with hot water. Addition of K_2CrO_4 gives the confirmatory yellow precipitate $PbCrO_4$.

PbS can be dissolved with hot HNO_3 (to distinguish it from HgS) and reprecipitated as white $PbSO_4$ on addition of H_2SO_4. To confirm, the $PbSO_4$ is dissolved in ammonium acetate and precipitated as $PbCrO_4$. The lead sulfate is dissolved in acetate solution because of relatively great stability of the acetate-complex ion, $Pb(C_2H_3O_2)_3^-$.

Important Concepts

alkali elements
metallic properties
preferred-ion formation
electrode potentials vs. ionization potentials
compounds of alkali elements
alkaline-earth elements
compounds of alkaline-earth elements
hard water
carbonate vs. noncarbonate hardness
limestone cave formation
ion exchange

cation vs. anion exchangers
analysis for group I and II elements
aluminum and compounds
Bayer process
Hall-Héroult process
thermite reaction
alums
alumina
tin and compounds
lead and compounds
analysis for Al, Sn, Pb

Exercises

*17.1 Alkali elements Give a brief general description of metallic properties and tell how they can be explained.

**17.2 Alkali elements Explain why the first ionization potential of the alkali elements decreases from top to bottom of the group. Suggest a reason why the second ionization potential of lithium is so much larger than the first.

**17.3 Alkali elements LiI has a NaCl structure. Assuming iodide ions are in contact, calculate the apparent radius of iodide ion from the observed fact that the unit cell of LiI has an edge length of 0.600 nm. From your calculated iodide radius, figure out the maximum permissible value of the lithium-ion radius that is consistent with it.

Ans. 0.212 nm, 0.088 nm

**17.4 Alkali elements Distinguish clearly between ionization potential and electrode potential. Why is there no simple correlation between the two?

***17.5 Alkali elements The ionization potentials of three elements X, Y, and Z are 5, 6, and 7 eV, respectively; hydration energies are 200, 300, and 400 kJ/mol. Assuming the sublimation energies are all the same, which of these elements will have the greatest electrode potential?

*17.6 Alkali elements Sodium makes up 2.6 weight % of the Earth's crust; potassium 2.4 weight %. How many more sodium atoms are there than potassium atoms in the Earth's crust?

*17.7 Alkali elements Draw a diagram for a cell that could be used to make sodium from NaOH.

Label anode and cathode. Give electrode reactions for each. Suggest some reasons why NaOH might be better than NaCl in such a cell.

**17.8 Alkali elements Write balanced net equations for each of the following reactions:
a Metallic sodium is added to water.
b Na_2O is made from a suitable salt.
c Metallic potassium is exposed to oxygen.
d An aqueous solution of NaCl is electrolyzed.

**17.9 Alkali elements Explain why an aqueous solution of Na_2CO_3 is more basic than an aqueous solution of $NaHCO_3$.

*17.10 Alkaline-earth elements State specifically what characteristics calcium has that justify the name alkaline-earth element.

**17.11 Alkaline-earth elements Helium has an s^2 configuration. Why is it not an alkaline-earth element?

**17.12 Alkaline-earth elements Explain qualitatively why calcium in aqueous solution is more likely to be $Ca^{2+}(aq)$ than $Ca^+(aq)$. Tell why $Ca^{3+}(aq)$ is improbable.

***17.13 Ion stability Given a metal X which has a sublimation energy of 100 kJ/mol, first ionization potential of 5 eV, second ionization potential of 12 eV, and respective hydration energies of -380 kJ and -1280 kJ/mol for X^+ and X^{2+}. Predict whether $X^+(aq)$ or $X^{2+}(aq)$ will be more stable. Justify your answer.

**17.14 Sizes Explain why the Ca^{2+} ion is smaller than the K^+ ion. How would you expect the size of Ca^+ to compare with that of K^+?

***17.15 Alkaline-earth elements Explain why the electrode potentials of group I and II elements are very similar at the bottom of the periodic table but quite different at the top.

**17.16 Alkaline-earth elements Explain why compounds of the alkaline-earth elements are likely to be less soluble than compounds of the alkali elements.

**17.17 Stoichiometry Dolomite is rarely composed of an exact 1:1 ratio of $MgCO_3$ and $CaCO_3$ but is more likely to be $MgCO_3 \cdot (1-x)CaCO_3$. What would be the value of x in a particular sample of dolomite which on being heated to drive off CO_2 loses 46.0% of its weight?

**17.18 Alkaline-earth elements Make a flow chart showing each step involved in the extraction of magnesium from seawater. Write a balanced chemical equation for each reaction involved. Seawater contains 1272 ppm (parts per million by weight) of magnesium. How many liters of seawater (density 1.043 g/ml) would you need to process to get 0.50 kg of magnesium? *Ans. 380 liters*

**17.19 Alkaline-earth compounds Suppose you have a 50-kg bag of lime that accidentally gets wet. How much heat will be liberated in the process?

***17.20 Alkaline-earth compounds Describe with the aid of balanced chemical equations how you might make each of the following conversions:
a CaH_2 to $Ca(OH)_2$ *b* $Ca(OH)_2$ to $CaCO_3$
c $MgCO_3$ to $MgSO_4$ *d* $MgSO_4$ to $MgCO_3$
e $CaCl_2$ to $CaCO_3$

*17.21 Limestone Explain how limestone caves are formed and how stalactites and stalagmites are produced in these caves. Write net balanced equations for all reactions.

*17.22 Hard water What is meant by the term "hard water"? What is the difference between temporary and permanent hardness? Suggest three ways (including equations) whereby temporary hard water can be softened.

**17.23 Hard water If a sample of hard water contains the equivalent of 400 ppm of $CaCO_3$ in a dissolved state, how many liters of 0.01 M $Ca(OH)_2$ should you add per liter of the above hard water to reduce the hardness to 40 ppm of $CaCO_3$ equivalent? Assume the density of the hard water is 1.00 g/ml. *Ans. 0.4 liter*

**17.24 Ion exchange What is an ion exchanger? What is the difference between a cation exchanger and an anion exchanger? Explain how a pair of cation exchanger plus anion exchanger can be used to remove NaCl from water. Why would this not be feasible as a way of getting drinking water from the sea?

***17.25 Ion exchange Suggest a mechanism which would explain why a cation exchanger might be a good membrane for purifying water by electrodialysis.

*17.26 Qualitative analysis Explain why flame tests rather than precipitation tests are generally used in qualitative analysis of alkali elements.

**17.27 Qualitative analysis Addition of CO_3^{2-} to a solution that may contain Mg^{2+}, Ca^{2+}, Sr^{2+}, and/or Ba^{2+} gives a white precipitate. Dissolving the white precipitate in an acetic acid buffer followed by treatment with K_2CrO_4 gives no precipitate, even on subsequent addition of NH_3 and alcohol. Which ions are definitely present, which definitely absent, and which may be there?

**17.28 Qualitative analysis Explain with the help of equations why addition of NH_3 facilitates the precipitation of Sr^{2+} and makes feasible its separation from Ba^{2+}.

**17.29 Aluminum In the Earth's crust, aluminum is almost twice as abundant as iron. Yet, iron is much cheaper. Suggest two reasons for this discrepancy.

**17.30 Aluminum Make a flow chart showing the procedure required to go from impure aluminum oxide to pure aluminum. What chemical reactions occur at each step of the way?

***17.31 Aluminum In the purification of $Al(OH)_3$ by the Bayer process, why should agitation with air help to precipitate $Al(OH)_3$?

**17.32 Aluminum Assuming the cost of electric energy is 1 cent/kWh (about 4×10^5 C), what would be the electric cost for preparing 0.5 kg of aluminum?   *Ans. 13 cents*

***17.33 Aluminum The first three ionization potentials of aluminum are 5.98, 18.82, and 28.44 eV, respectively. The hydration energy of $Al^{+3}(g)$ is -4640 kJ/mol. Recalling that $e^- + H^+(aq) \longrightarrow \frac{1}{2}H_2(g)$ liberates 465 kJ, estimate the enthalpy change for $Al(s) + 3H^+(aq) \longrightarrow Al^{+3}(aq) + \frac{3}{2}H_2(g)$. What assumptions have you made?

*17.34 Aluminum How much hydrogen (STP) can you get from 1 kg of Al and 1 liter of 6 M NaOH?   *Ans. 200 liters*

***17.35 Aluminum Assuming the Al goes to Al_2O_3, tell which of the following reactions probably releases more heat:

$$2Al + 3FeO \longrightarrow 3Fe + Al_2O_3$$
$$2Al + \tfrac{3}{4}Fe_3O_4 \longrightarrow \tfrac{9}{4}Fe + Al_2O_3$$
$$2Al + Fe_2O_3 \longrightarrow 2Fe + Al_2O_3$$

Explain your reasoning.

**17.36 Aluminum Write balanced net equations for each of the following:

a Aqueous NaOH is added to aqueous Al^{3+} until a white precipitate appears.
b Excess NaOH is added to the precipitate from (*a*) until the precipitate disappears.
c The clear solution from (*b*) is treated dropwise with NH_4Cl solution until a white precipitate reappears.

*17.37 Aluminum What percent of alum, $KAl(SO_4)_2 \cdot 12H_2O$, is water? To prepare a crystal of alum from K_2SO_4 and $Al_2(SO_4)_3 \cdot 18H_2O$, how much water must you add per mole of K_2SO_4?

***17.38 Qualitative analysis An unknown solution may contain Al^{3+}, Cr^{3+}, and/or Zn^{2+}. Addition of aqueous ammonia dropwise produces a white precipitate that dissolves in excess of ammonia. If the original solution is treated with excess sodium hydroxide and oxidized with hydrogen peroxide, the solution remains colorless and gives no precipitate with Pb^{2+}. What is the probable makeup of the original unknown?

*17.39 Tin A block of β tin (density 7.31 g/cm³) that has a volume of 10 cm³ is kept at 0°C. What volume should be allowed for when the β tin converts to α tin (5.75 g/cm³)?

**17.40 Tin Write balanced net equations for each of the following reactions:
a Metallic tin is dissolved in hydrochloric acid to give Sn^{2+}.
b Aqueous NaOH is added dropwise to the solution from (*a*) to form a white precipitate which redissolves in excess NaOH.
c Oxygen gas is bubbled through the solution from (*b*) until the tin is completely converted to stannate ion.
d Acid is added dropwise to the solution from (*c*) until a white precipitate forms.

**17.41 Tin Describe with the aid of equations how you would go about making the following conversions:
a Sn to $SnCl_2$ *b* $SnCl_2$ to $SnCl_4$
c $SnCl_4$ to SnO_2 *d* SnO_2 to Sn

**17.42 Tin When a small amount of Sn^{2+} is added to $HgCl_2$, a white precipitate is formed. When a large amount of Sn^{2+} is added to the same quantity of $HgCl_2$, a black precipitate is formed. Explain with the help of equations.

**17.43 Lead How do you account for the fact that lead has +2 and +4 compounds, but not +3? The electron configuration of Pb^0 is $6s^2 6p^2$.

**17.44 Lead The principal ore of lead is PbS. Tell how this can be converted to metallic Pb. Silver sulfide is a common impurity in PbS. Tell how the by-product silver is extracted.

*17.45 Lead Suggest a reason why children living in old tenements frequently show signs of lead poisoning.

*17.46 Qualitative analysis Set up a flow chart showing how $PbCl_2$ can be separated from AgCl and Hg_2Cl_2, all of which are white insoluble solids.

**17.47 Lead Diagram a lead storage bat-tery. Indicate essential components. Label anode and cathode, write electrode reactions for each, and indicate the migration directions of charged particles when the cell is discharging. Explain why the voltage of the cell decreases as discharge continues.

**17.48 Qualitative analysis Write equations for each of the following processes:

a $PbCl_2(s)$ dissolves in hot water.
b H_2S added to (*a*) gives a black precipitate.
c The black precipitate from (*b*) is oxidized by hot HNO_3 (to form NO_2 and S).
d Addition of sulfuric acid to the solution from (*c*) gives a white precipitate.

Chapter 18

TRANSITION ELEMENTS: SOME GENERAL ASPECTS

Intervening between groups II and III of the periodic table are subgroups of elements collectively referred to as transition elements. The term originally was applied to elements of intermediate character between the extreme left and the extreme right of the periodic table; it was later restricted to elements that use *d* electrons in bonding. Because the latter definition may lead to ambiguity as to whether a particular element, e.g., Ag, should be included, it is better to define transition elements solely on the basis of posi-

tion in the periodic table, i.e., the 10 subgroups between main groups II and III. As shown in Fig. 18.1, this gives us, in the fourth period, Sc, Ti, V, Cr, Mn, Fe, Co, Ni, Cu, and Zn. Each of these heads a subgroup named after itself. Thus, the scandium subgroup includes scandium (Sc), yttrium (Y), elements 57 through 71, and 89 through 103. In this chapter, we consider general aspects of transition-element chemistry. In the next chapter, we look in detail at some specific examples.

18.1
TRANSITION-ELEMENT PROPERTIES

Aside from the fact that transition elements are all metals, there is no single property that *all* the transition elements have in common. However, there are some features that appear in a large fraction of these elements, and so they are considered characteristic transition-element properties. These include: (1) a large tendency to form complex ions, (2) variety of oxidation states, (3) color in the compound state, and (4) paramagnetism of the compounds. Let us look briefly at each of these points. We consider them in greater detail in following sections.

Formation of Complex Ions

What do we mean by a complex ion? Strictly speaking the term should be applied to any charged particle that contains more than one atom, e.g., SO_4^{2-}. However, species such as sulfate are best considered as single units. What we have in mind here is a more limited definition of complex ion, viz., a complex species composed of pieces that have recognizable separate existence in solution. Thus, for example, the surprisingly stable unit $Fe(CN)_6^{4-}$ can be put

483

FIG. 18.1   **Periodic Table**   (Elements in color are transition elements.)

H 1																	He 2
Li 3	Be 4											B 5	C 6	N 7	O 8	F 9	Ne 10
Na 11	Mg 12											Al 13	Si 14	P 15	S 16	Cl 17	Ar 18
K 19	Ca 20	Sc 21	Ti 22	V 23	Cr 24	Mn 25	Fe 26	Co 27	Ni 28	Cu 29	Zn 30	Ga 31	Ge 32	As 33	Se 34	Br 35	Kr 36
Rb 37	Sr 38	Y 39	Zr 40	Nb 41	Mo 42	Tc 43	Ru 44	Rh 45	Pd 46	Ag 47	Cd 48	In 49	Sn 50	Sb 51	Te 52	I 53	Xe 54
Cs 55	Ba 56	* 57–71	Hf 72	Ta 73	W 74	Re 75	Os 76	Ir 77	Pt 78	Au 79	Hg 80	Tl 81	Pb 82	Bi 83	Po 84	At 85	Rn 86
Fr 87	Ra 88	† 89–103	Ku 104	Ha 105	? 106	? 107											

*	La 57	Ce 58	Pr 59	Nd 60	Pm 61	Sm 62	Eu 63	Gd 64	Tb 65	Dy 66	Ho 67	Er 68	Tm 69	Yb 70	Lu 71
†	Ac 89	Th 90	Pa 91	U 92	Np 93	Pu 94	Am 95	Cm 96	Bk 97	Cf 98	Es 99	Fm 100	Md 101	No 102	Lr 103

together from Fe^{2+} and CN^-, both of which are common species in aqueous solution. SO_4^{2-}, on the other hand, cannot be made by mixing solutions of S^{6+} and O^{2-}. S^{6+} is only a hypothetical species, and O^{2-} does not exist at all in aqueous solution. Other examples of complex ions are $Ag(NH_3)_2^+$, $Cu(NH_3)_4^{2+}$, $Zn(CN)_4^{2-}$, and $Cr(H_2O)_6^{3+}$. As noted in Sec. 15.6, complex ions are generally in equilibrium with their dissociation fragments, but the equilibrium constant for dissociation is often very small.

Variety of Oxidation States

There is a tremendous range in the variety of compounds formed by a typical transition element. For example, chromium forms oxides CrO, Cr_2O_3, CrO_2, and CrO_3, corresponding, respectively, to oxidation states +2, +3, +4, and +6. It also forms oxysalts such as K_2CrO_4 and $K_2Cr_2O_7$, corresponding to oxidation state +6; complex compounds such as $K_3Cr(CN)_6$, corresponding to oxidation state +3; and even strange compounds such as $Cr(CO)_6$, in which the oxidation state is 0. Diversity of oxidation state is a clear hallmark of transition-metal compounds.

Color in the Compound State

The characteristic color of group I, group II, Al, Sn, and Pb compounds is white (in powder form) or colorless (in the form of single crystals). In contrast, compounds of the transition elements are usually highly colored. Thus, for example, CrO is black, Cr_2O_3 is green, CrO_2 is brown, CrO_3 is red, K_2CrO_4 is yellow, etc. However, not all transition-metal compounds are colored. $CrCl_2$, for example, is white as a powder. Some complex ions such as $Ag(NH_3)_2^+$ are colorless in aqueous solution. What is it that decides color? Why are transition-element compounds more likely to be colored than not?

Paramagnetism of the Compounds

In Sec. 4.10, it was noted that there are three kinds of magnetism: ferromagnetism, paramagnetism, and diamagnetism. Ferromagnetism is strong attraction into a magnetic field; it is relatively rare. Fe, Co, and Ni of the transition elements show ferromagnetism in the elemental state. Some of the compounds are also ferromagnetic. Diamagnetism is a weak repulsion out of a magnetic field. Everything has some diamagnetism, but it is often obscured by other effects. It arises from induced electron motion in atoms. Sometimes, as in NaCl, it is the only kind of magnetism present. Paramagnetism, rather moderate attraction into a magnetic field, is especially characteristic of transition-metal compounds. It is especially noticeable for compounds containing unpaired electrons. As we see later in this chapter, there is a strong correlation between paramagnetism, oxidation state, and electron configuration. In the next section we take a closer look at what is special about the electron configurations of the transition elements.

18.2
ELECTRON CONFIGURATIONS

Transition elements owe their special place in the periodic table to belated filling of inner orbitals. As can be seen from the orbital-filling diagram of Fig. 4.25, the $3d$ orbital, for example, fills after the $4s$. Why should electrons go back into the third shell after the fourth shell has been started? The reason is shown in Fig. 18.2. There is a sharp drop of the $3d$ energy level relative to $4s$ around $Z = 21$. In atom buildup, s orbitals continuously drop in energy. They have finite probability density at the nucleus, and so increase in nuclear charge is unshielded and gradually pulls in the s electrons. Not so for $3d$. The d electrons have most of their probability density away from the nucleus, and so increase in Z does not immediately affect them. Stepwise increase in Z is

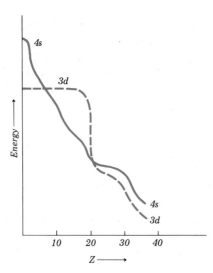

FIG. 18.2 Relative energies of $3d$ and $4s$ orbitals.

just about fully canceled by stepwise increase in number of screening electrons. It is only when electrons are added in outer regions that they do not screen well, and, hence, the $3d$ level then starts getting pulled in.

Figure 18.3 gives detailed electron configurations for each of the first-row transition elements. The overall change is that the third shell ($3s^2 3p^6 3d^1$ in Sc) is built up to 18 electrons ($3s^2 3p^6 3d^{10}$ in Zn). All the while, except for chromium and copper, two electrons remain in the fourth shell (the $4s$ subshell). The apparent anomaly for chromium is due to the fact that $3d$ and $4s$ subshells are very close in energy at this point, and there is a decrease in energy when one of the electrons of the $4s^2$ pair reduces electron-electron repulsion by moving over to a $3d$ orbital. In copper the dropping of a $4s$ electron to the $3d$ subshell does not relieve electron-electron repulsion, but by this time Z has increased so much that $3d$ gets pulled lower than $4s$.

When electrons are removed from transition-metal atoms, it is not always obvious from the electron configuration which electronic level will be

FIG. 18.3 Electron Configurations of First-Row Transition Elements

Element	Symbol	Z	Electron configuration
Scandium	Sc	21	$1s^2\ 2s^2\ 2p^6\ 3s^2\ 3p^6\ 3d^1\ 4s^2$
Titanium	Ti	22	_____(18) _____ $3d^2\ 4s^2$
Vanadium	V	23	_____(18) _____ $3d^3\ 4s^2$
Chromium	Cr	24	_____(18) _____ $3d^5\ 4s^1$
Manganese	Mn	25	_____(18) _____ $3d^5\ 4s^2$
Iron	Fe	26	_____(18) _____ $3d^6\ 4s^2$
Cobalt	Co	27	_____(18) _____ $3d^7\ 4s^2$
Nickel	Ni	28	_____(18) _____ $3d^8\ 4s^2$
Copper	Cu	29	_____(18) _____ $3d^{10}\ 4s^1$
Zinc	Zn	30	_____(18) _____ $3d^{10}\ 4s^2$

depleted. On the basis of orbital filling in buildup of the periodic table, it might seem that since $3d$ electrons are added after $4s$, they should on ionization be removed before the $4s$. However, this prediction is unwarranted because the two processes differ from each other in a major way. In the buildup of the periodic table, the number of electrons is increased at the same time that the nuclear charge is increased. On the other hand, in ionization the number of electrons is decreased while the nuclear charge stays constant. The experimental fact is that $4s$ electrons are removed before $3d$ electrons in the ionization of transition-element atoms.

For the second-row transition elements, yttrium through cadmium, electronic expansion involves $4d$ and $5s$. For the third-row transition elements, lanthanum through mercury, not only are $5d$ and $6s$ involved but also the $4f$ subshell is being filled to 14 electrons. The elements involved in the $4f$ expansion are called the *lanthanide* elements. A similar situation involving the $5f$ expansion occurs in the last row of the periodic table, giving rise to the *actinide* elements.

18.3
OXIDATION STATES

A typical transition element shows a great variety of oxidation states. Figure 18.4 lists the more common states found. Included also (in parentheses) are less common states, such as those which are unstable to disproportionation in aqueous solution or which have been prepared in only a few solid-state compounds. It should be noted that there is presently considerable research activity in attempting to prepare unusual oxidation states; so the listing in the table should not be considered complete.

In addition to showing the large number of possible states, Fig. 18.4 shows several noteworthy features. In each row there is a maximum in the highest state shown near the middle of the row. Thus, for the first row the maximum oxidation state increases regularly from $+3$ for Sc to $+4$ for Ti and so on up to $+7$ for Mn, after which there is a falloff to $+2$ for Zn. Connected with this trend is the fact that elements toward the center of the row generally show more oxidation states than those toward the ends. A careful look at the table discloses one other feature. In going down a subgroup, there is generally a trend toward higher oxidation states. For example, in the Fe subgroup the $+2$ and $+3$ states, which predominate for Fe, give way to $+4$, $+6$, and $+8$, which predominate for Os.

Before making too much fuss over probable reasons for the above trends, we should recall that the concept of oxidation state is artificial and rests on a rather arbitrary assignment of shared electrons to more electronegative atoms. Nevertheless, it is possible to associate increasing maximum oxidation state with increasing number of s and d electrons available for bonding. The falloff after the center of the row can be related to lowering of the d-shell energy relative to s, hence to decreasing availability of those d electrons for

FIG. 18.4 Oxidation States of Transition Elements (Less common states are given in parentheses.)

Sc	Ti	V	Cr	Mn	Fe	Co	Ni	Cu	Zn
+3	(+2)	+2	+2	+2	+2	+2	+2	+1	+2
	+3	+3	+3	(+3)	+3	+3	(+3)	+2	
	+4	+4	+4	+4	(+4)	(+4)			
		+5	+6	(+6)	(+6)				
				+7					

Y	Zr	Nb	Mo	Tc	Ru	Rh	Pd	Ag	Cd
+3	+4	+3	+3	+4	+2	+3	+2	+1	+2
		+5	+4	(+6)	+3	+4	(+3)	(+2)	
			+5	+7	+4	(+6)	+4	(+3)	
			+6		(+5)				
					+6				
					(+7)				
					(+8)				

La	Hf	Ta	W	Re	Os	Ir	Pt	Au	Hg
+3	+4	(+4)	(+2)	(+3)	(+2)	(+2)	+2	+1	+1
		+5	(+3)	+4	(+3)	+3	(+3)	+3	+2
			+4	(+5)	+4	+4	+4		
			+5	+6	+6	(+6)			
			+6	+7	+8				

bonding. Similarly, it is possible to connect increasing preference for higher states down a subgroup with increasing availability of d and s electrons as atomic size increases.

18.4
COMPLEX IONS

The idea of complex ions evolved from observations that a transition-metal compound such as $CrCl_3$ can add other compounds, e.g., ammonia, to form related, more complex materials which are known as *coordination compounds*. Thus, for example, it is possible to have $CrCl_3 \cdot 3NH_3$, $CrCl_3 \cdot 4NH_3$, $CrCl_3 \cdot 5NH_3$, and $CrCl_3 \cdot 6NH_3$. Little was known initially about the way these complexes were held together. The dot formulation simply stated that, in some unspecified way, 3, 4, 5, or 6 mol of NH_3 could be associated with 1 mol of $CrCl_3$. In the early 1900s, Alfred Werner, Swiss chemist, carried out extensive studies of these compounds and came up with the idea that they consist of a central metal atom (e.g., Cr) surrounded by a tightly bound first layer of atoms and molecules (the so-called "first coordination shell") and a second layer (the second coordination shell) that is loosely bound. As a result of his pioneering investigations, Werner has been called the father of coordination chemistry. A remarkable feature of his work is that most of his ideas were developed before much was known about chemical theories of bonding.

488

Clues as to the nature of coordination compounds came from observations such as the following:

$$CrCl_3 \cdot 6NH_3(aq) \xrightarrow{Ag^+} 3 \text{ mol of AgCl precipitated}$$

$$CrCl_3 \cdot 5NH_3(aq) \xrightarrow{Ag^+} 2 \text{ mol of AgCl precipitated}$$

$$CrCl_3 \cdot 4NH_3(aq) \xrightarrow{Ag^+} 1 \text{ mol of AgCl precipitated}$$

$$CrCl_3 \cdot 3NH_3(aq) \xrightarrow{Ag^+} \text{no AgCl precipitated}$$

All the chloride can be precipitated from the first compound, none from the last, and only part from the middle two. Clearly there are two kinds of chloride. One is bound in the inner coordination shell and is not precipitable; the other is in the outer coordination shell and is easily accessible for $Cl^- + Ag^+ \longrightarrow AgCl(s)$.

We now know that there is usually a limit to the number of atoms or molecules that can be accommodated in the first coordination shell. This is called the *coordination number*. It is 6 for an octahedral complex such as $Cr(NH_3)_6^{3+}$, 4 for a tetrahedral complex such as $Zn(NH_3)_4^{2+}$, and 2 for a linear complex such as $Ag(NH_3)_2^+$. How the geometry is related to the coordination number is discussed in the next section. Here we simply note that the above precipitation reactions can be explained as follows:

$CrCl_3 \cdot 6NH_3$ is composed of $[Cr(NH_3)_6]^{3+}$ and $3Cl^-$.
$CrCl_3 \cdot 5NH_3$ is composed of $[CrCl(NH_3)_5]^{2+}$ and $2Cl^-$.
$CrCl_3 \cdot 4NH_3$ is composed of $[CrCl_2(NH_3)_4]^+$ and Cl^-.
$CrCl_3 \cdot 3NH_3$ is composed of $[CrCl_3(NH_3)_3]$.

In other words, in this series chloride gradually replaces NH_3 in the grouping of six molecules clustered about the central Cr atom. The remaining chloride is outside and can be precipitated.

Before looking at structures of complex ions, we need to consider the proper way to name them. We use the word *ligand* (from the Latin *ligare*, meaning "to tie") to designate any group attached to the central metal atom. Typical ligands would be Cl^- (chloro), H_2O (aquo), NH_3 (ammine), CN^- (cyano). For complex *cations*, such as $Cr(NH_3)_6^{3+}$, we give the number and name of ligands followed by the name of the central atom with its oxidation number indicated by Roman numerals in parentheses. Thus, $Cr(NH_3)_6^{3+}$ is hexaamminechromium(III). If the attached ligands are not all alike, they are named in the same order in which they should be written in the formula; i.e., anion ligands precede neutral ligands. Thus, for $CrCl(NH_3)_5^{2+}$ we have monochloropentaamminechromium(III). Numbers of ligands are designated by Greek prefixes *mono-* (usually omitted), *di-, tri-, tetra-, penta-, hexa-, hepta-,* and *octa-*, except that *bis-* (twice), *tris-* (thrice), *tetrakis-* (four times) may be used, especially when the name of the ligand contains a numerical designation (e.g., ethylenediamine). Complex *anions,* such as $Cr(CN)_6^{3-}$, are named by giving the number and name of ligands followed by the name of the element with an *-ate* ending and its oxidation number in parentheses. $Cr(CN)_6^{3-}$ is hexa-cyanochromate(III). Further details on naming complex compounds are given in Appendix A2.1.

EXAMPLE 1

Name the following complexes: $CoCl_4^{2-}$, $Co(NH_3)_6^{2+}$, $Co(NH_3)_6^{3+}$, $CoCl_2(NH_3)_4^+$.

Solution

$CoCl_4^{2-}$ tetrachlorocobaltate(II)

$Co(NH_3)_6^{2+}$ hexaamminecobalt(II)

$Co(NH_3)_6^{3+}$ hexaamminecobalt(III)

$CoCl_2(NH_3)_4^+$ dichlorotetraamminecobalt(III)

18.5
STRUCTURE AND ISOMERISM

The most common geometry for a complex ion is octahedral. As shown in Fig. 18.5 for $Cr(NH_3)_6^{3+}$, the metal atom is at the center of the octahedron, and the ligands are attached at the corners. One of the problems with the representation of Fig. 18.5 is that it may suggest that the up and down ligands are different from the other four. Such is *not* the case. All six NH_3 positions are equivalent. Perhaps a better way to show an octahedral complex is as was done for $Al(H_2O)_6^{3+}$ in Fig. 9.2.

If we replace any one of the NH_3 molecules of $Cr(NH_3)_6^{3+}$ by Cl^-, we get $CrCl(NH_3)_5^{2+}$, as shown in Fig. 18.6. There are six possible drawings that can be made to correspond to different chloride positions, but all these are equivalent and correspond to the same ion.

If we replace two of the NH_3's by Cl^-, we get $CrCl_2(NH_3)_4^+$, as shown in Fig. 18.7. Here the placing of chlorides relative to each other can lead to two distinct arrangements. At the top of the figure, the two chloride atoms occupy adjacent ligand sites; at the bottom, they occupy nonadjacent sites. The top complex is referred to as *cis*- (Latin, meaning "on this side"); the bottom

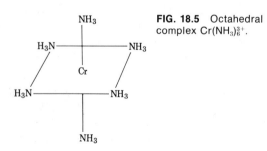

FIG. 18.5 Octahedral complex $Cr(NH_3)_6^{3+}$.

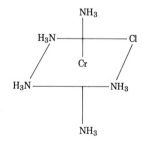

FIG. 18.6 Octahedral complex $CrCl(NH_3)_5^{2+}$.

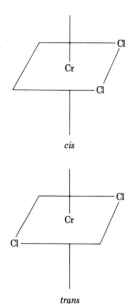

cis

trans

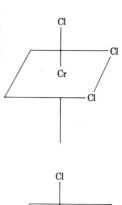

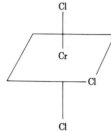

FIG. 18.7 Octahedral isomers of $CrCl_2(NH_3)_4^+$. (NH_3 molecules not shown.)

FIG. 18.8 Isomers of $CrCl_3(NH_3)_3$. (NH_3 not shown.)

complex is referred to as *trans-* (Latin, meaning "across"). They correspond to different compounds and have different sets of properties. Together, they represent an example of *isomerism,* different spatial arrangements of the same collection of atoms. The two isomers are called, respectively, *cis-* and *trans-*dichlorotetraamminechromium(III).

When we put three chloride and three ammonia groups on the same chromium, we again get two possible arrangements, as is shown in Fig. 18.8. At the top of the figure, all three chlorides are on the same triangular face of the octahedron. The chlorides are all adjacent to each other and therefore equivalent to each other. The other isomer, as shown at the bottom, has one chloride different from the other two.

Octahedral complexes are found throughout the periodic table, even for nontransition elements. Thus, for example, Mg^{2+} attaches six water molecules to itself to give $Mg(H_2O)_6^{2+}$, as in the compound $MgCl_2 \cdot 6H_2O$. However, it is among the transition elements that we find the widest variety of octahedral complexes. Typical examples from iron chemistry would be FeF_6^{3-}, $Fe(H_2O)_6^{3+}$, $Fe(NH_3)_6^{3+}$, and $Fe(CN)_6^{3-}$. Some of these, such as $Fe(NH_3)_6^{3+}$, are found only in crystals. In aqueous solution, they would convert to $Fe(H_2O)_6^{3+}$. $Fe(CN)_6^{3-}$, however, is quite stable to water replacement for the CN^-. Each of these complexes can be written as ML_6, where L stands for each of six identical ligands attached to the central M.

It is also possible to have complexes with fewer than six attached ligands but still end up with six near-neighbor atoms. An example of such a complex is $Fe(C_2O_4)_3^{3-}$. It has only three oxalate ($C_2O_4^{2-}$) groups per iron atom, but, as shown in Fig. 18.9, each oxalate group has to do double duty in providing two points of attachment to the central iron. Such a group is called a *bidentate group* (from the Latin *bi-,* meaning "two," and *dens,* meaning "tooth").

491

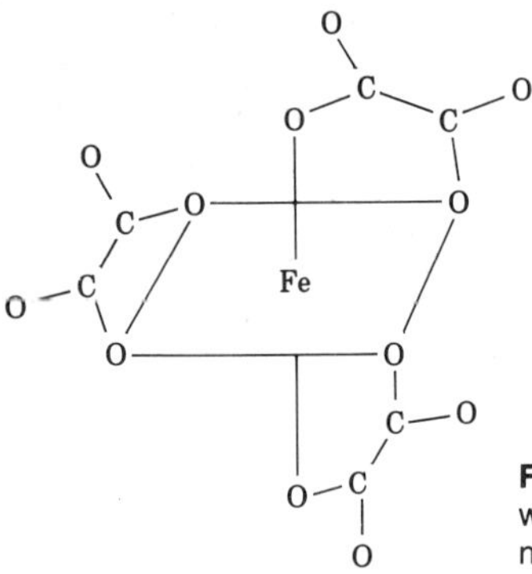

FIG. 18.9 Complex ion Fe(C₂O₄)₃³⁻ with octahedral arrangement of near-neighbor oxygen atoms.

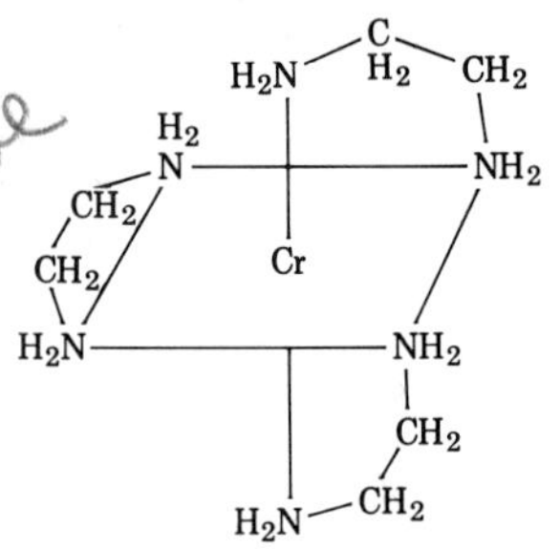

FIG. 18.10 Complex ion Cren₃³⁺ with octahedral arrangement of near-neighbor nitrogen atoms.

The Fe atom ends up with six near-neighbor oxygen atoms, essentially at the corners of an octahedron, but the oxygen atoms are linked together in pairs through the carbon atoms. Any group such as oxalate that bridges two or more coordination positions is called a *chelating agent,* and the resulting complex is called a *chelate.* (Chelate comes from the Greek word *chēlē,* meaning "claw.") The proper name for the chelate shown in Fig. 18.9 is trisoxalatoferrate(III).

Another example of a bidentate group is ethylenediamine, $H_2NCH_2CH_2NH_2$, usually abbreviated *en.* It attaches itself through the two nitrogen atoms, as is shown in Fig. 18.10 for trisethylenediaminechromium(III). Some chelating agents bind at more than two positions, in which case they may be tridentate, tetradentate, pentadentate, or even hexadentate. An example of the last of these is ethylenediaminetetraacetate (generally abbreviated EDTA):

It is hexacoordinate and can wrap itself around a central metal atom M so that binding occurs through the four negative oxygens and the two nitrogens.

Although they are not so common as octahedral complexes, there are many complexes in which the coordination number is not 6 but 4. Two kinds of geometry are then possible: tetrahedral or square planar. Tetrahedral complexes, such as $Zn(NH_3)_4^{2+}$, tend to be formed by elements toward the right end of the transition-element rows, especially the $3d$ sequence. Square-planar complexes also tend to show up at the right end of the transition-element rows, especially for palladium (element number 46) and platinum (element number 78). A typical square-planar complex would be $PtCl_4^{2-}$, tetrachloroplatinate(II). As can be seen in Fig. 18.11, tetrahedral and square-planar complexes differ from each other in the relation of ligands to one other. In the tetrahedral complex, each ligand position is adjacent to each of the other three; in the square-planar complex, each ligand has two near ligand neighbors and one further away. This does not affect the number of

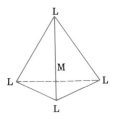

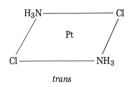

trans

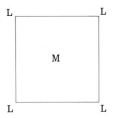

FIG. 18.11   Four-coordinate complexes
ML$_4$.

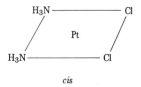

cis

FIG. 18.12 Two possible arrangements for the square-planar complex
PtCl$_2$(NH$_3$)$_2$ (seen in perspective).

all form are different

isomers when all the ligands are alike—there is only one isomer for ML$_4$, be it tetrahedral or square planar. However, if two of the ligands are replaced by two of another kind, e.g., MX$_2$Y$_2$, then we can have two isomers for the square-planar arrangement but only one isomer for the tetrahedral. Figure 18.12 shows the two possible arrangements for dichlorodiammineplatinum(II). In the trans isomer, the chlorine atoms are on opposite corners of the square; in the cis isomer, on adjacent corners. It is interesting to note that one of the early ways of telling whether a given compound was tetrahedral or square planar was to see how many isomers it could form. In a tetrahedron, all corners are adjacent to each other so that the equivalent of a trans complex is not possible.

In addition to the isomerism discussed above, which is called *geometric isomerism*, complex ions also may show what is called *optical isomerism*. Optical isomers occur when two species are related as mirror images of each other. The two species have practically all chemical properties identical, but they differ in that one turns the plane of polarized light clockwise and the other turns it counterclockwise. Figure 18.13 shows the relation between the two optical isomers of Co*en*$_3^{3+}$, trisethylenediaminecobalt(III). Although in both cases the central cobalt is bound to six nitrogen atoms, there are two different ways in which the *en* bridges are arranged. No amount of rotation can superimpose one complex on its mirror image. The pair of optical isomers are called *enantiomers*, or *enantiomorphs*. They may be individually designated *d* (for *dextro*) or *l* (for *levo*), depending on whether the plane of polarized light is rotated clockwise or counterclockwise. The requirement for having optical isomers is that there be no center of symmetry or plane of symmetry in the complex. (Optical isomers are discussed further in Sec. 20.3.)

The study of complexes is now one of the most thriving areas of chemical research (known as coordination chemistry). Why is it so important?

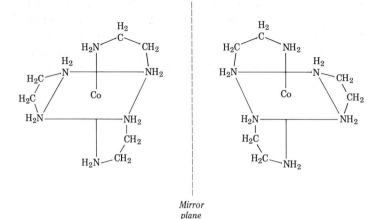

Mirror
plane

FIG. 18.13 Mirror-image relation of optical isomers of Coen_3^{3+}.

1 Complexes are structural units in many crystalline compounds.

2 They are often the most important species in aqueous solutions.

3 They are frequently involved as intermediates in the mechanisms of chemical reactions.

4 They are increasingly recognized as serving vital functions in the physiological processes of plant and animal growth. Chlorophyll, for example, the green matter in plants that absorbs light energy for photosynthetic conversion of CO_2 and H_2O into plant carbohydrates, is essentially a magnesium complex. As shown in Fig. 18.14 the magnesium is bound to four nitrogen atoms. Hemoglobin, the red coloring matter in blood, involved in carrying oxygen, is a closely related complex in which iron is similarly bound to four nitrogens. The precise geometry around the magnesium or iron is very important for controlling how the entire molecule carries out its function. Other metals that apparently have similar critical roles in establishing favorable geometry for physiological reactions are cobalt, copper, zinc, and manganese. Only trace amounts are required, but without them many enzymes would be unable to function. The problem of determining the specific geometry around active sites of enzymes is currently one of the most exciting challenges in chemistry.

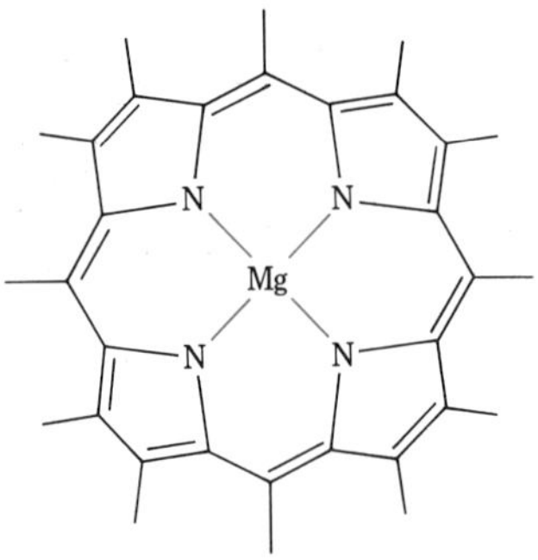

FIG. 18.14 Schematic representation of part of a chlorophyll molecule. (Intersections of lines are occupied by carbon atoms.)

Why should the geometry be so interesting? We have not mentioned it above, but, as was evident in early parts of this book, there is a close relation between the spatial arrangement of bonded atoms and the nature of chemical bonding. Indeed, early study of complexes was most important in clarifying the connection between chemical bonding and structure. In the next section, we take a closer look at this problem of bonding in complexes.

18.6
BONDING IN COMPLEXES

There are three ways in which we can approach the description of bonding in complexes: (1) a valence-bond approach, in which atoms sit side by side and electrons are exchanged between bonded atoms; (2) a molecular-orbital approach, in which the electrons are delocalized (spread out) over the whole complex; and (3) a strictly ionic approach, in which the complex is viewed as an assembly of individual ions. For some complexes such as FeF_6^{3-}, the ionic approach is not a bad model; for others, such as $Fe(CN)_6^{3-}$, the ionic model is not the best approximation.

Let us imagine the complex ion $Fe(CN)_6^{3-}$ as being assembled from Fe^{3+} and six CN^- ions. As shown on the left side of Fig. 18.15, the Fe^{3+} can be visualized to be at the center of an octahedron and the CN^- ions at the corners. It seems reasonable that the complex would stick together because of the electric attraction between the positive center and the negative CN^- ions. However, we have not explained the geometry. Why should there be six CN^- ions, and why should they occupy the corners of an octahedron? Why not tie on four CN^- ions at the corners of a tetrahedron? Why not form $Fe(CN)_4^-$ rather than $Fe(CN)_6^{3-}$?

This is where the valence-bond method proves more powerful than a simple ionic model. As was noted in Sec. 5.9, there is a connection between shapes of molecules and the orbitals used in bonding. Specifically, it was pointed out that if the orbitals used were two d, one s, and three p (the so-called "d^2sp^3 hybrids"), then we would have an octahedral geometry. On the right side of Fig. 18.15 is presented a schematic view of how use of hybrid orbitals can lead to octahedral geometry for $Fe(CN)_6^{3-}$. The d^2sp^3 hybrid orbitals of the iron point toward the corners of an octahedron; each cyanide shares its

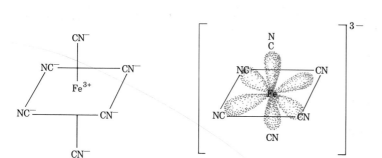

FIG. 18.15 Two models of $Fe(CN)_6^{3-}$ complex. Left shows an ionic model; right shows a valence-bond model, where electron pairs are shared between CN and d^2sp^3 hybrid orbitals of central Fe atom.

496

Chapter 18
Transition
Elements:
Some General
Aspects

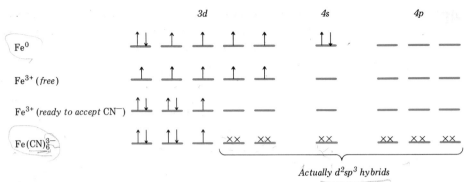

FIG. 18.16 Electron population for iron in its various stages on the way to being bound in $Fe(CN)_6^{3-}$. Arrows refer to electrons from the iron; x's refer to electrons contributed from the cyanide.

electrons with the iron by putting a pair of electrons at least part of the time in one of the d^2sp^3 hybrids. Schematically, the electron population can be worked out as shown in Fig. 18.16. Fe $(Z = 26)$ has electron configuration $1s^22s^22p^63s^23p^63d^64s^2$. To form Fe^{3+}, we strip off the two $4s$ electrons and one of the $3d$ electrons. This leaves five electrons in the five $3d$ orbitals. To bind on the CN^-, we will need to use two of the d orbitals, and so we crowd all the iron's electrons into the remaining three. We have six orbitals vacant ready to accept CN^-. Each CN^- puts a pair of electrons (shown by x's in Fig. 18.16) into an available iron orbital, and we thus establish a covalent bond. The whole complex sticks together because there is a sharing of electrons between a d^2sp^3 hybrid orbital of the iron and some orbital on the cyanide.

The beauty of the above picture is that it shows there should be one unpaired electron in the complex $Fe(CN)_6^{3-}$. This corresponds exactly with what is observed in the paramagnetism of the complex. Although Fe^{3+} has five unpaired electrons, $Fe(CN)_6^{3-}$ has but one. The former is called *high spin,* the latter is called *low spin.*

As a further illustration of how the above description works, we can apply it to $Cr(CO)_6$. Cr $(Z = 24)$ has electron configuration $1s^22s^22p^63s^23p^63d^54s^1$. To bind on six CO molecules, we crowd all the $3d^54s^1$ electrons of the chromium into three of the orbitals, as follows:

$$\underset{3d}{\underline{\uparrow\downarrow}\ \underline{\uparrow\downarrow}\ \underline{\uparrow\downarrow}\ \underline{xx}\ \underline{xx}}\quad \underset{4s}{\underline{xx}}\quad \underset{4p}{\underline{xx}\ \underline{xx}\ \underline{xx}}$$

and then use the rest of the available orbitals for electrons from CO.

Tetrahedral geometry results when we use sp^3 hybrids. Thus, the $Zn(NH_3)_4^{2+}$ complex can be represented this way:

$$\underset{3d}{\underline{\uparrow\downarrow}\ \underline{\uparrow\downarrow}\ \underline{\uparrow\downarrow}\ \underline{\uparrow\downarrow}\ \underline{\uparrow\downarrow}}\quad \underset{4s}{\underline{xx}}\quad \underset{4p}{\underline{xx}\ \underline{xx}\ \underline{xx}}$$

Zn^{2+} has a $3d^{10}$ configuration, and so all the d orbitals are filled. Electron pairs from NH_3 can be accommodated in $4s$ and $4p$.

Square-planar geometry results when we use dsp^2 hybrids (a set of metal orbitals made up of one d, one s, and two p). An example would be $Ni(CN)_4^{2-}$, which can be represented as follows:

$$\underline{\uparrow\downarrow}\ \underline{\uparrow\downarrow}\ \underline{\uparrow\downarrow}\ \underline{\uparrow\downarrow}\ \underline{\text{xx}} \qquad \underline{\text{xx}} \qquad \underline{\text{xx}}\ \underline{\text{xx}}\ \underline{}$$

$$\underset{3d}{} \qquad \underset{4s}{} \qquad \underset{4p}{}$$

Square-planar complexes are most prevalent for metal atoms that have d^8 configurations. As can be seen, all eight of these d electrons can be crowded into four of the d orbitals, leaving one d orbital for making the dsp^2 hybrids.

Molecular-orbital pictures are probably the most faithful representations of the overall electron distribution in complexes, but they are quite difficult to work out. To do so we would need to look in detail at how electrons behave under the attractive influence of several centers such as we would have in a complex. Such a full treatment is what is called *ligand-field theory*. In the following section we look at a simplified version of the problem.

18.7
CRYSTAL-FIELD THEORY

Crystal-field theory started as an attempt to answer the following questions: Given an atom in a crystal, how does the electric field produced by neighboring ions affect the electrons in the atom? How are the energy levels changed? How are the electron probability densities modified from what they would be in the free atom? These questions have relevance for us in describing complex ions since we can understand many properties of complex ions by considering what happens to d electrons of a central atom when surrounding ligands are brought up. Crystal-field theory simply looks at how the approaching charge clouds affect the various d electrons. (Ligand-field theory is a more elegant version in which the d electrons are not confined to the central atom alone but are allowed to wander over the whole complex.)

Let us start by considering the crystal-field problem of a Ti^{3+} ion in an oxide lattice (e.g., in Ti_2O_3, where the Ti^{3+} sits in an octahedral hole made by packing together oxide ions). What effect would the negative charge of the oxide ions have on the titanium? The electron configuration of Ti^{3+} is $1s^2 2s^2 2p^6 3s^2 3p^6 3d^1$. Except for $3d^1$, the electron subshells are filled and hence spherically symmetric. We can ignore them. So far as $3d^1$ is concerned, the electron can be accommodated in any one of the five d orbitals. These five have different shapes, as was shown in Fig. 4.22. Which of these orbitals is likely to be the one occupied? If Ti^{3+} were an isolated ion in the gas phase, it would make no difference, since all five of the d orbitals would be of equal energy. However, in a crystal the answer depends on the arrangement of neighbors.

Let us locate the Ti^{3+} ion at the origin of a set of axes, as shown in Fig. 18.17. For octahedral symmetry, there will be one ligand equally distant from the origin on each of the six axial directions. If we imagine the set of six ligands brought in uniformly toward the metal ion, what effect would there be

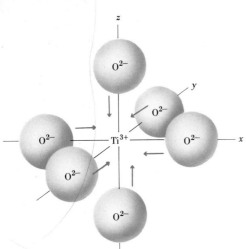

FIG. 18.17 Octahedral arrangement of oxide ions around Ti^{3+} in Ti_2O_3.

on the $3d$ electron? Specifically, which of the five $3d$ orbitals would most likely be occupied?

Two of the five orbitals, d_{z^2} and $d_{x^2-y^2}$, have high electron density along the axes. The other three, d_{xy}, d_{yz}, and d_{zx}, have high electron density between axes. Clearly, there will be strong repulsion between the ligand and the d electron when the d electron is concentrated along the axes (d_{z^2} or $d_{x^2-y^2}$). The repulsion will be less if the electron has minimum charge density on the axes (d_{xy}, d_{yz}, and d_{zx}). The result is that, as shown in Fig. 18.18, the set of five d orbitals is energetically split by the amount Δ into two subsets. The higher-energy subset (called e_g) has high electron density on the axes, and the lower-energy one (called t_{2g}) has high electron density away from the axes. For Ti^{3+} the lowest-energy configuration corresponds to having the $3d$ electron in one of the t_{2g} orbitals. As indicated in the next sections, the splitting of d orbitals is important in deciding the magnetic and spectral properties of complex ions.

tetrade

18.8
COLOR

When white light interacts with a substance giving rise to color, there is absorption of part of the visible spectrum. For example, if the blue portion of white light is absorbed, then the remainder appears red; conversely, if red frequencies are absorbed, the substance appears blue. Since most transition-element compounds are colored, there must be energy transitions which can use up some of the energy of visible light. Figure 18.19 lists some characteristic colors of first-row transition-element aquo ions and indicates approximate wavelengths for maximum light absorption. Because the eye is not equally sensitive to all wavelengths, it is not simple to go from absorption maximum to perceived color. Furthermore, the absorption does not occur

498

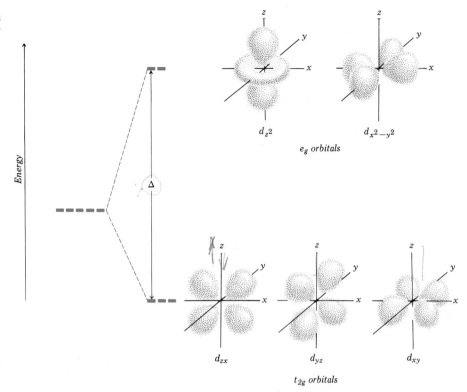

FIG. 18.18   Energy-level splitting of 3d orbitals by octahedral environment.

sharply at a single wavelength but instead is spread out over a *band* of the spectrum. The visible region of the light spectrum extends approximately from 400 to 700 nm, and so some of the maxima noted in the figure are in the ultraviolet (below 400 nm) and some are in the infrared (above 700 nm).

Absorption bands are characteristic of transition-element ions and are attributed to electronic transitions involving the d orbitals. In the crystal-field point of view, the d orbitals are separated into different levels of energy. For Ti^{3+} surrounded by an octahedron of H_2O molecules, the minimum energy state would correspond to having the $3d^1$ unpaired electron in one of the lower three orbitals (Fig. 18.18). As light is absorbed by the sample, the electron is raised from the lower t_{2g} set of orbitals to the upper e_g set. The absorption of energy Δ gives rise to the color.

In cases involving more than one d electron, the cause of absorption bands is qualitatively similar to the d^1 case. However, the situation can be more complicated. More than one electron can be excited, and interelectron repulsions between excited and nonexcited electrons can produce different transition energies. Hence, as noted in Fig. 18.19, several absorption bands may be observed.

In the case of $Fe(H_2O)_6^{3+}$ it is noted that the ion is colorless: no major absorption band is observed in the visible. The reason for this is the $3d^5$ configu-

500

Chapter 18
Transition
Elements:
Some General
Aspects

FIG. 18.19   Colors and Absorption Wavelengths

Ion	Configuration	Observed color	Absorption maxima, nm
$Ti(H_2O)_6^{3+}$	d^1	Violet	493
$V(H_2O)_6^{3+}$	d^2	Blue	389, 562
$V(H_2O)_6^{2+}$	d^3	Violet	358, 541, 910
$Cr(H_2O)_6^{3+}$	d^3	Violet	264, 407, 580
$Fe(H_2O)_6^{3+}$	d^5	Colorless	
$Fe(H_2O)_6^{2+}$	d^6	Pale green	962
$Co(H_2O)_6^{2+}$	d^7	Pink	515, 625, 1220
$Ni(H_2O)_6^{2+}$	d^8	Green	395, 741, 1176
$Cu(H_2O)_6^{2+}$	d^9	Blue	794

ration. Of the five electrons, three are in the t_{2g} set and two are in the e_g set. There is one electron in *each* of the d orbitals. The only way energy could be absorbed would be for an electron from a lower t_{2g} orbital to be raised to an e_g orbital. There are no empty orbitals, and so the promoted electron would have to pair up. Such a change where an electron goes from an unpaired to a paired situation rarely occurs during a light-absorption process.

What happens if we change the ligands around a central atom? How would the color be affected? In general, ligands differ quite markedly from each other in the extent to which they can separate the set of five d orbitals into subsets. In other words, the magnitude of the energy-splitting Δ, shown in Fig. 18.18, depends on the nature of the ligands. H_2O as a ligand produces a relatively moderate effect; CN^-, on the contrary, invariably produces a very large splitting. As a rule, the energy splitting for a given central atom increases with change of ligand in the sequence

$$I^- < Br^- < Cl^- < F^- < OH^- < H_2O < NH_3 < CN^-$$

The further to the right in the series, the bigger the Δ, and the more the absorption band moves from the low-energy red to the high-energy blue end of the spectrum. The observed color, of course, changes accordingly. $V(H_2O)_6^{4+}$, for example, absorbs in the red and appears blue; $V(CN)_6^{2-}$ absorbs in the violet and appears yellow. The above sequence of ligands is known as the *spectrochemical series*.

18.9
MAGNETIC PROPERTIES

Because of unpaired electron spin, paramagnetism is a likely possibility for transition-element compounds. There are two properties of electrons in atoms that can give rise to paramagnetism: One is the electron spin, and the other is the orbital motion of the electron. In the lanthanides and actinides (where f electrons deep in the atom are involved) both orbital and spin effects contribute to the observed magnetism. However, in the other transition elements the orbital magnetism is destroyed. Being more exposed to the environment, the d electrons get tumbled about by interaction with surrounding ligands. They lose much of their freedom of orbital motion and, as a result, the magnetism that comes from the orbital motion is quenched. We shall deal with magnetic moment* from electron spin only. This can be calculated from the formula

$$\text{Spin magnetic moment} = \sqrt{n(n+2)}$$

where n is the number of unpaired electrons. The units are *Bohr magnetons*. In these units, a single $1s$ electron has a magnetic moment of 1.73 Bohr magnetons.

Figure 18.20 summarizes the magnetic moments as calculated from the above spin-only formula and those actually observed for hydrated ions of the first-row transition elements. In general, there is quite good agreement. In certain cases the values turn out to exceed slightly those predicted on the basis of spin only. These small discrepancies are probably due to incomplete quenching of the orbital magnetic moment.

For some compounds of the transition elements the observed magnetic moments turn out to be considerably smaller than predicted by the spin-only formula. For example, $K_4Fe(CN)_6$ does not show the paramagnetic moment of 4.9 Bohr magnetons that would be predicted for Fe^{2+} (d^6—four unpaired electrons—as shown in Fig. 18.20). Instead, $K_4Fe(CN)_6$ is observed to be diamagnetic; it contains *no* unpaired electrons. How can this be accounted for? $K_4Fe(CN)_6$ solid consists of potassium cations, K^+, and ferrocyanide anions, $Fe(CN)_6^{4-}$. We can consider the $Fe(CN)_6^{4-}$ anion as being assembled from Fe^{2+} (d^6) and six ligand CN^- ions. If the six CN^- ions are brought up to

*The *magnetic moment* of a bar magnet is equal to the strength of the magnetic pole at each end times the distance between them. The magnetic moment is a magnetic analog of the electric dipole moment mentioned in Sec. 5.4.

**FIG. 18.20   Predicted and Observed Paramagnetic Moments of Various
Transition-Metal Ions (Units are Bohr magnetons.)**

Ion	e^- configuration	Unpaired e^-	Magnetic moment calculated	Experimental
Sc^{3+}	$3d^0$ _ _ _ _ _	0	0	0
Ti^{3+}	$3d^1$ ↑ _ _ _ _	1	1.73	1.75
Ti^{2+}	$3d^2$ ↑↑ _ _ _	2	2.84	2.76
V^{2+}	$3d^3$ ↑↑↑ _ _	3	3.87	3.86
Cr^{2+}	$3d^4$ ↑↑↑↑ _	4	4.90	4.80
Mn^{2+}	$3d^5$ ↑↑↑↑↑	5	5.92	5.96
Fe^{2+}	$3d^6$ ↑↓↑↑↑↑	4	4.90	5.0
Co^{2+}	$3d^7$ ↑↓↑↓↑↑↑	3	3.87	4.4
Ni^{2+}	$3d^8$ ↑↓↑↓↑↓↑↑	2	2.84	2.9
Cu^{2+}	$3d^9$ ↑↓↑↓↑↓↑↓↑	1	1.73	1.8
Zn^{2+}	$3d^{10}$ ↑↓↑↓↑↓↑↓↑↓	0	0	0

502

Chapter 18
Transition
Elements:
Some General
Aspects

surround the Fe^{2+} octahedrally, as in the process illustrated in Fig. 18.17, the e_g and t_{2g} orbitals of the iron atom are perturbed differently. With the approach of ligand charge clouds along the axes, the electrons of the iron are repelled out of the e_g set (d_{z^2} and $d_{x^2-y^2}$) into the t_{2g} set (d_{xy}, d_{yz}, and d_{zx}). The only way all six electrons can be accommodated in the t_{2g} set of three orbitals is for an electron pair to occupy each orbital. This forces the electron spins to pair up and cancel each other. Actually, there is considerable repulsion between two electrons in the same orbital. Only if the crystal-field effect is strong enough—i.e., if Δ in Fig. 18.18 is big enough—would it be energetically preferable to put two electrons in the same orbital (i.e., pay the repulsion energy) rather than go to a higher-energy orbital. Thus, whether pairing occurs depends critically on the magnitude of Δ compared with electron-electron repulsion energy. As we saw in Sec. 18.8, the magnitude of Δ depends on the nature of the ligand. CN^-, for example, produces a much bigger energy-splitting Δ than does H_2O. Hence, for a given transition element, CN^- complexing is more likely to force electron pairing than is H_2O. In addition, Δ increases with increasing central-atom charge and with increasing principal quantum number (that is, $5d > 4d > 3d$).

Complexes in which Δ is small have the largest number of unpaired electrons; they are called "high-spin" complexes. Complexes in which Δ is large have the fewest unpaired electrons; they are called "low-spin" complexes. From iron chemistry we have $Fe(H_2O)_6^{3+}$ as a high-spin complex (five unpaired electrons) and $Fe(CN)_6^{3-}$ as a low-spin complex (one unpaired electron). They both correspond to d^5. Figure 18.21 compares the relative energy splitting in the two complexes.

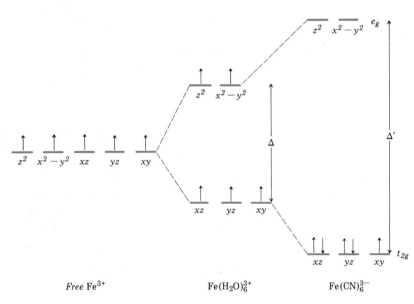

FIG. 18.21 Relative splitting of d orbitals and change in number of unpaired electrons for $Fe(H_2O)_6^{3+}$ and $Fe(CN)_6^{3-}$.

Important Concepts

transition elements
transition-element properties
complex-ion formation
diversity of oxidation states
color of complexes
magnetic properties
electron configurations
$3d$–$4s$ crossover
coordination chemistry
coordination number
naming of complexes
octahedral complexes
cis and trans isomers
ligands
chelation
bidentate

tetrahedral complexes
square-planar complexes
geometric vs. optical isomerism
bonding in complexes
valence-bond description
d^2sp^3 hybrids
crystal-field theory
splitting of d orbitals
t_{2g} vs. e_g orbitals
absorption bands
spectrochemical series
electron paramagnetism
spin-only magnetic moment
high-spin complexes
low-spin complexes

Exercises

*18.1 Transition elements Draw up a periodic table which includes only the transition elements. Specify the atomic number and indicate the electron population of the two outermost subshells for each element.

***18.2 Electronic configurations Explain why the energy of the $3d$ level abruptly falls below that of the $4s$ level around $Z = 21$.

**18.3 Electronic configurations Explain why for chromium ($Z = 24$) the configuration $3d^5\,4s^1$ might be lower in energy than $3d^4\,4s^2$.

***18.4 Oxidation states Explain why we might expect the transition elements to show more oxidation states per element on the average than the pretransition elements show.

*18.5 Complex-ion nomenclature Give the systematic name for each of the following complex ions: $Cr(NH_3)_6^{3+}$, $CrCl(NH_3)_5^{2+}$, $CrCl_2(NH_3)_4^+$, $CrCl_3(NH_3)_3$.

*18.6 Complex-ion nomenclature Write the correct formula for each of the following complex ions:

a Dichlorotetraamminecobalt(III)
b Dichlorotetraamminecobalt(II)
c Tetrachlorodiamminecobaltate(III)
d Hexacyanoferrate(II)
e Hexaaquoiron(III)

**18.7 Structures All the complexes in Exercise 18.6 are octahedral. Draw sketches showing the spatial arrangement of ligands. Indicate in which cases cis-trans isomerism would be expected.

*18.8 Complex ions Draw sketches of and tell how many geometric isomers one may have for each of the following:

a Octahedral $CrCl(NH_3)_5^{2+}$
b Octahedral $CrCl_5(NH_3)^{2-}$
c Octahedral $CrCl_2(NH_3)_4^+$
d Octahedral $CrCl_4(NH_3)_2^-$

e Octahedral $CrCl_3(NH_3)_3$

f Square planar $PtClF_3^{2-}$

g Square planar $PtCl_2F_2^{2-}$

h Tetrahedral $Zn(H_2O)_2(NH_3)_2^{2+}$

**18.9 Complex ions Given an octahedral complex of the type MX_2Y_4, you would like to know whether it is cis or trans. By reacting it with more X so as to convert MX_2Y_4 into MX_3Y_3, you find you can get but one product. What can you conclude about the identity of the original complex?

**18.10 Complex ions Suppose you had a complex ion ML_5 in the shape of a square pyramid with four ligands at the corners of a square base containing M in the center and a fifth ligand at the apex. Draw the structure and predict how may different isomers should result when L_5 is stepwise replaced by L_4X, L_3X_2, L_2X_3, and LX_4.

*18.11 Optical isomers Tell which of the following should exist as optical isomers:

a Square planar $PtCl_3Br$

b Tetrahedral $Zn(OH)IClBr^{2-}$

c Octahedral $CoCl_3(NH_3)_3$

d Octahedral complex with all six ligands different

e Square-planar complex with all four ligands different

*18.12 Complex ions Tell what orbitals of a metal atom are likely to be used to form a complex of each of the following geometries: (*a*) octahedral, (*b*) square planar, and (*c*) tetrahedral. Give a specific example of each type of complex.

**18.13 Complex ions Suppose you have two transition-metal ions having d^2 and d^8 configurations, respectively. Which of these ions is more likely to form an octahedral complex, say, with chloride ion? Which, square planar? Justify your answers.

**18.14 Complex ions Draw a structural diagram showing how the hexadentate ligand EDTA wraps itself around Cr^{3+}. Tell whether this complex should exist as optical isomers.

**18.15 Bonding in complexes Give a valence-bond description of the bonding in the complex $Cr(CN)_6^{3-}$. Tell what orbitals are used and how many unpaired electrons we expect for this complex.

**18.16 Bonding in complexes Nickel ($Z = 28$) combines with four CO molecules to give the tetrahedral complex $Ni(CO)_4$. Give a valence-bond description of this complex.

**18.17 Bonding in complexes When four cyanide groups attach to Ni^{2+} ($Z = 28$), we get a square-planar complex. When four cyanide groups attach to Zn^{2+} ($Z = 30$), we get a tetrahedral complex. Explain in terms of valence-bond theory how this comes about.

**18.18 Crystal-field theory Draw a clearly labeled splitting diagram showing how the energies of the five $3d$ orbitals change when an octahedral set of ions is brought up to a Ti^{3+} ion. What features of the d-electron distributions account for the splitting?

**18.19 Crystal-field theory Explain why the same set of octahedral ligands brought up to the same distance from a transition-metal ion produces a bigger splitting of d orbitals when they are $5d$ than when they are $3d$.

***18.20 Crystal-field theory Predict how the relative energy of the five d orbitals would be affected by bringing only four ligands on the x and y axes up to a transition-metal ion. Draw an energy diagram and justify your relative ordering.

**18.21 Spectral properties Why are transition-metal compounds so often colored? How does this fit in with the fact that compounds of pretransition elements are frequently colorless?

*18.22 Spectral properties What would you expect would happen to the color of $Ti(H_2O)_6^{3+}$ if you squeezed on the complex, as by putting it under pressure?

**18.23 Spectral properties How might you explain the fact that replacing the H_2O ligands in $Fe(H_2O)_6^{3+}$ by CN^- changes the color of the complex from almost colorless to deep red?

*18.24 Magnetic properties How many unpaired electrons would you expect in each of the following:

a Free Cr^{3+} ion (d^3)

b Free Mn^{2+} ion (d^5)

c $Cr(H_2O)_6^{3+}$ (weak crystal field)

d $Mn(H_2O)_6^{2+}$ (weak crystal field)

e $Cr(CN)_6^{3-}$ (strong crystal field)

f $Mn(CN)_6^{4-}$ (strong crystal field)

**18.25 Magnetic properties Using the data from Fig. 18.20, make a graph showing how the calculated magnetic moment varies with the number of d electrons. How would your graph be changed if each of your d^n ions were placed in a strong-field octahedral environment such as would be provided by cyanide ions?

Free atom no spilting

Chapter 19

TRANSITION ELEMENTS: SOME DETAILED CHEMISTRY

In the preceding chapter, we considered the general characteristics of transition elements and how they could be interpreted in terms of ligand interaction with d electrons. In this chapter we look at the special features that characterize individual transition elements. We cannot take time to look at all the elements, but we will consider only those that are most frequently encountered, have special technological importance, or uniquely illustrate some fundamental characteristic of chemical behavior.

We will essentially proceed from left to right across the transition-element sequence. We will start with the rare-earth elements, then examine the members of the 3d expansion (Ti, V, Cr, Mn, Fe, Co, Ni, Cu, Zn) with an occasional excursion down a subgroup to look at special cases such as silver and mercury. Throughout the following sections, it is advisable to keep in mind normal periodic trends and the general characteristics of transition elements as discussed in the preceding chapter.

19.1
RARE-EARTH ELEMENTS

In crossing the periodic table from left to right, the first subgroup we encounter is the scandium subgroup. It contains, besides scandium and yttrium, a place for 15 elements (lanthanum through lutetium), and below that another place for 15 elements (actinium through lawrencium). The 14 elements with lanthanum are called the *lanthanides* or, more frequently, *rare-earth elements*. The 14 elements with actinium are called the *actinides*. Figure 19.1 indicates the electron configurations characteristic of these elements.

FIG. 19.1   Electron Configurations in Scandium Subgroup

Element	Symbol	Z	Electron population
Scandium	Sc	21	2, 8, 8 + 3d^1, 4s^2
Yttrium	Y	39	2, 8, 18, 8 + 4d^1, 5s^2
Lanthanum	La	57	2, 8, 18, 18, 8 + 5d^1, 6s^2
−Lutetium	−Lu	−71	−32
Actinium	Ac	89	2, 8, 18, 32, 18, 8 + 6d^1, 7s^2
−Lawrencium	−Lr	−103	−32

The rare-earth elements correspond to electron filling in the $4f$ subshell. Since the $4f$ subshell is third outermost, changes in its electronic population are well screened from neighboring atoms by outer shells. Consequently, all the lanthanides have properties that are remarkably alike. They generally occur together and, because of the similarity in chemical properties, are difficult to separate. Separation can be accomplished by fractional crystallization or by ion-exchange techniques. Both separations rely on slight differences of properties (e.g., solubility, complex-ion formation, and hydration). In going through the sequence from La^{3+} to Lu^{3+}, the ionic radius gradually shrinks from 0.103 to 0.086 nm. This shrinkage, the *lanthanide contraction,* arises from increase of nuclear charge during progressive filling of an interior subshell.

The most successful technique for separating the lanthanide elements is ion exchange. An ion-exchange column is typically constructed from a glass tube about 1 m long and 1 cm in diameter which is packed with ion-exchange resin (Sec. 17.6). It is *fixed* with the ions to be separated and then *eluted* (i.e., washed out) with a complexing-agent solution. The fixing is accomplished by simply pouring a solution of the lanthanide salts onto the top of the column. The eluting solution, which might contain, for example, EDTA as a complexing agent, is slowly dribbled through so as to remove the ions. The rate at which M^{3+} ions come off the column depends on their relative tendency to complex with EDTA.

$$M^{3+} \text{ (on exchanger)} + EDTA^{4-} \rightleftharpoons M(EDTA)^{-}$$

Because the lanthanides are so similar to each other, there are only tiny differences in their tendency to form complexes. However, the geometry of having many tiny beads of resin in a long column ensures many successive adsorptions and dissolvings into the solution, and so small equilibrium differences are magnified. Figure 19.2 shows a typical column setup. Figure 19.3 gives a graphic record of the wash solution coming out of the column. As can be seen, the smaller ion, Lu^{3+}, washes out first, corresponding to tighter binding in the EDTA complex.

The lanthanide elements need to be protected from air since they react vigorously with moisture and oxygen. They are generally handled in closed inert-gas chambers, particularly when used for production of electronic and optical devices. They are of current interest for magnetic alloys and possibly as a way to store hydrogen in the solid state.

Because the lanthanide ions are generally characterized by an incomplete $4f$ subshell, paramagnetism due to unpaired electrons is expected. The magnetic moments *cannot* be calculated from the spin-only formula, which was found to work reasonably well for paramagnetism arising from d electrons. The reason for this difference is that, whereas d electrons are exposed to interaction with the ligands, f electrons are deep inside the ion and are shielded from the quenching effect of the environment on orbital motion.

The relatively good shielding of f electrons is important in determining the spectral characteristics of lanthanide ions. Instead of the broad absorption bands arising from d-electron transitions, the lanthanide ions have spectra consisting generally of many sharp lines. The sharpness is attributable to lack of interaction between the f levels and the environment.

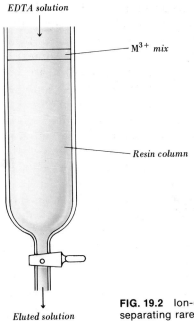

EDTA solution

M^{3+} *mix*

Resin column

Eluted solution

FIG. 19.2 Ion-exchange column for separating rare-earth elements.

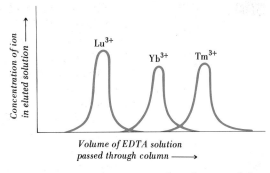

Concentration of ion in eluted solution ⟶

Lu^{3+}

Yb^{3+}

Tm^{3+}

Volume of EDTA solution passed through column ⟶

FIG. 19.3 Ion-exchange separation of rare-earth ions.

The 15 actinide elements correspond to electron filling involving the 5f subshell. Two important actinides are uranium and plutonium, used for nuclear energy. They are discussed further in Sec. 22.4.

19.2 TITANIUM

The second subgroup of transition elements is the titanium subgroup. It contains the elements titanium, zirconium, hafnium, and element 104. The most important member of the group is titanium which, in its elemental state, can be used as a structural material.

There was a flurry of excitement when the first moon-rock analyses indicated unusually high titanium content. However, this turned out to be a fluke of the sampling site and not representative of the entire moon. In the Earth's crust, titanium is quite abundant, but commercially useful deposits are scarce. The principal sources are rutile (TiO_2), ilmenite ($FeTiO_3$), and iron ores. The usual method for getting titanium is to convert the oxides with chlorine to $TiCl_4$, which is then reduced with magnesium:

$$TiO_2 \xrightarrow[C]{Cl_2} TiCl_4 \xrightarrow{Mg} \underset{99.3\%}{Ti} + MgCl_2$$

Pure titanium which has low carbon impurity and is free from hydrogen embrittlement is extremely strong (stronger than iron). Because it also has a high melting point and is resistant to corrosion, the metal is in great demand as a structural material, for example, in rocket engines and supersonic planes.

The Concorde was able to get away with a structural skin of aluminum (mp 660°C) by limiting itself to 2.2 times the velocity of sound, but it is believed that the second generation SSTs will almost inevitably have to use titanium (mp 1660°C). Until recently, the principal use of titanium has been for hardening and toughening steel.

In its compounds, titanium exhibits oxidation states of $+2$, $+3$, and $+4$. The most important state is $+4$; probably the most important compound is TiO_2, titanium dioxide, or titania. This compound is quite inert and has good covering power; it is used extensively as a pigment in both the paint industry and the cosmetic industry.

Zirconium and hafnium are remarkable because they have essentially identical chemical properties. Their atomic radii, 0.1454 and 0.1442 nm, respectively, are very close because the usual size increase down a group is canceled by the lanthanide contraction, which intervenes between these two elements. The resemblance of zirconium and hafnium is so marked that all naturally occurring zirconium minerals are contaminated with hafnium.

Zirconium and hafnium are both important for the production of nuclear energy. Zirconium has a particularly low probability of capturing neutrons, which, coupled with its high resistance to corrosion, makes it ideal for cladding uranium or plutonium fuel rods in nuclear reactors. Hafnium, on the other hand, has a very high neutron-capture probability; it is used to make control rods for regulating the free-neutron level in a nuclear reactor (Sec. 22.4).

In 1964 Soviet scientists reported a new element, kurchatovium, element 104, which was synthesized by bombardment of plutonium with neon nuclei:

$$^{242}_{94}Pu + {}^{22}_{10}Ne \longrightarrow {}^{260}_{104}Ku + 4{}^{1}_{0}n$$

As is traditionally true, they had the privilege of naming the new element, which they promptly did after Professor Kurchatov, Soviet physicist. However, the University of California (Berkeley) group, which was trying to make the same element, could not reproduce the Soviet work but came up with another isotope of the same element. They promptly christened it rutherfordium (Rf), after Ernest Rutherford, discoverer of the nucleus. At the present time, the name as well as the discovery remain in dispute.

19.3
VANADIUM

The elements of the vanadium subgroup are vanadium, niobium, tantalum, and hahnium ($Z = 105$). Hahnium is a recently discovered synthetic element made by heavy-ion bombardment; it is named after Otto Hahn, codiscoverer with Lise Meitner of the phenomenon of nuclear fission (Sec. 22.4).

The principal element of the group is vanadium. Its name comes from Vanadis, the Scandinavian goddess of beauty, and reflects the beautiful colors of various vanadium compounds. The pure metal is very hard to prepare, and since its main use is as an additive to steel, vanadium is usually made as ferrovanadium (solid solution of iron and vanadium). When added to steel, vanadium scavenges oxygen and nitrogen and thereby increases tensile strength, toughness, and elasticity.

In its compounds, vanadium shows oxidation states +2, +3, +4, and +5. Probably the most important compound is the pentoxide, V_2O_5. This is a red solid made by thermal decomposition of ammonium vanadate (NH_4VO_3). It is used as a catalyst in oxidation reactions in which O_2 is the oxidizing agent, e.g., conversion of SO_2 to SO_3 for making H_2SO_4.

Both niobium and tantalum are rather rare. The principal minerals are mixed oxides of the two metals along with those of iron and manganese. Although tantalum is rare, its desirable properties have led to rather extensive use. It is very ductile and preceded tungsten as filament material in light bulbs and electron tubes. It is also used as a rectifier for converting alternating current to direct current. This rectifier consists of an aqueous solution with two electrodes, one of which is Ta. When the Ta starts to act as an anode, it immediately forms an oxide coat, which cuts off the current. Tantalum can, however, easily act as a cathode and so permits flow of current in only one direction. Since Ta is very resistant to corrosion, it is used extensively for apparatus in chemical plants, especially equipment designed for handling acids. Since it is also compatible with human tissue, it finds use in surgery, as for bone pins.

19.4
CHROMIUM

The chromium subgroup contains the elements chromium, molybdenum, and tungsten. They are all metals of small atomic volume, extremely high melting point, great hardness, and excellent resistance to corrosion. Molybdenum is important as a trace element in the soil, where it plays an important role in nitrogen fixation by bacteria. Tungsten, also called *wolfram*, hence the symbol W, has a very high melting point and an extremely low vapor pressure.* The combination makes tungsten useful for lamp filaments. These filaments have an unusual property: when they wear out and get thin, they develop hot spots which apparently catalyze the redeposition of tungsten to heal the hot spots. Tungsten is also used in making high-speed steel for cutting tools. Addition of tungsten increases the ability of iron to hold hardness at high temperatures and slows down the tearing off of small particles that causes dulling of tools.

The most common element of the subgroup is chromium. Its principal mineral is chromite ($FeCr_2O_4$). Direct reduction by heating with C gives ferrochromium (solid solution of Cr in Fe) for addition to steel. High-chrome steel (up to 30% Cr), or *stainless steel*, is very resistant to corrosion. Most of the

---

*At room temperature, the vapor pressure of tungsten has been calculated to be one atom per universe!

510

Chapter 19
Transition
Elements:
Some Detailed
Chemistry

remaining chromite is converted to sodium chromate (Na_2CrO_4) by heating it with Na_2CO_3 in air:

$$8Na_2CO_3(s) + 4FeCr_2O_4(s) + 7O_2(g) \longrightarrow$$
$$2Fe_2O_3(s) + 8Na_2CrO_4(s) + 8CO_2(g)$$

The sodium chromate is leached out with acid to form $Na_2Cr_2O_7$, an important oxidizing agent.

Chromium metal in the massive form is quite resistant to corrosion. It takes a high polish, which lasts because of formation of an invisible, self-protective oxide coat. Consequently, chromium finds much use as a plating material. The plate is usually put on by electrolyzing the object in a bath made by dissolving $Na_2Cr_2O_7$ and H_2SO_4 in water.

Most compounds of chromium are colored, a fact which is reflected in the name "chromium," from the Greek word for color, *chroma*. The characteristic oxidation states are +2, +3, and +6. Representative species are shown in Fig. 19.4.

The chromous ion, Cr^{2+}, is a beautiful blue ion obtained by reducing either Cr^{3+} or $Cr_2O_7^{2-}$ with Zn metal. However, it is rapidly oxidized by air. When base is added, chromous hydroxide precipitates. It is not amphoteric. On exposure to air, light blue $Cr(OH)_2$ is oxidized by O_2 to give pale green $Cr(OH)_3$ (also written $Cr_2O_3 \cdot xH_2O$).

Many chromic salts, such as chromic nitrate, $Cr(NO_3)_3$, dissolve in water to give violet solutions. [The violet color is due to the hydrated chromic ion, $Cr(H_2O)_6^{3+}$. To show that it is really a hexaaquo ion, an ingenious method was developed, known as the *isotope-dilution method*. H_2O containing some oxygen-18 isotope is added to an ordinary solution containing Cr^{3+} and normal H_2O. The H_2O is then sampled at intervals and examined for its oxygen-18/oxygen-16 ratio. It is found by this method that six H_2O per Cr^{3+} do not participate in the instantaneous dilution of the added oxygen 18. The method works for Cr^{3+} because $Cr(H_2O)_6^{3+}$ is very slow in exchanging its bound H_2O for H_2O that comes from the solvent.]

Solutions of chromic salts can be kept indefinitely exposed to the air without oxidation or reduction. In general, they are slightly acid because of hydrolysis of the chromic ion. The hydrolysis reaction (Secs. 9.7 and 15.10) can be written in either of the following ways:

$$Cr^{3+} + H_2O \longrightarrow CrOH^{2+} + H^+ \qquad K_{hyd} = 1.5 \times 10^{-4}$$
$$Cr(H_2O)_6^{3+} \longrightarrow Cr(H_2O)_5OH^{2+} + H^+ \qquad K_{diss} = 1.5 \times 10^{-4}$$

When a base is gradually added to chromic solutions, a green, slimy precipitate, which can be formulated either as $Cr(OH)_3 \cdot xH_2O$ or

FIG. 19.4 Chromium Species in Various Oxidation States

	+2	+3	+6
Acidic solution	Cr^{2+}, chromous	Cr^{3+}, chromic	$Cr_2O_7^{2-}$, dichromate
Basic solution	$Cr(OH)_2(s)$	$Cr(OH)_4^-$, chromite	CrO_4^{2-}, chromate

$Cr_2O_3 \cdot xH_2O$, first forms but then disappears as excess OH^- is added. A deep green color characteristic of chromite ion, written as CrO_2^- or $Cr(OH)_4^-$, is produced. The precipitation and redissolving associated with this amphoteric behavior can be described as follows:

$$Cr^{3+} + 3OH^- \longrightarrow Cr(OH)_3(s)$$

$$Cr(OH)_3(s) + OH^- \longrightarrow Cr(OH)_4^-$$

When filtered off and heated, $Cr(OH)_3$ loses water to form Cr_2O_3, chromium sesquioxide. This is an inert, green powder, much used as the artist's pigment, chrome green.

Chromic ion forms a great number of complex ions in which Cr is surrounded by six other atoms arranged at the corners of an octahedron. Typical octahedral complexes are CrF_6^{3-}, $Cr(NH_3)_6^{3+}$, and $Cr(H_2O)_6^{3+}$. In potassium chrome alum, $KCr(SO_4)_2 \cdot 12H_2O$, the $Cr(H_2O)_6^{3+}$ complex occurs as a unit occupying some of the crystal lattice sites.

In the +6 oxidation state, chromium is known principally as the chromates and dichromates. The chromate ion, CrO_4^{2-}, can be made by oxidizing chromite ion, $Cr(OH)_4^-$, in basic solution with a moderately good oxidizing agent such as hydrogen peroxide. The reaction, written for basic solution, is

$$2Cr(OH)_4^- + 3HO_2^- \longrightarrow 2CrO_4^{2-} + 5H_2O + OH^-$$

Chromate ion is yellow and has a tetrahedral structure, as shown in Fig. 19.5. When chromate is acidified, the yellow color is replaced by orange, the result of formation of $Cr_2O_7^{2-}$, dichromate ion:

$$2CrO_4^{2-} + 2H^+ \rightleftharpoons Cr_2O_7^{2-} + H_2O$$

The change is reversed by adding base. The structure of the dichromate ion, as shown in Fig. 19.5, consists of two tetrahedra sharing an oxygen atom at a common corner. The dichromate ion is a very good oxidizing agent, especially in acid solution. The half-reaction

$$Cr_2O_7^{2-} + 14H^+ + 6e^- \longrightarrow 2Cr^{3+} + 7H_2O$$

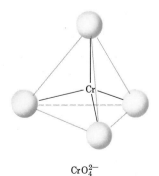

CrO_4^{2-}

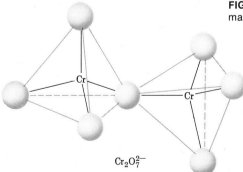

$Cr_2O_7^{2-}$

FIG. 19.5 Structures of chromate and dichromate.

has an electrode potential of $+1.33$ V; therefore, $Cr_2O_7^{2-}$ is among the strongest of the common oxidizing agents. It will, for example, oxidize hydrogen peroxide to form oxygen gas:

$$Cr_2O_7^{2-} + 3H_2O_2 + 8H^+ \longrightarrow 3O_2(g) + 2Cr^{3+} + 7H_2O$$

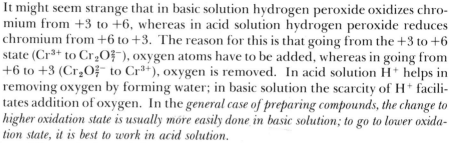

It might seem strange that in basic solution hydrogen peroxide oxidizes chromium from $+3$ to $+6$, whereas in acid solution hydrogen peroxide reduces chromium from $+6$ to $+3$. The reason for this is that going from the $+3$ to $+6$ state (Cr^{3+} to $Cr_2O_7^{2-}$), oxygen atoms have to be added, whereas in going from $+6$ to $+3$ ($Cr_2O_7^{2-}$ to Cr^{3+}), oxygen is removed. In acid solution H^+ helps in removing oxygen by forming water; in basic solution the scarcity of H^+ facilitates addition of oxygen. In the *general case of preparing compounds, the change to higher oxidation state is usually more easily done in basic solution; to go to lower oxidation state, it is best to work in acid solution.*

When solutions of dichromate are made very acid, especially in the presence of a dehydrating agent such as concentrated H_2SO_4, the uncharged species CrO_3 is formed. This deep-red solid is chromium trioxide, or, as it is sometimes called, chromic anhydride. It is a very powerful oxidizing agent and is used extensively in preparative organic chemistry. Suspensions of CrO_3 in concentrated H_2SO_4 are frequently used in laboratories as "cleaning solution" for degreasing glass equipment.

The presence of chromium in solution can often be detected by characteristic color changes. If chromium is present in an original unknown as chromate or dichromate, it will be reduced by H_2S in acid solution to Cr^{3+}:

$$8H^+ + Cr_2O_7^{2-} + 3H_2S \longrightarrow 2Cr^{3+} + 3S + 7H_2O$$

forming milky white, finely divided sulfur in the process. When the solution, still containing H_2S, is made basic, $Cr(OH)_3$ will precipitate. Subsequent addition of excess NaOH converts the chromium to soluble chromite ion. The appearance of a green color in the solution at this point is a strong indication of the presence of chromium. It can be confirmed by treating the green, basic solution with H_2O_2 to oxidize chromite to yellow chromate.

19.5
MANGANESE

The elements of the manganese subgroup are manganese, technetium, and rhenium. Manganese is by far the most important of the group; technetium is radioactive and does not occur in nature; rhenium is so rare as to constitute a chemical curiosity.

Manganese is not a very common element. The most important minerals are the oxides, such as MnO_2, pyrolusite. Since most metallic Mn goes into steel production, alloys of Mn with Fe are generally prepared. Two such alloys are ferromanganese (about 80% Mn) and spiegeleisen (about 30% Mn); they are made by reducing mixed oxides of Fe and Mn with C or CO as reducing agents. When added to steel, Mn acts as a scavenger in combining with O

FIG. 19.6 **Manganese Species in Various Oxidation States**

	+2	+3	+4	+6	+7
Acidic solution	Mn^{2+} manganous		$MnO_2(s)$		MnO_4^- permanganate
Basic solution	$Mn(OH)_2(s)$	$Mn(OH)_3(s)$	$MnO_2(s)$	MnO_4^{2-} manganate	MnO_4^- permanganate

and S to form easily removable substances. In high amounts (up to 14 percent), it imparts special hardness and toughness, such as is needed for resistance to battering abrasion.

In its chemical compounds manganese shows oxidation states +2, +3, +4, +6, and +7. Representative species are shown in Fig. 19.6. Most of these are colored and paramagnetic. Although Mn^{2+} solutions appear colorless, salts, such as manganous sulfate, $MnSO_4$, are pink. Mn^{2+} is a very poor reducing agent; neutral or acid solutions can be kept indefinitely. When base is added to Mn^{2+}, a white precipitate of $Mn(OH)_2$ is formed. This solid, unlike manganous salts, is promptly oxidized by air to the +3 state.

In the +3 state, manganese can exist briefly as the red manganic ion, Mn^{3+}. It is a very powerful oxidizing agent and can even oxidize water to liberate oxygen. The electrode potential of $Mn^{3+} + e^- \longrightarrow Mn^{2+}$ (+1.51 V) is high enough that manganic ion can oxidize itself to the +4 state:

$$Mn^{3+} + 2H_2O \longrightarrow MnO_2(s) + 4H^+ + e^-$$
$$\underline{Mn^{3+} + e^- \qquad \longrightarrow Mn^{2+}}$$
$$2Mn^{3+} + 2H_2O \longrightarrow Mn^{2+} + MnO_2(s) + 4H^+$$

This disproportionation can be prevented by (1) complexing the manganese, for example, with cyanide, CN^-, to give $Mn(CN)_6^{3-}$; (2) forming an insoluble salt, such as manganic hydroxide, variously written $Mn(OH)_3$, $MnOOH$, or even Mn_2O_3.

In the +4 state the principal compound of manganese is manganese dioxide, MnO_2. As mentioned in Sec. 11.8, it is the oxidizing agent in the flashlight cell.

When heated with basic substances in air, MnO_2 is oxidized from its original black color to deep green, the result of conversion to manganate ion, MnO_4^{2-}. Though stable in alkaline solution, this ion (which represents Mn in the +6 state) disproportionates when the solution is acidified:

$$3MnO_4^{2-} + 4H^+ \longrightarrow MnO_2(s) + 2MnO_4^- + 2H_2O$$

MnO_4^-, permanganate ion, is a very good oxidizing agent, especially in acid solution. The half-reaction

$$MnO_4^- + 8H^+ + 5e^- \longrightarrow Mn^{2+} + 4H_2O$$

has an electrode potential of +1.51 V, which means that MnO_4^- is one of the strongest common oxidizing agents. Permanganate solutions are frequently

514

Chapter 19
Transition
Elements:
Some Detailed
Chemistry

used in analytical chemistry to determine amounts of reducing agents. Disappearance of the deep violet color of MnO_4^- can be used as the end-point indicator. The usual procedure is to make the solutions acid, so as to ensure complete reduction to Mn^{2+}. In neutral or alkaline solutions, MnO_2 is formed instead; in very basic solutions, MnO_4^{2-}.

The chemistry of manganese well illustrates how the characteristics of compounds change in going from low oxidation state to high. In low oxidation states manganese exists as a cation (Mn^{2+}) which forms basic oxides and hydroxides. In higher oxidation states it exists as anions (MnO_4^{2-} and MnO_4^-) derived from acidic oxides.

Manganese in qualitative analysis of unknowns usually occurs as Mn^{2+} or MnO_4^-. If it occurs as permanganate, it will be reduced by H_2S in acid solution to Mn^{2+}:

$$2MnO_4^- + 5H_2S + 6H^+ \longrightarrow 2Mn^{2+} + 5S + 8H_2O$$

forming finely divided sulfur in the process. When the solution, still containing H_2S, is made basic, MnS will precipitate. Treatment with acid and an oxidizing agent (to separate Mn from Cr), followed by addition of NaOH, precipitates red manganic hydroxide, $Mn(OH)_3$. Very strong oxidizing agents such as sodium bismuthate, $NaBiO_3$, convert $Mn(OH)_3$ to the characteristic violet color of MnO_4^-.

19.6
IRON

Iron is the first of three closely related elements, Fe, Co, and Ni, occurring in horizontal sequence in the middle of the first transition period. These elements, the so-called "iron triad," have electron populations as shown in Fig. 19.7. Removal of the $4s$ electrons is relatively easy; hence a +2 state is formed by all. In Fe, additional removal of a $3d$ electron happens easily because a half-filled $3d$ level is left; in Co and Ni it does not happen so readily. The +3 state of Co must be stabilized as by formation of a complex ion; the +3 state of Ni is very rare, and compounds of Ni^{3+} are powerful oxidizing agents.

The properties of the iron-triad elements are very similar, as is shown in Fig. 19.8. Melting and boiling points are uniformly high; ionization potentials are nearly the same; the electrode potentials are moderately more negative

FIG. 19.7   Electron Population of Iron-Triad Elements

Sublevel	Iron	Cobalt	Nickel
$1s$	2	2	2
$2s$ and $2p$	8	8	8
$3s$ and $3p$	8	8	8
$3d$	⇅ ↑ ↑ ↑ ↑	⇅ ⇅ ↑ ↑ ↑	⇅ ⇅ ⇅ ↑ ↑
$4s$	⇅	⇅	⇅

FIG. 19.8   Some Properties of Elements of the Iron Triad

Property	Fe	Co	Ni
Melting point, °C	1535	1490	1450
Boiling point, °C	2700	2900	2700
Ionization potential, eV	7.90	7.86	7.63
Electrode potential (from M^{2+}), V	−0.44	−0.28	−0.25
Density, g/cm³	7.9	8.7	8.9

than the electrode potential of hydrogen. In addition, these elements are alike in that all are *ferromagnetic;* i.e., they are strongly attracted into a magnetic field and show permanent magnetization when removed from such a field. That these elements are magnetic is not surprising. The electronic configurations in Fig. 19.7 would lead us to expect that there would be unpaired electrons. However, it is surprising that the magnetization is so large and so persistent. The explanation is that in these metals there are regions or domains of magnetization, where all the individual magnetic dipoles line up in the same direction. Of all the elements at room temperature, only iron, cobalt, and nickel satisfy the conditions necessary for domain formation. These conditions are that the ions contain unpaired electrons, that the ions be the right size, and that the distance between ions be exactly right. Manganese metal has most of the properties needed to be ferromagnetic. However, its ions are too close. Addition of Cu to Mn increases the average spacing, and so the resulting alloy becomes ferromagnetic.

The element Fe has an industrial importance which exceeds that of any other element. It is very abundant, ranking fourth in the Earth's crust (after O, Si, and Al); it is very common, being an essential constituent of several hundred minerals; it is easy to make by simply heating some of its minerals with C; it has many desirable properties, especially when impure. For all these reasons, Fe has become such a distinctive feature of civilization that it marks the Iron Age of archaeological chronology.

Most of the iron in the Earth's crust is combined with oxygen, silicon, or sulfur. The important minerals are hematite (Fe_2O_3), magnetite (Fe_3O_4), and siderite ($FeCO_3$), usually contaminated with complex iron silicates from which these minerals have been produced by weathering. Iron sulfides, such as iron pyrites (FeS_2), also called *fool's gold,* are also quite abundant. They are not used as sources of iron because sulfur is an objectionable impurity in the final product.

Iron is practically never produced in the pure state, since it is difficult to make and too expensive for most purposes. Furthermore, impure iron, i.e., steel, has desirable properties, especially when the specific impurity is carbon in carefully controlled amounts. The industrial production of steel is carried out on a massive scale in the well-known blast furnace (Fig. 19.9). Iron ore, limestone, and carbon as coke are added at the top, and preheated air or oxygen is blown in at the bottom. As the charge settles through the furnace, molten iron forms and trickles down to a pit at the bottom, from which it is peri-

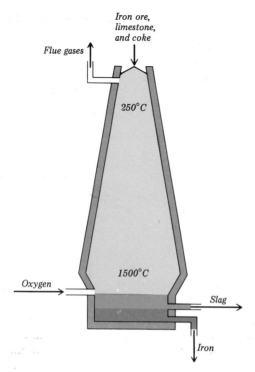

FIG. 19.9 Blast furnace.

odically drawn off. All told, it takes about 12 h for material to pass through the furnace.

The chemical processes that occur in the blast furnace are still obscure. It is generally agreed, however, that the active reducing agent is not carbon, but carbon monoxide. As the carbon moves down through the furnace, it is oxidized by incoming oxygen in the reaction

$$2C(s) + O_2(g) \longrightarrow 2CO(g) \qquad \Delta H^\circ = -220 \text{ kJ}$$

thus forming the reducing agent carbon monoxide and liberating large amounts of heat. The CO gas moves up the furnace and encounters oxides of iron in various stages of reduction, depending on the temperature of the particular zone. At the top of the furnace, where the temperature is lowest (250°C), the iron ore (mostly Fe_2O_3) is reduced to Fe_3O_4:

$$3Fe_2O_3(s) + CO(g) \longrightarrow 2Fe_3O_4(s) + CO_2(g)$$

As the Fe_3O_4 settles further, it gets reduced to FeO:

$$Fe_3O_4(s) + CO(g) \longrightarrow 3FeO(s) + CO_2(g)$$

Finally, toward the bottom of the furnace, FeO is eventually reduced to iron:

$$FeO(s) + CO(g) \longrightarrow Fe(s) + CO_2(g)$$

Since the temperature at the lowest part of the furnace (1500°C) is above the melting point of impure iron, the solid melts and drips down into the hearth at the very bottom. The net equation for the overall reduction is

$$Fe_2O_3(s) + 3CO(g) \longrightarrow 2Fe(l) + 3CO_2(g)$$

In addition, there is combination of carbon dioxide with hot carbon:

$$C(s) + CO_2(g) \longrightarrow 2CO(g)$$

and thermal decomposition of limestone:

$$CaCO_3(s) \longrightarrow CaO(s) + CO_2(g)$$

Both of these reactions are helpful: The former raises the concentration of CO, and the latter removes silica-containing contaminants present in the original ore. Lime (CaO), being a basic oxide, reacts with the acidic oxide SiO_2 to form calcium silicate ($CaSiO_3$). In the form of a lavalike *slag*, calcium silicate collects at the bottom of the furnace, where it floats on the molten iron and protects it from oxidation.

Pig iron, the crude product of the blast furnace, contains about 4% carbon, 2% silicon, a trace of sulfur, and up to 1% of phosphorus and manganese. Sulfur is probably the worst impurity since it makes steel break when worked; it must be avoided, as it is hard to remove in refining operations. When pig iron is remelted with scrap iron and cast into molds, it forms *cast iron*. This can be either *gray* or *white*, depending on the speed of cooling. When cooled slowly (as in sand molds), the carbon impurity separates out as tiny flakes of graphite, giving gray cast iron which is relatively soft and tough. When cooled rapidly (as in water-cooled molds), the carbon remains combined in the form of a compound iron carbide, Fe_3C, also called *cementite*. White cast iron is as much as 75% cementite and is extremely hard and brittle.

Most pig iron is refined into steel by burning out the impurities to leave small controlled amounts of carbon. In the *open-hearth* process, some carbon is removed by oxidation with air and iron oxide, the latter being added as hematite and rusted scrap iron. The process is usually carried out on a shallow hearth which is so arranged that a hot-air blast can play over the surface.

The *Bessemer process* is much more rapid (10 to 15 min) but gives a less uniform product. In this process, molten pig iron taken directly from the blast furnace is poured into a large pot having blowholes at the bottom, and a blast of air or oxygen is swept through the liquid to burn off most of the carbon and silicon. Frequently, the Bessemer and open-hearth processes are combined to take advantage of the good points of each: A preliminary blowing in the Bessemer converter gets rid of most of the carbon and silicon; a following burn-off in an open-hearth furnace gets rid of phosphorus.

The properties of Fe in the form of steel are very much dependent on the percentage of impurities present, on the heat treatment of the specimen, and even on the working to which the sample has been subjected. For these reasons, the following comments about Fe properties do not necessarily apply to every given sample. Compared with most metals, Fe is a fairly good reducing agent. With nonoxidizing acids it reacts to liberate H_2 by the reaction $Fe(s) + 2H^+ \longrightarrow Fe^{2+} + H_2(g)$. It also has the ability to replace less-active

518

Chapter 19
Transition
Elements:
Some Detailed
Chemistry

metals in their solutions. For example, a bar of Fe placed in a solution of $CuSO_4$ is immediately covered with a reddish deposit of Cu formed by the reaction $Fe(s) + Cu^{2+} \longrightarrow Fe^{2+} + Cu(s)$. In concentrated HNO_3, Fe becomes passive; i.e., it loses the ability to react with H^+ and Cu^{2+} and appears to be inert. It is believed that passivity is due to formation of a submicroscopic surface coating of oxide, which slows down rate of reaction. When the film is broken, reactivity is restored.

19.7
COMPOUNDS OF IRON

The two common oxidation states of iron are +2 (ferrous) and +3 (ferric). Compounds in which the oxidation state is fractional, as in Fe_3O_4, can be thought of as mixtures of the two oxidation states.

In the +2 state, iron exists as ferrous ion, Fe^{2+}. This in water is a pale green, almost colorless, ion. Except in acid solutions, it is rather hard to keep since it is easily oxidized to the +3 state by oxygen in the air. However, the rate of oxidation by oxygen is inversely proportional to H^+ concentration, and so acid solutions of ferrous salts can be kept for long periods. When base is added, a nearly white precipitate of ferrous hydroxide, $Fe(OH)_2$, is formed. On exposure to air, it turns brown, owing to oxidation to hydrated ferric oxide, $Fe_2O_3 \cdot xH_2O$. For convenience, the latter is often designated $Fe(OH)_3$, ferric hydroxide. The oxidation can be written as

$$4Fe(OH)_2(s) + O_2(g) + 2H_2O \longrightarrow 4Fe(OH)_3(s)$$

In the +3 state, iron exists as the colorless ferric ion, Fe^{3+}. Aqueous solutions are generally acid, indicating that appreciable hydrolysis must take place. This can be written as

$$Fe^{3+} + H_2O \rightleftharpoons FeOH^{2+} + H^+ \qquad K_{hyd} = 6.8 \times 10^{-3}$$

Apparently, the yellow-brown color so characteristic of ferric solutions is mainly due to $FeOH^{2+}$. By addition of an acid such as HNO_3, the color can be made to disappear. (However, it will not disappear on addition of HCl because yellow $FeCl^{2+}$ forms instead.) On addition of base, a slimy, red-brown, gelatinous precipitate forms; it may be written $Fe(OH)_3$. This can be dehydrated to form red or yellow Fe_2O_3.

In both the +2 and +3 states, iron shows a great tendency to form complex ions. For example, ferric ion combines with thiocyanate ion, NCS^-, to form $FeNCS^{2+}$. This complex has such a deep red color that it can be detected at concentrations as low as 10^{-5} M. The formation of this complex is the basis of one of the most sensitive tests for the presence of Fe^{3+}.

In the usual scheme of qualitative analysis, iron precipitates as black FeS, insoluble in basic solution. If ferric ion is present in the original unknown, it is reduced by H_2S in acid solution to ferrous ion:

$$2Fe^{3+} + H_2S \longrightarrow 2Fe^{2+} + 2H^+ + S(s)$$

FeS can be separated from CoS and NiS because it dissolves fairly quickly in Na_2SO_4–$NaHSO_4$ buffer, whereas CoS and NiS are slow to dissolve. Separation of iron from cobalt and nickel can also be achieved by using the fact that Fe^{2+} plus excess NH_3, when exposed to air, form insoluble ferric hydroxide, whereas Co^{2+} and Ni^{2+} form soluble ammonia complexes. The presence of iron can be confirmed by adding thiocyanate, to get the deep red color of $FeNCS^{2+}$.

19.8
CORROSION OF IRON

Corrosion is a general term applied to the process in which uncombined metals change over to compounds. In the special case of iron the corrosion process is called *rusting*. Economically, rusting is a serious problem. It has been estimated that one-seventh of the annual production of iron goes simply to replace that lost by rusting. Still, despite much study, corrosion is a mysterious process, and its chemistry is not well understood.

Rust appears to be hydrated ferric oxide, corresponding approximately to $2Fe_2O_3 \cdot 3H_2O$. However, since the water content is not fixed, it is preferable to write $Fe_2O_3 \cdot xH_2O$. Rust will not form in dry air or in water that is completely free of air, and so it appears that both oxygen and water are required for rust formation. Furthermore, it is observed that rusting is generally speeded up by acids, strains in the iron, contact with less-active metals, and presence of rust itself (autocatalysis).

In order to account for the observed facts, the following steps have been proposed as the mechanism by which rusting occurs:

$$Fe(s) \longrightarrow Fe^{2+} + 2e^- \tag{1}$$

$$e^- + H^+ \longrightarrow H \tag{2}$$

$$4H + O_2(g) \longrightarrow 2H_2O \tag{3}$$

$$4Fe^{2+} + O_2(g) + (4 + 2x)H_2O \longrightarrow 2(Fe_2O_3 \cdot xH_2O)(s) + 8H^+ \tag{4}$$

In step (1) ferrous ions are produced by loss of electrons from neutral Fe. This process cannot go very far unless there is some way to get rid of the electrons which accumulate on the residual Fe. One way to do this is by step (2), in which H^+ ions, either from water or from acid substances in the water, pick up electrons to form neutral H atoms. Fe is a good catalyst for hydrogenation reactions, and so it is believed that step (3) now occurs to use up the H atoms. In the meantime, the ferrous ion from step (1) reacts with O_2 gas by step (4) to form the rust and restore H^+. The net reaction, obtained by adding all four steps, is

$$4Fe(s) + 3O_2(g) + 2xH_2O \longrightarrow 2(Fe_2O_3 \cdot xH_2O)(s)$$

Since H^+ accelerates step (2) and is replenished in step (4), it is a true catalyst for the reaction and explains the observation that acids speed up the rate of rust formation.

520

Chapter 19
Transition
Elements:
Some Detailed
Chemistry

The above mechanism also accounts for a remarkable observation called *electrolytic corrosion*. When iron pipes are directly connected to copper pipes, the iron is observed to corrode much faster than normally. The explanation lies in step (1). Electrons from dissolution of the iron flow to the copper, where their energy is lower. This relieves excess negative charge on the iron and allows more Fe^{2+} to leave the metal. Electrolytic corrosion used to be very common, for example, when new copper plumbing was hooked up directly to an existing system composed of iron pipes. It can now be prevented by simply inserting a junction that does not conduct electric current very well.

One of the strongest supports for the above stepwise rusting mechanism comes from the observation that the most serious pitting of a rusting iron object often occurs where the oxygen supply is limited. If oxygen is abundant, step (4) promptly occurs to deposit rust on the iron before Fe^{2+} can move very far away. This makes it difficult for more iron to dissolve, and the reaction is self-stopping. However, if the oxygen supply is restricted, especially in an aqueous environment, Fe^{2+} may have a chance to diffuse quite some distance before encountering enough oxygen to form rust. This means that the rust may deposit at some point away from where pitting occurs. Common examples of this are observed around ill-fitted rivet heads. As shown in Fig. 19.10a, the rivet shank, although protected from air, is eaten away, but the rust forms where the rivet head overlaps the plate. Seeping moisture allows Fe^{2+} to diffuse out to the surface, where it can react with oxygen. Another example occurs with iron posts that are set in water. Pitting usually starts where there are strains in the metal, but the rust may form, as shown in Fig. 19.10b,

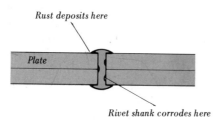

Rust deposits here

Plate

Rivet shank corrodes here

(a)

FIG. 19.10 Iron corrosion as enhanced by restricted oxygen supply.

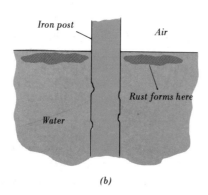

Iron post　　　*Air*

Rust forms here

Water

(b)

near the air-water line, where the dissolved oxygen is most plentiful. This makes the situation go from bad to worse since the waterline rust now acts as a curtain to keep oxygen away from the corroding iron. Self-stopping is no longer possible, and severe pitting can now occur where the oxygen supply is restricted. The common practice of coating car bottoms to prevent rusting can actually do more harm than good if the covering job is not done properly.

Although there are still many unanswered questions about rusting, it is clear what must be done to prevent it. The most direct approach is to shut off the reactants oxygen and water. This can be done by smearing grease over the iron to be protected, painting it with an ordinary paint or, better, with an oxidizing paint that makes the iron passive, or plating the iron with some other metal. Painting or greasing is probably cheapest, but it must be done thoroughly; otherwise rusting may only be accelerated by partial exclusion of oxygen. Plating with another metal is more common when appearance is a factor. Chrome plating, for example, is usually chosen because of its dressy look. Zinc plating, or galvanizing, though it does not look as good, is actually more permanent. Tin plating looks good and is relatively cheap, but it may not last.

The relative merits of metals used for plating depend on the activity of the metal relative to Fe and the ability of the metal to form a self-protective coat. Zinc, for example, reacts with O_2 and CO_2 in air to form an adherent coating of hydroxycarbonate, $Zn_2(OH)_2CO_3$, which prevents further corrosion. Also, Zn has a bigger electrode potential than Fe; if Zn plating is damaged so that both Zn and Fe are exposed, it is the Zn that is preferentially oxidized to form a compound that plugs the hole. Tin also forms a self-protective coat, but Sn has a smaller electrode potential than Fe; if a tin coating is punctured, it is the underlying Fe that is preferentially oxidized.

One of the most elegant ways to protect iron from corrosion is by *cathodic protection*. In this method, iron is charged to a negative voltage compared with its surroundings. This tends to make the iron act as a cathode instead of an anode (required for oxidation) and effectively stops corrosion. Actually, zinc plating is a method of cathodic protection, since zinc has a larger electrode potential than iron and forces electrons onto the iron. In practice, pipelines and standpipes can be protected by driving zinc or magnesium stakes into the ground and connecting them by wires to the object to be protected. In salt water, where rusting is unusually severe, steel ships can be protected by strapping blocks of magnesium to the hulls. These preferentially corrode (since they are acting as anodes), but they can easily be replaced, while the iron is essentially untouched.

19.9
COBALT AND NICKEL

The other elements of the iron triad, cobalt and nickel, are much less abundant than iron and are harder to extract from their minerals. The name "cobalt" reflects this difficulty, since it comes from the German word *Kobold*, meaning "goblin." Cobalt minerals look very much like copper minerals and were occasionally worked by mistake as sources of copper. Furthermore, arsenic often occurs with cobalt, and so poisonous fumes may be present, all obviously due to black magic. Similar troubles with nickel minerals led to their being named after *Nickel,* a mischievous underground spirit in German superstition.

Cobalt

The important minerals are sulfides and arsenides. Extraction of cobalt is very complex. It involves roasting in a blast furnace, dissolving with sulfuric acid, and precipitating by addition of sodium carbonate. The hydroxide so produced is dehydrated to the oxide, which can then be reduced with hydrogen.

The ferromagnetism of Co is very strong, higher than that of Fe, and accounts for its extensive use in magnets, especially in alloys such as the *alnico* alloys (Co, Ni, Al, and Cu). Other alloys such as stellite (55% Co, 15% W, 25% Cr, and 5% Mo) are important for their extreme hardness and resistance to corrosion. They are used, for example, in high-speed tools and surgical instruments.

In its compounds cobalt shows oxidation states of +2 (cobaltous) and +3 (cobaltic). Most cobaltous solutions are pink, due to the hydrated ion $Co(H_2O)_6^{2+}$. Addition of base precipitates dark blue insoluble hydroxide, $Co(OH)_2$. In the absence of oxygen, $Co(OH)_2$ can be dehydrated to give CoO, much used to produce blue color in pottery and enamel. Cobalt(II) forms many complex ions, which, however, are easily oxidized. An exception is the characteristic bright blue complex $CoCl_4^{2-}$, which can be used as a test for the presence of cobalt. The complex is stabilized in the presence of excess chloride ion.

In aqueous solution the cobaltic ion, Co^{3+}, is a very powerful oxidizing agent. The potential for $Co^{3+} + e^- \longrightarrow Co^{2+}$ is +1.84 V, which means that Co^{3+} is strong enough to oxidize water to oxygen. Only a few simple cobaltic salts such as CoF_3 and $Co_2(SO_4)_3 \cdot 18H_2O$ have been made, and these decompose in aqueous solution:

$$4Co^{3+} + 2H_2O \longrightarrow 4Co^{2+} + O_2(g) + 4H^+$$

Unlike the simple ion, the complex ions of Co(III) are quite stable to reduction. There are a tremendous number of these, ranging from simple octahedral $Co(CN)_6^{3-}$ and $Co(NH_3)_6^{3+}$ to complicated *polynuclear* complexes, in which several cobalt atoms are bridged together. Vitamin B_{12} is a complicated organic complex in which cobalt occurs in a distorted octahedral environment.

Nickel

The principal minerals of nickel are mixed sulfides of iron and nickel. Most ores are very poor in nickel content, and so they have to be concentrated before smelting, usually by *flotation*. In this process ore is ground up and agitated briskly with water to which oil and wetting agents have been added. Earthy particles (*gangue*) are wet by the water and hence sink, whereas the fine particles of minerals get carried off with the froth. The concentrate is then roasted in air to get rid of some of the sulfur as SO_2, burned in a furnace (smelted) to form oxide, and finally reduced with carbon. To get pure nickel, the final product is refined, either electrolytically or by forming $Ni(CO)_4$. In the Mond process, carbon monoxide is passed over impure nickel at 80°C to form volatile $Ni(CO)_4$, nickel tetracarbonyl. This is distilled, purified, and then heated to about 200°C, where it decomposes into nickel and carbon monoxide.

The properties of nickel metal are much like those of cobalt. More than 65 percent of nickel production goes into iron alloys to increase their strength and corrosion resistance. The rest goes into Ni–Cu alloys, e.g., nickel coinage, or is used as the pure metal. In the latter case it is used for plating steel and as a catalyst for hydrogenation reactions.

The chemistry of nickel compounds is mainly that of the +2 state. In aqueous solution it exists as green nickelous ion, Ni^{2+}, which, on treatment with base, precipitates light green nickelous hydroxide, $Ni(OH)_2$. On being heated, $Ni(OH)_2$ loses water to form black NiO. The complex ions of nickel(II) are almost as numerous as those of cobalt. However, unlike cobaltous complexes, they are quite stable to air oxidation. Some are octahedral, for example, blue $Ni(NH_3)_6^{2+}$; others are planar, for example, yellow $Ni(CN)_4^{2-}$; still others are tetrahedral, for example, blue-green $NiCl_4^{2-}$.

In basic solution, nickelous hydroxide can be oxidized by the powerful oxidizing agent hypochlorite, ClO^-. The product is a dark-colored oxide of indefinite composition variously described as NiO_2, Ni_2O_3, or Ni_3O_4. Whatever it is, it is a very good oxidizing agent and forms the cathode material in the Edison storage battery (Fig. 19.11). On discharge, the cathode reaction can be written as

$$Ni_2O_3(s) + 2e^- + 3H_2O \longrightarrow 2Ni(OH)_2(s) + 2OH^-$$

$Ni(OH)_2$ sticks to the cathode, and so the cell can be recharged by application of an external voltage. The anode reaction can be written as

$$Fe(s) + 2OH^- \longrightarrow Fe(OH)_2(s) + 2e^-$$

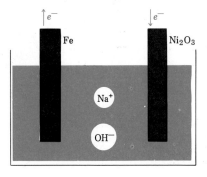

FIG. 19.11 Cell of Edison storage battery.

Although it has lower voltage than the lead storage cell, the Edison cell has an advantage because the OH^- produced at the cathode is used up at the anode; hence, there is no concentration change or change in voltage output as the cell runs down.

Qualitative analysis by the usual scheme puts Fe, Co, and Ni together as black sulfides, insoluble in basic solution. Treatment of FeS, CoS, and NiS with Na_2SO_4–$NaHSO_4$ buffer dissolves the FeS and leaves behind CoS and NiS. To distinguish cobalt from nickel, the sulfides can be dissolved in acid solution, boiled with bromine water to destroy H_2S, and treated with potassium nitrite. The appearance of insoluble yellow $K_3Co(NO_2)_6$ shows the presence of cobalt. Nickel can be identified by adding a special reagent, dimethylglyoxime. From basic solution this reagent precipitates reddish orange nickel dimethylglyoxime. The latter is a square-planar complex of the following structure:

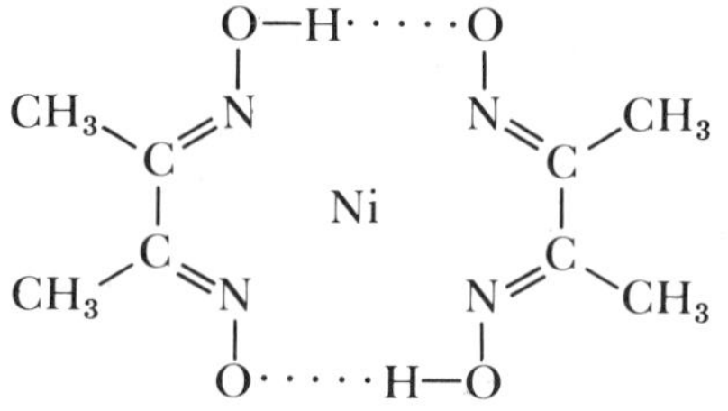

$$
\begin{array}{c}
O\!-\!H\cdots\cdots O \\
\vert \qquad\quad \vert \\
CH_3\!\diagdown_{C}\!\!=\!\!N \qquad N\!\!=\!\!_{C}\!\diagup CH_3 \\
\vert \qquad Ni \qquad \vert \\
CH_3\!\diagup^{C}\!\!=\!\!N \qquad N\!\!=\!\!^{C}\!\diagdown CH_3 \\
\vert \qquad\quad \vert \\
O\cdots\cdots H\!-\!O
\end{array}
$$

19.10
COPPER

Copper is the head element of a famous subgroup called *coinage metals.* Included in the subgroup are copper, silver, and gold. All have been known since antiquity because, unlike preceding elements, they are sometimes found in nature in the uncombined, or *native,* state. Besides native copper, which is often 99.9 percent pure, the element occurs as two principal classes of minerals: sulfide ores (accounting for about 90 percent of the occurrence) and oxide ores. The principal sulfide ores are chalcocite (Cu_2S) and chalcopyrite ($CuFeS_2$), also called copper pyrites. Principal oxide ores are cuprite (Cu_2O) and malachite [$CuCO_3 \cdot Cu(OH)_2$].

In order to make copper metal, the sulfide minerals are first concentrated by flotation, roasted in air, and then smelted. The overall equation can be written as

$$2CuFeS_2(s) + 5O_2(g) \longrightarrow 2Cu(s) + 2FeO(s) + 4SO_2(g)$$

It produces tremendous quantities of sulfur dioxide, which can be converted on the spot to sulfuric acid.

Raw copper product is about 97 to 99 percent pure and must be refined for most uses. This can be done best in a $CuSO_4$ electrolysis cell, such as that sketched in Fig. 19.12. The impure copper is made an anode, and a sheet of pure copper acts as cathode. By careful control of the voltage, only the cop-

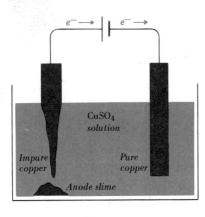

FIG. 19.12 Electrorefining of copper.

per is transferred from the anode to the cathode. Suppose, for example, we have iron and silver as typical impurities. The tendency to be oxidized decreases as follows:

$$Fe(s) \longrightarrow Fe^{2+} + 2e^- \qquad E° = +0.44 \text{ V}$$
$$Cu(s) \longrightarrow Cu^{2+} + 2e^- \qquad E° = -0.34 \text{ V}$$
$$Ag(s) \longrightarrow Ag^+ + e^- \qquad E° = -0.80 \text{ V}$$

By keeping the cell voltage at an appropriate value, only the iron and copper go into solution as Fe^{2+} and Cu^{2+} ions. Silver, oxidizing with more difficulty, simply drops off as the anode dissolves away. At the cathode, where reduction must occur, high concentration of Cu^{2+} combines with the fact that Cu^{2+} is more readily reduced than Fe^{2+} to bring about deposition of pure copper. Common impurities in crude copper are iron, nickel, arsenic, antimony, and bismuth (all of which, like iron, are oxidized and remain in the refining solution), and silver, gold, and traces of platinum (all of which, like silver, are not oxidized but collect at the bottom of the cell). With efficient operation, recovery of valuable metals from the anode residue (so-called "anode slime") pays for the whole refinery operation, leaving the copper as profit.

Metallic copper is malleable, ductile, and, especially when very pure, an excellent conductor of heat and electricity. Except for silver, it has the lowest electric resistance of any metal. It is used extensively in wires and switches.

Chemically, copper is a poorer reducing agent than hydrogen and does not dissolve in acids unless they contain oxidizing anions (e.g., nitric acid). When exposed to air, it slowly tarnishes, forming a green hydroxycarbonate. (The green patina often observed on bronze statues is copper hydroxycarbonate.) Copper is an important constituent of thousands of alloys such as *brass* (copper plus zinc), *bronze* (copper plus tin), and *Monel metal* (copper, nickel, iron, and manganese).

The compounds of copper correspond to oxidation states +1 (cuprous) and +2 (cupric). The +2 state is the one commonly observed in most situations, especially in an aqueous environment. The simple cuprous ion, Cu^+,

526

Chapter 19
Transition
Elements:
Some Detailed
Chemistry

cannot exist in aqueous solution since it oxidizes and reduces itself by the reaction

$$2Cu^+ \longrightarrow Cu^{2+} + Cu(s)$$

However, the cuprous condition can be stabilized by formation of complex ions, such as $CuCl_2^-$, or insoluble compounds, such as cuprous oxide, Cu_2O. The reddish color observed on metallic copper that has been heated in air is apparently due to a surface coating of Cu_2O. In a classic test for reducing sugars (e.g., glucose, but not sucrose), Cu_2O is formed when the reducing sugar is heated with an alkaline solution of a cupric salt.

Although many anhydrous cupric salts are white, hydrated salts are blue, owing to presence of hydrated cupric ion. This may be written $Cu(H_2O)_6^{2+}$, but two of the H_2O molecules are farther away than the other four. In general, aqueous solutions of cupric salts are acidic because of hydrolysis:

$$Cu^{2+} + H_2O \rightleftharpoons CuOH^+ + H^+$$

However, the hydrolysis is not very extensive ($K = 4.6 \times 10^{-8}$). When base is added to these solutions, light blue cupric hydroxide, $Cu(OH)_2$, is formed. When treated with aqueous ammonia, $Cu(OH)_2$ dissolves to give a deep blue solution. The color is usually attributed to a copper-ammonia complex ion, $Cu(NH_3)_4^{2+}$:

$$Cu(OH)_2(s) + 4NH_3 \longrightarrow Cu(NH_3)_4^{2+} + 2OH^-$$

Like most ammonia complexes, $Cu(NH_3)_4^{2+}$ can be destroyed by heat or by addition of acid. Heat is effective because it boils the NH_3 out of the solution:

$$Cu(NH_3)_4^{2+} \longrightarrow Cu^{2+} + 4NH_3(g)$$

Addition of acids results in neutralization of the NH_3:

$$Cu(NH_3)_4^{2+} + 4H^+ \longrightarrow Cu^{2+} + 4NH_4^+$$

It is interesting to note that addition *of an acid* to a solution of $Cu(NH_3)_4^{2+}$ precipitates a *hydroxide*. As acid is added and NH_3 is neutralized, the concentration of Cu^{2+} rises until the K_{sp} of $Cu(OH)_2$ is exceeded.

One of the least soluble of cupric compounds is black cupric sulfide, CuS. Its K_{sp} is very low (8×10^{-37}), which indicates that not even very concentrated H^+ can dissolve appreciable amounts of it. It is possible, however, to dissolve CuS by heating it with nitric acid. Dissolving occurs, not because H^+ reacts with S^{2-} to form H_2S, but because hot nitrate ion (especially in acid solution) is a very good oxidizing agent and oxidizes sulfide ion to elementary sulfur. The net reaction can be written as

$$3CuS(s) + 2NO_3^- + 8H^+ \longrightarrow 3Cu^{2+} + 3S(s) + 2NO(g) + 4H_2O$$

Probably the best-known cupric compound is copper sulfate pentahydrate, $CuSO_4 \cdot 5H_2O$, also known as *blue vitriol*. It is used extensively as a germicide and fungicide since cupric ion is toxic to lower organisms. Application to water supplies for controlling algae and its use on grapevines to control molds depend on this toxicity.

Qualitative analysis for copper ion is usually carried out by precipitating black CuS with H_2S in acid solution. The CuS can be separated from another black compound, HgS, which precipitates at the same time by treating both with hot HNO_3. The CuS dissolves, but the HgS does not. Cu^{2+} can be confirmed by adding NH_3 to form the deep blue complex $Cu(NH_3)_4^{2+}$.

19.11
SILVER

Silver is a rather rare element which occurs principally as native silver, argentite (Ag_2S), and horn silver ($AgCl$). Relatively little comes from silver ores; most comes as a by-product of copper and lead production. The main problem in extracting silver from its ores is to get the rather inert silver (or very insoluble silver compounds) to go into solution. This can be accomplished by blowing air through a suspension of ore in dilute aqueous sodium cyanide ($NaCN$). With native silver, the reaction can be written as

$$4Ag(s) + 8CN^- + 2H_2O + O_2(g) \longrightarrow 4Ag(CN)_2^- + 4OH^-$$

Were it not for the presence of cyanide ion, the oxygen would not be able to oxidize elemental silver to a higher oxidation state. In the presence of cyanide ion, Ag^+ forms a strongly associated complex ion and is thus stabilized. To recover the silver from the residual solution, it is necessary to use a rather strong reducing agent, such as aluminum or zinc in basic solution. A typical reaction would be

$$Zn(s) + 2Ag(CN)_2^- + 4OH^- \longrightarrow 2Ag(s) + 4CN^- + Zn(OH)_4^{2-}$$

Massive silver appears almost white because of its high luster. It is too soft to be used pure in jewelry and coinage and is usually alloyed with copper for these purposes. Because of expense it cannot be used much for its best property, electric and thermal conductivity, which are second to none. In the colloidal state, silver usually appears black.

The compounds of silver are mainly in the +1 state. The Ag^+ ion is sometimes called *argentous ion,* after the Latin word for silver, *argentum.* Ag^+ does not hydrolyze appreciably in aqueous solution: it is a relatively good oxidizing agent. It also forms many complex ions—for example, $Ag(NH_3)_2^+$ and $Ag(CN)_2^-$—which are linear. When base is added to solutions of silver salts, a brown oxide is formed:

$$2Ag^+ + 2OH^- \longrightarrow Ag_2O(s) + H_2O$$

This oxide is insoluble in excess OH^- but does dissolve in ammonia to form the colorless complex ion $Ag(NH_3)_2^+$, diamminesilver(I):

$$Ag_2O(s) + 4NH_3 + H_2O \longrightarrow 2Ag(NH_3)_2^+ + 2OH^-$$

Solutions containing $Ag(NH_3)_2^+$ are frequently used for silver plating. They have the advantage of providing a very low concentration of Ag^+ so that mild

528

Chapter 19
Transition
Elements:
Some Detailed
Chemistry

reducing agents, such as glucose, react very slowly to give a compact silver plate.

Probably the most interesting of all the silver compounds are the silver halides AgF, AgCl, AgBr, and AgI. AgF is very soluble in water, but the others are quite insoluble. The low solubility is rather surprising; salts of +1 cations and −1 anions are generally soluble. In this respect AgF is normal; it dissolves much like NaF or KF. The abnormal insolubility of the other silver halides is attributed to the fact that their lattice energies are higher than expected. Let us compare AgCl and NaCl. Ag^+ and Na^+ are about the same size, and so we might expect ionic attraction to Cl^- to be about the same. However, Ag^+ has 46 electrons, whereas Na^+ has only 10. In general, the more electrons, the stronger the van der Waals attraction to neighboring atoms (Sec. 6.12). Consequently, AgCl is held together more strongly than NaCl. In fact, the lattice energy of AgCl is 904 kJ/mol; that of NaCl, 770 kJ/mol.

Except for AgF, the silver halides are photosensitive, i.e., sensitive to light. For this reason, they find extensive use in photographic emulsions. The chemistry of the photographic process is complicated. It is usually described in three basic steps: (1) exposure, (2) development, and (3) fixing.

1   Exposure   When photographic film, consisting of a dispersion of silver bromide in gelatin, is exposed to light, grains of AgBr are somehow activated. This is not a visible change but is called formation of a *latent image*. According to one theory (the Gurney-Mott theory), it involves the following sequence of events (see Fig. 19.13):

a An incoming quantum of light $h\nu$ kicks an electron out of Br^- to form Br^0 and e^-.

b The e^- wanders through the crystal of AgBr and eventually gets trapped at a surface defect, which might, for example, be a speck of Ag_2S.

c An interstitial Ag^+, such as is commonly found in AgBr (Sec. 7.8), diffuses to the trap site, where the Ag^+ and the trapped e^- combine to give Ag^0.

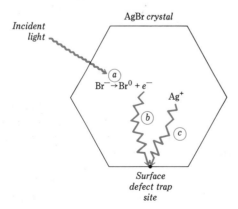

FIG. 19.13 Suggested steps for formation of latent image in the photographic process. Sequence shown is repeated about 50 times.

d A second quantum of light energy $h\nu$ comes along and ejects a second e^-, which then migrates to the Ag^0 and converts it to Ag^-.

e A second interstitial Ag^+ subsequently diffuses over and combines

$$Ag^+ + Ag^- \longrightarrow Ag_2$$

f The process repeats until a clump of about 50 silver atoms is built up. The AgBr grain is now "activated."

2 Development Grains that have been activated can now be preferentially reduced with a mild reducing agent. The half-reaction

for example, is just strong enough to bring about reduction of activated AgBr to elemental silver:

$$AgBr(s) + e^- \longrightarrow Ag(s) + Br^-$$

3 Fixing Since all AgBr eventually turns black when exposed to light, the whole film would turn black eventually. However, the photographic image can be permanently fixed by immediately washing out any nonactivated AgBr grains in the emulsion. This can be done by dissolving the AgBr in a solution containing a high concentration of thiosulfate ion, $S_2O_3^{2-}$:

$$AgBr(s) + 2S_2O_3^{2-} \longrightarrow Ag(S_2O_3)_2^{3-} + Br^-$$

The result is a fixed negative image of the exposure. To get a positive image, the whole process is repeated. By shining light through the negative onto another emulsion, the light and dark areas can be inverted.

Qualitative analysis for silver ion usually starts with precipitation of white AgCl on addition of HCl to the original unknown. The AgCl can be distinguished from another white precipitate, Hg_2Cl_2, which may form at the same time by adding NH_3. The AgCl dissolves to give $Ag(NH_3)_2^+$ and Cl^-, whereas the Hg_2Cl_2 turns black due to conversion to $HgNH_2Cl$ and Hg. Addition of HNO_3 to a filtrate containing $Ag(NH_3)_2^+$ and Cl^- reprecipitates AgCl.

19.12
ZINC AND CADMIUM

Zinc, cadmium, and mercury are the last subgroup of transition elements. Zinc occurs primarily as sphalerite (ZnS), also called zinc blende. The metal is prepared by roasting the sulfide in air to convert it to oxide and then reducing the oxide with carbon. The reactions are

$$2ZnS(s) + 3O_2(g) \longrightarrow 2ZnO(s) + 2SO_2(g)$$
$$ZnO(s) + C(s) \longrightarrow Zn(g) + CO(g)$$

Rapid condensation of the zinc vapor produces the fine powder known as zinc dust.

Massive zinc has fairly good metallic properties except that it is rather brittle. With acids, ordinary zinc gives the well-known evolution of H_2. Strangely enough, this is very rapid when the zinc is impure but almost too slow to be observed when the zinc is very pure. Impurities apparently serve as centers from which the H_2 gas can evolve.

In air, zinc tarnishes only slightly, probably because it forms an invisible, self-protective coat of oxide, hydroxide, or carbonate. Because it withstands corrosion well and because it gives cathodic protection to iron (Sec. 19.8), zinc is often used as a coating on iron to keep it from rusting. Iron protected in this way is called *galvanized iron;* it can be made by dipping iron into molten zinc or by electroplating. The other important use of zinc is to make copper-zinc alloys, known as brass.

In all its compounds zinc shows only a $+2$ oxidation state. The ion Zn^{2+} is colorless. In aqueous solution it hydrolyzes slightly:

$$Zn^{2+} + H_2O \rightleftharpoons Zn(OH)^+ + H^+ \qquad K_{hyd} = 2.0 \times 10^{-10}$$

When base is added, white zinc hydroxide, $Zn(OH)_2$, is precipitated. This hydroxide is amphoteric and further addition of base dissolves it to give zincate ion, $Zn(OH)_4^{2-}$. Like other transition elements, zinc has a great tendency to form stable complex ions. For example, $Zn(OH)_2$ easily dissolves in aqueous ammonia to form the tetrahedral complex, $Zn(NH_3)_4^{2+}$.

When hydrogen sulfide is bubbled through zinc salt solutions that are not too acidic, white zinc sulfide precipitates. The solubility product is 1×10^{-22}, and so ZnS is essentially insoluble in neutral solution. Addition of acid raises the solubility and gives a method for separating ZnS from other sulfides such as CuS, Ag_2S, and CdS.

Zinc sulfide is used extensively in the white pigment lithopone, an approximately equimolar mixture of ZnS and $BaSO_4$. Unlike white lead paints, it is not toxic. ZnS is also used in making fluorescent screens. Impure ZnS acts as a phosphor; i.e., it can convert energy such as that of an electron beam into visible light.

Zinc oxide is probably the most important of the zinc compounds. It can be made by oxidizing zinc vapor in air. ZnO has many specialized uses (e.g., filler and vulcanization activator in rubber tires, white pigment, ointment base, and cement), but its most interesting use is as a photoconductor in copying machines. It conducts electric current when illuminated; when the light is shut off, it reverts to an insulator. In the Xerox method of electrostatic printing (Fig. 19.14), a photoconductive coating on a rotating drum is electrically charged. The electric charge is then made to leak off by exposure to light, but only in those places where light is reflected from white parts of a document to be copied. The remaining electric image is then used to pick up negatively charged black powder (i.e., carrier plus resin pigment) for subsequent transfer to paper where it is fused by heat to fix a permanent image.

The properties of cadmium are so similar to those of zinc that the two elements invariably occur together. The principal source of cadmium is the flue

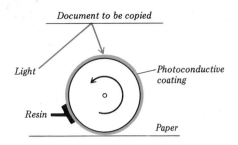

Document to be copied

Light

Photoconductive coating

Resin

Paper

FIG. 19.14 Schematic representation of photocopying.

dust from the purification of zinc by distillation. Cadmium is more volatile than zinc, and so it evaporates first and concentrates in the first distillate. The main use of cadmium is as a plating on other metals, such as steel. It is particularly good as a protective coat for alkaline conditions because, unlike zinc, it is not amphoteric and does not dissolve in base. The other principal use of cadmium is in making low-melting alloys, such as Wood's metal (mp 70°C), used in automatic fire sprinklers.

In compounds the usual oxidation state of cadmium is +2. It exists in aqueous solutions as colorless Cd^{2+} ion. With H_2S, it forms insoluble yellow CdS ($K_{sp} = 1.0 \times 10^{-28}$), used as the pigment *cadmium yellow*. Like zinc, cadmium forms a variety of complex ions, including $Cd(NH_3)_4^{2+}$, $Cd(CN)_4^{2-}$, $CdCl_4^{2-}$, and CdI_4^{2-}, all of which are tetrahedral. Some of the salts of cadmium are unusual in that they do not dissociate completely into ions in aqueous solution as practically all other salts do.

Although cadmium is a relatively rare element, it is now recognized as a major environmental contaminant. The most notorious example of cadmium poisoning comes from Japan in the form of *Itai-itai disease*. (An equivalent English name would be "ouch-ouch.") Symptoms include pain in the joints, bone deformation, and, in the later stages, susceptibility to multiple fractures by as little a disturbance as coughing. Contamination of water, food, and air all contribute as pathways for ingestion into the body. The cadmium accumulates in the liver and kidneys and apparently does its damage by inactivating sulfur-containing enzymes. Recently, tobacco smoke (which contains cadmium) has been implicated as a cadmium pollutant to smoker and non-smoker alike.

Qualitative analysis for zinc and cadmium depends on precipitation of their sulfides. Yellow CdS is precipitated when H_2S is added to an acidic solution containing Cd^{2+}. If the residual solution is then made basic with NH_3, white ZnS is formed. To separate CdS from HgS and CuS, which precipitate along with it in acidic solution, we can make use of the fact that CdS and CuS are soluble in boiling HNO_3, whereas HgS is not. Addition of NH_3 to a solution containing Cu^{2+} and Cd^{2+} gives the blue color characteristic of $Cu(NH_3)_4^{2+}$. Presence of cadmium can be detected by first precipitating out the Cu^{2+} with H_2S in acidic solution in the presence of a high concentration of chloride ion (which keeps Cd^{2+} in solution as $CdCl_4^{2-}$), and then adding

$NaC_2H_3O_2$ and H_2S. The added acetate ion serves to reduce the H^+ concentration, thereby raising the S^{2-} concentration sufficiently to precipitate yellow CdS.

19.13
MERCURY

The only common mineral of mercury is cinnabar (HgS), from which the element is produced by roasting in air:

$$HgS(s) + O_2(g) \longrightarrow Hg + SO_2(g)$$

Mercury is a liquid at room temperature, and its symbol emphasizes this, since it comes from the Latin *hydrargyrum,* meaning "liquid silver." The liquid is not very volatile (vapor pressure is 0.0000024 atm at 25°C), but the vapor is very poisonous, and *prolonged exposure even to the liquid should be avoided.*

Liquid mercury has high metallic luster, but it is not a very good metal. It has higher electric resistance than any other transition metal. However, for some uses, as in making electric contacts, its fluidity is such a great advantage that its mediocre conductivity can be tolerated. Furthermore, its inertness to air oxidation, relatively high density, and uniform expansion with temperature lead to special uses, as in barometers and thermometers.

Liquid mercury dissolves many metals, especially softer ones such as copper, silver, gold, and the alkali elements. The resulting alloys are called *amalgams.* Probably their most distinctive property is that the reactivity of the metal dissolved in the mercury is lowered. For example, the reactivity of sodium dissolved in mercury is so low that sodium amalgam can be kept in contact with water with only slow evolution of hydrogen.

In its compounds mercury shows both +1 (mercurous) and +2 (mercuric) oxidation states. The mercurous compounds are unusual because they all contain two mercury atoms bound together. In aqueous solutions, the ion is a double ion, corresponding to Hg_2^{2+}, in which there is a covalent σ bond between the two mercury atoms. Experimental evidence for this is the lack of paramagnetism of mercurous compounds. Hg^+ would have one unpaired electron, whereas Hg_2^{2+} has the two electrons paired in a σ-bonding molecular orbital.

Except for doubling, mercurous ion behaves much like Ag^+; for example, it reacts with chloride ion to precipitate white mercurous chloride, Hg_2Cl_2, also known as calomel. At one time, calomel was used in medicine as a purgative. However, when exposed to light, it may decompose to Hg and $HgCl_2$. Because of the possible hazard of forming $HgCl_2$, which is quite poisonous, calomel is no longer used in internal medicine. Unlike Ag^+, mercurous ion does not form an ammonia complex. When aqueous ammonia is added to Hg_2Cl_2, the solid turns black because of formation of finely divided mercury:

$$Hg_2Cl_2(s) + 2NH_3 \longrightarrow HgNH_2Cl(s) + Hg + NH_4^+ + Cl^-$$

The compound $HgNH_2Cl$, mercuric ammonobasic chloride, is white, but its color is obscured by the intense black of the mercury. This difference in behavior toward NH_3 provides a simple test for distinguishing $AgCl$ from Hg_2Cl_2.

In the $+2$ state, mercury is frequently represented as the simple ion Hg^{2+}. However, it is usually found in the form of complex ions, insoluble solids, or weak salts. For example, in a solution of the weak salt mercuric chloride, the concentration of Hg^{2+} is much smaller than the concentration of undissociated $HgCl_2$ molecules.

When H_2S is passed through a mercuric solution, a black precipitate of HgS is obtained. The solubility of HgS ($K_{sp} = 1.6 \times 10^{-54}$) is very low. Even boiling nitric acid, for example, will not dissolve it. Aqua regia, however, which supplies nitrate for oxidizing sulfide and chloride for complexing the mercuric, does take it into solution.

The electrode potentials

$$2Hg^{2+} + 2e^- \longrightarrow Hg_2^{2+} \qquad E° = +0.92 \text{ V}$$
$$Hg_2^{2+} + 2e^- \longrightarrow 2Hg(l) \qquad E° = +0.79 \text{ V}$$

are close enough that any reducing agent which is capable of reducing mercuric to mercurous is also able to reduce mercurous to mercury. Thus, if a limited amount of Sn^{2+} is added to a mercuric solution, only Hg_2^{2+} is formed; but if Sn^{2+} is added in excess, the reduction goes all the way to Hg.

Qualitative analysis for mercury depends on whether mercury is present as mercurous or mercuric. If present as mercurous ion, Hg_2Cl_2 will precipitate along with $AgCl$ when HCl is first added. Both Hg_2Cl_2 and $AgCl$ are white but can be distinguished from each other by the fact that, if NH_3 is added to Hg_2Cl_2, a black color appears, owing to formation of Hg and $HgNH_2Cl$. $AgCl$, on the other hand, is soluble in aqueous NH_3, where it forms $Ag(NH_3)_2^+$ and Cl^-. The presence of silver can be confirmed by reprecipitating white $AgCl$ on addition of HNO_3.

Mercuric mercury precipitates as HgS when H_2S is added under acidic conditions. The HgS can be confirmed by dissolving it in aqua regia and then reducing with $SnCl_2$ to give Hg_2Cl_2 and Hg.

19.14
MERCURY IN THE ENVIRONMENT

Although nature has tried to lock mercury away as very insoluble HgS, humans have opened Pandora's box by extracting the metal and by burning coal. The supposition until recently was that since elemental mercury is rather inert and eventually goes to the very insoluble sulfide, no great harm was done. We know now that the mercury threat to the environment was vastly underrated.

Metallic mercury is not highly toxic. The vapor, on the other hand, is a hazard, particularly over long exposure. The yellow sulfur that is sometimes seen sprinkled around in a laboratory where there is spilled mercury repre-

534

Chapter 19
Transition
Elements:
Some Detailed
Chemistry

sents an attempt to convert mercury to HgS and thereby prevent buildup of even small amounts of vapor.

Inorganic compounds of mercury, when soluble, are toxic, but only moderately so. Mercuric chloride, for example, when taken orally, damages the intestinal tract and the kidneys. The organic compounds, particularly dimethylmercury, $(CH_3)_2Hg$, are extremely toxic, and their uses as slimicides, fungicides, mildew killers, and germ sprays pose a threat to the environment.

The most famous case of environmental contamination by mercury occurred in Japan as *Minamata disease*. Minamata is a small fishing village on the southwest coast of Kyushu, and there in 1953 people began to show signs of a strange illness characterized progressively by soreness of hands and face, tunnel vision, dizziness, loss of control of body movements, mental disorder, and finally death. Health authorities correlated the symptoms with fish consumption and by analysis of silt in the harbor implicated a nearby factory which had used $HgCl_2$ as a catalyst for making polyvinyl chloride, a common plastic. The culprit was believed to be $(CH_3)_2Hg$ stored in the body tissue of fish and produced by bacterial action on mercury wastes in the mud slimes.

Where does most human-made mercury pollution come from? This is a difficult question to answer because it depends on (1) the amount of mercury a particular usage represents and (2) the leakage from that use into the environment. The biggest user of mercury in the United States is the chlor-alkali industry, which uses mercury as the cathode in electrochemical decomposition of aqueous sodium chloride. As can be seem from Fig. 19.15, the process is a continuous one in which aqueous sodium chloride flows into the cell and is electrolyzed between graphite anodes and a flowing-mercury-pool cathode. At the cathode, Na^+ is reduced to Na^0, which dissolves in the mercury to give sodium amalgam. The amalgam is siphoned off and allowed to react with water in an iron pan so as to form aqueous NaOH and regenerate the mercury. In principle, there is no mercury loss. However, waste mercury in many

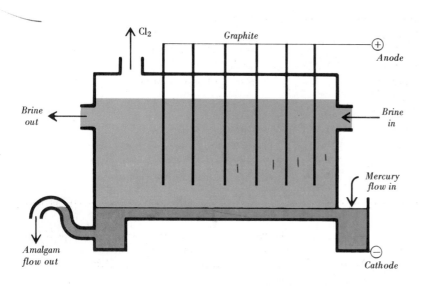

FIG. 19.15 Electrolytic cell for obtaining Cl_2 and NaOH from aqueous NaCl.

Cl₂

Graphite

⊕
Anode

Brine out

Brine in

Mercury flow in

Amalgam flow out

⊖
Cathode

cases was simply washed out to accumulate in river and lake bottoms. Bacteria in the mud feast on the mercury and generate dimethylmercury for absorption into the food chain. Marine diatoms with mercury compounds adsorbed on their surfaces get eaten by higher organisms, fish eat the organisms, people eat the fish, and so on up the food chain. Each step concentrates the mercury contamination.

Other sources of mercury pollution are the paper industry, commercial laundries, seed dressings in agriculture, and coal combustion. In the paper industry, PMA, or phenyl mercury acetate ($C_6H_5HgOCOCH_3$), has been used for slime control. In the United States the practice has been given up because of a government directive not to use PMA for paper that comes in contact with food. Since the paper use cannot be guaranteed, use of PMA for any paper has been abandoned. Commercial laundries, especially those with diaper service, sometimes use PMA to suppress mold. Mercury seed dressings, although suspended for interstate commerce in the United States, continue to produce tragedies elsewhere. In several cases, seed grains that had been mercury-treated and dyed red to show this were diverted from use as seed, washed to remove the dye, fed to animals, and eventually fed to people *via* sick animals that were rushed to the slaughterhouse.

Coal contains about 1 ppm Hg. Worldwide combustion of coal amounts to about 5×10^9 tons/yr, which means about 5000 tons of Hg into the atmosphere. This airborne contamination may be part of the explanation why fish in isolated mountain lakes have sometimes shown surprisingly large Hg content. Mercury is excreted by fish only very slowly. The half-life in freshwater fish is about 200 days; in humans it is about 70 days.

Important Concepts

rare-earth elements	manganese	flotation
lanthanides	iron	copper
lanthanide contraction	blast furnace	electrolytic refining
ion-exchange separation	steel	silver
titanium	rusting	photographic process
zirconium and hafnium	electrolytic corrosion	zinc and cadmium
vanadium	cathodic protection	mercury
chromium	cobalt and nickel	mercury pollution

Exercises

*19.1 **Rare-earth elements** How does the size of the rare-earth tripositive ion change in the sequence from La^{3+} to Lu^{3+}? Explain why this change in size occurs. How would you expect this size change to influence the complex-ion–forming ability of the rare-earth ions? Explain.

**19.2 Rare-earth elements How do the magnetic and optical properties of the $4f$ rare-earth ions generally differ from those of the $3d$ transition-metal ions? What is the main reason for the difference?

*19.3 Rare-earth elements The great chemical similarity between rare-earth elements makes separation quite difficult. Tell with the aid of a sketch how this separation may be achieved. What is the underlying reason why the separation works?

*19.4 Titanium Suppose as manager of a chemical plant you receive an order for titanium to be used as sheet metal for aircraft. What would you order for raw material? How would you make the element, and what pitfalls would you need to avoid?

***19.5 Titanium In the preparation of pure titanium from TiO_2, carbon is used as a reducing agent. Yet carbon is an objectionable impurity in titanium which is destined for use as a structural material. Resolve this apparent contradiction in terms of the specific chemistry involved.

*19.6 Zirconium and hafnium Zr and Hf are both important in the nuclear energy industry. What special attributes make them desirable for such use? Explain why the separation of Zr and Hf is so much more difficult than that of most other pairs of elements in a given group.

*19.7 Element 104 Predict the ground-state electron configuration for element 104. Comment critically on the use of the words "actinide" or "transition element" to describe this element.

*19.8 Vanadium How many d electrons and how many unpaired electrons would there be in V^0, V^{2+}, V^{3+}, V^{4+}, and V^{5+}?

**19.9 Chromium Write balanced chemical equations for each of the following conversions:
 a Oxidation of $FeCr_2O_4$ by O_2 in the presence of Na_2CO_3 to form Fe_2O_3, Na_2CrO_4, and CO_2
 b Reduction of dichromate ion by Zn in acid solution to give Cr^{2+} and Zn^{2+}
 c Oxidation of $Cr(OH)_2$ by O_2 to give $Cr(OH)_3$
 d Addition of excess base to a solution of Cr^{3+}
 e Reduction of $Cr_2O_7^{2-}$ in acidic solution by ethyl alcohol (C_2H_5OH) to give acetaldehyde (CH_3CHO) and Cr^{3+}

***19.10 Chromium In aqueous solution Cr^{3+} is bound to six near-neighbor water molecules rather firmly, whereas the bonding to more distant water molecules is considerably weaker. Tell how

mass spectrometric analysis, using water labeled with oxygen 18, can be used to show that there are but six water molecules in this inner set.

**19.11 Chromium How many isomers are there for each of the following:
 a $CrCl_2(H_2O)_4^+$ b $CrCl_3(H_2O)_3$
 c $CrCl_2(C_2O_4)_2^{3-}$ d $Cr(C_2O_4)_3^{3-}$
Draw possible structures for each. *Ans. 2, 2, 3, 2*

**19.12 Chromium Progressive addition of base to a solution of $Cr(H_2O)_6^{3+}$ first produces a precipitate, which then redissolves when excess base is added. Write equations for the stepwise reaction under the requirement that the coordination number of chromium to oxygen remain six in each species shown.

*19.13 Chromium How many grams of K_2SO_4 and of $Cr_2(SO_4)_3 \cdot 18H_2O$ would you need to grow a 5.00-g crystal of potassium chrome alum? What other ingredient is needed?

**19.14 Chromium Upon acidification with H_2SO_4, a *yellow* chromium-containing solution turns *orange*. Bubbling in SO_2 changes the orange to *green*. Subsequent stepwise addition of NaOH gives a *light green* precipitate and then a *deep green* solution. Final addition of H_2O_2 converts the deep green to *yellow*. Identify each of the colored species, and write balanced net equations for each of the changes.

**19.15 Manganese The common oxidation states of manganese are 0, +2, +3, +4, +6, and +7. Give the electron configuration for each of these states and tell how many unpaired electrons there are in each, both as an isolated species and in the presence of a strong octahedral crystal field.

**19.16 Manganese A *colorless solution* of manganous ion is treated with base to give a *white precipitate*. On standing in air, the white precipitate turns *red*. Addition of dilute sulfuric acid to the red precipitate gives a *colorless solution* plus particles of a *black compound*. When filtered off and heated with Na_2CO_3 in air, the black compound turns *green*. Upon addition of acid to the green compound, a *black material* and a *deep violet solution* are produced. Identify each of the italicized species, and write equations for the changes that occur.

***19.17 Manganese Manganese dioxide is an example of a nonstoichiometric compound, which can be written MnO_{2-x}. In a particular experiment to find out the value of x for a given sample, the MnO_{2-x} was boiled in HCl to give Cl_2 and Mn^{2+}. What value of x corresponds to evolution of 1.428 liters of $Cl_2(g)$ at 35°C and 0.955 atm from 5.00 g of MnO_{2-x}?   *Ans. 0.075*

**19.18   Iron   The burning of iron pyrites converts FeS_2 to Fe_2O_3. What would be the percent loss in weight shown by pure FeS_2 in such a burning? A given specimen composed of FeS_2 plus inert material shows a weight loss of 20.0 percent on being burned. What percent of the specimen is inert material?

*19.19   Iron   What is the chemical difference between iron, pig iron, cast iron, and steel?

**19.20   Iron   Describe with the help of chemical equations what is going on in the following experiment: An iron bar is thrust into a solution of concentrated nitric acid, where it reacts vigorously with the generation of brown fumes and then subsides. If now rinsed off and placed in a solution of $CuSO_4$, nothing happens until the surface is scratched, at which point a reddish deposit spreads over the iron bar.

*19.21   Iron   A piece of pure iron dissolves in dilute sulfuric acid to give a *pale green* solution. When base is added, a *whitish* precipitate forms but quickly turns *brown* on exposure to air. When the brown precipitate is dissolved in sulfuric acid, a colorless solution is formed, but it turns *yellow* when hydrochloric acid is added. Tell what the colored species are and write equations for the reactions that occur.

***19.22   Iron   A given solution, 25.0 ml of 0.30 *M* Fe^{3+}, shows an initial pH of 1.38. H_2S is bubbled through the solution until the pH falls to 1.28. What fraction of the Fe^{3+} has been reduced?
Ans. 3.6%

*19.23   Rusting   What is the chemical composition of rust? What ingredients are needed for rust formation? Indicate three different methods of rust prevention and tell why they work.

**19.24   Rusting   How does the mechanism proposed in Sec. 19.8 account for the following observations:

a Garbage cans rust faster when they contain acid waste.

b Iron pipes connected to copper pipes rust faster than iron pipes alone.

c Rust deposits frequently form at some distance from where the iron is eroded.

d Iron tools coated with grease generally show little rust formation.

e Wet steel wool rusts overnight, whereas dry steel wool stays shiny almost indefinitely.

f When the positive pole of a car battery is grounded to the chassis, an automobile rusts much faster than if the negative pole is so connected.

*19.25   Rusting   Tell what are the advantages and disadvantages of preventing rust formation by plating with zinc, with chromium, or with tin. Tell why each of these platings is effective.

*19.26   Cobalt and nickel   What is the chemical rationale for naming these elements after goblins and underground spirits? What are the key differences in the chemical behavior of cobalt and nickel as compared with that of iron?

*19.27   Cobalt   The element cobalt is famous for forming an enormous number of complex ions. Give the structure and the systematic name (Appendix A2.1) for each of the following:

a $Co(CN)_6^{3-}$ *b* $Co(NH_3)_6^{3+}$

c *trans*-$CoCl_2(NH_3)_4^+$ *d* $CoCl_3(NH_3)_3$

e *cis*-$Co(OH)(H_2O)(CO)_4^{2+}$

**19.28   Nickel   The cell reaction for the Edison cell is

$$Fe(s) + Ni_2O_3(s) + 3H_2O \longrightarrow$$
$$2Ni(OH)_2(s) + Fe(OH)_2(s)$$

Diagram a cell that makes use of this reaction. Label the anode and cathode, write appropriate electrode reactions, and tell which way positive and negative charges move in the internal and external circuit.

*19.29   Qualitative analysis   A black material may be one or more of the following: FeS, CoS, or NiS. What chemical tests can you perform to tell which it is? Write chemical equations for any reactions involved.

*19.30   Copper   Given a copper mineral that is mostly $CuFeS_2$, tell with the help of equations how you can convert this mineral to 99.999% copper.

**19.31   Copper   It has been estimated that copper smelters in the United States produce 4 million tons/yr of SO_2 by-product. How many tons of H_2SO_4 could this make? What weight of $CuFeS_2$ would have to be processed to produce this much SO_2?

*19.32   Copper   Explain the principle behind the electrorefining of copper. Why does the method work? Tell what happens to typical impurities Fe and Ag during the refining.

***19.33 Copper** A *blue solution* containing cupric ion is treated with aqueous ammonia until a *light blue precipitate* forms. Addition of excess ammonia gives a *deep blue solution*. When dilute nitric acid is added to the deep blue solution, a *light blue precipitate* appears and then dissolves in excess to restore the original *blue solution*. Tell what species are indicated by the italicized descriptions and write balanced equations for the chemical reactions that occur. Explain why on successive repetition of these manipulations the light blue precipitate eventually fails to reappear.

19.34 Copper Given that the hydrolysis constant of Cu^{2+} is 4.6×10^{-8}, what would be the pH of a solution made by dissolving 100 g of $Cu(NO_3)_2$ in enough water to make 0.250 liter of solution? *Ans. 3.5*

19.35 Copper Tell how you could make each of the following conversions. Give equations for each reaction.

a $Cu(s)$ to $CuS(s)$
b $CuS(s)$ to $Cu(OH)_2(s)$
c $CuS(s)$ to $CuSO_4 \cdot 5H_2O(s)$
d $Cu_2O(s)$ to $Cu(NH_3)_4^{2+}$

*19.36 Silver** Explain why addition of cyanide ion helps to bring about air oxidation of elemental silver.

19.37 Silver The brown-black tarnish often seen on silverware is Ag_2S. It can be removed by wrapping the silverware in aluminum foil and putting it in a solution of NaCl. Explain why the process works.

19.38 Silver Excess aqueous ammonia is added dropwise to a solution of $AgNO_3$ until a *brown precipitate* that first appears dissolves to give a *colorless solution*. Upon subsequent treatment with dilute HCl, the colorless solution gives a *white precipitate*. Tell what the italicized species are and write net equations for the interconversions.

***19.39 Silver** Given the following potentials:

$$Ag(s) \longrightarrow Ag^+ + e^- \qquad E^0 = -0.799V$$
$$Ag(s) + 2CN^- \longrightarrow Ag(CN)_2^- + e^- \qquad E^0 = +0.31V$$

Calculate the dissociation constant of $Ag(CN)_2^-$.
Ans. 1.8 × 10⁻¹⁹

19.40 Silver Write the electronic configuration for neutral silver ($Z = 47$). How would it be changed for Ag^+? By writing an electronic formula for $Ag(NH_3)_2^+$ and specifying which orbitals are used for bonding, show that a linear configuration for the molecule is not unexpected. What geometry would you expect if you could crowd four NH_3 molecules around the Ag^+?

19.41 Silver The solubility products for AgCl, AgBr, and AgI are 1.7×10^{-10}, 5.0×10^{-13}, and 8.5×10^{-17}, respectively. How might you account qualitatively for this trend? If you shook up a mixture of AgCl, AgBr, and AgI with water so as to saturate the solution, what would be the concentration of each ion in the final solution?

***19.42 Silver** Ultrapure AgBr in perfect crystalline form would probably not be very good for photography. Explain.

19.43 Silver Tell what goes on chemically in the three steps of photography: exposure, development, and fixing. Write balanced equations for any reactions that occur.

***19.44 Zinc** You are given a white material which may be any one of the following: ZnO, $ZnCO_3$, $Zn(OH)_2$, ZnS, $Zn(NH_3)_4Cl_2$. What tests would you run to identify your compound? Be specific. Tell what reagents you could use and what observations you would make.

19.45 Zinc Write balanced chemical equations for the following chain of reactions:

a A piece of zinc is dissolved in aqueous NaOH.
b HCl is added to the product from (*a*) until a white precipitate appears.
c Gentle heating of the white precipitate from (*b*) produces a loss of weight, but no loss of zinc.
d The product from (*c*) is thrown into dilute HCl, where it dissolves.

19.46 Zinc ZnO was an important ingredient in the early copying machines because of its photoconductivity. Explain how the copying process makes use of photoconductivity.

*19.47 Cadmium** A given colorless solution may contain Zn^{2+} or Cd^{2+}. What one reagent could you add to tell immediately which one it is?

*19.48 Mercury** Given that the vapor pressure of liquid mercury is 0.0000024 atm at 25°C, what weight of mercury would it take to pollute a room that is $4 \times 4 \times 2$ m³? The density of liquid mercury is 13.6 g/ml. What size droplet could infect the above room?

*19.49 **Mercury** What is the experimental evidence for believing that mercurous ion is a double ion Hg_2^{2+} rather than monatomic Hg^+? Justify your answer.

19.50 **Mercury Write balanced equations for the reactions that occur in the following experiments:

a Addition of HCl to a mercurous solution produces a white precipitate.

b Addition of aqueous NH_3 to the precipitate from (*a*) causes it to turn black.

c When the final residue from (*b*) is heated with hot HNO_3, it dissolves to give a clear colorless solution.

d Addition of H_2S to the solution from (*c*) gives a black precipitate.

19.51 **Mercury If a little bit of stannous chloride is added to mercuric chloride solution, a grayish precipitate is formed. If a lot of stannous chloride is added, the mixture turns black. Explain with the help of equations.

19.52 **Mercury How is mercury useful in the commercial production of Cl_2 and NaOH from aqueous NaCl? Trace the path of the mercury and show how it could get into the environment. How would you go about decontaminating a mud slime that contains mercury?

***19.53 **Mercury** The average mercury content of fish in particular contaminated waters was found to be 0.86 ppm. Suppose you had eaten a 100-g portion of such fish daily. How much mercury would you have ingested in a year? Assuming that mercury is eliminated from the body at a rate that corresponds to a half-life of 70 days, approximately how much mercury would have accumulated in your body at the end of 350 days? (*Hint:* Divide the 350 days into 5 half-lives. Assume the amount ingested in the first 70 days has 4.5 half-lives to be eliminated; the second 70 days, 3.5 half-lives; etc.)

Ans. 31 mg, 8 mg

Chapter 20

CHEMISTRY OF CARBON

The chemistry of carbon has two aspects: inorganic and organic. The dividing line is not sharp. Generally, the term *inorganic* applies to elemental carbon, compounds of carbon with metals, and compounds of carbon with nonmetals other than hydrogen. The term *organic* covers compounds of carbon with hydrogen, and derivatives of such hydrocarbons. At one time it was believed that organic compounds were essen-

tially different from inorganic in coming only from living organisms. This distinction is no longer valid. Indeed, the number of organic compounds coming from the synthetic chemist's laboratory exceeds the number that come from living things.

In this chapter we consider first the inorganic aspects of the chemistry of carbon. Then we turn to the organic.

20.1
ELEMENTAL CARBON

Carbon is the head element of group IV, which contains carbon, silicon, germanium, tin, and lead. There is a pronounced change in metallic character going down the group. The lightest member, carbon, forms a covalent solid of complex structure which exhibits no metallic properties; the two heaviest elements, tin and lead, are quite good metals. In between are silicon and germanium. There is a definite trend from acidic to basic character of the oxides as we go down the group.

Carbon is not very plentiful in the Earth's crust, but it is the second most abundant element in the human body. It occurs in all plant and animal tissue, combined with hydrogen and oxygen, and in their geologic derivatives, petroleum, coal, and natural gas, where it is combined mostly with hydrogen in the form of hydrocarbons. Combined with oxygen, carbon occurs as carbon dioxide and as carbonate in rocks such as limestone. In the free state, carbon occurs as diamond and graphite, the two allotropic forms of the element.

As shown in Fig. 20.1, the principal difference between diamond and graphite is that in diamond, each C atom has four nearest neighbors, while in graphite, each C has three nearest neighbors. In the diamond lattice the distance between C atoms is 0.154 nm and each atom is bonded by sp^3-hybrid orbitals to four other atoms at the corners of a tetrahedron. The result is an infinite interlocked structure extending in three dimensions. The giant molecule formed is very hard (the hardest known naturally occurring substance) and has a high melting point (3500°C). Furthermore, diamond is a noncon-

540

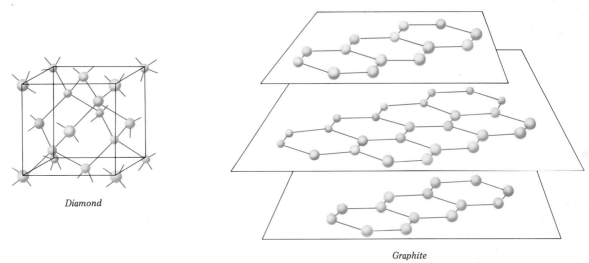

Diamond

Graphite

FIG. 20.1 Allotropic forms of carbon.

ductor of electricity.   All pairs of electrons are localized between specific pairs of carbon atoms and are not free to migrate through the crystal.

In graphite the structure consists of giant sheetlike molecules, 0.340 nm apart, held weakly to each other by van der Waals forces. Within the sheet each C atom is covalently bound by sp^2-hybrid orbitals to three C neighbors 0.142 nm away. Since each C has four valence electrons and only three C's to bond to, there are more than enough electrons to establish single bonds by use of sp^2 hybrids. The fourth electron goes into the p_z orbital perpendicular to the plane. However, since there is no preference as to which atom the last electron should bond to (all three neighbors being equivalent), it must be considered as forming a partial π bond to all three neighbors.

The electronic configuration of graphite can be represented as a resonance hybrid of the three formulas shown at the top of Fig. 20.2. Alternatively, the lower left of the figure gives another representation, where circles stand for the π electrons that are in molecular orbitals derived from the overlap of p_z orbitals (perpendicular to the sheet). Each C atom thus uses its s, p_x, and p_y orbitals to form three σ bonds within the sheet and its p_z orbital to form π bonds above and below the sheet. A portion of the π system is shown on the lower right of Fig. 20.2.

Massive graphite is a soft, gray, high-melting solid with a dull metallic luster and fairly good electric conductivity parallel to the sheet. The softness is attributed to weak sheet-to-sheet bonding, which permits adjacent layers to slide over each other. The high melting point is traceable to strong covalent bonding within the sheets. The conductivity and metallic luster presumably stem from the freedom of π electrons (one per carbon) to move from atom to atom. Because of its high melting point and its electric conductivity, graphite finds extensive use as electrode material.

541

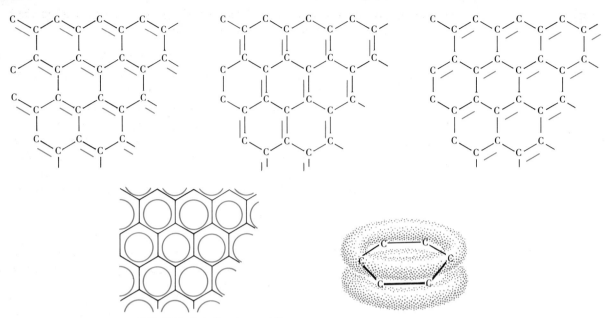

FIG. 20.2 Bonding in graphite.

Besides massive graphite, there are several porous forms of carbon which resemble graphite in character. These include coke (made by heating coal in the absence of air), charcoal (made from wood in the same way), and carbon black (soot). They all have tremendous surface areas (for example, 1 cm³ of charcoal has a surface area of approximately 50 m²). These forms of carbon have strong adsorption properties.

Under normal conditions graphite is the stable form of carbon. At high pressure, the principle of Le Châtelier predicts that diamond should become the more stable since its density (3.51 g/cm³) exceeds that of graphite (2.25 g/cm³). By raising the pressure to about 10^5 atm and the temperature to about 2000°K (to increase the rate), diamonds have been prepared synthetically. Transition metals such as chromium and platinum act as catalysts for the conversion. Although not usually of gem quality, synthetic diamonds are useful in industrial applications as abrasives.

20.2
INORGANIC COMPOUNDS OF CARBON

At room temperature carbon is rather inert, but at higher temperatures it reacts with a variety of other elements. With metals and semimetals, carbon forms solid carbides, such as silicon carbide (SiC), iron carbide (Fe_3C), and calcium carbide (CaC_2). Silicon carbide, formed by heating silica (SiO_2) with graphite, is the industrial abrasive carborundum. It has Si and C atoms occupying alternate positions in a diamond lattice. Iron carbide is the essential constituent of white cast iron. Calcium carbide, obtained by heating CaO with coke, reacts with water to liberate acetylene:

$$CaC_2(s) + 2H_2O \longrightarrow Ca^{2+} + 2OH^- + C_2H_2(g)$$

The formation of acetylene reflects the fact that the CaC_2 lattice contains Ca^{2+} and C_2^{2-} ions.

With nonmetals carbon forms molecular compounds which vary from simple carbon monoxide to complex hydrocarbons. With the nonmetal sulfur, carbon reacts at high temperature to form carbon disulfide (CS_2). Liquid CS_2 is a familiar solvent, particularly for such substances as rubber and sulfur. It is hazardous, however, because it is toxic and highly flammable. When carbon disulfide vapor is heated with chlorine gas, the following reaction occurs:

$$CS_2(g) + 3Cl_2(g) \longrightarrow CCl_4(g) + S_2Cl_2(g)$$

The carbon tetrachloride (CCl_4) thus formed is a good solvent for molecular solutes. As a cleaning fluid, it should be used with caution; although it is not flammable, the liquid can penetrate the skin, and both liquid and vapor are toxic.

With oxygen, carbon forms carbon monoxide (CO) and carbon dioxide (CO_2). These oxides are produced in the combustion of carbon or hydrocarbons, with carbon monoxide predominating when the supply of oxygen is limited. As previously indicated (Secs. 14.3 and 19.6), CO is an important industrial fuel and reducing agent. It is a colorless, odorless gas that is quite poisonous because it interferes with the normal oxygen-carrying function of hemoglobin in red blood cells. Instead of forming a complex with oxygen (oxyhemoglobin), hemoglobin forms a more stable complex with carbon monoxide (carboxyhemoglobin). The tissue cells are thus starved for lack of oxygen.

Unlike CO, CO_2 is not poisonous and, in fact, is necessary for various physiological processes, e.g., the maintenance of the proper pH of blood. Since it is *produced* by respiration and *used up* in photosynthesis, the concentration in the atmosphere remains fairly constant at about 0.04 percent by volume. Over the years, however, there appears to be a gradual rise in atmospheric CO_2 because of increased combustion of fossil fuels. It has been suggested that this might eventually produce significant climatic heating since CO_2 in the atmosphere acts by a greenhouse effect to trap infrared radiation.

Commercially, CO_2 is generally derived from the distilling industry, where fermentation of sugar to alcohol

$$C_6H_{12}O_6 \xrightarrow{\text{yeast}} 2C_2H_5OH + 2CO_2(g)$$

cheaply produces large amounts of by-product CO_2. The gas can also be formed by thermal decomposition of limestone or by reaction of carbonates with acid. The gas is approximately $1\frac{1}{2}$ times as dense as air and settles in pockets to displace air. Since it is not combustible itself, it acts as an effective blanket in fire fighting. The phase relations of CO_2 and the use of CO_2 as a refrigerant have been indicated in Sec. 7.15.

Compared with most gases, CO_2 is quite soluble in water; at 1 atm pressure and room temperature the solubility is 0.03 M. (It is twice as soluble in alcohol, where it has the peculiar physiological effect of increasing the rate of alcohol absorption. This may account for the extra exhilarating effect of champagne and combinations such as Scotch and soda.) Aqueous solutions of CO_2 are acid, with a pH of about 4. Although it has been suggested that this acidity arises from carbonic acid, H_2CO_3, this acid has never been isolated. In aqueous CO_2 solutions, more than 99 percent of the solute remains in the form of linear $:\overset{..}{O}::C::\overset{..}{O}:$ molecules. However, a small amount does indeed react to form H_2CO_3, which can dissociate to H^+ and bicarbonate ion. Thus there are two simultaneous equilibria:

$$CO_2 + H_2O \rightleftharpoons H_2CO_3$$
$$H_2CO_3 \rightleftharpoons H^+ + HCO_3^-$$

which can be combined to give

$$CO_2 + H_2O \rightleftharpoons H^+ + HCO_3^-$$

The constant for this last equilibrium, loosely called the first dissociation of carbonic acid, is 4.2×10^{-7}. The dissociation of bicarbonate ion into H^+ and carbonate ion, CO_3^{2-}, has a constant 4.8×10^{-11}.

The carbonate and bicarbonate ions are planar, with C bonded to three O atoms at the corners of an equilateral triangle. The situation is reminiscent of graphite, with more than enough electrons to form single bonds to all three oxygens; as a result, the electronic distribution is represented as a resonance hybrid. For carbonate ion, the contributing resonance forms are usually written as in Fig. 20.3.

Derived from carbonic acid are two series of salts: bicarbonates, such as $NaHCO_3$, and carbonates, such as Na_2CO_3. The compounds are made industrially by the *Solvay process,* in which CO_2 (from thermal decomposition of limestone) and NH_3 (recycled in the process) are dissolved in NaCl solution. NH_3 neutralizes CO_2 by the reaction

$$NH_3 + CO_2 + H_2O \longrightarrow NH_4^+ + HCO_3^-$$

and the HCO_3^- formed precipitates as $NaHCO_3$ if the temperature of the NaCl solution is kept at 15°C or lower. On thermal decomposition, $NaHCO_3$ gives Na_2CO_3:

$$2NaHCO_3(s) \longrightarrow Na_2CO_3(s) + CO_2(g) + H_2O(g)$$

Sodium carbonate and sodium bicarbonate are industrial chemicals of primary importance. Na_2CO_3, commonly called soda ash, is used, for ex-

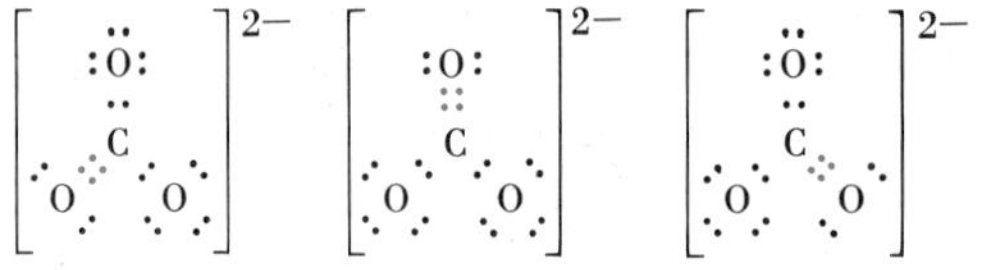

FIG. 20.3 Resonance formulas of carbonate ion.

ample, in making glass. $Na_2CO_3 \cdot 10H_2O$, called washing soda, is used in laundering because of its mild basic reaction resulting from hydrolysis of carbonate ion:

$$CO_3^{2-} + H_2O \rightleftharpoons HCO_3^- + OH^-$$

$NaHCO_3$, or baking soda, is a principal component of baking powders, used to replace yeast in baking. Yeast brings about fermentation of sugars, releasing CO_2 gas, which "raises" the dough; with baking powder, the CO_2 for leavening is obtained by reaction of $NaHCO_3$ with acid substances such as alum.

In addition to the compounds that carbon forms with oxygen, there are numerous compounds in which carbon is bonded to the nonmetal nitrogen. The simplest of these is cyanogen, C_2N_2. It is very poisonous. In many chemical reactions it behaves like a halogen. For example, in basic solution it disproportionates according to the equation

$$C_2N_2(g) + 2OH^- \longrightarrow CN^- + OCN^- + H_2O$$

The cyanide ion, CN^-, resembles chloride ion in giving an insoluble silver salt, $AgCN$ ($K_{sp} = 1.6 \times 10^{-14}$). Cyanide ion forms many complex ions with transition-metal ions, for example, $Fe(CN)_6^{3-}$. It also combines with H^+ to form a weak acid, HCN, which in solution is called hydrocyanic acid (prussic acid). Like cyanogen, HCN is poisonous. Death may result from a few minutes' exposure at 300 ppm.

The anion OCN^-, formed by disproportionation of cyanogen, is called cyanate ion. It exists in many salts, e.g., ammonium cyanate, NH_4OCN. This last compound is of special interest because on being heated it is converted to urea, $CO(NH_2)_2$, the principal end product of protein metabolism. The discovery of this reaction by Wöhler in 1828 was a milestone in chemistry. It represented the first time a person was able to synthesize in the laboratory a compound previously thought to be produced only in living organisms.

20.3
SATURATED HYDROCARBONS

There are a fantastic number of compounds containing carbon and hydrogen. Some of these are composed solely of carbon and hydrogen and are called *hydrocarbons;* others contain additional elements and are called hydrocarbon derivatives. It has been estimated that the hydrocarbons and their derivatives number more than a million. Why are there so many? In the first place, carbon atoms can bond to each other to form chains of varying length. Second, adjacent carbon atoms can share one, two, or three pairs of electrons; therefore, a carbon chain of given length can have different numbers of attached hydrogen atoms. Third, the more atoms a molecule contains, the more ways there are to join the atoms to each other (structural isomerism). Finally, different atoms or groups of atoms can be substituted for hydrogen atoms to yield a large number of derivatives.

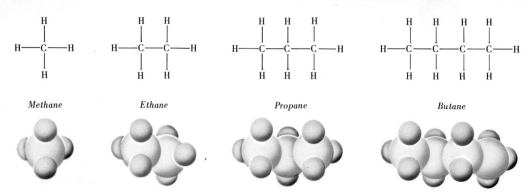

Methane   Ethane   Propane   Butane

FIG. 20.4 Some hydrocarbons.

The C atom has four valence electrons and can use its sp^3-hybrid orbitals to form four covalent bonds directed toward the corners of a tetrahedron. C and H are of about the same electronegativity, and the C—C bond strength is not far from that of C—H. This means that, instead of one simplest hydrocarbon, CH_4 (methane), a whole series of compounds is possible; examples are C_2H_6 (ethane), C_3H_8 (propane), and C_4H_{10} (butane). The structural formulas are shown in Fig. 20.4. The series does not stop with butane but continues almost indefinitely, with each member having the general formula C_nH_{2n+2}. In all these molecules, each C atom bonds to four other atoms, the maximum number it can attach. Hence, the compounds are referred to as being *saturated. Unsaturated* means that a compound contains less than the maximum amount of hydrogen. The *saturated hydrocarbons* are called *alkanes,* with the ending *-ane* appearing in all their names. They are also sometimes called *paraffins.* The first 10 members of the series are listed in Fig. 20.5. As can be seen from the formulas, the members increase stepwise in composition by addition of the CH_2 unit (called the *methylene unit*). Such a series of compounds incrementally related is called a *homologous series.*

For each of the first three entries in Fig. 20.5 there is but one way to arrange the atoms in space. For butane (C_4H_{10}) there are two possibilities:

FIG. 20.5 Saturated Hydrocarbons, C_nH_{2n+2}

Formula	Name	Melting point, °C	Boiling point, °C
CH_4	Methane	−182	−161
C_2H_6	Ethane	−183	−89
C_3H_8	Propane	−188	−42
C_4H_{10}	Butane	−138	−1
C_5H_{12}	Pentane	−130	36
C_6H_{14}	Hexane	−95	69
C_7H_{16}	Heptane	−91	98
C_8H_{18}	Octane	−57	126
C_9H_{20}	Nonane	−54	151
$C_{10}H_{22}$	Decane	−30	174

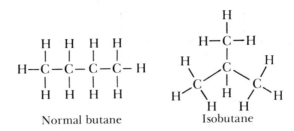

Normal butane   Isobutane

Both molecules correspond to composition C_4H_{10}, but normal butane is a linear sequence, $CH_3CH_2CH_2CH_3$, whereas isobutane consists of three CH_3 groups and one H attached to a central atom. The two molecules have different properties. Compounds differing in spatial arrangement of atoms are called *isomers*. The more atoms, the greater the number of isomers. For butane, it might seem possible that there would be isomers other than the two shown above. The problem arises because two-dimensional display formulas do not take into account possible rotation in space about single bonds. The spatial relations can best be seen by use of molecular models such as those diagrammed in Fig. 20.6. Of the five configurations shown, the first four correspond to the same molecule (*n*-butane) twisted into different shapes; the fifth corresponds to a different molecule (isobutane). No amount of twisting can convert the iso to the normal isomer.

For pentane (C_5H_{12}) there are three isomers:

$$CH_3CH_2CH_2CH_2CH_3 \qquad CH_3\underset{\underset{CH_3}{|}}{C}HCH_2CH_3 \qquad CH_3-\underset{\underset{CH_3}{|}}{\overset{\overset{CH_3}{|}}{C}}-CH_3$$

Normal pentane   2-Methylbutane   2,2-Dimethylpropane

They can be systematically distinguished by name by calling the parent skeleton the longest connected sequence of C atoms and then using numbers and prefixes to locate attachments to the parent alkane. (See Appendix A2.2 for systematic nomenclature.)

Hexane (C_6H_{12}) has five isomers; heptane, nine; octane, 18; nonane, 35; decane, 75; etc. By the time one reaches $C_{40}H_{82}$ (tetracontane), the number of possible isomers reaches more than 61 trillion.

For higher hydrocarbons there is, besides the *structural isomerism* just

FIG. 20.6 Various configurations of butane molecules.

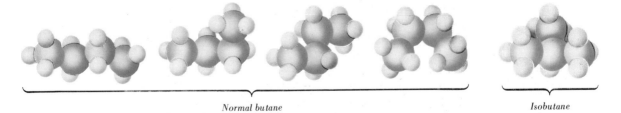

Normal butane   *Isobutane*

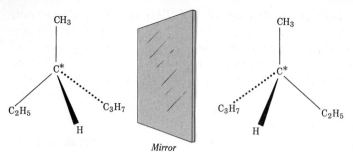

FIG. 20.7   Optical isomers of 3-methylhexane.

described, the possibility of optical isomerism. (Recall Sec. 18.5.) This happens, for example, with the hydrocarbon

$$C_2H_5-\overset{\overset{\displaystyle CH_3}{|}}{\underset{\underset{\displaystyle H}{|}}{C^*}}-CH_2CH_2CH_3$$

The carbon with the asterisk, called an asymmetric carbon atom, has attached to it four different groups (C_2H_5, ethyl; CH_3, methyl; C_3H_7, propyl; and H). These groups can be arranged in two ways, related to each other as mirror images, as shown in Fig. 20.7. The arrangements are called optical isomers or enantiomers (from the Greek word for opposite). (In these three-dimensional representations, a solid line represents a bond in the plane of the diagram, a wedge represents a bond coming in front of the diagram, and a dotted line represents a bond going in back of the diagram.) It is not possible to convert one optical isomer into another by rotation in space. For example, if the molecule on the left is rotated 180° about the CH_3—C^* bond, the C_3H_7 projects in front of the diagram, which is different from the mirror image on the right. The mirror images can be distinguished as *d* (from *dextro,* meaning "right") and *l* (from *levo,* meaning "left").

Optical isomers are so named because they rotate the plane of polarized light differently. As shown in Fig. 20.8, natural light has the electric field vibrating in all possible directions perpendicular to the beam direction. When

FIG. 20.8   Rotation of plane-polarized light.

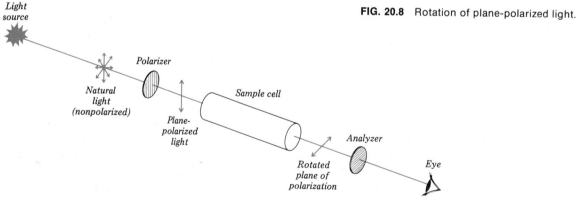

passed through a "polarizer," such as a crystal of calcite or a sheet of polaroid, the light vibrations are filtered out except for those in a single plane. This is called plane-polarized light. Passage through a sample cell containing but one kind of optical isomer rotates the plane of polarization. An "analyzer" (polaroid or calcite) transmits light only when lined up with the plane of polarization. Adjustment of the analyzer enables one to measure the degree of rotation. One optical isomer rotates the plane of polarization to the right; the other rotates the beam an equal amount to the left.

Optical isomers are very difficult to separate; they have identical chemical properties except when reacting with other molecules that show optical isomerism. As a consequence, they can sometimes be separated by slight solubility differences in a solvent that is composed of a single optical isomer. It is interesting to note that, in biological systems, chemical reactions frequently involve optically active molecules. In particular, reactions with optically active reagents, such as enzymes, often go faster with one enantiomer than with the other; hence, there is a preference in nature for that enantiomer.

Before leaving saturated hydrocarbons, we should note that it is also possible for carbon atoms to form *rings*, or *cycles*. The simplest cyclic hydrocarbon is cyclopropane, for which the molecular formula is C_3H_6. It consists of three C's at the corners of a triangle with two H's attached to each C. Other examples are cyclobutane (C_4H_8), cyclopentane (C_5H_{10}), and cyclohexane (C_6H_{12}). Cyclohexane shows the normal tetrahedral bond angle of $109.5°$, but the C—C—C bond angle in C_3H_6, C_4H_8, and C_5H_{10} is less than this. Consequently, the smaller cyclic compounds are said to be "strained" and have enhanced reactivity.

Saturated hydrocarbons are generally quite inert. However, they will react with halogens, especially when heated or exposed to light. A typical reaction would be

$$CH_4 + Cl_2 \xrightarrow{\text{light}} CH_3Cl + HCl$$

Substitution reactions with saturated hydrocarbons are generally slow and fairly hard to bring about.

Another characteristic reaction of hydrocarbons is that they all burn in air, usually to produce CO or CO_2, depending on the oxygen supply. This reaction is the basis for using hydrocarbons as fuels. In the Bunsen burner, a typical reaction would be

$$CH_4(g) + 2O_2(g) \longrightarrow CO_2(g) + 2H_2O(g) \qquad \Delta H = -887 \text{ kJ/mol}$$

For the burning of gasoline, a representative reaction would be

$$C_8H_{18}(g) + 12\tfrac{1}{2}O_2(g) \longrightarrow 8CO_2(g) + 9H_2O(g) \qquad \Delta H = -5460 \text{ kJ/mol}$$

Crude petroleum, our major source of hydrocarbons, consists of an almost infinite variety, e.g., straight chains, branched chains, unsaturated hydrocarbons, rings, etc. In the refining process, the various volatility fractions are separated from each other by distillation. The groups generally differ by the number of C's per molecule and can be classified as follows: gases (C_1 to C_4), gasoline (C_5 to C_{12}), kerosene (C_{10} to C_{16}), fuel and gas oil (C_{15} to

C_{22}), and lubricating oils (C_{19} to C_{35}). The residue of the distillation gives paraffin wax and asphalt (C_{36} to C_{90}).

The problem with petroleum is that the demand for gasoline exceeds that for other fractions. To enhance the supply of gasoline, four recourses are possible:

1 Cracking Higher hydrocarbons are thermally or catalytically decomposed to lower ones, for example, $C_{14}H_{30} \longrightarrow C_7H_{16} + C_7H_{14}$.

2 Polymerization Unsaturated hydrocarbons of the type C_4 and C_5 are polymerized to give C_8 and C_9 compounds.

3 Alkylation Small isoalkanes such as isobutane (C_4H_{10}) are made to combine with a small unsaturated hydrocarbon such as *butene* (C_4H_8) (see Sec. 20.4) to give a large isoalkane (C_8H_{18}).

4 Re-forming Straight-chain hydrocarbons are broken and reassembled as isohydrocarbons and partially dehydrogenated compounds. Approximately 40 percent of present fuels are made this way.

The isohydrocarbons and unsaturated hydrocarbons are better for low-knock combustion (Sec. 12.10) than are normal saturated hydrocarbons. References for low-knock combustion are isooctane (2,2,4-trimethylpentane), C_8H_{18}, which is assigned an *octane number* of 100, and *n*-heptane, C_7H_{16}, which knocks badly and is assigned an octane number of 0. Other fuels are rated by comparison with mixtures of the two reference compounds that give the same knock. Thus, for example, octane number 70 means a fuel that knocks like 70 parts (by volume) of iso-C_8H_{18} and 30 parts *n*-C_7H_{16}.

20.4
UNSATURATED HYDROCARBONS

Unsaturated hydrocarbons are characterized by sharing of two or three pairs of electrons between adjacent carbon atoms. The simplest examples are $H_2C{=}CH_2$ (ethylene) and $HC{\equiv}CH$ (acetylene). Ethylene is the first member of the ethylene, or *olefin*, *series*, general formula C_nH_{2n}, all of which contain but one double bond; acetylene is the first member of the acetylene series, general formula C_nH_{2n-2}, all of which contain but one triple bond. Unlike saturated hydrocarbons, these compounds are quite reactive and can add on hydrogen to become fully saturated. In systematic nomenclature, the olefins are sometimes called *alkenes*, and the acetylenes are called *alkynes*.

Ethylene is a planar molecule with a σ bond and a π bond between the two C atoms, as described in Sec. 5.9. The double bond prevents free rotation of the ends of the molecule around the C-to-C axis. Thus it is possible to get additional isomers when different groups are substituted for the H atoms. Thus, for example, for C_4H_8 (butene), there are four possible isomers:

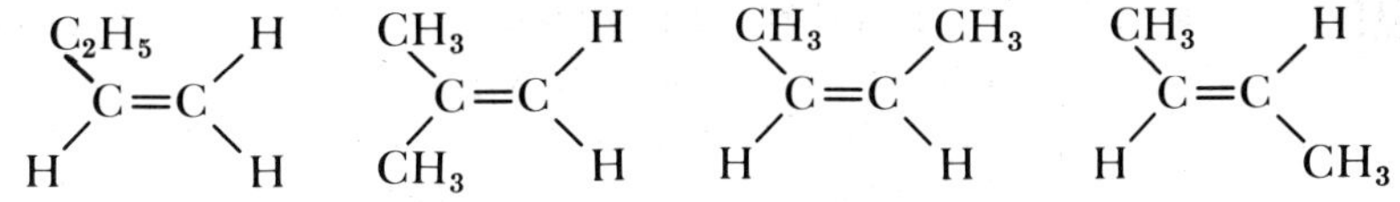

$$\underset{H}{\overset{C_2H_5}{>}}C=C\underset{H}{\overset{H}{<}} \qquad \underset{CH_3}{\overset{CH_3}{>}}C=C\underset{H}{\overset{H}{<}} \qquad \underset{H}{\overset{CH_3}{>}}C=C\underset{H}{\overset{CH_3}{<}} \qquad \underset{H}{\overset{CH_3}{>}}C=C\underset{CH_3}{\overset{H}{<}}$$

The first and second are *structural* isomers, they differ in which atoms are connected to which; the third and fourth are *geometrical* isomers, they differ in how the same set of connections is arranged in space. The third isomer has both methyl groups on the same side of the molecule (*cis*); the fourth has them across from each other (*trans*).

The olefins are quite reactive. Typical reactions are:

1 **Addition of halogen:**

$$H_2C=CH_2 + Cl_2 \longrightarrow CH_2ClCH_2Cl$$

2 **Addition of hydrogen in the presence of a catalyst:**

$$H_2C=CH_2 + H_2 \overset{Pt}{\longrightarrow} CH_3CH_3$$

3 **Oxidation:**

$$H_2C=CH_2 \overset{KMnO_4}{\longrightarrow} CH_2OHCH_2OH$$

4 **Combustion:**

$$C_2H_4 + 3O_2 \longrightarrow 2CO_2 + 2H_2O$$

Acetylene (C_2H_2) is a linear molecule $H-C\equiv C-H$ with one σ bond and two π bonds between the carbon atoms (Sec. 5.9). It is an important industrial chemical which can be made cheaply by heating lime and coke in an electric furnace:

$$CaO + 3C \longrightarrow CaC_2 + CO$$

and then adding water to the calcium carbide:

$$CaC_2 + 2H_2O \longrightarrow Ca(OH)_2 + C_2H_2$$

It is a colorless gas with a characteristic penetrating odor. Typical chemical reactions are:

1 **Addition of chlorine:**

$$H-C\equiv C-H + 2Cl_2 \longrightarrow CHCl_2CHCl_2$$

2 **Replacement of hydrogen by a metal:**

$$H-C\equiv C-H + 2Ag^+ \overset{alcohol}{\longrightarrow} Ag-C\equiv C-Ag + 2H^+$$

The product silver acetylide is highly explosive in the dry state.

3 **Combustion:**

$$2C_2H_2 + 5O_2 \longrightarrow 4CO_2 + 2H_2O$$

Two major uses of acetylene are as a fuel in the oxyacetylene torch (for welding and cutting metals) and as a raw material for making more complicated organic compounds.

The *dienes* or *diolefins* are unsaturated hydrocarbons with two double bonds per molecule. The first member of the series is $H_2C=CHCH=CH_2$, called butadiene. One of its important derivatives is *isoprene*, also called 2-methylbutadiene:

$$H_2C=\overset{\overset{\displaystyle CH_3}{|}}{C}-CH=CH_2$$

It is a normally colorless liquid, but if sodium is added, a sticky, rubberlike material results. A process called *polymerization* has occurred in which an individual, or monomer, unit C_5H_8 has joined to other similar units to build up long molecules of the type $(C_5H_8)_n$. These are called *polymers*. They may have molecular weights of the order of 100,000 or greater. The polymerization process for isoprene can be visualized as resulting from break up of the two double bonds to form one double bond in the middle and two unpaired electrons at the ends, as follows:

$$H-\overset{\overset{\displaystyle H}{|}}{C}-\overset{\overset{\displaystyle CH_3}{|}}{C}-\overset{\overset{\displaystyle H}{|}}{C}-\overset{\overset{\displaystyle H}{|}}{C}-H \longrightarrow H-\overset{\overset{\displaystyle H}{|}}{C}-\overset{\overset{\displaystyle CH_3}{|}}{C}=\overset{\overset{\displaystyle H}{|}}{C}-\overset{\overset{\displaystyle H}{|}}{C}-H$$

The ends then bond to other units to give a repeating sequence:

$$-CH_2-\overset{\overset{\displaystyle CH_3}{|}}{C}=CH-CH_2-CH_2-\overset{\overset{\displaystyle CH_3}{|}}{C}=CH-CH_2-CH_2-\overset{\overset{\displaystyle CH_3}{|}}{C}=CH-CH_2-$$

The long chains so formed help to explain the elastic properties of rubber. Because the atoms are in motion, the molecules tend to coil up on themselves and the end-to-end distance of the chain tends to shrink. Forces are required to stretch the molecule out to its maximum length. At lower temperatures, where the kinetic motion is diminished, the elasticity is less.

Natural rubber, or "caoutchouc," is found in the sap of the rubber tree (*hevea brasiliensis*). After evaporation of water, the rubber globules are coagulated to give a very sticky material. In the process of *vulcanization*, 4 to 5 percent sulfur is added to introduce cross-links between the molecules. The result is to give a substance that is not so sticky and keeps its shape. Other materials such as carbon can be added to serve as fillers, as in the manufacture of automobile tires. *Synthetic rubber* is a rubber substitute with a structure very similar to polymeric isoprene.

20.5
AROMATIC HYDROCARBONS

Aromatic hydrocarbons are a special class of compounds in which carbon atoms are arranged in loops, with double and single bonds alternating so as to endow the molecule with special properties. The most famous example of an aromatic hydrocarbon is benzene, C_6H_6. As described in Sec. 5.8, the molecule has six carbon atoms at the corners of a regular hexagon and a hydrogen atom attached to each carbon; it is best described as a resonance hybrid. The carbon-carbon distances do not alternate between short and long but are all equal to 0.140 nm, intermediate between a single (0.154 nm) and a double (0.133 nm) bond. The heat evolved on combustion has been determined to be 3300 kJ/mol, or 150 kJ less than would be anticipated for a cyclic compound consisting of alternating single and double bonds. The benzene grouping of atoms is an especially important one that shows up in a number of aromatic compounds. The simplest derivative of benzene is toluene ($C_6H_5CH_3$, methylbenzene) in which one of the hydrogens has been replaced by a methyl group. Like benzene, toluene can be obtained from destructive distillation of coal and petroleum. Destructive distillation simply means heating material in absence of air to drive off volatile products. Coal, when subject to destructive distillation, yields coke (mainly carbon), coal gas (light hydrocarbons), ammonia, and coal tar. Distillation of the coal tar gives benzene, toluene, and a variety of other aromatic compounds. Some of these are pictured in Fig. 20.9. They are all mildly fragrant, which explains the name "aromatic."

As can be seen for the three compounds at the right of the top row in Fig. 20.9, when there are two substituents on a benzene ring, three isomers result, depending on where the substituents are placed. If they are on adjacent carbon atoms, then the isomer is called *ortho;* if separated by another carbon, *meta;* if across the ring, *para.* Thus, for dimethylbenzene, $C_6H_4(CH_3)_2$, also called *xylene,* we have *o*-xylene, *m*-xylene, and *p*-xylene. Benzene, toluene, and the xylenes find extensive use as solvents and also as starting points for the synthesis of other organic compounds.

Phenol, C_6H_5OH, also shown in Fig. 20.9, has a hydroxyl group substi-

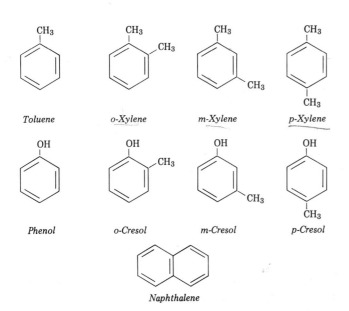

Toluene   *o-Xylene*   *m-Xylene*   *p-Xylene*

Phenol   *o-Cresol*   *m-Cresol*   *p-Cresol*

Naphthalene

FIG. 20.9 Some aromatic compounds. (In these representations, each corner of the hexagon represents a C atom and attached H atom. If an H atom has been replaced, only the substituent is shown. Only one resonance form is shown for each compound.)

tuted for one of the hydrogen atoms of benzene. It is famous as a disinfectant, sometimes under the name carbolic acid, and is important for the manufacture of many medical and industrial organic compounds. Cresols, which are also significant by-products of coal-tar distillation, have both a methyl and a hydroxyl group substituted on the same benzene ring. The three isomeric forms possible are shown in the second row of Fig. 20.9.

The last compound shown in Fig. 20.9 is naphthalene. It also comes from coal tar and can be visualized as consisting of two benzene rings fused together. In the process, two C atoms and four H atoms are lost. Hence, the C-to-H ratio increases as we go from benzene, C_6H_6, to naphthalene, $C_{10}H_8$. Naphthalene is just the first of a series of compounds which consist of networks of benzene rings. In the limit, as more and more rings are fused together, one reaches pure carbon in the form of graphite, as was shown in Fig. 20.2.

What special properties entitle a substance to be called aromatic? The presence of resonant benzene rings is one criterion. More important is diminished reactivity of the molecule. One way of rationalizing this is to note that the carbon p_z orbitals perpendicular to the plane of the benzene ring overlap each other so that the electrons are *delocalized* over the whole ring (see Fig. 20.2). As noted before (Sec. 16.2), spreading out an electron generally reduces its energy and hence its reactivity.

20.6
ALCOHOLS AND ETHERS

Alcohols are a class of organic compounds that can be represented by the shorthand notation ROH. R stands for any hydrocarbon residue, i.e., the part of a hydrocarbon molecule that is left after one H has been removed. Examples of hydrocarbon residues are CH_3 (methyl), C_2H_5 (ethyl), C_3H_7 (propyl), and C_4H_9 (butyl). Hydrocarbons can be thought of as compounds in which H is attached to a hydrocarbon residue. Thus, we have CH_3—H, C_2H_5—H, C_3H_7—H, and C_4H_9—H. If now the H is replaced by OH, we have another set of properties associated with presence of the hydroxyl grouping. Such a *grouping of substituent atoms* which bestows characteristic properties on an organic molecule is called a *functional group*. OH is the functional group that gives to molecules properties that are characteristic of alcohols. Figure 20.10 lists a few of the simpler alcohols.

Methyl alcohol, CH_3OH, also called *methanol,* can be considered to be a derivative of methane, with OH replacing one of the H atoms of CH_4. It is sometimes called wood alcohol, since it was originally obtained by destructive distillation of wood. Now it is usually made by combination of CO and H_2 in the presence of catalysts such as Cr_2O_3 and ZnO. Methanol is an important industrial solvent, a fuel, and a raw material for making organic compounds. It is quite poisonous and when ingested by breathing or drinking can lead to blindness and death. One of its uses is as a denaturing agent to make ethyl alcohol unfit for human consumption.

Ethyl alcohol, C_2H_5OH, also called *ethanol,* is derived from ethane by sub-

FIG. 20.10   Alcohols

Formula	Common name	Melting point, °C	Boiling point, °C
CH_3OH	Methyl alcohol	-98	65
CH_3CH_2OH	Ethyl alcohol	-117	79
$CH_3CH_2CH_2OH$	*n*-Propyl alcohol	-127	97
$CH_3CHOHCH_3$	*iso*-Propyl alcohol	-90	82
$CH_3CH_2CH_2CH_2OH$	*n*-Butyl alcohol	-90	117
$CH_3CH_2CHOHCH_3$	*sec*-Butyl alcohol	-115	100
$(CH_3)_3COH$	*tert*-Butyl alcohol	26	82

stituting an OH for one of the H atoms. The oldest and still used method of manufacturing ethyl alcohol is the fermentation of sugars by yeast. Typical is the reaction

$$C_6H_{12}O_6 \xrightarrow{\text{zymase}} 2CO_2 + 2C_2H_5OH$$
Glucose   Ethyl alcohol

as in the fermentation of grape juice to make wine. Yeast produces an enzyme zymase which acts as a catalyst for the conversion. Other carbohydrates, e.g., starch, can be used as raw material for making ethyl alcohol. In the making of beer, grain such as barley or oats is converted to "malt" by soaking in water and allowing to germinate. The germination produces an enzyme *diastase* which catalyzes conversion of starch into sugar. Subsequent fermentation gives ethyl alcohol. Beer generally has fairly low alcoholic content (3 to 6%), owing its characteristic flavor to addition of hops. Wine usually runs much stronger (10 to 14% C_2H_5OH). Distillation can be used to produce "distilled liquors," such as whiskey and brandy, which generally run from 40 to 50% alcohol. The maximum alcoholic content achievable by ordinary distillation is 95% (by volume). The last trace of water (8% by weight) has to be removed with the aid of chemical dehydrating agents. "Absolute alcohol" is 100% C_2H_5OH. Most ethyl alcohol is used in alcoholic beverages, but it is also important as a solvent and as a reagent for organic synthesis.

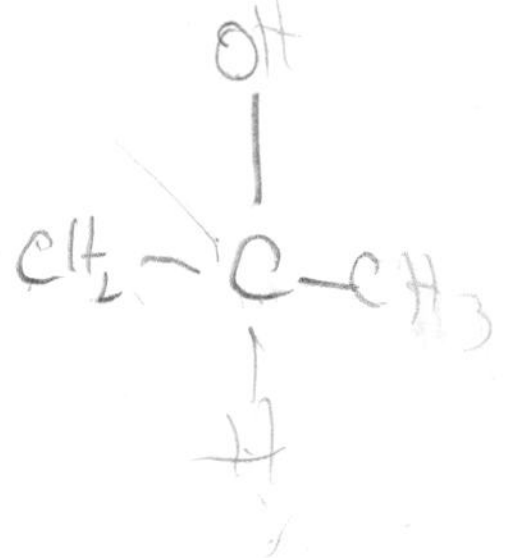

Polyalcohols are organic compounds that contain several OH groups per molecule. One example is $CH_2OHCHOHCH_2OH$, trihydroxypropane, commonly called *glycerol* or *glycerin*. It is produced when animal fat or vegetable oil is boiled with NaOH; therefore it is a by-product in the manufacture of soap. It is used as an antifreeze and also in making cosmetics, inks, glues, etc.

Propyl alcohol, C_3H_7OH, as is clear from Fig. 20.10, can exist in two isomeric forms, depending on whether the OH is attached to the middle C or the end C of the three-C chain. Normal propyl alcohol, designated *n*-C_3H_7OH, has the OH attached to a terminal C, and so it can be written $CH_3CH_2CH_2OH$. It is an example of a *primary alcohol,* i.e., one in which OH is bonded to a C with at least two H atoms attached. Isopropyl alcohol, designated iso-C_3H_7OH or $CH_3CHOHCH_3$, has the OH bonded to a C with but one H attached. It is an example of a *secondary alcohol*. Still another possibility would be to have a *tertiary alcohol*. This is one in which OH is bonded to a C that has no H attached. An example would be tertiary butyl alcohol, $(CH_3)_3COH$.

One of the most characteristic reactions of alcohols is their ability to be oxidized. For example,

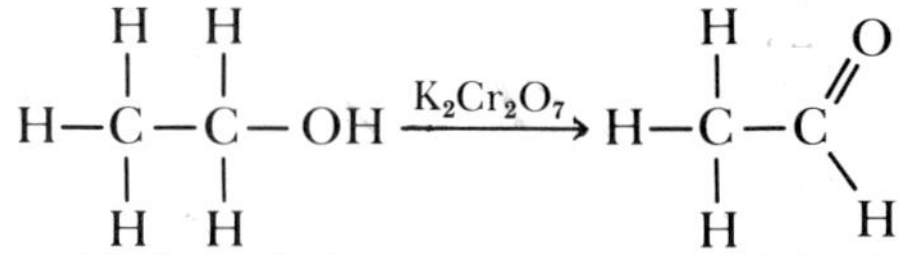

$$
\underset{\text{Ethyl alcohol}}{H-\overset{\overset{\displaystyle H}{|}}{\underset{\underset{\displaystyle H}{|}}{C}}-\overset{\overset{\displaystyle H}{|}}{\underset{\underset{\displaystyle H}{|}}{C}}-OH} \xrightarrow{K_2Cr_2O_7} \underset{\text{Acetaldehyde}}{H-\overset{\overset{\displaystyle H}{|}}{\underset{\underset{\displaystyle H}{|}}{C}}-\overset{\displaystyle O}{\underset{\displaystyle H}{C}}}
$$

The product, acetaldehyde, CH_3CHO, is two hydrogen atoms short of ethyl alcohol. This is typical of organic reactions in that "oxidation" frequently does not mean addition of oxygen, but instead means subtraction of hydrogen, i.e., dehydrogenation. In the above reaction it is seen that the essential change is

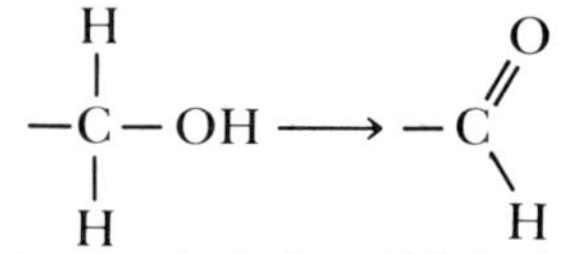

$$
\underset{\text{Primary alcohol}}{-\overset{\overset{\displaystyle H}{|}}{\underset{\underset{\displaystyle H}{|}}{C}}-OH} \longrightarrow \underset{\text{Aldehyde}}{-\overset{\displaystyle O}{\underset{\displaystyle H}{C}}}
$$

It illustrates the general rule that any primary alcohol can be oxidized to an aldehyde, for which the general formula is RCHO.

What happens if we start with a secondary alcohol? In such a case the essential change would be

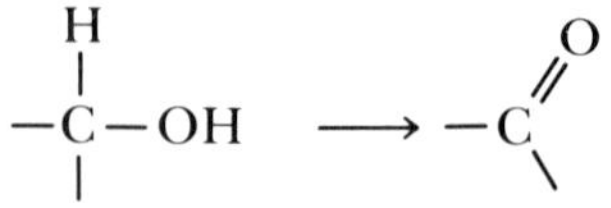

$$
\underset{\text{Secondary alcohol}}{-\overset{\overset{\displaystyle H}{|}}{\underset{\underset{\displaystyle |}{}}{C}}-OH} \longrightarrow \underset{\text{Ketone}}{-\overset{\displaystyle O}{C}}
$$

The CO, or *carbonyl,* group, when attached to other carbon atoms gives a class of compounds called ketones, for which the general formula is RCOR'. Tertiary alcohols, because of their structure, do not undergo corresponding dehydrogenation.

Another characteristic reaction of alcohols is dehydration, the elimination of a molecule of water, to form *ethers*. Ethers have the general formula R—O—R' and can be prepared by heating alcohols with a good dehydrating agent such as concentrated sulfuric acid. A typical reaction would be

$$
\underset{\text{Ethyl alcohol}}{2C_2H_5OH} \xrightarrow{H_2SO_4} \underset{\text{Diethyl ether}}{C_2H_5OC_2H_5} + H_2O
$$

The product, diethyl ether, is a colorless liquid with a rather pleasant odor. It is highly flammable and needs to be handled with great care. The liquid is quite volatile (bp 35°C), and the vapor, being fairly dense, can flow quite some distance along a bench top to an open flame. Diethyl ether is frequently used as an anesthetic and also, because of its high dissolving power for organic compounds, as an extracting solvent.

On prolonged standing in air, diethyl ether can become extremely hazardous because it may form an explosive peroxide. Before any ether is distilled, it should be tested for presence of peroxides. A simple test is to shake some of the ether with ferrous sulfate and potassium thiocyanate and look for the telltale red color of $FeNCS^{2+}$. Peroxides can be destroyed by addition of reducing agents such as $FeSO_4$. To prevent peroxide formation, ether is often stored with iron wire in it.

20.7
ALDEHYDES AND KETONES

Aldehydes have the general formula RCHO. They are prepared by oxidation of a primary alcohol. A typical reaction would be

$$2C_2H_5OH + O_2 \xrightarrow[300°C]{Cu} 2CH_3CHO + 2H_2O$$

Ethyl alcohol   Acetaldehyde

The lighter aldehydes have a sharp irritating odor; the higher aldehydes generally have pleasant odors. For example, $C_8H_{17}CHO$, nonyl aldehyde, smells like geraniums.

[Aldehydes act as mild reducing agents.] For example, acetaldehyde can be used to reduce silver-ammonia complex to a metallic silver mirror:

$$2Ag(NH_3)_2^+ + CH_3CHO + 3OH^- \longrightarrow$$

Acetaldehyde

$$2Ag(s) + 4NH_3 + 2H_2O + CH_3COO^-$$

Acetate

[The reducing power of many sugars is due to the presence of aldehyde groups.] A famous test for reducing sugars uses Fehling's solution, an alkaline solution containing Cu^{2+} complexed with sodium tartrate. Upon slight warming, a reddish deposit of Cu_2O is produced.

The simplest of the aldehydes is *formaldehyde*, HCHO. It is prepared by the oxidation of methanol, CH_3OH. "Formalin," an aqueous solution of HCHO, is much used as an antiseptic, since HCHO is toxic to lower forms of life. Polymerization of formaldehyde gas goes quite easily. It can be visualized as due to opening up of the double bond between C and O and then stringing together the fragments into a long repeating chain:

$$n \; \overset{H}{\underset{H}{\diagdown}}C{=}O \longrightarrow n \; \overset{H}{\underset{H}{\diagup}}C{-}O{-} \longrightarrow (-\overset{\overset{\displaystyle H}{|}}{\underset{\underset{\displaystyle H}{|}}{C}}{-}O{-})_n$$

The polymeric material is a white solid, called paraformaldehyde. Upon heating, it liberates HCHO gas, and so it is used in disinfection of sick rooms.

Ketones have the general formula $R-\overset{\overset{\displaystyle O}{\|}}{C}-R'$. They can be made by ox-idation of secondary alcohols. The simplest one is dimethyl ketone, CH_3COCH_3, commonly called acetone. It is used extensively as a solvent for fats, oils, waxes, etc. Unlike aldehydes which are fairly sensitive to oxidation, ketones are relatively inert to mild oxidizing agents. However, when heated with strong oxidizing agents such as $KMnO_4$, ketones get converted to acids (see following Sec. 20.8). The oxidation breaks the R-to-carbonyl bond ran-domly on the two sides of the carbonyl group so that a mixture of products is generally obtained:

$$RCOR' \xrightarrow{KMnO_4} RCOOH \text{ and } R'COOH$$

20.8
CARBOXYLIC ACIDS

Carboxylic acids have the general formula RCOOH, in which the carboxyl group COOH consists of a carbon with a doubly bonded oxygen and a singly bonded hydroxyl group. Dissociation of the hydroxyl group is what gives these compounds their acidic properties. The long-chain acids, where R con-tains 15 to 17 carbon atoms, are not very soluble in water; they are often called fatty acids since they are derived from animal fats. Stearic acid, for ex-ample, $C_{17}H_{35}COOH$, comes from beef tallow.

Figure 20.11 lists some of the simpler carboxylic acids. Formic acid, HCOOH, which occurs in nettles and in ants, gets its name from the Latin word *formica* for "ants." Early chemists used to make it by distillation of red ants. Acetic acid, CH_3COOH, the second member of the series, is the acid constituent of vinegar. It can be made by acetic acid fermentation, which occurs in the presence of a microorganism called *bacterium aceti*, or "mother of vinegar." As with the yeast that produces alcoholic fermentation, the *bac-terium aceti* that triggers acetic acid fermentation is often present as stray spores in the air. This explains why conversion of grape juice to wine frequently goes on to form vinegar.

To get concentrated acetic acid, the usual method is to air-oxidize ace-taldehyde, obtained either from the catalytic hydration of acetylene

$$H-C\equiv C-H + H_2O \xrightarrow[H_2SO_4]{HgSO_4} CH_3CHO$$

or from the catalytic dehydrogenation of ethyl alcohol. Pure acetic acid is called *glacial acetic acid.*

Butyric acid, $CH_3CH_2CH_2COOH$, has a very disagreeable smell. Its production in small amounts is what gives the characteristic smell to rancid butter. Palmitic acid ($C_{15}H_{31}COOH$), stearic acid ($C_{17}H_{35}COOH$), and oleic acid ($C_{17}H_{33}COOH$) do not occur in nature as such, but compounds derived from them do occur in various oils and fats. Oleic acid is related to stearic acid by having one double bond in the hydrocarbon chain instead of being fully saturated; it is obtained from olive oil.

FIG. 20.11 Carboxylic Acids

Formula	Common name	Boiling point, °C	K_{diss}
HCOOH	Formic acid	101	1.77×10^{-4}
CH_3COOH	Acetic acid	118	1.76×10^{-5}
C_2H_5COOH	Propionic acid	141	1.34×10^{-5}
C_3H_7COOH	Butyric acid	164	1.54×10^{-5}

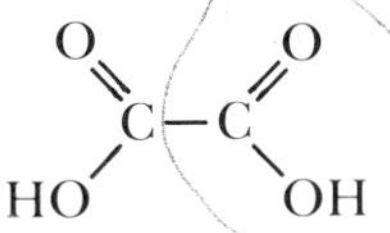

Some of the most important organic acids differ from the above in having two or more COOH groups per molecule or a combination of COOH and OH. The simplest is oxalic acid, which consists of two carboxyl groups joined together. It can be written as $(COOH)_2$ or $H_2C_2O_4$. The latter formulation emphasizes that there are two dissociable protons. The first one comes off with $K_{diss} = 5.36 \times 10^{-2}$; the second one, with $K_{diss} = 5.3 \times 10^{-5}$. Two types of salts can be produced, for example, $K_2C_2O_4$ (potassium oxalate) and KHC_2O_4 (potassium hydrogen oxalate). The latter occurs in the cell sap of many plants, e.g., rhubarb and sorrel. Oxalic acid itself is poisonous.

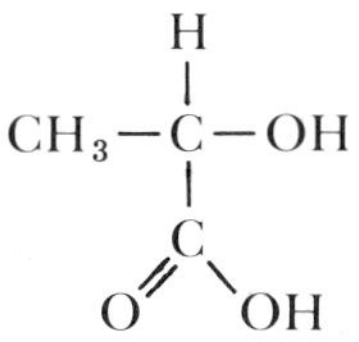

Lactic acid, $CH_3CHOHCOOH$, has both a carboxyl group ($K_{diss} = 1.4 \times 10^{-4}$) and a hydroxyl group. It occurs in sour milk, in muscles after activity, and in many other places where partial breakdown of sugars is occurring. It is interesting to note that the central carbon in lactic acid has four different groups attached to it, and so it is asymmetric. As a result, two possible configurations can exist, both of which are found in nature. The dextro form is found in muscle tissue; the levo form, in yeast. Sour milk has both.

Tartaric acid, shown in Fig. 20.12, has two carboxyl groups and two hydroxyl groups. It occurs as the monopotassium salt, $KHC_4H_4O_6$, also called potassium acid tartrate, in crystalline deposits in wine barrels. When purified, the potassium salt is known as cream of tartar. It finds extensive use in making baking powder.

Citric acid, also shown in Fig. 20.12, is widely distributed in nature, particularly in citrus fruits. Lemon juice, for example, contains from 5 to 8% of citric acid.

20.9 ESTERS

Esters have the general formula RCOOR′ and are derived from an acid RCOOH and an alcohol R′OH. In naming an ester, we give the R′ group

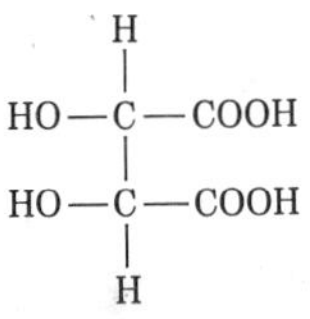

FIG. 20.12 Complex organic acids.

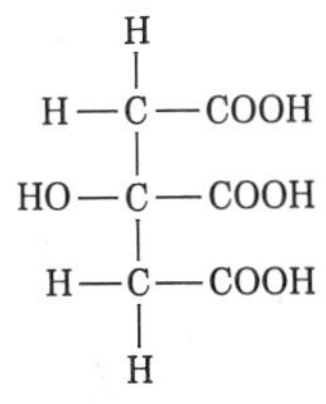

Tartaric acid

Citric acid

and then indicate the acid from which the rest of the molecule is derived. Thus, for $CH_3COOC_2H_5$ we have ethyl ester of acetic acid, or ethyl acetate.

[Esters are produced by interaction of an alcohol and an acid.] The reaction

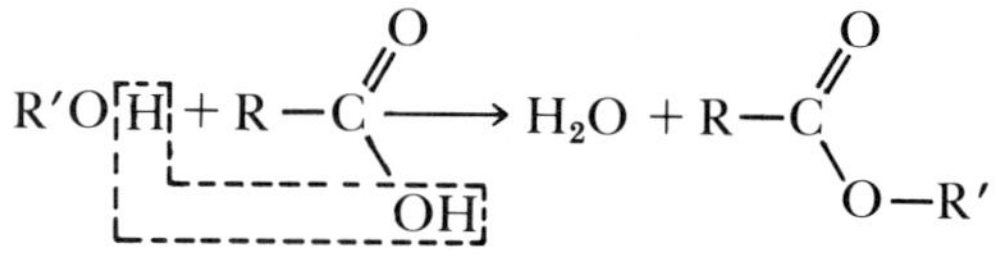

$$R'O[H] + R - C \overset{O}{\underset{[OH]}{\diagdown}} \longrightarrow H_2O + R - C \overset{O}{\underset{O-R'}{\diagdown}}$$

[formally looks like a neutralization leading to formation of water. However, it is much slower than normal acid-base neutralization. [Furthermore, as can be verified by tracer experiments, it is the OH of the acid which pairs with the H of the alcohol; this is opposite to what we might expect.]

Esters generally have pleasant odors and tastes. They are the principal flavor and odor constituents of fruits and flowers. For example, ethyl acetate ($CH_3COOC_2H_5$) smells like apples; amyl acetate ($CH_3COOC_5H_{11}$), like bananas. Ethyl acetate is an important solvent for varnishes and quick-drying automobile lacquers.

The natural fats and oils are also esters, principally of the polyalcohol glycerol. Animal fats, which tend to be solid, are generally rich in the glyceryl ester of palmitic acid, called *palmitin,* and the glyceryl ester of stearic acid, called *stearin.* Oils, on the other hand, such as olive oil and whale oil tend to be richer in the glyceryl ester of oleic acid, called *olein.* The formulas of these are shown in Fig. 20.13. Beef fat has about 75% palmitin plus stearin and 25% olein. Lard is about 40% palmitin plus stearin and 60% olein. Olive oil, a liquid, is about 75% olein; it presumably owes its liquidity to the presence of the double bond in the $C_{17}H_{33}$ side chain.

[Liquid fats often have disagreeable odors and tastes due to the unsaturated character of the side chains.] The amount of unsaturation, i.e., the extent of double bonding, can be reduced by catalytic hydrogenation. Thus, for example, when cottonseed oil is treated with hydrogen in the presence of nickel, it converts to a smooth creamy solid. Oleomargarine is a mixture of fats and oils and partially hydrogenated oils.

In addition to being important as foods (the energy liberated in the oxidation of 1 g of oleomargarine is about 40 kJ), fats and oils are raw materials

FIG. 20.13 Glyceryl esters of some fatty acids.

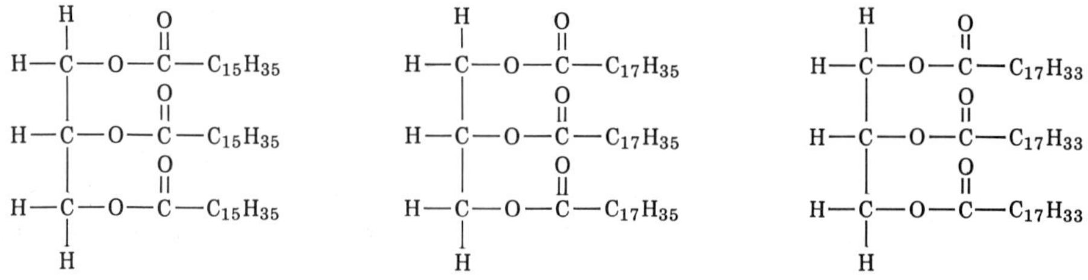

Palmitin   *Stearin*   *Olein*

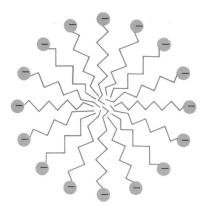

FIG. 20.14 Soap micelle.

for manufacture of soap and glycerol. For example, when stearin is boiled with NaOH, the reaction that occurs is

$$\begin{array}{lll}
C_{17}H_{35}COOCH_2 & & HOCH_2 \\
| & & | \\
C_{17}H_{35}COOCH & + 3NaOH \longrightarrow 3C_{17}H_{35}COONa + HOCH \\
| & & | \\
C_{17}H_{35}COOCH_2 & & HOCH_2 \\
\text{Stearin} & \text{Soap} & \text{Glycerol}
\end{array}$$

Products are glycerol and the sodium salt of stearic acid (soap). The cleansing action of soap is due to the dual nature of the soap anion, part hydrocarbon and part ionic. The hydrocarbon end is hydrophobic; i.e., it "hates water," and so the hydrocarbon parts cluster together. The charged part is hydrophilic; i.e., it "likes water," and so the carboxylate ends dissolve in water. The result is that, when placed in water, soap anions do not really dissolve. Instead, they form small colloidal particles called *micelles,* as shown schematically in Fig. 20.14. The negative charges are dissolved in water; the hydrocarbon chains are dissolved in each other. Cleansing action of soap is thought to stem from the dissolving of grease (hydrocarbon in nature) in the hydrocarbon clusters.

Soap substitutes were developed to get around the annoying fact that soap gives a curdy precipitate in hard water due to precipitation of insoluble calcium stearate. Alkylbenzenesulfonates (ABS) such as

$$\begin{array}{cccc}
CH_3 & CH_3 & CH_3 & CH_3 \\
| & | & | & | \\
CH_3CH-CH_2CH-CH_2CH-CH_2CH-
\end{array}\!\!\!\bigotimes\!\!-SO_3^-Na^+$$

do not precipitate with Ca^{2+} and make good soap substitutes, since they also lead to micelle formation. However, molecules of the type shown have one major flaw—they are not biodegradable. Apparently, bacteria are unable to eat branched chains. As a result, the branched alkylbenzenesulfonates tend to accumulate in the environment. Even as little as 1 ppm, for example, produces excessive foam formation in rivers.

The search for substitutes was based on the observation that ordinary straight-chain soaps are readily biodegradable. However, the problem was not solved until synthetic molecular sieves (see Sec. 21.2) were developed. These are ingenious silicate structures which allow molecules of small cross section to pass through them but exclude those of large cross section. With pore diameters of 0.5 nm, hydrocarbon straight chains of cross section 0.49 nm can be let through. Kerosene, for example, composed of C_{10} to C_{16} hydrocarbons, can have its n-alkanes separated out and converted to linear alkylbenzenesulfonates (LAS), which are completely biodegradable. Most detergent formulations now use LAS instead of ABS.

20.10 AMINES

Amines are derivatives of ammonia in which one or more of the hydrogen atoms of NH_3 have been replaced by R groups. Because of the unshared pair of electrons on the nitrogen, amines generally act as bases. Figure 20.15 shows some of the simpler amines. The K_b values listed are for base action in water, as follows:

$$CH_3NH_2 + H_2O \rightleftharpoons CH_3NH_3^+ + OH^-$$

As can be seen, the base strength of aniline (the amine which has a benzene ring attached to nitrogen) is considerably lower than that of the other amines, all of which are not very different from ammonia ($K_b = 1.8 \times 10^{-5}$). In the case of aniline, the unshared pair of electrons interacts with the aromatic ring so that it is less available for accepting a proton.

Closely related to the amines are the important compounds known as *amino acids*. Structurally, they are bifunctional; they contain two functional groups, an amine group and the organic-acid group. The simplest amino acid is glycine, NH_2CH_2COOH, also called aminoacetic acid. It differs from acetic acid (CH_3COOH) in that an NH_2 group has been substituted for one of the methyl's hydrogen atoms. Because glycine, like other amino acids, contains a basic NH_2 group and an acidic $COOH$ group, it has the ability to act both as a base and as an acid. This is illustrated by the following reactions:

$$H_2NCH_2COOH + H^+ \longrightarrow [H_3NCH_2COOH]^+$$
$$H_2NCH_2COOH + OH^- \longrightarrow [H_2NCH_2COO]^- + H_2O$$

FIG. 20.15 Amines

Formula	Name	Melting point, °C	Boiling point, °C	K_b, in H_2O
CH_3NH_2	Methylamine	−93.5	−6.3	5×10^{-4}
$(CH_3)_2NH$	Dimethylamine	−96	7.4	5.4×10^{-4}
$(CH_3)_3N$	Trimethylamine	−117	2.9	5.9×10^{-4}
$C_2H_5NH_2$	Ethylamine	−81	16.6	5.6×10^{-4}
$C_6H_5NH_2$	Aniline	−6.3	184	3.8×10^{-10}

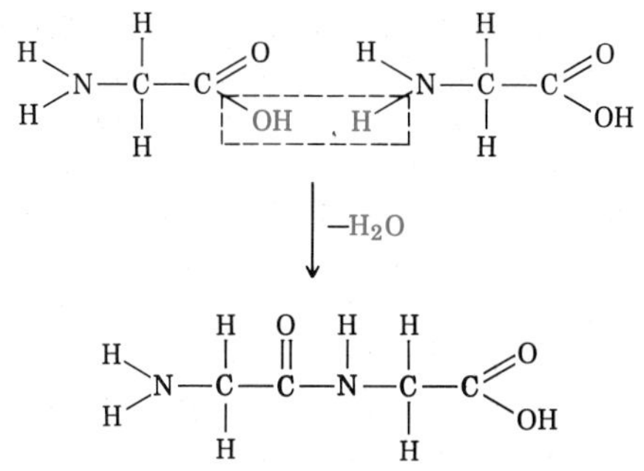

FIG. 20.16 Formation of a peptide link.

In the first reaction, the H^+ is neutralized by being joined to the amino acid through the unshared pair of electrons on the nitrogen atom. In the second reaction, OH^- is neutralized by interaction with a proton from the COOH group. Since the first reaction represents *addition of* H^+ to one end of the amino acid and the second reaction *removal of* H^+ from the other end, the two processes can occur together without intervention of external acid or base. The proton transfer can be represented as follows:

$$H_2NCH_2COOH \longrightarrow {}^+H_3NCH_2COO^-$$

The species produced, which is highly polar, is called a *zwitterion,* from the German word for "mongrel" or "hybrid." The net charge on the zwitterion is 0.

The most common source of amino acids is from the breakdown of proteins. These are extremely complex molecules of high molecular weight, which, on being boiled in acid or base, break up to form amino acids. More than 20 such amino acids have been identified. All proteins (hair, fingernails, skin, muscles, tendons, and blood) can be considered to be condensation products of two or more of these acids. The characteristic feature is the group $\begin{smallmatrix} H & O \\ | & \| \\ -N-C- \end{smallmatrix}$, called the *peptide link.* Figure 20.16 shows schematically how the peptide link might be established. We shall have more to say about proteins in Sec. 20.13.

20.11
ORGANIC REACTIONS

There are two important features that characterize reactions between organic compounds: One is the relative slowness compared with many familiar inorganic reactions. For example, whereas HCl reacts with NaOH practically instantaneously, the esterification between acetic acid and ethyl alcohol takes hours and even then occurs only if the reaction mixture is heated and a catalyst such as sulfuric acid is present. The other characteristic feature of organic reactions is that, in general, the greater part of the reacting molecule remains relatively unchanged during the course of the reaction.

Both the slowness of reaction and the retention of a major part of a molecule's identity can be exceedingly exasperating at times, but in general they prove very useful. For one thing, they allow the organic chemist to focus attention on a small portion of the molecule with the confident expectation that the rest of the molecule will not change much while the reaction is being carried out. For another thing, the slowness of reaction allows the changes to be stopped well before equilibrium is established. As a result, it is frequently possible to isolate compounds which if allowed to remain in the reaction mixture would react further to give different products.

Of the many types of organic reactions, some of the most commonly encountered are *addition, elimination, substitution,* and *polymerization.*

Addition

In *addition,* a reactant adds to an organic molecule containing a multiple bond. The multiple bond can be in a functional group, such as C=O of ketone, or in the carbon skeleton, such as C=C of an unsaturated hydrocarbon. Typical is the addition of hydrogen cyanide to aldehydes and ketones. To understand how it comes about, we note that in the C=O grouping oxygen, being more electronegative than carbon, carries a slightly negative charge (δ^-) and carbon is correspondingly positive (δ^+).

Step (1): $HC\equiv N \longrightarrow H^+ + : \overset{-}{C}\equiv N$

Step (2):

$$\overset{\delta^-}{O}=\underset{CH_3}{\overset{\overset{\displaystyle CH_3}{|}\,\delta^+}{C}}\quad :\overset{-}{C}\equiv N: \longrightarrow \overset{-}{O}-\underset{CH_3}{\overset{\overset{\displaystyle CH_3}{|}}{C}}-C\equiv N:$$

Step (3):

$$H^+ + \overset{-}{O}-\underset{CH_3}{\overset{\overset{\displaystyle CH_3}{|}}{C}}-C\equiv N: \longrightarrow H-O-\underset{CH_3}{\overset{\overset{\displaystyle CH_3}{|}}{C}}-C\equiv N:$$

In step (1) hydrogen cyanide dissociates. In step (2) the negatively charged CN^- seeks the positive end of the CO group. After the attack the bonds rearrange and the electronegative oxygen picks up an additional share of electrons. The negatively charged anion is then attacked by a proton [step (3)] to give the final product (called a *cyanohydrin*). Cyanohydrins are useful in organic synthesis, since they can be hydrolyzed to replace the CN group by COOH. In this way one can prepare organic acids with a hydroxyl group on the carbon that is adjacent to COOH.

Addition to a carbon-carbon double bond is typified by the reaction of HBr with ethylene. The overall change is

$$\underset{\substack{\\ \text{Ethylene}}}{\overset{\displaystyle H}{\underset{\displaystyle H}{>}}C=C\overset{\displaystyle H}{\underset{\displaystyle H}{<}}} + H-Br \longrightarrow \underset{\substack{\\ \text{Bromoethane}}}{H-\underset{\overset{|}{H}}{\overset{\overset{|}{H}}{C}}-\underset{\overset{|}{H}}{\overset{\overset{|}{H}}{C}}-Br}$$

The mechanism of the reaction is also believed to be stepwise. HBr dissociates to give a proton and a bromide ion. The proton, being positively charged, adds to the double bond, which is an electron-rich region. This can pull electrons away from one of the carbon atoms so that it looks positive, thereby attracting the negative bromide ion. The sequence may look like this:

$$\textbf{Step (1):} \quad CH_2{=}CH_2 + H^+ \longrightarrow CH_3-\overset{\displaystyle H}{\underset{\displaystyle H}{\overset{|}{\underset{|}{C}}}}{}^+$$

$$\textbf{Step (2):} \quad \left(CH_3-\overset{\displaystyle H}{\underset{\displaystyle H}{\overset{|}{\underset{|}{C}}}}{}^+ + Br^- \longrightarrow CH_3-CH_2Br \right)$$

In step (1) a positively charged ion is formed; it contains carbon with only six instead of the usual eight electrons. This is called a *carbonium ion*. It is highly reactive and frequently appears as an intermediate in organic reactions. Carbonium ions can be primary (1°), secondary (2°), or tertiary (3°), depending on the number of other carbon atoms joined to the positive carbon. Thus, we can have

$$\underset{\text{Primary (1°)}}{CH_3-\overset{\displaystyle H}{\overset{|}{\underset{+}{C}}}-H} \qquad \underset{\text{Secondary (2°)}}{CH_3-\overset{\displaystyle CH_3}{\overset{|}{\underset{+}{C}}}-H} \qquad \underset{\text{Tertiary (3°)}}{CH_3-\overset{\displaystyle CH_3}{\overset{|}{\underset{+}{C}}}-CH_3}$$

In general, the ease of formation of carbonium ions increases in the order 1° < 2° < 3°. This increasing order helps to explain a famous generalization known as Markovnikov's rule: *When HX adds to a double bond, the hydrogen atom goes to the carbon that already has more hydrogen.* To illustrate, addition of HCl to isobutylene proceeds as follows:

$$CH_3-\overset{\displaystyle CH_3}{\overset{|}{C}}{=}CH_2 + H^+ \longrightarrow CH_3-\overset{\displaystyle CH_3}{\underset{+}{\overset{|}{C}}}-CH_3$$

$$CH_3-\overset{\displaystyle CH_3}{\underset{+}{\overset{|}{C}}}-CH_3 + Cl^- \longrightarrow CH_3-\overset{\displaystyle CH_3}{\underset{\displaystyle Cl}{\overset{|}{\underset{|}{C}}}}-CH_3$$

The alternative path, where the proton adds to the other carbon of the double bond, would lead to a less stable carbonium ion.

Elimination

Elimination reactions are the reverse of addition reactions. They produce rather than consume unsaturated hydrocarbons. To illustrate, addition of HBr to $R-CH{=}CH_2$ to form $RCHBrCH_3$ can be reversed by treating the halogenated hydrocarbon with a concentrated solution of KOH in ethyl alcohol. Elimination seems to proceed by two mechanisms, depending on the specific reaction conditions. One mechanism is the exact reverse of the addition mechanism discussed above.

$$\text{Step (1):} \quad R-\underset{\underset{Br}{|}}{C}HCH_3 \longrightarrow Br^- + R\overset{+}{C}HCH_3$$

$$\text{Step (2):} \quad R\overset{+}{C}HCH_3 \longrightarrow RCH{=}CH_2 + H^+$$

The second is quite similar but differs in timing sequence. It is called a "concerted" mechanism in that steps occur simultaneously:

$$R-\underset{\underset{H}{|}}{C}\overset{Br\,\,H}{\underset{\underset{H}{|}}{C}}{-}H + OH^- \longrightarrow \underset{H}{\overset{R}{}}C{=}C\overset{H}{\underset{H}{}} + H_2O + Br^-$$

The attack here is by OH^- which extracts a proton, leaving the pair of electrons. Since the C attached to Br cannot accommodate five electron pairs, Br^- leaves, taking with it the fifth electron pair. Because the concerted mechanism is initiated by OH^-, it predominates over the two-step mechanism in highly basic solution.

Substitution

Substitution reactions are more complicated than either addition or elimination because they involve both removal of a group and addition of a different group. There are three general ways by which this can be accomplished: (1) attach the new group first, then remove the old group; (2) remove the old group, then attach the new group; (3) simultaneously attach the new group while breaking off the old group. Mechanism (1) is not very common since it means putting five bonds on a carbon; mechanisms (2) and (3) are preferred. Whether a reaction such as $RBr + Cl^- \longrightarrow RCl + Br^-$ goes by (2) or (3) depends on the nature of R. The more readily R forms a carbonium ion, the more likely the reaction goes by mechanism (2). Thus, for *tert*-butyl bromide, the reaction goes in two consecutive steps:

$$(CH_3)_3CBr \longrightarrow (CH_3)_3C^+ + Br^- \qquad \text{slow}$$
$$(CH_3)_3C^+ + Cl^- \longrightarrow (CH_3)_3CCl \qquad \text{fast}$$

However, for *sec*-butyl bromide, the reaction goes by mechanism (3) in a single concerted step. As shown in Fig. 20.17, the entering group (Cl^-) approaches one side of the molecule, while the leaving group (Br^-) departs from the other side. As indicated, the tetrahedral arrangement around the central carbon is inverted in the process (the H that went off to the left now goes off to the right). If, as in this case, the attached groups are all different so that the molecule is optically active, this type of substitution by displacement inverts dextro to levo, or vice versa. It is known as the *Walden inversion*.

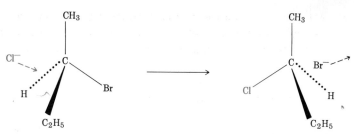

FIG. 20.17 Substitution of Cl for Br by displacement mechanism. (Solid bonds are in plane of diagram; wedge bond comes forward; dotted bond goes back.)

Another important class of substitution reactions involves benzene and its derivatives. We take as an example the chlorination of benzene:

$$\text{Benzene} \quad -\text{H} + \text{Cl}-\text{Cl} \longrightarrow \quad -\text{Cl} + \text{HCl} \quad \text{Chlorobenzene}$$

Although benzene appears to contain double bonds, it does not add molecules but instead undergoes substitution. To do otherwise would destroy the resonance stabilization of the ring system. When a positive ion adds to the C_6H_6 ring to form a carbonium ion, the following step is not the addition of a negative ion but the elimination of a proton so as to preserve maximum resonance possibility. For the above reaction, $FeCl_3$ is a catalyst, and the reaction is believed to occur as follows:

$$FeCl_3 + Cl_2 \longrightarrow FeCl_4^- + Cl^+ \qquad \text{fast}$$
$$Cl^+ + C_6H_6 \longrightarrow C_6H_6Cl^+ \qquad \text{very slow}$$
$$C_6H_6Cl^+ \longrightarrow C_6H_5Cl + H^+ \qquad \text{fast}$$

The function of the $FeCl_3$ is to form the cation Cl^+, which adds to the ring system and brings about elimination of H^+. The net effect is to substitute Cl for one of the six equivalent H atoms of C_6H_6. The overall reaction is fairly difficult to bring about. Long heating with the catalyst is required, and even then the yield of product is not very good.

Polymerization

The fourth type of reaction mentioned above, *polymerization*, can be of two very different kinds. In one of these, called *addition polymerization*, molecules add together to form giant molecules; in the other, called *condensation polymerization*, small molecules are split out as the giant molecule is built up. Addition polymerization, which is used for production of polyvinyl chloride, is thought to proceed in the following manner:

$$\text{R}-\text{O}-\text{O}-\text{R} \longrightarrow \text{R}-\text{O}\cdot + \cdot\text{O}-\text{R}$$
$$\text{Organic peroxide} \qquad \text{Peroxide free radicals}$$

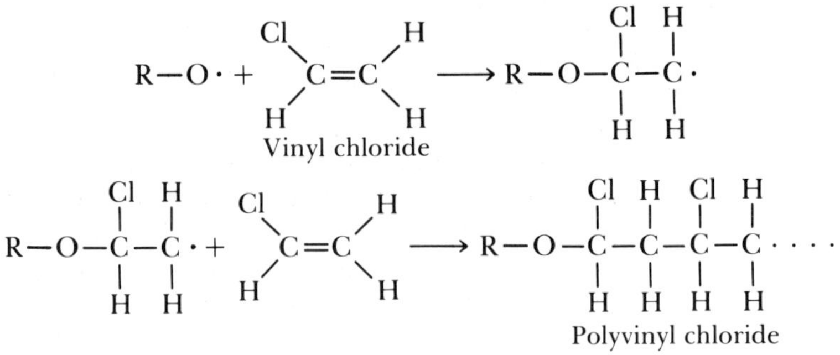

$$R-O\cdot + \underset{\underset{H}{|}}{\overset{\overset{Cl}{|}}{C}}=\underset{\underset{H}{|}}{\overset{\overset{H}{|}}{C} \longrightarrow R-O-\overset{\overset{Cl}{|}}{\underset{\underset{H}{|}}{C}}-\overset{\overset{H}{|}}{\underset{\underset{H}{|}}{C}}\cdot$$

Vinyl chloride

$$R-O-\overset{\overset{Cl}{|}}{\underset{\underset{H}{|}}{C}}-\overset{\overset{H}{|}}{\underset{\underset{H}{|}}{C}}\cdot + \overset{Cl}{\underset{H}{}}C=C\overset{H}{\underset{H}{}} \longrightarrow R-O-\overset{\overset{Cl}{|}}{\underset{\underset{H}{|}}{C}}-\overset{\overset{H}{|}}{\underset{\underset{H}{|}}{C}}-\overset{\overset{Cl}{|}}{\underset{\underset{H}{|}}{C}}-\overset{\overset{H}{|}}{\underset{\underset{H}{|}}{C}}\cdots$$

Polyvinyl chloride

In the first step peroxide added as *initiator* breaks a covalent bond so as to leave an unpaired electron (shown by the dot) on each R—O· residue. The residue is a free radical, is very reactive, and initiates a chain reaction. As shown in the second step, the free radical combines with a vinyl chloride molecule to form a new free radical which is one —CHClCH$_2$— unit longer than the original. Chain propagation thus continues, and a giant molecule is progressively built up until the free radical is destroyed, either by combining with another free radical or by reacting with some reagent added as an *inhibitor*. Such a mechanism accounts for the observation that the final product contains giant molecules of different molecular weights. It also explains the fact that a small amount of initiator can produce a great deal of polymerization. It should be noted that in the final giant molecule, the RO end group is an insignificant fraction of the molecule, and so the polymer is essentially an aggregate of CHClCH$_2$ units added together. Polyvinyl chloride (PVC) is one of our most common plastics, e.g., as in Saran wrap.

In condensation polymerization, buildup of the polymer occurs by splitting out a simpler species such as H$_2$O. In order that polymerization may occur, two conditions must be met: (1) Molecule A must be able to interact with molecule B. (2) Both molecules A and B must contain two functional groups so that, after A and B combine, the free ends can continue to react to extend the polymer. An example of condensation polymerization is the formation of Dacron polyester. Here the reaction is between methyl terephthalate, CH$_3$OOCC$_6$H$_4$COOCH$_3$, and ethylene glycol, HOCH$_2$CH$_2$OH. It involves elimination of methyl alcohol, CH$_3$OH. The reaction can be pictured as in Fig. 20.18. The step shown is then followed by similar reactions at both ends of the molecule.

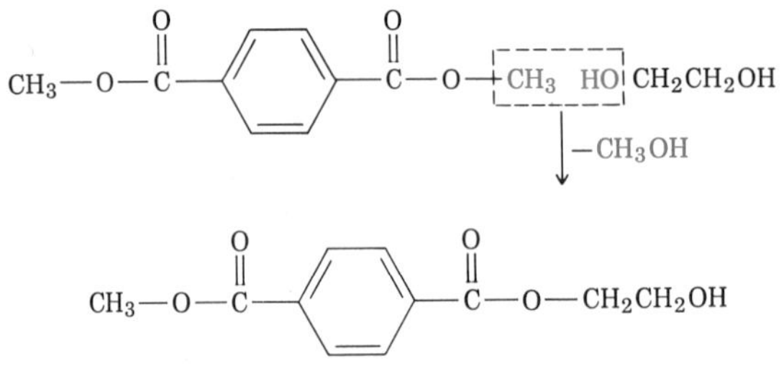

FIG. 20.18 Condensation polymerization.

The carbohydrates received their name because the ones that were first analyzed had the empirical formula $C_x(H_2O)_y$, in which carbon appears to be hydrated. Not only is the formula misleading, but further studies showed that not all carbohydrates have such a formula.

In a plant, carbohydrates form supporting tissues that make up its structure. The structural carbohydrates are called *polysaccharides* and, as we shall see, are high polymers composed of a tremendous number of simple monosaccharide units. The monosaccharides are *sugars*. However, not all sugars are monosaccharides; some of them contain more than one saccharide unit.

In addition to serving as building blocks for carbohydrates, two monosaccharides occur in significant quantities in nature. They are glucose and fructose. Both contain six carbon atoms (hence, they are called *hexoses*) and have the molecular formula $C_6H_{12}O_6$. There are a large number of isomers of this formula, but only two in addition to glucose and fructose occur naturally. All four of these hexoses are optically active, but only the D isomer (as opposed to the L isomer) occurs in nature. The four naturally occurring hexoses are shown in Fig. 20.19. As can be noted, all but fructose are in the form of a six-membered ring, with oxygen acting as one member of the ring. For fructose, the ring is five-membered, and two of the six carbon atoms are attached outside the ring.

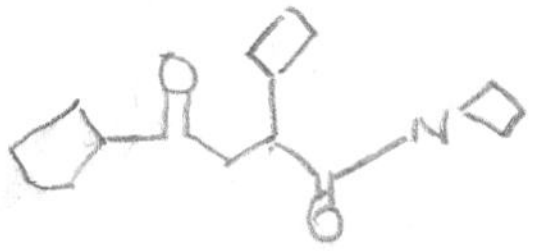

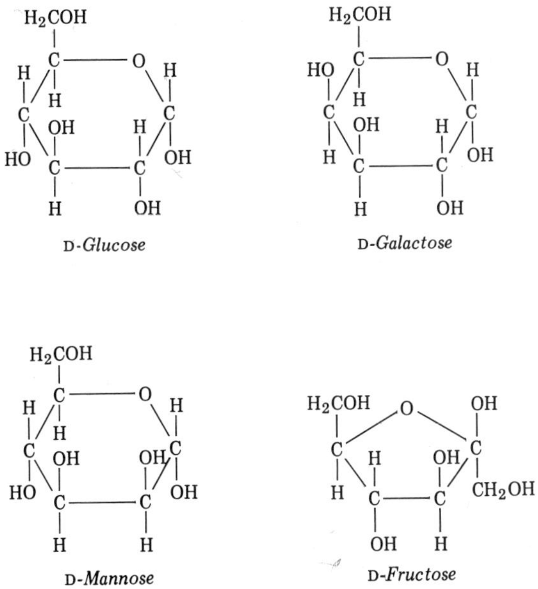

FIG. 20.19 Hexoses.

D-Glucose

D-Galactose

D-Mannose

D-Fructose

Besides the ring structures shown, the molecules may open up the rings as in the following rearrangement:

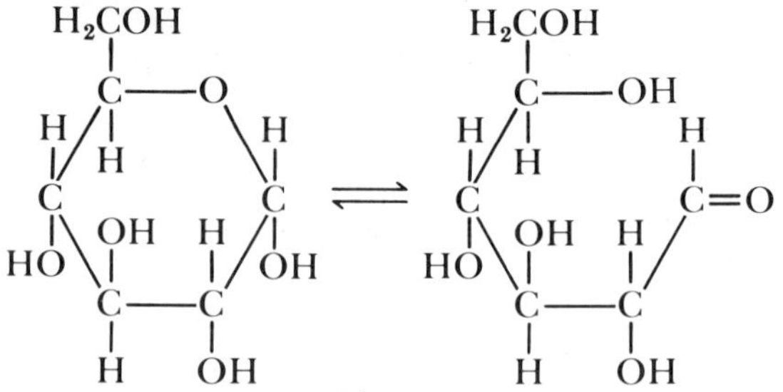

D-Glucose

When the ring is open, an aldehyde group (CHO) is present and consequently glucose as well as galactose and mannose give positive tests for an aldehyde group by reducing Fehling's solution. On the other hand, when the fructose ring opens, it forms a ketone, and so it does not show this test. Hence, fructose is not a "reducing sugar."

Glucose is also called *dextrose.* It is the principal sugar in blood, where it occurs at about 0.1 percent. When oral nutrition is not possible, glucose in saline solution is administered intravenously. As the most abundant of the monosaccharides, glucose occurs widely in fruits such as grapes and so is also called *grape sugar.* Fructose also occurs widely in fruits, as well as in honey, but the other hexoses generally occur only as polysaccharides.

In addition to the hexoses (6-carbon sugars), there are two pentoses (5-carbon sugars) that occur naturally in complex, high-molecular-weight compounds (e.g., nucleic acids). These pentoses are D-ribose and D-deoxyribose. Their formulas, corresponding to $C_5H_{10}O_5$ and $C_5H_{10}O_4$, respectively, are shown in Fig. 20.20. All living cells contain D-ribose, and so this sugar is of considerable interest to biological chemists.

Disaccharides are formed by linking together two monosaccharides. An example is *sucrose,* or ordinary table sugar. As shown in Fig. 20.21, it consists of D-glucose joined to D-fructose. Important natural sources of sucrose are sugar cane and sugar beets. Although both are about 15% sucrose, the sugar yield per acre from cane is higher than from beets. Because sugar cane grows only in tropical climates, its availability has been limited at various periods in history by political considerations. Indeed, the sugar-beet industry owes its origin in France to the British blockade during the Napoleonic Wars. Today, about half the world's supply of sucrose is from sugar beets.

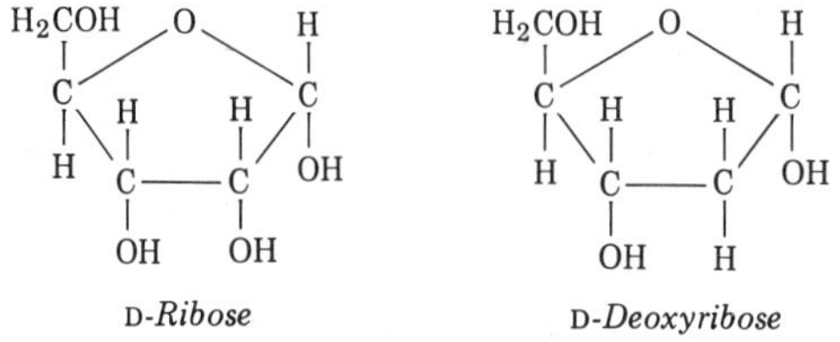

FIG. 20.20 Pentoses.

D-*Ribose* D-*Deoxyribose*

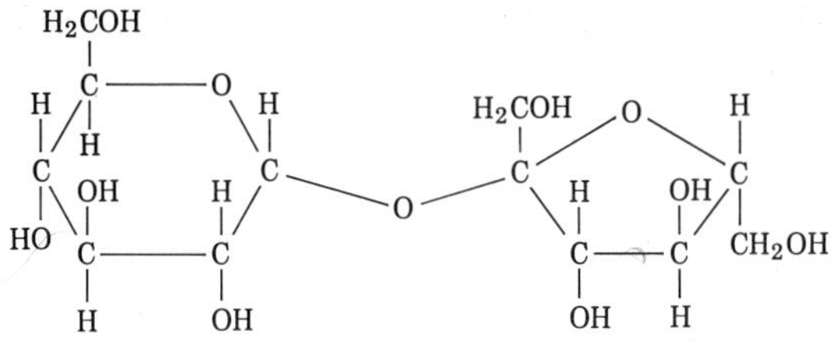

D-*Glucose* + D-*fructose* = *sucrose*

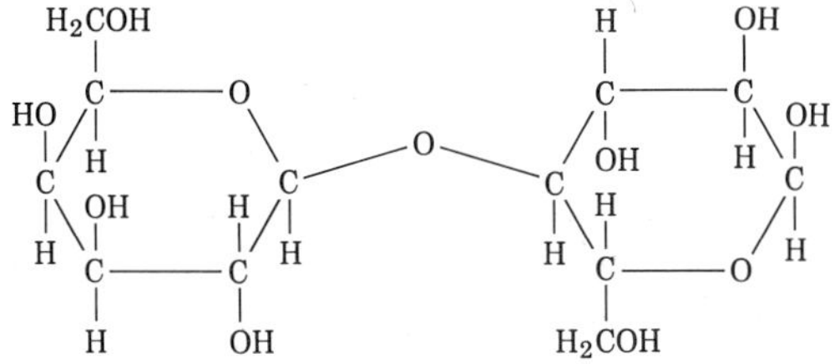

D-*Galactose* + D-*glucose* = *lactose*

FIG. 20.21 Disaccharides.

cellulose

Another familiar disaccharide is *lactose,* which makes up about 5 percent of milk. As shown in Fig. 20.21, it is a disaccharide of D-galactose and D-glucose.

Just as two monosaccharides can link to form a dissacharide, additional monosaccharides can be linked on to give *polysaccharides.* For example, one glucose unit can link to a second glucose unit to form the disaccharide called *maltose.* At either end of the maltose unit additional glucoses can link to give long, nearly infinite, chains of glucose units:

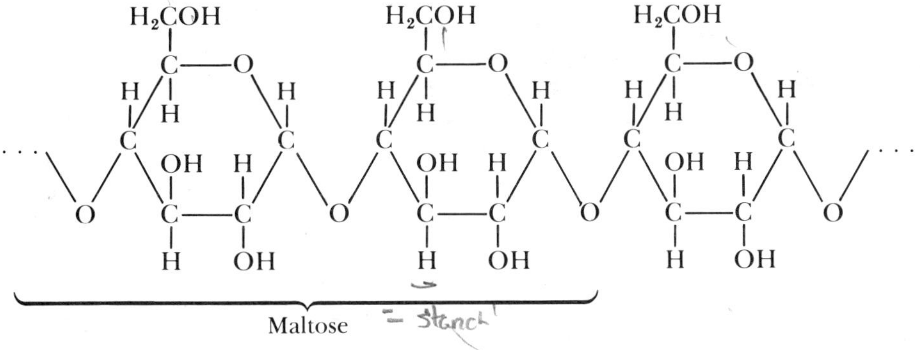

Maltose = *starch*

The huge linear polymer which is formed is one of the forms of *starch,* which represents a reserve energy store for plants. However, not all starch has this relatively simple structure. Chemical studies on the degradation of starch indicate that there are two types of starch, one, called *amylose,* composed of the end-to-end linkages of glucose units, and a second, called *amylopectin,* which contains both end-to-end linkages and side-chain linkages, resulting from joining CH_2OH groups to other glucose units. When this occurs, branches in

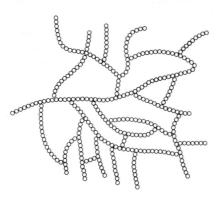

FIG. 20.22 Representation of glycogen structure. (Glucose units are shown as circles.)

the chain are formed. In the branched form of starch, one CH_2OH unit out of about 20 or 30 glucose units takes part in the branching.

It is estimated that the molecular weight of starch is very high, ranging in value to about a million. Even higher molecular weights are found in the related polysaccharide, glycogen, which is the form in which carbohydrate is stored in animals. It occurs in muscles and in the liver to be released under metabolic demand. Glycogen, like starch, consists of linked glucose units. It differs from starch in that not only branched chains are present but cross-links are formed which link one polymer chain to another. This kind of network structure is shown in Fig. 20.22. Molecular weights of glycogen are estimated to be as high as a hundred million or so.

A slight variant in the way glucose units are put together gives a long-chain polymer, known as cellulose. It is the most abundant organic compound in the world. Cotton, for example, is more than 90 percent pure cellulose, with polymers having molecular weights of the order of half a million.

20.13 PROTEINS

Proteins are high-molecular-weight organic compounds, composed mainly of C, H, O, and N, but also possibly containing some P and S. Some important examples are *keratin,* the primary constituent of skin, hair, and fingernails, and *collagen,* which makes up tendons and developing bones. Because of their function and extended network structure (molecular weight in the millions), these proteins are classified as *fibrous proteins.* Other proteins exist as smaller discrete molecules. These are roughly spherical in shape and are called *globular proteins.* Hemoglobin, the red matter in blood, with a molecular weight of 68,000 amu, is an example of a globular protein. Another is egg albumin; its molecular weight is 43,000. All proteins are polymers and consist of a large number of simple building units called *amino acids.* These acids have already been mentioned in Sec. 20.10 as bifunctional molecules which contain both an amine group and a carboxyl group. They are formed when a protein is boiled in acidic or basic solution so as to fragment the polymer.

[handwritten margin notes: peptide <70, amino acid, protein >70, enzymes, structures]

Although 26 amino acids have been found in nature, only 20 occur regularly in proteins. All but one can be considered to have the general formula

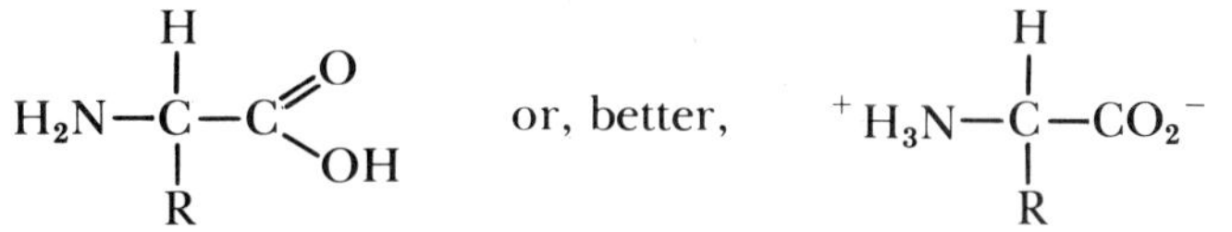

$$H_2N-\overset{\displaystyle H}{\underset{\displaystyle R}{C}}-C\overset{\displaystyle O}{\underset{\displaystyle OH}{}} \qquad \text{or, better,} \qquad {}^+H_3N-\overset{\displaystyle H}{\underset{\displaystyle R}{C}}-CO_2^-$$

The formula on the right represents the zwitterion form, as was mentioned in Sec. 20.10. Structures for all 20 of the protein-derived amino acids are shown in Fig. 20.23. In each case the structural factor in common is indicated in color. The last amino acid listed, proline, is not quite in the same general form as the others. It has the R group linked to the amino group.

Because, in all amino acids except glycine, the central carbon atom is bound to four different groups, an asymmetry results which manifests itself in optical activity. This is illustrated in Fig. 20.24 in which the two optical isomers of a general amino acid are shown. The two forms are designated as L and D. It is interesting to note that only the L-amino acids occur naturally in protein structures.

How are the amino acids bound together? As we have already seen in Sec. 20.10, reaction of the amine end of one amino acid with the carboxyl end of another, so as to split out water, forms the peptide link $-\overset{\displaystyle H}{\underset{}{N}}-\overset{\displaystyle O}{\underset{}{C}}-$ Further condensation polymerization produces longer chains. By convention, molecules formed from a relatively small number of amino acid residues are called *peptides;* the name *protein* is reserved for molecules of higher molecular weight, i.e., something greater than 70 amino acid residues. Peptides have biological importance, particularly as hormones. For example, the pituitary hormone *oxytocin* controls milk ejection in mammals. It has the following composition:

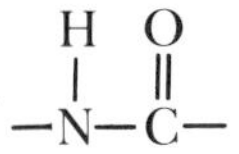

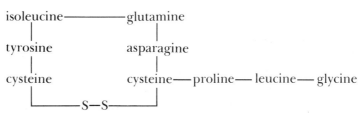

It consists of nine amino acid residues linked in a continuous chain which partly loops back on itself through a disulfide bridge involving the two cysteine residues.

The above structure does not tell us the actual shape of the molecule or its configuration in space. It does, however, tell us the sequence of amino acids so that we know all the covalent chemical bonds in the molecule. Such a structure is called a *primary structure.*

The biochemical function of a protein depends not only on the chemical sequence but also on the spatial configuration of the entire molecule. Each

$$H_3N^\pm - \overset{\overset{\displaystyle H}{|}}{\underset{\underset{\displaystyle H}{|}}{C}} - CO_2^-$$

Glycine

$$H_3N^\pm - \overset{\overset{\displaystyle H}{|}}{\underset{\underset{\displaystyle CH_3}{|}}{C}} - CO_2^-$$

Alanine

$$H_3N^\pm - \overset{\overset{\displaystyle H}{|}}{\underset{\underset{\displaystyle CH_3CHCH_3}{|}}{C}} - CO_2^-$$

Valine

$$H_3N^\pm - \overset{\overset{\displaystyle H}{|}}{\underset{\underset{\underset{\displaystyle CH_3CHCH_3}{|}}{\displaystyle CH_2}}{C}} - CO_2^-$$

Leucine

$$H_3N^\pm - \overset{\overset{\displaystyle H}{|}}{\underset{\underset{\displaystyle CH_3CH_2CHCH_3}{|}}{C}} - CO_2^-$$

Isoleucine

$$H_3N^\pm - \overset{\overset{\displaystyle H}{|}}{\underset{\underset{\displaystyle CH_2OH}{|}}{C}} - CO_2^-$$

Serine

$$H_3N^\pm - \overset{\overset{\displaystyle H}{|}}{\underset{\underset{\displaystyle CH_3CHOH}{|}}{C}} - CO_2^-$$

Threonine

$$H_3N^\pm - \overset{\overset{\displaystyle H}{|}}{\underset{\underset{\underset{\displaystyle COOH}{|}}{\displaystyle CH_2}}{C}} - CO_2^-$$

Aspartic acid

$$H_3N^\pm - \overset{\overset{\displaystyle H}{|}}{\underset{\underset{\underset{\displaystyle COOH}{|}}{\displaystyle (CH_2)_2}}{C}} - CO_2^-$$

Glutamic acid

$$H_3N^\pm - \overset{\overset{\displaystyle H}{|}}{\underset{\underset{\displaystyle (CH_2)_4NH_2}{|}}{C}} - CO_2^-$$

Lysine

$$H_3N^\pm - \overset{\overset{\displaystyle H}{|}}{\underset{\underset{\underset{\displaystyle NH}{||}}{\displaystyle (CH_2)_3NHC-NH_2}}{C}} - CO_2^-$$

Arginine

$$H_3N^\pm - \overset{\overset{\displaystyle H}{|}}{\underset{\underset{\displaystyle CH_2}{|}}{C}} - CO_2^-$$

Histidine

$$H_3N^\pm - \overset{\overset{\displaystyle H}{|}}{\underset{\underset{\underset{\displaystyle NH_2}{|}}{\displaystyle \underset{\displaystyle CO}{\underset{|}{CH_2}}}}{C}} - CO_2^-$$

Asparagine

$$H_3N^\pm - \overset{\overset{\displaystyle H}{|}}{\underset{\underset{\underset{\underset{\displaystyle NH_2}{|}}{\displaystyle CO}}{\displaystyle (CH_2)_2}}{C}} - CO_2^-$$

Glutamine

$$H_3N^\pm - \overset{\overset{\displaystyle H}{|}}{\underset{\underset{\displaystyle CH_2}{|}}{C}} - CO_2^-$$

Phenylalanine

$$H_3N^\pm - \overset{\overset{\displaystyle H}{|}}{\underset{\underset{\displaystyle CH_2}{|}}{C}} - CO_2^-$$

Tryptophan

$$H_3N^\pm - \overset{\overset{\displaystyle H}{|}}{\underset{\underset{\displaystyle CH_2}{|}}{C}} - CO_2^-$$

Tyrosine

$$H_3N^\pm - \overset{\overset{\displaystyle H}{|}}{\underset{\underset{\underset{\displaystyle SH}{|}}{\displaystyle CH_2}}{C}} - CO_2^-$$

Cysteine

$$H_3N^\pm - \overset{\overset{\displaystyle H}{|}}{\underset{\underset{\underset{\underset{\displaystyle CH_3}{|}}{\displaystyle S}}{\displaystyle (CH_2)_2}}{C}} - CO_2^-$$

Methionine

Proline

FIG. 20.23 Amino acids.

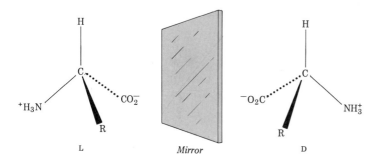

FIG. 20.24 Optical isomers of an amino acid.

amino acid segment can be set in space in a variety of possible orientations. The final three-dimensional disposition of the molecule is fixed not only by constraints produced by the covalent bonds but also by hydrogen bonding and other similar interactions between segments of the protein chain.

A particularly important spatial configuration is the *alpha helix*, diagrammed in Fig. 20.25, which seems to be the most stable single configuration for a protein chain. An important feature is that it is right-handed; i.e.,

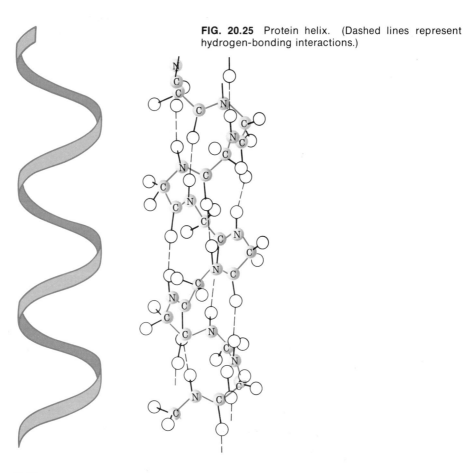

FIG. 20.25 Protein helix. (Dashed lines represent hydrogen-bonding interactions.)

575

the amino acid chain follows the pattern of the thread on a right-hand screw. There are 3.6 amino acid residues per turn of the helix, and each NH group forms a hydrogen bond to the carboxyl group of the third amino acid residue further along the chain. These hydrogen-bonding interactions give the alpha helix its stable configuration. However, in the presence of water, hydrogen bonds can be formed not only within the protein molecule itself but also with the solvent; hence, part of the helix may uncoil and the configuration may change in solution.

A further complication is that, even if we can establish the fraction of a protein molecule that is in an alpha-helix form (this can be done by a variety of physical measurements), we still do not have the whole story on its structure. The various helical segments of the protein molecule can hydrogen-bond with side groups and use other interactions to fold the helix into a more compact, often globular, form. Generally, the alpha-helix structure is referred to as the protein's *secondary structure,* and the folding of helical segments into additional structural configurations is called its *tertiary structure.*

In addition, some proteins have additional structure called *quaternary structure.* This arises if two or more protein molecules are bonded together to form a unit. For example, hemoglobin, the oxygen carrier in human blood, consists of four subunits bound together in a quasi-globular tetrahedral arrangement.

20.14
NUCLEIC ACIDS

Nucleic acids are complex nonamino acid portions of the nucleoproteins, the protein complexes that occur in all living cells and play vital roles in cell duplication and protein synthesis. Nucleic acids are of two types: ribonucleic acid (RNA) and deoxyribonucleic acid (DNA). They contain phosphoric acid, sugars, and two special nitrogen-containing bases, purine and pyrimidine. The formulas of the latter are

Purine   Pyrimidine

A typical example of a ribonucleic acid (RNA) is shown in Fig. 20.26. It is a polymer composed of sugar (ribose) rings hooked together through phosphate linkages. Each phosphate linkage carries a negative charge which is neutralized by a counter ion, either a metal ion, such as Na^+ or Mg^{2+}, or a more complex organic cation, which might be a substituted ammonium ion. The repeating unit phosphate-sugar-X is called a *nucleotide.* The X groups are generally derivatives of purines or pyrimidines. The major ones are shown in Fig. 20.27. Adenine and guanine are derivatives of purine; uracil and cy-

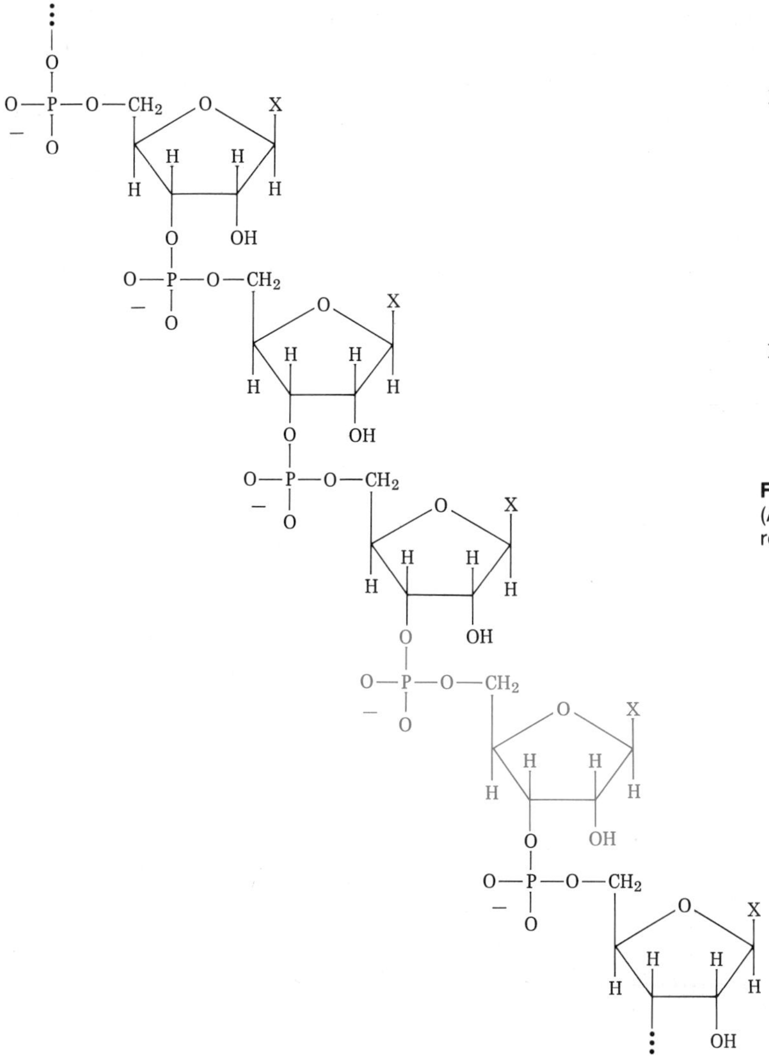

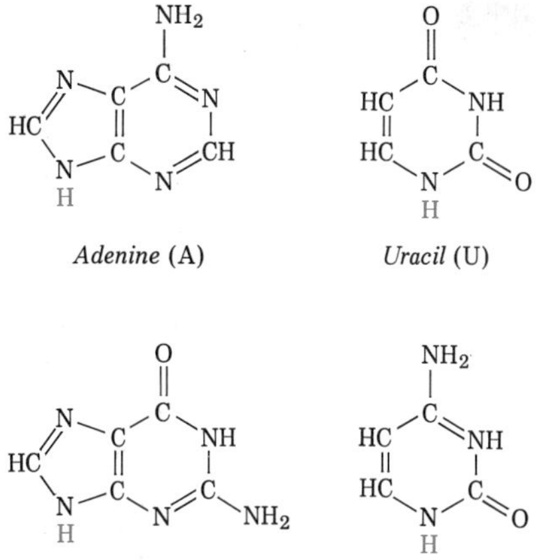

FIG. 20.27 RNA nucleotide substituents. (Attachment to the sugar chain is through replacement of the colored H.)

Adenine (A)

Uracil (U)

Guanine (G)

Cytosine (C)

FIG. 20.26 RNA polymer. (Repeat unit in color.)

tosine are derivatives of pyrimidine. In addition to these bases, a few other so-called "strange bases" as well as methyl-substituted or hydrogen-added variants are found.

Deoxyribonucleic acid (DNA) is structurally similar in many respects to ribonucleic acid. One principal difference is in the identity of the sugar unit. Instead of OH groups, there are H atoms. The absence of OH is reflected in the name *deoxy*ribonucleic acid. There is also a difference in the identity of the attached nitrogen bases. Again four bases predominate, with three of them being the same as before (adenine, cytosine, and guanine), but instead of uracil there is now a methyl derivative of uracil called thymine:

Thymine (T)

An important difference between DNA and RNA is the relationship between the relative numbers of the four nucleotides. Whereas in RNA the relative numbers may vary without apparent restraint, in DNA there is a definite restraint: the number of A (adenine) nucleotides equals the number of T (thymine), and the number of G (guanine) equals the number of C (cytosine). The equality conditions reflect the structural nature of the DNA molecule. It is a double helix, two representations of which are given in Fig. 20.28. The above nitrogen bases form cross-links between the two strands. The cross-links are thought to be held together by hydrogen bonds which are specific for binding A to T and for binding G to C. Thus, A is said to be the complement of T, and G the complement of C. In the final structure there must be one A for each T and one G for each C. It also follows that the arrangement of bases in one of the two chains fixes the arrangement of bases in the other.

DNA is the informational molecule of an organism. It contains coded within it, through the relative arrangement of A, T, C, and G, all the hereditary information that the organism needs for building each of the many pro-

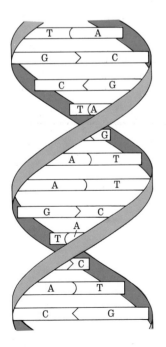

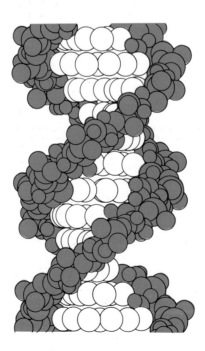

FIG. 20.28 DNA double helix composed of two sugar-phosphate chains held together by base pairs as shown.

teins that constitute the organism. Furthermore, the DNA molecule must be able to replicate itself as the organism grows. For this purpose the double-stranded nature of DNA is vital. The two strands are complementary to each other, so that each contains all the information. During cell division, the strands unzip and separately produce two molecules, each containing all the information that was coded into the original molecule.

What is the key to the information code stored in the DNA molecule? Ultimately the code must dictate the sequence in which 20 amino acids are to be used to build a particular protein. Clearly, only four DNA bases cannot code one-at-a-time for 20 amino acids. Furthermore, two adjacent nucleotides taken together would not suffice either. The number of possible combinations of two adjacent nucleotides would be 4×4, or 16. Since 16 possibilities are inadequate for the task, it is assumed that the code works through three adjacent nucleotides. For three adjacent nucleotides, there are $4 \times 4 \times 4$, or 64, possible combinations, more than adequate for coding 20 amino acids.

In using the information contained within the DNA molecule, it is RNA that actually does the work. Corresponding to the portion of a DNA molecule that contains the information for a specific protein chain, a complementary RNA chain is constructed. This copy, which is called *messenger RNA*, contains bases complementary to those of a portion of the DNA strand. For example, if the original DNA strand contains the sequence T—G—C, the corresponding messenger-RNA strand has A—C—G. Each triplet, which is called a *codon*, in the messenger RNA serves to specify the amino acid that is to be added next in the synthesis of the protein chain. Hence, messenger RNA serves as a template for protein synthesis. In reading the genetic code, another type of RNA, called *transfer RNA*, recognizes a codon and brings the indicated amino acid to the messenger-RNA template.

Important Concepts

diamond
graphite
carbides
oxides of carbon
carbonates and bicarbonates
saturated hydrocarbons
homologous series
structural isomers
optical isomers
petroleum and gasoline
unsaturated hydrocarbons
olefins

acetylenes
diolefins
aromatic hydrocarbons
functional groups
alcohols
ethers
aldehydes
ketones
carboxylic acids
esters
fats and oils
soap

amines
zwitterion
peptide link
addition reactions
carbonium ions
elimination reactions
substitution reactions
Walden inversion
polymerization
addition polymerization
condensation polymerization
carbohydrates
saccharides
polysaccharides
sugars

hexoses
pentoses
starch
cellulose
proteins
amino acids
protein structure
amino acid sequence
alpha helix
nucleic acids
DNA
RNA
nucleotide
genetic code

Exercises

*20.1 **Organic chemistry** Suggest reasons why in the early days of chemistry a compound such as ethyl alcohol would be classified as organic rather than inorganic.

*20.2 **Group IV elements** How might you explain the decreasing acidity of the dioxides of the group IV elements as one goes down the group?

20.3 **Elemental carbon Compare the two allotropic forms of carbon, graphite and diamond, with reference to each of the following: spatial arrangement of atoms, coordination number, bond strength, bond distances, description of the bonding.

20.4 **Diamond unit cell Given the cubic unit cell shown for diamond in Fig. 20.1, calculate the edge length of the unit cell, using 0.154 nm for the carbon-carbon distance. How many carbon atoms are there per unit cell? Calculate the theoretical density of diamond. *Ans. 0.356 nm*

*20.5 **Graphite** Explain why graphite is an excellent conductor parallel to the sheet direction but a poor conductor perpendicular to the sheet.

20.6 **Graphite A white, slippery insulating compound can be obtained by replacing the carbon atoms of the graphite sheet shown in Fig. 20.2 alternately by boron and nitrogen atoms. The com-

pound BN is very high melting but does not conduct electric current. Describe the probable bonding in BN and suggest reasons why it may or may not be like graphite.

20.7 **Charcoal How might you explain the strong adsorption properties of charcoal? The density of charcoal is about 0.3 g/cm^3; that of diamond is 3.5 g/cm^3. Into how small fragments should you powder diamond to get the same surface area as an equivalent weight of charcoal? *Ans. 1 × 10^{-6} cm*

*20.8 **Reactions of carbon** Suppose you had some radioactive carbon which you wanted to convert to radioactive acetylene. Sketch out the synthetic pathway you would use; include balanced equations. How many grams of radioactive carbon would you need to produce 1 liter of acetylene at STP?

*20.9 **Oxides of carbon** In the combustion of methane in air, how would the ratio of methane to air have to be changed to get the combustion products to span all the way from only $CO(g)$ to only $CO_2(g)$? Assume 100 percent efficiency.

20.10 **Carbon dioxide Tap water is almost never neutral but generally shows a pH of about 5. This deviation from neutrality is usually attributed to dissolving of CO_2 from the atmosphere. Assuming $4.2 × 10^{-7}$ for the first dissociation constant of carbonic acid, calculate the probable CO_2 concentration of a water sample showing a pH of 5.10.

**20.11 Carbon dioxide The solubility of CO_2 in water is given as 0.03 M when the CO_2 pressure is 1 atm. In a typical sample of air, the CO_2 pressure is about 0.03 atm. Calculate what pH would be expected for water in equilibrium with a typical sample of air. *Ans. 4.7*

**20.12 Sodium carbonate How would you proceed to convert carbon into sodium carbonate? Give equations. What pH would you expect for a solution that is 0.50 M Na_2CO_3?

***20.13 Sodium bicarbonate If $K_I = 4.2 \times 10^{-7}$ and $K_{II} = 4.8 \times 10^{-11}$ for CO_2 in H_2O, what pH would you expect for a solution that is 0.50 M $NaHCO_3$?

*20.14 Structural isomers By drawing the carbon skeletons, show that there are nine structural isomers for the saturated hydrocarbon heptane, C_7H_{16}.

**20.15 Optical isomers Two of the nine structural isomers of heptane can also show optical isomerism. Which of the isomers found in Exercise 20.14 exist as enantiomers?

**20.16 Optical isomers Suppose the sample cell shown in Fig. 20.8 is split vertically into two equal compartments. If one optical isomer is placed in one compartment and its mirror image in the other, what would you predict would be observed at the eye? What would probably be the result if one compartment were 3 times as long as the other?

**20.17 Cyclic compounds The C—C—C bond angles in cyclohexane are 109.5°; those in benzene are 120°. What does this tell you about the planarity of the respective carbon skeletons?

*20.18 Gasoline What is gasoline? How is it related to petroleum? Write a typical equation showing how the yield of gasoline-type hydrocarbons can be increased by each of the following: cracking, polymerization, alkylation, re-forming.

*20.19 Olefin series Write the structural formulas for the first four members of the olefin series, C_nH_{2n}. How many isomers are there of each? Draw structural formulas.

*20.20 Olefins Using structural formulas, write equations for the addition of chlorine to the four isomers of butene, C_4H_8. How many different products result?

**20.21 Diolefins Using structural formulas, show how the 2-methyl butadiene undergoes polymerization. Show how cis-trans isomerism is involved in the starting material and in the final product.

**20.22 Polymerization Suppose you make a synthetic rubber which is a copolymer of butadiene and styrene in alternating sequence. Styrene can be visualized as ethylene in which one H is replaced by a benzene ring C_6H_5. Using structural formulas, show how the polymerization might proceed and what the repeat unit of the final molecule might be. What will be the elemental percent composition of the final product?

***20.23 Aromatic compounds The ΔH for the reaction

$$C_6H_6(l) + \tfrac{15}{2}O_2(g) \longrightarrow 6CO_2(g) + 3H_2O(l)$$

is -3300 kJ, which is 150 kJ smaller in magnitude than would be expected for a cyclic compound containing alternating double or single bonds. Does this mean that $C_6H_6(l)$ is more or less stable than expected compared with the elements?

**20.24 Aromatic compounds Given the structures that are shown in Fig. 20.9, how many dihydroxybenzene types would you expect? How many trihydroxybenzenes? Draw their structures.

**20.25 Alcohols What are commercial ways of making methyl alcohol and ethyl alcohol? Suggest a mechanism by which H_2SO_4 acts as a catalyst for hydration of $H_2C\text{=}CH_2$ to CH_3CH_2OH.

*20.26 Alcohols Figure 20.10 shows formulas of three butyl alcohols. Draw structural formulas to show how they differ from each other. Which of these shows optical isomerism?

**20.27 Alcohols Using *n*-propyl alcohol, show with structural formulas what products would be formed on (*a*) oxidation, (*b*) reduction, and (*c*) dehydration.

**20.28 Ethers Why is it prudent to test an ether with ferrous sulfate and potassium thiocyanate before distilling it? Be specific.

***20.29 Aldehydes Write balanced net equations for each of the following reactions:

a Methanol is passed with oxygen over hot copper.

b The product from (*a*) is bubbled through an

alkaline solution of silver-ammonia complex to precipitate elemental silver.

c The product from (*a*) is passed through an alkaline solution of Cu(II) to precipitate a reddish solid.

*20.30 Ketones What secondary alcohol would be the precursor of acetone? If acetone were heated with $KMnO_4$, what acid would result?

***20.31 Organic acids Without making a numerical calculation, predict how the pH of a formic acid solution should compare with the pH of an equimolar acetic acid solution, given that K_{diss} of formic acid is 10 times greater.

**20.32 Acetic acid When cider ferments, the result is sometimes "hard cider," i.e., alcoholic, and sometimes vinegar. Show how the two are related.

***20.33 Oxalic acid Draw the structural formula of oxalic acid. Show how this formula helps to account for the fact that K_{II} is not approximately 10^{-5} times K_I as is normally observed for diprotic acids (for example, H_2CO_3 and H_2SO_3).

*20.34 Esters Esterification has been called the analog of acid-base neutralization. In what two important respects is it different?

*20.35 Soap What is soap? How is it related to fat? How can the cleansing action of soap be explained in terms of colloid formation? What advantage over soap is a detergent such as alkylbenzenesulfonate?

***20.36 Amines Explain why aqueous solutions of amines generally give high-pH solutions. Why is aniline so different from other amines?

**20.37 Zwitterion What is a zwitterion? Why does it form? What happens to the concentration of a zwitterion when the pH of a solution is raised? lowered?

**20.38 Peptide link What condition would likely enhance formation of a peptide link?

*20.39 Organic reactions What are the principal differences between organic reactions, such as that between methyl alcohol and acetic acid, and inorganic reactions, such as that between Fe^{2+} and MnO_4^-?

**20.40 Addition reactions Using structural formulas write equations for (*a*) the addition of HCN to acetone and (*b*) the addition of HCN to ethylene.

**20.41 Carbonium ions What is a carbonium ion? How might you rationalize that a tertiary carbonium ion forms more easily than a primary carbonium ion? How would this be involved in explaining how HBr adds to $(CH_3)_2C{=}CH_2$? What would probably be the final product when HBr adds to $CH_3CH_2CH{=}CH_2$?

**20.42 Elimination Show with structural formulas two different mechanisms by which HBr would be eliminated from $CH_3CHBrCH_2CH_3$. Which of the mechanisms would predominate in highly basic solution?

**20.43 Substitution Suppose you carry out the substitution reaction

$$OH^- + H{-}\underset{\underset{C_2H_5}{|}}{\overset{\overset{CH_3}{|}}{C}}{-}Cl \longrightarrow H{-}\underset{\underset{C_2H_5}{|}}{\overset{\overset{CH_3}{|}}{C}}{-}OH + Cl^-$$

What evidence would help you prove that it occurs by a Walden inversion? Draw a sketch showing the spatial relations involved.

***20.44 Substitution Explain how $FeCl_3$ acts as a catalyst in the substitution of Cl for H when C_6H_6 reacts with Cl_2 to form C_6H_5Cl.

**20.45 Addition polymerization Explain how organic peroxides $R{-}O{-}O{-}R$ can help to initiate the polymerization of ethylene to form polyethylene. How might you explain that the resulting polymer will generally consist of chains of various lengths instead of one giant polymer where all the $-CH_2CH_2-$ units are polymerized together?

*20.46 Carbohydrates Suggest two reasons why the name "carbohydrate" might be a misnomer.

***20.47 Carbohydrates What is the main difference between D-glucose and D-galactose? How might you tell if a given compound were one or the other?

*20.48 Carbohydrates Explain how D-glucose can give an aldehyde test with Fehling's solution even though there is no aldehyde group in the formula shown on page 569.

*20.49 Carbohydrates Show how glucose, fructose, and sucrose are related to each other. Write empirical formulas for each. What difference is there in the percent carbon in these compounds?

**20.50 Starch How is starch related structurally to the sugar glucose? What is glycogen, and how is it related to starch and to glucose?

20.51 Proteins Take 0.9 g/cm³ as the typical density of a protein. Assuming spherical shape, what would be the radius of a hemoglobin molecule if it had roughly this density? *Ans. 3.1 × 10⁻⁷ cm*

20.52 Amino acids Which of the 20 amino acids shown in Fig. 20.23 has the highest percent nitrogen? (Try to answer this one *without* calculating the percent nitrogen of all the amino acids.)

20.53 Amino acids The formulas presented in Fig. 20.23 are set up so as to show the common feature in all these amino acids. In which of the cases are other zwitterion formulations possible? How would they differ from the ones shown?

20.54 Optical activity Which of the amino acids shown in Fig. 20.23 have an asymmetric carbon other than the one shown in color?

20.55 Proteins What is the essential structural feature that characterizes all proteins? What is meant by primary, secondary, tertiary, and quaternary structure as applied to proteins?

20.56 Nucleic acids Tell briefly what is meant by each of the following: nucleic acid, nucleotide, DNA, messenger RNA.

20.57 Nucleic acids Tell specifically how sugars are related to nucleic acids. Show how the sugars shown in Fig. 20.20 are related to nucleic acids.

20.58 Nucleic acids Explain how the term "base" can be applied to the structure features of nucleic acids. What special relations exist between some of the bases in DNA? What is the reason for these special relations?

***20.59 DNA** Tell how genetic information is believed to be chemically coded in the structure of DNA.

20.60 DNA Suppose there were only three kinds of DNA bases instead of four. What difference would this make in the number of bases needed for a codon? Justify your answer.

Chapter 21

CHEMISTRY OF SOME NONMETALS OTHER THAN CARBON

Unlike the chemistry of the metals, where a few general principles can suffice to cover a large part of the chemistry, the behavior of nonmetals is so characteristic of each element that we need to consider them individually. In the preceding chapter, we considered the chemistry of carbon; in this chapter, we take up boron, silicon, nitrogen, phosphorus, sulfur, and the halogens. For the halogens, group properties vary so systematically that we shall be able to summarize some of the chemistry through group relations; each of the other elements will be treated as an important special case.

21.1
BORON

Boron is the head element of group III, which contains boron, aluminum, gallium, indium, and thallium. Some representative properties of these elements are shown in Fig. 21.1. Except for boron, which is sometimes classed as a semimetal, these elements show typical metallic properties. The special character of boron stems principally from the small size of its atom. As shown by the ionization potential in Fig. 21.1, boron gives up electrons less readily than do other atoms of the group. Since low ionization potential favors metallic properties, it is not surprising that boron is the least metallic of the group. However, ionization potential is not the only factor which determines whether an element is metallic. For example, gold (9.22 eV) has a higher first ionization potential than boron (8.30 eV) and yet is a typical metal. The detailed structure of the solid is also important. Gold has 12 atoms as nearest neighbors; boron has 6 or less. In general, good metals have their atoms characterized by a large number of nearest neighbors; nonmetals, by relatively few.

From the ns^2np^1 electronic configurations given in Fig. 21.1, it might be expected that all the group III elements would form +3 ions. However, boron has such a firm hold on its three valence electrons that it does not exist as B^{3+} cations in its compounds but takes part in chemical combination only through covalent binding. The other members of group III give up their electrons more readily; hence, formation of a +3 ion becomes progressively easier down the group.

As discussed in Sec. 9.7, a highly charged cation in water tends to pull electrons to itself and thereby weaken the O—H bonds in water. The larger

FIG. 21.1 Properties of Group III Elements

Symbol	Z	Electronic configuration	Melting point, °C	Boiling point, °C	Ionization potential, eV	Electrode potential, V
B	5	(2) $2s^2 2p^1$	2040	4100	8.30	-0.87 (from H_3BO_3)
Al	13	(10) $3s^2 3p^1$	659.7	2300	5.98	-1.66 (from Al^{3+})
Ga	31	(28) $4s^2 4p^1$	29.8	2430	6.00	-0.53 (from Ga^{3+})
In	49	(46) $5s^2 5p^1$	155	2170	5.79	-0.34 (from In^{3+})
Tl	81	(78) $6s^2 6p^1$	304	1460	6.11	-0.34 (from Tl^{+})

the cation, the smaller is the effect. In going down group III the effect of this change is well illustrated. Boron is so small that if a B^{3+} ion were placed in water, it would pull electrons to itself from water strongly enough to rupture the O—H bond and release H^+. As a result, $B(OH)_3$ and the corresponding oxide B_2O_3 are acidic. Al^{3+} and Ga^{3+} are larger than B^{3+}, and they hydrolyze less; $Al(OH)_3$, Al_2O_3, and the corresponding compounds of gallium are amphoteric. In^{3+} and Tl^{3+} are still larger. Their interactions with water are so small that the O—H bond of water is essentially unperturbed; i.e., the ions are but slightly hydrolyzed. Their hydroxides are basic. Thus, in going down group III there is a pronounced change from acidic behavior to basic behavior.

In nature, boron is moderately rare. It occurs principally as the borates (oxyboron anions), e.g., borax, $Na_2B_4O_7 \cdot 10H_2O$. The element may be produced by reducing the oxide B_2O_3 with a metal such as magnesium, or reducing boron trichloride, BCl_3, with hydrogen. The purest boron is made by thermal decomposition of $BI_3(g)$ on a heated tantalum filament. It is extremely strong and is now being used in composite materials, e.g., boron fibers in an epoxy matrix, for specialty applications such as rocket-motor casings.

Massive boron is very hard. It has a dull metallic luster but is a poor conductor of electricity. When its temperature is raised, the conductivity increases. This is unlike metallic behavior; therefore, boron and substances like it (silicon and germanium) are called *semiconductors*. The explanation of semiconductivity is that at room temperature, electrons are bound rather tightly to local centers, but as the temperature is raised, they are freed and are able to move through the crystal. The higher the temperature, the greater the number of electrons freed; hence, the conductivity rapidly increases.

At room temperature, boron is inert to all except the most powerful oxidizing agents, such as fluorine and concentrated nitric acid. However, when fused with alkaline oxidizing mixtures, such as NaOH and $NaNO_3$, it reacts to form borates. Boron also combines at higher temperature with metallic elements to give a bewildering variety of borides, for example, Mn_4B, FeB, Cr_3B_4, ZrB_2, CaB_6. Most of the borides are chemically quite inert, but some of them, particularly the alkaline-earth borides such as MgB_2, react with acids to produce boron-hydrogen compounds that are of unusual interest. A whole series of hydrides can be obtained, ranging from B_2H_6 (diborane) to $B_{18}H_{22}$

585

586 Chapter 21
Chemistry
of Some
Nonmetals
Other than
Carbon

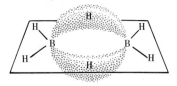

FIG. 21.2 Molecular-orbital representation of diborane.

(octadecaborane). All the compounds are surprising since there seem to be too few electrons to hold them together. Diborane, B_2H_6, for example, has only 12 valence electrons (three from each boron and one from each hydrogen) for what appears to be seven bonds (three bonds in each BH_3 unit and one bond between them).

There is no simple valence-bond structure which can be written for B_2H_6. A relatively simple molecular-orbital description has been worked out and is shown in Fig. 21.2. Four of the H atoms are in the same plane as the two B atoms; the other two H atoms are above and below this plane. Whereas the four outer H's are bonded to B by four conventional σ bonds, the other two H's are bound by *three-center* molecular orbitals. Each three-center bond, sometimes called a "banana bond," extends over three atoms, the two B's and the bridging H. One three-center molecular orbital (most simply visualized as an sp^3 orbital of one B plus a $1s$ of a bridge H plus an sp^3 of the other B) is above the plane of the rest of the molecule, the other is below; each of these molecular orbitals accommodates one pair of electrons. Such a molecular-orbital scheme not only gives a proper electron count but also is consistent with the observation that B_2H_6 has no paramagnetism and that two of the six H atoms are structurally and chemically different from the other four.

All the boron hydrides, ranging from gaseous B_2H_6 to solid $B_{18}H_{22}$, catch fire in air to form dark-colored products of unknown composition. In the absence of air, they decompose on heating to boron and hydrogen. They react with water to form hydrogen and boric acid.

The only important oxide of boron is B_2O_3, boric oxide. As already mentioned, it is acidic, dissolving in water to form H_3BO_3, boric acid. Boric acid is an extremely weak acid, for which K_I is 6.0×10^{-10}. Because its acidity is so slight, it can safely be used as an eyewash to take advantage of its antiseptic properties.

The borates, formed either by neutralization of boric acid or reaction of B_2O_3 with basic oxides, are extremely complicated compounds. A few, such as $LaBO_3$, contain discrete BO_3^{3-} ions. As shown in Fig. 21.3, the simple BO_3^{3-}, or orthoborate ion, is a planar ion with the three oxygen atoms at the corners of an almost equilateral triangle. In more complex anions, such as the one also shown in Fig. 21.3, there are three oxygen atoms about each boron atom, but some of the oxygens are bridges to other boron atoms. Other borates are even more complex and may have, in addition to triangular BO_3 units, tetrahedral BO_4 units. This is true of borax, the most common of the borates. It is extensively used in water softening, partly because it reacts with Ca^{2+} to form insoluble calcium borate and partly because it hydrolyzes to give an alkaline solution (Sec. 17.5). Because borax dissolves many metal oxides to

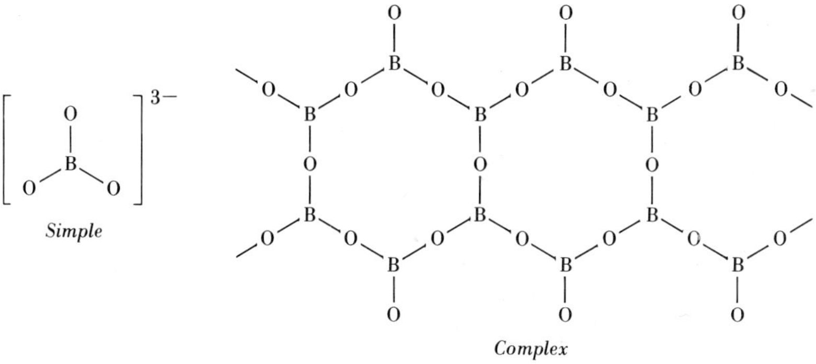

FIG. 21.3 Borate ions.

form easily melted borates, it is widely used as a flux in soldering operations. By removing oxides such as Cu_2O from the surface of hot brass, the flux allows fresh metal surfaces to fuse together.

The boron halides (BF_3, BCl_3, BBr_3, and BI_3) are unusual in that the boron atom in these molecules has only a sextet of electrons; hence, it can accommodate another pair of electrons. This occurs, for example, in the reaction

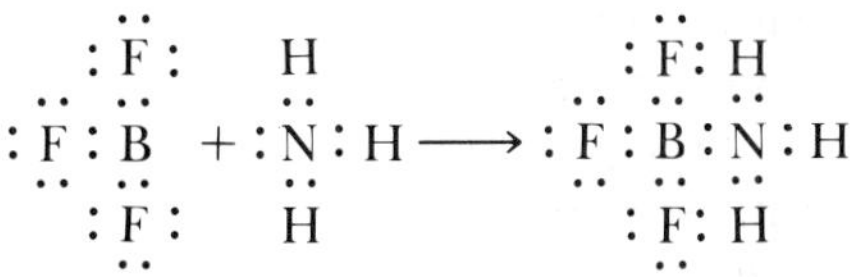

$$
\begin{array}{ccc}
:\!\ddot{F}\!: & H & :\!\ddot{F}\!:\ H \\
:\!\ddot{F}\!:\!B\ +:\!N\!:\!H \longrightarrow & :\!\ddot{F}\!:\!B\!:\!N\!:\!H \\
:\!\ddot{F}\!: & H & :\!\ddot{F}\!:\ H
\end{array}
$$

The product BF_3NH_3 is sometimes called an *addition compound.* The action of BF_3 as a Lewis acid, i.e., its ability to draw a pair of electrons to itself, makes it useful as a catalyst. BF_3 is one of the strongest Lewis acids known, though apparently not so strong as either BCl_3 or BBr_3.

21.2 SILICON

Silicon is the second member of group IV. This group contains carbon, silicon, germanium, tin, and lead. Like the members of group III, they show a pronounced change from acidic behavior for the light elements to more basic behavior for the heavy elements. Carbon and silicon are especially important because between them their compounds account for all living material and practically all the earth's minerals. In addition, silicon and germanium are of special interest because they are used in solid-state electronics.

Figure 21.4 gives some of the properties of group IV elements. Each of the elements has four electrons in its outermost shell. Since the outermost shell can usually accommodate eight electrons, it becomes questionable whether the atom would find it energetically favorable to lose electrons or gain electrons. For C and Si, and to some extent for Ge, the compromise is to share electrons in all compounds; for Sn and Pb, the formation of cations is favored.

587

FIG. 21.4 Properties of Group IV Elements

Symbol	Z	Electronic configuration	Melting point, °C	Boiling point, °C	Ionization potential, eV	Electrode potential, V
C	6	(2) $2s^2 2p^2$	3500	4200	11.26	+0.20 (from CO_2)
Si	14	(10) $3s^2 3p^2$	1420	2400	8.15	−0.86 (from SiO_2)
Ge	32	(28) $4s^2 4p^2$	937	2800	8.13	−0.1 (from GeO_2)
Sn	50	(46) $5s^2 5p^2$	232	2260	7.33	−0.14 (from Sn^{2+})
Pb	82	(78) $6s^2 6p^2$	327	1700	7.42	−0.13 (from Pb^{2+})

The chemistry of silicon resembles that of carbon in several respects. For example, silicon forms tetrahedral SiH_4 and a few higher hydrosilicons which contain chains of silicon atoms. However, Si—Si and Si—H bonds are relatively weak compared with Si—O bonds, and so the chemistry of silicon is primarily concerned with oxygen compounds rather than with hydrosilicons. Furthermore, unlike the smaller carbon atom which forms double and triple bonds, silicon invariably forms single bonds. As a result, oxygen-silicon compounds usually contain Si—O—Si bridges in which oxygen is bonded by single bonds to two silicon atoms instead of being bonded by a double bond to one silicon atom. This is unlike the case of carbon, where oxygen is frequently found bonded to a single carbon atom, as in the carbonyl group, C=O.

Silicon is the second most abundant element in the Earth's crust (25.8 wt %). It is as important in the mineral world as carbon is in the organic. As SiO_2 (which is called silica) it accounts for most beach sands, quartz, flint, and opal; as complex oxysilicates, it accounts for practically all rocks, clays, and soils.

The preparation of elemental Si is quite difficult. It can be accomplished by reduction of SiO_2 with Mg or by reduction of $SiCl_4$ with Zn. Since it is mainly used for addition to steel, it is usually prepared as ferrosilicon by reduction of mixtures of SiO_2 and iron oxides with coke.

For the electronics industry, the need is for ultrapure silicon. As mentioned in Sec. 7.8, some solids such as silicon and germanium are electric insulators in the pure state but become good conductors when doped with group III or group V elements. The enhanced conductivity increases with the amount of impurity and rapidly increases with increasing temperature. Many important devices such as transistors and solar batteries depend on such electrical properties. To get ultra-high-purity silicon, the starting materials such as $SiCl_4$ and zinc have to be prepurified by the best techniques available, e.g., vacuum distillation and selective adsorption. As a last step, an ingot of high-purity silicon is subject to a *zone-refining* process. In this process, the high-purity ingot is slowly drawn through a long quartz tube filled with inert gas. Heating coils wound in strips around the tube, as shown in Fig. 21.5, melt the silicon in narrow zones. At each solid-liquid interface there is an equilibrium $Si(s) \rightleftharpoons Si(l)$. Impurities, which are generally more soluble in the liquid phase, remain in the molten zone and get swept to the end of the ingot, where they can be cut off and discarded. Zone refining has made it possible to reduce impurity concentrations to one part per billion (i.e., 1×10^{-7} percent).

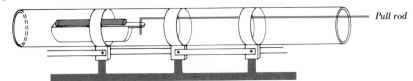

Heater loops

Pull rod

FIG. 21.5 Zone-refining technique to purify silicon.

For use in *solar cells,* a thin wafer of ultrapure silicon is doped with boron (a group III element), giving it *p*-type character (recall Sec. 7.8). One side of the wafer then has a group V element such as phosphorus diffused into it, giving it *n*-type character. When the *n*-type surface is exposed to the sun, photons of solar energy penetrate to the *p/n* junction and give up their energy by creating an electron and a "hole." A single cell can generate only about 0.4 V, but with thousands of them placed in series, large power production can be achieved. Unfortunately, the cells are rather expensive.

Most of the compounds of silicon are oxy compounds. Other compounds tend to be unstable with respect to conversion to the oxy compounds. Thus, hydrosilicons, prepared by reaction of silicides (for example, Mg_2Si) with acid, are unstable in oxygen with respect to rapid conversion to SiO_2. Silane (SiH_4), for example, is oxidized as follows:

$$SiH_4(g) + 2O_2(g) \longrightarrow SiO_2(s) + 2H_2O(g) \qquad \Delta G° = -1220 \text{ kJ}$$

which is to be compared with

$$CH_4(g) + 2O_2(g) \longrightarrow CO_2(g) + 2H_2O(g) \qquad \Delta G° = -800 \text{ kJ}$$

Disilane (Si_2H_6), trisilane (Si_3H_8), and tetrasilane (Si_4H_{10}) have been prepared, but they are progressively even less stable as the silicon-silicon chain length increases.

The silicates (oxy compounds of silicon) have been extensively investigated. With few exceptions, the silicon is tetrahedrally bonded to four oxygen atoms. As shown in Fig. 21.6, four valence electrons from silicon ($3s^2 3p^2$) and six valence electrons from each oxygen are insufficient to complete the octets of all the atoms. Consequently, to produce a stable compound, the oxygen atoms may obtain electrons from other atoms and become negative in the process. This produces the discrete orthosilicate anion, SiO_4^{4-}, found, for example, in the mineral zircon ($ZrSiO_4$). Alternatively, the oxygen atoms may complete their octets by sharing electrons with other silicon atoms. Since one, two, three, or four of the oxygen atoms can thus bridge to other silicon atoms, increasingly complex structures are possible.

FIG. 21.6 Tetrahedral SiO_4 unit.

590

Chapter 21
Chemistry
of Some
Nonmetals
Other than
Carbon

One bridge oxygen per silicon atom gives $Si_2O_7^{6-}$, which is analogous to $Cr_2O_7^{2-}$. Two bridge oxygens per silicon lead to formation of extended chains called *pyroxene* chains. The pyroxenes are a class of minerals—spodumene, $LiAl(SiO_3)_2$, is an example—which are second only to feldspars as the most common constituents of igneous rock. As shown by the two views of Fig. 21.7, the pyroxene chains are strings of corner-sharing tetrahedra where the repeat unit is SiO_3^{2-}. Each of the oxygen atoms that is not a bridge oxygen picks up an electron to complete its octet; so each pyroxene chain is a negatively charged anion. In the compounds, cations such as Li^+ and Al^{3+} hold the anions together by ionic attractions. Pyroxene chains of varying length occur in the material called *water glass,* made by dissolving silica in aqueous NaOH:

$$SiO_2 + 2NaOH \longrightarrow \text{``}Na_2SiO_3\text{''} + H_2O$$

The quotation marks indicate that the ratio of sodium to silicon can be highly variable. Water glass is used as an egg preservative (where it presumably acts by sealing the pores), fabric fireproofer, and adhesive for cardboard cartons.

If two pyroxene strands are laid parallel to each other and cross-linked to share oxygen bridges, the result is to form a band anion known as an *amphibole* chain (Fig. 21.8). Amphibole chains are found in a variety of common minerals, including, for example, the hornblendes, which are hydroxysilicates containing Ca, Mg, and Fe as cations that balance the charge of the anion.

If the SiO_4 tetrahedra are cross-linked so that there are three bridge oxygens per silicon, the result is to form infinite sheets. A portion of such a sheet is shown in Fig. 21.9. The open circles represent oxygen atoms slightly below the plane of the paper; the black circles represent silicon atoms in the

FIG. 21.7 Two views of the pyroxene chain. In (*b*) the view is down along the Si—O bond sticking out of the plane of the paper.

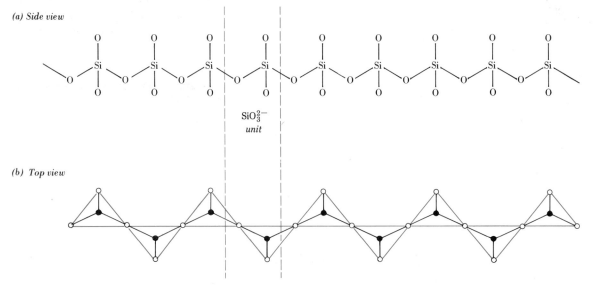

(a) Side view

SiO_3^{2-}
unit

(b) Top view

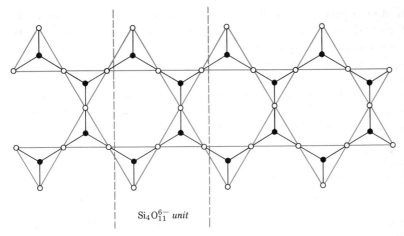

$Si_4O_{11}^{6-}$ *unit*

FIG. 21.8 Amphibole chain.

plane of the paper, with oxygen atoms slightly above the plane. The oxygen atoms sticking out above the plane are negatively charged and are attracted to positive ions, which in turn are attracted to other, similar sheet-silicate ions turned upside down. If the sheets were perfect, they would have the general formula $(Si_2O_5^{2-})_n$. With a counter ion such as Mg^{2+}, the compound would be $MgSi_2O_5$. It occurs in the compound $2MgSi_2O_5 \cdot Mg(OH)_2$, known as talc. Talc, when viewed edge-on, appears as in Fig. 21.10a. It consists of $Si_2O_5^{2-}$ layers with Mg^{2+} and OH^- ions sandwiched in between. The whole sandwich is electrically neutral. Therefore, talc is very soft. It is one of the softest minerals known.

The Si in the SiO_4 tetrahedra is commonly replaced in part by Al, usually in the ratio of about 1 out of 4. Since Al is tripositive whereas Si is tetrapositive, substitution of Al for Si would result in a net negative charge for the sandwich shown in Fig. 21.10a. This attracts positive ions such as K^+. Thus, for example, we can explain the compound $KMg_3AlSi_3O_{10}(OH)_2$, a form of mica. It comes from $2MgSi_2O_5 = Mg_2Si_4O_{10}$ with replacement of one

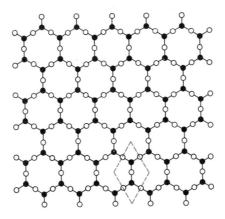

FIG. 21.9 Silicate sheet. Repeat unit, outlined in color, is $Si_2O_5^{2-}$.

591

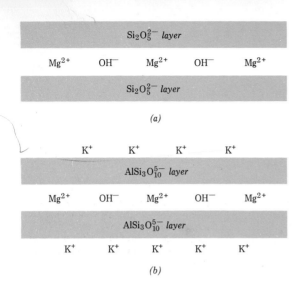

$Si_2O_5^{2-}$ *layer*

Mg^{2+} OH^- Mg^{2+} OH^- Mg^{2+}

$Si_2O_5^{2-}$ *layer*

(a)

K^+ K^+ K^+ K^+

$AlSi_3O_{10}^{5-}$ *layer*

Mg^{2+} OH^- Mg^{2+} OH^- Mg^{2+}

$AlSi_3O_{10}^{5-}$ *layer*

K^+ K^+ K^+ K^+ K^+

(b)

FIG. 21.10 (*a*) Edge view of layer stacking in talc, $Mg_3(Si_2O_5)_2(OH)_2$. (*b*) Edge view of layer stacking in mica, $KMg_3AlSi_3O_{10}(OH)_2$.

Si out of four by Al, K and combination with $Mg(OH)_2$. As shown in Fig. 21.10*b*, rather strong binding to adjacent sandwiches is expected because of the K^+ ions on the sides of the sandwich. Clay minerals have layer structures that are quite similar to those in talc and mica.

In the limit there can be four bridge O atoms per Si. This leads to three-dimensional structures such as those found in quartz (SiO_2), feldspars (e.g., orthoclase, $KAlSi_3O_8$), and zeolites (e.g., ultramarine, $Na_3Al_2Si_6O_{12}S$). In *feldspars,* the common constituents of igneous rocks (e.g., granite), replacement of part of the Si by Al gives the framework a negative charge, which must be compensated for by cations, usually K^+, Na^+, or Ca^{2+}. Orthoclase, $KAlSi_3O_8$, can be considered to be derived from $4SiO_2=Si_4O_8$ with replacement of one out of four Si by Al, K. The Al sits in a tetrahedron of O atoms, just as does the Si, but the K^+ is found in cavities in the framework. In the *zeolites* the cavities are very large so that besides cations, H_2O can be accommodated too. Figure 21.11 shows a portion of the framework that characterizes the zeolite structure in ultramarine. Action of zeolites as ion exchangers was mentioned in Sec. 17.6, but their use in that application has been largely supplanted by synthetic resins. Synthetic zeolites have recently been tailor-made so as to create cavities and passages of desired dimensions for use as *molecular sieves.* Figure 21.12 shows a portion of the structure of a synthetic zeolite known as *Linde A.* Each of the globular clusters at the corners of the cube is like the entire cluster shown in Fig. 21.11. Thus there are eight small cavities at the cube corners and a very large cavity in the cube center. The large cavities are connected by openings of 0.42 nm. Molecules that can squeeze through them can be absorbed in the cavities and can be held there by various attractive forces such as van der Waals forces. Even hydrocarbons can be absorbed. As mentioned on page 562, molecular sieves can be used to separate straight-chain from branched hydrocarbons for

592

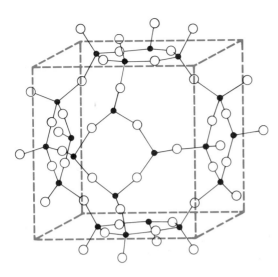

FIG. 21.11 Ultramarine. (Filled circles represent Si or Al atoms. Open circles represent O atoms. The central cavity can accommodate Na^+ and various anions, such as S_2^{2-}, Cl^-, or even SO_4^{2-}. For clarity, atoms on the back face of the cube have not been shown. Four oxygen atoms projecting toward the reader from the front face have also been omitted.)

making biodegradable detergents. In a recent development, Union Carbide has synthesized a molecular sieve that lets water through but absorbs organic molecules below a certain size. Such a sieve could find application in the cleaning of waste water.

Derived from SiO_2 are other silicate systems of great practical importance, e.g., glass and cement. *Glass* is made by fusing SiO_2 (sand) with basic substances such as CaO and Na_2CO_3. Special glasses such as Pyrex contain other acidic oxides (B_2O_3) substituted for some of the SiO_2. Like SiO_2, glass is slowly etched by basic solutions. As a consequence, glass stoppers frequently stick fast in reagent bottles containing basic solutions such as NaOH and Na_2CO_3. *Cement,* a complex aluminum silicate, is made by sintering limestone and clay at high temperature and grinding the product to a fine powder. When mixed with water and allowed to stand, it sets to a hard, rigid solid. The reactions involve slow hydration of silicates to form a complex interlock-

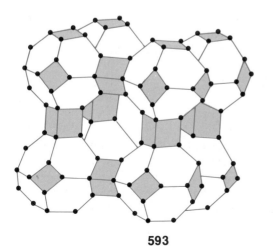

FIG. 21.12 Linde A. (Filled circles represent positions of Si or Al atoms. O bridges are not shown.)

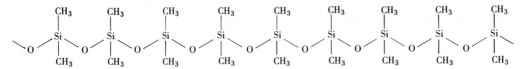

FIG. 21.13 Methyl silicone chain.

ing structure. The hydration evolves considerable heat, which may cause cracking unless provision is made for its removal.

The high thermal stability of Si—O—Si chains has been exploited in the *silicones*, compounds in which organic residues are bonded to Si atoms in place of negatively charged silicate oxygens. A typical example of a silicone is the chainlike methyl silicone shown in Fig. 21.13. Thanks to the methyl groups, this silicone has lubricating properties characteristic of hydrocarbon oils; thanks to the oxysilicon backbone, it is unreactive even at high temperatures. More complicated silicone polymers are made possible by having oxygen or hydrocarbon bridges between chains. The materials are rubbery and are used as electrical insulators at elevated temperatures.

21.3 NITROGEN

$-1 + 1 + 5 - 7$

Nitrogen is the top element of group V, which also contains phosphorus, arsenic, antimony, and bismuth. Collectively, the elements are called "pnicogens," and their compounds are "pnictides." The name comes from the Greek word *pnigmos,* meaning "suffocation." The lighter members of the group, N and P, are typical nonmetals and form only acidic oxides; the middle members, arsenic and antimony, are semimetals and form amphoteric oxides; the heaviest member, bismuth, is a metal and forms mostly basic oxides. Figure 21.14 summarizes some of the properties of the elements. As shown, each of the atoms has five valence electrons (ns^2np^3) in its outermost shell. Sharing electrons with more-electronegative atoms corresponds to a maximum oxidation state of $+5$; sharing with less-electronegative atoms, to a minimum oxidation state of -3. In addition, a $+3$ state corresponding to not sharing the pair of s electrons (so-called "inert pair") is common to all. Nitrogen and phosphorus are unusual in that they show all the oxidation states from -3 to $+5$, inclusive.

Nitrogen occurs principally as diatomic N_2 in the atmosphere. It is also found as Chile saltpeter ($NaNO_3$) and in plants and animals in the form of proteins. The average composition of proteins is 51% by weight C, 25% O, 16% N, 7% H, 0.4% P, and 0.4% S.

Elemental nitrogen is usually obtained by fractional distillation of liquid air. Since N_2 has a lower boiling point (77.4°K) than O_2 (90.2°K), it is more volatile and evaporates preferentially in the first fractions of gas (Sec. 14.6). Very pure N_2 can be made by thermal decomposition of ammonium nitrite, NH_4NO_2:

$$NH_4NO_2(s) \longrightarrow N_2(g) + 2H_2O(g)$$

FIG. 21.14   Properties of Group V Elements

Symbol	Z	Electronic configuration	Melting point, °C	Boiling point, °C	Ionization potential, eV	Electrode potential, V
N	7	(2) $2s^2 2p^3$	−210.0	−195.8	14.5	+1.25 (from NO_3^-)
P	15	(10) $3s^2 3p^3$	44.1	280	11.0	−0.50 (from H_3PO_3)
As	33	(28) $4s^2 4p^3$	Sublimes	Sublimes	10	+0.23 (from As_4O_6)
Sb	51	(46) $5s^2 5p^3$	631	1380	8.6	+0.21 (from $Sb(OH)_2^+$)
Bi	83	(78) $6s^2 6p^3$	271	1500	8	+0.32 (from $Bi(OH)_2^+$)

It is interesting to note that pure nitrogen obtained from decomposition of compounds was the key that led to the discovery of the noble gases. Lord Rayleigh, in 1894, noted that N_2 from the decomposition of compounds was of slightly lower density (1.2505 g/liter at STP) than the residual gas obtained from the atmosphere by removal of oxygen, carbon dioxide, and water (1.2572 g/liter at STP). In collaboration with Sir William Ramsay, Rayleigh removed the nitrogen from air residue by various reactions, such as combining N with hot Mg to form solid Mg_3N_2. After removal of the nitrogen, there was still some gas remaining. Unlike any gas known at the time, it was completely unreactive and was christened "argon" from the Greek word *argos,* meaning "lazy." Later spectroscopic investigations showed that crude argon, and hence the atmosphere, contains the other noble-gas elements helium, neon, krypton, and xenon. Including the noble gases, the average composition of the Earth's atmosphere is as shown in Fig. 21.15. As can be seen from the data, nitrogen is by far the predominant constituent of the atmosphere.

The N_2 molecule contains a triple bond. Although very stable with respect to dissociation into single atoms, N_2 is thermodynamically unstable with respect to the reaction

$$2N_2(g) + 5O_2(g) + 2H_2O \longrightarrow 4H^+ + 4NO_3^-$$

It is fortunate that this reaction is very slow; otherwise, atmospheric N_2 and O_2 would combine with the oceans to form solutions of dilute nitric acid.

FIG. 21.15   Composition of Dry Air

Component	Percent by volume	Boiling point, °K
Nitrogen (N_2)	78.09	77.4
Oxygen (O_2)	20.95	90.2
Argon (Ar)	0.93	87.4
Carbon dioxide (CO_2)	0.023–0.050	Sublimes
Neon (Ne)	0.0018	27.2
Helium (He)	0.0005	4.2
Krypton (Kr)	0.0001	121.3
Hydrogen (H_2)	0.00005	20.4
Xenon (Xe)	0.000008	163.9

596

Chapter 21
Chemistry
of Some
Nonmetals
Other than
Carbon

The principal compound of nitrogen is probably ammonia, NH_3. It occurs to a slight extent in the atmosphere, primarily as a product of putrefaction of nitrogen-containing animal or vegetable matter. Commercially it is important as the most economic pathway for nitrogen *fixation*, i.e., the conversion of atmospheric N_2 into useful compounds. In the Haber process, synthetic ammonia is made by passing a nitrogen-hydrogen mixture through a bed of catalyst consisting of iron plus oxides such as Al_2O_3. At about 500°C (a compromise between making the reaction go faster and making the NH_3 less stable) and 1000 atm, there is about 50 percent conversion of N_2 to NH_3:

$$N_2(g) + 3H_2(g) \longrightarrow 2NH_3(g) \qquad \Delta H° = -92 \text{ kJ}$$

NH_3 is a polar molecule, pyramidal in shape, with the three H atoms occupying the base of the pyramid and an unshared pair of electrons, the apex. It is easily condensed (liquefaction temperature, $-33°C$) to a liquid of great solvent power. Like water, it can dissolve a great variety of salts. In addition, it has the rather unique property of dissolving alkali and alkaline-earth metals to give solutions which contain solvated electrons (Fig. 17.6).

Ammonia gas is very soluble in water, consistent with the fact that both NH_3 and H_2O are polar molecules. Not so easy to explain is the basic character of the aqueous solutions formed. At one time it was thought that NH_3 reacts with H_2O to form the weak base ammonium hydroxide, NH_4OH. However, it has not been possible to prove the existence of such a species in aqueous solution, and so the basic nature of aqueous NH_3 is best represented by the equilibrium

$$NH_3 + H_2O \rightleftharpoons NH_4^+ + OH^- \qquad K = 1.8 \times 10^{-5}$$

Upon neutralization of NH_3 with acids, ammonium salts can be formed; these contain the tetrahedral NH_4^+ ion. They resemble potassium salts, except that they give slightly acid solutions. This can be interpreted as a dissociation of an acid:

$$NH_4^+ \rightleftharpoons NH_3 + H^+ \qquad K = 5.5 \times 10^{-10}$$

Some ammonium salts, such as ammonium nitrate, NH_4NO_3, are thermally unstable because they undergo autooxidation. As an illustration, NH_4NO_3 decomposes (sometimes explosively) when heated to produce nitrous oxide, N_2O, by the reaction

$$NH_4NO_3(s) \longrightarrow N_2O(g) + 2H_2O(g)$$

Whereas ammonia and ammonium salts represent nitrogen in its lowest oxidation state (-3), the highest oxidation state of nitrogen $(+5)$ appears in the familiar compounds nitric acid (HNO_3) and nitrate salts. Nitric acid is one of the most important industrial acids, and large quantities of it are produced, principally by the catalytic oxidation of ammonia. In the process, called the *Ostwald process*, the following steps are important:

$$4NH_3(g) + 5O_2(g) \xrightarrow{\text{Pt}} 4NO(g) + 6H_2O(g)$$
$$2NO(g) + O_2(g) \longrightarrow 2NO_2(g)$$
$$3NO_2(g) + H_2O \longrightarrow 2H^+ + 2NO_3^- + NO(g)$$

In the first step a mixture of ammonia and air is passed over a platinum catalyst heated to about 800°C. The product nitric oxide (NO) is then oxidized to nitrogen dioxide (NO_2). When dissolved in H_2O, NO_2 disproportionates to form nitric acid and NO. To get 100 percent acid, it is necessary to distill off volatile HNO_3.

Pure nitric acid is a colorless liquid which on exposure to light turns brown because of slight decomposition to NO_2:

$$4HNO_3 \longrightarrow 4NO_2(g) + O_2(g) + 2H_2O$$

It is a strong acid in that it is 100 percent dissociated in dilute solutions to H^+ and nitrate ion, NO_3^-. Like carbonate (Fig. 20.3), nitrate ion is planar; it is sometimes represented as a resonance hybrid of three contributing formulas. The ion is colorless and forms a great variety of nitrate salts, most of which are quite soluble in aqueous solutions.*

In acid solution, nitrate ion is a good oxidizing agent. By proper choice of concentration and reducing agent, it can be reduced to compounds of nitrogen in all the other oxidation states. The possible half-reactions and their electrode potentials are

$$NO_3^- + 2H^+ + e^- \longrightarrow NO_2(g) + H_2O \qquad E° = +0.79 \text{ V}$$
$$NO_3^- + 3H^+ + 2e^- \longrightarrow HNO_2 + H_2O \qquad E° = +0.94 \text{ V}$$
$$NO_3^- + 4H^+ + 3e^- \longrightarrow NO(g) + 2H_2O \qquad E° = +0.96 \text{ V}$$
$$2NO_3^- + 10H^+ + 8e^- \longrightarrow N_2O(g) + 5H_2O \qquad E° = +1.12 \text{ V}$$
$$2NO_3^- + 12H^+ + 10e^- \longrightarrow N_2(g) + 6H_2O \qquad E° = +1.25 \text{ V}$$
$$NO_3^- + 8H^+ + 6e^- \longrightarrow NH_3OH^+ + 2H_2O \qquad E° = +0.73 \text{ V}$$
$$2NO_3^- + 17H^+ + 14e^- \longrightarrow N_2H_5^+ + 6H_2O \qquad E° = +0.83 \text{ V}$$
$$NO_3^- + 10H^+ + 8e^- \longrightarrow NH_4^+ + 3H_2O \qquad E° = +0.88 \text{ V}$$

Since all the electrode potentials are quite positive, nitrate ion is a better oxidizing agent than H^+. This explains why metals such as copper and silver, which are too poor as reducing agents to dissolve in HCl, for example, will dissolve in HNO_3. Both of these acids contain the oxidizing agent H^+, but only HNO_3 has the additional oxidizing agent NO_3^-. Some metals, such as gold, are insoluble in HCl and also in HNO_3, but they are soluble in a mixture of the two. This mixture, called *aqua regia*, normally contains one part of concentrated HNO_3 to three parts of concentrated HCl. Its enhanced dissolving power is due to oxidizing ability of nitrate ion in strong acid plus complexing ability of chloride ion.

Since the various electrode potentials shown above are very roughly the same, reduction of NO_3^- may yield any of several species. The actual composition of the product depends on the rates of the different reactions. These rates are influenced by the concentration of NO_3^-, the concentration of H^+,

---

*Because of the solubility of nitrates, it is not usual to find solid nitrates occurring naturally as minerals. The extensive deposits of $NaNO_3$ in Chile occur in a desert region where there is insufficient rainfall to wash them away. These deposits probably originated from decomposition of nitrogenous marine organisms which were cut off from the sea.

598

Chapter 21
Chemistry
of Some
Nonmetals
Other than
Carbon

the temperature, and the reducing agent used. As can be seen, compounds are possible in the +4, +3, +2, +1, −1, and −2 states, as well as +5 and −3. Some of the more common representative compounds are discussed below.

The +5 State

In addition to nitric acid and the nitrates, nitrogen corresponding to the +5 state is found in nitrogen pentoxide, N_2O_5. At room temperature, N_2O_5 is a white solid which decomposes slowly into NO_2 and oxygen. At slightly elevated temperatures, it may explode. With water it reacts quite vigorously to form HNO_3.

The +4 State

When concentrated nitric acid is reduced with metals, brown fumes are evolved. The brown gas is NO_2, nitrogen dioxide. The molecule contains an odd number of valence electrons (five from the nitrogen and six from each of the oxygens); so it should be, and is, paramagnetic. When the brown NO_2 gas is cooled, its color fades, and the paramagnetism diminishes. These observations are explained by assuming that two NO_2 molecules pair up (dimerize) to form a single molecule of N_2O_4, nitrogen tetroxide:

$$2NO_2(g) \rightleftharpoons N_2O_4(g) \qquad \Delta H° = -61.1 \text{ kJ}$$

At 60°C and 1 atm pressure, half the nitrogen is present as NO_2 and half as N_2O_4. As the temperature is raised, decomposition of N_2O_4 is favored. The NO_2–N_2O_4 mixture is poisonous and is a strong oxidizing agent.

The +3 State

The most common representatives of the +3 state are the salts called nitrites. Nitrites such as $NaNO_2$ can be made by heating sodium nitrate above its melting point:

$$2NaNO_3(l) \longrightarrow 2NaNO_2(l) + O_2(g)$$

Nitrites are important industrially in the manufacture of azo dyes, which contain the $-N=N-$ group, as in azobenzene, $C_6H_5-N=N-C_6H_5$. The azo dyes are intensely colored and account for more than half of synthetic dyes.

The +2 State

NO, nitric oxide, is, like NO_2, an "odd" molecule; it contains an uneven number of electrons. However, unlike NO_2, NO is colorless and does not dimerize appreciably in the gas phase. It can be made in several ways:

$$4NH_3(g) + 5O_2(g) \xrightarrow{\text{Pt}} 4NO(g) + 6H_2O$$
$$3Cu(s) + 8H^+ + 2NO_3^- \longrightarrow 3Cu^{2+} + 2NO(g) + 4H_2O$$
$$N_2(g) + O_2 \longrightarrow 2NO(g)$$

The first of these reactions is the catalytic oxidation that is the first step of the Ostwald process for making nitric acid. The second occurs when copper is warmed with dilute nitric acid. The third, which is extremely endothermic (by 180 kJ), occurs only at high temperatures or when large amounts of energy are added. The third reaction occurs during electric storms and is one of the paths by which atmospheric nitrogen is made available to plants. It also occurs in the internal-combustion engine. The higher the temperature at which an engine operates, the more NO is formed in the exhaust gases. In air, NO is rapidly oxidized to brown NO_2:

$$2NO(g) + O_2(g) \longrightarrow 2NO_2(g)$$

Nitric oxide combines with many transition-metal cations to form complex ions. The most familiar of these is $FeNO^{2+}$, the nitroso ferrous ion, which forms in the "brown-ring" test for nitrates. When concentrated sulfuric acid is carefully poured into a solution containing ferrous ion and nitrate, a brown layer appears at the junction of the H_2SO_4 and the nitrate-containing solution. The NO for the complex is formed by reduction of NO_3^- by Fe^{2+}.

The +1 State

When solid ammonium nitrate is gently heated, it melts and undergoes autooxidation according to the following equation:

$$NH_4NO_3(l) \longrightarrow N_2O(g) + 2H_2O(g)$$

The compound formed, N_2O, is called nitrous oxide. Its electronic formula can be written $:N::N::O:$. Compared with the other oxides of nitrogen, nitrous oxide is considerably less poisonous. Small doses are mildly intoxicating, hence the common name "laughing gas." Large doses produce general anesthesia and in dentistry are frequently used for this purpose. Nitrous oxide has appreciable solubility in fats, a property exploited in making self-whipping cream. Cream is packaged with nitrous oxide under pressure to increase its solubility. When the pressure is released, the nitrous oxide escapes and forms tiny bubbles.

The −1 State

Hydroxylamine, NH_2OH, is representative of nitrogen with oxidation number −1. It can be considered to be derived from NH_3 by substituting a hydroxyl group for one of the hydrogen atoms. Like NH_3, NH_2OH has an unshared pair of electrons and so can pick up a proton to form NH_3OH^+:

$$H:\overset{..}{\underset{..}{N}}:\overset{..}{\underset{..}{O}}:H + H_2O \rightleftharpoons \left[H:\overset{\overset{\textstyle H}{|}}{\underset{\underset{\textstyle H}{|}}{N}}:\overset{..}{\underset{..}{O}}:H \right]^+ + OH^- \qquad K = 6.6 \times 10^{-8}$$

Thus, hydroxylamine solutions are slightly basic, but less so than NH_3 solutions. Analogous to ammonium salts (for example, NH_4Cl), there are hydroxylammonium salts (for example, NH_3OHCl). Since hydroxylamine

600

Chapter 21
Chemistry
of Some
Nonmetals
Other than
Carbon

corresponds to nitrogen in an intermediate oxidation state, it can act both as oxidizing agent and as reducing agent.

The −2 State

The compound hydrazine, N_2H_4, is similar to ammonia in many ways. This compound can be made by bubbling chlorine through a solution of ammonia:

$$Cl_2(g) + 4NH_3 \longrightarrow N_2H_4 + 2NH_4^+ + 2Cl^-$$

When pure, N_2H_4 is a colorless liquid at room temperature. It is unstable with respect to disproportionation:

$$2N_2H_4(l) \longrightarrow N_2(g) + 2NH_3(g) + H_2(g)$$

and is violently explosive in the presence of air or other oxidizing agents. It is quite poisonous. In aqueous solution it acts as a base, since it can add one or two protons to the unshared pairs of electrons:

$$
\begin{array}{c}
H\ \ H \\
\overset{\cdot\cdot}{H:N:N:H} \\
\overset{\cdot\cdot}{}\ \ \overset{\cdot\cdot}{}
\end{array}
+ H_2O \longrightarrow
\left[
\begin{array}{c}
H\ \ H \\
\overset{\cdot\cdot}{H:N:N:H} \\
\overset{\cdot\cdot}{}\ \ \overset{\cdot\cdot}{} \\
H
\end{array}
\right]^+
+ OH^- \qquad K = 9.8 \times 10^{-7}
$$

Hydrazine has become important as a rocket propellant. For example, the reaction

$$N_2H_4(l) + 2H_2O_2(l) \longrightarrow N_2(g) + 4H_2O(g) \qquad \Delta H° = -643 \text{ kJ}$$

takes place in the presence of Cu^{2+} ion as catalyst. It is strongly exothermic and is accompanied by a large increase in volume.

The −3 State

In addition to ammonia and the ammonium salts, nitrogen forms other compounds in which it is assigned an oxidation state of −3. These include the nitrides, such as Mg_3N_2 and TiN. Mg_3N_2 is quite reactive and combines with water to liberate ammonia. TiN, on the other hand, is inert and can be used to make containers for high-temperature reactions. The compound nitrogen triiodide (NI_3) should also be included with the −3 oxidation state of nitrogen, since nitrogen is more electronegative than iodine. At room temperature, NI_3 is a solid which is violently explosive; it is notorious for the fact that even a fly's landing on it can set it off.

The above list of nitrogen compounds is by no means exhaustive. We should add to it the proteins and amino acids, as discussed in Sec. 20.13. In nature there is constant interconversion between the various forms of nitrogen. The *nitrogen cycle,* which traces the interconversion, is shown in simplified form in Fig. 21.16. When plant and animal proteins break down, as in digestion and decay, the principal end products are NH_3 and N_2, released to the atmosphere, and various nitrogen-containing ions, added to the soil.

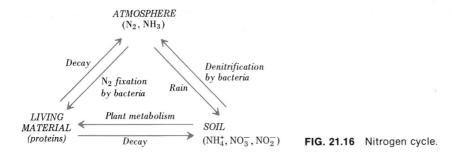

ATMOSPHERE
(N_2, NH_3)

Decay

N_2 fixation
by bacteria

Rain

Denitrification
by bacteria

LIVING
MATERIAL
(proteins)

Plant metabolism

Decay

SOIL
(NH_4^+, NO_3^-, NO_2^-)

FIG. 21.16 Nitrogen cycle.

NH_3 in the atmosphere can be returned to the soil by being dissolved in rain. Elemental nitrogen can be returned by two paths: (1) Nitrogen-fixing bacteria which live on the roots of leguminous plants, such as clover, convert N_2 to proteins and other nitrogen compounds. (2) Lightning discharges convert N_2 and O_2 to NO, which in turn is oxidized to NO_2. The NO_2 dissolves in rainwater to form nitrates and nitrites, which are washed into the soil. As a final step of the cycle, plants absorb the nitrogen compounds from the soil and convert them to plant proteins. The plant proteins are ingested by animals as food and are then broken down and reassembled as animal proteins or excreted as waste to the soil.

The nitrogen cycle as outlined above is in precarious balance. Frequently, the balance is locally upset, as, for example, by intensive cultivation and removal of crops. In such cases, it is necessary to replenish the nitrogen by addition of synthetic fertilizers such as NH_3, NH_4NO_3, and KNO_3.

21.4
PHOSPHORUS

The second element of group V, phosphorus, occurs principally as *phosphate rock*. This is mostly $Ca_5(OH, F)(PO_4)_3$, *apatite,* where the OH^- and F^- substitute freely for each other, giving either hydroxyapatite or fluorapatite. Like nitrogen compounds, phosphorus compounds are essential constituents of all animal and vegetable matter. Bones, for example, contain about 60% $Ca_3(PO_4)_2$; nucleic acids such as DNA and RNA contain polyester chains of sugars and phosphates (Sec. 20.14).

Elemental phosphorus can be made by reduction of calcium phosphate with coke in the presence of sand. The reaction can be represented by the equation

$$Ca_3(PO_4)_2(s) + 3SiO_2(s) + 5C(s) \longrightarrow 3CaSiO_3 + 5CO(g) + P_2(g)$$

Since the reaction is carried out at high temperature, the phosphorus formed is a diatomic gas. It can be condensed to a solid by running the product gases through water. The condensation serves not only to separate the phosphorus from the carbon monoxide but also to protect it from reoxidation by air.

602

Chapter 21
Chemistry
of Some
Nonmetals
Other than
Carbon

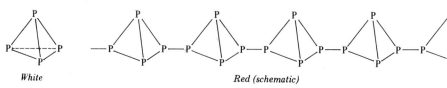

White   *Red (schematic)*

FIG. 21.17 White and red phosphorus.

There are several forms of solid phosphorus. White phosphorus consists of discrete tetrahedral P_4 molecules, as shown on the left of Fig. 21.17. Red phosphorus is polymeric, consisting of chains of P_4 tetrahedra linked together, perhaps in the manner shown in Fig. 21.17. At room temperature the stable modification of elemental phosphorus is the red form. It is less volatile, less soluble (especially in nonpolar solvents), and less reactive than white phosphorus. The white form must be handled with care because it ignites spontaneously in air and is extremely poisonous.

At room temperature ordinary red phosphorus is not especially reactive, but at higher temperatures it reacts with many other elements to form a variety of compounds. For example, when heated with calcium, it forms solid calcium phosphide, Ca_3P_2. With chlorine it can form either liquid phosphorus trichloride, PCl_3, or solid phosphorus pentachloride, PCl_5, depending on the relative amount of chlorine supplied. The three compounds just mentioned illustrate the three most important oxidation states of phosphorus, -3, $+3$, and $+5$.

When Ca_3P_2 is placed in water, it reacts vigorously to form <u>phosphine</u>, PH_3, a toxic gas that smells like dead fish:

$$Ca_3P_2(s) + 6H_2O \longrightarrow 2PH_3(g) + 3Ca^{2+} + 6OH^-$$

In structure PH_3 resembles NH_3 in being a pyramidal molecule. Compared with NH_3, PH_3 is practically insoluble in water and is much less basic. In air, PH_3 usually bursts into flame, apparently because it is ignited by spontaneous oxidation of the impurity P_2H_4.[*]

When phosphorus is burned in a limited supply of oxygen, it forms the oxide P_4O_6 (phospho<u>rous</u> oxide). (Note the spelling: The name for the element ends in -*us*, and the name for $+3$ compounds ends in -*ous*.) The structure, shown in Fig. 21.18, can be visualized as derived from a P_4 tetrahedron by insertion of an oxygen bridge between each pair of phosphorus atoms. P_4O_6 is the anhydride of phosphorous acid, and when cold water is added to it, H_3PO_3 is formed. Phosphorous acid is peculiar because only two of the three hydrogen atoms in the molecule can dissociate:

$$H_3PO_3 \rightleftharpoons H^+ + H_2PO_3^- \qquad K_I = 1.6 \times 10^{-2}$$
$$H_2PO_3^- \rightleftharpoons H^+ + HPO_3^{2-} \qquad K_{II} = 7 \times 10^{-7}$$

[*]The will-o'-the-wisp, or faint, flickering light, sometimes observed in marshes, may be due to spontaneous ignition of impure PH_3. The PH_3 might be formed by reduction of naturally occurring phosphorus compounds.

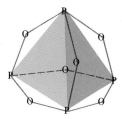

FIG. 21.18 Phosphorous oxide (P_4O_6). Shaded part highlights the tetrahedral arrangement of P atoms. The dashed oxygen bridge is in back.

The reason the third hydrogen does not dissociate is that it is attached directly to the phosphorus instead of to an oxygen. The structure of H_3PO_3 can be indicated by writing $HPO(OH)_2$ instead of $P(OH)_3$. Neutralization of H_3PO_3 by bases can produce two series of salts, the dihydrogen phosphites, for example, NaH_2PO_3, and the monohydrogen phosphites, for example, Na_2HPO_3. The phosphites, especially in basic solutions, are very strong reducing agents. Even in acid solution (where they immediately are converted to H_3PO_3) they are moderately good reducing agents.

In the +5 state, phosphorus exists as several oxy compounds of varying complexity. The least complicated is the oxide, P_4O_{10}, called phosphoric oxide, phosphorus pentoxide, or phosphoric anhydride. It is a white solid formed when phosphorus is burned in an unlimited supply of oxygen. Though called a pentoxide (because of its simplest formula, P_2O_5), this material consists of discrete P_4O_{10} molecules. The structure can be visualized as derived from the molecule shown in Fig. 21.18 by sticking an additional oxygen atom on each phosphorus.

When exposed to moisture, P_4O_{10} turns gummy as it picks up water. The affinity for water is so great that P_4O_{10} is frequently used as a dehydrating agent. With a large amount of water, the acid H_3PO_4, or orthophosphoric acid, is formed. This is a triprotic acid for which the stepwise dissociation is as follows:

$$H_3PO_4 \rightleftharpoons H^+ + H_2PO_4^- \qquad K_I = 7.5 \times 10^{-3}$$
$$H_2PO_4^- \rightleftharpoons H^+ + HPO_4^{2-} \qquad K_{II} = 6.2 \times 10^{-8}$$
$$HPO_4^{2-} \rightleftharpoons H^+ + PO_4^{3-} \qquad K_{III} = 10^{-12}$$

(Like SO_4^{2-} and CrO_4^{2-}, PO_4^{3-} is tetrahedral in structure.) Three series of salts are possible: the dihydrogen phosphates (for example, NaH_2PO_4), the monohydrogen phosphates (for example, Na_2HPO_4), and the normal phosphates (for example, Na_3PO_4). When dissolved in water, NaH_2PO_4 gives slightly acid solutions. Solutions of $NaHPO_4$ are slightly basic. Solutions of Na_3PO_4 are quite basic because there is no acid dissociation to counterbalance the strong hydrolysis of PO_4^{3-}. $Ca(H_2PO_4)_2$ is used with $NaHCO_3$ in baking powders to produce carbon dioxide. The reaction may be written as

$$H_2PO_4^- + HCO_3^- \longrightarrow CO_2(g) + H_2O + HPO_4^{2-}$$

but it does not occur until water is added. Na_3PO_4 has been used in water softening, but it is under vigorous attack by environmentalists for contributing to water pollution *via* eutrophication. Phosphates are essential nutrients for

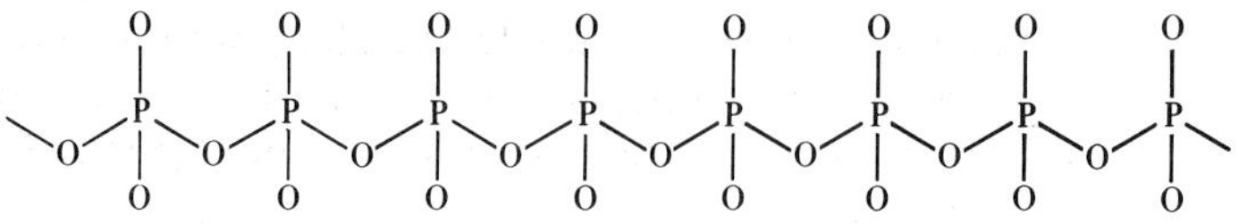

604

Chapter 21
Chemistry
of Some
Nonmetals
Other than
Carbon

FIG. 21.19 Metaphosphate chain.

growth, and their excessive presence in domestic waste water can nourish biologic processes beyond desirable rates.

H_3PO_4 is only one of a series of phosphoric acids that may be formed by the hydration of P_4O_{10}. To distinguish it from other phosphoric acids, H_3PO_4 is called *orthophosphoric* acid, and its salts are called *orthophosphates*. Among the other phosphoric acids are *pyrophosphoric* acid, $H_4P_2O_7$, and *metaphosphoric* acid, HPO_3, both of which can be made by heating H_3PO_4. On standing in water, all the phosphoric acids convert to orthophosphoric acid.

The metaphosphates, with simplest formula MPO_3, exist in a bewildering variety of complex salts. They are all polymeric in structure and, as shown in Fig. 21.19, are built up of PO_3^- units connected to each other so that each phosphorus atom remains tetrahedrally associated with four oxygen atoms. Of the many metaphosphates reported, we might mention the trimetaphosphate $Na_3P_3O_9$. This material is a white, crystalline solid which is produced by heating NaH_2PO_4. The reaction can be written as

$$3NaH_2PO_4 \longrightarrow Na_3P_3O_9(s) + 3H_2O(g)$$

When melted and chilled suddenly, $Na_3P_3O_9$ does not crystallize but instead forms a glass (sometimes called Graham's salt). The glass is soluble in water and in solution can precipitate Ag^+ and Pb^{2+} but not Ca^{2+}. In fact, it seems to form a complex with Ca^{2+} which makes it impossible to precipitate Ca^{2+} with the usual reagents such as carbonate. Because of this so-called "sequestering" action on Ca^{2+}, the material has been used extensively in water softening under the trade name Calgon. Recent investigations suggest that it is a high polymer of the type $(NaPO_3)_n$, where n can be as high as 1000.

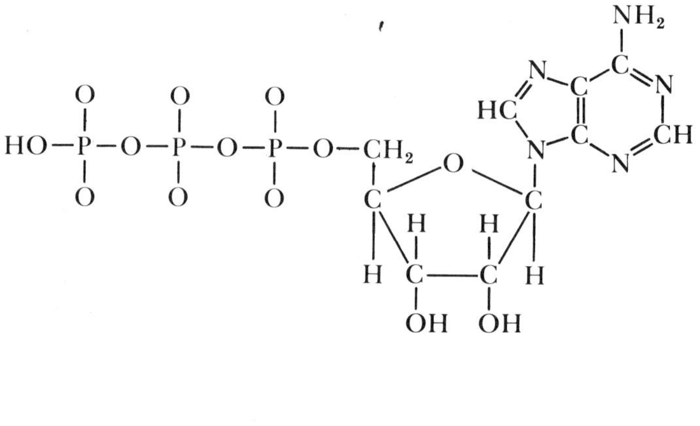

Like nitrogen, phosphorus is an essential constituent of living cells. One principal function is as phosphate groups that provide a means for storing energy in the cells. For example, when water splits a phosphate group off adenosine triphosphate (ATP), to form adenosine diphosphate (ADP), approximately 33 kJ of heat is liberated per mole. This energy can be used for the mechanical work of muscle contraction. Further discussions of this interesting subject are found in textbooks on biochemistry.

21.5
SULFUR

Sulfur is the second element of group VI. Other members of the group besides oxygen are selenium, tellurium, and polonium. The group is sometimes known as the *chalcogens*, from the Greek word *chalkos* for "copper" and *genes* for "born." Most copper minerals are either oxygen or sulfur compounds and frequently contain the other members of the group. Some of the group properties are given in Fig. 21.20.

Oxygen stands alone from the group in being a diatomic gas at room temperature. The other elements are solids with more complex structural units. There is, in going down the group, increasing tendency toward formation of long strings of atoms held together by covalent bonds. The bottom element, polonium, appears to be typically metallic. The increasing complexity of structure is due to increasing atomic size down the group. In general, the larger the atom, the less the tendency to form multiple bonds, and the greater the tendency of each atom to be bound to more than one other atom.

Perhaps the most striking variation in these elements is the decreasing oxidizing strength from oxygen to polonium. As the electrode potentials in the last column of Fig. 21.20 indicate, there is much greater tendency for elemental oxygen to form H_2O than for elemental polonium to form H_2Po. In fact, unlike H_2O and H_2S, the compounds H_2Se, H_2Te, and H_2Po are quite good reducing agents, better than hydrogen.

When bound to more electronegative atoms, the elements of group VI show positive oxidation states. Such states of oxygen are found only in compounds with fluorine (for example, OF_2) since fluorine is the only element more electronegative than oxygen. All the other elements of group VI form oxy compounds in which the elements, being less electronegative than oxy-

FIG. 21.20   Properties of Group VI Elements

Symbol	Z	Electronic configuration	Melting point, °C	Boiling point, °C	Ionization potential, eV	Electrode potential, V (X to H_2X)
O	8	(2) $2s^2 2p^4$	−219	−183.0	13.61	+1.23
S	16	(10) $3s^2 3p^4$	119	444.6	10.36	+0.14
Se	34	(28) $4s^2 4p^4$	220	685	9.75	−0.40
Te	52	(46) $5s^2 5p^4$	450	1390	9.01	−0.72
Po	84	(78) $6s^2 6p^4$	....	....	8.43	−1.0

606   Chapter 21
Chemistry
of Some
Nonmetals
Other than
Carbon

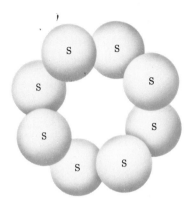

FIG. 21.21 S_8 molecule.

gen, are assigned positive oxidation numbers. An examination of these compounds shows that +4 and +6 are the most common.

Although sulfur is not very abundant, it is readily available because of its
occurrence in large beds of the free element. These beds, usually located several hundred feet underground, are thought to be due to bacterial decomposition of calcium sulfate. They are exploited by the *Frasch process* in which
superheated water (at about 170°C) is pumped down to the beds to melt the
sulfur, and the sulfur is then blown to the surface with compressed air.
Besides being found as the free element, sulfur occurs naturally in many sulfide and sulfate minerals, such as $CuFeS_2$, Cu_2S, and $CaSO_4 \cdot 2H_2O$. It is also
important in some biological molecules such as the amino acid cysteine,
$HSCH_2CH(NH_2)COOH$. S—S links also occur as bridges between various
parts of protein molecules (e.g., insulin).

There are several allotropic modifications of sulfur, the most important
being *rhombic* (also known as α) and monoclinic (β) sulfur. In the rhombic
form, which is the stable one at room temperature, sulfur atoms are linked to
each other, as shown in Fig. 21.21. The puckered, eight-membered rings are
stacked, as in Fig. 21.22, to form pipes that line up parallel in one plane but
are criss-cross relative to pipes in an adjacent plane. Above 96°C, monoclinic

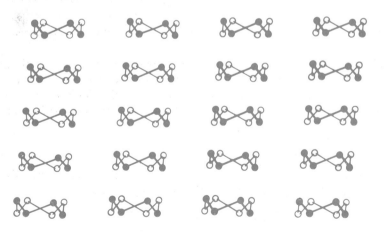

FIG. 21.22 Stacks of S_8 rings as arranged in rhombic sulfur.

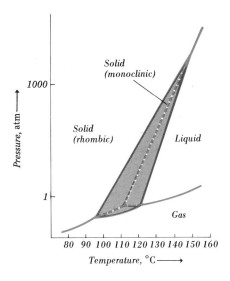

FIG. 21.23 Phase diagram for sulfur. (Pressure axis is distorted.)

sulfur is stable; it also apparently consists of S_8 rings, but the arrangement is less ordered.

When heated above the melting point, sulfur goes through a variety of changes. Starting as a mobile, pale yellow liquid, it gradually thickens above 160°C and then becomes less viscous as the boiling point is approached. The change in viscosity has been attributed to opening up of S_8 rings, joining the fragments together to form less mobile long chains, and then breaking the chains into fragments as the average kinetic energy per atom is increased.

The phase relations of sulfur are shown in Fig. 21.23. Because sulfur can exist in two solid modifications, the diagram contains four regions, corresponding to two solid states, one liquid, and one gas. At any temperature and pressure lying within the shaded triangle, monoclinic sulfur is the stable form. Thus, if rhombic sulfur is heated at 1 atm pressure to about 110°C and held there, it slowly converts to monoclinic sulfur. This is a very slow process, and under usual conditions is not observed. For the usual rapid heating of sulfur, the effective phase diagram is that corresponding to the dashed lines. Solid rhombic sulfur superheats without changing to monoclinic and melts at a temperature (112.8°C) which is below the melting point of monoclinic sulfur (119.25°C).

Although much of the sulfur produced is used directly in insecticides, fertilizers, paper and pulp fillers, and rubber, most of it is converted to industrially important oxy compounds, especially sulfuric acid. Sulfuric acid is made from sulfur dioxide, SO_2. This is usually made by burning sulfur in air:

$$S(s) + O_2(g) \longrightarrow SO_2(g)$$

or comes as a by-product of the preparation of various metals from their sulfide ores. For example, SO_2 is formed in the roasting of Cu_2S:

$$Cu_2S(s) + O_2(g) \longrightarrow 2Cu(s) + SO_2(g)$$

608

Chapter 21
Chemistry
of Some
Nonmetals
Other than
Carbon

The SO_2 is oxidized in the presence of catalysts such as vanadium pentoxide (V_2O_5) or platinum:

$$2SO_2(g) + O_2(g) \longrightarrow 2SO_3(g)$$

The product, SO_3, or sulfur trioxide, is the anhydride of H_2SO_4. We might expect the final step to be the dissolving of SO_3 in water; however, SO_3 reacts with water to form a fog of H_2SO_4, and the uptake of SO_3 by water is extremely slow. The usual method for circumventing this difficulty is to dissolve the SO_3 in pure H_2SO_4 and then dilute the product $H_2S_2O_7$, pyrosulfuric acid, with water:

$$SO_3(g) + H_2SO_4 \longrightarrow H_2S_2O_7$$
$$H_2S_2O_7 + H_2O \longrightarrow 2H_2SO_4$$

Pure H_2SO_4 is a liquid at room temperature; it has a tremendous affinity for water and forms several compounds, or hydrates, such as $H_2SO_4 \cdot H_2O$ and $H_2SO_4 \cdot 2H_2O$. Ordinary commercially available, concentrated sulfuric acid is approximately 93% H_2SO_4 by weight and can be thought of as a solution of H_2SO_4 and $H_2SO_4 \cdot H_2O$. Frequently, concentrated H_2SO_4 is used as a dehydrating agent, as, for example, in desiccators to keep substances dry. It is also used in reactions to favor splitting out of water. An example of the latter is the manufacture of ethers from alcohols:

$$2C_2H_5OH \xrightarrow{H_2SO_4} C_2H_5OC_2H_5 + H_2O$$

In dilute aqueous solutions, H_2SO_4 is a strong acid, but only for dissociation of one proton. The dissociation constant for the second proton is 1.26×10^{-2}. Because of the second dissociation, solutions of HSO_4^-, as from $NaHSO_4$, are acidic.

For the half-reaction

$$HSO_4^- + 3H^+ + 2e^- \longrightarrow SO_2 + 2H_2O$$

the electrode potential is +0.11 V. This means that HSO_4^- at 1 m concentration is a mild oxidizing agent. However, the action at room temperature is generally not observed because reaction is so slow. With hot, concentrated solutions, the situation is different. For example, sodium bromide plus hot H_2SO_4 produces not only HBr but also some bromine by oxidation of Br^- to Br_2. Furthermore, some of the less active metals such as copper are soluble in hot, concentrated sulfuric acid, presumably because of oxidation by sulfate.

In addition to the oxy compounds in the +6 oxidation state, there are important oxy compounds corresponding to the +4 state. The simplest of these is the dioxide, SO_2. At room temperature, it is a gas, but it is quite easily liquefied. The easy liquefaction reflects the fact that the molecule is polar because it has a nonlinear arrangement of atoms. Sulfur dioxide has a disagreeable, choking odor and is somewhat poisonous. Its emission from copper and zinc smelters and from power plants that burn fossil fuels (e.g., coal, on the average, contains about 3% by weight sulfur) makes SO_2 a serious environmental pollutant. It is especially toxic to lower organisms such as fungi. For this reason it is used for sterilizing dried fruit and wine barrels.

With water, SO_2 dissolves to give acid solutions, which contain about 5% of the sulfur as sulfurous acid, H_2SO_3. The compound H_2SO_3 has never been isolated pure; any attempt to concentrate it, as by heating, simply expels SO_2. H_2SO_3 is a weak diprotic acid, for which the principal equilibria are

$$SO_2 + H_2O \rightleftharpoons H^+ + HSO_3^- \qquad K_I = 1.25 \times 10^{-2}$$
$$HSO_3^- \rightleftharpoons H^+ + SO_3^{2-} \qquad K_{II} = 5.6 \times 10^{-8}$$

It forms two series of salts: sulfites, for example, Na_2SO_3, and hydrogen sulfites, for example, $NaHSO_3$. Addition of acid to sulfites or hydrogen sulfites liberates SO_2 and is a convenient way of making sulfur dioxide in the laboratory. Sulfites, hydrogen sulfites, and sulfurous acid are mild reducing agents and are relatively easily oxidized to sulfates, though sometimes the reaction is quite slow.

When solutions containing sulfite ion are boiled with elemental sulfur, the solid sulfur dissolves according to the reaction

$$S(s) + SO_3^{2-} \rightleftharpoons S_2O_3^{2-}$$

The ion formed, $S_2O_3^{2-}$, is called *thiosulfate* ion, where the prefix *thio-* indicates substitution of a sulfur atom for an oxygen atom. $S_2O_3^{2-}$ contains two different kinds of sulfur atoms, as can be shown by the following experiment: Solid sulfur containing a radioactive isotope is boiled with a solution containing nonradioactive sulfite. The thiosulfate formed is found to be radioactive, but after acid is added so as to reverse the above reaction, all the radioactivity is recovered as precipitated solid sulfur. The implication is that the same sulfur atom which adds to SO_3^{2-} to form $S_2O_3^{2-}$ is dropped off when acid is added. This can be true only if the added sulfur atom is bound in a way that is unlike the binding of the sulfur atom already in SO_3^{2-}. The structure proposed for $S_2O_3^{2-}$ has a sulfur atom at the center of a tetrahedron and another sulfur with three oxygens at the corners (Fig. 21.24). Thiosulfate acts as a mild reducing agent:

$$2S_2O_3^{2-} \longrightarrow S_4O_6^{2-} + 2e^- \qquad E° = -0.08 \text{ V}$$

It has, for example, the ability to reduce I_2 to I^-. The reaction, which produces *tetrathionate ion,* $S_4O_6^{2-}$, is frequently used in determining the amount of iodine in a solution. It thus makes possible the quantitative analysis

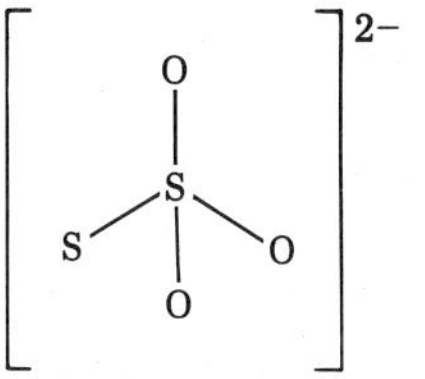

FIG. 21.24 Structure of thiosulfate ion.

610

Chapter 21
Chemistry
of Some
Nonmetals
Other than
Carbon

of many oxidizing agents. The oxidizing agent is treated with excess I$^-$, and the liberated I$_2$ is titrated with thiosulfate solution. Thiosulfate ion also has the ability to form complex ions. The silver-thiosulfate complex, $Ag(S_2O_3)_2^{3-}$, is so stable that thiosulfate solutions can be used to dissolve insoluble silver halides in the photographic fixing process.

Besides positive oxidation states, sulfur also shows negative states, especially -2. The most familiar such compound is probably hydrogen sulfide, H$_2$S, notorious for its rotten-egg odor. Not so well known is the fact that hydrogen sulfide is as poisonous as hydrogen cyanide. The pure compound can be made by interaction of a sulfide such as FeS with acid:

$$FeS(s) + 2H^+ \longrightarrow Fe^{2+} + H_2S(g)$$

or by warming a solution of thioacetamide:

$$\underset{\text{Thioacetamide}}{CH_3-\overset{\overset{\displaystyle S}{\|}}{C}-NH_2} + H_2O \longrightarrow \underset{\text{Acetamide}}{CH_3-\overset{\overset{\displaystyle O}{\|}}{C}-NH_2} + H_2S(g)$$

This latter reaction is important as an easily controlled laboratory source of hydrogen sulfide for qualitative analysis.

Like water, hydrogen sulfide has a bent molecule and is polar. It is a mild reducing agent and can, for example, reduce ferric ion to ferrous ion:

$$2Fe^{3+} + H_2S(g) \longrightarrow 2Fe^{2+} + S(s) + 2H^+$$

During the course of the reaction, the solution becomes milky from the production of colloidal sulfur. In aqueous solution, hydrogen sulfide is a weak diprotic acid for which the dissociation constants are $K_I = 1.1 \times 10^{-7}$ and $K_{II} = 1 \times 10^{-14}$. A detailed consideration of the equilibria in aqueous hydrogen sulfide solutions was given in Sec. 15.9. The sulfides are derived from H$_2$S. Their solubility in water varies widely from those which, like Na$_2$S, are quite soluble in water to those which, like HgS, require drastic treatment to be brought into solution. Figure 21.25 lists various representative sulfides and methods required to dissolve them. The alkali-metal and alkaline-earth-metal sulfides dissolve readily. Because they are so soluble, they cannot be precipitated by bubbling H$_2$S through solutions of their salts.

As already discussed in Sec. 15.9, some sulfides that are insoluble in water can be dissolved by raising the H$^+$ concentration. ZnS, for example, is soluble in 0.3 M H$^+$ because the H$^+$ lowers the concentration of sulfide ion by combining with it to form H$_2$S. The net equation can be represented as

$$ZnS(s) + 2H^+ \longrightarrow Zn^{2+} + H_2S$$

FIG. 21.25   Solubilities of Sulfides

Soluble in water	Na$_2$S, K$_2$S, (NH$_4$)$_2$S, BaS
Soluble in 0.3 M H$^+$	ZnS, FeS, MnS, CoS
Soluble in hot HNO$_3$	CuS, Ag$_2$S PbS, SnS
Soluble in aqua regia	HgS

The sulfides in the third row of Fig. 21.25 are so insoluble that they cannot be dissolved by H^+ alone. However, hot nitric acid oxidizes sulfide to sulfur and hence lowers the sulfide-ion concentration sufficiently to permit solubility. For CuS, the net reaction can be written as

$$3CuS(s) + 8H^+ + 2NO_3^- \longrightarrow 3Cu^{2+} + 3S(s) + 2NO(g) + 4H_2O$$

The least soluble of the sulfides shown in Fig. 21.25, mercuric sulfide, is not appreciably soluble in hot HNO_3. In order to "dissolve" it, aqua regia must be used in order that oxidation of the sulfide ion be accompanied by complexing of the mercuric ion. The net reaction may be written as

$$HgS(s) + 2NO_3^- + 4Cl^- + 4H^+ \longrightarrow HgCl_4^{2-} + 2NO_2(g) + S(s) + 2H_2O$$

The differences in solubility behavior of metal sulfides can be used to great advantage in the separation and identification of various elements.

In addition to sulfides, sulfur forms polysulfides, in which two or more sulfur atoms are bound together in a chain. When Na_2S is boiled with sulfur, the product formed is Na_2S_x and is thought to consist of Na^+ ions and S_x^{2-} ions. The polysulfide chains are of varying length, formed by progressive addition of sulfur atoms to sulfide ion:

$$\left[:\overset{..}{\underset{..}{S}}:\right]^{2-} + :\overset{..}{\underset{..}{S}}: \longrightarrow \left[:\overset{..}{\underset{..}{S}}:\overset{..}{\underset{..}{S}}:\right]^{2-} \longrightarrow \left[:\overset{..}{\underset{..}{S}}:\overset{..}{\underset{..}{S}}:\overset{..}{\underset{..}{S}}:\right]^{2-} \longrightarrow$$

$$\left[:\overset{..}{\underset{..}{S}}:\overset{..}{\underset{..}{S}}:\overset{..}{\underset{..}{S}}:\overset{..}{\underset{..}{S}}:\right]^{2-} \text{, etc.}$$

The simplest of the polysulfide chains is the disulfide, S_2^{2-}; it is found in the mineral FeS_2, iron pyrites, or fool's gold. Solid FeS_2 has an NaCl-like structure consisting of an array of alternating Fe^{2+} and S_2^{2-} ions. In acid solution, disulfides (and other polysulfides) break down to form solid sulfur and H_2S. In some respects, disulfides resemble peroxides. They are, for example, oxidizing agents, especially for metal sulfides. Thus, a solution of Na_2S_2 can oxidize stannous sulfide, SnS (+2 state of tin), to SnS_3^{2-} (+4 state of tin):

$$SnS(s) + S_2^{2-} \longrightarrow SnS_3^{2-}$$

In addition to the above compounds, which are mainly inorganic, there are many important organic compounds that contain sulfur. These include, for example, the sulfonates, which are of interest as synthetic detergents. Actually, the term "detergent" includes soap, but by common usage it has come to mean a soap substitute. Prime among these are the benzenesulfonates such as

$$CH_3CH_2CH_2CH_2CH_2CH_2CH_2CH_2CH_2CH_2 - \underset{\underset{CH_3}{|}}{\overset{\overset{H}{|}}{C}} - \underset{}{\overset{}{\bigcirc}} - \underset{\underset{O}{\|}}{\overset{\overset{O}{\|}}{S}} - O^- Na^+$$

They resemble soap in having a long hydrocarbon tail and a charged anionic group at the end. The above molecule is an example of a linear sulfonate and is to be contrasted with a sulfonate of the type

612 Chapter 21
Chemistry
of Some
Nonmetals
Other than
Carbon

$$CH_3-CH-CH_2-CH-CH_2-CH-CH_2-C\underset{CH_3}{\overset{CH_3}{|}}\cdots$$

The branched sulfonates are also very good detergents; they reduce surface tension and are good wetting agents, but because of the chain branching, they are not biodegradable. Hence, they accumulate in the water environment and, even at 1 ppm levels, cause serious foaming problems. Although they are not harmful even up to 50 ppm, there is a certain lack of appeal in drinking water with a "head" on it.

21.6
THE HALOGENS

The elements of group VII, fluorine, chlorine, bromine, iodine, and astatine, are collectively called *halogens*. Their name comes from the Greek *halos*, meaning "salt," and *genes*, meaning "born," and reflects the fact that they are "salt producers." They all have high electronegativity and form negative halide ions such as are found in ionic salts.

Although the chemistry of the group VII elements is somewhat complex, similarities within the group are more pronounced than in any of the other groups except I and II. They all form diatomic molecules and are good oxidizing agents. Figure 21.26 shows some of the other properties. At room temperature fluorine and chlorine are gases, bromine is a liquid, and iodine is a solid; all are volatile, however, so even for bromine and iodine, vapors are present at room temperature.

As indicated by the relatively high values of the ionization potentials, it is fairly difficult to remove an electron from a halogen atom. In fact, more energy is required to remove an electron from a halogen atom than from any other atom in the same period except for the noble gas. Within the group itself there is a decrease going down the group; the larger the halogen atom, the less firmly bound are the outermost electrons.

FIG. 21.26 Properties of Group VII Elements

Symbol	Z	Electronic configuration	Melting point, °C	Boiling point, °C	Ionization potential, eV	Electrode potential, V (X_2 to X^-)
F	9	(2) $2s^2 2p^5$	−223	−187	17.42	+2.87
Cl	17	(10) $3s^2 3p^5$	−102	−34.6	13.01	+1.36
Br	35	(28) $4s^2 4p^5$	−7.3	58.78	11.84	+1.09
I	53	(46) $5s^2 5p^5$	114	183	10.44	+0.54
At	85	(78) $6s^2 6p^5$	....	....	....	+0.2

Of greater significance chemically are the electrode potentials, given in the last column of Fig. 21.26. The potentials show that fluorine gas is the best oxidizing agent of the group. The reason, however, is not so simple as it appears. The overall half-reaction

$$\tfrac{1}{2}X_2(g) + e^- \longrightarrow X^-(aq)$$

can be constructed from the set of consecutive steps

$$\tfrac{1}{2}X_2(g) \xrightarrow{(1)} X(g) \xrightarrow{(2)} X^-(g) \xrightarrow{(3)} X^-(aq)$$

Step (1) corresponds to breaking up $\tfrac{1}{2}$ mol of diatomic X_2 molecules to give monatomic gas; in step (2) each gaseous X atom picks up one electron to form a mononegative X^- ion; in step (3) the X^- ions go from the gaseous state to the aqueous state; i.e., they become hydrated. The free-energy changes for the various steps are listed in Fig. 21.27. In the case of fluorine, for example, the sum of the three processes has a net free-energy change of -730 kJ* for $\tfrac{1}{2}F_2(g) + e^- \longrightarrow F^-(aq)$.

Comparing the halogens with each other, we see from Fig. 21.27 that the most significant difference between the elements is in the third step, the hydration step. Except for chlorine, the first step requires about the same amount of energy to break up the X_2 molecule. The second step is about equally favorable for adding an electron to any gaseous halogen atom [actually, most favorable in the case of $Cl(g)$]. Hence, if only the sum of the first two steps were considered, there would be relatively little difference between the various halogens. The biggest single factor in making the overall tendencies differ greatly from each other is the hydration free energy. In going from F^- to I^- there is a large increase in ionic radius, and this makes for a large decrease in affinity for water. The main reason why F_2 is such a good oxidizing agent [and $F^-(aq)$ is such a poor reducing agent] is that the fluoride ion is strongly hydrated.

*To convert this to electrode potential, we need to add $+465$ kJ for the hydrogen-electrode reference reaction $e^- + H^+(aq) \longrightarrow \tfrac{1}{2}H_2(g)$ and use $\Delta G^\circ = -n\mathfrak{F}E^\circ$.

FIG. 21.27 Free-Energy Changes in Kilojoules per Mole for Each of the Halogens

Step	Fluorine	Chlorine	Bromine	Iodine
$\tfrac{1}{2}X_2(g) \longrightarrow X(g)$	+63	+105	+79	+59
$X(g) + e^- \longrightarrow X^-(g)$	−333	−348	−324	−295
$X^-(g) \longrightarrow X^-(aq)$	−460	−348	−318	−279
$\tfrac{1}{2}X_2(g) + e^- \longrightarrow X^-(aq)$	−730	−591	−563	−515

Fluorine is about half as abundant as chlorine and is widely distributed in nature. It occurs principally as the minerals fluorspar, CaF_2; cryolite, Na_3AlF_6; and fluorapatite, $Ca_5F(PO_4)_3$. Because none of the ordinary chemical oxidizing agents is capable of extracting electrons from fluoride ions, elemental fluorine is prepared by electrolytic oxidation of molten fluorides. At room temperature fluorine is a pale yellow gas which is extremely corrosive and reactive. With hydrogen it forms violently explosive mixtures because of the reaction

$$H_2(g) + F_2(g) \longrightarrow 2HF(g) \qquad \Delta H° = -537 \text{ kJ}$$

On the skin it causes severe "burns," which are quite slow to heal.

Hydrogen fluoride is usually made by the action of sulfuric acid on fluorspar. Because of hydrogen bonding, liquid HF has a higher boiling point (19.5°C) than any of the other hydrogen halides. Hydrogen bonding is also present in the gas phase and accounts for polymeric species, $(HF)_x$, where x is some small number such as 6 or less. In aqueous solutions HF is called hydrofluoric acid and is unique among the hydrogen halides in being a weak, rather than a strong, acid ($K_{diss} = 6.7 \times 10^{-4}$). It is also able to dissolve glass. The latter reaction is attributed to the formation of fluorosilicate ions, as in the equation

$$SiO_2(s) + 6HF \longrightarrow SiF_6^{2-} + 2H^+ + 2H_2O$$

where glass is represented for simplicity as SiO_2.

In general, most simple fluoride salts formed with +1 cations are soluble (for example, KF and AgF) and give slightly basic solutions because of the hydrolysis of F^- to HF. With +2 cations, however, the fluorides are usually insoluble (for example, CaF_2 and PbF_2). The formation of insoluble, inert fluorides as surface coatings is apparently the reason why fluorine and its compounds can be stored in some metal containers such as copper.

Most amazing of the fluorine compounds are the fluorocarbons. These materials are derived from the hydrocarbons by substitution of fluorine atoms for hydrogen atoms. Thus, the fluorocarbon corresponding to methane, CH_4, is tetrafluoromethane, CF_4. This compound is typical of the saturated (i.e., containing no double bonds) fluorocarbons in being extremely inert. For example, unlike methane, it can be heated in air without burning. Furthermore, it can be treated with boiling nitric acid, concentrated sulfuric acid, and strong oxidizing agents such as potassium permanganate with no change. Reducing agents such as hydrogen and carbon do not affect it even at temperatures as high as 1000°C. Because of their inertness, the fluorocarbons find application for special uses. For example, $C_{12}F_{26}$ is an ideal insulating liquid for heavy-duty transformers that operate at high temperature. Just as ethylene, C_2H_4, can polymerize to form polyethylene, so tetrafluoroethylene, C_2F_4, can polymerize to form polytetrafluoroethylene. The polymerization can be imagined to proceed by the opening up of the double bond to form an unstable intermediate, which joins with other molecules to produce a high polymer:

$$n\left(\begin{array}{c}F\\ \\F\end{array}C=C\begin{array}{c}F\\ \\F\end{array}\right) \longrightarrow n\left[\begin{array}{ccc}&F&F\\ &|&|\\ -&C&-C-\\ &|&|\\ &F&F\end{array}\right] \longrightarrow \begin{array}{cccccc}F&F&F&F&F&F\\ |&|&|&|&|&|\\ -C&-C&-C&-C&-C&-C-\\ |&|&|&|&|&|\\ F&F&F&F&F&F\end{array}$$

The high polymer is a plastic known commercially as Teflon and, like other saturated fluorocarbons, is inert to chemical attack. It is unaffected even by boiling aqua regia or ozone. Though rather expensive, fluorocarbon polymers find use as structural materials where corrosive conditions are extreme, as in chemical plants. They are also familiar for coating "greaseless" frying pans—a boon to amateur cooks and calorie counters.

Finally, it should be noted that fluorine was the key element in cracking the problem of making the noble-gas elements combine chemically with other elements. The search for compounds of the group 0 elements (helium, neon, argon, krypton, xenon, and radon) had been carried on for many years. The synthetic breakthrough came as a result of an observation by Neil Bartlett that O_2 reacts with PtF_6 to form a solid compound O_2PtF_6. X-ray diffraction indicated that the orange-colored compound consists of O_2^+ and PtF_6^-. Recognizing that the ionization potential of the O_2 molecule (12.2 eV) is very close to that of xenon (12.1 eV), Bartlett tried the reaction between xenon and PtF_6. Reaction occurred readily at room temperature to produce a red solid, $XePtF_6$, the first real compound of a noble-gas element.

Since this discovery in 1962, a number of other noble-gas compounds have been prepared. These are principally compounds of xenon, but a few compounds of krypton and radon have also been prepared. They are all rather unstable.

The shapes of the xenon compounds are quite interesting. As shown in Fig. 21.28, XeF_2 is linear and XeF_4 is square planar. Neither the bond lengths nor the bond energies are particularly unusual, and so it must be concluded that xenon forms normal chemical bonds. The bonding can be accounted for by a simple valence-bond scheme that utilizes the normally empty $5d$ orbitals of xenon. In XeF_2, for example, one of the $5d$ orbitals is hybridized with the $5s$ and the three $5p$ orbitals to form a set of five sp^3d hybrids. These are directed, as in Fig. 21.28 (left), toward the corners of a trigonal bipyramid. (This can be visualized as two triangular pyramids sharing a common face.)

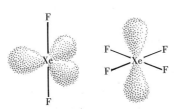

FIG. 21.28 Xenon fluorides, XeF_2 and XeF_4.

616

Chapter 21
Chemistry
of Some
Nonmetals
Other than
Carbon

There are 10 electrons to accommodate (one from each of two fluorine atoms plus eight from xenon.) We put these in the five sp^3d orbitals to get covalent bonds and three unshared pairs. If the two fluorine atoms are as far apart as possible, we get the structure shown on the left in Fig. 21.28. For XeF_4 we need to use the combination (one $5s$, three $5p$, and two $5d$) to get six sp^3d^2-hybrid orbitals. Twelve electrons (one from each of four fluorines and eight from xenon) are disposed as four covalent bonds and two unshared pairs. With the four fluorine atoms in a plane, the structure is square planar, as is shown on the right in Fig. 21.28.

21.8
CHLORINE

Chlorine is the most abundant of the halogens. It occurs as chloride ion in seawater, salt wells, and salt beds, where it is combined with Na^+, K^+, Mg^{2+}, and Ca^{2+}. On a small scale, the element can be made by chemical oxidation, as with MnO_2:

$$MnO_2(s) + 2Cl^- + 4H^+ \longrightarrow Mn^{2+} + Cl_2(g) + 2H_2O$$

On a commercial scale, it is more economically prepared by electrolytic oxidation of either aqueous or molten NaCl (see Sec. 11.3). The element is a greenish yellow gas (in fact, it gets its name from the Greek, *chloros*, meaning "green") and has a choking odor. Although not so reactive as fluorine, it is a good oxidizing agent and explodes with hydrogen when mixtures of H_2 and Cl_2 are exposed to ultraviolet light. Most of the commercial chlorine is used as a bleach for paper and wood pulp and for large-scale disinfecting of public water supplies. Both of these uses depend on its oxidizing action.

The most important compounds of chlorine are those which correspond to the oxidation states -1, $+1$, $+5$ and $+7$. The -1 state is familiar as the one assigned in HCl and chloride salts. Although HCl can be produced by direct combination of the elements, a more convenient method of preparation is the heating of NaCl with concentrated H_2SO_4:

$$NaCl(s) + H_2SO_4 \longrightarrow NaHSO_4(s) + HCl(g)$$

Hydrogen chloride gas is very soluble in water, and it is the aqueous solution that is properly referred to as hydrochloric acid. Commercially available, concentrated hydrochloric acid is 37% by weight HCl, or 12 M. Unlike HF, HCl is a strong acid and is practically 100 percent dissociated into ions in 1 M solution. Why is HCl stronger than HF? The high hydration energy of fluoride ion would seem to favor dissociation of HF. The fact that HCl is the more highly dissociated arises because the bond in HCl (427 kJ/mol) is appreciably weaker than the bond in HF (575 kJ/mol).

The $+1$ oxidation state of chlorine is represented by hypochlorous acid, HOCl, and its salts, the hypochlorites. Hypochlorous acid is produced to a limited extent when chlorine gas is dissolved in water:

$$Cl_2 + H_2O \rightleftharpoons Cl^- + H^+ + HOCl$$

The yield can be greatly increased by tying up the Cl^- and H^+, as by adding silver oxide (silver ion to precipitate AgCl and oxide ion to neutralize H^+). The formula of hypochlorous acid is usually written HOCl, instead of HClO, to emphasize the fact that the proton is bonded to the oxygen and not directly to the chlorine. The acid is weak, with a dissociation constant 3.2×10^{-8}, and exists only in aqueous solution. It is a powerful oxidizing agent, stronger, for example, than MnO_4^-. Hypochlorites, such as NaClO, can be made by neutralization of HOCl solutions. They are produced more economically by disproportionation of chlorine in basic solution:

$$Cl_2 + 2OH^- \longrightarrow Cl^- + ClO^- + H_2O$$

Commercially, the process is carried out by electrolyzing cold, aqueous sodium chloride solution and stirring vigorously. The stirring serves to mix chlorine produced at the anode:

$$2Cl^- \longrightarrow Cl_2 + 2e^-$$

with hydroxide ion produced at the cathode:

$$2e^- + 2H_2O \longrightarrow H_2(g) + 2OH^-$$

so that the above reaction can occur. Solutions of hypochlorite ion are sold as laundry bleaches, e.g., Clorox.

In aqueous solution, hypochlorite ion is unstable with respect to self-oxidation and, when warmed, disproportionates by the equation

$$3ClO^- \longrightarrow 2Cl^- + ClO_3^-$$

The product chlorate ion (ClO_3^-) contains chlorine in oxidation state +5. It is pyramidal in structure, with the three oxygen atoms forming the base of a pyramid and chlorine the apex. Probably the most important chlorate salt is $KClO_3$, used as an oxidizing agent in matches, fireworks, and some explosives. Since $KClO_3$ is only moderately soluble in water, it can be precipitated by addition of KCl to chlorate-containing solution. The chlorate solutions can be produced by electrolyzing hot chloride solutions that are vigorously stirred. Steps in the production can be summarized as follows:

$$2Cl^- + 2H_2O \xrightarrow{\text{electrolyze}} Cl_2 + 2OH^- + H_2(g)$$

$$3Cl_2 + 6OH^- \xrightarrow{\text{stir, heat}} 5Cl^- + ClO_3^- + 3H_2O$$

$$K^+ + ClO_3^- \longrightarrow KClO_3(s)$$

As seen from the equation for the second step, only one-sixth of the chlorine is converted to ClO_3^-, which makes the process seem rather inefficient. However, on continued electrolysis, the chloride produced in the second step is reoxidized in the first step.

Unlike hypochlorite, chlorate ion is the anion of a strong acid. The parent acid, $HClO_3$, chloric acid, has not been prepared pure since it is unstable. When attempts are made to concentrate it, as by evaporation, violent explosions occur. The principal reaction is

$$4HClO_3 \longrightarrow 4ClO_2(g) + O_2(g) + 2H_2O(g)$$

618 Chapter 21
Chemistry
of Some
Nonmetals
Other than
Carbon

Chlorate ion, like hypochlorite ion, is a good oxidizing agent. When $KClO_3$ is heated, it can decompose by two reactions:

$$2KClO_3(s) \longrightarrow 2KCl(s) + 3O_2(g)$$

$$4KClO_3(s) \longrightarrow 3KClO_4(s) + KCl(s)$$

The first is catalyzed by surfaces, such as powdered glass or MnO_2, from which oxygen can readily escape. In the absence of such catalysts, especially at lower temperatures, the formation of potassium perchlorate ($KClO_4$) is favored. The perchlorate ion has a tetrahedral configuration, with the chlorine at the center and the four oxygens at the corners. In aqueous solutions perchlorate is potentially a good oxidizing agent, but its reactions are slow. For example, a solution containing ClO_4^- and the very strong reducing agent Cr^{2+} (chromous ion) can be kept for weeks without any appreciable oxidation to Cr^{3+} (chromic ion).

Like chlorate, perchlorate ion is the anion of a strong acid. Consequently, in aqueous solution there is practically no association of H^+ and ClO_4^-. However, when perchlorate salts are treated with sulfuric acid, pure hydrogen perchlorate ($HClO_4$) may be distilled off under reduced pressure. The anhydrous compound is a liquid at room temperature and is extremely dangerous because it may explode spontaneously. The danger is especially great in the presence of reducing agents such as organic material (e.g., wood, cloth, etc.). Dilute aqueous solutions, on the other hand, are safe. They are useful reagents for the chemist since perchlorate ion has less tendency than any other anion to form complex ions with metal cations. Furthermore, perchloric acid is a very strong acid ($K_{diss} = 550$).

Why is perchloric acid so strong? The dissociation of an oxyacid involves breaking a hydrogen-oxygen bond to form a hydrated hydrogen ion and a hydrated anion. The bigger the anion, the less its hydration energy. Consequently, since ClO_4^- is obviously bigger than ClO^- we might expect $HClO_4$ to be less dissociated than $HOCl$. Since the reverse is true, it must be that the bond holding the proton to OCl^- is stronger than the bond holding H^+ to ClO_4^-. That this is reasonable can be seen by noting that oxygen is more electronegative than chlorine. In the series $HOCl$, $HOClO$, $HOClO_2$, $HOClO_3$, we are adding more oxygen atoms to the chlorine. The more oxygens we add, the more we pull electrons away from the H—O bond and the more we tend to weaken it. Hence, $HOCl$ is the weakest and $HOClO_3$ is the strongest acid of the series. In general, for any series of oxyacids the acid corresponding to highest oxidation number (i.e., the one containing most oxygen) is the most highly dissociated. Thus, HNO_3 is stronger than HNO_2, and H_2SO_4 is stronger than H_2SO_3.

21.9 BROMINE

Bromine, from the Greek word *bromos*, meaning "stink," occurs as bromine ion in seawater, brine wells, and salt beds. It is less than a hundredth as abundant as chlorine. The element is usually prepared by chlorine oxidation of bromide solutions, as by sweeping chlorine gas through seawater. Since chlorine is a stronger oxidizing agent than bromine, the reaction

$$Cl_2(g) + 2Br^- \longrightarrow Br_2 + 2Cl^-$$

occurs as indicated. Removal of the bromine from the resulting solution can be accomplished by sweeping the solution with air because bromine is quite volatile. At room temperature pure bromine is a mobile but dense red liquid of pungent odor. It is a dangerous substance since it attacks the skin to form slow-healing sores.

Although less powerful as an oxidizing agent than is chlorine, bromine readily reacts with other elements to form bromides. Hydrogen bromide is a strong acid, like HCl, but is more easily oxidized. Whereas HCl can be made by heating NaCl and H_2SO_4, HBr cannot be made by heating NaBr and H_2SO_4. Hot H_2SO_4 easily oxidizes HBr to Br_2, and so a nonoxidizing acid such as H_3PO_4 must be used instead.

In basic solution, bromine disproportionates to give bromide ion and hypobromite ion (BrO^-). The reaction is quickly followed by further disproportionation:

$$3BrO^- \longrightarrow 2Br^- + BrO_3^-$$

to give bromate ion, BrO_3^-. It is a good oxidizing agent, slightly stronger than chlorate and considerably faster.

One of the most important uses of bromine is in making silver bromide for photographic emulsions (Sec. 19.11). However, the principal use has been in making dibromoethane ($C_2H_4Br_2$) for addition to gasoline. Tetraethyllead, $(C_2H_5)_4Pb$, added to gasoline as an antiknock agent, decomposes on burning to form lead deposits. The dibromoethane is added to prevent accumulation of lead deposits in the engine, but, of course, it increases the amount of lead exhausted into the air, principally as PbClBr.

21.10
IODINE

Of all the halogens, iodine is the only one that occurs naturally in a positive oxidation state. In addition to I^- in seawater and salt wells, it is found as sodium iodate ($NaIO_3$) mixed with $NaNO_3$ in Chile saltpeter. The Chilean ore is processed by reducing the $NaIO_3$ with controlled amounts of $NaHSO_3$. The principal reaction is

$$5HSO_3^- + 2IO_3^- \longrightarrow I_2 + 5SO_4^{2-} + 3H^+ + H_2O$$

Excess hydrogen sulfite must be avoided, for it would reduce I_2 to I^-. In the United States most iodine is produced by chlorine oxidation of I^- from salt wells.

At room temperature elemental iodine crystallizes as black leaflets with metallic luster. As shown by X-ray analysis, the solid consists of discrete I_2

620

Chapter 21
Chemistry
of Some
Nonmetals
Other than
Carbon

molecules, but its properties are different from those of usual molecular solids. For example, its electric conductivity, though small, increases with increasing temperature like that of a semiconductor. Furthermore, liquid iodine also has perceptible conductivity, which decreases with increasing temperature like that of a metal. Thus, feeble as they are, metallic properties do appear even in the halogen group.

When heated, solid iodine readily sublimes to give a violet vapor which consists of I_2 molecules. The violet color is the same as that observed in many iodine solutions, such as in CCl_4 or hydrocarbons. However, in water and in alcohol, iodine solutions are brown, probably because of specific interactions between I_2 and the solvent. When iodine is brought in contact with starch, a characteristic deep blue color results. The color, which has been attributed to a starch-I_2 complex, is the basis of a test using starch–potassium iodide paper to look for presence of oxidizing agents. Oxidizing agents convert I^- to I_2, and the I_2 combines with the starch to form the blue complex. With very strong oxidizing agents the color may fade because of oxidation of I_2 to a higher oxidation state.

Iodine is only slightly soluble in water (0.001 M), but the solubility can be vastly increased by the presence of iodide ion. The color changes from brown to deep red because of formation of triiodide ion, $I_2 + I^- \longrightarrow I_3^-$. The triiodide ion is also known in solids such as NH_4I_3. X-ray investigations indicate that it is linear, $I—I—I^-$. No electronic formula conforming to the octet rule can be written for this ion. Apparently, the central iodine atom, perhaps because of its large size, can accommodate more than eight electrons in its valence shell. A possible model using normally empty $5d$ orbitals would be like that shown in Fig. 21.29. Hybridization of the $5d$ with the $5s$ and three $5p$ orbitals of the central iodine would give sp^3d orbitals directed to the corners of a trigonal bypyramid. Two iodines could then form ordinary σ bonds with the central iodine along the molecular axis, leaving three unshared pairs of electrons in the plane perpendicular to it. In basic solutions, I_2 disproportionates to form iodide ion and hypoiodite ion (IO^-):

$$I_2 + 2OH^- \longrightarrow I^- + IO^- + H_2O$$

Further disproportionation to give iodate ion (IO_3^-) is hastened by heating or by addition of acid. Iodate ion in acid solution is a weaker oxidizing agent than either bromate ion or chlorate ion. Furthermore, iodate salts are not quite so explosive as chlorates or bromates. The greater stability of iodates is also evident in the fact that HIO_3, unlike $HClO_3$ and $HBrO_3$, can be isolated pure (as a white solid). The latter acids detonate when attempts are made to concentrate them.

FIG. 21.29 Possible structure of triiodide ion.

In the +7 state, the oxysalts of iodine are called periodates (pronounced per-eye-oh-dates). There are several kinds: those derived from HIO_4 (metaperiodic acid), those derived from H_5IO_6 (paraperiodic acid), and possibly others. In the metaperiodates the iodine is bonded tetrahedrally to four oxygen atoms (this ion is the analog of ClO_4^-); in the paraperiodates there are six oxygen atoms bound octahedrally to the iodine atom. The fact that there are paraperiodates is apparently due to the large size of the iodine atom. As in I_3^-, it is necessary to assume that the valence shell of iodine is expanded to contain more than eight electrons. Paraperiodic acid, H_5IO_6, is moderately weak ($K_I = 5.1 \times 10^{-4}$), but metaperiodic acid, HIO_4, is strong.

In going down the halogen group the atoms of the elements get progressively larger, and it becomes easier to oxidize the halide ion (X^-) to free halogen (X_2). This shows up as an instability of iodide solutions to air oxidation. The oxidation is slow for basic and neutral solutions but becomes appreciably faster for acid solutions.

So far as uses are concerned, iodine is less widely used than other halogens. It finds limited use for its antiseptic properties, both as tincture of iodine (solution of I_2 in alcohol) and as iodoform (CHI_3). Small amounts of iodine are required in the human diet, and so traces of sodium iodide (10 ppm) are frequently added to table salt.

Important Concepts

boron
diborane
borates
silicon
zone refining
oxysilicates
pyroxenes
amphiboles
layered silicates
feldspars
zeolites
molecular sieves
glass
silicones
nitrogen
Haber process
ammonia
Ostwald process
nitric acid
aqua regia
reduction products of nitric acid
nitrogen cycle
phosphorus

phosphorous oxide
phosphoric oxide
phosphoric acid
phosphates
sulfur
sulfuric acid
sulfur dioxide
thiosulfate
hydrogen sulfide
sulfonates
halogens
relative oxidizing strength of halogens
fluorine
fluorocarbons
noble-gas compounds
chlorine
hypochlorites
chlorates
perchlorates
strength of oxyacids
bromine
iodine

Exercises

**21.1 Boron How can you account for the fact that boron, with an ionization potential less than that of gold, is nonmetallic whereas gold is metallic?

*21.2 Boron Write the balanced half-reaction in acidic solution for which the electrode potential shown in Fig. 21.1 applies in the case of boron.

**21.3 Boron Explain why B_2O_3 is an acidic oxide whereas the other M_2O_3 of group III are more basic.

**21.4 Boron Write balanced net equations for each of the following reactions:

a Boron is heated with magnesium to form magnesium diboride.

b The product from (*a*) is treated with acid to generate diborane and boron.

c The diborane from (*b*) burns in air to form B_2O_3.

**21.5 Boron Describe the spatial arrangement of atoms and the nature of the chemical bonding in B_2H_6.

***21.6 Boron The B_4H_{10} molecule contains one direct B—B bond and four BHB bonds. In addition there is an H attached by a σ bond to two of the B's and two H's attached by σ bonds to the other two. Suggest a possible structure that fits this description. Each boron forms four bonds.

***21.7 Boron How might you rationalize the fact that BF_3 is not so strong a Lewis acid as is BCl_3 or BBr_3?

*21.8 Silicon In what ways is the chemistry of silicon like that of carbon? In what ways is it different?

*21.9 Silicon Make a flow sheet showing how ultrapure silicon can be made, starting with SiO_2.

*21.10 Silicon Describe the atomic arrangement and the nature of the bonding in elemental silicon. Explain why partial substitution of Si atoms by B or by P would make elemental silicon a good conductor.

*21.11 Silicon What is the usual arrangement of oxygen atoms about silicon? Show how the arrangement is achieved in each of the following: $LiAl(SiO_3)_2$, talc, $KAlSi_3O_8$, SiO_2.

*21.12 Silicon What would be the essential difference between silicate minerals that contain pyroxene chains and those that have amphibole chains?

**21.13 Silicon Talc is $2MgSi_2O_5 \cdot Mg(OH)_2$; mica is $KMg_3AlSi_3O_{10}(OH)_2$. Tell how the properties of these differ and show how the structural features account for the difference.

**21.14 Silicon What is meant by the term "molecular sieve"? Give an example and indicate the essential structural features. Explain why molecular sieves are useful in the making of biodegradable detergents.

**21.15 Silicon Explain why glass stoppers frequently stick in reagent bottles containing aqueous Na_2CO_3. What would you do to get around this problem?

*21.16 Silicon Why is provision for heat removal important in the setting of concrete?

**21.17 Silicon How many ppm (parts per million) of phosphorus impurity would you have to incorporate in pure silicon to get 1.5×10^{18} electron current carriers per cubic centimeter? The density of silicon is 2.32 g/cm³.

*21.18 Nitrogen Why is the name *pnicogen* appropriate for nitrogen? Be specific in giving illustrative examples.

*21.19 Nitrogen Given the average composition of proteins as 51% C, 25% O, 16% N, and 7% H, deduce a simplest formula for a compound having this weight percent composition. Compare the atom ratio with that in the simplest amino acid, glycine.

**21.20 Nitrogen Why might you expect $N_2(l)$ to have a lower boiling point than $O_2(l)$?

**21.21 Nitrogen If all the components but nitrogen were removed from the standard atmosphere, what would be the density (in grams per liter) of the remaining "atmosphere" at 0°C? Assume the pressure of nitrogen is not changed by removal of the other components.

***21.22 Nitrogen Write a balanced net equation for the conversion of N_2, O_2, and H_2O to HNO_3. Under standard conditions (25°C;

$P_{N_2} = P_{O_2} = 1$ atm; $[HNO_3] = 1\ M$), the $\Delta G°$ for this reaction is $+8$ kJ/mol of HNO_3. In a situation where $P_{N_2} = 0.78$ atm and $P_{O_2} = 0.21$ atm, at what concentration of HNO_3 would the ΔG for the above reaction be zero? Assume the HNO_3 is 100 percent dissociated. *Ans. 0.074 M*

*21.23 Nitrogen Explain why raising the pressure in the Haber synthesis of ammonia increases the yield. Explain also why raising the temperature decreases the yield.

**21.24 Ammonia At 1000°K the equilibrium constant for the reaction $N_2(g) + 3H_2(g) \rightleftharpoons 2NH_3(g)$ expressed in partial pressures is 3.52×10^{-7}. The ΔH for this reaction is -92 kJ. Assuming $\Delta H°$ does not change with temperature, calculate K_p at 500°C. *Ans. 9.1 × 10⁻⁶*

***21.25 Ammonia At 500°C, the equilibrium constant expressed in terms of partial pressures has a value of 9.1×10^{-6} for the reaction $N_2(g) + 3H_2(g) \rightleftharpoons 2NH_3(g)$. What would be the value of K expressed in terms of concentration (moles per liter)? How many moles per liter of NH_3 would be in equilibrium with $[N_2] = 10\ M$ and $[H_2] = 1.0\ M$ at this temperature? *Ans. 0.60 M*

***21.26 Ammonia In the NH_3 molecule, the N—H bond length is 0.1015 nm and the H—N—H bond angle is 106.6°. Calculate the distance from the N atom to the plane defined by the three hydrogen atoms. *Ans. 0.038 nm*

***21.27 Ammonia The dipole moment for the N—H bond is 1.31 Debye units. In Exercise 21.26, it is determined that the N atom is 0.038 nm above the plane of the three hydrogen atoms in NH_3. Calculate the net dipole moment for the NH_3 molecule. *Ans. 1.47 Debye units*

**21.28 Ammonia What is the pH of each of the following solutions: (a) 0.25 M NH_3, (b) 0.25 M NH_4Cl, and (c) a solution made by mixing equal volumes of (a) and (b)?

*21.29 Nitrogen Calculate the increase in volume that accompanies the explosion of 1 g of $NH_4NO_3(s)$ at 100°C and 1 atm pressure. The density of the solid is 1.725 g/cm³.

**21.30 Nitrogen Suppose you wanted to make a sample of radioactive nitric acid from a sample of elemental nitrogen that contains the radioisotope N^{13}. Write chemical equations for the reactions. Given that the lifetime of N^{13} is so short (9.97 min), what difficulties do you foresee? How might you do tracer experiments with nitric acid and not run into these difficulties?

**21.31 Nitric acid Why are solutions of concentrated nitric acid generally yellow? How might you decolorize them?

**21.32 Nitric acid Given the half-reactions for the reduction of nitrate in acid solution, as shown is Sec. 21.3, what regularity can you discover in the number of H^+ needed per nitrate ion? How do you account for departures from the above regularity?

**21.33 Nitrogen Suggest a reason why the base strength of NH_3 which is the reaction $NH_3 + H_2O \longrightarrow NH_4^+ + OH^-$ is greater than that of N_2H_4 for the analogous reaction.

*21.34 Nitrogen cycle How might the nitrogen cycle be kept in balance locally without having recourse to addition of commercial fertilizers?

*21.35 Phosphorus An early recipe for making elemental phosphorus goes as follows: evaporate some fresh urine to dryness; let it stand in the cellar for a few months until it putrefies; heat with sand. Explain the chemical basis for this recipe and indicate key ingredients involved.

**21.36 Phosphorus What feature of the structures shown for red and white phosphorus in Fig. 21.17 would help to account for the observed fact that red phosphorus is less reactive than white phosphorus?

*21.37 Phosphorus Write balanced net equations showing how a sample of radioactive elemental phosphorus could have the radioactivity incorporated in each of the following: PH_3, H_3PO_3, H_3PO_4.

**21.38 Phosphorus Given two solutions 0.1 M H_3PO_3 and 0.1 M H_3PO_4, sketch very roughly what the pH titration curve should look like for progressive addition of 0.1 M NaOH to each of these solutions.

***21.39 Phosphorus What would be the pH of 10.0 ml of 0.100 M H_3PO_4 solution? How much 0.100 M NaOH would you have to add to raise the pH by one unit? *Ans. 7.2 ml*

*21.40 Phosphorus What structural features do P_4O_6 and P_4O_{10} have in common? What structural feature does P_4O_{10} have in common with $(NaPO_3)_n$?

*21.41 Phosphorus Trisodium orthophosphate, Na_3PO_4, and sodium metaphosphate, $NaPO_3$, are both used in water softening. Explain briefly their chemical action. What is a major disadvantage that accompanies this use?

21.42 Group VI Write balanced half-reactions for the action of group VI elements as oxidizing agents to conform with the electrode potentials given in Fig. 21.20. How do the elements compare with respect to oxidizing strength? Suggest two possible reasons for the special place of oxygen in this sequence?

21.43 Sulfur What features of the chemistry of sulfur result in its being found in relatively concentrated deposits rather than being uniformly spread throughout nature? Two major aspects are involved.

21.44 Sulfur Show how the electron configuration of the sulfur atom leads to the molecular shape shown in Fig. 21.21. What bond angle would you expect? Compare with the observed value of $105°$. How do you explain the difference?

21.45 Sulfur The gaseous phase of sulfur contains diatomic molecules. Describe the probable nature of the bonding in S_2 using molecular-orbital ideas.

***21.46 Sulfur** Using 0.102 nm for the radius of a neutral sulfur atom, what diameter sphere could you pass through the center of the S_8 molecule shown in Fig. 21.21? The S—S bond length is 0.207 nm, and the S—S—S bond angle is $105°$.

Ans. 0.26 nm

21.47 Sulfur Using the phase diagram shown in Fig. 21.23, draw curves for what the temperature-vs.-time heating curve should look like at 1 atm (*a*) if the heat addition is uniformly very fast; (*b*) if the heat addition is uniformly very slow.

*21.48 Sulfur** When sulfur is heated at a normal rate, the yellow solid goes first to a clear yellow, mobile liquid; then to a dark, very viscous form; and finally to a dark, less viscous liquid. Explain what structural changes could account for these observations.

21.49 Sulfur When dark plastic sulfur is allowed to sit at room temperature for a long time, it eventually goes back to a brittle, yellow solid. How might you explain this observation?

*21.50 Sulfur** Make a flow sheet showing how the sulfur in the mineral ZnS can be converted to sulfuric acid. Indicate critical conditions and write balanced equations for all the reactions.

21.51 Sulfuric acid What concentration of sulfuric acid should you take to have a pH of 1.00?

*21.52 Sulfur** Explain why H_2SO_4 is a stronger acid than H_2SO_3.

*21.53 Sulfur** How would you expect the pH of $0.10\ M\ NaHSO_3$ to compare with that of $0.10\ M\ NaHSO_4$? Explain your reasoning.

21.54 Sulfur Given a solution that contains Zn^{2+}, Cu^{2+}, and Hg^{2+}, how could you use H_2S to separate these ions from each other and end up with each of them still in solution?

21.55 Halogens How might you account for the fact that F_2 has a weaker bond strength than does Cl_2?

21.56 Halogens A new halogen, Z, has a bond strength in Z_2 of 200 kJ/mol, an electron affinity of 3.10 eV, and a hydration free-energy change of -300 kJ/mol of $Z^-(g)$. Where would this halogen fit in the electrode-potential listing of Fig. 21.26?

21.57 Halogens What is the main reason the elemental halogens are quite volatile species whereas, except for oxygen, the group VI elements are relatively nonvolatile?

*21.58 Fluorine** In what ways is the chemistry of fluorine quite different from that of the other halogens? Be specific.

*21.59 Fluorine** Aqueous solutions of HF cannot be stored in glass bottles. Explain with the help of a chemical equation. What container might you use instead of glass?

21.60 Fluorine The fluorocarbons C_nF_{2n+2} are more like stiff, rodlike molecules than are the normal hydrocarbons, C_nH_{2n+2}, which tend to be somewhat globular. How might you account for this?

*21.61 Noble-gas compounds** Show how valence-bond theory accounts for the fact that XeF_2 is linear and XeF_4 is square planar. What objection might be raised to this model?

21.62 Chlorine Diagram a suitable electrolytic cell that could be used to produce elemental chlorine from aqueous NaCl. Write electrode reactions. What change would occur in the solution concentration after passage of 1.00 A for 1.00 h through 1.00 liter of 1.00 M NaCl?

21.63 Chlorine Given commercial hydrochloric acid that is 37% by weight HCl (12 M), what is the density (in grams per milliliter) of the solution?

*21.64 Chlorine Given some elemental chlorine that is labeled with radioactive tracer ^{36}Cl, how could you get the radioactivity into the form of $KClO_3$?

**21.65 Chlorine A common reagent in the qualitative analysis laboratory is an aqueous solution of chlorine. Usually the solution has to be fresh to be any good. Why should it deteriorate in time? Old solutions can often be somewhat rejuvenated by putting in a few drops of $6 M$ HCl. Explain why this might help.

**21.66 Chlorine Explain why a solution of sodium hypochlorite would be expected to be basic. What should be the pH of 0.5 M NaClO?

Ans. 10.6

*21.67 Chlorine Explain why the acids HClO, $HClO_2$, $HClO_3$, and $HClO_4$ show increasing dissociation constants in the order given.

*21.68 Chlorine How would you explain the following observation? Gentle heating of $KClO_3(s)$ gives a clear liquid. Vigorous bubbling appears when powdered glass is added to the liquid.

**21.69 Chlorine A simple preparation for oxygen is to heat $KClO_3$ to which the black powder MnO_2 has been added. When carrying out the prep, it is always recommended that the decomposition first be tried on a pinch of the mixture. Why the precaution?

*21.70 Bromine Bromine can be prepared by sweeping chlorine gas through seawater. How might the bromine be separated from excess chlorine?

*21.71 Bromine How would you proceed to convert the bromine in NaBr to HBr?

*21.72 Bromine Leaded gasoline contains about 1 ml of $Pb(C_2H_5)_4$ per liter. The density of tetraethyllead is 1.65 g/ml. How many milliliters of $C_2H_4Br_2$ would you need to add to an average auto tank (80 liters) to take care of this lead as $PbClBr$? The density of $C_2H_4Br_2$ is 2.18 g/ml.

*21.73 Iodine Explain why the reduction of Chile saltpeter, $NaIO_3$, to elemental iodine by $NaHSO_3$ has to be carefully controlled as to amounts. What would be the ideal ratio by weight of $NaHSO_3$ to $NaIO_3$?

*21.74 Iodine When starch–potassium iodide paper is used to test an aqueous solution of chlorine, a blue color first appears but then quickly fades away. Explain with the help of equations. Would acid help or hinder the fading?

*21.75 Iodine What is special about the triiodide ion? How might you account for its formation?

**21.76 Iodine Explain why aqueous solutions of sodium iodide start to discolor on standing in air. Why should this discoloration become more pronounced at lower pH?

18   19

Chapter 22

THE ATOMIC NUCLEUS

In the preceding chapters, the emphasis has been on interpretation of chemical behavior in terms of the electronic structure of atoms. Little has been said of the nucleus because, apart from its charge, which controls the electronic configuration, the nucleus has relatively little influence on chemical behavior. In this final chapter we consider some aspects of nuclear behavior.

22.1 NUCLEAR STABILITY

A nucleus $_Z^A X$, where Z is the atomic number and A is the mass number, is usually pictured as containing Z protons and $A - Z$ neutrons crowded together into a small volume. Some nuclei are stable; they seem to persist indefinitely. Others are unstable; on a random basis individuals of a given collection may suddenly emit a simpler particle and change into a different nucleus. Such spontaneous change of one nucleus into another, accompanied by particle emission, is called *radioactivity*.

Radioactivity can be of two types: *natural* (occurring in nature) or *induced* (brought about by particle bombardment). Natural radioactivity was discovered in the uranium mineral pitchblende by Becquerel in 1896 only a few months after Röntgen had discovered X rays: induced radioactivity was found by Rutherford in 1919.

What kinds of particles are emitted in radioactivity? There are five main types: alpha, beta, gamma, neutrons, and protons. They are all highly energetic with kinetic energies per particle of the order of a million electronvolts (MeV). (This is to be compared with chemical binding energies, which are typically of the order of 3 to 4 eV.) They differ from each other in mass and charge. Figure 22.1 summarizes some of the characteristics. Alpha particles, which are emitted from elements such as uranium and radium, can be considered to be energetic helium nuclei. Beta particles are like electrons; since they can be positive or negative, they may be distinguished as β^+ and β^-. The β^+ is often called a positron. Gamma particles, more usually called gamma rays, are equivalent to very energetic X rays; they are essentially bundles of electromagnetic energy. Neutrons and protons are the fundamental particles that we have seen before in accounting for the mass and charge of the nucleus.

626

FIG. 22.1   Particles Produced in Radioactivity [Charge is given in elementary charge units (1.60 × 10⁻¹⁹ C); mass is given in atomic mass units (1.66 × 10⁻²⁴ g)]

Particle	Symbol*	Description	Charge	Mass
Alpha	α or ^4_2He	Helium nucleus	+2	4.0015
Beta	β^- or $^0_{-1}e$	Negative electron	−1	0.000549
	β^+ or ^0_1e	Positive electron (positron)	+1	0.000549
Gamma	γ	Electromagnetic radiation	0	~0
Neutron	n or ^1_0n	Neutral particle	0	1.00866
Proton	p or ^1_1H	Hydrogen nucleus	+1	1.00728

*In writing the systematic symbols, such as ^4_2He, the subscript indicates the charge and the superscript indicates the mass rounded off to the nearest whole number.

What decides which particular kind of particle will be emitted by a radioactive nucleus? This is not an easy question to answer. We get a clue by looking at the pattern that characterizes stable nuclei. Any nucleus is very small. It has a radius of about 10^{-13} cm (r is approximately equal to 1×10^{-13} times the cube root of the mass number A). If we crowd Z protons and $A - Z$ neutrons into this small space, the difficult thing to understand is how only positive and neutral charges can be packed together to give a stable nucleus without its flying apart as a result of the enormous electrical repulsions. The neutrons must be at least partly responsible for the binding. First, there is no nucleus that consists only of two or more protons with no neutrons. Second, the more protons there are in a nucleus, the more neutrons per proton we need for stability. This latter point is demonstrated by the "belt of stability," shown in Fig. 22.2, where a plot is made of the stable (nonradioactive) nuclei. Each point corresponds to a known nucleus containing a given number of protons and a given number of neutrons. The dashed line represents the direction along which nuclei would lie if they all contained an equal number of neutrons and protons. As can be seen, for the light elements the nonradioactive nuclei fall along this line; they contain

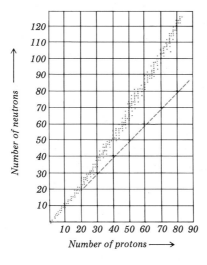

FIG. 22.2 Stable nuclei. Each dot represents a known nucleus. Dashed line indicates where number of protons equals number of neutrons.

Number of neutrons →

120 110 100 90 80 70 60 50 40 30 20 10

10 20 30 40 50 60 70 80 90
Number of protons ⟶

approximately equal numbers of neutrons and protons (for example, $^{14}_{7}N$ has 7 protons and 7 neutrons; $^{16}_{8}O$ has 8 protons and 8 neutrons; $^{23}_{11}Na$ has 11 protons and 12 neutrons). However, for the heavier nuclei there are significantly more neutrons than protons (for example, $^{138}_{56}Ba$ has 56 protons and 82 neutrons; $^{202}_{80}Hg$ has 80 protons and 122 neutrons; $^{208}_{82}Pb$ has 82 protons and 126 neutrons). In other words, the more protons there are in a stable nucleus, the higher will be its neutron-to-proton ratio.

Nuclei which do not fall within the belt of stability are radioactive; i.e., their neutron-to-proton ratios are either too high or too low. Some kind of radioactive process, as discussed in the following section, must occur in order to bring the nucleus back to stability. The bigger question as to how neutrons act to hold protons together is still essentially unanswered.

At the present time there are two models used for the nucleus: the liquid-drop model and the shell model. The two are not mutually exclusive but emphasize different aspects. The liquid-drop model, first proposed by Bohr (1936), imagines the nucleus as consisting of neutrons and protons aggregated in a random, disordered fashion somewhat like that of molecules in a drop of liquid. The strongest support for the model comes from the fact that the nuclear density is approximately constant at about 10^{14} g/cm^3 for all atoms of the periodic table. The nuclear-shell model, due mainly to Mayer (1950), considers nuclear particles to be arranged in energy levels in the nucleus just as electrons are arranged outside the nucleus. There are special complications arising from the fact that two kinds of particles (neutrons and protons) must be accommodated, but the model gains considerable support from the special stability of certain nuclei. These especially stable nuclei occur whenever the number of neutrons or the number of protons equals one of the so-called "magic numbers," 2, 8, 20, 50, 82, or 126, a situation reminiscent of closed shells of electrons. Nuclei which contain simultaneously a magic number of protons and a magic number of neutrons are the most stable, for example $^{4}_{2}He$, $^{16}_{8}O$, $^{40}_{20}Ca$, and $^{208}_{82}Pb$.

22.2
TYPES OF RADIOACTIVITY

If a nucleus lies outside the belt of stability, it tends to reach a stable configuration by a radioactive process which changes the number of neutrons and/or protons to a more favorable value. How does a nucleus get outside the belt of stability? Induced radioactivity results when stable nuclei are subjected to bombardment by other particles. If the energy of the incoming particles is of the proper magnitude, bombarded nuclei combine with incident particles to form new nuclei which, if unstable, undergo radioactive decay. An example of such a process occurs when $^{12}_{6}C$ nuclei are bombarded with protons which have been accelerated to high energies in a cyclotron. The process can be described by the nuclear equation

$$^{12}_{6}C + ^{1}_{1}H \longrightarrow ^{13}_{7}N$$

The equation is balanced: it shows conservation of charge in that the subscripts on the left of the equation add up to the subscript on the right; it shows conservation of mass in that superscripts on the left add up to that on the right. The nucleus produced, $^{13}_{7}N$, is unstable. It lies below the belt of stability and has too few neutrons (six) for the number of protons (seven). Radioactivity occurs in which there is emission of a positron. The positron has a +1 charge and essentially no mass, and so its emission has the effect of lowering the nuclear charge from +7 to +6 while leaving the total mass at 13. In other words, $^{13}_{7}N$ converts to $^{13}_{6}C$. The radioactive conversion can be written

$$^{13}_{7}N \longrightarrow {}^{0}_{1}e + {}^{13}_{6}C$$

where $^{0}_{1}e$ represents a positron.

The rate at which radioactive disintegration occurs gives a measure of the stability of a nucleus and is usually expressed in terms of the *half-life* of the nucleus. This is the time required for half a given number of atoms or molecules to disappear. For the above decay of $^{13}_{7}N$, the half-life, usually designated as $t_{1/2}$, is 10.1 min. This means that, of any aggregation of $^{13}_{7}N$ nuclei, half will have disintegrated to $^{13}_{6}C$ in 10.1 min; in another 10.1 min, half the remainder will have disintegrated; and so forth. Each succeeding interval of 10.1 min cuts the nuclei remaining by a factor of $\frac{1}{2}$. Thus, we would have $1, \frac{1}{2}, \frac{1}{4}, \frac{1}{8}, \frac{1}{16}, \frac{1}{32}$, etc., of the original nuclei left at successive 10.1-min time intervals. Since radioactive decay is a statistical process which is practically unaffected by changes in temperature or chemical binding, no one can predict which specific $^{13}_{7}N$ nucleus of a collection will disintegrate next. Only the probability of decay can be stated, and this is done by specifying the half-life. The shorter the half-life, the more probable the decay. Hence, the shorter the half-life, the more intense is the radioactivity from a given sample.

Radioactive decay is a first-order rate process (Sec. 12.2) in which the rate of decay —i.e., the number of nuclear disintegrations per unit time—is proportional to the number of unstable nuclei present. This can be written as

$$\frac{\Delta N}{\Delta t} = - kN$$

where k is the radioactive decay constant and N is the number of radioactive nuclei. The minus sign comes from the fact that Δ is defined as an increase, but the number of nuclei is *decreasing* with time—in other words, $\Delta N/\Delta t$ is a negative number. At any specific time t the number of nuclei present can be found by solving the above equation to give

$$\log \frac{N_0}{N} = \frac{kt}{2.303}$$

where N is the number of nuclei left at time t and N_0 is the number of nuclei at time 0. To solve for the half-life, we can set N_0/N equal to 2, and so we get

$$t_{1/2} = \frac{2.303}{k} \log 2 = \frac{0.693}{k}$$

As with chemical reactions, k and hence $t_{1/2}$ can be determined from experiment by plotting log N versus t. The slope of the line so obtained is equal to $-k/2.303$.

EXAMPLE 1

A given sample of a radioactive isotope shows an activity of 8640 counts/min at one time and 7620 counts/min 1 h later. What is its half-life?

Solution

Assume counts per minute is proportional to number of nuclei present.

$$\log \frac{N_0}{N} = \log \frac{8640}{7620} = \frac{(k)\,(1\text{ h})}{2.303}$$

$$k = \frac{2.303 \log (8640/7620)}{1\text{ h}} = 0.126\text{ h}^{-1}$$

$$t_{1/2} = \frac{0.693}{k} = \frac{0.693}{0.126\text{ h}^{-1}} = 5.52\text{ h}$$

At present there are more than 800 radioactive isotopes known. No two have exactly the same half-life. Thus, precise measurement of half-lives can be used to identify isotopes. Half-lives vary from fractions of a second to billions of years. When half-lives are extremely short, the nuclei do not survive very long. Thus, for $^{28}_{15}$P where $t_{1/2} = 0.28$ sec, 99.9 percent of the nuclei would be gone in 2.8 sec. During this time, the radioactivity is, of course, very intense. When half-lives are very long, survival chances of a particular nucleus are correspondingly great. For $^{232}_{90}$Th, the half-life is so long, 1.4×10^{10} yr, that it is not easy to tell that radioactive decay is occurring at all.

Natural radioactivity, as the name implies, refers to the decay of naturally occurring unstable isotopes. These are of long half-life, or else they are the result of other radioactive disintegrations which can be traced back to some long-lived isotope. For example, there are no stable nuclei with atomic numbers higher than 83. Still, there are appreciable amounts of $^{234}_{90}$Th in nature even though it has a half-life of only 24.1 days. With such a half-life, it would be expected that after a couple of years the $^{234}_{90}$Th in nature would have disappeared. The fact is, however, that $^{234}_{90}$Th is constantly being regenerated by the decay of $^{238}_{92}$U. This disintegration

$$^{238}_{92}\text{U} \longrightarrow\ ^{234}_{90}\text{Th} + ^{4}_{2}\text{He}$$

has a half-life of 4.5×10^9 yr. If we take the age of the Earth as roughly 5 billion years, then approximately half the $^{238}_{92}$U originally present at the creation is still with us and continues to replenish the $^{234}_{90}$Th supply.

Although consideration of radioactivity as artificially induced or natural is convenient, it may be more useful to classify radioactive reactions according to the types of particles which unstable nuclei eject. In terms of the belt of stability shown in Fig. 22.2, three sorts of unstable nuclei can be considered:

(1) above the belt of stability, (2) below the belt of stability, or (3) beyond ($Z > 83$) the belt of stability.

1   In case 1 the nucleus has too high a neutron-proton ratio. It can remedy matters either by ejecting a neutron or by forming and ejecting a beta particle (electron). Simple neutron ejection is rarely observed because it usually occurs so rapidly. For example, the decay

$$^{5}_{2}\text{He} \longrightarrow {}^{4}_{2}\text{He} + {}^{1}_{0}n$$

to produce an alpha particle and a neutron ($^{1}_{0}n$) has been calculated to have a half-life of 2×10^{-21} sec, much too short to be observed.*

Beta emission is much more common. It can be visualized as a process in which a neutron emits an electron and converts to a proton: $n \longrightarrow p^{+} + e^{-}$. It corrects a too-high neutron-proton ratio by emitting one unit of negative charge and thereby increasing the positive charge of the residual nucleus. Since the beta particle has very little mass, its emission does not change the mass number of the emitting nucleus. A few examples of beta decay are

$$^{14}_{6}\text{C} \longrightarrow {}^{14}_{7}\text{N} + {}^{0}_{-1}e \qquad t_{1/2} = 5570 \text{ yr}$$
$$^{90}_{38}\text{Sr} \longrightarrow {}^{90}_{39}\text{Y} + {}^{0}_{-1}e \qquad t_{1/2} = 28 \text{ yr}$$
$$^{137}_{55}\text{Cs} \longrightarrow {}^{137}_{56}\text{Ba} + {}^{0}_{-1}e \qquad t_{1/2} = 30 \text{ yr}$$

The first of these examples is used in radiocarbon age dating (see below); the second and third represent important fission products from uranium-bomb explosions.

2   If, as in case 2, an unstable nucleus lies below the belt of stability, it has too low a neutron-proton ratio. It must increase the number of neutrons, decrease the number of protons, or do both simultaneously. One device is to absorb into the nucleus one of the orbital electrons, usually an electron of the K shell. Such a process is called K capture (also EC, or electron capture). It reduces the nuclear charge by one unit, leaving the mass number unchanged, as in the following example:

$$^{90}_{42}\text{Mo} \xrightarrow{\text{K capture}} {}^{90}_{41}\text{Nb} \qquad t_{1/2} = 5.7 \text{ h}$$

K capture frequently occurs in fission products. Invariably, it is accompanied by X-ray emission as an outer-shell electron drops into the K shell to fill the vacancy, thus liberating energy.

Another way to raise an n/p ratio is for the nucleus to emit a positron ($^{0}_{1}e$, a positive electron). This process, typified by

$$^{11}_{6}\text{C} \longrightarrow {}^{11}_{5}\text{B} + {}^{0}_{1}e \qquad t_{1/2} = 20.5 \text{ min}$$

---

*However, there are some neutron emissions which are delayed long enough that they are observable. These occur, for instance, in some of the products resulting from the fission of $^{235}_{92}\text{U}$ nuclei. A specific case of this is found in the decay of high-energy $^{87}_{36}\text{Kr}$ by the reaction

$$^{87}_{36}\text{Kr} \longrightarrow {}^{86}_{36}\text{Kr} + {}^{1}_{0}n$$

for which the half-life appears to be about 1 min.

decreases the nuclear charge by one unit and leaves the mass essentially unchanged. Positrons do not survive very long; they usually combine with electrons to annihilate each other by making gamma rays.

3 In case 3 the nuclei lie beyond the belt of stability; they have too many protons crammed into one nucleus for stability, no matter how many neutrons are present. This happens for all nuclei with 84 or more protons; they all lie beyond the belt of stability. No one of the above steps by itself can lead to a stable nucleus. Instead, it is necessary to split off larger pieces, and even then a series of steps may be required. Most commonly, the piece split off is an alpha particle ($^4_2$He), and, in fact, most of the heavy nuclei are alpha emitters. With $^{212}_{84}$Po, a single step is sufficient to reach stability:

$$^{212}_{84}\text{Po} \longrightarrow \, ^{208}_{82}\text{Pb} + \, ^4_2\text{He} \qquad t_{1/2} = 3 \times 10^{-7} \text{ sec}$$

With $^{234}_{92}$U, many steps involving a combination of α and β^- decays are required, as in the following:

$$^{234}_{92}\text{U} \xrightarrow{\alpha} \, ^{230}_{90}\text{Th} \xrightarrow{\alpha} \, ^{226}_{88}\text{Ra} \xrightarrow{\alpha} \, ^{222}_{86}\text{Rn} \xrightarrow{\alpha} \, ^{218}_{84}\text{Po} \xrightarrow{\alpha} \, ^{214}_{82}\text{Pb} \xrightarrow{\beta^-} \, ^{214}_{83}\text{Bi} \xrightarrow{\alpha}$$

$$^{210}_{81}\text{Tl} \xrightarrow{\beta^-} \, ^{210}_{82}\text{Pb} \xrightarrow{\beta^-} \, ^{210}_{83}\text{Bi} \xrightarrow{\beta^-} \, ^{210}_{84}\text{Po} \xrightarrow{\alpha} \, ^{206}_{82}\text{Pb}$$

Other steps leading to the same stable nucleus, $^{206}_{82}$Pb, are also possible.

Finally, to complete the discussion of types of radioactive decay, it should be noted that gamma rays frequently accompany other kinds of radioactivity, especially beta emission and positron emission. Gamma rays represent the principal way in which a nucleus in an excited energy state can get rid of excess energy.

Until fairly recently it was thought that the periodic table was limited to only 92 elements. However, as radioactive processes were studied in greater detail, it was observed that nuclei of atomic number higher than 92 could be produced artificially. These transuranium elements have now been extended to $Z = 107$ and will probably go even higher.

The first transuranium element to be prepared (1940) was neptunium, made by irradiation of $^{238}_{92}$U with neutrons. The nucleus formed, $^{239}_{92}$U, decays by beta emission:

$$^{239}_{92}\text{U} \longrightarrow \, ^{239}_{93}\text{Np} + \, ^0_{-1}e \qquad t_{1/2} = 23.5 \text{ min}$$

It produces $^{239}_{93}$Np, which is also beta-active. It decays as follows:

$$^{239}_{93}\text{Np} \longrightarrow \, ^{239}_{94}\text{Pu} + \, ^0_{-1}e \qquad t_{1/2} = 2.3 \text{ days}$$

It produces plutonium, the second transuranium element, which is alpha-active. Higher elements have been produced by similar irradiation of transuranium elements with neutrons or even with nuclei of lighter elements such as helium or carbon. The transuranium elements have different chemical properties from each other and can be separated by chemical means. The making and separation of plutonium is an important industry, because the element is used as a source of nuclear energy.

Radioactive nuclei are useful because they are easy to follow. Their presence even in trace amounts can be detected by darkening of photographic plates or by use of various devices such as the Geiger counter. Figure 22.3 shows a simplified sketch of such a counter. It is filled with argon gas and is designed to count individually the energetic particles given off in radioactivity. A high voltage is imposed between the central wire and the surrounding copper cylinder, but no current flows until an ionizing particle, such as an alpha particle, passes through the chamber. Nuclear emissions differ in their penetrating power. In general, gammas are most penetrating, betas are less, and alphas are the least. A thin mica window on the face of the counter makes it easier for particles to enter.

There are numerous ways in which radioactivity has been applied in industry, in medicine, and in the research laboratory. Following are a few examples:

   1   Flow in a pipe system, even when it is underground, can be monitored by injecting a bit of radioactive material into the fluid and surveying with a counter from the outside.

   2   Wear on a cutting tool can be measured by doping the tool with a small amount of radioactive tracer and watching how fast the radioactivity appears in the lubricant.

   3   The course of industrial wastes dumped into streams can be followed by doping the effluent with a radioactive tracer.

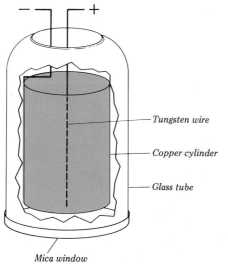

FIG. 22.3 Geiger counter.

Tungsten wire

Copper cylinder

Glass tube

Mica window

4   Large-scale mixing, as when a small amount of gasoline additive is added to a million-liter tank in an oil refinery, can be checked for completeness by doping the additive with a radioactive tracer and sampling the mix at intervals until the radioactivity becomes uniformly distributed.

5   Radioactive tracers added to the blood stream enable us to monitor movement of biochemical components in the human body.

6   Cobalt 60, an intense gamma emitter with a half-life of 5.26 yr, is often used in cancer therapy.

7   Iodine 131, a β^- emitter with a half-life of 8.0 days, also gives off gamma rays. It is often fed as KI "cocktail" for treatment of overactive thyroids. The iodine is selectively taken up by the thyroid gland.

8   Phosphorus 32, another beta emitter with a half-life of 14.3 days, is often injected intravenously for treatment of leukemia.

9   Plutonium 238, an alpha emitter with an 86-yr half-life, has been used as a long-lasting, low-level power source for pacemakers to regulate heart action.

10   One of the most useful chemical applications of radioactive tracers has been in elucidation of mechanisms of complex organic reactions. For example, by feeding radioactive CO_2 (labeled with $^{14}_{6}C$) to plants, it has been possible to work out the course of photosynthetic conversion of CO_2 to carbohydrates. The CO_2 is apparently first converted to an organic phosphate and then stepwise to sugar.

11   The fact that $^{14}_{6}C$ is radioactive has led to development of a rather novel method for dating archaeological discoveries. The basic ideas are as follows: Carbon dioxide in the atmosphere contains mostly $^{12}_{6}C$ and a little $^{13}_{6}C$, both of which are nonradioactive. In addition, there is a trace of radioactive $^{14}_{6}C$ which, even though it is constantly decaying, remains rather uniform in abundance. Cosmic rays act on $^{14}_{7}N$ of the atmosphere to form $^{14}_{6}C$ at about the same rate as $^{14}_{6}C$ decays. Because decay balances production, the ratio of $^{14}CO_2$ to $^{12}CO_2$ in the atmosphere does not change with time. Now, it is well known that plants absorb CO_2 from the atmosphere in the process of photosynthesis. So long as the plant is alive, the ratio of $^{14}_{6}C$ to $^{12}_{6}C$ in the plant is the same as that in the atmosphere. However, once the plant has been removed from the life cycle, as, for instance, when a tree is chopped down, the ratio of $^{14}_{6}C$ to $^{12}_{6}C$ begins to diminish as the $^{14}_{6}C$ atoms undergo radioactive decay. Thus, we can use the observed isotope ratio as a kind of clock that measures the time since a plant stopped living. The half-life of $^{14}_{6}C$ is 5570 years; therefore, at the end of 5570 years the ratio $^{14}_{6}C$ to $^{12}_{6}C$ becomes half as great as it would be in the atmosphere. There are problems with the method, as, for example, in assuming that the cosmic-ray flux stays constant for thousands of years. Given that assumption, however, we can determine the age of a wooden relic or, for that matter, of any once-living material by burning a sample to CO_2 and measuring its ratio of $^{14}_{6}C$ to $^{12}_{6}C$.

EXAMPLE 2

A piece of charcoal from Lascaux Cave in France has been found to have a ^{14}C-to-^{12}C ratio that is 13 percent that of the atmosphere. How old is the specimen?

Solution

Let N_0 = relative number of $^{14}_6$C at time $t = 0$

$\qquad N$ = relative number of $^{14}_6$C t years later $= 0.13 N_0$

Then

$$\log \frac{N_0}{N} = \frac{kt}{2.303}$$

where $k = \dfrac{0.693}{t_{1/2}} = \dfrac{0.693}{5570 \text{ yr}}$

$$\log \frac{1}{0.13} = \left(\frac{0.693}{5570}\right)\left(\frac{t}{2.303}\right)$$

$t = 16{,}000$ yr

---

12 Perhaps the most impressive application of tracer techniques has been in elucidation of chemical properties of almost-zero amounts of artificially produced elements. In the case of mendelevium, element 101, with a count of only 17 disintegrations it was possible to decide that its properties are similar to those of thulium ($Z = 69$), which is the element just above it in the periodic table.

22.4
NUCLEAR ENERGY

Besides ordinary radioactive decay, there is another kind of nuclear instability which is based on interconversion of mass and energy. The observed fact is that the mass of a nucleus is always less than the sum of its component neutrons and protons. By the Einstein relation $E = mc^2$ (where E is energy in joules, m is mass in kilograms, and c is the speed of light, 2.9979×10^8 m/sec), a deficiency in mass is equivalent to a deficiency in energy. Stated another way, the assembled nucleus is lower in energy than the isolated component particles by an amount equal to the missing mass. Hence, the missing mass gives a measure of the binding energy of the neutrons and protons in the particular nucleus. How this works is illustrated below for the case of $^{56}_{26}$Fe. For convenience, we introduce the term *nucleon* to refer to either a neutron or a proton in the nucleus.

635

EXAMPLE 3

Calculate the binding energy per nucleon in $_{26}^{56}$Fe. The mass of a proton is 1.00728 amu; the mass of a neutron is 1.00866 amu; the observed mass of one $_{26}^{56}$Fe atom is 55.9349 amu. Nuclear binding energies are usually expressed in million electronvolts, where 1 MeV $= 1.602 \times 10^{-13}$ J.

Solution

26 protons $= 26(1.00728)$ $= 26.1893$ amu

30 neutrons $= 30(1.00866)$ $= 30.2598$ amu

$\qquad\qquad\qquad\qquad\qquad\qquad$ 56.4491 amu

$+ 26$ electrons $= 26(0.0005486) = $ 0.01426 amu

Expected mass of atom $\qquad = 56.4634$ amu

$-$ Observed mass of atom $\qquad = 55.9349$ amu

Missing mass $\qquad\qquad\qquad\qquad$ 0.5285 amu

Nucleus contains $26 + 30 = 56$ nucleons

$$\frac{0.5285 \text{ amu}}{56 \text{ nucleons}} = 0.009438 \text{ amu per nucleon}$$

$(0.009438 \text{ amu})(1.6605 \times 10^{-24} \text{ g/amu}) = 1.567 \times 10^{-26}$ g

$(1.567 + 10^{-29} \text{ kg})(2.9979 \times 10^8 \text{ m/sec})^2 = 1.408 \times 10^{-12}$ J

$$\frac{1.408 \times 10^{-12} \text{J}}{1.602 \times 10^{-13} \text{ J/MeV}} = 8.79 \text{ MeV}$$

Thus, for $_{26}^{56}$Fe we have a nuclear binding energy of 8.79 MeV per nucleon. A similar calculation for $_{1}^{2}$H gives 1.12 MeV per nucleon; for $_{92}^{238}$U, 7.6 MeV per nucleon. Figure 22.4 shows how the binding energy compares for the various nuclei. As can be seen, intermediate elements of mass number about 60 have the highest binding energies. The other elements have smaller binding energies and are unstable with respect to conversion to them. This means, for example, that if a heavy nucleus, such as uranium, is converted to iron, the difference in binding energy per nucleon should be liberated. Similarly, if a light nucleus, such as hydrogen, is converted to iron, energy should also be liberated.

Such energy-releasing conversions are the bases for utilization of nuclear energy. Breakup of large nuclei to intermediate ones is called *nuclear fission;* merging of small nuclei to intermediate ones is called *nuclear fusion.* The most important fissionable isotopes are ^{235}U and ^{239}Pu.

A typical fission process is the following:

$$_{0}^{1}n + _{92}^{235}U \longrightarrow [_{92}^{236}U] \longrightarrow _{56}^{141}Ba + _{36}^{92}Kr + 3_{0}^{1}n + Q$$

It is represented schematically in Fig. 22.5. A neutron impinging on a ^{235}U

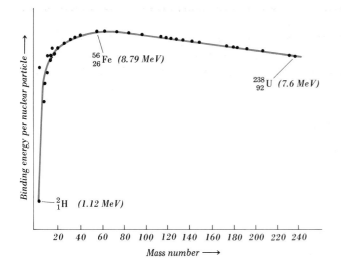

$^{56}_{26}$Fe *(8.79 MeV)*

$^{238}_{92}$U *(7.6 MeV)*

$^{2}_{1}$H *(1.12 MeV)*

Binding energy per nuclear particle →

20 40 60 80 100 120 140 160 180 200 220 240

Mass number →

FIG. 22.4 Binding energy per nucleon in the various nuclei.

nucleus* gets absorbed by it to produce temporarily a compound nucleus ^{236}U. This almost immediately breaks up into two approximately equal fragments, Ba and Kr, as well as three neutrons. Emitted at the same time is a large burst of energy *Q*, which is mostly kinetic and amounts to about 200 MeV per fission.

The number of neutrons emitted per fission is variable. It is usually two or three but may go as high as six. The weighted average for the above fission, which was a closely guarded secret during World War II, is 2.43.

The nuclei resulting from a fission generally have high neutron-proton ratios, much higher than they should have for stable nuclei of that *Z*. Hence, they are beta-active. For the above case, we would have the following subsequent decays:

$$^{141}_{56}\text{Ba} \xrightarrow[\text{18 min}]{\beta^-} {}^{141}_{57}\text{La} \xrightarrow[\text{3.9 h}]{\beta^-} {}^{141}_{58}\text{Ce} \xrightarrow[\text{33 days}]{\beta^-} {}^{141}_{59}\text{Pr (stable)}$$

$$^{92}_{36}\text{Kr} \xrightarrow[\text{3 sec}]{\beta^-} {}^{92}_{37}\text{Rb} \xrightarrow[\text{5 sec}]{\beta^-} {}^{92}_{38}\text{Sr} \xrightarrow[\text{2.7 h}]{\beta^-} {}^{92}_{39}\text{Y} \xrightarrow[\text{3.5 h}]{\beta^-} {}^{92}_{40}\text{Zr (stable)}$$

*Natural uranium is 99.3 percent $^{238}_{92}$U and 0.7 percent $^{235}_{92}$U. The more abundant isotope requires high-energy neutrons to initiate fission; the less abundant one undergoes fission with readily available low-energy neutrons. Therefore, in practice it is desirable to have uranium enriched in ^{235}U. Separation of uranium 235 from uranium 238 can be achieved by gaseous diffusion of UF_6, as described on page 183; it was the basis of an enormous secret project during World War II and continues to be an important industry. At present, fissionable material in bombs is more likely to be $^{239}_{94}$Pu. Plutonium 239 also undergoes fission with slow neutrons and, being chemically different from uranium, it can be rather easily separated from uranium 238.

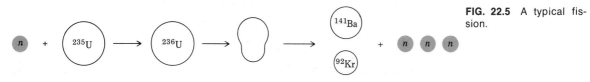

FIG. 22.5 A typical fission.

637

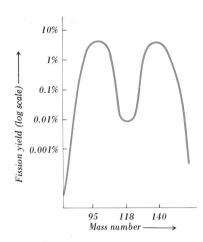

FIG. 22.6 Fission yield vs. mass number. (Note that the vertical scale is not linear.)

As can be seen, the half-lives of the fission products gradually get longer as the consecutive decay products approach stability. Actually, the fission shown is just one of the ways in which ^{235}U undergoes fission. Figure 22.6 shows what the statistical distribution of fission products looks like.

About 90 percent of the energy release in fission occurs in the first 1.0 μsec. Delayed energy release comes from the fission products—gamma rays, β^- particles, and delayed neutron emission. This delayed energy release is what constitutes the *fallout* problem. When fissionable material such as ^{235}U is used in a nuclear explosion, the fission products that have short half-lives disappear rather rapidly. Some of the decay products, however, such as ^{90}Sr and ^{137}Cs, have appreciable half-lives; so they persist—long enough to "fall out" over the environment, get into the food chain, be absorbed in the human system, and possibly create a radiation hazard.

An important feature of the fission process is that more neutrons are produced in the fission than are needed to initiate it. This means that the fission process can become self-sustaining as a chain reaction. When the chain propagates so that more neutrons are generated per unit time than are absorbed or lost to the outside, then the chain reaction will go on at an ever-increasing rate. This is what happens when a system passes what is called *critical size*. Neutron loss to the outside becomes relatively less important, and the chain reaction tends to go into a runaway condition. In a nuclear bomb, subcritical masses are kept separate from each other until a triggering mechanism fires them together so they exceed criticality.

In a nuclear reactor, such as used in electricity-generating power stations, design requirements are quite the opposite. The idea is to keep the system just short of criticality so that the chain reaction continues but not in a runaway fashion. As shown in Fig. 22.7, which is a schematic version of a "boiling-water reactor," control is achieved by interspersing uranium fuel material with cadmium rods. Cadmium has a high neutron capture ability and acts as a control by keeping the neutron number down to some small, desirable level. Fission goes on, but at a slow, useable rate. By adjusting the

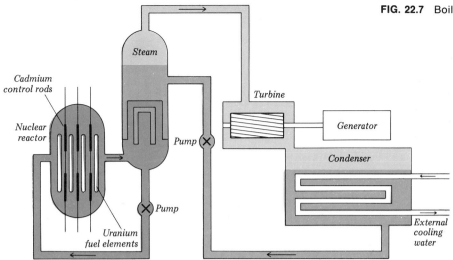

FIG. 22.7   Boiling-water nuclear reactor.

relative position of the control rods, the nuclear reactor can be set to operate at any desired power level. Because nuclear reactors are designed so differently from nuclear bombs, the often voiced fear of some people that nuclear power plants might blow up like a bomb is not justified. There may be a radiation hazard from an accident in a nuclear power plant, but not a nuclear-bomb hazard.

To get a better idea of what is involved in nuclear-fission energy, it is instructive to calculate how much uranium 235 is needed for criticality. We can estimate this from the published information that the first atomic bomb, which was dropped on Hiroshima, was equivalent to 20,000 tons of TNT. One ton of TNT releases 4.18×10^9 J; so 20 kilotons would correspond to 8.4×10^{13} J. If each nuclear fission liberates 200 MeV, we have (200 MeV) $\times (1.6 \times 10^{13}$ J/MeV), per fission, or 3.2×10^{-11} J. One bomb then would be equal to 8.4×10^{13} J divided by 3.2×10^{-11} J per fission, or 2.6×10^{24} atoms. Dividing this by the Avogadro number gives 4.3 mol of ^{235}U, or 1.0 kg. (Actually, bombs are not pure uranium 235 nor is the efficiency 100 percent; so this figure is a lower limit.) Given that the density of uranium is 19.05 g/cm^3, we can calculate that the above critical mass would correspond to 52 cm^3, or a sphere about 5 cm in diameter.

Actually the extraction of energy from uranium is currently rather limited because the supply of ^{235}U is limited. New reactor technology is being developed, however, to create *breeder* reactors. These operate with ^{235}U or ^{239}Pu as regular fuels, but in addition they are charged with hard-to-fission materials such as ^{238}U and ^{232}Th. Under proper conditions of neutron irradiation, these can be converted to ^{239}Pu and ^{233}U by the following reactions:

$$\,_{0}^{1}n + \,_{92}^{238}\mathrm{U} \longrightarrow [\,_{92}^{239}\mathrm{U}] \xrightarrow{\beta^-} \,_{93}^{239}\mathrm{Np} \xrightarrow{\beta^-} \,_{94}^{239}\mathrm{Pu}$$

$$\,_{0}^{1}n + \,_{90}^{232}\mathrm{Th} \longrightarrow [\,_{90}^{233}\mathrm{Th}] \xrightarrow{\beta^-} \,_{91}^{233}\mathrm{Pa} \xrightarrow{\beta^-} \,_{92}^{233}\mathrm{U}$$

Both ^{239}Pu and ^{233}U are easy to fission. The idea, of course, is have our cake and eat it. As we consume nuclear fuel, we generate more to replace it. Since ^{238}U and ^{232}Th are relatively abundant, their exploitation as nuclear fuels would put off the energy crisis by several hundred years.

A more dramatic possibility lies in exploiting nuclear *fusion* as a source of controllable energy. Nuclear fusion seems to be the way the sun and other stars generate their energy. It has been achieved in the hydrogen bomb, but its controlled application has turned out to be very difficult. The goal is to release energy by binding nucleons of very light elements into nuclei of heavier elements. Feasible reactions, as demonstrated in the hydrogen bomb, are as follows:

$$\begin{aligned}
{}_1^2\text{H} + {}_1^2\text{H} &\longrightarrow {}_2^3\text{He} + {}_0^1n + 3.2 \text{ MeV} \\
{}_1^2\text{H} + {}_1^2\text{H} &\longrightarrow {}_1^1\text{H} + {}_1^3\text{H} + 4.0 \text{ MeV} \\
\underline{{}_1^3\text{H} + {}_1^2\text{H}} &\underline{\longrightarrow {}_2^4\text{He} + {}_0^1n + 17 \text{ MeV}} \\
5{}_1^2\text{H} &\longrightarrow {}_2^3\text{He} + {}_1^1\text{H} + {}_2^4\text{He} + 2{}_0^1n + 24 \text{ MeV}
\end{aligned}$$

In the first reaction, two deuterium nuclei combine to give helium 3 and a neutron; in the second, two deuterium nuclei combine to give a protium and a tritium; in the third reaction, tritium and deuterium combine to give ordinary helium 4 plus a neutron. All three reactions occur together, and the net result is the conversion of five deuterium nuclei to helium 3, helium 4, protium, and two neutrons. Energy emission is 24 MeV for 10 nucleons, or 2.4 MeV per nucleon. This is about 3 times as great as the 200 MeV for 235 nucleons, or 0.85 MeV per nucleon, obtainable from uranium via nuclear fission.

The abundance ratio of deuterium to protium in nature is 1:7000. This may not seem very much, but there is such a fantastic amount of protium in the world that the amount of deuterium is quite appreciable. It has been calculated that there is enough deuterium in a liter of water to be equivalent in energy to 300 liters of gasoline. The problem in getting out the energy is that nuclear fusion, as given by the equations above, has an enormous energy of activation. It occurs only at very high nuclear velocities (corresponding to temperatures of the order of 100 million degrees) and at very high nuclear densities. These are contradictory requirements. To get high nuclear velocities we need high temperatures, but high temperatures mean expansion, therefore, fewer nuclei per cubic centimeter. Current efforts for control of nuclear fusion are concentrated on use of plasmas (ionized gases) to attain the high temperatures needed. Containment of the plasmas, however, is a major problem. It is being tried by strong magnetic fields.

Nuclear fusion in bombs (i.e., so-called "thermonuclear weapons") is not so difficult to bring about. All we need is a fission bomb to generate the high temperature needed. Particularly convenient for a fusion bomb is the reaction

$$\text{{}_3^6Li} + {}_0^1n \longrightarrow {}_2^4\text{He} + {}_1^3\text{H}$$

which can be carried out in lithium deuteride. Neutrons from a fission bomb can act on the lithium to generate tritium, and this in turn interacts with the deuterium to release 17 MeV as indicated above. The reaction appears to be the middle phase of an especially destructive three-phase bomb; fission-fusion-fission. An ordinary ^{235}U fission bomb sets off nuclear fusion in ^6LiD, which then sets off nuclear fission in a ^{238}U jacket on the warhead. ^{238}U normally does not undergo fission, but with fast neutrons it can be done. The diabolic part comes from the fact that ^{238}U costs only about \$25/kg whereas ^{235}U costs more than \$30,000/kg. Fusion bombs without the U-238 blanket are less destructive, but they emit relatively more neutrons. This is the concept behind the so-called "neutron bomb," more properly called "enhanced radiation weapon." It is a fission-fusion device which emits relatively large amounts of high-energy neutrons. Neutron irradiation of people produces cell damage due to ionizing effects of the neutrons colliding with protons in the cells. Massive doses are generally fatal, if not immediately, then shortly thereafter.

22.5
ENERGY CRISIS

One of civilization's most pressing current problems is energy. Where does it come from? Where does it go? How long before we run out of it?

Practically all our energy comes ultimately from the sun. The sun, as a nuclear fusion reactor, pours 3.8×10^{26} J/sec into space. We on Earth get about 3.5×10^{24} J/yr to play with. Some of this gets converted into plants by the process of photosynthesis:

$$6CO_2 + 6H_2O \longrightarrow C_6H_{12}O_6 + 6O_2 \qquad \Delta H = +2800 \text{ kJ}$$

Plant material so produced in the past was converted to the fossil fuels petroleum and coal. When we burn these now, we are using sunlight of the past. At present, it is estimated that only about 0.1 percent of the solar flux goes into plant growth.

Where else does the energy from the sun go? Lest we think we might ultimately be able to use it all, we should note that approximately one-third of the total goes to power the rain cycle. It is estimated there is about 5×10^{17} liters of rain per year. To evaporate this much water requires (5×10^{17} liters) $\times$ (55 mol/liter)(41 kJ/mol) $= 1.1 \times 10^{24}$ J. Other weather cycles also have their energy requirements.

Still it seems as if there is plenty of solar energy to spare. Why should we worry about an "energy crisis"? Part of the anxiety comes from the fact that the rate of energy consumption is going up faster than the population. For instance, world energy consumption is increasing at an annual rate of 3.6 percent per year. World population is growing only at 2 percent per year. This means that per capita consumption of energy is going up at 1.6 percent per year. For the United States, the growth rate in energy consumption is smaller (i.e., 2.8 percent per year), but it starts from a higher base. It is a fact that, although the United States accounts for only $6\frac{1}{2}$ percent of the world's population, it accounts for 30 percent of the energy consumption. The other part of the anxiety comes from a realization that direct utilization of solar energy is going to be an expensive business. It is true that water exposed to the sun can easily be heated to useful temperatures, but this does not work very well when the air temperature drops below 0°C or when there is cloud cover. Devices will have to be made to convert and store the solar energy for when it is needed. Silicon solar cells (Sec. 21.2) and cells for photoelectrolysis of water (Sec. 11.7) use quite elegant materials that demand lots of energy for their fabrication. Devices for direct utilization of solar energy will not be cheap.

Figure 22.8 shows a schematic representation of the present energy flow from source to use in the United States. As can be seen, most of the energy comes from fossil fuels (natural gas, petroleum, and coal). Relatively little comes from nuclear energy. About a quarter of it goes into electricity, and three-quarters into fuel. It may come as a shock that half the energy is eventually wasted and only half ends up as useful work. However, losses in generation and transmission of power are inevitable.

More interesting than representative figures for energy consumption are the reserve figures. How long will our energy sources last? We can get an idea of this by assuming all the United States consumption is furnished by a single energy source. If it all came from coal, proven United States reserves would last 125 yr. If it all came from petroleum, reserves would last only 5 yr. Natural gas is worth only 5 yr. Natural uranium, if used in nuclear reactors of

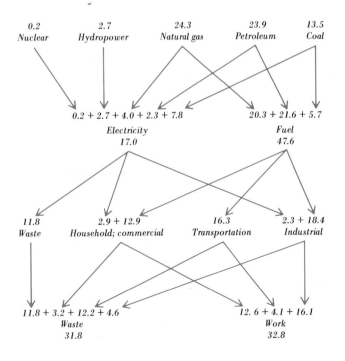

FIG. 22.8　Energy flow in United States in units of 10^{18} J/yr.

the conventional fission type, would carry us along for only 2.3 yr, possibly for 15 yr. If breeder reactors were operational, then we could go for 115 yr, possibly 750 yr. The spectacular source is nuclear fusion. If we could make it work, it would last us 10^6 yr, maybe 10^9 yr. In other words, nuclear fusion offers the possibility of an essentially unlimited supply of energy which would last us "forever." Any rosy prognosis, however, must be severely tempered. How will we protect ourselves from radiation? How will we manage the nuclear waste disposal?

At present, practical energy sources are severely limited. Already, local crises of supply are beginning to occur. Inevitably, they will become more frequent in the near future.

Important Concepts

nucleus
radioactivity
natural radioactivity
induced radioactivity
alpha particles
beta particles
gamma rays
neutrons
protons
neutron/proton ratio
belt of stability
types of radioactivity
half-life
positron emission
beta decay

K capture
transuranium elements
applications of radioactivity
tracers
carbon dating
nuclear energy
nucleon binding energy
nuclear fission
nuclear fusion
fallout
critical size
nuclear reactor
breeder reactor
energy crisis

Exercises

**22.1 Particles How do the particles α, β, and γ differ from each other in charge and mass? Given a beam that consists of one of these particles, how could you tell experimentally which it is? (*Hint:* magnetic field)

*22.2 Energy Given an alpha particle which has energy 1 MeV, How many H_2 bonds could you break with this energy? Dissociation energy of H_2 is 432 kJ/mol.

**22.3 Nuclear equations In each of the following, one nuclear particle has been left out. Name the missing particle and balance each equation:

a $^{14}_{7}N + ^{4}_{2}He. \longrightarrow ^{17}_{8}O$
b $^{23}_{11}Na \longrightarrow ^{26}_{12}Mg + ^{1}_{1}H$
c $^{27}_{13}Al + ^{4}_{2}He \longrightarrow ^{30}_{15}P$
d $^{40}_{19}K \longrightarrow ^{0}_{-1}e$
e $^{10}_{5}B \longrightarrow ^{7}_{3}Li + ^{4}_{2}He$

*22.4 Nuclear density Calculate the density in grams per cubic centimeter of the nuclear region for each of the following atoms: $^{16}_{8}O$, $^{56}_{26}Fe$, $^{238}_{92}U$.

**22.5 Nuclear forces The repulsive force between two charges goes as $q_1 q_2 / r^2$. Compare the magnitude of the repulsion between two protons in a nucleus with that between two protons at the distance characteristic of the H_2 molecule.

**22.6 Nuclear stability $^{23}_{11}Na$ is the stable isotope of sodium. What is its neutron-to-proton ratio? Compare it with that of the unstable isotopes $^{22}_{11}Na$ and $^{24}_{11}Na$ and predict the types of radioactivity of the unstable isotopes.

**22.7 Radioactivity $^{31}_{15}P$ is the stable nucleus of phosphorus. Predict what would be the result of bombarding $^{31}_{15}P$ with protons in a cyclotron. Write a nuclear equation for the process.

*22.8 Radioactivity Tell what is meant by each of the following and indicate under what condition each is likely to be obtained: (*a*) positron emission, (*b*) K capture, (*c*) alpha-particle emission, (*d*) beta-particle emission.

*22.9 Half-life Given that the half-life of $^{13}_{7}N$ is 10.1 min, what fraction of a given sample will be left after 1 h?

**22.10 Half-life The half-life of $^{38}_{17}Cl$ for β^- emission is 37.3 min. Given a 1.00-g sample at time zero, how long will it be before the sample is reduced to 0.25 g?   *Ans. 75 min*

**22.11 Half-life Given radioactive isotope X. When measured at 10-min intervals, a given sample decays as follows: 0.0562, 0.0466, 0.0386, 0.0320, 0.0266, 0.0221, 0.0183, 0.0150, 0.0126 in units of micromoles. What is the half-life of the isotope?

**22.12 Radioactivity How can one account for the fact that there is still ^{234}Th in nature even though its half-life of 24 days is so much shorter than the age of the Earth?

**22.13 Radioactivity Make a square grid showing number of protons and number of neutrons as the coordinates. Mark on this grid each of the nuclei in the decay chain from $^{234}_{92}U$ to $^{206}_{82}Pb$ given on page 632. Draw arrows for each step. What distinguishes an α step from a β^- step? Starting with $^{238}_{92}U$, draw arrows for the following sequence: α, β^-, β^-, α, α, α, α, β^-, α, α, β^-, β^-, β^-, and α. What is the identity of the final nucleus?

***22.14 Radioactivity $^{239}_{94}Pu$ is an alpha emitter with a half-life of 24,400 yr. Given a 1.00-g sample of this plutonium, how many alpha particles will it emit per second?   *Ans. 2.27×10^9*

**22.15 Radioactivity Suppose you are given two identically appearing pieces of two different metals, one of which is radioactive. How would you proceed to tell which is which? Be specific.

*22.16 Radioactivity Astatine is a very radioactive element with short half-lives for all its isotopes. Its activity is so intense, it is hard to work with. Given a very dilute aqueous solution of sodium astatide, how might you proceed to show that the chemistry of astatine resembles that of the halogens?

*22.17 Radiocarbon Explain how radiocarbon dating can be used to estimate the age of charcoal paintings in caves inhabited by prehistoric man. What key assumption is the basis of the method?

**22.18 Radiocarbon A wooden club found in a troglodyte cave shows a $^{14}C/^{12}C$ ratio which is 10 percent of the ratio shown by the trees outside. How old would you date the club?

**22.19 Nuclear energy The observed mass of the $^{16}_{8}O$ atom is 15.99491 amu. Calculate the average binding energy in million electron volts per nuclear particle in this isotope.   *Ans. 7.99 MeV*

*22.20 Nuclear energy What is the main difference between nuclear fission and nuclear fusion? What makes one so much easier than the other?

**22.21 Nuclear energy On the basis of Fig. 22.4, what is the ratio of the energy to be obtained from fission of ^{238}U to ^{56}Fe compared with the energy to be obtained from fusion of an equal mass of ^2H to ^{56}Fe?

***22.22 Fallout Among the fallout products from fission is $^{90}_{38}Sr$ which has a half-life for beta decay of 28.1 yr. On the basis of its position in the periodic table, why might this fission product be a particularly dangerous one?

**22.23 Nuclear energy Why might it be particularly dangerous to pour a solution containing ^{239}Pu from a flat tray into a spherical flask?

**22.24 Nuclear energy Natural uranium is 99.3% ^{238}U (which is not fissionable with slow neutrons) and 0.7% ^{235}U (which is fissionable). How might these isotopes be separated from each other? What is meant by the term *breeder reactor* and what is its main advantage? How would you set up a breeder reactor?

**22.25 Nuclear fusion What are the main obstacles to peaceful extraction of nuclear energy from fusion? Why is it relatively simpler to power a bomb with nuclear fusion than to get out the energy in controlled fashion?

*22.26 Energy crisis In what sense is it true that all our energy comes ultimately from the sun?

**22.27 Energy crisis Enthusiasts in favor of utilizing solar energy point to the very tiny fraction we now are using. What problems do you foresee in trying to solve the energy crisis by exploiting greater use of sunlight?

*22.28 Energy crisis According to Fig. 22.8, what is the chief culprit in inefficient utilization of energy? How might the fuel cells discussed in Sec. 11.9 contribute significantly in solving the energy problem? What applications might be specifically recommended?

Appendix 1

SI UNITS

Most of the units used in this text are SI units, as recommended by the International Committee on Weights and Measures. The International System of Units (usually designated SI, after Système International) is constructed from seven base units. These are the following:

Name of unit	Physical quantity	Symbol for unit
Meter	Length	m
Kilogram	Mass	kg
Second	Time	s (sec)
Ampere	Electric current	A
Kelvin	Thermodynamic temperature	K (°K)
Candela	Light intensity	cd
Mole	Amount of substance	mol

The symbols given in parentheses, although not officially recommended, are used in this text for the sake of clarity.

Decimal fractions or multiples of these units are indicated as follows:

Fraction	Prefix	Symbol	Multiple	Prefix	Symbol
10^{-1}	Deci-	d	10^{1}	Deka-	da
10^{-2}	Centi-	c	10^{2}	Hecto-	h
10^{-3}	Milli-	m	10^{3}	Kilo-	k
10^{-6}	Micro-	μ	10^{6}	Mega-	M
10^{-9}	Nano-	n	10^{9}	Giga-	G
10^{-12}	Pico-	p	10^{12}	Tera-	T
10^{-15}	Femto-	f			
10^{-18}	Atto-	a			

Derived from the base units are various specially named units. These include the following:

Name of unit	Physical quantity	Symbol for unit	Definition of unit
Newton	Force	N	$kg\ m/sec^2$
Pascal	Pressure	Pa	$N/m^2 = kg\ m^{-1}\ sec^{-2}$
Joule	Energy	J	$kg\ m^2/sec^2$
Watt	Power	W	$J/sec = kg\ m^2/sec^3$
Coulomb	Electric charge	C	A sec
Volt	Electric potential difference	V	$J\ A^{-1}\ sec^{-1} = kg\ m^2\ sec^{-3}\ A^{-1}$
Ohm	Electric resistance	Ω	$V/A = kg\ m^2\ sec^{-3}\ A^{-2}$
Siemens	Electric conductance	S	$\Omega^{-1} = A/V = sec^3\ A^2\ kg^{-1}\ m^{-2}$
Farad	Electric capacitance	F	$A\ sec/V = A^2\ sec^4\ kg^{-1}\ m^{-2}$
Hertz	Frequency	Hz	sec^{-1} (cycle per second)

Use of the above units is recommended. There are also certain decimal fractions and multiples of SI units, having special names, which do not belong to the International System of Units; their use is to be *progressively discouraged*. Among such are the following:

Name of unit	Physical quantity	Symbol for unit	Definition of unit
Angstrom	Length	Å	$10^{-10}\ m = 10^{-8}\ cm$
Dyne	Force	dyn	$10^{-5}\ N$
Bar	Pressure	bar	$10^5\ N/m^2$
Erg	Energy	erg	$10^{-7}\ J$

There are, in addition, other units that are not simple fractions or multiples of SI units. They do not belong to the International System of Units and are recommended to be *abandoned*. These include the following:

Name of unit	Physical quantity	Symbol for unit	Definition of unit
Inch	Length	in	$2.54 \times 10^{-2}\ m$
Pound	Mass	lb	0.453502 kg
Atmosphere*	Pressure	atm	$101,325\ N/m^2$
Torr	Pressure	Torr	$(101,325/760)\ N/m^2$
Millimeter of mercury	Pressure	mmHg	$13.5951 \times 980.665 \times 10^{-2}\ N/m^2$
Calorie	Energy	cal	4.184 J

*Use of this unit is sanctioned for a limited period of time.

Besides the above, there are certain natural units tied directly to the properties of microscopic constituents of matter. Use of these as natural units is acceptable. Important examples are the following:

Unit or physical quantity	Symbol	Conversion factor
Atomic mass unit	amu	1.6605×10^{-24} g
Avogadro number	N	6.0222×10^{23} molecules per mole
Boltzmann constant	k	1.3806×10^{-23} J/deg
Electron charge	e	1.6022×10^{-19} C
Electron mass	m	9.1096×10^{-28} g
Electronvolt	eV	1.6022×10^{-19} J
Faraday constant	$\tilde{\mathfrak{F}}$	9.6487×10^{4} C/equiv
Gas constant	R	8.2057×10^{-2} l atm mol^{-1} deg^{-1}
		8.3143 J mol^{-1} deg^{-1}
Planck's constant	h	6.6262×10^{-34} J sec
Speed of light	c	2.9979×10^{8} m/sec

Appendix 2

CHEMICAL NOMENCLATURE

A2.1
INORGANIC COMPOUNDS

Compounds composed of but two elements have names derived directly from the elements. Usually the more electropositive element is named first, and the other element is given an *-ide* ending. Thus, we have sodium chloride, NaCl; calcium oxide, CaO; and aluminum nitride, AlN. If more than one atom of an element is involved, prefixes such as *di-* (for 2), *tri-* (3), *tetra-* (4), *penta-* (5), and *sesqui-* ($1\frac{1}{2}$) are used. For example, AlF_3 is aluminum trifluoride, Na_3P is trisodium phosphide, and N_2O_4 is dinitrogen tetroxide. When the same two elements form more than one compound, the compounds can be distinguished as in the following example:

$FeCl_2$	$FeCl_3$
(1) Iron dichloride	Iron trichloride
(2) Ferrous chloride	Ferric chloride
(3) Iron(II) chloride	Iron(III) chloride

In (1) distinction is made through use of prefixes; in (2) the endings *-ous* and *-ic* denote the lower and higher oxidation states, respectively, of iron; in (3), the Stock system, Roman numerals in parentheses indicate the oxidation states. In a given series of compounds, the suffixes *-ous* and *-ic* may not be sufficient for complete designation but may need to be supplemented by one of the other methods of nomenclature. For example, the oxides of nitrogen are usually named as follows:

649

N_2O	nitrous oxide
NO	nitric oxide
N_2O_3	dinitrogen trioxide, or nitrogen sesquioxide
NO_2	nitrogen dioxide
N_2O_4	dinitrogen tetroxide, or nitrogen tetroxide
N_2O_5	dinitrogen pentoxide, or nitrogen pentoxide

Compounds containing more than two elements are named differently, depending on whether they are bases, acids, or salts. Since most bases contain hydroxide ion (OH^-), they are generally called hydroxides, e.g., sodium hydroxide (NaOH), calcium hydroxide [$Ca(OH)_2$], and arsenic trihydroxide [$As(OH)_3$]. The naming of acids and of salts derived from them is more complicated, as can be seen from the following series:

Acid	Sodium salt
$HClO$, hypochlorous acid	$NaClO$, sodium hypochlorite
$HClO_2$, chlorous acid	$NaClO_2$, sodium chlorite
$HClO_3$, chloric acid	$NaClO_3$, sodium chlorate
$HClO_4$, perchloric acid	$NaClO_4$, sodium perchlorate
H_2SO_3, sulfurous acid	Na_2SO_3, sodium sulfite
H_2SO_4, sulfuric acid	Na_2SO_4, sodium sulfate

When there are only two common oxyacids of a given element, the one corresponding to lower oxidation state is given the *-ous* ending, and the other the *-ic* ending. If there are more than two oxyacids of different oxidation states, the prefixes *hypo-* and *per-* may also be used. As indicated in the above example, the prefix *hypo-* indicates an oxidation state lower than that of an *-ous* acid, and the prefix *per-*, an oxidation state higher than that of an *-ic* acid. For salts derived from oxyacids the names are formed by replacing the ending *-ous* by *-ite* and *-ic* by *-ate*. Salts derived from polyprotic acids (for example, H_3PO_4) are best named so as to indicate the number of hydrogen atoms left unneutralized. For example, NaH_2PO_4 is monosodium dihydrogen phosphate, and Na_2HPO_4 is disodium monohydrogen phosphate. Frequently, the prefix *mono-* is left off. For monohydrogen salts of diprotic acids, such as $NaHSO_4$, the presence of hydrogen may also be indicated by the prefix *bi-*. Thus, $NaHSO_4$ is sometimes called sodium bisulfate, though the name "sodium hydrogen sulfate" is preferred.

Complex cations, such as $Cr(H_2O)_6^{3+}$, are named by giving the number and name of the groups attached to the central atom followed by the name of the central atom with its oxidation number indicated by Roman numerals in parentheses. Thus, $Cr(H_2O)_6^{3+}$ is hexaaquochromium(III). Complex anions, such as $PtCl_6^{2-}$, are named by giving the number and name of attached groups followed by the name of the element with an *-ate* ending and its oxidation number in parentheses. Thus, $PtCl_6^{2-}$ is hexachloroplatinate(IV). If the attached groups (*ligands*) are not all alike, it is customary to name the ligands in the same order in which they should be written in the formula—i.e., anion ligands generally precede neutral ligands. If more than one kind of anion ligand is present, the order is H⁻ (hydrido), O^{2-} (oxo), OH⁻ (hydroxo), other monatomic anions (in order of increasing electronegativity of the elements—for example, F⁻, fluoro, last), polyatomic anions (in order of increasing number of atoms), and organic anions (in alphabetical order). If more than one kind of neutral ligand is present, the order is H_2O (aquo), NH_3 (ammine), other inorganic ligands (in order of increasing electronegativity of their central atom—for example, CO, carbonyl, precedes NO, nitrosyl), and organic ligands (in alphabetical order). To indicate the numbers of each kind of ligand, Greek prefixes are used: *mono-* (usually can be omitted), *di-, tri-, tetra-, penta-, hexa-, hepta-,* and *octa-.* Instead of these prefixes, *bis-* (twice), *tris-* (thrice), *tetrakis-* (four times), etc., may be used, especially when the name of the ligand itself contains a numerical designation (e.g., ethylenediamine, frequently abbreviated *en*). Some examples of the application of the above rules follow:

$CrCl_2(H_2O)_4^+$	dichlorotetraaquochromium(III)
$CrCl_4(H_2O)_2^-$	tetrachlorodiaquochromate(III)
$Cr(H_2O)(NH_3)_5^{3+}$	aquopentaamminechromium(III)
$Ga(OH)Cl_3^-$	hydroxotrichlorogallate(III)
cis-$PtBrCl(NO_2)_2^{2-}$	*cis*-bromochlorodinitroplatinate(II)
trans-$Co(OH)Cl en_2^+$	*trans*-hydroxochlorobisethylenediaminecobalt(III)
$Mn(CO)_3(C_6H_6)^+$	tricarbonylbenzenemanganese(I)

In case of complex-ion isomerism, the names *cis-* or *trans-* may precede the formula or the complex-ion name to indicate the spatial arrangement of the ligands. Cis means the ligands occupy adjacent coordination positions; trans means opposite positions.

A2.2
ORGANIC COMPOUNDS

The key rules recommended by the International Union of Pure and Applied Chemistry (IUPAC) are summarized as follows:

1 Choose as the parent carbon skeleton the longest sequence of C atoms that contains the principal functional group.

2 Name the parent structure using the name of the alkane that contains the same number of C atoms as the chosen structure. Replace *-ane* by *-ene* for double bond or *-yne* for triple bond. If a functional group is present, drop the final *-e* and add suffixes as follows:

-ol for alcohol (OH)
-al for aldehyde (CHO)
-one for ketone (CO)
-oic acid for acid (COOH)

3 Use prefixes in alphabetic order to denote other substituents.

4 Locate substituents and points of unsaturation by numbering the C atoms of the parent skeleton with the following criteria used in decreasing order of priority:

a Assign the C atom of the principal functional group the number 1 if it is terminal.

b Assign numbers so that the location of the principal functional group is as low as possible if the group is nonterminal.

c Assign numbers so that substituents are located by lowest possible numbers. If there are two kinds of substituents, give low-number preference to the first named.

5 If an attached side chain bears substituents, it too must be numbered, starting with the C atom which is attached to the parent carbon skeleton. Names of substituents on the side chain and numbers locating them are enclosed in parentheses with the name of the side chain:

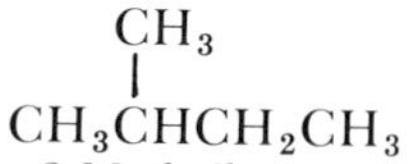

$$
\begin{array}{cc}
CH_3 & CH_3 \\
| & | \\
CH_3CHCH_2CH_3 & CH_3CH_2CHCH=CH_2 \\
\text{2-Methylbutane} & \text{3-Methyl-1-pentene}
\end{array}
$$

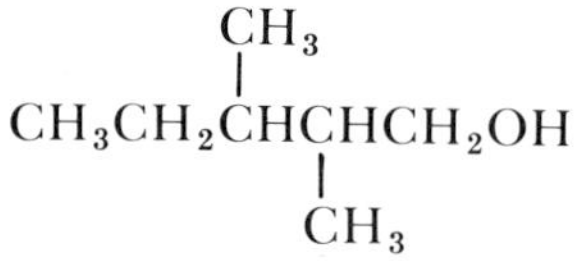

$$
\begin{array}{cc}
CH_3 & O \\
| & \| \\
CH_3CH_2CHCHCH_2OH & CH_3CH_2CH=CHCCH_3 \\
| & \\
CH_3 & \\
\text{2,3-Dimethyl-1-pentanol} & \text{3-Hexene-2-one}
\end{array}
$$

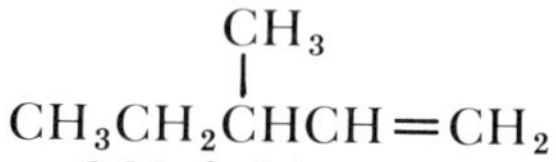

Esters are named by replacing the suffix of the parent acid *-oic acid* by *-oate:*

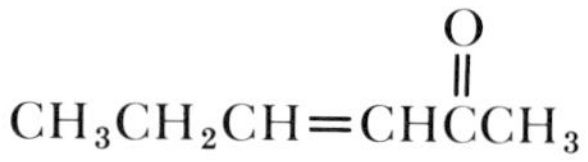

$$
\begin{array}{cc}
& CH_3 \\
& | \\
CH_3CH_2CH_2COOCH_3 & CH_3COOCH_2CH=CCH_3 \\
\text{Methylbutanoate} & \text{3-Methyl-2-butenylethanoate}
\end{array}
$$

Cyclic aliphatic hydrocarbons are named by prefixing *cyclo-* to the name of the corresponding open-chain hydrocarbon having the same number of C atoms as the ring:

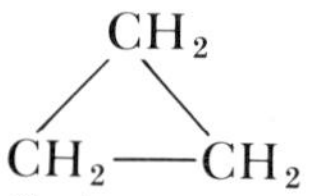

$$
\begin{array}{ccc}
CH_2 & CH_2-CH_2 & CH_2 \\
/\ \backslash & | \quad\quad | & CH_2 \quad\quad CH_2 \\
CH_2-CH_2 & CH_2-CH_2 & CH=CH \\
\text{Cyclopropane} & \text{Cyclobutane} & \text{Cyclopentene}
\end{array}
$$

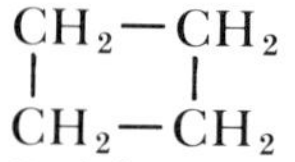

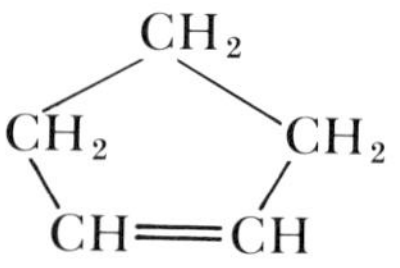

Rings containing atoms other than C (heterocycles) as well as aromatic rings are usually designated by trivial (nonsystematic) names:

Pyridine   Benzene   Naphthalene

To locate substituents, rings are numbered clockwise around the periphery as shown for naphthalene.

Definitive rules for nomenclature of organic chemistry are given in the *Handbook of Chemistry and Physics,* Chemical Rubber Co., Cleveland, Ohio.

Appendix 3

VAPOR PRESSURE OF WATER

Temperature, °C	Pressure, atm	Pressure, mmHg	Temperature, °C	Pressure, atm	Pressure, mmHg
0	0.00603	4.58	23	0.0277	21.07
1	0.00648	4.93	24	0.0294	22.38
2	0.00697	5.29	25	0.0313	23.76
3	0.00748	5.69	26	0.0332	25.21
4	0.00803	6.10	27	0.0352	26.74
5	0.00861	6.54	28	0.0373	28.35
6	0.00923	7.01	29	0.0395	30.04
7	0.00989	7.51	30	0.0419	31.82
8	0.0106	8.04	35	0.0555	42.18
9	0.0113	8.61	40	0.0728	55.32
10	0.0121	9.21	45	0.0946	71.88
11	0.0130	9.84	50	0.1217	92.51
12	0.0138	10.52	55	0.1553	118.04
13	0.0148	11.23	60	0.1966	149.38
14	0.0158	11.99	65	0.2468	187.54
15	0.0168	12.79	70	0.3075	233.7
16	0.0179	13.63	75	0.3804	289.1
17	0.0191	14.53	80	0.4672	355.1
18	0.0204	15.48	85	0.5705	433.6
19	0.0217	16.48	90	0.6918	525.8
20	0.0231	17.54	95	0.8341	633.9
21	0.0245	18.65	100	1.0000	760.0
22	0.0261	19.83	105	1.1922	906.1

Appendix 4

MATHEMATICAL OPERATIONS

Multiplication by a positive power of 10 corresponds to moving the decimal point to the right; multiplication by a negative power of 10 corresponds to moving the decimal point to the left:

1.23×10^4 is 12,300.
1.23×10^{-4} is 0.000123.

Numbers expressed with powers of 10 can be added or subtracted directly only if the powers of 10 are the same:

$$1.23 \times 10^4 + 1.23 \times 10^5 = 1.23 \times 10^4 + 12.3 \times 10^4$$
$$= 13.5 \times 10^4$$
$$1.23 \times 10^{-4} - 1.23 \times 10^{-5} = 1.23 \times 10^{-4} - 0.123 \times 10^{-4}$$
$$= 1.11 \times 10^{-4}$$

When powers of 10 are multiplied, exponents are added; when divided, exponents are subtracted:

$$(1.23 \times 10^4)(1.23 \times 10^5) = (1.23 \times 1.23)(10^4 \times 10^5)$$
$$= 1.51 \times 10^9$$
$$\frac{1.23 \times 10^{-4}}{1.23 \times 10^{-5}} = \frac{1.23}{1.23} \times \frac{10^{-4}}{10^{-5}} = 1.00 \times 10$$

655

In taking square roots of powers of 10, the exponent is divided by 2; in taking cube roots, by 3:

Square root of 9×10^4 is 3×10^2.
Cube root of 8×10^{-12} is 2×10^{-4}.

A4.2 LOGARITHMS

A *logarithm* of a given number is the power to which a base number must be raised to equal the given number. There are in common usage two bases for logarithms: the base 10 and the base e ($e = 2.71828\ldots$). These can be distinguished by writing "log" for the base-10 system and "ln" for the base-e system. The latter is derived from the name "natural logarithm" for reference to base e. The two systems are related by the equality

$$2.303 \log x = \ln x$$

For numerical calculations it is usually more convenient to use the base-10 logarithms because of the decimal nature of our number system. When needed, natural logarithms can be derived from a table of base-10 logarithms by use of the multiplier 2.303. The table on pages 658 and 659 gives the base-10 logarithms.

One principal use of logarithms in this text is in connection with pH, defined as the negative of the logarithm of the hydrogen-ion concentration. For a hydrogen-ion concentration of 0.00036 M the pH is found as follows:

$$
\begin{aligned}
\log 0.00036 &= \log (3.6 \times 10^{-4}) \\
&= \log 3.6 + \log 10^{-4} \\
&= 0.556 - 4 \\
&= -3.444 \\
\mathrm{pH} &= +3.444
\end{aligned}
$$

Sometimes, the reverse procedure is required. For example, if a solution has a pH of 8.50, its hydrogen-ion concentration can be found as follows:

$$
\begin{aligned}
\mathrm{pH} &= 8.50 \\
\log [H^+] &= -8.50 = 0.50 - 9 \\
[H^+] &= 3.2 \times 10^{-9}
\end{aligned}
$$

The number 3.2 is the antilogarithm of 0.50 (the number whose logarithm is 0.50). Antilogarithms are obtained by using the table in reverse, i.e., by looking up the logarithm in the body of the table and then finding the number which corresponds to it.

A4.3 QUADRATIC EQUATIONS

A *quadratic equation* is an algebraic equation in which a variable is raised to the second power but no higher and which can be written in the form

$$ax^2 + bx + c = 0$$

The solution of such an equation is

$$x = \frac{-b \pm \sqrt{b^2 - 4ac}}{2a}$$

where the plus-or-minus sign indicates that there are two roots. Thus, the equation obtained in Example 2 of Sec. 15.2,

$$1.8 \times 10^{-5} = \frac{(1.00 - y)(1.00 - y)}{y}$$

when rewritten gives

$$y^2 + (-2.00 - 1.8 \times 10^{-5})y + 1.00 = 0$$

for which the roots are

$$y = \frac{-(-2.000018) \pm \sqrt{(-2.000018)^2 - 4(1)(1.00)}}{2(1)}$$

$$= +1.004 \quad \text{or} \quad 0.996$$

The first root $y = 1.004$ is inadmissible from the nature of the problem (y cannot be greater than 1.00, which represents all the acid present). The second root $y = 0.996$ must be the correct one. It might be noted that the usual rules for carrying through significant figures do not apply when we operate with the quadratic formula.

A4.4
SOLVING EQUATIONS BY SUCCESSIVE APPROXIMATIONS

Complicated algebraic equations can often be solved by the method of successive approximation. To use this method, the assumption is made that one or more terms can be neglected so as to give a simple approximate equation, which can quickly lead to a first approximate answer. This answer is then substituted into the terms that were neglected to give a better approximate equation, which is solved to give a second approximate answer (presumably better than the first). This second approximate answer is then fed back. The sequence continues until two successive trials give the same self-consistent value for the unknown.

EXAMPLE

Try to solve $4x^3 - 0.800x^2 + 0.0500x - 0.00060 = 0$ by a method of successive approximation in which only the linear term is retained. (If you have a hunch that x is going to be larger than 1, you should throw away the low powers of x; if you think x is going to be smaller than 1, throw away the high powers.)

Logarithms

	0	1	2	3	4	5	6	7	8	9
10	0000	0043	0086	0128	0170	0212	0253	0294	0334	0374
11	0414	0453	0492	0531	0569	0607	0645	0682	0719	0755
12	0792	0828	0864	0899	0934	0969	1004	1038	1072	1106
13	1139	1173	1206	1239	1271	1303	1335	1367	1399	1430
14	1461	1492	1523	1553	1584	1614	1644	1673	1703	1732
15	1761	1790	1818	1847	1875	1903	1931	1959	1987	2014
16	2041	2068	2095	2122	2148	2175	2201	2227	2253	2279
17	2304	2330	2355	2380	2405	2430	2455	2480	2504	2529
18	2553	2577	2601	2625	2648	2672	2695	2718	2742	2765
19	2788	2810	2833	2856	2878	2900	2923	2945	2967	2989
20	3010	3032	3054	3075	3096	3118	3139	3160	3181	3201
21	3222	3243	3263	3284	3304	3324	3345	3365	3385	3404
22	3424	3444	3464	3483	3502	3522	3541	3560	3579	3598
23	3617	3636	3655	3674	3692	3711	3729	3747	3766	3784
24	3802	3820	3838	3856	3874	3892	3909	3927	3945	3962
25	3979	3997	4014	4031	4048	4065	4082	4099	4116	4133
26	4150	4166	4183	4200	4216	4232	4249	4265	4281	4298
27	4314	4330	4346	4362	4378	4393	4409	4425	4440	4456
28	4472	4487	4502	4518	4533	4548	4564	4579	4594	4609
29	4624	4639	4654	4669	4683	4698	4713	4728	4742	4757
30	4771	4786	4800	4814	4829	4843	4857	4871	4886	4900
31	4914	4928	4942	4955	4969	4983	4997	5011	5024	5038
32	5051	5065	5079	5092	5105	5119	5132	5145	5159	5172
33	5185	5198	5211	5224	5237	5250	5263	5276	5289	5302
34	5315	5328	5340	5353	5366	5378	5391	5403	5416	5428
35	5441	5453	5465	5478	5490	5502	5514	5527	5539	5551
36	5563	5575	5587	5599	5611	5623	5635	5647	5658	5670
37	5682	5694	5705	5717	5729	5740	5752	5763	5775	5786
38	5798	5809	5821	5832	5843	5855	5866	5877	5888	5899
39	5911	5922	5933	5944	5955	5966	5977	5988	5999	6010
40	6021	6031	6042	6053	6064	6075	6085	6096	6107	6117
41	6128	6138	6149	6160	6170	6180	6191	6201	6212	6222
42	6232	6243	6253	6263	6274	6284	6294	6304	6314	6325
43	6335	6345	6355	6365	6375	6385	6395	6405	6415	6425
44	6435	6444	6454	6464	6474	6484	6493	6503	6513	6522
45	6532	6542	6551	6561	6571	6580	6590	6599	6609	6618
46	6628	6637	6646	6656	6665	6675	6684	6693	6702	6712
47	6721	6730	6739	6749	6758	6767	6776	6785	6794	6803
48	6812	6821	6830	6839	6848	6857	6866	6875	6884	6893
49	6902	6911	6920	6928	6937	6946	6955	6964	6972	6981
50	6990	6998	7007	7016	7024	7033	7042	7050	7059	7067
51	7076	7084	7093	7101	7110	7118	7126	7135	7143	7152
52	7160	7168	7177	7185	7193	7202	7210	7218	7226	7235
53	7243	7251	7259	7267	7275	7284	7292	7300	7308	7316
54	7324	7332	7340	7348	7356	7364	7372	7380	7388	7396

	0	1	2	3	4	5	6	7	8	9
55	7404	7412	7419	7427	7435	7443	7451	7459	7466	7474
56	7482	7490	7497	7505	7513	7520	7528	7536	7543	7551
57	7559	7566	7574	7582	7589	7597	7604	7612	7619	7627
58	7634	7642	7649	7657	7664	7672	7679	7686	7694	7701
59	7709	7716	7723	7731	7738	7745	7752	7760	7767	7774
60	7782	7789	7796	7803	7810	7818	7825	7832	7839	7846
61	7853	7860	7868	7875	7882	7889	7896	7903	7910	7917
62	7924	7931	7938	7945	7952	7959	7966	7973	7980	7987
63	7993	8000	8007	8014	8021	8028	8035	8041	8048	8055
64	8062	8069	8075	8082	8089	8096	8102	8109	8116	8122
65	8129	8136	8142	8149	8156	8162	8169	8176	8182	8189
66	8195	8202	8209	8215	8222	8228	8235	8241	8248	8254
67	8261	8267	8274	8280	8287	8293	8299	8306	8312	8319
68	8325	8331	8338	8344	8351	8357	8363	8370	8376	8382
69	8388	8395	8401	8407	8414	8420	8426	8432	8439	8445
70	8451	8457	8463	8470	8476	8482	8488	8494	8500	8506
71	8513	8519	8525	8531	8537	8543	8549	8555	8561	8567
72	8573	8579	8585	8591	8597	8603	8609	8615	8621	8627
73	8633	8639	8645	8651	8657	8663	8669	8675	8681	8686
74	8692	8698	8704	8710	8716	8722	8727	8733	8739	8745
75	8751	8756	8762	8768	8774	8779	8785	8791	8797	8802
76	8808	8814	8820	8825	8831	8837	8842	8848	8854	8859
77	8865	8871	8876	8882	8887	8893	8899	8904	8910	8915
78	8921	8927	8932	8938	8943	8949	8954	8960	8965	8971
79	8976	8982	8987	8993	8998	9004	9009	9015	9020	9025
80	9031	9036	9042	9047	9053	9058	9063	9069	9074	9079
81	9085	9090	9096	9101	9106	9112	9117	9122	9128	9133
82	9138	9143	9149	9154	9159	9165	9170	9175	9180	9186
83	9191	9196	9201	9206	9212	9217	9222	9227	9232	9238
84	9243	9248	9253	9258	9263	9269	9274	9279	9284	9289
85	9294	9299	9304	9309	9315	9320	9325	9330	9335	9340
86	9345	9350	9355	9360	9365	9370	9375	9380	9385	9390
87	9395	9400	9405	9410	9415	9420	9425	9430	9435	9440
88	9445	9450	9455	9460	9465	9469	9474	9479	9484	9489
89	9494	9499	9504	9509	9513	9518	9523	9528	9533	9538
90	9542	9547	9552	9557	9562	9566	9571	9576	9581	9586
91	9590	9595	9600	9605	9609	9614	9619	9624	9628	9633
92	9638	9643	9647	9652	9657	9661	9666	9671	9675	9680
93	9685	9689	9694	9699	9703	9708	9713	9717	9722	9727
94	9731	9736	9741	9745	9750	9754	9759	9763	9768	9773
95	9777	9782	9786	9791	9795	9800	9805	9809	9814	9818
96	9823	9827	9832	9836	9841	9845	9850	9854	9859	9863
97	9868	9872	9877	9881	9886	9890	9894	9899	9903	9908
98	9912	9917	9921	9926	9930	9934	9939	9943	9948	9952
99	9956	9961	9965	9969	9974	9978	9983	9987	9991	9996

First approximation

Assume $x = 0$ in first two terms.

The equation becomes $0.0500x - 0.00060 = 0$, for which the solution is $x = 0.012$.

Second approximation

Assume $x = 0.012$ in first two terms.

The equation becomes $4(0.012)^3 - 0.800(0.012)^2 + 0.0500x - 0.00060 = 0$, which reduces to $0.0500x - 0.00071 = 0$, for which the solution is $x = 0.014$.

Third approximation

Assume $x = 0.014$ in first two terms.

The equation becomes $4(0.014)^3 - 0.800(0.014)^2 + 0.0500x - 0.00060 = 0$, which reduces to $0.0500x - 0.00075 = 0$, for which the solution is $x = 0.015$.

Fourth approximation

Assume $x = 0.015$ in first two terms.

The equation becomes $4(0.015)^3 - 0.800(0.015)^2 + 0.0500x - 0.00060 = 0$, which reduces to $0.0500x - 0.00077 = 0$, for which the solution is $x = 0.016$.

Fifth approximation

Assume $x = 0.016$ in first two terms.

The equation becomes $4(0.016)^3 - 0.800(0.016)^2 + 0.0500x - 0.00060 = 0$, which reduces to $0.0500x - 0.00078 = 0$, for which the solution is $x = 0.016$.

Since two successive trials lead to the same answer $x = 0.016$, we assume we have a self-consistent answer.

Normally, it takes no more than two or three successive trials to come up with a self-consistent answer, provided the first assumption is a reasonable one. If the first assumption is a bad one, then in general the succeeding steps will not converge on an answer, and the calculation using that assumption should be abandoned.

Appendix 5

DEFINITIONS FROM PHYSICS

A5.1
VELOCITY AND ACCELERATION

When an object changes its position, it is said to undergo a *displacement*. The rate at which displacement changes with time is called the *velocity* and has the dimensions of distance divided by time (e.g., centimeters per second). *Acceleration* is the rate at which velocity changes with time and has the dimensions of velocity divided by time (e.g., centimeters per second per second, or cm/sec²).

A5.2
FORCE AND MASS

Force can be thought of as a push or pull on an object which tends to change its motion, to speed it up or slow it down or cause it to deviate from its path. Mass is a quantitative measure of the inertia of an object to having its motion changed. Thus, mass determines how difficult it is to accelerate an object. Quantitatively, force and mass are related by the equation

$$F = ma$$

where F is the force which produces acceleration a in mass m. If m is in kilograms and a is in meters per second per second, then F is in kilogram meters per second per second, or newtons. (For reference, 1 N is approximately the force exerted by an apple in the earth's gravity.) If m is in grams and a is in centimeters per second per second, then F is in gram centimeters per second

661

per second, or dynes. The recommended unit for force is the newton, which is equal to 10^5 dyn. Weight is an expression of force and arises because every object has mass and is being accelerated by gravity.

A5.3
MOMENTUM AND IMPULSE; ANGULAR MOMENTUM

In dealing with collision problems it is useful to have terms for describing the combined effect of mass and velocity and its change with time. Mass times velocity mv, called the *momentum*, determines the length of time required to bring a moving body to rest when decelerated by a constant force. Thus, for a particle of momentum mv to be stopped by a constant force F the time required t is mv/F.

The *impulse* is defined for the case of a constant force as Ft, where t is the time during which the force F acts. Thus, for the stopping of a particle originally of momentum mv by force F in time t the impulse is just

$$Ft = F\frac{mv}{F} = mv$$

This is true if the particle comes to a complete rest. If, however, the particle bounces back, as it would on collision with a rigid wall, the particle is reflected from the wall with momentum $-mv$ (the minus sign indicating that the velocity is now in the opposite direction). The total impulse, counting the time for deceleration to zero and acceleration to $-mv$, is twice what it was before, or $2mv$.

In considering the pressure exerted by a gas, impulse comes in as follows: The pressure, or force per unit area, is the rate of collision per unit area times the effect of each collision:

$$\text{Pressure} = \frac{\text{force}}{\text{area}} = \frac{\text{number of collisions}}{(\text{time}) (\text{area})} \times ?$$

$$? = \frac{(\text{force}) (\text{time})}{\text{number of collisions}} = \text{impulse per collision}$$

In contrast to *linear momentum*, which measures mv along a straight line, there is also *angular momentum*, which describes an analogous quantity for spinning movement or movement along a curved path. For motion along a curved path, angular momentum is defined as mvr, where r is the radius of curvature of the path.

A5.4
WORK AND ENERGY

When a force F operates on (e.g., pushes) an object through a distance d, work W is done:

$$W = Fd$$

If force is expressed in newtons (kilogram meters per second per second) and distance in meters, then work has the dimensions newton meters (kg m²/sec²), or joules. One joule is thus the work done in moving one kilogram through one meter so as to increase its velocity by one meter per second all in one second. If force is expressed in dynes (gram centimeters per second per second) and distance in centimeters, then work has the dimensions dyne centimeters (g cm²/sec²), or ergs. One erg is thus the work done in moving one gram through one centimeter so as to increase its velocity by one centimeter per second all in one second. (For reference, 1 erg is approximately the work a fly does in one push-up.)

Energy is the ability to do work, and the dimensions of energy are the same as those of work. Kinetic energy is the energy a body possesses because of its motion and mass. It is equal to one-half the mass times the square of its velocity. Potential energy is the energy a body possesses because of its position or arrangement with respect to other bodies.

A5.5
ELECTRIC CHARGE AND ELECTRIC FIELD

Electric charge is a property assigned to objects to account for certain observed attractions or repulsions which cannot be explained in terms of gravitational attraction between masses. Electric charge can be of two types, positive and negative. Objects which have the same type of electric charge repel each other; objects with opposite charges attract each other. Originally, a unit of charge was defined as the quantity of electric charge which at a distance of one centimeter from another identical charge produced a repulsive force of one dyne in a vacuum. This unit of charge was called the electrostatic unit (esu). An electron has a negative charge of 4.80×10^{-10} esu. The unit of electric charge is now defined as the coulomb, which is the amount of charge transferred by a current flow of one ampere (see below) for one second. In coulombs, the charge of an electron is 1.60×10^{-19} C.

An electric field is said to exist at a point if a force of electric origin is exerted on any charged body placed at that point. The intensity of an electric field is defined as the magnitude of the electric force exerted on a unit charge. Any electrically charged body placed in an electric field moves unless otherwise constrained. The direction of a field is usually defined as the direction in which a positive charge would move.

A5.6
VOLTAGE AND CAPACITANCE

An electric capacitor is a device for storing electric charge. In its simplest form, a capacitor consists of two parallel, electrically conducting plates separated by some distance. The capacitor can be charged by making one plate positive and the other plate negative. In order to transfer a unit positive charge from the negative plate to the positive plate, work must be done against the electric field which exists between the charged plates. Therefore, the potential energy of the unit charge is increased in the process. In other words, there is a change in potential energy in going from one plate to the other. This difference in potential energy for a unit charge moved from one plate to the other is called the potential difference, or the voltage, of the capacitor. Voltage, or potential difference, is not restricted to capacitors but may exist between any two points so long as work must be done in transferring an electric charge from one point to the other. The potential difference between two points is said to be one volt if one joule (that is, 10^7 ergs) is required to move one coulomb of charge from one point to the other. To move an electron through a potential difference of one volt requires an amount of energy, called the electron volt, equal to 1.6×10^{-19} J.

Capacitance is the term used to describe quantitatively the amount of charge that can be stored on a capacitor. It is equal to the amount of charge that can be stored on the plates when the voltage difference between the plates is one volt. In general, the amount of charge a capacitor can hold is directly proportional to the voltage; the capacitance is simply the proportionality constant

$$Q = CV$$

If Q, the charge, is one coulomb and if V, the voltage, is one volt, then C, the capacitance, is one farad. The capacitance of a capacitor depends on the capacitor design (e.g., area of the plates and distance between them) and on the nature of the material between the plates. For a parallel-plate capacitor the capacitance is given approximately by the following equation:

$$C = \frac{KA}{4\pi d}$$

where A is the area of the plates, d is the distance between the plates, and K is the dielectric constant of the material between the plates. For a vacuum the dielectric constant K is exactly equal to 1; for all other substances K is greater than 1. Some typical dielectric constants are 1.00059 for air at STP, 1.00026 for hydrogen gas at STP, 1.0046 for HCl gas at STP, 80 for liquid water at 20°C, 28.4 for ethyl alcohol at 0°C, 2 for petroleum, and 4 for solid sulfur.

A5.7
ELECTRIC CURRENT

A collection of moving charges is called an electric current. The unit of current is the ampere, which is defined as the constant current which if maintained in two straight, parallel conductors of infinite length and negligible circular cross section that are placed one meter apart in a vacuum would produce between these conductors a force equal to 2×10^{-7} newton per meter of length. One ampere corresponds to a flow of one coulomb of charge past a point in one second. Since current specifies the rate at which charge is transferred, the current multiplied by time gives the total amount of charge transferred:

$$Q = It$$

If the current I is in amperes (coulombs per second) and the time t is in seconds, the charge Q is in coulombs.

The current that a wire carries is directly proportional to the voltage difference between the ends of the wire. The proportionality constant, called the conductance of the wire, is equal to the reciprocal of the resistance of the wire:

$$I = \frac{1}{R}V \qquad \text{or} \qquad V = IR$$

If V is the potential difference in volts and I is the current in amperes, R is the resistance in ohms.

There are two important kinds of current, direct and alternating. Direct current implies that the charge is constantly moving in the same direction along the wire. Alternating current implies that the current reverses its direction at regular intervals of time. The usual house current is 60-cycle alternating current; i.e., it goes through 60 complete back-and-forth oscillations per second.

Appendix 6

DISSOCIATION CONSTANTS K_c AND SOLUBILITY PRODUCTS K_{sp} AT 25°C

Dissociation constants (first step only)*

H_3BO_3	6.0×10^{-10}	HPO_4^{2-}	10^{-12}
$CO_2 + H_2O$	4.2×10^{-7}	H_2O	1.0×10^{-14}
HCO_3^-	4.8×10^{-11}	H_2S	1.1×10^{-7}
$HC_2H_3O_2$	1.8×10^{-5}	HS^-	1×10^{-14}
HCN	4.0×10^{-10}	H_2SO_3	1.3×10^{-2}
$NH_3 + H_2O$	1.8×10^{-5}	HSO_3^-	5.6×10^{-8}
HNO_2	4.5×10^{-4}	HSO_4^-	1.3×10^{-2}
H_3PO_3	1.6×10^{-2}	HF	6.7×10^{-4}
$H_2PO_3^-$	7×10^{-7}	$HOCl$	3.2×10^{-8}
H_3PO_4	7.5×10^{-3}	$HClO_2$	1.1×10^{-2}
$H_2PO_4^-$	6.2×10^{-8}		

*In order of appearance in the periodic table.

Solubility products*

$Mg(OH)_2$	8.9×10^{-12}	NiS	3×10^{-21}
MgF_2	8×10^{-8}	PtS	8×10^{-73}
MgC_2O_4	8.6×10^{-5}	$Cu(OH)_2$	1.6×10^{-19}
$Ca(OH)_2$	1.3×10^{-6}	CuS	8×10^{-37}
CaF_2	1.7×10^{-10}	$AgCl$	1.7×10^{-10}
$CaCO_3$	4.7×10^{-9}	$AgBr$	5.0×10^{-13}
$CaSO_4$	2.4×10^{-5}	AgI	8.5×10^{-17}
CaC_2O_4	1.3×10^{-9}	$AgCN$	1.6×10^{-14}
$Sr(OH)_2$	3.2×10^{-4}	Ag_2S	5.5×10^{-51}
$SrSO_4$	7.6×10^{-7}	ZnS	1×10^{-22}
$SrCrO_4$	3.6×10^{-5}	CdS	1.0×10^{-28}
$Ba(OH)_2$	5.0×10^{-3}	Hg_2Cl_2	1.1×10^{-18}
$BaSO_4$	1.5×10^{-9}	Hg_2Br_2	1.3×10^{-22}
$BaCrO_4$	8.5×10^{-11}	Hg_2I_2	4.5×10^{-29}
$Cr(OH)_3$	6.7×10^{-31}	HgS	1.6×10^{-54}
$Mn(OH)_2$	2×10^{-13}	$Al(OH)_3$	5×10^{-33}
MnS	7×10^{-16}	SnS	1×10^{-26}
FeS	4×10^{-19}	$Pb(OH)_2$	4.2×10^{-15}
$Fe(OH)_3$	6×10^{-38}	$PbCl_2$	1.6×10^{-5}
CoS	5×10^{-22}	PbS	7×10^{-29}

*In order of appearance in the periodic table.

Appendix 7
STANDARD ELECTRODE POTENTIALS AT 25 °C

Half-reaction	$E°$, V
$F_2(g) + 2H^+ + 2e^- \longrightarrow 2HF$	$+3.06$
$F_2(g) + 2e^- \longrightarrow 2F^-$	$+2.87$
$O_3(g) + 2H^+ + 2e^- \longrightarrow O_2(g) + H_2O$	$+2.07$
$Ag^{2+} + e^- \longrightarrow Ag^+$	$+1.98$
$Co^{3+} + e^- \longrightarrow Co^{2+}$	$+1.82$
$H_2O_2 + 2H^+ + 2e^- \longrightarrow 2H_2O$	$+1.77$
$MnO_4^- + 4H^+ + 3e^- \longrightarrow MnO_2(s) + 2H_2O$	$+1.70$
$Au^+ + e^- \longrightarrow Au(s)$	ca. $+1.7$
$HClO_2 + 2H^+ + 2e^- \longrightarrow HClO + H_2O$	$+1.64$
$HClO + H^+ + e^- \longrightarrow \frac{1}{2}Cl_2(g) + H_2O$	$+1.63$
$Ce^{4+} + e^- \longrightarrow Ce^{3+}$	$+1.61$
$H_5IO_6 + H^+ + 2e^- \longrightarrow IO_3^- + 3H_2O$	$+1.6$
$MnO_4^- + 8H^+ + 5e^- \longrightarrow Mn^{2+} + 4H_2O$	$+1.51$
$Mn^{3+} + e^- \longrightarrow Mn^{2+}$	$+1.51$
$BrO_3^- + 6H^+ + 5e^- \longrightarrow \frac{1}{2}Br_2 + 3H_2O$	$+1.50$
$Au^{3+} + 3e^- \longrightarrow Au(s)$	$+1.50$
$Cl_2(g) + 2e^- \longrightarrow 2Cl^-$	$+1.36$
$NH_3OH^+ + 2H^+ + 2e^- \longrightarrow NH_4^+ + H_2O$	$+1.35$
$Cr_2O_7^{2-} + 14H^+ + 6e^- \longrightarrow 2Cr^{3+} + 7H_2O$	$+1.33$
$2HNO_2 + 4H^+ + 4e^- \longrightarrow N_2O(g) + 3H_2O$	$+1.29$
$Tl^{3+} + 2e^- \longrightarrow Tl^+$	$+1.25$
$MnO_2(s) + 4H^+ + 2e^- \longrightarrow Mn^{2+} + 2H_2O$	$+1.23$
$O_2(g) + 4H^+ + 4e^- \longrightarrow 2H_2O$	$+1.23$
$ClO_3^- + 3H^+ + 2e^- \longrightarrow HClO_2 + 6H_2O$	$+1.21$
$IO_3^- + 6H^+ + 5e^- \longrightarrow \frac{1}{2}I_2 + 3H_2O$	$+1.20$
$ClO_4^- + 2H^+ + 2e^- \longrightarrow ClO_3^- + H_2O$	$+1.19$
$PuO_2^+ + 4H^+ + e^- \longrightarrow Pu^{4+} + 2H_2O$	$+1.15$
$Br_2 + 2e^- \longrightarrow 2Br^-$	$+1.09$
$N_2O_4(g) + 2H^+ + 2e^- \longrightarrow 2HNO_2$	$+1.07$
$Br_2(l) + 2e^- \longrightarrow 2Br^-$	$+1.07$
$PuO_2^{2+} + 4H^+ + 2e^- \longrightarrow Pu^{4+} + 2H_2O$	$+1.04$

Half-reaction	$E°$, V
$N_2O_4(g) + 4H^+ + 4e^- \longrightarrow 2NO(g) + 2H_2O$	+1.03
$V(OH)_4^+ + 2H^+ + e^- \longrightarrow VO^{2+} + 3H_2O$	+1.00
$HNO_2 + H^+ + e^- \longrightarrow NO(g) + H_2O$	+1.00
$Pu^{4+} + e^- \longrightarrow Pu^{3+}$	+0.97
$NO_3^- + 4H^+ + 3e^- \longrightarrow NO(g) + 2H_2O$	+0.96
$2Hg^{2+} + 2e^- \longrightarrow Hg^{2+}$	+0.92
$2NO_3^- + 4H^+ + 2e^- \longrightarrow N_2O_4(g) + 2H_2O$	+0.80
$Ag^+ + e^- \longrightarrow Ag(s)$	+0.80
$Hg_2^{2+} + 2e^- \longrightarrow 2Hg(l)$	+0.79
$Fe^{3+} + e^- \longrightarrow Fe^{2+}$	+0.77
$O_2(g) + 2H^+ + 2e^- \longrightarrow H_2O_2$	+0.68
$UO_2^+ + 4H^+ + e^- \longrightarrow U^{4+} + 2H_2O$	+0.62
$MnO_4^- + e^- \longrightarrow MnO_4^{2-}$	+0.56
$H_3AsO_4 + 2H^+ + 2e^- \longrightarrow HAsO_2 + 2H_2O$	+0.56
$I_2 + 2e^- \longrightarrow 2I^-$	+0.54
$Cu^+ + e^- \longrightarrow Cu(s)$	+0.52
$VO^{2+} + 2H^+ + e^- \longrightarrow V^{3+} + H_2O$	+0.36
$Fe(CN)_6^{3-} + e^- \longrightarrow Fe(CN)_6^{4-}$	+0.36
$Cu^{2+} + 2e^- \longrightarrow Cu(s)$	+0.34
$UO_2^{2+} + 4H^+ + 2e^- \longrightarrow U^{4+} + 2H_2O$	+0.33
$Cu^{2+} + e^- \longrightarrow Cu^+$	+0.15
$Sn^{4+} + 2e^- \longrightarrow Sn^{2+}$	+0.15
$S(s) + 2H^+ + 2e^- \longrightarrow H_2S(g)$	+0.14
$HSO_4^- + 3H^+ + 2e^- \longrightarrow SO_2 + 2H_2O$	+0.11
$P(s) + 3H^+ + 3e^- \longrightarrow PH_3(g)$	+0.06
$UO_2^{2+} + e^- \longrightarrow UO_2^+$	+0.05
$2H^+ + 2e^- \longrightarrow H_2(g)$	Zero
$Pb^{2+} + 2e^- \longrightarrow Pb(s)$	−0.13
$Sn^{2+} + 2e^- \longrightarrow Sn(s)$	−0.14
$Mo^{3+} + 3e^- \longrightarrow Mo(s)$	ca. −0.2
$Ni^{2+} + 2e^- \longrightarrow Ni(s)$	−0.25
$V^{3+} + e^- \longrightarrow V^{2+}$	−0.26
$H_3PO_4 + 2H^+ + 2e^- \longrightarrow H_3PO_3 + H_2O$	−0.28
$Co^{2+} + 2e^- \longrightarrow Co(s)$	−0.28
$Tl^+ + e^- \longrightarrow Tl(s)$	−0.34
$In^{3+} + 3e^- \longrightarrow In(s)$	−0.34
$Cd^{2+} + 2e^- \longrightarrow Cd(s)$	−0.40
$Cr^{3+} + e^- \longrightarrow Cr^{2+}$	−0.41
$Eu^{3+} + e^- \longrightarrow Eu^{2+}$	−0.43
$Fe^{2+} + 2e^- \longrightarrow Fe(s)$	−0.44
$Ga^{3+} + 3e^- \longrightarrow Ga(s)$	−0.53
$U^{4+} + e^- \longrightarrow U^{3+}$	−0.61
$Cr^{3+} + 3e^- \longrightarrow Cr(s)$	−0.74
$Zn^{2+} + 2e^- \longrightarrow Zn(s)$	−0.76
$TiO^{2+} + 2H^+ + 4e^- \longrightarrow Ti(s) + H_2O$	ca. −0.9
$V^{2+} + 2e^- \longrightarrow V(s)$	ca. −1.2

Half-reaction	$E°$, V
$Mn^{2+} + 2e^- \longrightarrow Mn(s)$	-1.18
$Zr^{4+} + 4e^- \longrightarrow Zr(s)$	-1.53
$Al^{3+} + 3e^- \longrightarrow Al(s)$	-1.66
$Hf^{4+} + 4e^- \longrightarrow Hf(s)$	-1.70
$U^{3+} + 3e^- \longrightarrow U(s)$	-1.80
$Be^{2+} + 2e^- \longrightarrow Be(s)$	-1.85
$Th^{4+} + 4e^- \longrightarrow Th(s)$	-1.90
$Pu^{3+} + 3e^- \longrightarrow Pu(s)$	-2.07
$Sc^{3+} + 3e^- \longrightarrow Sc(s)$	-2.08
$\frac{1}{2}H_2(g) + e^- \longrightarrow H^-$	-2.25
$Y^{3+} + 3e^- \longrightarrow Y(s)$	-2.37
$Mg^{2+} + 2e^- \longrightarrow Mg(s)$	-2.37
$Ce^{3+} + 3e^- \longrightarrow Ce(s)$	-2.48
$La^{3+} + 3e^- \longrightarrow La(s)$	-2.52
$Na^+ + e^- \longrightarrow Na(s)$	-2.71
$Ca^{2+} + 2e^- \longrightarrow Ca(s)$	-2.87
$Sr^{2+} + 2e^- \longrightarrow Sr(s)$	-2.89
$Ba^{2+} + 2e^- \longrightarrow Ba(s)$	-2.90
$Ra^{2+} + 2e^- \longrightarrow Ra(s)$	-2.92
$Cs^+ + e^- \longrightarrow Cs(s)$	-2.92
$Rb^+ + e^- \longrightarrow Rb(s)$	-2.93
$K^+ + e^- \longrightarrow K(s)$	-2.93
$Li^+ + e^- \longrightarrow Li(s)$	-3.05

Appendix 8

ATOMIC AND EFFECTIVE IONIC RADII

Atomic* and effective ionic† radii, nm

Ac^{3+}	0.112	C^0	0.077	Er^0	0.158	In^0	0.150
Ag^0	0.134	C^{4+}	0.016	Er^{3+}	0.089	In^{3+}	0.080
Ag^+	0.115	Ca^0	0.174	Eu^0	0.185	Ir^0	0.127
Ag^{2+}	0.094	Ca^{2+}	0.100	Eu^{2+}	0.117	Ir^{4+}	0.0625
Ag^{3+}	0.075	Cd^0	0.141	Eu^{3+}	0.0947	K^0	0.196
Al^0	0.130	Cd^{2+}	0.095	F^0	0.071	K^+	0.138
Al^{3+}	0.0535	Ce^0	0.165	F^-	0.133	La^0	0.169
Am^{3+}	0.0975	Ce^{3+}	0.101	Fe^0	0.117	La^{3+}	0.103
Am^{4+}	0.085	Ce^{4+}	0.087	Fe^{2+}	0.078	Li^0	0.134
As^0	0.122	Cf^{3+}	0.095	Fe^{3+}	0.0645	Li^+	0.076
As^{3+}	0.058	Cf^{4+}	0.0821	Ga^0	0.12	Lu^0	0.156
As^{5+}	0.046	Cl^0	0.099	Ga^{3+}	0.062	Lu^{3+}	0.086
Au^0	0.134	Cl^{-1}	0.181	Gd^0	0.162	Mg^0	0.145
Au^+	0.137	Cl^{7+}	0.027	Gd^{3+}	0.0938	Mg^{2+}	0.072
Au^{3+}	0.085	Cm^{3+}	0.097	Ge^0	0.122	Mn^0	0.117
B^0	0.090	Cm^{4+}	0.085	Ge^{2+}	0.073	Mn^{2+}	0.083
B^{3+}	0.027	Co^0	0.116	Ge^{4+}	0.053	Mn^{3+}	0.065
Ba^0	0.198	Co^{2+}	0.065	H^0	0.037	Mn^{4+}	0.053
Ba^{2+}	0.135	Co^{3+}	0.0545	H^-	0.144	Mn^{7+}	0.046
Be^0	0.125	Cr^0	0.118	Hf^0	0.144	Mo^0	0.130
Be^{2+}	0.045	Cr^{2+}	0.080	Hf^{4+}	0.071	Mo^{4+}	0.065
Bi^0	0.15	Cr^{3+}	0.0615	Hg^0	0.144	Mo^{6+}	0.059
Bi^{3+}	0.103	Cr^{6+}	0.044	Hg^+	0.119	N^0	0.075
Bi^{5+}	0.076	Cs^0	0.235	Hg^{2+}	0.102	N^{3-}	0.146
Bk^{3+}	0.096	Cs^+	0.167	Ho^0	0.158	N^{3+}	0.016
Bk^{4+}	0.083	Cu^0	0.117	Ho^{3+}	0.0901	N^{5+}	0.013
Br^0	0.114	Cu^+	0.077	I^0	0.133	Na^0	0.154
Br^-	0.196	Cu^{2+}	0.073	I^-	0.220	Na^+	0.102
Br^{3+}	0.059	Dy^0	0.160	I^{5+}	0.095	Nb^0	0.134
Br^{5+}	0.031	Dy^{3+}	0.0912	I^{7+}	0.053	Nb^{3+}	0.072
Br^{7+}	0.025						

*Atomic radii (M^0) are determined as half the observed interatomic distance between identical single-bonded atoms.

†Effective ionic radii (M^{n+}), taken from R. D. Shannon, *Acta Crystallographica*, **A32:** 751 (1976), are determined by computer fitting of observed distances in oxide and fluoride crystals. Values given are for octahedral coordination.

Nb^{4+}	0.068	Pt^0	0.130	Si^0	0.118	Tl^{3+}	0.089
Nb^{5+}	0.064	Pt^{2+}	0.080	Si^{4+}	0.040	Tm^0	0.158
Nd^0	0.164	Pt^{4+}	0.0625	Sm^0	0.162	Tm^{3+}	0.088
Nd^{3+}	0.0983	Pu^{3+}	0.100	Sm^{2+}	0.117	U^0	0.142
Ni^0	0.115	Pu^{4+}	0.086	Sm^{3+}	0.0958	U^{3+}	0.103
Ni^{2+}	0.069	Ra^{2+}	0.126	Sn^0	0.140	U^{4+}	0.089
Np^{3+}	0.101	Rb^0	0.216	Sn^{4+}	0.069	U^{6+}	0.073
Np^{4+}	0.087	Rb^+	0.152	Sr^0	0.191	V^0	0.122
Np^{5+}	0.075	Re^0	0.128	Sr^{2+}	0.118	V^{2+}	0.079
O^0	0.073	Re^{4+}	0.063	Ta^0	0.134	V^{3+}	0.064
O^{2-}	0.140	Re^{6+}	0.055	Ta^{3+}	0.072	V^{4+}	0.058
Os^0	0.126	Re^{7+}	0.053	Ta^{4+}	0.068	V^{5+}	0.054
Os^{4+}	0.063	Rh^0	0.125	Ta^{5+}	0.064	W^0	0.130
P^0	0.110	Rh^{3+}	0.0665	Tb^0	0.161	W^{4+}	0.066
P^{3+}	0.044	Rh^{4+}	0.060	Tb^{3+}	0.0923	W^{5+}	0.062
P^{5+}	0.038	Ru^0	0.125	Tb^{4+}	0.076	W^{6+}	0.060
Pa^{3+}	0.104	Ru^{3+}	0.068	Tc^0	0.127	Y^0	0.162
Pb^0	0.154	Ru^{4+}	0.062	Tc^{4+}	0.0645	Y^{3+}	0.090
Pb^{2+}	0.119	S^0	0.102	Tc^{7+}	0.056	Yb^0	0.170
Pb^{4+}	0.078	S^{2-}	0.184	Te^0	0.135	Yb^{2+}	0.102
Pd^0	0.128	S^{4+}	0.037	Te^{2-}	0.221	Yb^{3+}	0.0868
Pd^{2+}	0.086	S^{6+}	0.012	Te^{4+}	0.097	Zn^0	0.125
Pd^{4+}	0.062	Sb^0	0.143	Te^{6+}	0.056	Zn^{2+}	0.074
Pm^0	0.163	Sb^{3+}	0.076	Th^0	0.165	Zr^0	0.145
Pm^{3+}	0.097	Sb^{5+}	0.060	Th^{4+}	0.094	Zr^{4+}	0.072
Po^0	0.153	Sc^0	0.144	Ti^0	0.132		
Po^{4+}	0.094	Sc^{3+}	0.0745	Ti^{2+}	0.086		
Po^{6+}	0.067	Se^0	0.117	Ti^{3+}	0.067		
Pr^0	0.164	Se^{2-}	0.198	Ti^{4+}	0.061		
Pr^{3+}	0.099	Se^{4+}	0.050	Tl^0	0.155		
Pr^{4+}	0.085	Se^{6+}	0.042	Tl^+	0.150		

Appendix 9

REFERENCES

References of general utility covering many of the topics in this text are J. E. Huheey, *Inorganic Chemistry*, Harper & Row; G. Hägg, *General Inorganic Chemistry*, Wiley; R. T. Morrison and R. N. Boyd, *Organic Chemistry*, Allyn and Bacon; G. M. Barrow, *Physical Chemistry*, McGraw-Hill; E. J. King, *Qualitative Analysis and Electrolytic Solutions*, Harcourt, Brace; and *McGraw-Hill Encyclopedia of Science and Technology*, McGraw-Hill.

Additional information and background material can be found in the following references, which are listed by the chapters to which they most apply:

Chapter 1

S. J. Baum and C. W. J. Scaife, *Chemistry: A Life Science Approach*, Macmillan. W. F. Kieffer, *Chemistry: A Cultural Approach*, Harper & Row. H. G. Jerrard and D. B. McNeill, *A Dictionary of Scientific Units*, Chapman & Hall.

Chapter 2

M. J. Sienko, *Chemistry Problems*, Benjamin. R. Kiesel and E. S. Gore, *Problems for Chemistry*, Allyn and Bacon. E. I. Peters, *Problem Solving for Chemistry*, Saunders.

Chapter 3

R. T. Sanderson, *Chemical Periodicity*, Reinhold. R. Rich, *Periodic Correlations*, Benjamin.

Chapter 4

M. Born, *Atomic Physics*, Hafner. J. C. Slater, *Modern Physics*, McGraw-Hill.

Chapter 5

A. L. Companion, *Chemical Bonding,* McGraw-Hill. J. W. Linnett, *The Electronic Structure of Molecules,* Methuen. L. Pauling, *The Nature of the Chemical Bond,* Cornell. C. A. Coulson, *Valence,* Oxford. R. J. Gillespie, *Molecular Geometry,* Van Nostrand.

Chapter 6

W. Kauzmann, *Kinetic Theory of Gases,* Benjamin. M. A. Paul, *Physical Chemistry,* Heath. W. F. Sheehan, *Physical Chemistry,* Allyn and Bacon.

Chapter 7

F. H. MacDougall, *Physical Chemistry,* Macmillan. W. J. Moore, *Physical Chemistry,* Prentice-Hall. G. W. Castellan, *Physical Chemistry,* Addison-Wesley.

Chapter 8

E. J. King, *Qualitative Analysis and Electrolytic Solutions,* Harcourt, Brace. W. Ostwald, *Solutions,* Longmans. A. G. Williamson, *An Introduction to Nonelectrolyte Solutions,* Wiley.

Chapter 9

H. W. Willard and N. H. Furman, *Elementary Quantitative Analysis,* Van Nostrand. R. P. Bell, *Acids and Bases: Their Quantitative Behavior,* Methuen. J. N. Butler, *Solubility and pH Calculations,* Addison-Wesley.

Chapter 10

W. G. Davies, *Introduction to Chemical Thermodynamics,* Saunders. L. K. Nash, *Chemthermo: A Statistical Approach to Classical Chemical Thermodynamics,* Addison-Wesley.

Chapter 11

D. A. MacInnes, *The Principles of Electrochemistry,* Reinhold. W. M. Latimer, *Oxidation Potentials,* Prentice-Hall.

Chapter 12

E. L. King, *How Chemical Reactions Occur,* Benjamin. J. O. Edwards, *Inorganic Reaction Mechanisms,* Benjamin. E. F. Obert, *Internal Combustion Engines,* International Textbook.

Chapter 13

F. H. MacDougall, *Physical Chemistry,* Macmillan. M. J. Sienko, *Equilibrium,* Benjamin.

Chapter 14

Barnstead Company, *The Barnstead Basic Book on Water,* Barnstead. M. Ardon, *Oxygen,* Benjamin. R. P. Bell, *The Proton in Chemistry,* Cornell.

Chapter 15

T. Moeller, *Qualitative Analysis,* McGraw-Hill. M. J. Sienko, *Equilibrium,* Benjamin. J. Wasser, *Quantitative Chemistry,* Benjamin.

Chapter 16

G. M. Barrow, *The Structure of Molecules,* Benjamin. H. B. Gray, *Electrons and Chemical Bonding,* Benjamin.

Chapter 17

W. M. Latimer, and J. H. Hildebrand, *Reference Book of Inorganic Chemistry,* Macmillan. T. Moeller, *Inorganic Chemistry,* Wiley.

Chapter 18

E. M. Larsen, *Transitional Elements,* Benjamin. L. E. Orgel, *An Introduction to Transition-Metal Chemistry,* Methuen.

Chapter 19

E. M. Larsen, *Transitional Elements,* Benjamin. F. A. Cotton and G. Wilkinson, *Basic Inorganic Chemistry,* Wiley.

Chapter 20

M. H. O'Leary, *Contemporary Organic Chemistry,* McGraw-Hill. W. Herz, *The Shape of Carbon Compounds,* Benjamin. C. H. DePuy and K. L. Rinehart, *Introduction to Organic Chemistry,* Wiley. L. Mandelkern, *An Introduction to Macromolecules,* Heidelberg.

Chapter 21

R. Steudel, *Chemistry of the Nonmetals,* De Gruyter. W. M. Latimer and J. H. Hildebrand, *Reference Book of Inorganic Chemistry,* Macmillan.

Chapter 22

R. S. Shankland, *Atomic and Nuclear Physics,* Macmillan. G. Friedlander and J. W. Kennedy, *Nuclear and Radiochemistry,* Wiley. G. Bryerton, *Nuclear Dilemma,* Ballantine.

Excellent references for qualitative analysis are E. J. King, *Qualitative Analysis and Electrolytic Solutions,* Harcourt, Brace; and T. R. Hogness and W. C. Johnson, *Qualitative Analysis and Chemical Equilibrium,* Holt.

Useful sources of data are *Handbook of Chemistry and Physics,* Chemical Rubber Co.; *The Merck Index of Chemicals and Drugs,* Merck & Co.; *Stability Constants,* London Chemical Society; *Interatomic Distances,* London Chemical Society; and K. S. Pitzer and L. Brewer, *Thermodynamics,* McGraw-Hill.

INDEX

Periodic Table of the Elements

1 H 1.008																	2 He 4.003
3 Li 6.941	4 Be 9.012											5 B 10.81	6 C 12.011	7 N 14.007	8 O 15.999	9 F 18.998	10 Ne 20.179
11 Na 22.990	12 Mg 24.305											13 Al 26.982	14 Si 28.086	15 P 30.974	16 S 32.06	17 Cl 35.453	18 Ar 39.948
19 K 39.098	20 Ca 40.08	21 Sc 44.956	22 Ti 47.90	23 V 50.941	24 Cr 51.996	25 Mn 54.938	26 Fe 55.847	27 Co 58.933	28 Ni 58.70	29 Cu 63.546	30 Zn 65.38	31 Ga 69.72	32 Ge 72.59	33 As 74.922	34 Se 78.96	35 Br 79.904	36 Kr 83.80
37 Rb 85.468	38 Sr 87.62	39 Y 88.906	40 Zr 91.22	41 Nb 92.906	42 Mo 95.94	43 Tc 98.906	44 Ru 101.07	45 Rh 102.906	46 Pd 106.4	47 Ag 107.868	48 Cd 112.41	49 In 114.82	50 Sn 118.69	51 Sb 121.75	52 Te 127.60	53 I 126.904	54 Xe 131.30
55 Cs 132.905	56 Ba 137.33	57–71 *	72 Hf 178.49	73 Ta 180.948	74 W 183.85	75 Re 186.207	76 Os 190.2	77 Ir 192.22	78 Pt 195.09	79 Au 196.966	80 Hg 200.59	81 Tl 204.37	82 Pb 207.2	83 Bi 208.980	84 Po (210)	85 At (210)	86 Rn (222)
87 Fr (223)	88 Ra 226.025	89–103 †	104 Ku (257)	105 Ha (260)	106 ?	107 ?											

* | 57 La 138.906 | 58 Ce 140.12 | 59 Pr 140.908 | 60 Nd 144.24 | 61 Pm (145) | 62 Sm 150.4 | 63 Eu 151.96 | 64 Gd 157.25 | 65 Tb 158.925 | 66 Dy 162.50 | 67 Ho 164.930 | 68 Er 167.26 | 69 Tm 168.934 | 70 Yb 173.04 | 71 Lu 174.97 |

† | 89 Ac (227) | 90 Th 232.038 | 91 Pa 231.036 | 92 U 238.029 | 93 Np 237.048 | 94 Pu (242) | 95 Am (243) | 96 Cm (247) | 97 Bk (249) | 98 Cf (251) | 99 Es (254) | 100 Fm (253) | 101 Md (256) | 102 No (254) | 103 Lr (257) |

acid → Base →

acid →
Base →

(n-1) d

(n-2) f

55-2